S0-CDO-856

THE METAMATHEMATICS OF ALGEBRAIC SYSTEMS

STUDIES IN LOGIC

AND

THE FOUNDATIONS OF MATHEMATICS

VOLUME 66

Editors

A. HEYTING, *Amsterdam*
H. J. KEISLER, *Madison*
A. MOSTOWSKI, *Warszawa*
A. ROBINSON, *New Haven*
P. SUPPES, *Stanford*

Advisory Editorial Board

Y. BAR-HILLEL, *Jerusalem*
K. L. DE BOUVÈRE, *Santa Clara*
H. HERMES, *Freiburg i. Br.*
J. HINTIKKA, *Helsinki*
J. C. SHEPHERDSON, *Bristol*
E. P. SPECKER, *Zürich*

NORTH-HOLLAND PUBLISHING COMPANY
AMSTERDAM · LONDON

THE METAMATHEMATICS OF ALGEBRAIC SYSTEMS

COLLECTED PAPERS: 1936-1967

ANATOLIĬ IVANOVIČ MAL'CEV

translated, edited, and provided with supplementary notes by

BENJAMIN FRANKLIN WELLS, III

1971

NORTH-HOLLAND PUBLISHING COMPANY
AMSTERDAM · LONDON

© North-Holland Publishing Company, 1971

All rights reserved. No part of this book may be reproduced, stored in a retrieval system or transmitted, in any form or by any means, electronic, mechanical, photocopying, recording or otherwise without the prior permission of the copyright owner.

Library of Congress Catalog Card Number 73-157020

International Standard Book Number 0 7204 2266 3

PUBLISHERS:

NORTH-HOLLAND PUBLISHING COMPANY – AMSTERDAM

NORTH-HOLLAND PUBLISHING COMPANY, LTD. – LONDON

PRINTED IN THE NETHERLANDS

to the many miles on the Volga in a simple rowboat

(from a reminiscence of
P.S. Aleksandrov [B1])

TRANSLATOR'S FOREWORD

Almost half of Soviet mathematician Anatoliĭ Ivanovič Mal'cev's published research contributes to the metamathematics of algebraic systems or employs its techniques to obtain algebraic results. The purpose of this book is to offer the English reader convenient access to most of this material, as well as to three important surveys (Chapters 18, 26, 34). The book's contents pervade the theory of models, that broad region on the boundary of logic and algebra, but lean more toward metamathematics than toward universal algebra. The title suggests the breadth of Mal'cev's study, for algebraic systems generalize models (relational structures), algebras (algebraic structures), and partial algebras. Briefly, an *algebraic system* consists of a nonempty base set and a number of basic notions defined on it of four possible kinds: predicate (relation), operation, partial operation, distinguished element; in practice the last three are special forms of predicates. Please consult the *Index* and (I), (II) below for more information; Mal'cev's last publication [M16], soon to appear in English, provides a detailed introduction to the general theory of algebraic systems.

Several of the articles presented here have already seen print in English (*Bibliography,* Part II); you will find, I trust, that the efforts to be mathematically clear, smooth, and accurate have justified a "freer" translation: while 110 supplementary notes have been provided, many small improvements have been made without notice. In such work the impulses to be uniform in notation and definition and to conform with both original sources and ordinary usage must be balanced, as Mal'cev himself recognized in [M14]. For instance, certain terms, such as *variety, compactness theorem, model,* are used here before they appear in the Russian. In a conflict the mathematics gets translated literally, the language freely.

Among other abuses: use-mention distinctions in the metalanguage are ignored, so that symbols, for example, are mentioned without quotation marks; we do, however, strive to differentiate (usually by means of boldface type) the notation of a predicate (relation) symbol, operation (function) symbol, or individual constant symbol from the notation of the predicate, operation, or distinguished element it might designate in a particular algebraic system: thus, $\boldsymbol{P}$, P, $\boldsymbol{f}$, f, $\boldsymbol{a}$, a. The same goes for logical *vs.* mathematical variables. In particular, $\approx$ is the logical equality symbol, and = sees mathematical and metalogical service. You will often meet the standard trope of using the name of an algebraic system to refer to its base set, as well; thus if $\mathfrak{A} = \langle A, +\rangle$ is an algebraic system, we write $x \in \mathfrak{A}$, meaning $x \in A$.

A *factor algebra* (or *model,* or *system*) is more commonly known as a quotient algebra, etc. An indexed set (or *sequence,* or *system*) may be denoted by either $\langle a_\alpha \colon \alpha \in J\rangle$ or $\{a_\alpha \colon \alpha \in J\}$, and the index set J may have additional structure such as a partial or linear ordering. Sometimes the name of a formula will appear with a mathematical argument symbol, e.g., Pos(u), $\Phi(a)$. This always refers to the semantical interpretation of the formula in a given algebraic system. Thus "$\boldsymbol{x} \boldsymbol{\leqslant} \boldsymbol{y}$", "$x \leqslant y$", "$x \leqslant y$ holds (or is valid, or is true)" would be interchangeable in a fixed semantic context consisting of a set to which x, y belong and a binary relation $\leqslant$ between elements of the set.

For the metalanguage Mal'cev employs a naive set theory with the axiom of choice. In discussing sets we render the membership relation as "belongs to" and its converse as "contains"; the inclusion relation comes out "includes". "Iff" is short for "if and only if". Mal'cev most frequently uses first-order predicate logic (FOPL) as his formal language, but propositional calculus (PC) and second-order predicate logic (SOPL) also occur. The details of the formulation, proof structure, and semantic interpretation are not specified and are usually not pertinent as long as we can apply the standard structural classification of formulas, etc. By *formula* we always mean a well-formed formula, possibly with free variables. *Sentence* and *axiom* refer synonymously to closed formulas, i.e., those without free variables. Thus Mal'cev's term *axiomatizability* leads to confusion, because he does not require a set of axioms to be recursive. Therefore, this term is always introduced as an abbreviation for *first-order axiomatizability*; the other notion, rarely encountered here, is called *recursive axiomatizability.*

A *class* of algebraic systems is *abstract* (closed under isomorphism) most of the time – usually explicitly, but sometimes not. It always consists of algebras of the same similarity type.

The closely related notions of *similarity type* and *signature* present a methodological, if not conceptual, difficulty. Roughly speaking, the former specifies the number of basic operations, partial operations, predicates, and distinguished elements in an algebraic system and the rank (number of arguments) of each; the signature specifies the symbols of matching rank used to designate these mathematical objects in logical contexts. When Mal'cev commits himself to a complete definition of these two concepts and that of algebraic system, he follows one of two paths:

(I). An *algebraic system* $\mathfrak{A}$ is a sequence consisting of a nonempty *base set* A and *basic notions* (operations, predicates, etc.) defined on the base A and grouped according to the kind of notion. The *type* τ of $\mathfrak{A}$ is the corresponding sequence of the ranks of the basic notions, and its *signature* Σ is a similar sequence of distinct logical symbols of the kind and rank determined by τ. This is Grätzer's approach in [47].

(II). A *signature* Σ is a set of (distinct) logical symbols of various kinds (operation, predicate, etc.) together with an *arity map* α from Σ into the natural numbers which gives the ranks of the symbols. An *algebraic system* $\mathfrak{A}$ *with this signature* is specified by choosing a nonempty *base set* A and a 1–1 correspondence υ (the *valuation*) between the symbols in Σ and *basic notions* defined on A that preserves the kind and rank. While the *type* is not mentioned, it is clear that the arity map serves the same purpose. This method is in the style of Cohn [22].

Generally speaking, Mal'cev stuck to (I) through 1961 ([XXVI] probably dates from then) and switched to (II) in 1962. Although these two approaches could lead to different pictures of algebraic systems, or at least to different notations, they do not: Mal'cev uses the sequence notation for systems and makes no reference to the map υ (*N.B.:* the notation used in this translation recaptures the map υ in the transition from boldface to regular type as discussed above.) But doesn't (II) make the algebra depend on the logic? This may be a crucial point, but Mal'cev avoids it here by implying that an algebraic system always determines a definite signature. The drawback is that we may want several logical symbols to designate the same predicate, say, in the same passage. Luckily, this happens only once: in Chapter 26, §5.3. Discounting this, we can eliminate the undercover relation between systems and signatures by explicitly agreeing that we always think of "algebraic-system-with-signature", and that a class is composed of algebraic-systems-with-signatures that have the same signature and *a fortiori* the same type. One more convention: although a FOPL formula has a definite "signature" consisting of the extralogical symbols occurring in it, we shall say that it is a *formula of* any *signature* to which all these symbols belong.

The *Topic Table* indicates five aspects of A.I. Mal'cev's work in the metamathematics of algebraic systems. Other papers rate at least one mark in that table: in [M1] Mal'cev proves the semigroups embeddable in groups form a quasivariety that is not finitely axiomatizable; in [M2] he applies the methods of [II] to prove that every torsion-free nilpotent group is freely orderable; in [M13] he gives a simple formula defining the relation "x and y generate the group" in a free group with two (distinguished) free generators; [M15] can be viewed as a completeness result for n-ary propositional logic. There are several papers [M5, M10–M12] in universal algebra that might fall in the intended scope of this book.

On the other hand, not every chapter of this book is outstanding. In some cases they lay the groundwork on which others will have to build; some seem now to be dead-end or dated. But a few are simply jewels, any time, any place.

Thus this anthology is neither consistent nor complete. It will, I hope, prove both interesting and useful.

I would like to express my deep gratitude to Professor Ju. L. Eršov for his proposal (mostly adopted) of which papers to include, to Professor Alfred Tarski for his many helpful conversations and suggestions (mostly followed), and to Avatar Meher Baba for his inspiration, guidance, and (most of all) love.

B.F. Wells, III
Carmel Highlands, California
January 27, 1971.

BIOGRAPHICAL NOTE

Academician Anatoliĭ Ivanovič Mal'cev was born November 27, 1909, to the family of a glass-blower in Mišeronsk, Krivandinsk region, Moscow province, USSR. On finishing secondary school at Mineral'nye Vody in 1927, young Mal'cev entered the mathematics department of Moscow University. In 1931 he completed his university work and continued his teaching at a college in Ivanovo, which he had begun in 1930. From 1932 to 1960 he worked at the Ivanovo State Pedagogical Institute (ISPI), first as assistant, then as docent, and from 1943 as professor, holding the chair in higher algebra. Without interrupting his teaching at ISPI, Mal'cev spent 1934–37 as a graduate student at Moscow University. During 1939–41 he was a doctoral candidate at the Steklov Mathematical Institute, Academy of Science USSR, which awarded him a doctorate in 1941 and a professorship in 1944. From 1942 to 1960 he held the position of senior scientist at the Steklov Institute in addition to his post at ISPI.

Mal'cev won the State Prize in 1946 for his research in the theory of Lie groups. He was made an associate member of the Academy of Science in 1953 and elected to full membership in 1958. The title of Honored Scientist of the Russian SFSR was bestowed on him in 1956. From 1960 Mal'cev simultaneously headed the algebra section of the Mathematical Institute of the Academy's Siberian Division and held the chair of algebra and mathematical logic at Novosibirsk University. In 1964 A.I. Mal'cev won the Lenin Prize for his cycle of papers on the application of mathematical logic to algebra and the thoery of algebraic systems. This interaction, which motivates the present anthology, was an important focus of intense effort for Mal'cev throughout his mathematical life: his first and last publications – and many others – bear on it. Just as mathematics was not his whole life, this field was not his whole mathematics. He is widely and deeply respected for his contributions to pedagogy as well as to pure and topological algebra, for his service to his country as well as to the world of science, for his humanity as well as his responsibility. But one may say that mathematics was his vocation, and the metamathematics of algebraic systems, his specialty.

A.I. Mal'cev died July 7, 1967, while participating in the All-Union Conference on Topology in Novosibirsk. More details of his life and work can be found in the obituaries listed in the *Bibliography*, Part V.

CONTENTS

Translator's foreword vii

Biographical note xi

Chapter 1. Investigations in the realm of mathematical logic 1

Chapter 2. A general method for obtaining local theorems in group theory 15

Chapter 3. Representations of models 22

Chapter 4. Quasiprimitive classes of abstract algebras 27

Chapter 5. Subdirect products of models 32

Chapter 6. Derived operations and predicates 37

Chapter 7. Classes of models with an operation of generation 44

Chapter 8. Defining relations in categories 51

Chapter 9. The structural characterization of certain classes of algebras 56

Chapter 10. Certain classes of models 61

Chapter 11. Model correspondences 66
- Introduction 66
- §1. Multibase models 67
  - 1.1. Multibase predicates 67
  - 1.2. Axiomatizable and projective correspondences 70
  - 1.3. Some examples 72
- §2. Fundamental properties of projective correspondences 74
  - 2.1. Equality relations; unification of quantifiers 74
  - 2.2. Extrinsic local theorem 76
  - 2.3. Boundedness and extendability for correspondences 78
- §3. Quasiuniversal subclasses 80
  - 3.1. Stability 80
  - 3.2. The intrinsic local theorem 83
  - 3.3. Applications 89

Chapter 12. Regular products of models 95
Introduction 95
§1. Splitting correspondences 96
§2. Regular products 101

Chapter 13. Small models 114

Chapter 14. Free solvable groups 119

Chapter 15. A correspondence between rings and groups 124
§1. The direct mapping 125
§2. Groups with distinguished elements 126
§3. The inverse mapping 127
§4. The reciprocity of the correspondences σ and τ 129
§5. Some special cases 131
§6. Reductions and interpretations of classes of models 133
§7. The undecidability of sundry classes of metabelian groups 135
§8. Nilpotent groups 136

Chapter 16. The undecidability of the elementary theories of certain fields 138
§1. The field of rational functions 139
§2. Fields of formal power series 142

Chapter 17. A remark concerning "The undecidability of the elementary theories of certain fields" [XVI] 147

Chapter 18. Constructive algebras. I 148
Introduction 148
§1. Algebraic systems 151
1.1. Functions, operations, predicates 151
1.2. Generating sets. Terms 153
1.3. Primitive and quasiprimitive classes 157
1.4. Defining relations 159
1.5. Algebras of recursive functions 161
§2. Numbered sets 165
2.1. Mappings of numbered sets 165
2.2. Unreducibility for numberings 171
2.3. Equivalent numberings 178
§3. Numbered algebraic systems 187
3.1. **R**-numberings of algebraic systems 187
3.2. Subsystems 192
3.3. Homomorphisms and congruence relations 196

§4; Finitely generated algebras 201
4.1. General finitely generated algebras 201
4.2. Finitely presented algebras 206

Chapter 19. The undecidability of the elementary theory of finite groups 215

Chapter 20. Elementary properties of linear groups 221
Introduction 221
§1. The elementary nature of the Segre characteristic 223
1.1. The elementariness of diagonizability 223
1.2. The elementariness of the characteristics of diagonizable matrices 227
1.3. The elementariness of the Segre characteristic in the general case 230
1.4. Projective groups 234
§2. Elementary (arithmetic) types of linear and projective groups 238
2.1. The relative elementariness of certain subgroups 238
2.2. The group $GL(2, \mathfrak{K})/\mathfrak{Z}$ 241
2.3. The fundamental theorems 244
2.4. Concluding remarks 246

Chapter 21. The effective inseparability of the set of valid sentences from the set of finitely refutable sentences in several elementary theories 248

Chapter 22. Closely related models and recursively perfect algebras 255
§1. Closely related models 255
§2. Recursively perfect algebras 256
§3. Linear groups 258

Chapter 23. Axiomatizable classes of locally free algebras of various types 262
§1. Locally absolutely free algebras 262
§2. Ordered groupoids 265
§3. $\mathfrak{S}$-algebras 266
§4. Special formulas 267
§5. Standard formulas 271
§6. The reduction of negations of standard formulas 273
§7. The reduction of closed formulas 276

Chapter 24. Recursive abelian groups 282

Chapter 25. Sets with complete numberings 287
§1. Complete numberings 288
§2. Isomorphism. Factor numberings 291
§3. Enumerable families of elements 293
§4. Completely numbered sets whose every family of nonspecial elements is totally enumerable 295
§5. Universal series of sets 298
§6. Totally enumerable families of partial recursive functions 300
§7. Projective families of functions. Computable numberings 305
§8. Quasiordered families 308
§9. Intrinsically productive families 309

Chapter 26. Problems in the theory of classes of models 313
Introduction 313
§1. Fundamental concepts 313
1.1. Algebras and models 313
1.2. Classes of models 315
1.3. The first-order language 317
§2. Axiomatizable classes of models 319
2.1. General properties 319
2.2. Small models 321
2.3. Completeness and categoricy 322
2.4. A set-theoretical characterization of axiomatizable classes 323
2.5. Categories of models 326
§3. Some special axiomatizable classes 327
3.1. Homomorphically closed classes 327
3.2. Universal and Skolem subclasses 328
3.3. Direct products of models 331
3.4. Subdirect products 332
3.5. Convex and quasiaxiomatizable classes 333
§4. Ultraproducts 335
4.1. Basic definitions 335
4.2. Direct limits and ultralimits 338
4.3. Conditions for the axiomatizability of classes. Elementary relations 339
§5. A few second-order classes of models 342
5.1. Axioms of the second order 342
5.2. Projective classes 343
5.3. Reductive classes 344
5.4. Quasiuniversal classes 347
5.5. Formulas with quantifiers on unary predicates 350

Chapter 27. Toward a theory of computable families of objects 353
§1. Principal and complete numberings 353
§2. The α-order and α-topology 356
§3. Normal and subnormal numberings 359
§4. Effectively principal numberings 363
§5. Standard families and precomplete numberings 365
§6. Special and subspecial numberings 369
§7. Totally enumerable families 374

Chapter 28. Positive and negative numberings 379

Chapter 29. Identical relations in varieties of quasigroups 384
§1. The problem of identical relations 384
§2. Algebras with unary operations 386
§3. Partial quasigroups 390
§4. Varieties of quasigroups 392

Chapter 30. Iterative algebras and Post varieties 396
§1. Iterative algebras 398
§2. Iterative algebras of partial functions 399
§3. Congruences on $\mathfrak{P}_A^*$ and $\mathfrak{Q}_{A_w}^*$ 400
§4. Automorphisms 404
§5. Representations of iterative algebras 407
§6. Post variables 408
§7. Selector representations of pre-iterative algebras 411
§8. Subalgebras 413

Chapter 31. A few remarks on quasivarieties of algebraic systems 416
§1. Identities and quasidentities 416
§2. Quasivarieties of algebraic systems 418

Chapter 32. Multiplication of classes of algebraic systems 422
§1. The basic definition 422
§2. Products and axiomatizable classes 429
§3. Multiplication in special classes of systems 435
§4. Additional observations 440

Chapter 33. Universally axiomatizable subclasses of locally finite classes of models 447
§1. Conditions for universal axiomatizability 447
§2. Independent axiomatizability 449
§3. Graphs of finite degree 454
§4. Uniformly locally finite classes 455

Chapter 34. Problems on the border between algebra and logic 460
§1. The algorithmic nature of theories 464
1.1. E-theories and theories of total classes 464
1.2. Number theory 466
1.3. Field theory 466
1.4. The theory of groups and semigroups 467
1.5. The identity problem 469
1.6. Degrees of unsolvability of theories 469
§2. Varieties and quasivarieties 470
2.1. Lattices of subvarieties 470
2.2. Groupoids of quasivarieties 472

Bibliography 474
Part I. The articles translated in this collection 475
Part II. Previous English translations 477
Part III. Other works of A.I. Mal'cev 478
Part IV. Reviews cited in editor's notes 479
Part V. Obituaries with bibliographies 480
Part VI. General references 481

Topic table 489

Index 490

CHAPTER 1

INVESTIGATIONS IN THE REALM OF MATHEMATICAL LOGIC

This article is devoted to generalizing two theorems, one for propositional calculus (PC) and the other for first-order predicate logic (FOPL).

The first theorem is due to Gödel [46] and can be formulated as follows:

For any countable system of formulas of PC *to be consistent (cf.* § 1*), it is sufficient that every finite part of the system be consistent.*

In § 1 it is shown that this theorem holds not only for countable systems, but for systems of any power, as well.

The second theorem was obtained in its hitherto most general form by Skolem [152]. He showed that there is no way to construct a countable system of formulas of FOPL that completely characterizes the structure of the natural numbers.

In § 6 we prove the following more general statement:

Every infinite domain (cf. § 2*) for any system of* FOPL *formulas can be extended* (cf. [188]).

This implies that every system of formulas which has an infinite domain has domains of every power, that every infinite algebraic field has extensions, etc.

Sections 2–5 are devoted to an exposition of auxiliary concepts and theorems. In particular, several well-known results of Löwenheim, Skolem [151], and Gödel [46] are rederived.

§1. Let us consider a set S (generally infinite) of PC formulas. We say the set S is *consistent* iff it is possible to assign truth-values, T or F, to all of the elementary propositions (or *elementals*) from which the various formulas in S are built so that each formula in S has the value T according to the rules of PC.

Theorem 1: *In order that a system* S *of* PC *formulas be consistent, it is necessary and sufficient that every finite subsystem of* S *be consistent.*

Necessity is obvious. Since for finite S the theorem is trivial, it is enough with an application of induction to prove that if sufficiency holds for all systems of power less than $\aleph_\alpha$, then it holds for systems of power $\aleph_\alpha$. Let S be a system of power $\aleph_\alpha$, all of whose finite subsets are consistent.

We well-order the members of S in a transfinite sequence of the least possible type ω_α:

$$\mathsf{S} = \{A_0(\mathfrak{a}_0^1, ..., \mathfrak{a}_0^{n_0}), A_1(\mathfrak{a}_1^1, ..., \mathfrak{a}_1^{n_1}), ..., A_\omega(\mathfrak{a}_\omega^1, ..., \mathfrak{a}_\omega^{n_\omega}), ... \} .$$

(The $\mathfrak{a}_\nu^k$ are the elementary propositions from which the proposition A_ν is constructed.) Let us consider all possible initial segments of the sequence S. Since the type of the sequence is minimal, every initial segment has power less than $\aleph_\alpha$ and, consequently, is consistent by assumption. Thus, for any segment

$$\mathsf{S}^\lambda = \{A_\nu : \nu < \lambda\} \qquad (\lambda < \omega_\alpha) ,$$

it is possible to assign truth-values to the $\mathfrak{a}_\nu^k$ so that every formula in S^λ becomes true.

Every assignment of truth-values to the elementary propositions is called a *model*. Hence, for every segment S^λ there is at least one model $\mathfrak{M}^\lambda$ for which S^λ is true.

We now consider the sequence of models

$$\mathfrak{M}^0, \mathfrak{M}^1, ..., \mathfrak{M}^\omega, \mathfrak{M}^{\omega+1}, ... \tag{1}$$

corresponding to all possible initial segments of the sequence S. The elementary propositions $\mathfrak{a}_0^1, ..., \mathfrak{a}_0^{n_0}$ in the first member A_0 of S are given a system of truth-values by each model $\mathfrak{M}^\lambda$ $(\lambda > 0)$. Since only a finite number of different value-systems exist for the elementals $\mathfrak{a}_0^1, ..., \mathfrak{a}_0^{n_0}$, we can find a value-system which is assigned to these elementals by $\aleph_\alpha$ of the models in the sequence (1). We construct a formula A_0^* by negating those elementals with the value F in the chosen value-system and taking the conjunction of these propositions and the remaining elementals. The new formula has the form

$$A_0^* = \mathfrak{b}_0^1 \,\&\, \mathfrak{b}_0^2 \,\&\, ... \,\&\, \mathfrak{b}_0^{n_0} ,$$

where $\mathfrak{b}_0^k$ is either $\mathfrak{a}_0^k$ or $\neg\mathfrak{a}_0^k$.

Now consider the sequence

$$\mathsf{S}_1 = \{A_0^*, A_1, ..., A_\omega, ...\};$$

any initial segment S_1^λ of this sequence is consistent, for we can take as a model any of the old models $\mathfrak{M}^\mu$ ($\mu > \lambda$) that gives the elementals $a_0^1, \ldots, a_0^{n_0}$ the values required by the formula A_0^*. With segments of S_1 and their models in mind, we can define an analogous formula A_1^* and sequence

$$S_2 = \{A_0^*, A_1^*, A_2, \ldots, A_\omega, \ldots\}.$$

Suppose that for all $\kappa < \lambda$, we have already constructed a sequence S_κ (with $S_0 = S$) of the form

$$S_\kappa = \{A_\nu^* : \nu < \kappa\} \cup \{A_\kappa, A_{\kappa+1}, \ldots\},$$

all of whose initial segments are consistent. If λ is a successor, $\lambda = \lambda^- + 1$, then

$$S_{\lambda^-} = \{A_\nu^* : \nu < \lambda^-\} \cup \{A_{\lambda^-}, A_\lambda, \ldots\}.$$

We find $A_{\lambda^-}^*$ and conclude, as in the definition of S_1, that every segment of the sequence

$$S_\lambda = \{A_\nu^* : \nu < \lambda^-\} \cup \{A_{\lambda^-}^*, A_\lambda, \ldots\}$$

is consistent. If λ is a limit ordinal, then we put

$$S_\lambda = \{A_\nu^* : \nu < \lambda\} \cup \{A_\lambda, A_{\lambda+1}, \ldots\}.$$

We now wish to prove that every finite subsystem of S_λ is consistent, so we take some finite subset T of propositions from S_λ:

$$T = \{A_{\lambda_1}^*, A_{\lambda_2}^*, \ldots, A_{\lambda_k}^*, A_{\lambda_{k+1}}, \ldots, A_{\lambda_m}\},$$

where $\lambda_1 < \lambda_2 < \ldots < \lambda_m < \lambda$. Since λ is a limit ordinal, $\lambda_k + 1 < \lambda$. If we consider the sequence S_{λ_k+1}, we see that its segment $S_{\lambda_k+1}^{\lambda_m+1}$ contains all the formulas in T and is consistent, because every segment of a sequence S_κ for $\kappa < \lambda$ is consistent; thus, T is consistent. Therefore, every segment of S_λ is consistent by the induction hypothesis.

Finally, we form the set

$$S^* = \bigcup_{\lambda < \omega_\alpha} S_\lambda \sim S = \{A_\nu^* : \nu < \omega_\alpha\}.$$

Every finite set of the A^*_ν is consistent for it is included in a segment of one of the S_λ. Making use of this fact, we easily construct a model $\mathfrak{M}^*$ for the set S*. Indeed, each formula in S* has the form

$$\mathfrak{b}_\nu^1 \,\&\, \mathfrak{b}_\nu^2 \,\&\, \dots \,\&\, \mathfrak{b}_\nu^{n_\nu} \, ,$$

where $\mathfrak{b}_\nu^k$ is $\mathfrak{a}_\nu^k$ or $\neg\mathfrak{a}_\nu^k$. While the elemental $\mathfrak{a}_\nu^k$ can occur in various formulas in S*, by the finite consistency of S* it appears everywhere either negated or unnegated uniformly. In the first case we assign it the true-value T, and in the second, the value F. The model obtained in this fashion satisfies every proposition A^*_ν and, consequently, the original formula A_ν, as well. It follows that the system S is consistent. ∎

§ 2. As is well known, every formula of FOPL can be replaced with an equivalent formula in the Skolem normal form for satisfiability, that is, with a closed formula (*sentence*) of the form

$$(x_1) \dots (x_m)(\exists y_1) \dots (\exists y_n)\, \Psi(x_1, \dots, x_m, y_1, \dots, y_n) \, ,$$

where Ψ contains no quantifiers. For what follows, we assume in advance that every formula under consideration has already been put into normal form unless otherwise noted.

Let us consider a system S (infinite, in general) of formulas of FOPL and a set B of arbitrary objects. With respect to the sets S and B, we can construct other sets, called *configurations* on B. All these configurations are merely subsets of the universal set U, which is constructed as follows. Let

$$P_i(x), \neg P_i(x), Q_j(x,y), \neg Q_j(x,y), \dots$$

be all the elementary predicates occurring in members of S, along with their negations. We take one of these formulas and put some choice of elements of B – viewed as individual constant symbols – in place of the variables $x, y, \dots, z$. The expression so obtained is, by definition, an element of the universal set U. As we substitute all possible choices of objects from B in all the above formulas we get a set of such expressions, which is the set we call U.

As already mentioned, every subset of the universal set U is called a configuration on B. To distinguish between members of B and of U, we call the latter *terms*. Two terms in U are called *opposed* iff one arises from an atomic formula and the other from its negation by the substitution of the same elements of B. Grouping opposed formulas together, we decompose the set U into pairs.

A configuration is called *complete* iff it contains at least one term of each opposed pair in U. A configuration is called *consistent* iff it contains no two opposed elements. The set B is called a *domain* for S iff all the formulas of S can be satisfied by defining interpretations of the elementary predicates P_i, Q_j, etc., on the set B. The system S is called *consistent* iff it has a domain.

To decide whether B is a domain for S, we must first define interpretations of the elementary predicates, which is always done as follows: we pick a complete consistent configuration R on the set B and give the predicates the truth-value T for those values of the variables which give elements of R; for the remaining values of the variables in B we assign F. If some model so obtained satisfies the system S, then B is a domain for S. ([1]) On the other hand, if B is a domain for S, then for any interpretations satisfying S, we recover the corresponding configuration on B by taking the elementary predicates from S and their negations with those values of the variables that make them true.

Let us look at another interpretation of the elements of the universal set – one on which all further considerations will be based. We shall view the terms in U as distinct, ordinary elementary propositions of PC along with their negations, so that opposed terms become opposed propositions. From the terms in U we can now construct various PC propositions; let Φ be one of these. We consider the formula:

$$(\exists)\Phi = (\exists \boldsymbol{b}_1) \ldots (\exists \boldsymbol{b}_p)\Phi ,$$

where the $\boldsymbol{b}_i$ are all the elements of B that appear in the terms in U from which Φ is constructed. The formula $(\exists)\Phi$ is a sentence of FOPL. At the same time, Φ and $(\exists)\Phi$ are equivalent, for the former is satisfiable (in the sense of PC) iff the later is satisfiable (in the FOPL sense).

§3. So far we have not been assuming that the equality predicate occurs in the system S. Now we examine the general case, allowing equality to appear. For the results of the last section to be generally valid, we must correct a shortcoming in the definition of a consistent configuration: e.g., by the definition in §2 the configuration

$$P(\boldsymbol{a}, \boldsymbol{b}), \neg P(\boldsymbol{a}, \boldsymbol{c}), \boldsymbol{b} \approx \boldsymbol{c} ,$$

where $\boldsymbol{a}, \boldsymbol{b}, \boldsymbol{c}$ are distinct elements of B, is consistent, while according to the usual interpretation of equality it must be deemed inconsistent. The desired notion of consistency for a configuration R with equality on the set B can be obtained as follows. We establish a correspondence between B and any other set B' such that: (i) each element of B corresponds to one and only one element of B'; (ii) two different elements a, b of B correspond to the same element of B' iff the term $\boldsymbol{a} \approx \boldsymbol{b}$ belongs to R. Now we replace each element of B in every term in R with the corresponding element of B'. This results in a configuration R' on B'. The equality relation can occur in this configuration, but with the vital limitation that in R' there are no terms of the form $\boldsymbol{a}' \approx \boldsymbol{b}'$ for different elements $\boldsymbol{a}', \boldsymbol{b}'$ of B'. Consequently, the previous definition of consistency can be applied to R', whence we obtain the desired definition for R: the configuration R is called *consistent* iff the corresponding configuration R' is consistent in the sense of § 2.

A configuration with equality can be consistent from two points of view: (a) in the sense of §2, (b) in the sense just described. In the first case we say that the configuration is consistent with respect to *relative* equality, and in the second, with respect to *absolute* equality. It is obvious that an absolutely consistent configuration is also relatively consistent. The converse is not true, as the example at the beginning of this section shows. The following lemma gives conditions under which absolute and relative equalities are equivalent.

Lemma: *If the configuration R is consistent with respect to relative equality and satisfies the supplementary axioms:*

$$\left.\begin{array}{l}
(\boldsymbol{x})(\boldsymbol{x} \approx \boldsymbol{x}), \\
(\boldsymbol{x})(\boldsymbol{y})(\boldsymbol{x} \approx \boldsymbol{y} \rightarrow \boldsymbol{y} \approx \boldsymbol{x}), \\
(\boldsymbol{x})(\boldsymbol{y})(\boldsymbol{z})(\boldsymbol{x} \approx \boldsymbol{y} \,\&\, \boldsymbol{y} \approx \boldsymbol{z} \rightarrow \boldsymbol{x} \approx \boldsymbol{z}), \\
(\boldsymbol{x})(\boldsymbol{y})(\boldsymbol{P}(\boldsymbol{x}) \,\&\, \boldsymbol{x} \approx \boldsymbol{y} \rightarrow \boldsymbol{P}(\boldsymbol{y})), \\
\quad\vdots \\
(\boldsymbol{x})(\boldsymbol{y}) \ldots (\boldsymbol{z})(\boldsymbol{u})(\boldsymbol{v}) \ldots (\boldsymbol{w})(\boldsymbol{Q}(\boldsymbol{x}, \boldsymbol{y}, \ldots, \boldsymbol{z}) \,\&\, \boldsymbol{x} \approx \boldsymbol{u} \,\&\, \\
\qquad \&\, \boldsymbol{y} \approx \boldsymbol{v} \,\&\, \ldots \,\&\, \boldsymbol{z} \approx \boldsymbol{w} \rightarrow \boldsymbol{Q}(\boldsymbol{u}, \boldsymbol{v}, \ldots, \boldsymbol{w})), \\
\quad\vdots
\end{array}\right\} \quad \text{I}$$

where $\boldsymbol{P}(\boldsymbol{x}), \ldots, \boldsymbol{Q}(\boldsymbol{x}, \boldsymbol{y}, \ldots, \boldsymbol{z}), \ldots$ *are all the elementary predicates occurring in the system* S, *then R is also absolutely consistent.*

To prove this we divide the underlying set B into classes, gathering in each class those elements which are connected by the equality sign in R. By virtue of the system **I** such a decomposition is possible, and the resulting classes are pairwise disjoint. As the set B' we take the collection of all these classes; an element of B then corresponds to the class that contains it. It is easy to see that the conditions of the definition of absolute consistency are fulfilled, and, therefore, the configuration R is consistent with respect to absolute equality.∎

This lemma leads to the next theorem (an analogous theorem is found in [151] and [46]):

Theorem 2: *Let* S *be a system of formulas of* FOPL *with equality. We combine the systems* S *and* **I**, *and replace the equality sign throughout* S ∪ **I** *with an auxiliary predicate* $\boldsymbol{E}(\boldsymbol{x}, \boldsymbol{y})$. *The resulting system* S^r, *in which the equality sign does not occur, is equivalent to* S.

It is sufficient to show that if S^r is consistent, then S is absolutely consistent (i.e., is satisfiable by an absolutely consistent configuration), as the converse is obvious. Let R be a complete consistent configuration on a set B satisfying S^r. Let R^1 be the configuration obtained by replacing each term in R of the form $\boldsymbol{E}(\boldsymbol{a}, \boldsymbol{b})$ with the term $\boldsymbol{a} \approx \boldsymbol{b}$. The configuration R^1 is relatively consistent and satisfies S ∪ **I**. Hence, it is absolutely consistent by the lemma. The system S is thus seen to be absolutely consistent. ∎

Remark: In what follows we shall not replace the equality sign by a new predicate in passing from S to S^r, but shall be content with saying that S^r has *relativized* equality.

§4. The purpose of this section is to construct for every system S of FOPL formuals, a system of PC formulas equivalent to S with regard to being satisfiable. (²)

To begin with, suppose S is finite. Then we can replace it with a single sentence Φ of the form

$$(\boldsymbol{x}_1) \dots (\boldsymbol{x}_m)\,(\exists \boldsymbol{y}_1) \dots (\exists \boldsymbol{y}_n)\Psi\,.$$

Let B be some infinite set. We shall construct a subset B_ω of B and a set of propositions T_ω. T_ω will be consistent in the PC sense (and lead directly to a complete consistent configuration R_ω on B_ω satisfying Φ) iff Φ is consistent in the FOPL sense.

To do this, we first take an element $\boldsymbol{b}_0$ of B and put it in place of all the variables $\boldsymbol{x}_1, \dots, \boldsymbol{x}_m$ in Ψ. Then the expression so derived from Φ asserts the existence of certain $y_1, \dots, y_n$ that, with $\boldsymbol{b}_0$, are in the relation

$$\Psi(\boldsymbol{b}_0, \boldsymbol{b}_0, \ldots, \boldsymbol{b}_0, y_1, y_2, \ldots, y_n) .$$

As such y_i we choose n elements $\boldsymbol{b}_1, \ldots, \boldsymbol{b}_n$ of B different from $\boldsymbol{b}_0$. We put $\boldsymbol{B}_1 = \{\boldsymbol{b}_0, \boldsymbol{b}_1, \ldots, \boldsymbol{b}_n\}$ and let T_1 be the set consisting of the single expression

$$\Psi(\boldsymbol{b}_0, \boldsymbol{b}_0, \ldots, \boldsymbol{b}_0, \boldsymbol{b}_1, \boldsymbol{b}_2, \ldots, \boldsymbol{b}_n) .$$

Next we form all possible sequences (with repetitions) of m elements from B_1 and substitute each sequence not previously considered for the $\boldsymbol{x}_i$ in Ψ. For every such substitution of a new sequence, we choose n new elements as the y_i from the untouched portion of B, which remains infinite. The collection of all the elements so chosen together with the elements of B_1 is known as B_2. Each step in this selection procedure produces an expression of the form

$$\Psi(\boldsymbol{b}_{i_1}, \ldots, \boldsymbol{b}_{i_m}, \boldsymbol{b}_{j_1}, \ldots, \boldsymbol{b}_{j_n}) ,$$

where $\boldsymbol{b}_{i_1}, \ldots, \boldsymbol{b}_{i_m}$ are the elements of B_1 substituted for the $\boldsymbol{x}_i$ in Ψ, and $\boldsymbol{b}_{j_1}, \ldots, \boldsymbol{b}_{j_n}$ are the corresponding new elements chosen. Let T_2 be the collection of all the expressions so produced along with those in T_1. The process whereby we obtained B_2, T_2 from B_1, T_1 can be employed under wider circumstances and will be called an *application* of Φ. In particular, we can apply Φ to B_2, T_2 and obtain sets B_3, T_3, etc. By iteration we get two infinite sequences of sets:

$$B_1, B_2, \ldots, B_k, \ldots,$$

$$\mathrm{T}_1, \mathrm{T}_2, \ldots, \mathrm{T}_k, \ldots .$$

We let

$$B_\omega = B_1 \cup B_2 \cup \ldots \cup B_k \cup \ldots,$$

$$\mathrm{T}_\omega = \mathrm{T}_1 \cup \mathrm{T}_2 \cup \ldots \cup \mathrm{T}_k \cup \ldots .$$

Applying Φ to $B_\omega, \mathrm{T}_\omega$ produces no new elements, as all possible m-tuples have already been considered.

Let us take a closer look at T_ω. It consists of expressions of the form:

$$\Psi(\boldsymbol{c}_1, \ldots, \boldsymbol{c}_m, \boldsymbol{d}_1, \ldots, \boldsymbol{d}_n) ,$$

where the $\boldsymbol{c}_i$, $\boldsymbol{d}_i$ are elements of B_ω. Every such expression is a formula of PC built up from elementary propositions having the form of elementary predicates from Φ with members of B_ω in their argument places. Thus, in the terminology of § 2, T_ω is a system of PC propositions whose elementals are members of the universal set U_ω associated with B_ω and Φ. If T_ω is a consistent system of propositions, then any model satisfying T_ω in the sense of § 1 yields a consistent configuration on B_ω, to wit, the subset of U_ω consisting of those elementals from T_ω assigned the truth-value T and the negations of those assigned F by the model; if this configuration is extended arbitrarily to be complete as well as consistent, then the resulting configuration R_ω satisfies Φ in the sense of § 2. Thus we have proved that if T_ω is PC-consistent, then Φ is consistent (and, in particular, has B_ω as a domain). The validity of the converse is seen from the following remarks.

Every formula in T_ω has the form:

$$\Psi(\boldsymbol{c}_1, ..., \boldsymbol{c}_m, \boldsymbol{d}_1, ..., \boldsymbol{d}_n) .$$

Let us consider the set $(\exists)\mathrm{T}_\omega$ of the corresponding existentially closed FOPL formulas of the form:

$$(\exists \boldsymbol{c}_1) ... (\exists \boldsymbol{c}_m)(\exists \boldsymbol{d}_1) ... (\exists \boldsymbol{d}_n) \Psi(\boldsymbol{c}_1, ..., \boldsymbol{c}_m, \boldsymbol{d}_1, ..., \boldsymbol{d}_n) .$$

Every sentence in $(\exists)\mathrm{T}$ is a consequence of Φ (cf. [46]). If Φ is assumed to be consistent, then every finite subset of $(\exists)\mathrm{T}_\omega$ is satisfiable. As in § 2, this implies the PC-satisfiability of every finite subset of T_ω. By Theorem 1 the whole system T_ω is consistent.

We have proved the equivalence of the FOPL formula Φ and the system T_ω of PC formulas. ■

If the system S contains an infinite number of sentences, then the iteration is performed in the following manner. Again we choose an element $\boldsymbol{b}_0$ of an infinite set B, but now we apply every formula in S to this single element. Each formula demands the choice of a certain number of elements from B. The set B_1 is the collection of all these elements; we put $\boldsymbol{b}_0$ in B_1, too. The set T_1 of propositions is formed analogously. Now the sentences of S are applied successively to B_1, T_1. The application of each sentence Φ_ν produces sets B_2^ν, T_2^ν; we let B_2, T_2 be the corresponding unions of all of these sets, etc. As in the finite case, we obtain sets B_ω, T_ω. The arguments for the earlier case can be carried over to the general without change. ■

Remark: In the case of a finite system S a countable set can be chosen for B. If S is infinite, then a set of the same power as S can be taken. From this we obtain the following generalization of Löwenheim's theorem:

Every domain for an infinite system S *of* FOPL *sentences includes a subdomain whose power does not exceed that of* S. ∎

§ **5**. In many situations one must begin the iterated application of a sentence not with a single element $\boldsymbol{b}_0$, but with a whole set B_0.

E.g., suppose we have a set B_0, a sentence

$$\Phi = (\boldsymbol{x}_1) \dots (\boldsymbol{x}_m)(\exists \boldsymbol{y}_1) \dots (\exists \boldsymbol{y}_n)\, \Psi\,,$$

and a configuration R_0 on B_0 that does not satisfy Φ. We can ask whether it is possible to extend the configuration R_0 to a complete consistent configuration on a superset of B_0 that does satisfy Φ.

To solve this problem we can take an infinite set B from which to choose new elements and apply Φ iteratively, beginning with B_0, R_0. As before, we get two sequences of sets:

$$B_0, B_1, B_2, \dots, B_k, \dots,$$

$$R_0, \mathrm{T}_1, \mathrm{T}_2, \dots, \mathrm{T}_k, \dots;$$

we let

$$B_\omega = B_0 \cup B_1 \cup \dots \cup B_k \cup \dots\,,$$

$$\mathrm{T}_\omega = R_0 \cup \mathrm{T}_1 \cup \dots \cup \mathrm{T}_k \cup \dots\,.$$

If T_ω is consistent, then it leads to a configuration R_ω on B_ω extending R_0 and satisfying Φ. If, however, T_ω is inconsistent, then the problem has no solution. The proof is essentially the same as that in § 4.

The sets B_ω and T_ω, and the application of a sentence Φ in general, can be represented more intuitively with the help of certain finite sequences of divers ranks. The elements of B_0 are called Φ-*sequences* of rank 0. An $(m+1)$-tuple of the form:

$$\langle \boldsymbol{b}_1, \boldsymbol{b}_2, \dots, \boldsymbol{b}_m, j \rangle\,,$$

where the $\boldsymbol{b}_i$ belong to B_0 and $1 \leqslant j \leqslant n$, is called a Φ-sequence of rank 1. In general, if $\sigma_1, \sigma_2, \dots, \sigma_m$ are Φ-sequences, and the greatest of their ranks is k, then

$$\langle \sigma_1, \sigma_2, ..., \sigma_m, j\rangle \qquad (1 \leqslant j \leqslant n)$$

is a Φ-sequence of rank $k+1$. Those elements of B_0 occurring at any level in the formation of a particular Φ-sequence are called the *ground elements* of the sequence. It is clear that each Φ-sequence has only a finite number of ground elements.

We now take advantage of the Φ-sequences to describe the iterated application process. After substituting arbitrary elements $\boldsymbol{b}_1, \boldsymbol{b}_2, ..., \boldsymbol{b}_m$ of B_0 for $\boldsymbol{x}_1, \boldsymbol{x}_2, ..., \boldsymbol{x}_m$ in Ψ, instead of choosing n elements for the $\boldsymbol{y}_j$ from an arbitrary set we can take the Φ-sequences:

$$\langle \boldsymbol{b}_1, ..., \boldsymbol{b}_m, 1\rangle, \langle \boldsymbol{b}_1, ..., \boldsymbol{b}_m, 2\rangle, ..., \langle \boldsymbol{b}_1, ..., \boldsymbol{b}_m, n\rangle .$$

In the general case, if Φ-sequences $\sigma_1, \sigma_2, ..., \sigma_m$, the maximum of whose ranks is k, are substituted for the $\boldsymbol{x}_i$, then as the elements whose existence is asserted we take the rank $k+1$ Φ-sequences:

$$\langle \sigma_1, ..., \sigma_m, 1\rangle, \langle \sigma_1, ..., \sigma_m, 2\rangle, ..., \langle \sigma_1, ..., \sigma_m, n\rangle .$$

Selecting elements in this way, we find that B_k coincides with the set of all Φ-sequences of rank $\leqslant k$, B_ω coincides with the set of all Φ-sequences, and T_ω is the set of all expressions of the form:

$$\Psi(\sigma_1, ..., \sigma_m, \langle \sigma_1, ..., \sigma_m, 1\rangle, ..., \langle \sigma_1, ..., \sigma_m, n\rangle) ,$$

where the σ_i range independently over B_ω.

When the sentence Φ under consideration has the form:

$$(\boldsymbol{x})(\exists \boldsymbol{y}_1) ... (\exists \boldsymbol{y}_n) \Psi ,$$

we can get by with simpler Φ-sequences. For when we substitute an element $\boldsymbol{b}_0$ of B_0 for $\boldsymbol{x}$, we can take

$$\langle \boldsymbol{b}_0, 1\rangle, \langle \boldsymbol{b}_0, 2\rangle, ..., \langle \boldsymbol{b}_0, n\rangle$$

as the new elements. Replacing $\boldsymbol{x}$ with one of these in turn, we can write the new Φ-sequences in the simpler form:

$$\langle \boldsymbol{b}_0, j, 1\rangle, \langle \boldsymbol{b}_0, j, 2\rangle, ..., \langle \boldsymbol{b}_0, j, n\rangle ,$$

and so on. In this case, the set of propositions is just

$$\mathrm{T}_{\omega} = \{\Psi(\sigma, \langle\sigma, 1\rangle, ..., \langle\sigma, n\rangle) \colon \sigma \text{ is a } \Phi\text{-sequence}\}\ .$$

§6. In this section we prove the following theorem:

Theorem 3: *Let* S *be a system of* FOPL *formulas with equality; let R be a configuration on an infinite set B such that B and R form a model for* S *with absolute equality.* (3) *Then there are a set $B_{\omega} \supset B$ and a configuration $R_{\omega} \supset R$ on it that form a model for* S. *In other words, every infinite model for* S *has at least one proper extension that also satisfies* S.

Assuming S to be finite for the time being, we can replace it with an equivalent sentence

$$\Phi = (\boldsymbol{x}_1) \dots (\boldsymbol{x}_m)(\exists \boldsymbol{y}_1) \dots (\exists \boldsymbol{y}_n)\Psi(\boldsymbol{x}_1, ..., \boldsymbol{x}_m, \boldsymbol{y}_1, ..., \boldsymbol{y}_n)\ ,$$

in which equality is relativized according to the methods of §3. Note that the complete, absolutely consistent configuration R satisfies Φ, as well as S, by the remarks at the end of § 3. Let $\boldsymbol{b}^*$ be any element not contained in B, and put

$$B_0 = B \cup \{\boldsymbol{b}^*\}\ ,$$

$$\mathrm{T}_0 = R \cup \{\boldsymbol{b}^* \approx \boldsymbol{b}^*\} \cup \{\boldsymbol{b}^* \not\approx \boldsymbol{b} \colon \boldsymbol{b} \in B\}\ .$$

T_0 is a configuration on B_0, but, in general, no longer satisfies Φ. We shall show it is possible to extend B_0, T_0 to a model for Φ. Moreover, this new model will become a model for the original system S when elements connected by the equality sign are identified; it will then be an extension of B, R be virtue of the extra expressions in R', and the extension will be proper since $\boldsymbol{b}^*$ cannot coincide with any of the old elements because of the added conditions in T_0.

Starting with B_0, T_0, we apply Φ iteratively, using the Φ-sequences of the previous section as new elements. As usual we let

$$B_{\omega} = B_0 \cup B_1 \cup \dots \cup B_k \cup \dots\ ,$$

$$\mathrm{T}_{\omega} = \mathrm{T}_0 \cup \mathrm{T}_1 \cup \dots \cup \mathrm{T}_k \cup \dots\ .$$

If we succeed in proving that $\mathbf{T}_\omega$ is PC-consistent, then it will yield a configuration R_ω on B_ω extending $\mathbf{T}_0$ and satisfying Φ. Thus, by Theorem 1 it is sufficient to show that every finite subset of $\mathbf{T}_\omega$ is consistent.

Let V be a finite subset of $\mathbf{T}_\omega$. The members of V are PC formulas whose elementals are elementary predicates from Φ with Φ-sequences in B_ω as arguments. Since V is finite, in the construction of all of its members only a finite number of Φ-sequences are involved; let k be the greatest of their ranks. Furthermore, these Φ-sequences have only a finite number of ground elements all together. Let D be the finite subset of B_k consisting of all Φ-sequences on B_0 of rank $\leqslant k$ whose ground elements are included among those mentioned. In particular, D contains all the Φ-sequences occurring in V, as well as all their ground elements.

We define a map from D onto a subset of the domain B in the following way:

(a) each element of $D \cap B$ corresponds to itself;

(b) if $\boldsymbol{b}^*$ belongs to D, then it is mapped onto any element $\overline{\boldsymbol{b}}^*$ of $B \sim D$, which is non-empty because B is infinite;

(c) if $0 \leqslant l < k$ and the map has been defined on all members of D of rank $\leqslant l$, and if

$$\langle \sigma_1, ..., \sigma_m, 1\rangle, \langle \sigma_1, ..., \sigma_m, 2\rangle, ..., \langle \sigma_1, ..., \sigma_m, n\rangle \tag{2}$$

are Φ-sequences of rank $l+1$ belonging to D, and $\sigma_1, \sigma_2, ..., \sigma_m$ are mapped onto $\boldsymbol{b}_1, \boldsymbol{b}_2, ..., \boldsymbol{b}_m$, respectively, then the sequences (2) are mapped respectively onto any elements $\boldsymbol{c}_1, \boldsymbol{c}_2, ..., \boldsymbol{c}_n$ of B such that

$$\Psi(\boldsymbol{b}_1, ..., \boldsymbol{b}_m, \boldsymbol{c}_1, ..., \boldsymbol{c}_n)$$

is satisfied by R (such $\boldsymbol{c}_j$ must exist because B, R are a model for Φ).

Replacing the Φ-sequences occurring in V by the corresponding elements of B according to this map transforms the propositions in V into ones whose elementals belong to the universal set associated with B and Φ. Indeed, $\boldsymbol{b}^* \approx \boldsymbol{b}^*$ becomes $\overline{\boldsymbol{b}}^* \approx \overline{\boldsymbol{b}}^*$, an expression of the form $\boldsymbol{b}^* \not\approx \boldsymbol{b}$ becomes $\overline{\boldsymbol{b}}^* \not\approx \boldsymbol{b}$ ($\boldsymbol{b} \in B \cap D$), while members of R are unchanged; finally, a proposition of the form:

$$\Psi(\sigma_1, ..., \sigma_m, \langle \sigma_1, ..., \sigma_m, 1\rangle, ..., \langle \sigma_1, ..., \sigma_m, n\rangle)$$

becomes

$$\Psi(\boldsymbol{b}_1, ..., \boldsymbol{b}_m, \boldsymbol{c}_1, ..., \boldsymbol{c}_n)$$

for certain $\boldsymbol{b}_i$, $\boldsymbol{c}_j$ in B. Moreover, R satisfies each of these transformed propositions by the definition (a)–(c). Thus, the set of transformed propositions is consistent, and R induces a PC model satisfying the original system V, as well. ■

In this and the preceding section we have assumed S is finite for simplicity, as this permits us to replace it with a single sentence Φ. If S is infinite, we may no longer be able to do this. All the arguments, however, remain valid if we make the following changes: (i) the iteration process is carried out in its general form, as described at the end of §4; (ii) in representing the new elements as finite sequences, we must include in each new sequence (rank $\geqslant 1$) an index for the formula whose application produced it. The remaining details can be repeated almost literally.

For his manifold valuable advice, my heartfelt thanks are hereby expressed to Prof. A.N. Kolmogorov.

NOTES

(1) If B is a domain for S, and R is a complete consistent configuration on B satisfying S, then we shall say that B, R are themselves a *model* satisfying S, or simply, for S.

(2) The discerning reader may note that this construction, together with Theorem 1, yields the FOPL analogue to Theorem 1: the so-called compactness or local theorem for FOPL (admitting equality and arbitrary sets of predicate and individual constant symbols). In what follows, there is some difficulty in replacing a set of sentences with the set of corresponding sentences in Skolem normal form – as well as in dealing with free variables (or individual constants) – in the case of an infinite set of formulas (or sentences). The reader is also directed to [R1].

(3) I.e., for all $\boldsymbol{b}, \boldsymbol{c}$ in B, the term $\boldsymbol{b} \approx \boldsymbol{c}$ belongs to R iff $\boldsymbol{b} = \boldsymbol{c}$, in fact. Every model in the statement of Theorem 3 is assumed to have absolute equality.

CHAPTER 2

A GENERAL METHOD FOR OBTAINING LOCAL THEOREMS IN GROUP THEORY

In algebra, and especially in group theory, there are quite a few theorems of the form: if a certain property C holds for all subalgebras of a given algebra (group, ring, etc.) generated by a finite set of elements of this algebra, then C holds for the whole algebra. The purpose of the present note is to show that such propositions are not specifically algebraic in the majority of cases and can be obtained as immediate consequences of a general proposition of mathematical logic. This general approach to local theorems does not, of course, give the solutions to any difficult algebraic problems. In many cases, however, it makes the algebraic proofs of these theorems redundant, and sometimes permits one to see immediately that a theorem holds under somewhat broader assumptions. Thus, the propositions introduced below on solvable groups and groups with Sylow sequences were proved by S.N. Černikov only for the case of locally finite groups; Theorem 4 on the extension lattice isomorphisms was first established by Baer for countable groups, and only recently did L. Sadovskiĭ give a proof valid for uncountable groups as well.

The proposition of mathematical logic that interests us can be formulated in the following manner.

Let there be given a set $\{\boldsymbol{F}_i(\boldsymbol{x}_1, \ldots, \boldsymbol{x}_{n_i})\colon i \in I\}$ of function (predicate) symbols. These symbols represent functions F_i defined on a set M of arbitrary elements and taking at most two values, which will be denoted by 0, 1. An equation $F_i = 1$ is abbreviated as F_i, and equation $F_i = 0$ as $\sim F_i$; we write these symbolically as $\boldsymbol{F}_i$ and $\neg \boldsymbol{F}_i$. Besides the symbols for the functions, a set M_0 of individual symbols for certain elements of M will be considered as given. Both sets of function and individual symbols may be infinite. A property of the functions F_i is said to be *formulatable* in first-order predicate logic (FOPL) – i.e., Hilbert's restricted calculus – iff it is expressible by means of the notions "and", "or", "not", "if ... then", "equals", "for every element x in M", and "there exists an element x in M such that" (symbolized by &, $\vee$, $\neg$, $\rightarrow$, $\approx$, $(\boldsymbol{x})$, $(\exists \boldsymbol{x})$, respectively), in terms of a finite number of the $\boldsymbol{F}_i$ and a finite number of the individual constant symbols from M_0. An arbitrary system S of such FOPL sentences (or axioms) is called *consistent* iff there exists a set M on which functions F_i can be defined and individual elements chosen so that all sentences in S are satisfied.

Basic theorem (compactness or local theorem for FOPL): *If every finite part of an infinite system of* FOPL *sentences is consistent, then the entire system is consistent.* ■ ([1])

Now we consider a number of applications of this principle in the theory of groups.

§1. We shall call a property of groups *elementary* and *hereditary* (E-H) iff it satisfies the two requirements:

(i) It is a conjunction of properties, each of which is expressible in a finite manner by means of the words "and", "or", "not", "equals", "product of elements", "for every element of the group", etc. – in other words, properties formulatable in FOPL;

(ii) Every subgroup of a group having the property must itself have the property. ([2])

Examples of E-H group properties are: "the group is abelian", "the orders of all elements of the group are contained in a fixed finite set of numbers", "the orders of all elements of the group do not belong to a fixed set of numbers", "the order of the group does not exceed a fixed number", etc.

Let $E_1, E_2, \ldots, E_k$ be a finite number of E-H group properties. A group $\mathfrak{G}$ is of *type* $[E_1, E_2, \ldots, E_k]$ iff $\mathfrak{G}$ possesses a normal series of length k, $\mathfrak{G} = \mathfrak{G}_0 \supseteq \mathfrak{G}_1 \supseteq \ldots \supseteq \mathfrak{G}_k = \{e\}$, with the factor groups $\mathfrak{G}/\mathfrak{G}_1, \ldots, \mathfrak{G}_{k-1}/\mathfrak{G}_k$ having the properties $E_1, \ldots, E_k$, respectively.

Theorem 1: *In order that the group* $\mathfrak{G}$ *be of type* $[E_1, \ldots, E_k]$, *it is necessary and sufficient that every finitely generated subgroup of* $\mathfrak{G}$ *be of this type.*

Proof: In order to use the compactness theorem, we have to formulate the concept of a given type in FOPL. Toward this end we introduce symbols $\boldsymbol{x}_g^i$, where g runs over all elements of the given group $\mathfrak{G}$ and $i = 1, 2, \ldots, k$. These will be the individual symbols in the projected axiom system S. The basic predicate symbols in this system are $\boldsymbol{F}_1(\boldsymbol{x}, \boldsymbol{y}, \boldsymbol{z}), \ldots, \boldsymbol{F}_k(\boldsymbol{x}, \boldsymbol{y}, \boldsymbol{z})$. From the semantical point of view, the individual variables $\boldsymbol{x}^i, \boldsymbol{y}^i$, etc. and constants $\boldsymbol{x}_g^i$ will designate elements of the factor group $\mathfrak{G}_{i-1}/\mathfrak{G}_i$, and the relation $F_i(x^i, y^i, z^i)$ is given by the equation $x^i \cdot y^i = z^i$. One part of the collection S consists of axioms stating that the elements x^i form a group relative to F_i, and that this group has the property E_i ($i = 1, \ldots, k$). The other part of S consists of the sentences

$$F_1(x^1_g, x^1_h, x^1_{gh}),$$

$$x^{i-1}_g \approx x^{i-1}_h \approx x^{i-1}_e \to F_i(x^i_g, x^i_h, x^i_{gh}) \qquad (i = 2, ..., k),$$

$$x^{k-1}_g \approx x^{k-1}_h \approx x^{k-1}_e \to x^k_g \not\approx x^k_h \qquad (g \neq h),$$

where e is the identity of the group $\mathfrak{G}$, while g and h vary independently over the whole group. These last axioms mean that the group $\mathfrak{X}_i$ generated by the distinguished elements x^i_g is a homomorphic image of $\mathfrak{G}_{i-1}$ for $i = 1, ..., k$, and that the group $\mathfrak{X}_k$ is isomorphic to $\mathfrak{G}_{k-1}$. In order to suggest the dependence of this collection of sentences on the group $\mathfrak{G}$, we denote it by $S_{\mathfrak{G}}$. It is easy to see that the consistency of $S_{\mathfrak{G}}$ means that $\mathfrak{G}$ is of the type $[E_1, ..., E_k]$.

Theorem 1 now becomes obvious. In fact, if $\mathfrak{H}$ is a subgroup of $\mathfrak{G}$, then $S_{\mathfrak{H}}$ is included in $S_{\mathfrak{G}}$, and if $S_{\mathfrak{G}}$ is consistent, then so is $S_{\mathfrak{H}}$. That is, every subgroup of a group of type $[E_1, ..., E_k]$ is also of this type. Conversely, let every finitely generated subgroup of a group $\mathfrak{G}$ be of type $[E_1, ..., E_k]$. Let S^* be a finite subset of $S_{\mathfrak{G}}$. Only a finite number of the individual symbols x^i_g occur in S^*. The lower indices of these symbols are elements of $\mathfrak{G}$ and generate a finitely generated subgroup $\mathfrak{H}$. Clearly, S^* is included in $S_{\mathfrak{H}}$. Since $\mathfrak{H}$ is of type $[E_1, ..., E_k]$ by assumption, $S_{\mathfrak{H}}$ is consistent, and so is S^*. Thus every finite part of $S_{\mathfrak{G}}$ is consistent, and $\mathfrak{G}$ is of type $[E_1, ..., E_k]$. ∎

We mention a few particular cases of this theorem. First of all, let $E_1, ..., E_k$ all be the property of being abelian. Then the groups of type $[E_1, ..., E_k]$ are k-step solvable groups, and Theorem 1 becomes the proposition: *the group $\mathfrak{G}$ has a solvable normal series of length k iff every finitely generated subgroup of $\mathfrak{G}$ has a solvable normal series of length k* [15]. ∎

Suppose now that $k = 2$, E_1 is "the group contains no element of order p_0", and E_2 is "the group contains no element with its order in $P \sim \{p_0\}$", where P is the set of all prime numbers and $p_0 \in P$. Applying Theorem 1, we immediately find that *locally special groups are direct products of their Sylow subgroups* [14]. ∎

Now suppose $k = 2$, E_1 is "the order of the group does not exceed n", and E_2 is "the group is abelian". Then Theorem 1 reduces to the proposition: *a group $\mathfrak{G}$ has an abelian normal divisor with index $\leqslant n$ iff every finitely generated subgroup of $\mathfrak{G}$ has an abelian normal divisor with index $\leqslant n$.* ∎

According to Jordan [68], every finite group admitting an isomorphic representation by matrices of order r over the field of complex numbers includes an abelian normal divisor with index $\leqslant n$, where n depends only on r. Now let $\mathfrak{G}$ be an infinite periodic group of matrices of order r. Every finitely generated subgroups of $\mathfrak{G}$ is finite, and so, by Jordan's result, includes an abelian normal divisor of index $\leqslant n(r)$. Hence $\mathfrak{G}$ has an abelian normal divisor of finite index; this brings us to Schur's theorem: *every periodic group of matrices over a field of characteristic* 0 *has an abelian normal divisor of finite index* [146]. ■

§ 2. As a second example of the application of this general method we consider the concept of the so-called Sylow sequence. By definition a collection $\mathcal{N}$ of normal divisors of a periodic group $\mathfrak{G}$ is a *Sylow sequence* for $\mathfrak{G}$ iff it satisfies the conditions:

(I) Of any two subgroups in $\mathcal{N}$, one includes the other. The whole group $\mathfrak{G}$ $\mathfrak{G}$ and the identity subgroup are in $\mathcal{N}$.

(II) If $\mathfrak{N}_1, \mathfrak{N}_2 \in \mathcal{N}$, and $\mathfrak{N}_1 \supset \mathfrak{N}_2$, and the factor group $\mathfrak{N}_1/\mathfrak{N}_2$ contains elements whose orders are relatively prime, then there is a group $\mathfrak{N}_3$ in $\mathcal{N}$ which is intermediate: $\mathfrak{N}_1 \supset \mathfrak{N}_3 \supset \mathfrak{N}_2$.

(III) If $\mathfrak{N}_1, \mathfrak{N}_2, \mathfrak{N}_3 \in \mathcal{N}$, and $\mathfrak{N}_1 \supset \mathfrak{N}_2 \supset \mathfrak{N}_3$, then the order of any element of $\mathfrak{N}_1/\mathfrak{N}_2$ is relatively prime to the order of any element of $\mathfrak{N}_2/\mathfrak{N}_3$.

We now show that the property "the group has a Sylow sequence" is elementary; we shall obtain the result of Černikov: if every finite subgroup of a locally finite group $\mathfrak{G}$ has a Sylow sequence, then so does $\mathfrak{G}$ itself [16].

As the basic predicate symbols of the axiom system $S_{\mathfrak{G}}$ we take $>$ and $A_g(x)$ ($g \in \mathfrak{G}$). First of all, we put axioms in $S_{\mathfrak{G}}$ that state that the elements x of the basic domain M form a linearly ordered system with respect to $>$. Let $\boldsymbol{x}_0$ and $\boldsymbol{x}_1$ be individual constant symbols. The predicate $A_g(x)$ is intended to mean that g belongs to the set x. We now rewrite the conditions for a Sylow sequence in the language of FOPL (universal quantifiers governing the whole formula have been dropped for clarity):

(0^a) For all g, h in $\mathfrak{G}$: $A_g(x) \mathbin{\&} A_h(x) \to A_h(x) \to A_{gh^{-1}}(x)$; (i.e., x is a subgroup).

(0^b) For all g, h in $\mathfrak{G}$: $A_g(x) \to A_{hgh^{-1}}(x)$; (i.e., x is invariant).

(1^a) For all $g \neq e$, the identity element of $\mathfrak{G}$: $\neg A_g(\boldsymbol{x}_0)$, $A_g(\boldsymbol{x}_1)$; $A_e(\boldsymbol{x}_0)$, $A_e(\boldsymbol{x}_1)$; (i.e., x_0 is the subgroup $\{e\}$, x_1 is $\mathfrak{G}$).

(1^b) For all g in $\mathfrak{G}$: $x > y \mathbin{\&} A_g(y) \to A_g(x)$; (i.e., if $x > y$, then x includes y).

(2) For all g, h in $\mathfrak{G}$ and all distinct primes p, q: $x > y \mathbin{\&} A_g(x) \mathbin{\&} \neg A_g(y) \mathbin{\&}$ $\mathbin{\&} A_h(x) \mathbin{\&} \neg A_h(y) \mathbin{\&} A_{gp}(y) \mathbin{\&} A_{hq}(y) \to (\exists z)(x > z > y \mathbin{\&} ((A_g(z) \mathbin{\&}$ $\mathbin{\&} \neg A_h(z)) \vee (\neg A_g(z) \mathbin{\&} A_h(z))))$; (this clearly corresponds to (II) above).

(3) For all g, h in $\mathfrak{G}$ and all primes p: $\boldsymbol{x} > \boldsymbol{y} > \boldsymbol{z}$ & $A_g(\boldsymbol{x})$ & $\neg A_g(\boldsymbol{y})$ & & $A_h(\boldsymbol{y})$ & $\neg A_h(\boldsymbol{z})$ & $A_{g^p}(\boldsymbol{y}) \to \neg A_{h^p}(\boldsymbol{z})$; (i.e., if a prime power of an element in x, but not in y, lies in y, then this power of any element in y, but not in z, will not lie in z).

Thus, we see that the property "the group has a Sylow sequence of normal subgroups" is a conjunction of properties formulatable in FOPL.

Therefore, by use of the basic theorem we obtain the result of Černikov cited above. But it is easy to see that the notion of Sylow sequence makes sense not only for locally finite groups, but also for non-periodic groups. Furthermore, local finiteness was never used in our proof. Consequently, we can formulate a more general proposition: *if every finitely generated subgroup of a group* $\mathfrak{G}$ *has a Sylow sequence, then so does* $\mathfrak{G}$. ∎

One can go further in the construction of generalizations of the Sylow sequence for which the local theorem remains true. E.g., we can proceed as follows. We linearly order the set P of all prime numbers in an arbitrary fashion, letting $>_Q$ denote the irreflexive order relation so obtained, which we call a Q-order. We call a collection $\mathcal{H}$ of subgroups of a given group $\mathfrak{G}$ a *Sylow Q-sequence* iff it fulfills the following conditions:

(I′) Of any two subgroups in $\mathcal{H}$, one includes the other; also, $\mathfrak{G}$, $\{e\} \in \mathcal{H}$.

(II′) If $\mathfrak{H}_1, \mathfrak{H}_2 \in \mathcal{H}$, $\mathfrak{H}_1 \supset \mathfrak{H}_2$, and $g, h \in \mathfrak{H}_1 \sim \mathfrak{H}_2$, while $g^m, h^n \in \mathfrak{H}_1$ with $(m, n) = 1$, then there is a subgroup $\mathfrak{H}_3$ in $\mathcal{H}$ that lies properly between $\mathfrak{H}_1$ and $\mathfrak{H}_2$.

(III′) Suppose $\mathfrak{H}_1, \mathfrak{H}_2, \mathfrak{H}_3 \in \mathcal{H}$ with $\mathfrak{H}_1 \supset \mathfrak{H}_2 \supset \mathfrak{H}_3$, and let $g \in \mathfrak{H}_1 \sim \mathfrak{H}_2$, $h \in \mathfrak{H}_2 \sim \mathfrak{H}_3$. If p and q are primes, and $g^p \in \mathfrak{H}_2$, $h^q \in \mathfrak{H}_3$, then $p >_Q q$.

It is easy to convince one's self that, as in the case of Sylow sequences, the property "the group has a Q-sequence" is elementary. Applying the basic theorem we immediately obtain:

Theorem 2: *If every finitely generated subgroup of a group* $\mathfrak{G}$ *has a* Q-*sequence, then so does* $\mathfrak{G}$. ∎

Without leaving this round of ideas, we could indicate many other possibilities. We shall pause at only one of these.

We call a collection $\mathcal{N}$ of normal subgroups of a group $\mathfrak{G}$ a *commutator sequence* iff it satisfies the following conditions:

(A) $\mathcal{N}$ is linearly ordered by inclusion and contains $\mathfrak{G}$ and $\{e\}$, the identity subgroup.

(B) Let g, h be any two distinct elements of $\mathfrak{G}$. Then there is a normal subgroup $\mathfrak{N}$ in $\mathcal{N}$ such that either $\mathfrak{N}$ contains one of g, h, but not the other, or $\mathfrak{N}$ contains neither g nor h, but contains their commutator $ghg^{-1}h^{-1}$.

The property "the group has a commutator sequence" is easily seen to be an elementary property. Consequently, the following theorem holds:

Theorem 3: *If every finitely generated subgroup of a group $\mathfrak{G}$ has a commutator sequence, then so does $\mathfrak{G}$.* ■

This theorem implies several corollaries, E.g., let $\mathfrak{G}$ be a solvable group; then the series of successive commutator subgroups $\mathfrak{G} \supseteq \mathfrak{G}' \supseteq \ldots \supseteq \mathfrak{G}^{(n)} = \{e\}$ obviously satisfies the conditions for a commutator sequence for $\mathfrak{G}$. Thus, a locally solvable group always has a commutator sequence.

A second example: let us call a group *generalized-solvable* iff the intersection of the terms of its higher commutator series – transfinite, in general – is the identity subgroup. Clearly, this series will have the properties of a commutator sequence. Consequently, Theorem 3 holds for these groups. In particular, it is well known that finitely generated free groups are generalized-solvable. Applying Theorem 3, we conclude that *every locally free group has a commutator sequence and is, therefore, not simple.* ■

That locally free groups are not simple was first discovered by D. Fuks-Rabinovič with the aid of highly specialized calculations.

§3. As a last application of the general method we consider a theorem of Baer connected with the theory of lattices. Two groups $\mathfrak{G}$, $\mathfrak{H}$ are called *lattice isomorphic* iff it is possible to establish between their sets of subgroups a 1–1 correspondence under which the intersection of two subgroups of $\mathfrak{G}$ is mapped onto the intersection of the two corresponding subgroups of $\mathfrak{H}$. Baer's theorem can be formulated as follows:

Theorem 4: *If a lattice isomorphism between two groups $\mathfrak{G}$, $\mathfrak{H}$ is such that when restricted to any finitely generated subgroup, the restriction is induced by an ordinary group isomorphism from this subgroup into $\mathfrak{H}$, then the given lattice isomorphism is induced by a group isomorphism between $\mathfrak{G}$ and $\mathfrak{H}$.*

To prove this, we take a predicate symbol $\boldsymbol{A}(\boldsymbol{x}, \boldsymbol{y})$ and distinct individual constants $\bar{g}, \bar{h}$ $(g \in \mathfrak{G}, h \in \mathfrak{H})$ and consider the system S_1 consisting of the axioms

$$A(\bar{g}, \bar{h}_1) \,\&\, A(\bar{g}, \bar{h}_2) \rightarrow \bar{h}_1 \approx \bar{h}_2 \,,$$

$$A(\bar{g}_1, \bar{h}) \,\&\, A(\bar{g}_2, \bar{h}) \rightarrow \bar{g}_1 \approx \bar{g}_2 \,,$$

$$A(\bar{g}_1, \bar{h}_1) \,\&\, A(\bar{g}_2, \bar{h}_2) \rightarrow A(\overline{g_1 g_2}, \overline{h_1 h_2})$$

for every $g, g_1, g_2 \in \mathfrak{G}$ and every $h, h_1, h_2 \in \mathfrak{H}$. The system S_1 is meant to express that A establishes an isomorphism between $\mathfrak{G}$ and $\mathfrak{H}$. Let u be an element of $\mathfrak{G}$, and let $\mathfrak{U}$ be the cyclic subgroup it generates in $\mathfrak{G}$. The given lattice isomorphism maps $\mathfrak{U}$ onto some subgroup $\mathfrak{V}$ of $\mathfrak{H}$. Since on finitely generated subgroups the lattice isomorphism is induced by a group isomorphism, the subgroup $\mathfrak{V}$ is also cyclic. Let $v_1, \ldots, v_k$ be the primitive elements of $\mathfrak{V}$, finite in number. By Φ_u we denote the FOPL sentence

$$A(\bar{u}, \bar{v}_1) \vee \ldots \vee A(\bar{u}, \bar{v}_k) .$$

We take S_2 to be the set of all Φ_u for $u \in \mathfrak{G}$.

The consistency of S_1 means that $\mathfrak{G}$ and $\mathfrak{H}$ are isomorphic. The consistency of $S_1 \cup S_2$ means there exists an isomorphism from $\mathfrak{G}$ onto $\mathfrak{H}$ that maps every element u of $\mathfrak{G}$ onto a generator of the image $\mathfrak{V}$, under the induced isomorphism, of the cyclic subgroup $\mathfrak{U}$ generated in $\mathfrak{G}$ by u. By hopothesis such an isomorphism does exist when we restrict our attention to any finitely generated subgroup of $\mathfrak{G}$. Consequently, every finite subset of $S_1 \cup S_2$ is consistent. By the basic theorem $S_1 \cup S_2$ is itself consistent. Therefore, $\mathfrak{G}$ and $\mathfrak{H}$ are isomorphic in a manner that induces the original lattice isomorphism on their sets of subgroups. ■

NOTES

([1]) In a footnote the author refers to a "precise formulation and proof of this proposition" in [I]. Apparently, the above is the very first formulation of this result, although the essential argument can be discerned in the previous article; the reader is also referred to [R1].

([2]) A property satisfying (i) and (ii) is called *elementary* in the original. This translation adopts the more common usage that any property satisfying (i) is *elementary* (in the broad sense, for the conjunction need not be finite, and the property may not be formulatable in FOPL).

CHAPTER 3

REPRESENTATIONS OF MODELS

A significant number of local algebraic theorems can be deduced from the following local theorem for logic (the compactness theorem): *for the consistency (i.e., satisfiability) of an infinite system of formulas of first-order predicate logic* (FOPL) *with equality and arbitrary sets of symbols for individuals and predicates, it is necessary and sufficient that every finite subset of the given system be consistent.* ■

In 1941 I presented the compactness theorem in this form in [II], based on results in [I]. In the former article this theorem was used to solve several previously open questions in group theory. These results later appeared in the survey [81] and the monograph [80], where further consequences were indicated. A.A.Vinogradov [181] and I [M2] used the same method in the theory of ordered groups. The early article [II], however, fell into obscurity, and several years ago the possibility of applying local theorems of mathematical logic to algebra was rediscovered by a series of authors [54], although the application to ordered groups and the above compactness theorem (but not the local theorems on solvable groups in [II]) were only recently rediscovered by B. Neumann [113] and A. Robinson [129]. ([1])

To prove concrete local theorems one usually has to introduce auxiliary constructions. The purpose of the present article is to point out several sorts of local theorems whose concrete applications do not require these auxiliary constructions. As an example a new theorem on ordered groups is indicated.

Let $R_1, R_2, \dots$ be predicates defined on a set A, with each $R_i(x_1, \dots, x_{n_i})$ being defined by a FOPL formula $\Phi_i(\boldsymbol{x}_1, \dots, \boldsymbol{x}_{n_i})$ with free variables $\boldsymbol{x}_1, \dots, \boldsymbol{x}_{n_i}$ and predicate symbols from among $\boldsymbol{P}_1, \boldsymbol{P}_2, \dots$. Let $P_1, P_2, \dots$ be predicates defined on some set B with the same ranks as $\boldsymbol{P}_1, \boldsymbol{P}_2, \dots$, respectively, and let $x \to x^\sigma$ be a map from A onto B. We shall say that σ is a *representation* of the model $\mathfrak{A} = \langle A; R_1, R_2, \dots \rangle$ in the model $\mathfrak{B} = \langle B; P_1, P_2, \dots \rangle$ of type $R_i \to \Phi_i$ iff for every natural number i and every sequence $\langle a_1, \dots, a_{n_i} \rangle$ of elements of A, $R_i(a_1, \dots, a_{n_i})$ is true (in $\mathfrak{A}$) iff $\Phi_i(a_1^\sigma, \dots, a_{n_i}^\sigma)$ is true in $\mathfrak{B}$. We shall work with a given type of representation, seeking the model $\mathfrak{B}$ and the map σ.

Theorem 1: *If every finite submodel of a fixed model $\mathfrak{A}$ admits a representation in a submodel of some model in a given class arithmetical in the broad sense (i.e., first-order axiomatizable – see [163]) then $\mathfrak{A}$ admits such a representation in this class.*

As an example of an application of Theorem 1 we can point out the theorem on the isomorphic representability of a group by matrices of a given order, in case all finitely generated subgroups of this group are representable by matrices of this order. Theorem 1 follows immediately from the compactness theorem.■

The representations so far considered can be called *direct* since they represent elements as elements. We can also construct representations in which elements are represented by predicates.

Again let $\mathfrak{A} = \langle A; R_1, R_2, \ldots\rangle$ be a model. Let there correspond to each element a of A a predicate symbol $\boldsymbol{P}_a(\boldsymbol{x}_1, \ldots, \boldsymbol{x}_m)$ (2), and to each predicate $R_i(x_1, \ldots, x_{n_i})$, a FOPL formula $\Phi_i(\boldsymbol{P}_{x_1}, \ldots, \boldsymbol{P}_{x_{n_i}})$ with no free individual variables (i.e., a sentence), among whose predicate symbols appear $\boldsymbol{P}_{x_1}, \ldots, \boldsymbol{P}_{x_{n_i}}$, each of rank m; in the following definition these will be viewed as second-order variables. Let $\mathfrak{B}$ be a model among whose basic predicates are included a predicate P_a of rank m for each a in A, and whatever else the Φ_i may require. We shall say that the correspondence $a \to P_a$ is a *predicate representation* of the model $\mathfrak{A}$ in the model $\mathfrak{B}$ of type $R_i \to \Phi_i$ iff for every i and every sequence $\langle a_1, \ldots, a_{n_i}\rangle$ of elements of A, $R_i(a_1, \ldots, a_{n_i})$ is true (in $\mathfrak{A}$) iff $\Phi_i(P_{a_1}, \ldots, P_{a_{n_i}})$ is true in $\mathfrak{B}$. It is easy to see that the compactness theorem implies

Theorem 2: *If every finite submodel of a given model $\mathfrak{A}$ admits a predicate representation of type $R_i \to \Phi_i$, in some member of a fixed axiomatizable class of models, then $\mathfrak{A}$ admits a representation of the same type in a member of this class.* ■

Stone's theorem on the representability of infinite boolean algebras serves as an example for this theorem. To see this, let $\mathfrak{A}$ be a boolean algebra, take the predicate symbols $\boldsymbol{P}_a$ to be unary, and let the formulas

$$(\boldsymbol{x})(\boldsymbol{P}_u(\boldsymbol{x}) \leftrightarrow \boldsymbol{P}_v(\boldsymbol{x}) \vee \boldsymbol{P}_w(\boldsymbol{x})) \quad \text{and} \quad (\boldsymbol{x})(\boldsymbol{P}_u(\boldsymbol{x}) \leftrightarrow \neg \boldsymbol{P}_v(\boldsymbol{x}))$$

correspond to the relations $u = v+w$ and $u = v'$ on $\mathfrak{A}$. Representations of this type are then sought in the class defined by the ordinary FOPL sentences:

$$(\exists \boldsymbol{x})((\boldsymbol{P}_a(\boldsymbol{x}) \,\&\, \neg \boldsymbol{P}_b(\boldsymbol{x})) \vee (\neg \boldsymbol{P}_a(\boldsymbol{x}) \,\&\, \boldsymbol{P}_b(\boldsymbol{x}))),$$

where a, b are distinct elements of A.

As a second example we consider the so-called algebras of relations. Tarski [163] has shown that the class of representable relation algebras can be defined by a certain system of identities, and so, has the local property. This local property also follows immediately from Theorem 2; we take $\mathfrak{A}$ to be a relational algebra, the predicate symbols $\boldsymbol{P}_a$ to be binary, and let the formulas

$$(x)(y)(P_u(x,y) \leftrightarrow \neg P_v(x,y)),$$

$$(x)(y)(P_u(x,y) \leftrightarrow P_v(y,x)),$$

$$(x)(y)(P_u(x,y) \leftrightarrow P_v(x,y) \vee P_w(x,y)),$$

$$(x)(y)(P_u(x,y) \leftrightarrow (\exists z)(P_v(x,z) \,\&\, P_w(z,y)))$$

correspond respectively to the predicates $u = v'$, $u = v^{\cup}$, $u = v + w$, and $u = v \bullet w$ on $\mathfrak{A}$.

In an analogous manner Theorem 2 can be employed to deduce the local property for the representable projective algebras of Everett and Ulam [35] and, generally speaking, for all predicate algebras.

There is a marked interest in algebra in local theorems concerning decompositions of subgroups, ideals, and other systems of elements. Neither direct nor predicate representations are immediately suitable for the derivation of these theorems. It is, therefore, appropriate to introduce yet another type of representation.

Suppose in the model $\mathfrak{A} = \langle A; R_1, R_2, \ldots \rangle$ we want to determine a family of subsets $p_1, p_2, \ldots$ of A with certain properties. We segregate these properties into two classes. Into the first class S_1 go those properties expressing relations among the subsets $p_1, p_2, \ldots$; we assume these are described by FOPL sentences with individual variables ranging over $\mathcal{P} = \{p_1, p_2, \ldots\}$. E.g., to this class would belong the properties of $\mathcal{P}$ being an ordered system, a lattice, etc. In the second class S_2 we put properties connecting the p_i with elements of A. These we assume are described by FOPL sentences in prenex form, some of whose quantified variables range over $\mathcal{P}$, the others over A; the relation $a \in p$ is designated by the predicate symbol $\boldsymbol{\varepsilon}(\boldsymbol{a}, \boldsymbol{p})$. By constructing a *model* for the system $\mathrm{S}_1 \cup \mathrm{S}_2$ over $\mathfrak{A}$ we shall mean finding an auxiliary set $\mathcal{P} = \{p_1, p_2, \ldots\}$ and defining interpretations of $\boldsymbol{\varepsilon}$ and the other predicate symbols from S_1 and S_2 so that all sentences in S_1 and S_2 become true. Clearly, for such a general problem there can be no local theorem. There is, however, the more specialized

Theorem 3: *Suppose for a given mixed system* $\mathrm{S}_1 \cup \mathrm{S}_2$ *the sentences in* S_2 *have no existentially quantified variables ranging over the base of the model.*

If every finite subset of the base A of the infinite model $\mathfrak{A}$ is included in some submodel of $\mathfrak{A}$ over which there exists a model for $S_1 \cup S_2$, then there exists a model for $S_1 \cup S_2$ over $\mathfrak{A}$.

The proof proceeds as follows. We introduce an infinite set of unary predicate symbols $\boldsymbol{P}_a$ $(a \in A)$, and in the sentences of S_2 we replace each occurrence of $\boldsymbol{\varepsilon}(\boldsymbol{a},\boldsymbol{p})$ with $\boldsymbol{P}_a(\boldsymbol{p})$. In the manner of [I] we then reduce the sentences in S_2 to normal form for satisfiability, in which the universal quantifiers precede the existential. Since none of the existential quantifiers referred to A in the original sentences, neither do they in the sentences in normal form. In the latter we now drop the universal quantifiers ranging over A, and replace the free variables so created with all possible combinations of elements of A as individual constants. As a result, the system $S_1 \cup S_2$ is transformed into an equivalent system S* of ordinary FOPL sentences, to which the compactness theorem applies (3). ■

As an example we offer the local theorems of [II], according to which a group $\mathfrak{G}$ has a solvable normal (respectively, central) sequence of subgroups if every finitely generated subgroup has such a sequence. Theorem 4 can serve as another example:

Theorem 4: *Every partially ordered and locally nilpotent torsion-free group has a central sequence consisting of convex normal subgroups.*

For finitely generated groups this theorem is found in [M2], and Theorem 3 gives the extension to the general case. ■

We note that the system $S_1 \cup S_2$ is in essence a set of formulas of higher-order predicate logic, for predicates of predicates and quantifiers over predicates can occur in it. The method indicated reduces it to an infinite system S* of FOPL sentences whose variables range over an intermediary set of "ordinary" predicates. A compactness theorem for second-order sentences of another form has been proved by Henkin [54].

NOTES

(1) For further historical and mathematical information please see [R1].

(2) In the present context these should all have the same rank.

(3) *Some plausible clarifications:* The predicate symbols permissible in S_2 are those associated with the basic predicates of $\mathfrak{A}$ (operation symbols can appear if $\mathfrak{A}$ is an algebra), arbitrary symbols with arguments restricted to the auxiliary set, equality, and $\boldsymbol{\varepsilon}$, which alone has mixed arguments. A special (cf. [R1]), two-sorted form of Skolem's theorem is required. The second sentence in the proof should follow the fifth. Then in the reduc-

tion process all other subformulas involving individual constants from A can be eliminated in favor of T or F as determined by the diagram of $\mathfrak{A}$. The effort to construct a family $\mathcal{P}$ of subsets (cf. Theorem 3 as reformulated in [R1]) is thwarted by the occasional loss of extensionality for $\in$. But similar results, yielding an alternate proof of Theorem 4 below, are obtained in [XI, § 3].

CHAPTER 4

QUASIPRIMITIVE CLASSES OF ABSTRACT ALGEBRAS

The sequence $\mathfrak{A} = \langle A; f_1, f_2, \ldots \rangle$ consisting of a base set A and basic operations $f_i(x_1, \ldots, x_{n_i})$ defined on A and taking values in A, is called an (*abstract*) *algebra* [9]. When we speak of elements and subsets of an algebra, we invariably refer to elements and subsets of the base. We assume each operation f_i has a finite number n_i of arguments, but the number of operations may be infinite. Algebras $\mathfrak{A} = \langle A; f_1, f_2, \ldots \rangle$ and $\mathfrak{B} = \langle B; g_1, g_2, \ldots \rangle$ are *similar* (or of the same *type*) iff they have the same number of operations and corresponding operations have the same number of arguments. In what follows a *class of algebras* will always be a collection of similar algebras; moreover, the corresponding operations will be denoted identically. Algebras are assumed to be specified only up to isomorphism; that is, if an algebra belongs to the class under consideration, then so do all algebras isomorphic to it.

A class of algebras is called *primitive* (or *equational*, or a *variety*) iff it consists of exactly those algebras which satisfy a given set of identities of the form

$$(\boldsymbol{x}_1) \ldots (\boldsymbol{x}_m)(\varphi(\boldsymbol{x}_1, \ldots, \boldsymbol{x}_m) \approx \psi(\boldsymbol{x}_1, \ldots, \boldsymbol{x}_m)), \tag{1}$$

where φ, ψ are formal polynomials in $\boldsymbol{x}_1, \ldots, \boldsymbol{x}_m$ formed by superposing the symbols $\boldsymbol{f}_i$ designating the basic operations of the class a finite number of times. An important property of varieties is the presence in them of algebras with prescribed defining relations. In the literature, however, one meets other classes of algebras, e.g. cancellative semigroups or rings of characteristic zero, in which there are algebras for arbitrary (finite or infinite) systems of defining relations.

Let $\mathfrak{K}$ be an arbitrary class of algebras. We shall consider a system S of formal equations of the form

$$\varphi(\alpha_1, \ldots, \alpha_s) \approx \psi(\alpha_{s+1}, \ldots, \alpha_t),$$

where φ, ψ are polynomials and the α_j are elements of an auxiliary index set J.

An algebra $\mathfrak{A}$ in $\mathfrak{K}$ (a $\mathfrak{K}$-algebra) with generators a_α enumerated by means of the elements of J, is called an *algebra with defining relations* S in $\mathfrak{K}$ iff for any $\mathfrak{K}$-algebra $\mathfrak{B}$ and any map $\alpha \to b_\alpha$ of J into $\mathfrak{B}$ such that all the equations $\varphi(b_{\alpha_1}, ..., b_{\alpha_s}) = \psi(b_{\alpha_{s+1}}, ..., b_{\alpha_t})$ obtained from those in S by means of this map in fact hold in $\mathfrak{B}$, there is a homomorphism of $\mathfrak{A}$ into $\mathfrak{B}$ carrying a_α onto b_α $(\alpha \in J)$. ([1]) In case S is empty, the algebra $\mathfrak{A}$ will be *free* in $\mathfrak{K}$ in the sense of Sikorski [150]. Obviously, if an algebra with defining relations S in $\mathfrak{K}$ exists, it is unique apart from isomorphisms over J.

We shall call a class $\mathfrak{K}$ *finitely free* iff there are algebras in it for arbitrary finite sets of defining relations (with finite index sets). We shall call $\mathfrak{K}$ *free* iff it contains algebras for arbitrary sets of defining relations. In a free class every algebra is a factor algebra of a free algebra. In finitely free classes this holds for finitely generated algebras.

Theorem 1: *A class $\mathfrak{K}$ of algebras is free iff $\mathfrak{K}$ contains all subalgebras and direct products of its members.* ([2])

We show, e.g., that the free class $\mathfrak{K}$ necessarily contains the direct product of any two of its members, $\mathfrak{A}$ and $\mathfrak{B}$. Consider the $\mathfrak{K}$-algebra $\mathfrak{C}$ generated by the set $J = \{\langle a, b\rangle : a \in \mathfrak{A}, b \in \mathfrak{B}\}$ and defined in $\mathfrak{K}$ by the equations

$$f_i(\langle a_1, b_1\rangle, ..., \langle a_n, b_n\rangle) \approx \langle f_i(a_1, ..., a_n), f_i(b_i, ..., b_n)\rangle \ , \tag{2}$$

where the f_i are the basic operations of $\mathfrak{K}$. Supplementing the equations (2) with the relations $\langle a, b\rangle \approx \langle a, b'\rangle$ (U), we obtain a system of defining relations for $\mathfrak{A}$ adding all the equations $\langle a, b\rangle \approx \langle a', b\rangle$ (V) to (2) gives defining relations for $\mathfrak{B}$. If the generators $\langle a, b\rangle$, $\langle c, d\rangle$ are equal in $\mathfrak{C}$, then they are equal in the $\mathfrak{K}$-algebra with index set J and defining relations (2) and (U), i.e., in $\mathfrak{A}$, whence $a = c$; we find $b = c$ similarly. It follows that $\mathfrak{C}$ is the direct product of $\mathfrak{A}$ and $\mathfrak{B}$. ■

In McKinsey's paper [98] there emerged the importance for the theory of algebras not only of identities of the form (1), but also of so-called *conditional identities* having the form

$$(\boldsymbol{x}_1) \dots (\boldsymbol{x}_m)(\varphi_1 \approx \psi_1 \ \& \dots \& \ \varphi_n \approx \psi_n \to \varphi_{n+1} \approx \psi_{n+1}) , \tag{3}$$

where $\varphi_1, \psi_1, ..., \varphi_{n+1}, \psi_{n+1}$ are polynomials inthe variables $\boldsymbol{x}_1, ..., \boldsymbol{x}_m$. A class $\mathfrak{K}$ consisting of exactly those algebras satisfying some fixed system of conditional identities is called *quasiprimitive* (or a *quasivariety*). If the number of basic operations is finite, and $\mathfrak{K}$ can be characterized by a finite system of conditional identities, then it is called *quasiprimitive in the narrow sense* (or a *strict quasivariety*).

Let θ be a congruence on the algebra $\mathfrak{A}$. If $\mathfrak{A}$ belongs to a variety, then so does the factor algebra $\mathfrak{A}/\theta$. If $\mathfrak{A}$ belongs to a quasivariety $\mathcal{K}$, then $\mathfrak{A}/\theta$ belongs to $\mathcal{K}$ iff $\mathfrak{A}$ and θ satisfy the conditions

$$(\forall x_1 \ldots x_m \in \mathfrak{A})(\varphi_1 \equiv \psi_1 \,\&\, \ldots \,\&\, \varphi_n \equiv \psi_n \Rightarrow \varphi_{n+1} \equiv \psi_{n+1} (mod\, \theta))$$

corresponding to the conditional identities of the form (3) that define $\mathcal{K}$.

Since subalgebras and direct products of members of a quasivariety are themselves in it [59], *quasivarieties are free.* From this it quickly follows by a theorem of Birkhoff [9] that a *quasivariety is a variety if it contains all factor algebras of its members.* ∎

A base set A with a sequence of predicates defined on it will be called a *model,* as in [163]. In the usual manner, instead of operations one can take the corresponding predicates and consider algebras to be models. Let us agree to say that a class $\mathcal{K}$ of algebras is *locally definable* (or just *local*) iff from the fact that every finite submodel of an arbitrary algebra $\mathfrak{A}$ is isomorphic to a submodel of some $\mathcal{K}$-algebra it follows that $\mathfrak{A}$ itself belongs to $\mathcal{K}$. In case the basic operations of $\mathcal{K}$ are infinite in number, in this definition we take submodels relative to arbitrary finite subsystems of the basic operations.

Theorem 2: *In order for a class $\mathcal{K}$ of algebras to be a quasivariety, it is necessary and sufficient that it be local and finitely free.*

Necessity is obvious. To prove sufficiency, let S be some finite collection of formal equations of the form $\varphi(\boldsymbol{x}_1, \ldots, \boldsymbol{x}_s) \approx \psi(\boldsymbol{x}_{s+1}, \ldots, \boldsymbol{x}_t)$, where φ and ψ are polynomials, and the $\boldsymbol{x}_j$ are elements of some fixed countable set J. By assumption there is a $\mathcal{K}$-algebra $\mathfrak{A}$ with generators $a_{\boldsymbol{x}}$ $(\boldsymbol{x} \in J_0)$ and defining relations S in $\mathcal{K}$, where J_0 is a finite subset of J containing all the $\boldsymbol{x}_j$ occurring in S. If for $\boldsymbol{x}, \boldsymbol{x}'$ in J_0 it turns out that $a_{\boldsymbol{x}} = a_{\boldsymbol{x}'}$ in $\mathfrak{A}$, then we write the conditional identity

$$(\forall)(\& \, \mathrm{S} \rightarrow \boldsymbol{x} \approx \boldsymbol{x}'), \tag{4}$$

taking the universal closure with respect to the elements of J_0, now viewed as individual variables. Let T be the collection of all the conditional identities (4) so obtained. Since $\mathcal{K}$ is local, we see that it is completely characterized by the axioms T. ∎

Theorem 3: *In order for a class $\mathcal{K}$ of algebras to be a quasivariety, it is necessary and sufficient that $\mathcal{K}$ be local and contain all direct products of its members.*

Only sufficiency needs to be proved. Since $\mathfrak{K}$ is local, it contains all subalgebras of its members. By virtue of Theorems 1 and 2 this implies that $\mathfrak{K}$ is a quasivariety. ■

We can analogously prove

Theorem 4: *The class of all algebras of type $\langle n_1, \ldots, n_s\rangle$ that are isomorphically embeddable in members of a fixed quasivariety of type $\langle n_1, \ldots, n_s, n_{s+1}, \ldots, n_t\rangle$ is itself a quasivariety.* ■

Those semigroups which are embeddable in groups can serve as an illustration. It is well known [M1] that the class of such semigroups is defined by an infinite set of conditional identities that is not equivalent to any finite set of conditional identities. Thus, in particular, Theorem 4 does not hold for strict quasivarieties.

A single-valued function taking values in a set A, but not necessarily defined for all sequences of arguments from A, is called a *partial operation* on A. A base set A with a given sequence of partial operations on it is called a *partial algebra* [34]. A map σ from one partial algebra into another of the same type is called a *homomorphism* iff for every i and any elements $a, a_1, \ldots, a_{n_i}$ of the first partial algebra, the definition and truth of the equation $a = f_i(a_1, \ldots, a_{n_i})$ in the first implies the definition and validity of $a^\sigma = f_i(a_1^\sigma, \ldots, a_{n_i}^\sigma)$ in the second. A 1–1 homomorphism of a partial algebra $\mathfrak{A}$ into an algebra $\mathfrak{B}$ is called an *embedding* of $\mathfrak{A}$ in $\mathfrak{B}$. We shall say that a conditional identity of the form (3) is satisfied in a partial algebra $\mathfrak{A}$ iff when for given elements $a_1, \ldots, a_m$ of $\mathfrak{A}$ the polynomials $\varphi_1, \psi_1, \ldots, \varphi_n, \psi_n, \varphi_{n+1}$ are defined and the relations $\varphi_1 = \psi_1, \ldots, \varphi_n = \psi_n$ hold, then ψ_{n+1} is defined and $\varphi_{n+1} = \psi_{n+1}$. Thus, we can introduce the notion of a quasivariety of partial algebras specified by a given set of conditional identities. Also introducing, as in the earlier case, the concept of a partail algebra determined by indexed generators and formal equations of "defined" polynomials, we easily convince ourselves that such objects exist in the new sort of quasivariety for arbitrary systems of defining relations.

Theorem 5: *In a quasivariety $\mathfrak{K}$ of partial algebras, the partial algebras with finite systems of defining relations (with finite index sets) are finite. The existence of an algorithm for recognizing the embeddability of finite partial $\mathfrak{K}$-algebras in ordinary $\mathfrak{K}$-algebras is equivalent to the existence of an algorithm for deciding the word problem in ordinary $\mathfrak{K}$-algebras.*

For varieties this theorem was proved by Evans [34]. The proof of the more general case uses the same techniques. ■

There is also the following generalization of Theorem 4:

In order that a partial algebra $\mathfrak{A}$ be isomorphically embeddable in an ordinary algebra in a quasivariety $\mathfrak{K}$, it is necessary and sufficient that all of the conditional identities holding in all ordinary $\mathfrak{K}$-algebras be satisfied by $\mathfrak{A}$ in the weak sense. ■

NOTES

([1]) It should be added that $\alpha \rightarrow a_\alpha$ must satisfy the equations of S in the algebra $\mathfrak{A}$; this map might not be 1–1.

([2]) According to [R2], the author has pointed out that the results of this article are valid only if the class considered contains a one-element algebra.

CHAPTER 5

SUBDIRECT PRODUCTS OF MODELS

The theorem of Birkhoff [11] on the decomposability of a given abstract algebra into a subdirect product of indecomposable factors refers to the class of all algebras of a fixed similarity type; in some cases, however, it is desirable to have an analogous theorem for narrower or broader classes. E.g., in the study of rings with no zero divisors or of rings embeddable in skewfields, it is natural to consider only decompositions into subdirect products of rings in the same class. These cases are not immediately covered by Birkhoff's theorem, since factor rings of embeddable rings, e.g., may not be embeddable. In this article we survey an extensive range of classes in which an analogue of Birkhoff's theorem is seen to hold. It is appropriate for this to pass from algebras to models. For particular classes of models, subdirect decompositions have been studied by Pickert [120], Foster [37], and Fuchs [42]. The terminology follows [163].

In what follows we shall consider only classes of those models among whose basic relations the equality relation $\equiv$ is found ([1]). A class of similar models will be called *abstract* iff it contains all models isomorphic to any of its members. Corresponding relations in similar models will be denoted similarly.

Let $\mathfrak{A} = \langle A; P_1, P_2, \ldots \rangle$ and $\mathfrak{B} = \langle B; P_1, P_2, \ldots \rangle$ be similar models (with $P_1 = \equiv$). A binary relation σ with domain A and range B is called a *homomorphism* of $\mathfrak{A}$ onto $\mathfrak{B}$ – and $\mathfrak{B}$ is a *homomorphic image* of $\mathfrak{A}$ – iff for every s and all elements $a_1, \ldots, a_{n_s}$ of A and $b_1, \ldots, b_{n_s}$ of B, if $P_s(a_1, \ldots, a_{n_s})$ holds in $\mathfrak{A}$, and $a_1 \sigma b_1, \ldots, a_{n_s} \sigma b_{n_s}$, then $P_s(b_1, \ldots, b_{n_s})$ is true in $\mathfrak{B}$. Essential to the concept of homomorphism is exactly which relations are considered fundamental. E.g., if in place of a basic relation in a given model its negation were taken, the model might then have different homomorphic images. Let us agree to call an abstract class an *H-class* iff it contains all homomorphic images of its members.

Theorem 1: *In order that an arithmetic (i.e., (first-order) axiomatizable) class of models be an H-class, it is necessary and sufficient that it consist of all models satisfying a system of first-order axioms of the form*

$$(\mathfrak{O}_1 x_1) \ldots (\mathfrak{O}_m x_m)\, \Phi(x_1, \ldots, x_m)\,, \tag{1}$$

where the $\mathfrak{O}_i$ are arbitrary quantifiers, and $\Phi(x_1, \ldots, x_m)$ is a quantifier-free formula constructed from expressions of the form $P_s(x_{i_1}, \ldots, x_{i_{n_s}})$ with the aid of the connectives & and $\vee$ only.

Sufficiency follows from the arguments of Horn [59]; necessity is readily discovered when one replaces the axioms with systems of formulas of the propositional calculus ([2]) and remembers that any homomorphism from a submodel can be extended to a homomorphism of the whole model. ∎

Let $\mathcal{R}(M)$ be the collection of all possible models of a fixed similarity type defined on a base M. For $\mathfrak{M}_1$ and $\mathfrak{M}_2$ in $\mathcal{R}(M)$ let us write $\mathfrak{M}_1 \leqslant \mathfrak{M}_2$ iff the identity map from $\mathfrak{M}_1$ onto itself is a homomorphism of $\mathfrak{M}_1$ onto $\mathfrak{M}_2$. Clearly, $\leqslant$ lattice-orders the collection $\mathcal{R}(M)$, and, moreover, the lattice is complete. Let us consider a map σ from M onto some model $\mathfrak{A}$ of the chosen type. If for every s and all elements $a_1, \ldots, a_{n_s}$ of M, we define $P_s(a_1, \ldots, a_{n_s})$ to be equivalent to $P_s(a_1^\sigma, \ldots, a_{n_s}^\sigma)$, then we thereby turn M into a model $\mathfrak{M}_\sigma \in \mathcal{R}(M)$ isomorphic to $\mathfrak{A}$. If $\mathfrak{M}$ is any model in $\mathcal{R}(M)$, and σ is a homomorphism of $\mathfrak{M}$ onto $\mathfrak{A}$, then $\mathfrak{M} \leqslant \mathfrak{M}_\sigma$.

Let $\mathfrak{A}^\alpha = \langle A^\alpha; P_1, P_2, \ldots\rangle$ $(\alpha \in J)$ be similar models. We let A be the cartesian product of the sets A^α $(\alpha \in J)$; if for every s and all elements $a_1, \ldots, a_{n_s}$ of A, we take $P_s(a_1, \ldots, a_{n_s})$ to be true iff $P_s(a_1^\alpha, \ldots, a_{n_s}^\alpha)$ is true for every $\alpha \in J$, we have thereby constructed a model $\mathfrak{A} = \langle A; P_1, P_2, \ldots\rangle$ called the *direct product* of the system $\{\mathfrak{A}^\alpha \colon \alpha \in J\}$ [59]. Assume for each $\alpha \in J$ we have a map σ^α from some fixed model $\mathfrak{M}$ onto $\mathfrak{A}^\alpha$. The maps σ^α naturally induce a map σ from $\mathfrak{M}$ into $\mathfrak{A}$. If σ is an isomorphism of $\mathfrak{M}$ onto the corresponding submodel $\mathfrak{A}_0$ of $\mathfrak{A}$, then we say that the model $\mathfrak{M}$ is *decomposed into the subdirect product* $\mathfrak{A}_0$ of the models $\mathfrak{A}^\alpha$ with projections σ^α. Let $\mathcal{K}$ be an abstract class of models. A model $\mathfrak{M}$ is called *$\mathcal{K}$-indecomposable* iff in any decomposition of $\mathfrak{M}$ into a subdirect product of $\mathcal{K}$-models (i.e., members of $\mathcal{K}$) at least one of the projections is an isomorphism.

Theorem 2: *An abstract class $\mathcal{K}$ of models contains all subdirect products of its members iff for every set M, the set $\mathcal{K}(M)$ of $\mathcal{K}$-models with base M forms a complete lower subsemilattice of the lattice $\mathcal{R}(M)$ of all models with base M of the type of $\mathcal{K}$ (in other words, $\mathcal{K}(M)$ is closed under arbitrary products – in the sense of $\mathcal{R}(M)$ – of its members).*

Indeed, if the model $\mathfrak{M}$ with base M is a submodel of the direct product of $\mathcal{K}$-models $\mathfrak{A}^\alpha$ with the natural projections $\alpha\colon \mathfrak{M} \to \mathfrak{A}^\alpha$, then in accord with the remark above, each model $\mathfrak{A}^\alpha$ can be viewed as a model $\mathfrak{M}_\alpha \in \mathcal{R}(M)$; we find that $\mathfrak{M}$ is a subdirect product of the $\mathfrak{A}^\alpha$ iff $\mathfrak{M}$ is the lattice product of the models $\mathfrak{M}_\alpha$, proving the sufficiency of the conditions in Theorem 2. Necessity is easily seen. ■

Theorem 3: *In order that a model* $\mathfrak{A} = \langle A; P_1, P_2, \dots \rangle$ *in an abstract class* $\mathcal{K}$ *be a subdirect product of* $\mathcal{K}$*-indecomposable* $\mathcal{K}$*-models, it is necessary and sufficient that for every* s *and any sequence* $\mathfrak{a} = \langle a_1, \dots, a_{n_s} \rangle$ *of elements of* A *making* $P_s(a_1, \dots, a_{n_s})$ *false in* $\mathfrak{A}$, *there are* t *and* $\mathfrak{b} = \langle b_1, \dots, b_{n_t} \rangle$ $(b_i \in A)$ *such that the subset of* $\mathcal{K}(A)$ *consisting of all* $\mathcal{K}$*-models* $\geqslant \mathfrak{A}$ *that falsify* $P_t(b_1, \dots, b_{n_t})$ *has at least one maximal member* $\mathfrak{A}(s, \mathfrak{a})$ *in which* $P_s(a_1, \dots, a_{n_s})$ *is false.*

To prove sufficiency let us assume the maximality condition is fulfilled. We denote by $\mathfrak{A}_0$ the subdirect product of all the models $\mathfrak{A}(s, \mathfrak{a})$ with the identity map on $\mathfrak{A}$ as each projection. The induced map from $\mathfrak{A}$ onto $\mathfrak{A}_0$ is, in fact, an isomorphism, so we only have to verify the $\mathcal{K}$-indecomposability of each particular $\mathfrak{A}(s, \mathfrak{a})$. The base of $\mathfrak{A}(s, \mathfrak{a})$ is A, and every model $\mathfrak{A}' \in \mathcal{K}(A)$ greater than $\mathfrak{A}(s, \mathfrak{a})$ will also be greater than $\mathfrak{A}$. In view of the maximality of $\mathfrak{A}(s, \mathfrak{a})$, $P_t(b_1, \dots, b_{n_t})$ must be true in $\mathfrak{A}'$. In other words, in any subdirect product of $\mathcal{K}$-images of $\mathfrak{A}(s, \mathfrak{a})$ under proper homomorphisms, the expression $P_t(b_1, \dots, b_{n_t})$ will be true; thus the subdirect product will not be isomorphic to $\mathfrak{A}(s, \mathfrak{a})$.

Conversely, let the $\mathcal{K}$-model $\mathfrak{A}$ be decomposed into a subdirect product of $\mathcal{K}$-indecomposable $\mathcal{K}(A)$-models $\mathfrak{A}^\alpha \geqslant \mathfrak{A}$ with projections: the identity map on $\mathfrak{A}$. Suppose $P_s(a_1, \dots, a_{n_s})$ is false in $\mathfrak{A}$. Then it is false in one of the $\mathfrak{A}^\alpha$, as well, say $\mathfrak{A}^\lambda$. If for every $P_t(b_1, \dots, b_{n_t})$ false in $\mathfrak{A}^\lambda$, there were a model $\mathfrak{A}(t, \mathfrak{b}) \in \mathcal{K}(A)$ greater than $\mathfrak{A}^\lambda$ in which $P_t(b_1, \dots, b_{n_t})$ remained false, then $\mathfrak{A}^\lambda$ could be decomposed into the subdirect product of all the $\mathfrak{A}(t, \mathfrak{b})$ so obtained, contradicting the supposed indecomposability of $\mathfrak{A}^\lambda$. Therefore, some $P_t(b_1, \dots, b_{n_t})$ is found to be false in $\mathfrak{A}^\lambda$, while true in all proper homomorphic $\mathcal{K}$-images of $\mathfrak{A}^\lambda$; that is, $\mathfrak{A}^\lambda$ is maximal among $\mathcal{K}(A)$-models in which $P_t(b_1, \dots, b_{n_t})$ is false. ■

A subset of a partially ordered system is called a *chain* iff it is linearly ordered by the given order (i.e., of any two elements, one is greater than or equal to the other). Theorem 3 and Zorn's lemma imply the

Remark: *Let* $\mathcal{K}$ *be an abstract class of models. A* $\mathcal{K}$*-model* $\mathfrak{M}$ *is automatically decomposable into a subdirect product of* $\mathcal{K}$*-indecomposable* $\mathcal{K}$*-models, if the sum of any chain in* $\mathcal{K}(M)$ *is again a* $\mathcal{K}$*-model.* ■

With the help of this remark it is easy to prove the fundamental

Theorem 4: *If a class $\mathcal{K}$ of models is characterized by a system of first-order axioms of the form* (1) *or*

$$(x_1) \ldots (x_m)\, \Psi(x_1, \ldots, x_m)\,, \tag{2}$$

where $\Psi(x_1, \ldots, x_m)$ is a quantifier-free formula constructed from expressions of the form $P_s(x_{i_1}, \ldots, x_{i_{n_s}})$ with the aid of the signs &, ∨, ¬, then every $\mathcal{K}$-model can be decomposed into a subdirect product of $\mathcal{K}$-indecomposable $\mathcal{K}$-models.

The class $\mathcal{K}$ can be represented as the intersection of a class $\mathcal{K}_1$, characterized by the axioms of the form (1), and a class $\mathcal{K}_2$, characterized by axioms (2). Let M be an arbitrary set. By Theorem 1 the sum of a chain of models in $\mathcal{K}_1(M)$ is a $\mathcal{K}_1$-model. In particular, the sum $\mathfrak{A}$ of a chain of $\mathcal{K}(M)$-models is a $\mathcal{K}_1$-model. On the other hand, if for some sequence $\langle a_1, \ldots, a_m \rangle$ $(a_i \in M)$, one of the expressions $\Psi(a_1, \ldots, a_m)$ should prove false in $\mathfrak{A}$, then there would be a $\mathcal{K}$-model $\mathfrak{M}_0$ in the chain in which $\Psi(a_1, \ldots, a_m)$ failed; but this would contradict the validity of the axioms (2) in $\mathfrak{M}_0$. Therefore, the sum $\mathfrak{A}$ is a $\mathcal{K}_2$-model, as well; hence, $\mathfrak{A} \in \mathcal{K}$. It now follows from the Remark that the decomposition theorem holds in $\mathcal{K}$. ■

In particular, the decomposition theorem is valid for algebras satisfying some system of universal axioms of the form (2). Examples of such classes are the classes of rings embeddable in skewfields, rings without zero divisors, semigroups embeddable in groups, etc.

A second example of where this theorem holds is in the class of *directed sets* – those partial orderings that are characterized by the two axioms

$$(x)(y)(\exists z)(x < z \;\&\; y < z)\,,$$

$$(x)(y)(z)((x \nless y \lor y \nless z \lor x < z) \;\&\; (x \nless y \lor y \nless x))\,,$$

where the fundamental predicates are $\equiv$ and $<$. If, however, $\equiv$ and $\nless$ are considered fundamental, then every model in this calss will admit a proper decomposition into a subdirect product of models greater in the lattice ordering. Indeed, suppose we have $a < b < c$ in the directed set $\mathfrak{M}$. Let us extend the relation $\nless$ in two ways to obtain new models, $\mathfrak{M}_1$ and $\mathfrak{M}_2$. In $\mathfrak{M}_1$: for elements a_1, a_2 of M, put $a_1 < a_2$ iff $a_1 < a_2$ holds in $\mathfrak{M}$, but $a \leqslant a_1 < a_2 \leqslant b$ does not. We define $\mathfrak{M}_2$ similarly, replacing a, b with b, c, respectively. Clearly, with $\equiv$ and $\nless$ as the fundamental predicates, the model $\mathfrak{M}$ is the subdirect product of $\mathfrak{M}_1$ and $\mathfrak{M}_2$ with the identity map on $\mathfrak{M}$ as both projections.

NOTES

(1) Throughout this article the author intends this to be *relative equality* (cf. [I], § 3); some provision must be made to guarantee that $\equiv$ is an equivalence relation respecting the basic predicates). In particular, this relation is used in defining the notion of a function on a model, e.g., a homomorphism is, in fact, a mapping according to the definition below. Every map is a function with respect to an equivalence relation in the domain (and in the range), either given by the model context, or arbitrarily fixed in the absence of a model.

(2) This surely refers to the techniques of [I]; cf. [R1]. The *H*-class must be non-empty.

CHAPTER 6

DERIVED OPERATIONS AND PREDICATES

Let $\mathfrak{A} = \langle A; f_1, f_2, ...\rangle$ be an algebra with base A and basic operations $f_i(x_1, ..., x_{m_i})$ $(i = 1, 2, ...)$. Basic terms – polynomials involving the basic operations – define the fundamental type of *derived operations* on $\mathfrak{A}$. Each operation f_i can be viewed as an m_i+1-ary predicate $P_i(a_1, ..., a_{m_i}, b)$ defined on A and signifying $f_i(a_1, ..., a_{m_i}) = b$. Every well-formed formula $\varphi(x_1, ..., x_n)$ of first-order predicate logic with equality (FOPL) with free individual variables $\boldsymbol{x}_1, ..., \boldsymbol{x}_n$ and predicate symbols from among the $\boldsymbol{P}_i$, which designate the predicates just defined, is considered to determine a *derived predicate* on $\mathfrak{A}$. It may happen that φ represents an operation on A in the above sense. Thus we can get new operations on $\mathfrak{A}$ by this means as well as by forming terms.

In §1 we establish the general form of operations obtained by means of FOPL formulas in a class of algebras characterized by universal axioms; in §2 we give an abstract characterization of predicates representable as conjunctions of universal FOPL formulas; in §3, these results are used to determine the general form of derived operations satisfying various additional conditions.

§1. For an arbitrary formula φ of FOPL with free variables $\boldsymbol{x}_1, ..., \boldsymbol{x}_m, \boldsymbol{y}$, let $\Phi(\varphi)$ denote the expression

$$(\boldsymbol{x}_1) \ldots (\boldsymbol{x}_m)(\boldsymbol{u})(\boldsymbol{v})(\exists \boldsymbol{y})(\varphi(\boldsymbol{x}_1, ..., \boldsymbol{x}_m, \boldsymbol{y})\ \& $$

$$\& \ (\varphi(\boldsymbol{x}_1, ..., \boldsymbol{x}_m, \boldsymbol{u})\ \&\ \varphi(\boldsymbol{x}_1, ..., \boldsymbol{x}_m, \boldsymbol{v}) \rightarrow \boldsymbol{u} \approx \boldsymbol{v}))\ .$$

A formula of the form $(\boldsymbol{y}_1) \ldots (\boldsymbol{y}_p)\ \psi(\boldsymbol{x}_1, ..., \boldsymbol{x}_n, \boldsymbol{y}_1, ..., \boldsymbol{y}_p)$, where ψ is open (i.e., quantifier-free), is called *universal,* while a formula with no free variables is called a *sentence* (or *axiom*).

Theorem 1: *Let* T *be a system of universal sentences, possibly infinite in number; let* φ *be a* FOPL *formula with free variables* $\boldsymbol{x}_1, ..., \boldsymbol{x}_m, \boldsymbol{y}$; $\boldsymbol{P}_1, ..., \boldsymbol{P}_s$

(and $\approx$) are assumed to be the only predicate symbols occurring in φ and the members of T. *Suppose, too, that the sentence $\Phi(\varphi)$ is a logical consequence of the system* $\mathrm{S} = \mathrm{T} \cup \{\Phi(\boldsymbol{P}_1), \dots, \Phi(\boldsymbol{P}_s)\}$. *Then for some natural number t, there are formulas $\varphi_j(\boldsymbol{x}_1, \dots, \boldsymbol{x}_m)$ and terms $\mathfrak{g}_j(\boldsymbol{x}_1, \dots, \boldsymbol{x}_m)$ $(j = 1, 2, \dots, t)$ – the φ_j involving the $\boldsymbol{P}_i$ only, and the $\mathfrak{g}_j$ involving only new operation symbols $\boldsymbol{f}_i$ corresponding to $\boldsymbol{P}_i$ $(i = 1, \dots, s)$ (*[1]*) – such that the formulas*

$$\varphi_1(\boldsymbol{x}_1, \dots, \boldsymbol{x}_m) \vee \dots \vee \varphi_t(\boldsymbol{x}_1, \dots, \boldsymbol{x}_m), \tag{1}$$

$$\varphi_j(\boldsymbol{x}_1, \dots, \boldsymbol{x}_m) \,\&\, \varphi_k(\boldsymbol{x}_1, \dots, \boldsymbol{x}_m) \leftrightarrow$$
$$\leftrightarrow \mathfrak{g}_j(\boldsymbol{x}_1, \dots, \boldsymbol{x}_m) \approx \mathfrak{g}_k(\boldsymbol{x}_1, \dots, \boldsymbol{x}_m) \quad (j, k = 1, \dots, t), \tag{2}$$

$$\varphi(\boldsymbol{x}_1, \dots, \boldsymbol{x}_m, \boldsymbol{y}) \leftrightarrow ((\varphi_1(\boldsymbol{x}_1, \dots, \boldsymbol{x}_m) \leftrightarrow$$
$$\leftrightarrow \boldsymbol{y} \approx \mathfrak{g}_1(\boldsymbol{x}_1, \dots, \boldsymbol{x}_m)) \,\&\, \dots \,\&\, (\varphi_t \leftrightarrow \boldsymbol{y} \approx \mathfrak{g}_t)) \tag{3}$$

are consequences of S.

To prove this we let $H = \{\mathfrak{h}_\alpha : \alpha \in J\}$ be the set of all possible terms involving the $\boldsymbol{f}_i$ and the variables $\boldsymbol{x}_1, \dots, \boldsymbol{x}_m$. Now let

$$\mathrm{S}_1 = \mathrm{S} \cup \{\neg\varphi(\boldsymbol{x}_1, \dots, \boldsymbol{x}_m, \mathfrak{h}_\alpha) : \alpha \in J\}.$$

Assume this system has some model $\mathfrak{A}$. In view of the axioms $\Phi(\boldsymbol{P}_1), \dots, \Phi(\boldsymbol{P}_s)$, the model $\mathfrak{A}$ is an algebra with respect to the operations $f_1, \dots, f_s$ corresponding to the predicates $P_1, \dots, P_s$. Let $\langle a_1, \dots, a_m \rangle$ be any sequence of elements in $\mathfrak{A}$ (i.e., of the base in $\mathfrak{A}$); the subalgebra $\mathfrak{B}$ generated in $\mathfrak{A}$ by the elements $a_1, \dots, a_m$ also satisfies S_1 since the axioms in T are universal. Inasmuch as $\mathfrak{B}$ satisfies S, there is an element b of $\mathfrak{B}$ such that $\varphi(a_1, \dots, a_m, b)$ is true in $\mathfrak{B}$, but $b = \mathfrak{h}_\lambda(a_1, \dots, a_m)$ for some λ in J, a contradiction. The inconsistency of S_1 is thus proved; this implies the inconsistency of some finite subset of S_1, i.e., the axioms S imply the formula

$$\varphi(\boldsymbol{x}_1, \dots, \boldsymbol{x}_m, \boldsymbol{y}) \to \boldsymbol{y} \approx \mathfrak{g}_1 \vee \dots \vee \boldsymbol{y} \approx \mathfrak{g}_t \tag{4}$$

for appropriate terms $\mathfrak{g}_1 = \mathfrak{h}_{\alpha_1}, \dots, \mathfrak{g}_t = \mathfrak{h}_{\alpha_t}$ in H.

Taking as $\varphi_j(\boldsymbol{x}_1, \dots, \boldsymbol{x}_m)$ the formula

$$(\boldsymbol{z})\,(\varphi(\boldsymbol{x}_1, \dots, \boldsymbol{x}_m, \boldsymbol{z}) \leftrightarrow \mathfrak{g}_j(\boldsymbol{x}_1, \dots, \boldsymbol{x}_m))$$

for $j = 1, ..., t$, we obtain (1), (2), and (3) as consequenses of S. Indeed, let $a_1, ..., a_m$ be arbitrary elements of a model $\mathfrak{A}$ satisfying S. By assumption there is a b in $\mathfrak{A}$ for which $\varphi(a_1, ..., a_m, b)$ holds. It follows from (4) that $b = \mathfrak{g}_k(a_1, ..., a_m)$ for some $1 \leqslant k \leqslant t$. Hence, $\varphi_k(a_1, ..., a_m)$ is true; thus (1) is valid in $\mathfrak{A}$. Suppose for certain $a_1, ..., a_m, b$ in $\mathfrak{A}$ the expression $\varphi(a_1, ..., a_m, b)$ is false. There are then c and k for which $\varphi(a_1, ..., a_m, c)$ is true, $c = \mathfrak{g}_k(a_1, ..., a_m)$, and $c \neq b$, and thus $\varphi_k(a_1, ..., a_m)$ is true, as well. So the relations $\varphi_k(a_1, ..., a_m)$ and $b = \mathfrak{g}_k(a_1, ..., a_m)$ are not equivalent in $\mathfrak{A}$. Thus whenever the left-hand part of (3) is false in $\mathfrak{A}$, so is the right-hand part. The converse and the formulas (2) are analogously shown to hold in $\mathfrak{A}$ ([2]). ■

§2. Let $\mathfrak{K}$ be a class of models of the form $\langle M; P_1^{n_1}, ..., P_s^{n_s}\rangle$. Let us suppose that on every member of $\mathfrak{K}$ there is an additional predicate $P(x_1, ..., x_n)$ prescribed in some manner, not necessarily by a FOPL formula, not necessarily by the same means on each $\mathfrak{K}$-model. In the class of ordered groups, e.g., we can take the predicate "x is incommensurably smaller than y", which makes sense in any ordered group, even though it cannot be expressed by a FOPL formula. We say P is *invariant with respect to passage to* $\mathfrak{K}$*-submodels* iff for any elements $a_1, ..., a_n$ of a $\mathfrak{K}$-model $\mathfrak{M}$, the truth of $P(a_1, ..., a_n)$ in $\mathfrak{M}$ implies the truth of $P(a_1, ..., a_n)$ in any $\mathfrak{K}$-submodel of $\mathfrak{M}$ containing $a_1, ..., a_n$. Analogously, the predicate P is called *invariant with respect to passage to* $\mathfrak{K}$*-supermodels* iff for any elements $a_1, ..., a_n$ of a $\mathfrak{K}$-model $\mathfrak{M}$, the truth of $P(a_1, ..., a_n)$ in $\mathfrak{M}$ implies the truth of $P(a_1, ..., a_n)$ in every $\mathfrak{K}$-model including $\mathfrak{M}$ as a submodel. The predicate P is called *formular* (in $\mathfrak{K}$) if there is a FOPL formula φ with free variables $\boldsymbol{x}_1, ..., \boldsymbol{x}_n$ such that $\boldsymbol{P}(\boldsymbol{x}_1, ..., \boldsymbol{x}_n) \leftrightarrow \varphi(\boldsymbol{x}_1, ..., \boldsymbol{x}_n)$ is valid in every $\mathfrak{K}$-model. Obviously, the invariance of P with respect to passage to submodels is equivalent to the invariance of $\sim P$ with respect to passage to supermodels. A.Robinson [131] calls a formular predicate invariant with respect to passage to supermodels *persistent* in $\mathfrak{K}$.

A predicate $P(x_1, ..., x_n)$ whose truth or falsity is not necessarily defined for all sequences of elements from some set M is called a *partial* predicate on M. A map σ from a set M with a partial predicate P into a set with a (total) predicate by the same name is called a *P-homomorphism* of the partial model $\langle M, P\rangle$ into $\langle N, P\rangle$ iff for all elements $a_1, ..., a_n$ of M, the definition and truth of $P(a_1, ..., a_n)$ imply the truth of $P(a_1^\sigma, ..., a_n^\sigma)$. We say that σ *P-embeds* $\langle M, P\rangle$ in $\langle N, P\rangle$ iff σ maps M 1–1 onto M^σ, and for all $a_1, ..., a_n$ in M, if $P(a_1, ..., a_n)$ is defined, $P(a_1, ..., a_n)$ in $\langle M, P\rangle$ is equivalent to $P(a_1^\sigma, ..., a_n^\sigma)$ in $\langle N, P\rangle$.

Theorem 2: *Let $\mathcal{K}^*$ be a class of models with basic (total) predicates $P_1, ..., P_s, P$. In order that some formula of the form*

$$P(x_1, ..., x_n) \leftrightarrow \underset{\alpha \in J}{\&} (x_{n+1}) ... (x_{p_\alpha}) \psi_\alpha(x_1, ..., x_n, x_{n+1}, ..., x_{p_\alpha}) \qquad (5)$$

(the conjunction can be infinite, but the ψ_α are ordinary open FOPL *formulas in which* **P** *does not occur) be valid in all $\mathcal{K}^*$-models, it is necessary and sufficient that $\mathcal{K}^*$ possess the following property: if every finite submodel of the partial model $\mathfrak{M} = \langle M; P_1, ..., P_s; P\rangle$ (both with $P_1, ..., P_s$ ordinary and P partial) is $(P_1, ..., P_s, P)$-embeddable in some $\mathcal{K}^*$-model, then every $(P_1, ..., P_s)$-isomorphism σ of $\mathfrak{M}$ onto a $\mathcal{K}^*$-model $\mathfrak{M}^*$ is also a P-homomorphism.*

Necessity. Suppose the formula Ψ of the form (5) is valid in all the members of $\mathcal{K}^*$, σ is a map of the indicated sort, and for some $a_1, ..., a_n$ in $\mathfrak{M}$, $P(a_1, ..., a_n)$ is defined, while $P(a_1^\sigma, ..., a_n^\sigma)$ is false in $\mathfrak{M}^*$. Inasmuch as σ is a $(P_1, ..., P_s)$-isomorphism, and **P** does not occur in the formulas ψ_α, the right-hand part of Ψ is false in $\mathfrak{M}$ for $a_1, ..., a_n$. Therefore, there are λ in J and $a_{n+1}, ..., a_{p_\lambda}$ in $\mathfrak{M}$ such that $\psi_\lambda(a_1, ..., a_{p_\lambda})$ is false in $\mathfrak{M}$. By assumption, the finite submodel of $\mathfrak{M}$ with base $\{a_1, ..., a_{p_\lambda}\}$ is $(P_1, ..., P_s, P)$-embeddable by a map τ in some $\mathcal{K}^*$-model $\mathfrak{N}$, in which Ψ is valid. Since $\psi_\lambda(a_1^\tau, ..., a_{p_\lambda}^\tau)$ is false in $\mathfrak{N}$, $P(a_1^\tau, ..., a_n^\tau)$ is also false, and thus $P(a_1, ..., a_n)$ is false in $\mathfrak{M}$.

Sufficiency. Assume $\mathcal{K}^*$ has the indicated property. Let $\{\varphi_\alpha : \alpha \in J\}$ be the set of all universal formulas φ_α with free variables $x_1, ..., x_n$ such that the sentence

$$(x_1) ... (x_n) (P(x_1, ..., x_n) \to \varphi_\alpha(x_1, ..., x_n)) \qquad (6)$$

holds in all $\mathcal{K}^*$-models. Suppose there is a $\mathcal{K}^*$-model $\mathfrak{M}^*$ in which the system of formulas $\{\neg P(x_1, ..., x_n), \varphi_\alpha(x_1, ..., x_n) : \alpha \in J\}$ is satisfied by some sequence $\langle a_1, ..., a_n\rangle$ of elements of $\mathfrak{M}^*$. Let $\mathfrak{M}$ be the same model as $\mathfrak{M}^*$ except that P is taken to be the partial predicate true for $\langle a_1, ..., a_n\rangle$, but undefined for all other possible values of the arguments. The identity map σ of $\mathfrak{M}$ onto $\mathfrak{M}^*$ is a $(P_1, ..., P_s)$-isomorphism. If every finite submodel of $\mathfrak{M}$ were $(P_1, ..., P_s, P)$-embeddable in some $\mathcal{K}^*$-model, then by assumption σ would be a P-homomorphism, as well; this cannot be, since $P(a_1, ..., a_n)$ is true in $\mathfrak{M}$, but false in $\mathfrak{M}^*$. Therefore, $\mathfrak{M}$ includes a finite submodel $\mathfrak{M}_0$ – with base $\{a_1, ..., a_n, a_{n+1}, ..., a_q\}$ – that is not $(P_1, ..., P_s, P)$-embeddable in any $\mathcal{K}^*$-model. Let $P(a_1, ..., a_n) \& \psi(a_1, ..., a_q)$ (here, ψ is an open FOPL formula in which **P** does not occur) be the conjunction of the diagram of $\mathfrak{M}_0$ (cf. [130]) in terms of $a_1, ..., a_q$, which are distinct designations for possibly indistinct elements. Since $\mathfrak{M}_0$ is not embeddable, the formula

$$P(x_1, ..., x_n) \to (x_{n+1}) \dots (x_q) \neg \psi(x_1, ..., x_n, x_{n+1}, ..., x_q)$$

is valid in every $\mathcal{K}^*$-model; thus for some $\lambda \in J$,

$$\varphi_\lambda(x_1, ..., x_n) = (x_{n+1}) \dots (x_q) \neg \psi(x_1, ..., x_n, x_{n+1}, ..., x_q) . \quad (7)$$

But we know $\varphi_\lambda(a_1, ..., a_n)$ and $\psi(a_1, ..., a_n, a_{n+1}, ..., a_q)$ are both true in $\mathfrak{M}^*$, contradicting (7). This shows that there is no such model $\mathfrak{M}^*$; consequently,

$$\mathop{\&}_{\alpha \in J} \varphi_\alpha(x_1, ..., x_n) \to P(x_1, ..., x_n)$$

is valid in every $\mathcal{K}^*$-model; together with (6) this shows that

$$P(x_1, ..., x_n) \leftrightarrow \mathop{\&}_{\alpha \in J} \varphi_\alpha(x_1, ..., x_n)$$

is valid throughout $\mathcal{K}^*$. ■

If we take P to be a 0-ary predicate in Theorem 2 and $\mathcal{L}$ to be the collection of $\mathcal{K}^*$-models in which P is true, then we obtain the theorem of A.Tarski [163]: *a subclass $\mathcal{L}$ of a class $\mathcal{K}$ of models can be distinguished in $\mathcal{K}$ by means of a system of universal axioms iff for every $\mathcal{K}$-model $\mathfrak{M}$, the embeddability of every finite submodel of $\mathfrak{M}$ in an $\mathcal{L}$-model implies that $\mathfrak{M}$ is an $\mathcal{L}$-model.* ■

If we assume the predicate P in Theorem 2 to be formular, then we get as a second corollary the theorem of A.Robinson [131]: *in order that the formular predicate P on a (first-order) axiomatizable class $\mathcal{K}$ of models be definable by a universal formula, it is necessary and sufficient that P be invariant with respect to passage to $\mathcal{K}$-submodels.* ■

§ 3. The simplest formular predicates invariant with respect to passage to sub- and supermodels in any class $\mathcal{K}$ are the predicates defined by open formulas. If the class $\mathcal{K}$ is characterizable by universal axioms, then the open formulas define the only formular predicates with this property. In a class of models which are algebras, however, this ceases to be the case, since all formulas representing equations of basic terms give predicates with the desired invariance.

Theorem 3: *Let $\varphi(x_1, ..., x_n)$ be a* FOPL *formula representing a predicate invariant with respect to passage to sub- and superalgebras in a universally* *axiomatizable class $\mathfrak{K}$ of algebras. Then φ is equivalent in $\mathfrak{K}$ to an open formula constructed from expressions of the form $\mathfrak{g} \approx \mathfrak{h}$, where $\mathfrak{g}$ and $\mathfrak{h}$ are basic terms with variables among $x_1, \ldots, x_n$.*

By virtue of the second corollary to Theorem 2 there is an open formula ψ such that

$$\varphi(x_1, \ldots, x_n) \leftrightarrow (y_1) \ldots (y_p)\, \psi(x_1, \ldots, x_n, y_1, \ldots, y_p) \tag{8}$$

is valid in all $\mathfrak{K}$-algebras. Let $\{\langle \mathfrak{g}_1^\alpha, \ldots, \mathfrak{g}_p^\alpha \rangle : \alpha \in J\}$ be the set of all p-sequences of terms $\mathfrak{g}_i^\alpha(x_1, \ldots, x_n)$ constructed from the basic operations of $\mathfrak{K}$ and the variables $x_1, \ldots, x_n$. Let $\langle a_1, \ldots, a_n \rangle$ be any sequence of elements of an arbitrary $\mathfrak{K}$-algebra $\mathfrak{A}$; let $\mathfrak{B}$ be the subalgebra of $\mathfrak{A}$ generated by the a_i; since $\mathfrak{K}$ is universally axiomatizable, $\mathfrak{B}$ is itself a $\mathfrak{K}$-algebra, and (8) is valid in it. By assumption, the value of $\varphi(a_1, \ldots, a_n)$ is the same in $\mathfrak{B}$ as in $\mathfrak{A}$, so $\langle a_1, \ldots, a_n \rangle$ satisfies

$$\underset{\alpha \in J}{\&}\, \psi(x_1, \ldots, x_n, \mathfrak{g}_1^\alpha, \ldots, \mathfrak{g}_p^\alpha) \leftrightarrow \varphi(x_1, \ldots, x_n) \tag{9}$$

in $\mathfrak{A}$. Since the a_i are arbitrary, (9) is valid in $\mathfrak{A}$; since $\mathfrak{A}$ is arbitrary, (9) is valid throughout $\mathfrak{K}$. Inasmuch as (9) follows from the axioms characterizing $\mathfrak{K}$, a finite conjunction must already work. ∎

A predicate P defined on the models in a class $\mathfrak{K}$ is called *multiplicatively invariant* (in $\mathfrak{K}$) iff for all $\mathfrak{K}$-models $\mathfrak{A}$ and $\mathfrak{B}$ such that their direct product $\mathfrak{A} \times \mathfrak{B}$ belongs to $\mathfrak{K}$, and for all elements $\langle a_i, b_i \rangle$ of $\mathfrak{A} \times \mathfrak{B}$ ($a_i \in \mathfrak{A}, b_i \in \mathfrak{B}$, $i = 1, \ldots, n$):

$P(\langle a_1, b_1 \rangle, \ldots, \langle a_n, b_n \rangle)$ is true in $\mathfrak{A} \times \mathfrak{B}$ iff $P(a_1, \ldots, a_n)$ and $P(b_1, \ldots, b_n)$ hold in $\mathfrak{A}$ and $\mathfrak{B}$, respectively.

Theorem 4: *In a quasivariety $\mathfrak{K}$ (cf. [IV]) of algebras, a formular predicate P is a multiplicatively invariant operation iff it can be represented by a term.*

Sufficiency is obvious, so assume P is a multiplicatively invariant operation. Let P be defined in $\mathfrak{K}$ by the formula $\varphi(x_1, \ldots, x_m, y)$. According to the proof of Theorem 1, the formula (4) for φ and appropriate terms $\mathfrak{g}_1, \ldots, \mathfrak{g}_t$ is valid in every $\mathfrak{K}$-algebra. If none of the formulas

$$\varphi(x_1, \ldots, x_m, y) \leftrightarrow y \approx \mathfrak{g}_j(x_1, \ldots, x_m) \tag{10}$$

is valid throughout $\mathfrak{K}$, then there are $\mathfrak{K}$-algebras $\mathfrak{A}_j$ containing elements $a_1^j, \ldots, a_m^j, b^j$ such that $\varphi(a_1^j, \ldots, a_m^j, b^j)$ holds in $\mathfrak{A}_j$, but $b^j \neq \mathfrak{g}_j(a_1^j, \ldots, a_m^j)$ ($j = 1, \ldots, t$). Putting $a_i = \langle a_i^1, \ldots, a_i^t \rangle$ ($i = 1, \ldots, m$) and $b = \langle b^1, \ldots, b^t \rangle$, we have $\varphi(a_1, ..., a_m, b)$ and $b \neq \mathfrak{g}_j(a_1, ..., a_m)$ $(j = 1, ..., t)$ in $\mathfrak{A}_1 \times ... \times \mathfrak{A}_t$; this contradicts the validity of (4) in this algebra, which belongs to the quasi-variety $\mathcal{K}$. Therefore, one of the formulas (10) holds in all $\mathcal{K}$-algebras. ∎

Theorem 5: *In order that a derived operation P represented in a quasi-variety $\mathcal{K}$ by a formula $\varphi(x_1, ..., x_m, y)$ be definable by a term in $\mathcal{K}$, it is necessary and sufficient that it be homomorphically invariant, i.e., for any homomorphism σ of a $\mathcal{K}$-algebra $\mathfrak{A}$ onto a $\mathcal{K}$-algebra $\mathfrak{B}$ and any elements $a_1, ..., a_m, b$ of $\mathfrak{A}$, the truth of $\varphi(a_1, ..., a_m, b)$ in $\mathfrak{A}$ implies the truth of $\varphi(a_1^\sigma, ..., a_m^\sigma, b^\sigma)$ in $\mathfrak{A}$.*

Necessity is obvious, so suppose P is homomorphically invariant. Let us consider the $\mathcal{K}$-free algebra $\mathfrak{A}$ with free generators $a_1, ..., a_m, a_{m+1}, ...$ (cf. [IV]). By assumption, there is a term $\mathfrak{g}(x_1, ..., x_m)$ such that $\varphi(a_1, ..., a_m, \mathfrak{g}(a_1, ..., a_m))$ is true in $\mathfrak{A}$ ([3]). Since a map σ of the a_i onto arbitrary, possibly indistinct elements a_i^σ of an arbitrary $\mathcal{K}$-algebra $\mathfrak{B}$ extends to a homomorphism of $\mathfrak{A}$ into $\mathfrak{B}$, $\varphi(a_1^\sigma, ..., a_m^\sigma, \mathfrak{g}(a_1^\sigma, ..., a_m^\sigma))$ is true in $\mathfrak{B}$; hence, $\varphi(x_1, ..., x_m, \mathfrak{g})$ is valid throughout $\mathcal{K}$, i.e., in every $\mathcal{K}$-algebra the formula

$$\varphi(x_1, ..., x_m, y) \leftrightarrow y \approx \mathfrak{g}(x_1, ..., x_m)$$

is valid. ∎

NOTES

([1]) Universal axioms defining each f_i in terms of P_i should be added to S.

([2]) As it stands, (2) is impossible; the best correction is probably $\varphi_j \to (\varphi_k \leftrightarrow \mathfrak{g}_j \approx \mathfrak{g}_k)$ for $j, k = 1, ..., t$.

([3]) The subsequent argument can be applied to show that $\mathfrak{g}$ depends only on the first m of the indefinite number of free generators of $\mathfrak{A}$. Theorem 5 is also a corollary of Theorem 4.

CHAPTER 7

CLASSES OF MODELS WITH AN OPERATION OF GENERATION

In the theory of algebras an essential role is played by the concept of a set of generating elements. This notion is naturally transferable to the theory of models: if $\mathfrak{K}$ is a class of similar models, $\mathfrak{M}$ is a member of $\mathfrak{K}$ (a $\mathfrak{K}$-model), and B is a subset of $\mathfrak{M}$ (i.e., of the base of $\mathfrak{M}$), then we say *B generates* the submodel $\mathfrak{N}$ of $\mathfrak{M}$ in $\mathfrak{K}$ iff $\mathfrak{N}$ is the intersection of all $\mathfrak{K}$-submodels of $\mathfrak{M}$ that include B, and is itself a $\mathfrak{K}$-model (for terminology see [V], [163]). Because another sort of generation will appear later, we shall call this one *natural generation.* Correspondingly, a class $\mathfrak{K}$ of models is called a class with *natural generation* iff the intersection of any collection of $\mathfrak{K}$-submodels of a $\mathfrak{K}$-model $\mathfrak{M}$ either is empty or is itself a $\mathfrak{K}$-submodel of $\mathfrak{M}$. Some properties of classes with natural generation are studied below; also considered is another kind of generation springing from the study of algebras with partial operations.

All classes of models considered are assumed to be abstract (i.e., closed under isomorphisms). The number of basic predicates and distinguished elements in each class can be infinite, but each predicate is assumed to depend on only a finite number of arguments. A class $\mathfrak{K}$ of models is called *pseudo-axiomatizable* iff it possesses the following two properties: (a) for every system of sentences of first-order predicate logic (FOPL) involving only the predicate and individual constant symbols associated with $\mathfrak{K}$, if each finite subset of the system is satisfiable in some $\mathfrak{K}$-model, then the whole system is simultaneously satisfiable in some $\mathfrak{K}$-model; (b) for every cardinal number $\mathfrak{m}$, there is a cardinal number $\mathfrak{n}(\mathfrak{m})$ such that if B is a subset of power = $\mathfrak{m}$ of a $\mathfrak{K}$-model $\mathfrak{M}$, then $\mathfrak{M}$ has a $\mathfrak{K}$-submodel including B of power = $\mathfrak{n}(\mathfrak{m})$. ([1]) By the compactness and Löwenheim-Skolem theorems for FOPL, every (first-order) axiomatizable class of models is pseudoaxiomatizable. The PC_Δ-classes investigated by Tarski [163], which are constructed from axiomatizable classes by deleting a number of the basic predicates, are also pseudo-axiomatizable. In both cases, $\mathfrak{n}(\mathfrak{m}) = \mathfrak{m}$ works for sufficiently large $\mathfrak{m}$. The collection of all linearly ordered sets whose powers are limit cardinals serves as an example of a pseudoaxiomatizable class that does not admit an axiomatization and is not a PC_Δ-class.

§ **1.** (²) Let $\mathfrak{K}$ be a pseudoaxiomatizable class of models with natural generation. Let $\mathfrak{A} = \langle A; P_\gamma; e_\delta : \gamma \in \Gamma, \delta \in \Delta \rangle$ be a $\mathfrak{K}$-model. In logical contexts we use $\boldsymbol{P}_\gamma, \boldsymbol{e}_\delta$ to designate the predicates P_γ and distinguished elements e_δ of arbitrary $\mathfrak{K}$-models, but we shall take greater pains to distinguish between the notations for elements ($a, b, c, \ldots$) of models and for constant symbols ($\boldsymbol{u}, \boldsymbol{v}, \boldsymbol{w}, \ldots$) that may designate these elements in constructions like the diagram. Suppose $\mathfrak{A}$ is $\mathfrak{K}$-generated by a subset $B \subseteq A$. Let $C = A \sim B$, and let $U = \{\boldsymbol{u}_b : b \in B\}$ and $V = \{\boldsymbol{v}_c : c \in C\}$ be sets of new distinct individual constants. Interpretation in $\mathfrak{A}$ of the constants in U, V is obtained via the maps $\boldsymbol{u}_b \to b$, $\boldsymbol{v}_c \to c$. The diagram of $\mathfrak{A}$ (cf. [132]) can be thought of as the set $D_{\mathfrak{A}}(U, V)$ consisting of all FOPL formulas of the form $\boldsymbol{P}_\gamma(\mathfrak{c}_1, \ldots, \mathfrak{c}_{n_\gamma})$, $\neg \boldsymbol{P}_\gamma(\mathfrak{c}_1, \ldots, \mathfrak{c}_{n_\gamma})$, $\mathfrak{c} \approx \mathfrak{c}'$, or $\mathfrak{c} \not\approx \mathfrak{c}'$, true in $\mathfrak{A}$ under the indicated interpretation, where $\mathfrak{c}, \mathfrak{c}', \mathfrak{c}_1, \ldots$ range independently over $U \cup V \cup \{\boldsymbol{e}_\delta : \delta \in \Delta\}$. In fact, the correspondences used for interpretation need not be 1–1, as is frequently the case with $\boldsymbol{e}_\delta \to e_\delta$, and will happen in § 2 with the map from U onto a generating set. In particular, a *finite subdiagram* is specified for $\mathfrak{A}$ by choosing finite subsets of the index sets Γ, Δ, B, C, and is the conjunction of the finite number of members of $D_{\mathfrak{A}}(U, V)$ in which only predicates and constants with the chosen indices occur ($\approx$ may always occur, as well).

Let $X = \{\boldsymbol{x}_c : c \in C\}$ be a set of new distinct individual constants, in natural 1–1 correspondence with V (and C). Let c^* be an element of C, and consider the system of formulas

$$\mathbf{T}^{c^*} = D_{\mathfrak{A}}(U, V) \cup D_{\mathfrak{A}}(U, X) \cup \{\boldsymbol{v}_{c^*} \not\approx \boldsymbol{x}_c : c \in C\}.$$

If every finite subset of $\mathbf{T}^{c^*}$ were satisfiable in some $\mathfrak{K}$-model, then by assumption, there would be a $\mathfrak{K}$-model $\mathfrak{M}$ in which the whole of $\mathbf{T}^{c^*}$ would be satisfiable. But then $\mathfrak{M}$ would have two different submodels isomorphic to $\mathfrak{A}$, the isomorphisms coinciding on the generating set B, clearly a contradiction. It follows that there are finite subsets $\Gamma' \subseteq \Gamma$, $\Delta' \subseteq \Delta$, $\{b_1, \ldots, b_s\} \subseteq B$, and $\{c_1, \ldots, c_t\} \subseteq C$ specifying a finite subdiagram $D_{\mathfrak{A}}^{c^*}(\boldsymbol{u}_{b_1}, \ldots, \boldsymbol{u}_{b_s}, \boldsymbol{v}_{c_1}, \ldots, \boldsymbol{v}_{c_t})$ for $\mathfrak{A}$ such that the formula

$$D_{\mathfrak{A}}^{c^*}(\boldsymbol{u}_{b_1}, \ldots, \boldsymbol{u}_{b_s}, \boldsymbol{v}_{c_1}, \ldots, \boldsymbol{v}_{c_t}) \mathbin{\&} D_{\mathfrak{A}}^{c^*}(\boldsymbol{u}_{b_1}, \ldots, \boldsymbol{u}_{b_s}, \boldsymbol{x}_{c_1}, \ldots, \boldsymbol{x}_{c_t}) \to$$

$$\to \boldsymbol{v}_{c^*} \approx \boldsymbol{x}_{c_1} \vee \ldots \vee \boldsymbol{v}_{c^*} \approx \boldsymbol{x}_{c_t} \qquad (1)$$

is satisfied in any $\mathfrak{K}$-model by any assignment whatever of elements to the symbols $u_{b_i}, v_{c*}, v_{c_j}, x_{c_j}$, these all being distinct from the e_δ. Hence, the universal closure $\Phi_{\mathfrak{A}}^{c^*}$ of (1) with respect to the $u_{b_i}, v_{c*}, v_{c_j}, x_{c_j}$ (now viewed as free individual variables), is valid in every $\mathfrak{K}$-model. Using this we easily prove

Theorem 1: *Let $\mathfrak{K}$ be a pseudoaxiomatizable class with natural generation; let $\mathfrak{A}$ be a $\mathfrak{K}$-model generated by a subset B. If $\mathfrak{G}$ is the group of all automorphisms of $\mathfrak{A}$ that leave each element of B fixed, then for every element a in $\mathfrak{A}$, $a\mathfrak{G}$ is finite.*

For $a \in B$, the theorem is trivial. Suppose a belongs to C, the complement of B. Then $D_{\mathfrak{A}}^{a}(u_{b_1}, \ldots, u_{b_s}, v_{c_1}, \ldots, v_{c_t})$ is true in $\mathfrak{A}$ under the standard interpretation; thus, so is $D_{\mathfrak{A}}^{a}(u_{b_1}, \ldots, u_{b_s}, v_{c_1 g}, \ldots, v_{c_t g})$ for every $g \in \mathfrak{G}$. Since $\Phi_{\mathfrak{A}}^{a}$ holds in $\mathfrak{A}$, we have either $ag^{-1} = c_1$, or $ag^{-1} = c_2$, ..., or $ag^{-1} = c_t$ for each $g \in \mathfrak{G}$, i.e., $a\mathfrak{G} \subseteq \{c_1, \ldots, c_t\}$. ∎

In the class of all algebras of a fixed similarity type the stronger relation $a\mathfrak{G} = \{a\}$ holds for every element a in any member of the class. An example of a class of models where $a\mathfrak{G} \supset \{a\}$ quite frequently is the class of partially ordered sets in which each element has exactly two immediate successors ([3]).

§2. As in §1, let $\mathfrak{K}$ be a pseudoaxiomatizable class with natural generation; let $U = \{u_b : b \in B\}$ be a set of new distinct individual constants indexed (without repetitions) by a set B. Consider a pair $\langle f_\sigma, \mathfrak{A}_\sigma \rangle$ consisting of a $\mathfrak{K}$-model $\mathfrak{A}_\sigma = \langle A_\sigma; P_\gamma; e_\delta : \gamma \in \Gamma, \delta \in \Delta \rangle$ and a map f_σ (not necessarily 1–1) from U onto a subset $B_\sigma \subseteq A_\sigma$, which generates $\mathfrak{A}_\sigma$ in $\mathfrak{K}$ (don't worry now about the index σ). Let $C_\sigma = A_\sigma \sim B_\sigma$, and let $V = \{v_c : c \in C_\sigma\}$ be a set of new distinct constants in 1–1 correspondence with C_σ via the map $v_c \to c$. It makes sense to talk about the diagram $D_{\mathfrak{A}_\sigma}(U, V)$ of $\mathfrak{A}_\sigma$ with respect to these maps. Let

$$D_{(\sigma)}^{\lambda}(u_{b_1^\lambda}, \ldots, u_{b_s^\lambda}, v_{c_1^\lambda}, \ldots, v_{c_t^\lambda}) \qquad (\lambda \in \Lambda_\sigma)$$

be an indexing of all its finite subdiagrams (s, t depend on λ, of course). Then for every $\lambda \in \Lambda_\sigma$, the formula

$$(\exists z_1) \ldots (\exists z_t)\, D_{(\sigma)}^{\lambda}(u_{b_1^\lambda}, \ldots, u_{b_s^\lambda}, z_1, \ldots, z_t),$$

or more simply,

$$E_{(\sigma)}^{\lambda}(u_{b_1^\lambda}, \ldots, u_{b_s^\lambda}),$$

is true in $\mathfrak{A}_\sigma$ under the given interpretation. In this situation we have the

Lemma: *Let* $\mathfrak{M}$ *be an arbitrary model of the similarity type of* $\mathcal{K}$, $\mathfrak{M} = \langle M; P_\gamma; e_\delta : \gamma \in \Gamma, \delta \in \Delta \rangle$. *Suppose* (i) *for every* $c^* \in C_\sigma$, $\Phi^{c^*}_{\mathfrak{A}_\sigma}$ *is true in* $\mathfrak{M}$; (ii) *there is a map* h: $U \to M$ *such that for every* $\lambda \in \Lambda_\sigma$, $\mathrm{E}^{\lambda}_{(\sigma)}$ *is satisfied in* $\mathfrak{M}$ *by this interpretation. Then* $\mathfrak{A}_\sigma$ *is canonically isomorphic to a submodel of* $\mathfrak{M}$ *including the image* Uh.

Let $W = \{\boldsymbol{w}_d : d \in M\}$ be a set of new distinct constants in 1–1 correspondence with M. By (ii), for every $c^* \in C_\sigma$, there are elements $d_1(c^*), \ldots, d_t(c^*)$ of M such that the formula

$$\mathrm{D}^{c^*}_{\mathfrak{A}_\sigma}(\boldsymbol{u}_{b_1(c^*)}, \ldots, \boldsymbol{u}_{b_s(c^*)}, \boldsymbol{v}_{c_1(c^*)}, \ldots, \boldsymbol{v}_{c_t(c^*)})$$

is satisfied in $\mathfrak{M}$ by the maps h and $\boldsymbol{v}_{c_j(c^*)} \to d_j(c^*)$ (s, t depend on c^*, of course). This formula occurs in the hypothesis in $\Phi^{c^*}_{\mathfrak{A}_\sigma}$ and appears among the $\mathrm{D}^{\lambda}_{(\sigma)}$. Let $\mathrm{D}_{\mathfrak{M}}(W)$ be the diagram of $\mathfrak{M}$ in terms of the obvious map, and consider the system of formulas

$$\mathbf{R} = \mathrm{D}_{\mathfrak{M}}(W) \cup \{\mathrm{D}^{\lambda}_{(\sigma)} : \lambda \in \Lambda\} \cup$$

$$\cup \{\boldsymbol{v}_{c^*} \approx \boldsymbol{w}_{d_1(c^*)} \vee \ldots \vee \boldsymbol{v}_{c^*} \approx \boldsymbol{w}_{d_t(c^*)} : c^* \in C_\sigma\}.$$

By (i) and (ii), every finite subset of **R** is consistent, i.e., is satisfiable in some model, namely $\mathfrak{M}$. By the compactness theorem for FOPL, **R** itself is consistent. If $\mathfrak{N}$ is a model satisfying **R** (considered to be a set of sentences), then $\mathfrak{N}$ can be viewed as an extension of isomorphs of both $\mathfrak{M}$ and $\mathfrak{A}_\sigma$ according to the first two pieces of **R**. By the third, the isomorph of $\mathfrak{A}_\sigma$ is included in that of $\mathfrak{M}$; the desired isomorphism from $\mathfrak{A}_\sigma$ into $\mathfrak{M}$ is easily constructed. ∎

With the help of this lemma we can prove the following basic theorem:

Theorem 2: *Every pseudoaxiomatizable class* $\mathcal{K}$ *with natural generation is axiomatizable by means of first-order axioms in Skolem form (prenex sentences in which all universal quantifiers precede all existential).*

Let $\mathcal{K}$ be a pseudoaxiomatizable class of models with natural generation; let **S** be the set of all sentences of the form

$$(\boldsymbol{x}_1) \ldots (\boldsymbol{x}_m)(\exists \boldsymbol{y}_1) \ldots (\exists \boldsymbol{y}_n)\varphi \tag{2}$$

that are true in all $\mathcal{K}$-models, φ being an open formula with predicate symbols and constants among the $\boldsymbol{P}_\gamma$, $\boldsymbol{e}_\delta$ (and $\approx$). Let $\mathfrak{A}$ be a model of the similarity type of $\mathcal{K}$ satisfying $\mathbf{S}$, $\mathfrak{A} = \langle A; P_\gamma; e_\delta : \gamma \in \Gamma, \delta \in \Delta \rangle$. We must show $\mathfrak{A}$ is isomorphic to a $\mathcal{K}$-model. Suppose the cardinality of A is $\mathfrak{m}$; let I be a fixed set of power $\mathfrak{n}(\mathfrak{m})$. Let $\mathcal{K}'$ be the subset of the class $\mathcal{K}$ consisting of those $\mathcal{K}$-models whose bases are included in I. Taking A as the index set B in the earlier construction in this section, we let $\mathcal{F} = \{\langle f_\sigma, \mathfrak{A}_\sigma \rangle : \sigma \in \Sigma\}$ be the set of all pairs $\langle f_\sigma, \mathfrak{A}_\sigma \rangle$ such that $f_\sigma : U \to I$ and the image $B_\sigma = Uf_\sigma$ is a $\mathcal{K}$-generating subset of the $\mathcal{K}'$-model $\mathfrak{A}_\sigma$. As above, for each $\sigma \in \Sigma$, we can construct the formulas $\mathrm{E}^\lambda_{(\sigma)}$ $(\lambda \in \Lambda_\sigma)$. It follows from (b) in the definition of pseudoaxiomatizability that any map g from U into a $\mathcal{K}$-model $\mathfrak{N}$ is equivalent to a member of $\mathcal{F}$ in the sense that if $\mathfrak{N}'$ is the $\mathcal{K}$-submodel generated by Ug in $\mathfrak{N}$, then there is a $\tau \in \Sigma$ such that $\mathfrak{N}'$ is isomorphic to $\mathfrak{A}_\tau$, and f_τ is the composite of g and this isomorphism..From this it follows that any map from U into a $\mathcal{K}$-model $\mathfrak{N}$ satisfies the infinite formula

$$\bigvee_{\sigma \in \Sigma} \Big(\mathop{\&}_{\lambda \in \Lambda_\sigma} \mathrm{E}^\lambda_{(\sigma)} \Big) \tag{3}$$

i.e., this formula is valid in all $\mathcal{K}$-models. Hence, so is the equivalent infinite formula

$$\mathop{\&}_{\lambda^* \in \Lambda^*} \Big(\bigvee_{\sigma \in \Sigma} \mathrm{E}^{\sigma\lambda^*}_{(\sigma)} \Big), \tag{4}$$

where Λ^* is the set of all appropriate choice functions, i.e., functions $\lambda^* : \Sigma \to \bigcup\{\Lambda_\sigma : \sigma \in \Sigma\}$ such that $\sigma\lambda^* \in \Lambda_\sigma$. Thus for each $\lambda^* \in \Lambda^*$, the formula

$$\bigvee_{\sigma \in \Sigma} \mathrm{E}^{\sigma\lambda^*}_{(\sigma)} \tag{5}$$

is valid throughout $\mathcal{K}$. By (a) in the definition of pseudoaxiomatizability, there is a finite subset $\{\sigma_1, \ldots, \sigma_\nu\} \subseteq \Sigma$ such that $\mathrm{E}^{\sigma_1\lambda^*}_{(\sigma_1)} \vee \ldots \vee \mathrm{E}^{\sigma_\nu\lambda^*}_{(\sigma_\nu)}$ is valid in all $\mathcal{K}$-models. This means the universal closure

$$(\forall)\,(\mathrm{E}^{\sigma_1\lambda^*}_{(\sigma_1)} \vee \ldots \vee \mathrm{E}^{\sigma_\nu\lambda^*}_{(\sigma_\nu)}) \tag{6}$$

of this formula with respect to the $\boldsymbol{u}_b$ occurring in it is valid in all $\mathcal{K}$-models and thus is true in $\mathfrak{A}$, for it is a sentence equivalent to one in Skolem form (2). Therefore, $\mathfrak{A}$ satisfies (6), and (5) is valid in $\mathfrak{A}$. As λ^* was arbitrary, (4) and the equivalent (3) are satisfied in $\mathfrak{A}$ by every assignment of elements to the $\boldsymbol{u}_b$. In particular, if $h : U \to A$ is the map sending $\boldsymbol{u}_b$ to b $(b \in A)$, then there is a $\tau \in \Sigma$ such that all the formulas $\mathrm{E}^\lambda_{(\tau)}$ $(\lambda \in \Lambda_\tau)$ are satisfied in $\mathfrak{A}$ under this interpretation. For every $c^* \in C$, $\Phi^{c^*}_{\mathfrak{A}_\tau}$ is a member of $\mathbf{S}$, and thus true in $\mathfrak{A}$. Applying the lemma with $\mathfrak{A}$ in the place of $\mathfrak{M}$, we find that $\mathfrak{A}_T$ is isomorphic to a submodel of $\mathfrak{A}$ including Uh, but $Uh = A$, so $\mathfrak{A}$ is itself isomorphic to the $\mathcal{K}$-model $\mathfrak{A}_T$. ∎

Corollary: *Let $\mathcal{K}$ be an axiomatizable class of models such that the intersection of any collection of $\mathcal{K}$-submodels of given $\mathcal{K}$-model is either empty or a $\mathcal{K}$-model. Then every model of the type of $\mathcal{K}$ in which each finite subset is included in some $\mathcal{K}$-submodel is itself a $\mathcal{K}$-model.*

This follows immediately from the axiomatizability of $\mathcal{K}$ by means of axioms in Skolem form (cf. [91], part IV). ∎

§**3.** Theorem 2 does not provide an explicit description of those systems of Skolem axioms which specify classes of models with natural generation. The simplest sufficient test can be formulated as follows. A FOPL sentence of the form

$$(x_1) \dots (x_m)(\exists y_1) \dots (\exists y_n)\varphi(x_1, \dots, x_m, y_1, \dots, y_n) \,\&$$

$$\& \,(\forall)(\varphi(x_1, \dots, x_m, y_1, \dots, y_n) \,\&$$

$$\& \,\varphi(x_1, \dots, x_m, z_1, \dots, z_n) \to y_1 \approx z_1 \,\& \dots \&\, y_n \approx z_n)$$

($\varphi(\mathfrak{x}, \mathfrak{y})$ is quantifier-free, $\mathfrak{x} = \langle x_1, \dots, x_m \rangle, \mathfrak{y} = \langle y_1, \dots, y_n \rangle$) is called *explicitly functional.*

Theorem 3: *Every class $\mathcal{K}$ of models which is characterizable by a system of explicitly functional axioms has natural generation. Every automorphism of a $\mathcal{K}$-model $\mathfrak{M}$ that leaves each element of a generating subset of $\mathfrak{M}$ fixed also fixes the other elements of $\mathfrak{M}$, i.e., is the identity map.*

The proof is the same as in the case of algebras. ∎

More general than the explicitly functional axioms, at least in appearance, are the sentences of the form (φ open):

$$(\mathfrak{x}_1)(\exists \mathfrak{y}_1)(\mathfrak{x}_2)(\exists \mathfrak{y}_2) \dots (\mathfrak{x}_p)(\exists \mathfrak{y}_p)\, \varphi(\mathfrak{x}_1, \dots, \mathfrak{x}_p, \mathfrak{y}_1, \dots, \mathfrak{y}_p) \,\&$$

$$\& \,(\forall)(\varphi(\mathfrak{x}_1, \dots, \mathfrak{x}_p, \mathfrak{y}_1, \dots, \mathfrak{y}_p)\, \varphi(\mathfrak{x}_1, \dots, \mathfrak{x}_p, \mathfrak{z}_1, \dots, \mathfrak{z}_p) \to$$

$$\to \mathfrak{y}_1 \approx \mathfrak{z}_1 \,\& \dots \&\, \mathfrak{y}_p \approx \mathfrak{z}_p). \tag{7}$$

It is easy to see, however, that every sentence of the form (7) is equivalent to the conjunction of an explicitly functional sentence and a few universal sentences. Thus, the conclusions of Theorem 3 hold for classes characterizable by means of axioms of the form (7).

§ **4.** The sort of generation considered in topological algebras and algebras with partial operations is not always natural. This leads us to make the following definition. Let $\mathcal{K}$ be a class of models with fundamental predicates P_γ ($\gamma \in \Gamma$). Let $\mathfrak{M}$ be a model of the type of $\mathcal{K}$; let Π be a subset of Γ. A submodel $\mathfrak{N}$ of the model $\mathfrak{M}$ is said to be Π-*closed* in $\mathfrak{M}$ iff for every $\gamma \in \Pi$, every sequence $\langle a_1, \ldots, a_{n_\gamma - 1}\rangle$ of elements of $\mathfrak{N}$, and every element $\mathfrak{M}$, and every element b of $\mathfrak{M}$, $P_\gamma(a_1, \ldots, a_{n_\gamma - 1}, b)$ implies b belongs to $\mathfrak{N}$. It is easy to see that the intersection of any collection of Π-closed submodels of $\mathfrak{M}$ is either empty or a submodel Π-closed in $\mathfrak{M}$. If $\mathfrak{M}$ is a $\mathcal{K}$-model, and B is a subset of $\mathfrak{M}$, we say that B Π-*generates* the submodel $\mathfrak{N}$ of $\mathfrak{M}$ in $\mathcal{K}$ iff $\mathfrak{N}$ is the smallest Π-closed $\mathcal{K}$-submodel of $\mathfrak{M}$ including B. $\mathcal{K}$ is said to be a class with Π-*generation* iff the intersection of every collection of Π-closed $\mathcal{K}$-submodels of a given $\mathcal{K}$-model is either empty or a $\mathcal{K}$-submodel.

Theorem 4: *Let $\mathfrak{M}$ be a model with basic predicates $\boldsymbol{P}_\gamma$ ($\gamma \in \Gamma$). Suppose $\Pi \subseteq \Gamma$, and $\mathfrak{M}$ satisfies a sentence of the form $(\eth_1 \boldsymbol{x}_1) \ldots (\eth_m \boldsymbol{x}_m)\varphi$, where φ is an open positive* FOPL *formula, and $\eth_i = \exists$ implies $\boldsymbol{x}_i$ occurs only in the final (n_γth) argument places of the predicate symbols $\boldsymbol{P}_\gamma$, and then only for $\gamma \in \Pi$. Then every Π-closed submodel of $\mathfrak{M}$ satisfies this sentence.* ∎

In particular, it follows that every class of models axiomatizable by means of a system of universal axioms and axioms of the form indicated in Theorem 4 has Π-generation for the appropriate Π.

NOTES

(1) It becomes clear from what follows that in (a) "sentences" should be changed to "formulas (possibly with free variables)", and in (b) = should be replaced with $\leqslant$ at both occurrences.

(2) This section and the next have been edited in translation to improve the flow of argument and notation.

(3) This example is dubious; in particular, the class does not have natural generation. A better example is the class of algebras with a single unary operation in which each element in the range of the operation is the image of exactly three elements. Neither class contradicts the converse of Theorem 2 below, but the class of dense linear orders does.

CHAPTER 8

DEFINING RELATIONS IN CATEGORIES

The immediate purpose of the present article is the transfer of the theory of defining relations to classes of models. However, the basic structural notions of the theory of models – submodels, direct products, etc. – can be expressed by homomorphisms in the sense of [V] ; moreover, this can be done in the general theory of categories. Therefore, defining relations are introduced and studied from the outset in general categories. Finally, we indicate their interpretation in categories of models.

§ **1.** In agreement with Eilenberg and MacLane [28], a *category* $\mathfrak{K}$ is a class of elements on which are defined a partial binary operation, written juxtapositively, and a unary predicate determining the neutral elements; both are subject to the axioms: (1) if ab, bc are defined, then $a(bc)$ and $(ab)c$ are defined and equal; (2) if $a(bc)$ or $(ab)c$ is defined, then ab and bc are, too; (3) if e is a neutral element, then ee is defined and equal to e; the definition of ae or ea implies $ae = a$ or $ea = a$, respectively; (4) for every a in $\mathfrak{K}$ there are neutral elements e and e' such that $ea = ae' = a$. The neutral elements are also called the *objects* of the category $\mathfrak{K}$. If for two objects e, e', we have $ea = ae' = a$, then a is called a *homomorphism* of e into e'. A homomorphism a is called an *isomorphism* iff there is a homomorphism b such that ab and ba are neutral elements of $\mathfrak{K}$. In this case the objects ab, ba are said to be isomorphic.

In agreement with MacLane [100], an object $\mathfrak{A}$ of a category $\mathfrak{K}$ is called a *direct composition* of the system of objects $\mathfrak{A}_\alpha$ $(\alpha \in \Gamma)$ iff there are homomorphisms π_α of $\mathfrak{A}$ into $\mathfrak{A}_\alpha$ $(\alpha \in \Gamma)$ such that for any system of homomorphisms σ_α of an object $\mathfrak{B}$ into $\mathfrak{A}_\alpha$ $(\alpha \in \Gamma)$, there is a unique homomorphism $\xi: \mathfrak{B} \to \mathfrak{A}$ satisfying the relation $\sigma_\alpha = \xi\pi_\alpha$ for $\alpha \in \Gamma$. Dually, an object $\mathfrak{A}$ of $\mathfrak{K}$ is called a *$\mathfrak{K}$-free composition* of a collection $\{\mathfrak{A}_\alpha: \alpha \in \Gamma\}$ iff there are homomorphisms $\pi_\alpha: \mathfrak{A}_\alpha \to \mathfrak{A}$ $(\alpha \in \Gamma)$ such that for any collection of homomorphisms σ_α of the $\mathfrak{A}_\alpha$ into an arbitrary $\mathfrak{K}$-object $\mathfrak{B}$ $(\alpha \in \Gamma)$, there is a unique homomorphism ξ satisfying the relation $\sigma_\alpha = \pi_\alpha\xi$ for $\alpha \in \Gamma$. Direct and $\mathfrak{K}$-free compositions may not exist, but if they do then they are defined uniquely up to isomorphism.

A *subcategory* $\mathcal{L}$ of a category $\mathfrak{K}$ is a subclass of the class of $\mathfrak{K}$-objects together with all homomorphisms of $\mathfrak{K}$ associated with all possible pairs of $\mathcal{L}$-objects. The notions of isomorphs and direct and free compositions will generally change in meaning on passage from $\mathfrak{K}$ to $\mathcal{L}$. However, isomorphism and direct composition will be taken, in what follows, to be in the sense of the basic category, although free composition in subcategories will be considered in the sense of the subcategories.

§ 2. In what follows we shall need categories of a more specialized form; these we shall call *categories of structured sets* or simply, categories of structures. Analogous notions are considered by Isbell [61] and MacLane himself [100]. To begin with, there is a class $\mathfrak{K}$ of objects called *structures* (structured sets, spaces). Associated with every structure $\mathfrak{A}$ is a uniquely determined set $\mu(\mathfrak{A})$ called the *basic set* or *base* of $\mathfrak{A}$. In addition there is a collection H of single-valued unary mappings from bases of structures to bases of structures; it is demanded that the identity transformation of each structure onto itself belongs to H, and that for a map ρ from the base of the structure $\mathfrak{A}$ (or more simply, from $\mathfrak{A}$) into the structure $\mathfrak{B}$ and a map σ from $\mathfrak{B}$ into $\mathfrak{C}$, if $\rho, \sigma \in H$, then $\rho\sigma \in H$. The maps in H are called *homomorphisms*. We consider two structures to be *identical* iff they have a common base and the identity map of the base onto itself is a homomorphism of the first structure onto the second, and also of the second onto the first. We shall consider that on an arbitrary set A a $\mathfrak{K}$-structure has been defined, if some 1–1 correspondence of A with the base of a $\mathfrak{K}$-structure $\mathfrak{B}$ is added to H and required to be an isomorphism in the obvious sense. A class of structures is *abstract* iff whenever it contains a structure it contains as well all possible isomorphs of that structure. In the following we shall consider only abstract classes of structures. The collection H of all homomorphisms of $\mathfrak{K}$-structures is a category in the stated sense. Objects of this category will be identified with the structures of which they in fact are the identity maps; in this way, the class $\mathfrak{K}$ of structures can be viewed as a category.

In a category $\mathfrak{K}$ of structures, a structure $\mathfrak{B}$ is called a *substructure* of the $\mathfrak{K}$-structure $\mathfrak{A}$ iff: (1) $\mu(\mathfrak{B}) \subseteq \mu(\mathfrak{A})$; (2) a homomorphism of $\mathfrak{A}$ into an arbitrary $\mathfrak{K}$-structure $\mathfrak{C}$ becomes a homomorphism of $\mathfrak{B}$ into $\mathfrak{C}$ when restricted to $\mu(\mathfrak{B})$; (3) any map of a structure $\mathfrak{C}$ into $\mathfrak{B}$ which is a homomorphism of $\mathfrak{C}$ into $\mathfrak{A}$ is also a homomorphism of $\mathfrak{C}$ into $\mathfrak{B}$. The substructure $\mathfrak{B}$ is uniquely determined in $\mathfrak{A}$ by $\mu(\mathfrak{B})$. A $\mathfrak{K}$-structure $\mathfrak{B}$ will be called a *strong substructure* of $\mathfrak{A}$ iff in addition to (1)–(3) the following holds: (4) every homomorphism into $\mathfrak{B}$ is also a homomorphism into $\mathfrak{A}$.

The direct composition $\mathfrak{A}$ of the structures $\mathfrak{A}_\alpha$ $(\alpha \in \Gamma)$ is called *separable* iff whenever $a, b \in \mu(\mathfrak{A})$ (or simply, $a, b \in \mathfrak{A}$), if $a\pi_\alpha = b\pi_\alpha$ for every canonical homomorphism π_α $(\alpha \in \Gamma)$, then $a = b$. This composition is called *complete* iff for every system $\{a_\alpha \in \mathfrak{A}_\alpha : \alpha \in \Gamma\}$, there is an element $a \in \mathfrak{A}$ such that $a_\alpha = a\pi_\alpha$ $(\alpha \in \Gamma)$. A direct composition is called a *direct product* iff it is separable and complete. In this case the elements of the direct composition will be identified with the elements of the cartesian product of the bases of the factors.

We call a structure $\mathfrak{A}$ of the category $\mathcal{K}$ a *unit* iff $\mathfrak{A}$ contains but one element, and any map of an arbitrary $\mathcal{K}$-structure into $\mathfrak{A}$ is a homomorphism. $\mathfrak{A}$ is called a *null* iff $\mathfrak{A}$ has but one element, and any map of $\mathfrak{A}$ into an arbitrary $\mathcal{K}$-structure is a homomorphism. It is easy to see that if the category $\mathcal{K}$ has a null then every direct composition is complete and separable. The following is also easy to prove:

Theorem 1: *Suppose that in a category $\mathcal{K}$ all substructures are strong and all direct compositions are separable; suppose further that for every $\alpha \in \Gamma$, $\mathfrak{B}_\alpha$ is a substructure of the $\mathcal{K}$-structure $\mathfrak{A}_\alpha$, and that the direct compositions $\mathfrak{A}$ of $\{\mathfrak{A}_\alpha : \alpha \in \Gamma\}$ and $\mathfrak{B}$ of $\{\mathfrak{B}_\alpha : \alpha \in \Gamma\}$ exists. Then $\mathfrak{B}$ is a substructure of $\mathfrak{A}$.* ■

Let us agree to call a category $\mathcal{K}$ *multiplicatively closed* iff it contains the direct composition of an arbitrary collection of $\mathcal{K}$-structures.

Theorem 2: *Let the category $\mathcal{K}$ be multiplicatively closed, contain a unit, and have strong substructures and separable direct compositions. In order that all canonical homomorphisms of $\mathcal{K}$-structures into their $\mathcal{K}$-free compositions be isomorphims onto the corresponding substructures, it is necessary and sufficient that every $\mathcal{K}$-structure be isomorphically embeddable in a $\mathcal{K}$-structure with a unit substructure.* ■

In our further examination of categories of structures we shall always assume the following *axiom of definiteness* to be satisfied: the collection of all $\mathcal{K}$-structures defined on an arbitrary fixed set can be viewed as a set of definite power. A category $\mathcal{K}$ of structures is called *bounded* iff for every cardinal number $\mathfrak{m}$ one can find a cardinal $\mathfrak{n} = \mathfrak{n}(\mathfrak{m})$ such that in every $\mathcal{K}$-structure a subset of the base of power $\leqslant \mathfrak{m}$ is included in some $\mathcal{K}$-substructure of power $\leqslant \mathfrak{n}$.

§**3.** Let $\mathcal{K}_0$ be a general category, and $\mathcal{K}, \mathcal{L}$, subcategories with $\mathcal{K} \supseteq \mathcal{L}$. We say that an $\mathcal{L}$-object $\mathfrak{B}$ is a *replica* of a $\mathcal{K}$-object $\mathfrak{A}$ in the category $\mathcal{L}$ (an $\mathcal{L}$-

replica) iff there is a homomorphism $\pi: \mathfrak{A} \to \mathfrak{B}$ such that for any homomorphism σ of $\mathfrak{A}$ into an arbitrary $\mathcal{L}$ object $\mathfrak{C}$, there is a unique homomorphism $\xi: \mathfrak{B} \to \mathfrak{C}$ which satisfies the relation $\sigma = \pi\xi$. The homomorphism π is termed the canonical homomorphism of the replica. It is easy to see that if an $\mathcal{L}$-replica of $\mathfrak{A}$ replica exists, then it is determined uniquely up to isomorphism; in particular the $\mathcal{L}$-replica of an $\mathcal{L}$-object always coincides with the object itself.

Theorem 3: *Let $\mathcal{K} \supseteq \mathcal{L} \supseteq \mathcal{M}$ be subcategories of a general category $\mathcal{K}_0$. Let $\mathfrak{A}^{\mathcal{L}}, \mathfrak{A}^{\mathcal{M}}$ be replicas of a $\mathcal{K}$-object $\mathfrak{A}$ in $\mathcal{L}$ and $\mathcal{M}$, respectively. Then $\mathfrak{A}^{\mathcal{M}}$ is the $\mathcal{M}$-replica of $\mathfrak{A}^{\mathcal{L}}$. If the $\mathcal{K}$-object $\mathfrak{A}$ is the $\mathcal{L}$-free composition of the system of $\mathcal{K}$-objects $\mathfrak{A}_\alpha$ $(\alpha \in \Gamma)$ and the $\mathcal{L}$-replicas $\mathfrak{A}^{\mathcal{L}}, \mathfrak{A}_\alpha^{\mathcal{L}}$ $(\alpha \in \Gamma)$ of these $\mathcal{K}$-objects exist, then $\mathfrak{A}^{\mathcal{L}}$ is the $\mathcal{L}$-free composition of $\{\mathfrak{A}_\alpha^{\mathcal{L}}: \alpha \in \Gamma\}$.* ∎

We remark that if categories of structures are considered and $\mathfrak{B}$ is the $\mathcal{L}$-replica of a $\mathcal{K}$-structure $\mathfrak{A}$ with canonical homomorphism π, then $\mathfrak{B}$ has no substructure other than itself which includes $\mathfrak{A}\pi$, that is, $\mathfrak{B}$ is generated by the elements of $\mathfrak{A}\pi$.

A collection S of elements of a structure $\mathfrak{A}$ in a category $\mathcal{K}$ is called $\mathcal{K}$-*dense* in $\mathfrak{A}$ iff for any homomorphisms ρ, σ of $\mathfrak{A}$ into an arbitrary $\mathcal{K}$-structure, if $a\rho = a\sigma$ for all $a \in S$, then $\rho = \sigma$.

It is easy to see that if π_α $(\alpha \in \Gamma)$ are the canonical homomorphisms of the $\mathcal{K}$-free composition $\mathfrak{A}$ of $\{\mathfrak{A}_\alpha: \alpha \in \Gamma\}$, then the set $\cup\{(\mathfrak{A}_\alpha)\pi_\alpha: \alpha \in \Gamma\}$ is $\mathcal{K}$-dense in $\mathfrak{A}$. Furthermore, if $\Gamma_1 \subseteq \Gamma$, and if $\mathfrak{B}$ is a $\mathcal{K}$-substructure of $\mathfrak{A}$ and includes the set $\cup\{(\mathfrak{A}_\beta)\pi_\beta: \beta \in \Gamma_1\}$ as a $\mathcal{K}$-dense subset, then $\mathfrak{B}$ is the $\mathcal{K}$-free composition of $\{\mathfrak{A}_\beta: \beta \in \Gamma_1\}$. In addition, it follows from the definition that the canonical image of a $\mathcal{K}$-structure in its $\mathcal{L}$-replica (where $\mathcal{K} \supseteq \mathcal{L}$) is an $\mathcal{L}$-dense subset of the latter.

A subcategory $\mathcal{L}$ of a category $\mathcal{K}$ is called R-*complete* in $\mathcal{K}$ iff every $\mathcal{K}$-object has an $\mathcal{L}$-replica. Theorem 3 shows that if $\mathcal{K} \supseteq \mathcal{L} \supseteq \mathcal{M}$, and $\mathcal{M}$ is *R*-complete in $\mathcal{L}$, and $\mathcal{L}$ is R-complete in $\mathcal{K}$, then $\mathcal{M}$ is R-complete in $\mathcal{K}$.

Theorem 4: *If a category $\mathcal{K}$ is multiplicatively closed, then so is every* R-*complete subcategory. If $\mathcal{K}$ contains a unit, then so does every* R-*complete subcategory.* ∎

A structure $\mathfrak{A}$ in a category $\mathcal{K}$ is called *regular* iff for every subset S of $\mathfrak{A}$ there is a substructure of $\mathfrak{A}$ including S as a $\mathcal{K}$-dense subset. $\mathcal{K}$ is a *regular category* iff every $\mathcal{K}$-structure is regular.

Theorem 5: *If a subcategory $\mathcal{L}$ of a category $\mathcal{K}$ of structures contains a unit, and is multiplicatively closed, regular, and bounded, then $\mathcal{L}$ is* R-*complete in $\mathcal{K}$, and the $\mathcal{L}$-free composition of any collection of $\mathcal{L}$-structures exists.* ∎

The proof is analogous to the proof of the theorem on the existence of a topological algebra with given defining relations and generating topological space [M5].

§4. Let $\mathcal{K}_\tau$ (or $\mathcal{K}$, for short) be the class of all models of a fixed similarity type $\tau = \langle r_1, r_2, \ldots \rangle$, and let $\boldsymbol{P}_1^{r_1}, \boldsymbol{P}_2^{r_2}, \ldots$ be predicate symbols for the basic predicates of this class.

In agreement with [V], a single-valued transformation σ of a $\mathcal{K}$-model $\langle A; Q_1, Q_2, \ldots \rangle$ into a $\mathcal{K}$-model $\langle B; R_1, R_2, \ldots \rangle$ is called a *homomorphism* iff for all i and all sequences $\langle a_1, a_2, \ldots \rangle$ of elements from A, if $Q_i(a_1, a_2, \ldots)$ holds, then $R_i(a_1\sigma, a_2\sigma, \ldots)$ is also true.

The collection H of all such homomorphisms of $\mathcal{K}$-models satisfies the requirements of § 1, and the pair $\langle \mathcal{K}, H \rangle$ is a category of structures: the models of the given type τ. It is easy to verify that the notions of submodel and direct product of models coincide with the notions of substructure and direct composition, the unit structure is the one-element model in which all basic predicates are true, and the null structure is the one-element model with all predicates false.

Suppose $\mathcal{L}$ is a subclass of the class $\mathcal{K}$, A is a set of individual symbols, and $\mathbf{F}$ is a collection of formulas of the form $\boldsymbol{P}_i(a_1, a_2, \ldots)$, where the $a_j \in A$. It is natural to call an $\mathcal{L}$-model $\mathfrak{B} = \langle B; R_1, R_2, \ldots \rangle$ a *model with generators A and defining relations* $\mathbf{F}$ in the class $\mathcal{L}$ iff there is a map $\pi: A \to B$ with the properties: (1) $A\pi$ is $\mathcal{L}$-dense in $\mathfrak{B}$; (2) if $\boldsymbol{P}_i(a_1, a_2, \ldots) \in \mathbf{F}$, then $R_i(a_1\pi, a_2\pi, \ldots)$ holds. (3) let σ be a map of A into an arbitrary $\mathcal{L}$-model $\mathfrak{C} = \langle C; R_1', R_2', \ldots \rangle$ such that whenever $\boldsymbol{P}_i(a_1, a_2, \ldots) \in \mathbf{F}$, $R_i'(a_1\sigma, a_2\sigma, \ldots)$ holds. Then there is a map $\xi: \mathfrak{B} \to \mathfrak{C}$ satisfying the relation $\sigma = \pi\xi$.

Letting $\mathfrak{A} = \langle A; Q_1, Q_2, \ldots \rangle$ be the model with base A and predicates Q_i $(i = 1, 2, \ldots)$, where $Q_i(a_1, a_2, \ldots)$ holds iff $\boldsymbol{P}_i(a_1, a_2, \ldots) \in \mathbf{F}$, we see that $\mathfrak{B}$ *is the* $\mathcal{L}$*-replica of* $\mathfrak{A}$.

One can also take $\mathcal{K}$ to be the class of all possible models of given similarity type, their bases being endowed with topologies, and take H to be the class of continuous maps which are simultaneously homomorphisms in the model-theoretic sense. If $\mathcal{L}$ is the subclass of topological algebras, then the hypotheses of Theorem 5 are satisfied, and the $\mathcal{L}$-replica of a $\mathcal{K}$-structure $\mathfrak{A}$ is the topological algebra determined by the topological space of $\mathfrak{A}$ in the sense of [M5] and the positive diagram of $\mathfrak{A}$ as indicated above.

CHAPTER 9

THE STRUCTURAL CHARACTERIZATION OF CERTAIN CLASSES OF ALGEBRAS

MacLane [100] has found conditions under which a general category is isomorphic to a category of abelian semigroups with zero. In the more specialized theory of categories of structures, it is natural to consider, along with the usual isomorphism, a more specialized form: the structural equivalence. The basic purpose of this note is the determination of necessary and sufficient conditions for a category of structures to be structurally equivalent to a subclass of the class of all algebras of a fixed similarity type, this subclass being multiplicatively closed and containing all subalgebras of its members. The terminology and results of [VIII] are used throughout what follows.

§ **1.** Let $\mathcal{L}$ be a subcategory of a category $\mathcal{K}$ of structures, and let $\mathfrak{A} \in \mathcal{K}$. A subset of the base of $\mathfrak{A}$ is called *$\mathcal{L}$-free* iff every map of this set into an $\mathcal{L}$-structure $\mathfrak{B}$ can be extended to a homomorphism of $\mathfrak{A}$ into $\mathfrak{B}$. The structure $\mathfrak{A}$ is called *$\mathcal{L}$-free* iff $\mathfrak{A} \in \mathcal{L}$, and $\mathfrak{A}$ includes an $\mathcal{L}$-dense $\mathcal{L}$-free subset. It follows that $\mathcal{L}$-free structures possessing $\mathcal{L}$-dense $\mathcal{L}$-free subsets of identical cardinality are isomorphic. We note that if $\mathcal{L}$ is regular, bounded, and multiplicatively closed, then in $\mathcal{L}$ there exist $\mathcal{L}$-free structures with $\mathcal{L}$-dense $\mathcal{L}$-free subsets of arbitrary power.

A subcategory $\mathcal{L}$ of a category $\mathcal{K}$ of structures is called *homomorphically closed* in $\mathcal{K}$ iff every homomorphic image of an $\mathcal{L}$-structure in a $\mathcal{K}$-structure is an $\mathcal{L}$-substructure of the latter. We shall say that a subset S of an $\mathcal{L}$-structure $\mathfrak{A}$ is an *$\mathcal{L}$-generating* set for $\mathfrak{A}$ iff $\mathfrak{A}$ includes no $\mathcal{L}$-substructures that include S other than $\mathfrak{A}$ itself. From now on we consider only categories of structures with strong substructures, that is, only those categories in which a homomorphism into a substructure of a structure $\mathfrak{A}$ is a homomorphism into $\mathfrak{A}$ itself.

Theorem 1: *Let the subcategory $\mathcal{L}$ of a category of structures be homomorphically closed in itself and contain $\mathcal{L}$-free structures with $\mathcal{L}$-free $\mathcal{L}$-dense* *subsets of every power. Then $\mathcal{L}$ is bounded and regular, and the intersection of any collection of $\mathcal{L}$-substructures of an $\mathcal{L}$-structure is either empty or again an $\mathcal{L}$-substructure. In addition, every generating set of an $\mathcal{L}$-structure is $\mathcal{L}$-dense in it, and the pre-image of an $\mathcal{L}$-substructure of an $\mathcal{L}$-structure under a homomorphic mapping of an $\mathcal{L}$-structure $\mathfrak{B}$ into $\mathfrak{A}$ is an $\mathcal{L}$-substructure of $\mathfrak{B}$.* ■

A structure $\mathfrak{A}$ is called *$\mathcal{L}$-freely cyclic* iff $\mathfrak{A}$ is in $\mathcal{L}$ and has an $\mathcal{L}$-free $\mathcal{L}$-dense subset consisting of a single element. So, if $\mathcal{L}$ contains an $\mathcal{L}$-freely cyclic structure, every $\mathcal{L}$-free structure is an $\mathcal{L}$-free composition of $\mathcal{L}$-freely cyclic structures.

We single out the following cases where dense sets are also generating sets. The canonical image of a $\mathcal{K}$-structure in its $\mathcal{L}$-replica is $\mathcal{L}$-generating. If a $\mathcal{K}$-structure $\mathfrak{A}$ includes an $\mathcal{L}$-free subset S, then $\mathfrak{A}$ can include no more than one $\mathcal{L}$-substructure in which S is $\mathcal{L}$-dense. Therefore, an $\mathcal{L}$-free structure is $\mathcal{L}$-generated by an $\mathcal{L}$-free $\mathcal{L}$-dense subset.

§ **2.** By analogy with the theory of groups, a collection $\{M_\alpha\colon \alpha\in\Gamma\}$ of subsets of a given set is called *local* iff each finite subset of $\bigcup\{M_\alpha\colon \alpha\in\Gamma\}$ is included in an M_β for appropriate $\beta\in\Gamma$. The category $\mathcal{K}$ of structures is called *additive* iff the union of any local collection of $\mathcal{K}$-substructures of an arbitrary $\mathcal{K}$-structure is itself a $\mathcal{K}$-substructure. A $\mathcal{K}$-dense subset of some $\mathcal{K}$-structure $\mathfrak{A}$ is called *finitely $\mathcal{K}$-dense* iff every element of $\mathfrak{A}$ lies in a $\mathcal{K}$-substructure in which some finite subset of S is dense.

Theorem 2: *If a category $\mathcal{K}$ of structures is homomorphically closed and contains $\mathcal{K}$-free structures with $\mathcal{K}$-free $\mathcal{K}$-dense subsets of arbitrary power, then $\mathcal{K}$ is additive. In a regular additive category $\mathcal{K}$, a $\mathcal{K}$-free $\mathcal{K}$-dense subset of a $\mathcal{K}$-structure is finitely $\mathcal{K}$-dense.* ■

Since $\mathcal{K}$-free structures with $\mathcal{K}$-free $\mathcal{K}$-dense subsets of identical power are isomorphic, it follows that if there is a $\mathcal{K}$-free structure with a $\mathcal{K}$-free $\mathcal{K}$-finitely dense subset of power $\mathfrak{m}$, then every $\mathcal{K}$-free $\mathcal{K}$-dense set of power $\mathfrak{m}$ is finitely $\mathcal{K}$-dense.

§ **3.** Let us suppose that the category $\mathcal{K}$ of structures contains $\mathcal{K}$-free structures with $\mathcal{K}$-free $\mathcal{K}$-dense subsets of arbitrary finite power. For $n=1,2,\ldots$, let $\mathfrak{W}_n$ be a $\mathcal{K}$-structure with distinct elements $v_{n\alpha}$ ($\alpha\in\Gamma_n \supseteq \{1,2,\ldots,n\}$), where $\{v_{n1}, v_{n2}, \ldots, v_{nn}\}$ is a $\mathcal{K}$-free $\mathcal{K}$-dense subset of $\mathfrak{W}_n$. In each $\mathcal{K}$-structure $\mathfrak{A}$ we define operations $\Phi_{n\alpha}^{\mathfrak{A}}(x_1,\ldots,x_n)$, for $\alpha\in\Gamma_n$, $n=1,2,\ldots$, in the following manner: let $\langle a_1,\ldots,a_n\rangle$ be a sequence of elements of $\mathfrak{A}$; by assumption, there is one and only one homomorphism $\sigma\colon \mathfrak{W}_n \to \mathfrak{A}$ such that $v_{ni}\sigma = a_i$ $(i = 1, 2, \ldots, n)$; we let $\Phi^{\mathfrak{A}}_{n\alpha}(a_1, \ldots, a_n) = v_{n\alpha}\sigma$. We note that if $\mathfrak{A}$ and $\mathfrak{B}$ are two structures with the same base M, it may happen that for some $a_1, \ldots, a_n \in M$, $\Phi^{\mathfrak{A}}_{n\alpha}(a_1, \ldots, a_n) \neq \Phi^{\mathfrak{B}}_{n\alpha}(a_1, \ldots, a_n)$. It follows from the definition, however, that if $\mathfrak{B}$ is a $\mathcal{K}$-substructure of $\mathfrak{A}$, and $a_1, \ldots, a_n \in \mathfrak{B}$, then $\Phi^{\mathfrak{A}}_{n\alpha}(a_1, \ldots, a_n) = \Phi^{\mathfrak{B}}_{n\alpha}(a_1, \ldots, a_n)$, i.e., the Φ-operations $\Phi_{n\alpha}$ are *stable* with respect to passage to $\mathcal{K}$-sub- and superstructures. Furthermore, the $\Phi_{n\alpha}$ are *preserved under homomorphism*, i.e., if $\sigma\colon \mathfrak{A} \to \mathfrak{B}$ is a homomorphism between $\mathcal{K}$-structures, then

$$(\Phi^{\mathfrak{A}}_{n\alpha}(a_1, \ldots, a_n))\,\sigma = \Phi^{\mathfrak{B}}_{n\alpha}(a_1\sigma, \ldots, a_n\sigma) \quad (a_1, \ldots, a_n \in \mathfrak{A}).$$

It follows that the $\Phi_{n\alpha}$ are invariant with respect to passage to direct products.

Theorem 3: *Suppose the category $\mathcal{K}$ is multiplicatively closed in itself and contains for every finite cardinal a $\mathcal{K}$-free structure with a $\mathcal{K}$-free, finitely $\mathcal{K}$-dense subset of this power. Then a subset of a $\mathcal{K}$-structure $\mathfrak{A}$ is the base of a $\mathcal{K}$-substructure of $\mathfrak{A}$ iff it is Φ-closed.*

Here, a subset B of $\mathfrak{A}$ is *Φ-closed* iff for all numbers n, all $\alpha \in \Gamma_n$, and all n-sequences $\langle a_1, \ldots, a_n\rangle$ over B, $\Phi^{\mathfrak{A}}_{n\alpha}(a_1, \ldots, a_n) \in B$. ■

§ 4. A category $\mathcal{K}$ of structures is called a category with *divisible homomorphisms* iff whenever $\rho\colon \mathfrak{A} \to \mathfrak{B}$ is a homomorphism between $\mathcal{K}$-structures, and $\sigma\colon \mathfrak{B} \to \mathfrak{C}$ is an arbitrary mapping between the bases of $\mathcal{K}$-structures such that $\rho\sigma\colon \mathfrak{A} \to \mathfrak{C}$ is a homomorphism, then σ is also a homomorphism. It is easy to see that any class of algebras is a category with divisible homomorphisms.

Theorem 4: *Let $\mathcal{K}$ be a category of structures with divisible homomorphisms containing $\mathcal{K}$-free structures with $\mathcal{K}$-free, finitely $\mathcal{K}$-dense subsets of every power. Then every mapping of a $\mathcal{K}$-structure $\mathfrak{A}$ into a $\mathcal{K}$-structure $\mathfrak{B}$ which preserves the Φ-operations is a homomorphism of $\mathfrak{A}$ into $\mathfrak{B}$.* ■

Two categories $\mathcal{K}_1$ and $\mathcal{K}_2$ are *isomorphic* [28] iff it is possible to establish a 1–1 correspondence Ω between the elements (homomorphisms) of $\mathcal{K}_1$ and $\mathcal{K}_2$ which is an isomorphism between $\mathcal{K}_1$ and $\mathcal{K}_2$ viewed as partial semigroups. In the case of categories of structures this means that the rule Ω permits one to construct from any $\mathcal{K}_1$-structure a corresponding $\mathcal{K}_2$-structure with a generally different base, and to construct from every $\mathcal{K}_1$-homomorphism a $\mathcal{K}_2$-homomorphism between the corresponding $\mathcal{K}_2$-structures that satisfies the appropriate conditions, etc. The following stricter notion of structural equivalence will be important. Let $\mathcal{K}_1$ and $\mathcal{K}_2$ be categories of structures; we say that $\mathcal{K}_1$ is *structurally equivalent* to $\mathcal{K}_2$ iff there is a rule Ω that uniquely describes for every $\mathcal{K}_1$-structure $\mathfrak{A}$ a $\mathcal{K}_2$-structure $\mathfrak{A}^{\Omega}$ with the same base as $\mathfrak{A}$ such that every $\mathcal{K}_1$-homomorphism between $\mathcal{K}_1$-structures $\mathfrak{A}$ and $\mathfrak{B}$ is also a $\mathcal{K}_2$-homomorphism between $\mathfrak{A}^{\Omega}$ and $\mathfrak{B}^{\Omega}$, and there is a rule inverse to Ω with the corresponding properties. From Theorem 4 we obtain the immediate corollary:

Corollary: *If $\mathcal{K}$ satisfies the hypotheses of Theorem 4, then it is structurally equivalent to a subcategory of the category of all algebras of an appropriate similarity type.*

Indeed, above we saw how to turn a $\mathcal{K}$-structure into an algebra with basic operations $\Phi_{n\alpha}$ ($\alpha \in \Gamma_n; n = 1, 2, \dots$). Let $\mathcal{K}_1$ be the class of all the algebras obtainable from $\mathcal{K}$-structures by this means. Theorem 4 shows that the correspondence between objects of $\mathcal{K}$ and $\mathcal{K}_1$ is 1–1 and satisfies the condition of coincidence of homomorphisms. ∎

§ **5.** A subcategory $\mathcal{L}$ of a category $\mathcal{K}$ of structures is called *quasifree* in $\mathcal{K}$ iff $\mathcal{L}$ contains a unit and is multiplicatively closed in $\mathcal{K}$, and $\mathcal{K}$-substructures of $\mathcal{L}$-structures are $\mathcal{L}$-structures. A subcategory $\mathcal{L}$ of $\mathcal{K}$ is *free* in $\mathcal{K}$ iff it is quasifree and homomorphically closed in $\mathcal{K}$. According to Birkhoff's theorem, every free subcategory of the category of all algebras of a fixed similarity type is a class of algebras characterized by a system of identities, i.e., a variety in the sense of [IV]. Quasivarieties [IV] are a special case of quasifree subcategories. If the basic category $\mathcal{K}$ is bounded and regular, then every quasifree subcategory is R-complete. In particular, every quasifree subclass of the class of all algebras of a fixed type is R-complete. These subclasses can be characterized also by their purely structural properties.

Theorem 5: *In order that a category $\mathcal{K}$ of structures be structurally equivalent to a quasifree subcategory of the category of all algebras of some fixed type, it is necessary and sufficient that $\mathcal{K}$ contain a unit and be regular, bounded, additive, and multiplicatively and homomorphically closed in itself.*

Necessity follows from the elementary properties of algebras; sufficiency is implied by the previous theorems. ∎

§ **6.** A class $\mathcal{L}_1$ of algebras with fundamental operations $f_\alpha(x_1, \dots, x_{m(\alpha)})$ ($\alpha \in \Gamma_1$) is said to be *rationally equivalent* to a class $\mathcal{L}_2$ of algebras with fundamental operations $g_\beta(x_1, \dots, x_{n(\beta)})$ ($\beta \in \Gamma_2$) iff there are $\mathcal{L}_2$-polynomials $\varphi_\alpha(\boldsymbol{x}_1, \dots, \boldsymbol{x}_{m(\alpha)})$ and $\mathcal{L}_1$-polynomials $\chi_\beta(\boldsymbol{x}_1, \dots, \boldsymbol{x}_{n(\beta)})$ ($\alpha \in \Gamma_1, \beta \in \Gamma_2$) such that every $\mathcal{L}_1$-algebra, viewed relative to the χ-operations, is an $\mathcal{L}_2$-algebra, and every $\mathcal{L}_2$-algebra, viewed relative to the φ-operations, is an $\mathcal{L}_1$-algebra, and this correspondence is an involution [M5]. Rational equivalence generally differs from structural equivalence, but can coincide with it; we note the simplest case of this:

Theorem 6: *If quasifree subclasses of classes of all algebras of fixed types (although the type may differ between classes) are structurally equivalent, then they are rationally equivalent.*

Let the prescribed classes be $\mathcal{L}_1$ and $\mathcal{L}_2$, and let $f_\alpha(x_1, \ldots, x_m)$ be one of the fundamental operations of the class $\mathcal{L}_1$. Consider the $\mathcal{L}_1$-free algebra $\mathfrak{W}$ with $\mathcal{L}_1$-free generators $v_1, \ldots, v_m$. From the structural equivalence of $\mathcal{L}_1$ and $\mathcal{L}_2$, it follows that $\mathfrak{W}$ is an $\mathcal{L}_2$-free structure with $\mathcal{L}_2$-free generators $v_1, \ldots, v_m$. Therefore, the element $f_\alpha(v_1, \ldots, v_m)$ of $\mathfrak{W}$ must be representable by means of some $\mathcal{L}_2$-polynomial: $\varphi_\alpha(v_1, \ldots, v_m)$. From the equation $f_\alpha(v_1, \ldots, v_m) = \varphi_\alpha(v_1, \ldots, v_m)$ in $\mathfrak{W}$, it follows that $\boldsymbol{f}_\alpha(\boldsymbol{x}_1, \ldots, \boldsymbol{x}_m) \approx \varphi_\alpha(\boldsymbol{x}_1, \ldots, \boldsymbol{x}_m)$ is valid in every $\mathcal{L}_1$-algebra, the polynomial φ_α being interpreted via the structural equivalence. Analogously, we find that for any fundamental $\mathcal{L}_2$-operation $g_\beta(x_1, \ldots, x_n)$, there is an $\mathcal{L}_1$-polynomial $\chi_\beta(\boldsymbol{x}_1, \ldots, \boldsymbol{x}_n)$ such that $\boldsymbol{g}_\beta(\boldsymbol{x}_1, \ldots, \boldsymbol{x}_n) \approx \chi_\beta(\boldsymbol{x}_1, \ldots, \boldsymbol{x}_n)$ is valid in every $\mathcal{L}_2$-algebra. Thus, the classes $\mathcal{L}_1$ and $\mathcal{L}_2$ are rationally equivalent. ■

CHAPTER 10

CERTAIN CLASSES OF MODELS

A structural characterization of quasifree classes of algebras was given in [IX], §5. Using this result, we state below structural characterizations for universally axiomatizable classes of models and for quasiprimitive classes of algebras (quasiequational classes or quasivarieties). At the same time we resolve the question of an intrinsic, purely algebraic characterization of quasiprimitive classes of algebraic systems – left open in [IV]. Finally, we show that, up to structural equivalence, quasiprimitive classes of algebraic systems are the only elementary (i.e., first-order axiomatizable) classes of models, homomorphically closed in themselves, which admit a theory of defining relations in the sense of [VIII].

In what follows all categories of structures will be assumed to have strong substructures, while direct compositions, in the cases when they exist, will be assumed to coincide with the direct products [VIII].

§1. We agree to say a category $\mathfrak{K}$ of structures has *finitary homomorphisms* iff for any two $\mathfrak{K}$-structures $\mathfrak{A}$ and $\mathfrak{B}$, no matter what local system $\{\mathfrak{A}_\alpha : \alpha \in \Gamma\}$ of $\mathfrak{K}$-substructures covering $\mathfrak{A}$ is chosen, any mapping from $\mathfrak{A}$ into $\mathfrak{B}$ that is a homomorphism of $\mathfrak{A}_\alpha$ into an appropriate $\mathfrak{K}$-substructure of $\mathfrak{B}$ for all $\alpha \in \Gamma$ is a homomorphism of $\mathfrak{A}$ into $\mathfrak{B}$. Corresponding to the usual group-theoretic terminology, a $\mathfrak{K}$-structure $\mathfrak{A}$ is called *locally finite* iff every finite subset of $\mathfrak{A}$ lies in some finite $\mathfrak{K}$-substructure. Clearly, all categories of models have finitary homomorphisms; it is also easy to prove the following theorem:

Theorem 1: *Every category $\mathfrak{K}$ of structures with finitary homomorphisms and locally finite structures is structurally equivalent to an appropriate class of models.* ∎

A category $\mathfrak{K}$ of structures is *locally compatible* iff whenever every finite subset of an arbitrary collection $\mathcal{S}$ of $\mathfrak{K}$-structures, defined on subsets of a given set, is embeddable in some $\mathfrak{K}$-structure as a set of $\mathfrak{K}$-substructures, then the whole of $\mathcal{S}$ can also be simultaneously so embedded in some $\mathcal{K}$-structure. From this definition it follows, in particular, that in a locally compatible category $\mathcal{K}$ every increasing chain of $\mathcal{K}$-structures, each embedded in the next, can be embedded in some embracing $\mathcal{K}$-structure. The compactness theorem for first-order predicate logic (FOPL) shows that every elementary class of models is locally compatible.

We recall that a class $\mathcal{K}$ of models is called *universally axiomatizable* iff it can be characterized by a collection of universal FOPL sentences, i.e., sentences of the form $(x_1) \dots (x_n)\, \varphi(x_1, \dots, x_n)$, where the expression φ contains no quantifiers.

Theorem 2: *For a category $\mathcal{K}$ of structures to be structurally equivalent to some universally axiomatizable class of models, it is necessary and sufficient that $\mathcal{K}$ be locally compatible and have finitary homomorphisms, and any subset of a $\mathcal{K}$-structure be a $\mathcal{K}$-substructure.* ∎

Theorem 3: *If a universally axiomatizable class of models with fundamental predicates $P_1^{r_1}, \dots, P_k^{r_k}$ is structurally equivalent to a class of models with fundamental predicates $Q_1^{s_1}, \dots, Q_l^{s_l}$, then throughout the classes there are equivalences of the following form holding: $P_i(x_1, \dots, x_{r_i}) \leftrightarrow \varphi_i(x_1, \dots, x_{r_i})$, $(i = 1, \dots, k)$, $Q_j(x_1, \dots, x_{s_j}) \leftrightarrow \chi_j(x_1, \dots, x_{s_j})$ $(j = 1, \dots, l)$, where the φ_i, χ_j are open formulas constructed with the aid of the equality sign and the predicate symbols $Q_1, \dots, Q_l$ and $P_1, \dots, P_k$, respectively.* ∎

In case the number of fundamental predicates is infinite, Theorem 3 still holds, but infinite expressions must be admissible as the φ_i, χ_j.

§2. A model $\mathfrak{A}$ with predicates $P_1, P_2, \dots,$ whose ranks are $n_1, n_2, \dots,$ respectively, is called an *algebraic system* of type $\tau = \langle I; n_1, n_2, \dots \rangle$, where I is a subset of the index set for the predicates such that for $i \in I$, P_i is the predicate of an operation on the base of $\mathfrak{A}$ (or simply, on $\mathfrak{A}$). The class $\mathcal{K}_\tau$ of all algebraic systems of type τ is bounded, multiplicatively and homomorphically closed in itself, regular, and contains a unit structure. The notions of quasifree and free subclasses of a category $\mathcal{K}$ of structures were introduced in [IX]. If $\mathcal{K}$ is a category of models, then a quasifree or free subclass, distinguishable in $\mathcal{K}$ by means of some system of axioms (i.e., first-order axiomatizable relative to $\mathcal{K}$), is called *quasiprimitive* or *primitive* in $\mathcal{K}$, respectively. Quasiprimitive (primitive) subclasses of a class $\mathcal{K}_\tau$ are called, simply, *quasiprimitive (primitive) classes of algebraic systems* of the specified type. From the theorems of Tarski-Łoś [163, 89] and Bing [8] it follows that a subclass $\mathcal{L}$ of a class $\mathcal{K}$ of models is quasiprimitive in $\mathcal{K}$ iff it can be distinguished in

$\mathcal{K}$ by axioms of the form $(x_1) \dots (x_n)(\psi_1 \,\&\, \dots \,\&\, \psi_s \to \psi_{s+1})$, where the ψ_i are expressions of the forms $P_j(x_{i_1}, \dots, x_{i_{n_j}})$ or $x_k \approx x_l$.

Suppose the category $\mathcal{K}$: (a) is multiplicatively closed, and (b) contains a unit. Then the intersection of any collection of quasifree (free) subclasses of $\mathcal{K}$ is again a quasifree (free) subclass. Thus, for every class $\mathcal{S}$ of $\mathcal{K}$-structures, there is a smallest quasifree (free) subclass $\mathcal{T}$ of $\mathcal{K}$ that includes $\mathcal{S}$. The class $\mathcal{T}$ is called the *quasifree (free) closure* of $\mathcal{S}$ in $\mathcal{K}$ and is written $\mathcal{T} = \mathcal{S}^q$ ($\mathcal{T} = \mathcal{S}^f$). It is easy to see that $\mathcal{S}^q$ consists of all possible $\mathcal{K}$-substructures of direct products of $\mathcal{S}$-structures. In order to obtain an analogous characterization for $\mathcal{S}^f$, we lay these additional demands on $\mathcal{K}$: (c) $\mathcal{K}$ is homomorphically closed in itself, and (d) if $\mathfrak{A}, \mathfrak{B} \in \mathcal{K}$, and σ is a homomorphism of $\mathfrak{A}$ onto $\mathfrak{B}$, then the pre-image under σ of any $\mathcal{K}$-substructure of $\mathfrak{B}$ is a $\mathcal{K}$-substructure of $\mathfrak{A}$. Then the free closure $\mathcal{L}^f$ of a quasifree subclass $\mathcal{L}$ consists of all possible $\mathcal{K}$-structures which are homomorphic images of $\mathcal{L}$-structures. From this it follows that if $\mathcal{K}$ and $\mathcal{S}$ are elementary classes of models, then $\mathcal{S}^q$ and $\mathcal{S}^f$ are elementary ([1]). Furthermore, if a category $\mathcal{K}$ satisfying (a)–(d) is regular, and $\mathcal{L}$ is a quasifree subcategory, then every $\mathcal{L}^f$-free $\mathcal{K}$-structure belongs to $\mathcal{L}$. In particular, if $\mathcal{L}^f$ contains free structures with any number of $\mathcal{L}^f$-free generators, which are dense under these conditions, then the supply of free structures does not change on passing from $\mathcal{L}$ to its free closure.

Theorem 4: *Let the regular category* $\mathcal{K}$, *satisfying* (a)–(d), *contain a finite structure* $\mathfrak{A}$. *Then:* (i) $\mathcal{K}$*-free structures with different finite numbers of free generators are not isomorphic;* (ii) *in the minimal quasifree subclass* $\{\mathfrak{A}\}^q$ *and free subclass* $\{\mathfrak{A}\}^f$ *containing* $\mathfrak{A}$, *every structure with a finite generating set is finite;* (iii) *if the number of non-isomorphic* $\mathcal{K}$*-structures of finite power is finite, then in* $\{\mathfrak{A}\}^f$ *are included only a finite number of minimal quasifree and free subclasses containing more than units.* ∎

The statements (i) and (iii) are generalizations of theorems of Fujiwara [43] and Scott [148], proved for varieties of algebras.

§ **3.** Let $\langle \Gamma, < \rangle$ be a partial ordering in which any two elements have a common greater element, and let $\mathcal{K}$ be a general category. Suppose that with every $\alpha \in \Gamma$ is associated an object $\mathfrak{A}_\alpha \in \mathcal{K}$, and with every pair $\langle \alpha, \beta \rangle$ ($\alpha < \beta$; $\alpha, \beta \in \Gamma$) is associated a homomorphism $\pi_{\alpha\beta}\colon \mathfrak{A}_\alpha \to \mathfrak{A}_\beta$ such that $\alpha < \gamma < \beta$ implies $\pi_{\alpha\beta} = \pi_{\alpha\gamma}\pi_{\gamma\beta}$. We say that Γ and the mappings $\alpha \to \mathfrak{A}_\alpha$ and $\langle \alpha, \beta \rangle \to \pi_{\alpha\beta}$ constitute a *direct spectrum.* An object $\mathfrak{A}$ of $\mathcal{K}$ with specified maps $\pi_\alpha\colon \mathfrak{A}_\alpha \to \mathfrak{A}$ ($\alpha \in \Gamma$) is called the *limit of the spectrum* [28] iff $\pi_\alpha = \pi_{\alpha\beta}\pi_\beta$ ($\alpha < \beta$), and for any $\mathfrak{B} \in \mathcal{K}$ and any system of homomorphisms $\sigma_\alpha\colon \mathfrak{A}_\alpha \to \mathfrak{B}$

$(\alpha \in \Gamma)$ satisfying the conditions $\sigma_\alpha = \pi_{\alpha\beta}\sigma_\beta$ $(\alpha < \beta)$, there is one and only one homomorphism ξ: $\mathfrak{A} \to \mathfrak{B}$ for which $\sigma_\alpha = \pi_\alpha \xi$ $(\alpha \in \Gamma)$. In case $\mathcal{K}$ is a category of structures, it will be assumed without further mention that Γ has a least element 0, and the mappings $\pi_{\alpha\beta}$ are homomorphisms of $\mathfrak{A}_\alpha$ onto $\mathfrak{A}_\beta$ $(\alpha < \beta)$. Then it is possible to consider the $\mathfrak{A}_\alpha$ $(\alpha \in \Gamma)$ and $\mathfrak{A} = \mathit{lim}\ \mathfrak{A}_\alpha$ to be defined on $\mathfrak{A}_0$ with appropriately chosen equality relations (cf. [V]). If $\mathcal{K}$ is the category of all models of a fixed similarity type, then for any direct spectrum under the stated conditions the limit model exists, and its construction is given in [V]. There it is also shown that if a universal or positive sentence holds in all models of the spectrum, then it holds as well in the limit model. Inasmuch as all universally axiomatizable classes of algebraic systems are characterized by positive or universal axioms, every such class contains spectral limits of its systems. The possibility of inverting this gives the following theorem:

Theorem 5: *A multiplicatively closed class of algebraic systems of type τ containing all $\mathcal{K}_\tau$-subsystems of its members is elementary iff it contains limits of all direct spectra of its members.* ∎

In particular, a quasifree class of algebraic systems is quasiprimitive iff it contains limits of all spectra of its systems. Taking into account Theorem 5 of [IX], we get: in order that a category $\mathcal{K}$ of structures be structurally equivalent to an elementary class of algebras, multiplicatively closed and containing all subalgebras of its members, it is necessary and sufficient that $\mathcal{K}$ be multiplicatively and homomorphically closed in itself, bounded, regular, and additive, and have divisible homomorphisms and limits of direct spectra of its structures. Adding to these conditions the demand for existence of a unit structure, we obtain a structural characterization of quasivarieties of algebras.

Let $\mathcal{K}$ be an arbitrary category of structures, and let $\mathfrak{A} \in \mathcal{K}$. An equivalence relation θ defined on the base of $\mathfrak{A}$ is called a *congruence* on $\mathfrak{A}$ (cf. [61]) iff θ is associated with some homomorphism of $\mathfrak{A}$ onto an appropriate $\mathcal{K}$-structure. An equivalence on $\mathfrak{A}$ is called an *outer congruence* iff for any two homomorphisms ρ, σ of an arbitrary $\mathcal{K}$-structure $\mathfrak{A}$, generated by $\{b_\alpha\colon \alpha \in \Gamma\}$, into $\mathfrak{A}$, if $b_\alpha\rho \equiv b_\alpha\sigma(\theta)$ for all $\alpha \in \Gamma$, then $b\rho \equiv b\sigma(\theta)$ for all $b \in \mathfrak{B}$.

Obviously, in order that a quasifree class $\mathcal{K}$ of algebras be free, it is necessary and sufficient that every outer congruence in a $\mathcal{K}$-algebra be a congruence.

Thus, in order to get an intrinsic structural characterization for varieties of algebras, it suffices to adjoin to the above collection of structural properties characterizing quasivarieties of algebras, the demand that outer congruences be congruences.

§4. A class $\mathcal{K}$ of models is said to have *local embeddability* iff whenever every finite submodel of an arbitrary model $\mathfrak{M}$ of appropriate type is isomorphically embeddable in some $\mathcal{K}$-model, then $\mathfrak{M}$ itself is embeddable in some $\mathcal{K}$-model.

Lemma 1: *Let the class $\mathcal{K}$ of models with local embeddability contain $\mathcal{K}$-free models with free dense generating sets of arbitrary finite powers. Then for each Φ-operation $\Phi_{n\alpha}$* (cf. [IX], §3), *a formula of the form*

$$\Phi_{n\alpha}(x_1, ..., x_n) \approx x_{n+1} \leftrightarrow (\exists x_{n+2}) \dots (\exists x_s) \varphi_{n\alpha}(x_1, ..., x_s) \quad (1)$$

(where $\varphi_{n\alpha}$ is an appropriate conjunction of formulas of the form $x_i \approx x_l$, $P_j(x_{i_1}, ..., x_{i_k})$) is valid in all $\mathcal{K}$-models. ∎

Lemma 2: *If a class $\mathcal{K}$ of models with local embeddability is homomorphically closed in itself and contains $\mathcal{K}$-free models with $\mathcal{K}$-free, finitely $\mathcal{K}$-dense subsets of every cardinality, then by augmenting the fundamental predicates with the Φ-operations, we turn $\mathcal{K}$ into a structurally equivalent universally axiomatizable class of algebraic systems.* ∎

On the basis of these lemmas the following can be proved:

Theorem 6: *Suppose $\mathcal{K}$ is an elementary, homomorphically closed in itself class of models which is also* R*-complete in the class of all models of the type of $\mathcal{K}$. Then $\mathcal{K}$ is structurally equivalent to a quasiprimitive class of algebraic systems.*

Indeed, assuming $\mathcal{K}$ is all of the above and non-trivial, we find that R-completeness implies $\mathcal{K}$ contains models with $\mathcal{K}$-free $\mathcal{K}$-dense generating sets of arbitrary cardinality [VIII]. Since $\mathcal{K}$ is homomorphically closed it follows from Theorem 1 of [IX] that $\mathcal{K}$ is regular, and the non-empty intersection of $\mathcal{K}$-submodels of a $\mathcal{K}$-model is a $\mathcal{K}$-submodel. By virtue of the basic result of [VII], $\mathcal{K}$ is additive, in view of its first-order axiomatizability. Theorems 1 and 2 of [IX] now show that every free generating subset of a $\mathcal{K}$-model is finitely dense. Local embeddability follows immediately from axiomatizability by way of the compactness theorem for FOPL. Finally, by Lemma 2, the expansion of $\mathcal{K}$-models by the Φ-operations as defined in (1) yields a quasi-primitive class of algebraic systems structurally equivalent to $\mathcal{K}$. ∎

NOTE

(1) $\mathcal{S}^q$ is not necessarily elementary; cf. [XXXI], second corollary.

CHAPTER 11

MODEL CORRESPONDENCES

Introduction

In this article particular correspondences between models, the so-called projective correspondences, are singled out, and their basic properties are established. For correspondences of a more complicated sort, an intrinsic local theorem is proved; fundamental local theorems of the theory of groups are shown to be special cases of this theorem.

In the study of properties of classes of models, besides considering properties of individual models, usually expressed through relations among elements of a given model, it is fitting to examine relations among models in the large, like, e.g., the relations "the model $\mathfrak{M}$ is a homomorphic image of the model", "the model $\mathfrak{M}$ is isomorphic to a submodel of the model $\mathfrak{N}$", "$\mathfrak{M}$ is the direct product of $\mathfrak{M}_1$ and $\mathfrak{M}_2$", etc. The fundamental purpose of the present article is to single out those model relations which are most closely connected with first-order predicate logic (FOPL), and to study the basic properties of such correspondences. These correspondences, called projective, are introduced in § 1; their properties are studied in § 2.

In § 3 we consider correspondences and classes of models of a more complicated type, the descriptions of which requires the apparatus of second-order predicate logic (SOPL). For these classes we prove an intrinsic local theorem, the central result of this paper. Finally, we establish the first-order axiomatizability of the classes of $\overline{\text{RN}}$-, $\overline{\text{RI}}$-, and $\overline{\text{Z}}$-groups, and show that local theorems for these classes – and for the more complex classes of RN-, RI-, Z-, and $\tilde{\text{N}}$-groups, and freely orderable groups – are special cases of the intrinsic local theorem. The local theorem for freely orderable groups is new, apparently. Combining the properties of $\overline{\text{RN}}$, $\overline{\text{RI}}$, etc. with the demand that subgroups be convex, we can, by the same method, get a series of new local theorems for partially ordered groups as well.

In the theory of models one usually considers predicates defined on a single fundamental set, or base. In the study of model correspondences it proves necessary to systematically examine predicates and models with several bases. In formulas relating to such multibase models, the quantifiers on individual variables are regarded as relativized, or specialized, to the bases. The usual process of "unifying" variables permits, on the whole, the reduction of the study of multibase models to that of single-base ones; this method is used in §2 in the deduction of model correspondence properties from well-known properties of classes of usual models.

By analogy with specialized individual quantifiers, specialized predicate quantifiers can be introduced: $(\forall_\Pi \boldsymbol{P})$ and $(\exists_\Pi \boldsymbol{P})$ are understood as symbolic expressions for the phrases, "for every predicate P with the property Π", and "there is a predicate P with the property Π such that". Just the use of specialized predicate quantifiers enables us to formulate the fundamental intrinsic local theorem.

The axiomatization of predicate logic with many-sorted (multibase) predicates was investigated by A. Schmidt [144, 145]. For terminological unity and the reader's comfort, a short summary of the necessary concepts and results is given in §1.

Some of the ideas and results of this article were published in [III].

§1. Multibase models

§1.1. *Multibase predicates*

Let $\{M_\alpha : \alpha \in \mathrm{A}\}$ be a collection of sets, not necessarily distinct, but nonempty. We say that $P(x_1, ..., x_n)$ is an *n-place predicate of sort* $\langle i_1, ..., i_n \rangle$ $(i_k \in \mathrm{A}, k = 1, ..., n)$ in the given collection iff every n-sequence $\langle u_1, ..., u_n \rangle$ $(u_i \in M_{i_k}, k = 1, ..., n)$ is put into correspondence with either T (truth) or F (falsity). A predicate of sort $\emptyset$ constantly has one of the two truth-values T or F. In addition to the basic predicates there are the equality relations. Each of these will be denoted logically by the single symbol $\approx$; in particular, $\approx$ can link individual symbols of diverse pairs of sorts $i, j \in \mathrm{A}$.

In formulas the sort of individual variables will be either stipulated separately or signified by upper indices; thus, $\boldsymbol{x}^i$ and $\boldsymbol{y}^i$ are individual symbols for the elements of the base M_i, when $\boldsymbol{P}^{i_1, ..., i_n}$, $\boldsymbol{Q}^{i_1, ..., i_n}$ are predicate symbols for predicates of the sort $\langle i_1, ..., i_n \rangle$.

All quantifiers are assumed to be *specialized* in the sense that the expressions $(\boldsymbol{x}^i)$, $(\exists \boldsymbol{x}^i)$, and $(\forall \boldsymbol{P}^i)$ mean "for every x^i in M_i", "there is an x^i in M_i such that", and "for every predicate P^i of sort $\langle i \rangle$". For example, the sentence $(\boldsymbol{x}^i)(\exists \boldsymbol{y}^j)(\boldsymbol{x}^i \approx \boldsymbol{y}^j)$ means $M_i \subseteq M_j$.

More generally, if $\mathfrak{x}$ is a predicate or individual variable, and Π is some property, then the expressions $(\forall_\Pi \mathfrak{x})$, $(\exists_\Pi \mathfrak{x})$ respectively mean: "for every

$\mathfrak{x}$ with the property Π" and "there exists an $\mathfrak{x}$ with the property Π such that".

The usual definitions of (well-formed) formulas of predicate logic with equality are naturally generalized to the case of multibase predicates and specialized quantifiers; it is important that all of the usual identically true formulas and the usual equivalences remain valid in the multibase case.

A well-ordered system $\{M_\alpha: \alpha \in \mathrm{A}\}$ of sets together with a well-ordered system $\{P_\gamma(x_1, ..., x_{n_\gamma}): \gamma \in \Gamma\}$ of multibase predicates defined in it and a well-ordered system $\{a_\delta : \delta \in \Delta\}$ of fixed elements (A, Γ, Δ are sets) is called a *multibase model.* The sequence of the number of sets and the sorts of the predicates and fixed elements is called the *type* of such a model. The sets M_α, the predicates P_γ, and the elements a_δ are called the *fundamental* (or *basic*) sets, predicates, and distinguished elements of the model. As mentioned above, equality is always included, albeit not explicitly among the P_γ. "An element of the model" is short for "an element of a base of the model", etc.

The notions of homomorphism and isomorphism for multibase models will be used in the same sense as for the single-base ones. It is necessary to generalize the notion of submodel a bit.

Let M_α ($\alpha \in$ A) be the bases of a multibase model $\mathfrak{M}$, and B some subset of A. $\mathfrak{M}'$ is a B-*submodel* of $\mathfrak{M}$ iff it is a collection of subsets $M'_\alpha \subseteq M_\alpha$ ($\alpha \in$ B), $M'_\alpha = M_\alpha$ ($\alpha \in$ A$\sim$B), together with the predicates of $\mathfrak{M}$ restricted to the M'_α and with the fixed elements of $\mathfrak{M}$. Note that the same distinguished elements, if there are any, must belong to both $\mathfrak{M}$ and $\mathfrak{M}'$. In case B = A, $\mathfrak{M}'$ is called simply a *submodel* of $\mathfrak{M}$, written as $\mathfrak{M}' \subseteq \mathfrak{M}$ ([1]).

A collection of models of the same type together with all of their isomorphs is called an (abstract) *class of models*. $M_\alpha, P_\gamma, a_\delta$ are used as generic notations for the basic sets, predicates, and distinguished elements of the models in a class.

A formula containing no free variables, although it may have individual and predicate constants, which are not quantified, is called a *sentence, axiom,* or *closed formula.* We can think of all formulas as being in prenex form, and when we speak of quantifiers, we shall have prenex quantifiers in mind unless otherwise specified.

Universal formulas are those FOPL formulas containing only universal quantifiers.

A subclass $\mathcal{L}$ of a class $\mathcal{K}$ of models is called (*first-order*) *axiomatizable* (or *elementary*) *in* $\mathcal{K}$ iff there is a collection S – possibly infinite – of FOPL sentences such that $\mathcal{L}$ consists of all models in $\mathcal{K}$ that satisfy all the sentences in S. If S contains only universal sentences, then $\mathcal{L}$ is *universally axiomatizable* (or *universal*) *in* $\mathcal{K}$.

A.Tarski [163] and J.Łoś [89] obtained simple characterizations of universally axiomatizable subclasses. In order to present these in the form we need, we introduce the following definition (cf. [132]).

Let $\mathfrak{M}$ be a model with bases M_α ($\alpha \in \mathrm{A}$), predicates P_γ ($\gamma \in \Gamma$), and distinguished elements a_δ ($\delta \in \Delta$). The *diagram* $\mathrm{D}(\mathfrak{M})$ of the model $\mathfrak{M}$ is the collection of all sentences of the form $\boldsymbol{P}_\gamma(\boldsymbol{c}_1, \ldots, \boldsymbol{c}_n)$, $\neg \boldsymbol{P}_\gamma(\boldsymbol{c}_1, \ldots, \boldsymbol{c}_n)$, $\boldsymbol{c} \approx \boldsymbol{c}'$, $\boldsymbol{c} \not\approx \boldsymbol{c}'$ that are true in $\mathfrak{M}$, where $\boldsymbol{c}, \boldsymbol{c}', \boldsymbol{c}_1, \ldots, \boldsymbol{c}_n$ lies among new distinct individual constants in fixed 1–1 correspondence with the elements of $\mathfrak{M}$ or among the constants $\boldsymbol{a}_\delta$ designating the distinguished elements of $\mathfrak{M}$. In the case of several bases these new constants must have appropriate sorts specified ([2]).

A *finite subdiagram* of $\mathfrak{M}$ is specified by choosing finite subsets of Γ, Δ, and the set of new constants, and is the conjunction of the finite number of members of $\mathrm{D}(\mathfrak{M})$ in which only the chosen predicates and constants occur ($\approx$ may always occur, as well). If no confusion is likely to result, a finite subdiagram of $\mathfrak{M}$ will be denoted generically by $\mathrm{D}_f(\mathfrak{M})$. We say that such a $\mathrm{D}_f(\mathfrak{M})$ is *realizable* in a model $\mathfrak{N}$ iff $\mathfrak{N}$ satisfies the sentence

$$(\exists \boldsymbol{b}_1) \ldots (\exists \boldsymbol{b}_n)\, \mathrm{D}_f(\mathfrak{M}),$$

where the $\boldsymbol{b}_i$ are the new constants appearing in $\mathrm{D}_f(\mathfrak{M})$, now viewed as variables.

Theorem 1 (cf. Tarski [163], Łoś [89]): *In order that a subclass $\mathcal{K}^*$ of a class $\mathcal{K}$ of multibase models without distinguished elements be universally axiomatizable in $\mathcal{K}$, it is necessary and sufficient that for every $\mathcal{K}$-model $\mathfrak{M}$, if every finite subdiagram of $\mathfrak{M}$ is realizable in some $\mathcal{K}^*$-model, then $\mathfrak{M} \in \mathcal{K}^*$.*

The necessity follows immediately from the compactness theorem (cf. §2.2 below). For completeness we shall prove sufficiency. Let $\mathbf{U}$ be the collection of all universal sentences valid in every $\mathcal{K}^*$-model, and suppose the $\mathcal{K}$-model $\mathfrak{M}$ satisfies all these sentences. We must show that $\mathfrak{M} \in \mathcal{K}^*$. Suppose not. Then there is a finite subdiagram $\mathrm{D}_f(\mathfrak{M})$ that is not realizable in any $\mathcal{K}^*$-model. This means that in every $\mathcal{K}^*$-model the universal sentence

$$(\boldsymbol{b}_1) \ldots (\boldsymbol{b}_n) \;\neg \mathrm{D}_f(\mathfrak{M})$$

is valid and, therefore, belongs to $\mathbf{U}$ and holds in $\mathfrak{M}$. But $\mathfrak{M}$ satisfies the negation of this sentence – a contradiction. ■

§1.2. *Axiomatizable and projective correspondences*

We consider two classes $\mathcal{K}$ and $\mathcal{L}$ of models whose bases and predicates are respectively denoted by M_α, P_γ ($\alpha \in \mathrm{A}, \gamma \in \Gamma$), and N_β, Q_δ ($\beta \in \mathrm{B}, \delta \in \Delta$). We shall say that a *correspondence* σ is established between the elements of the classes $\mathcal{K}$ and $\mathcal{L}$ iff with every pair of models $\mathfrak{M} \in \mathcal{K}$, $\mathfrak{N} \in \mathcal{L}$ is associated one of the truth-values T or F. Moreover, we assume that if $\mathfrak{M}\sigma\mathfrak{N} = \mathrm{T}$ and $\mathfrak{M}, \mathfrak{N}$ are isomorphic to $\mathfrak{M}_1, \mathfrak{N}_1$, respectively, then $\mathfrak{M}_1 \sigma \mathfrak{N}_1 = \mathrm{T}$.

The correspondence σ is said to be *(first-order) axiomatizable* (or *elementary*) iff the truth of $\mathfrak{M}\sigma\mathfrak{N}$ is equivalent to the satisfiability in $\mathfrak{M}, \mathfrak{N}$ of a fixed collection S of FOPL sentences constructed in the following fashion: choose some set of new predicate symbols $\boldsymbol{S}_\lambda^{i_1, \ldots, i_k}$ ($i_1, \ldots, i_k \in \mathrm{A} \cup \mathrm{B}, \lambda \in \Lambda$), and write FOPL sentences with predicate symbols only of the form $\boldsymbol{P}_\gamma$, $\boldsymbol{Q}_\delta$, or $\boldsymbol{S}_\lambda$. Furthermore, we say that S is satisfiable in $\mathfrak{M}, \mathfrak{N}$ iff it is possible to define predicates S_λ ($\lambda \in \Lambda$) on the sets M_α, N_β ($\alpha \in \mathrm{A}, \beta \in \mathrm{B}$) such that with the given predicates of $\mathfrak{M}$ and $\mathfrak{N}$ on the bases M_α, N_β all the sentences of S are are true; one can consider all these bases and predicates to form a single multi-base model.

In case S is finite, the satisfiability of S in $\mathfrak{M}, \mathfrak{N}$ is equivalent to the truth in $\mathfrak{M}, \mathfrak{N}$ of an appropriate second-order sentence of the form

$$(\exists \boldsymbol{S}_{\lambda_1}) \ldots (\exists \boldsymbol{S}_{\lambda_\nu})\, \Phi(\boldsymbol{P}, \boldsymbol{Q}, \boldsymbol{S})\,.$$

Therefore, axiomatizable correspondences of this particular form can be called $\exists$-*correspondences*; by analogy it is possible to define $\forall$-, $\exists\forall$-, $\forall\exists$-correspondences, etc. In §3 we shall return to a special case of these; for the time being we define one more class of correspondences, which includes the class of axiomatizable correspondences and is as convenient to study as the latter case.

The correspondence σ between the classes $\mathcal{K}$ and $\mathcal{L}$ is called *projective* iff the truth of $\mathfrak{M}\sigma\mathfrak{N}$ ($\mathfrak{M} \in \mathcal{K}$, $\mathfrak{N} \in \mathcal{L}$) is equivalent to the satisfiability in $\mathfrak{M}, \mathfrak{N}$ of a fixed system S of sentences with the following structure: we take an auxiliary set E of indices and a set of predicate symbols $\boldsymbol{S}_\lambda^{i_1, \ldots, i_k}$ ($i_1, \ldots, i_k \in$ $\in \mathrm{A} \cup \mathrm{B} \cup \mathrm{E}, \lambda \in \Lambda$), and consider only the FOPL sentences all of whose predicate symbols are found among the $\boldsymbol{P}_\gamma$, $\boldsymbol{Q}_\delta$, and $\boldsymbol{S}_\lambda$. In this connection S is satisfiable in $\mathfrak{M}, \mathfrak{N}$ iff it is possible to find non-empty sets T_ϵ ($\epsilon \in \mathrm{E}$) and to define predicates S_λ on the sets $M_\alpha, N_\beta, T_\epsilon$, such that all of the sentences of S are true.

These notions of axiomatizable and projective correspondences between two classes of models are extended in an obvious and unambiguous way to the case of a correspondence $\sigma(\mathfrak{M}_1, \ldots, \mathfrak{M}_s)$ among elements of classes

$\mathcal{K}_1, ..., \mathcal{K}_s$. From the definitions we immediately obtain two corollaries.

Corollary 1: *The disjunction of a finite number and the conjunction of any number of axiomatizable (projective) correspondence among models from given classes* $\mathcal{K}_1, ..., \mathcal{K}_s$ *is again an axiomatizable (projective) correspondence.* ■

Corollary 2: *If* $\sigma(\mathfrak{M}_1, ..., \mathfrak{M}_s)$ *is a projective correspondence on the classes* $\mathcal{K}_1, ..., \mathcal{K}_s$, *and* $\mathcal{K}_s$ *is axiomatizable, then the correspondence* τ *given by*

$$\tau(\mathfrak{M}_1, ..., \mathfrak{M}_{s-1}) = (\exists \mathfrak{M}_s)\, \sigma(\mathfrak{M}_1, ..., \mathfrak{M}_s)$$

is also projective. ■

The formulation of Corollary 2 makes sense for $s > 2$. For $s = 2$ the expression

$$\tau(\mathfrak{M}_1) = (\exists \mathfrak{M}_2)\, \sigma(\mathfrak{M}_1, \mathfrak{M}_2)$$

gives a property τ of $\mathcal{K}_1$-models which is also called *projective,* and the collection of all $\mathcal{K}_1$-models possessing this property is called a *projective subclass* of $\mathcal{K}_1$. In other words, a subclass $\mathcal{K}^*$ of a class $\mathcal{K}$ of models is called *projective in* $\mathcal{K}$ iff it consists exclusively of $\mathcal{K}$-models which are related by a fixed projective correspondence to at least one model in a fixed axiomatizable class; note that a subclass axiomatizable in $\mathcal{K}$ is also projective in $\mathcal{K}$.

From Corollary 1 it follows that *the union of a finite number and the intersection of any number of projective subclasses of a given class of models are again projective subclasses of this class.* ■

If the underlying class $\mathcal{K}$ consists of all models of a given type, then its projective (axiomatizable) subclasses are called, simply, *projective (axiomatizable) classes of models.*

It is easy to see that projective subclasses of projective classes of models are projective classes. Analogously, *the collection* $\mathcal{K}_1^{**}$ *of models in a projective class* $\mathcal{K}_1^*$ *which are related by a projective correspondence* σ *to models in a projective class* $\mathcal{K}^*$ *is itself a projective class.*

To see this, let $\mathcal{K}_1$ and $\mathcal{K}_2$ be the classes of all models of the types of $\mathcal{K}_1^*$ and $\mathcal{K}_2^*$; let $\mathbf{R}_i$ be an axiom system characterizing a projective correspondence ρ_i between $\mathcal{K}_i$ and some auxiliary axiomatizable class $\mathcal{L}_i$ (with axioms $\mathbf{L}_i$) such that $\mathcal{K}_i^*$ consists of just those $\mathcal{K}_i$-models ρ_i-related to $\mathcal{L}_i$-models ($i = 1, 2$). Let the axiom system S characterize the correspondence σ. Each of the systems $\mathbf{R}_1$, $\mathbf{R}_2$, and S involves auxiliary predicate symbols $\boldsymbol{S}_\lambda^{i_1,\ldots,i_k}$; let us exchange these for new symbols in each of the axiom systems and consider the correspondence τ between $\mathcal{K}_1$ and $\mathcal{K}_2$ determined by $\mathrm{T} = \mathrm{L}_1 \cup \mathrm{L}_2 \cup \mathrm{R}_1 \cup \mathrm{R}_2 \cup \mathrm{S}$. We see immediately that the class $\mathcal{K}_1^{**}$ consists of just those $\mathcal{K}_1$-models which τ-correspond to $\mathcal{K}_2$-models. ∎

§1.3. *Some examples.*

Let $\mathcal{K}$ and $\mathcal{L}$ be classes of single-base models. We introduce an auxiliary predicate $S(x, y)$ ($x \in \mathfrak{M}$, $y \in \mathfrak{N}$), which is viewed as a relation (as yet undefined) establishing a correspondence between elements of the bases of the models $\mathfrak{M} \in \mathcal{K}$, $\mathfrak{N} \in \mathcal{L}$. By means of a FOPL axiom system S, written with the aid of the predicate symbol $\boldsymbol{S}$ and the basic predicate symbols for $\mathcal{K}$ and $\mathcal{L}$, we can specify a property of the correspondence S. Then $\mathfrak{M}$ S-corresponds to $\mathfrak{N}$ iff it is possible to establish a correspondence S between elements of $\mathfrak{M}, \mathfrak{N}$ that has the property S.

Examples of this simple form of axiomatizable correspondence are the relations "the model $\mathfrak{M}$ is isomorphic to the model $\mathfrak{N}$", "$\mathfrak{M}$ is a homomorphic (strong homomorphic) image of $\mathfrak{N}$", "$\mathfrak{M}$ is isomorphic to a submodel of $\mathfrak{N}$", "$\mathfrak{M}$ is a homomorphic image of a submodel of $\mathfrak{N}$", etc.

Thus we observe that *the collection of all factor models of models in an axiomatizable class is a projective class.* ∎

Instead of the relation S between elements of two models, one can consider a relation among elements of several models, and, in particular, obtain the correspondence "$\mathfrak{M}$ is the direct product of $\mathfrak{M}_1, \ldots, \mathfrak{M}_n$".

Let us examine a more complicated example. A *finitely complete subdirect product* of models $\mathfrak{M}_\alpha$ ($\alpha \in \mathrm{A}$), single-based and of the same type, is a submodel of the direct product of these models such that for any choice of a finite number of distinct indices $\alpha_1, \ldots, \alpha_m$ from A, and any elements $v_i \in \mathfrak{M}_{\alpha_i}$ ($i = 1, \ldots, m$), the submodel contains an element u whose α_ith projection is v_i for $i = 1, \ldots, m$.

We shall show that *the property that a model $\mathfrak{M}$ is a finitely complete subdirect product of models in a fixed axiomatizable class $\mathcal{K}$ is projective.*

Let $P_\gamma(z_1, \ldots, z_{n_\gamma})$ ($\gamma \in \Gamma$) be the basic predicates of the class $\mathcal{K}$ defined by the axiom system K. We also deal with the following three classes of all models of the indicated types:

$\mathcal{K}_1$: no predicates; individual variables $\boldsymbol{\alpha}, \boldsymbol{\alpha}_1, \ldots$ have this sort.

$\mathcal{K}_2$: predicates $Q_\gamma(x_1, \ldots, x_{n_\gamma})$ ($\gamma \in \Gamma$); individual variables $\boldsymbol{x}, \boldsymbol{x}_1, \ldots$ have this sort.

$\mathcal{K}_3$: predicates $R_\gamma(x_1, \ldots, x_{n_\gamma})$ ($\gamma \in \Gamma$); individual variables $\boldsymbol{y}, \boldsymbol{y}_1, \ldots$ have this sort.

We introduce a new predicate $S(\alpha, x, y)$ to be read as "y is the αth projection of x"; we let S denote the following axiom system (universal quantifiers governing the whole formula have been dropped for clarity):

(i) $(\exists y)\, S(\boldsymbol{\alpha}, x, y)$;

(ii) $S(\boldsymbol{\alpha}, x, y) \,\&\, S(\boldsymbol{\alpha}, x, y_1) \to y \approx y_1$;

(iii) $(\boldsymbol{\alpha})\,(\exists y)(S(\boldsymbol{\alpha}, x, y) \,\&\, S(\boldsymbol{\alpha}, x_1, y)) \to x \approx x_1$;

(iv) $Q_\gamma(x_1, ..., x_{n_\gamma}) \leftrightarrow (\boldsymbol{\alpha})(y_1) \dots (y_{n_\gamma})\,(S(\boldsymbol{\alpha}, x_1, y_1) \,\&$
$\&\, S(\boldsymbol{\alpha}, x_{n_\gamma}, y_{n_\gamma}) \to R_\gamma(y_1, ..., y_{n_\gamma}))$ $\qquad (\gamma \in \Gamma)$;

(v) $T(\boldsymbol{\alpha}, y) \leftrightarrow (\exists x)\, S(\boldsymbol{\alpha}, x, y)$;

(vi) $\boldsymbol{\alpha}_1 \not\approx \boldsymbol{\alpha}_2 \,\&\, \boldsymbol{\alpha}_1 \not\approx \boldsymbol{\alpha}_3 \,\&\, \dots \,\&\, \boldsymbol{\alpha}_{m-1} \not\approx \boldsymbol{\alpha}_m \,\&\, T(\boldsymbol{\alpha}_1, y_1) \,\&\, \dots \,\&\, T(\boldsymbol{\alpha}_m, y_m)$ -
$\to (\exists x)(S(\boldsymbol{\alpha}_1, x, y_1) \,\&\, \dots \,\&\, S(\boldsymbol{\alpha}_m, x, y_m))$ $\quad (m = 1, 2, 3, \dots)$.

(vii) FOPL sentences expressing that each axiom in K holds in T_α, the set of y such that $T(\alpha, y)$ is true.

These last axioms are written with universal quantifiers over $\boldsymbol{\alpha}$ and result from specializing the quantifiers in the axioms in K and replacing the $\boldsymbol{P}_i$ with $\boldsymbol{R}_i$.

It is clear that if for a model $\mathfrak{M} \in \mathcal{K}_2$ one can find models $A \in \mathcal{K}_1$ and $\mathfrak{N} \in \mathcal{K}_3$ such that S is satisfiable in A, $\mathfrak{M}$, $\mathfrak{N}$, then $\mathfrak{M}$ is a finitely complete subdirect product of models – based on the T_α – that belong to $\mathcal{K}$ in view of (vii). The converse is clear. ∎

If the axioms (vi) are deleted from S then the satisfiability of the remainder in models A, $\mathfrak{M}$, $\mathfrak{N}$ means that $\mathfrak{M}$ is a submodel of a direct product of $\mathcal{K}$-models. Therefore, *the property of a model being a submodel of a direct product of models from an axiomatizable (or projective) class is projective.* ∎

By similar means one can prove the projectiveness of groups being RN-, RI-, or Z-groups, or being partially ordered with an RN-, RI-, or Z-system of convex subgroups. Indeed, just this was proved in [II] and [M4]. In §3 below, however, a stronger theorem is proved on the simple axiomatizability of all the indicated group properties.

§2. Fundamental properties of projective correspondences

§2.1. *Equality relations; unification of quantifiers*

Let $\mathcal{K}$ be an axiomatizable class of multibase models with bases M_α ($\alpha \in \mathrm{A}$) and basic predicates P_γ ($\gamma \in \Gamma$), characterized by a system S of FOPL sentences. As already noted, in the expressions in S one may encounter the equality sign $\approx$ as well as other predicate symbols. The well-known device of relativizing equations permits consideration of systems with (absolute) equality to be reduced to the study of systems with a predicate of equivalence (cf. [I], §3; [56]). In the case of multibase models the relativation of equations can sometimes be carried out separately as follows.

Let us assume the set A of indices of the bases is divided into disjoint, nonempty subsets $\mathrm{A}_0, \mathrm{A}_1, \ldots, \mathrm{A}_t$, such that in S there are no equality signs linking individual variables whose sorts are different A_i. We now introduce new relations $\theta_1, \ldots, \theta_t$, with $x\theta_i y$ defined for all $x, y \in U_i$, where

$$U_i = \bigcup\{M_\alpha \colon \alpha \in \mathrm{A}_i\} \quad (i = 1, 2, \ldots, t)\,.$$

The new and old relations are connected by the axiom schemes:

$$\boldsymbol{x}\,\boldsymbol{\theta}_i\,\boldsymbol{x}, \quad \boldsymbol{x}\boldsymbol{\theta}_i\boldsymbol{y} \rightarrow \boldsymbol{y}\boldsymbol{\theta}_i\boldsymbol{x}, \quad \boldsymbol{x}\boldsymbol{\theta}_i\boldsymbol{y}\,\&\,\boldsymbol{y}\boldsymbol{\theta}_i\boldsymbol{z} \rightarrow \boldsymbol{x}\boldsymbol{\theta}_i\boldsymbol{z} \qquad (i = 1, \ldots, t)\,, \tag{1}$$

$$\boldsymbol{x}_1\boldsymbol{\eta}_1\boldsymbol{y}_1\ \&\ \ldots\ \&\ \boldsymbol{x}_n\boldsymbol{\eta}_n\boldsymbol{y}_n\ \&\ \boldsymbol{P}_\gamma(\boldsymbol{x}_1, \ldots, \boldsymbol{x}_n) \rightarrow \boldsymbol{P}_\gamma(\boldsymbol{y}_1, \ldots, \boldsymbol{y}_n) \qquad (\gamma \in \Gamma)\,, \tag{2}$$

where the $\boldsymbol{x}, \boldsymbol{y}, \boldsymbol{z}, \boldsymbol{x}_k, \boldsymbol{y}_k$ occur as variables of all possible sorts consistent with the predicates, and the $\boldsymbol{\eta}_k$ are all possible meaningful symbols among the $\boldsymbol{\theta}_1, \ldots, \boldsymbol{\theta}_t$. Finally, in every S-axiom each expression of the form $\boldsymbol{x} \approx \boldsymbol{y}$ (with sorts in A_i) is replaced by $\boldsymbol{x}\boldsymbol{\theta}_i\boldsymbol{y}$; the axioms so obtained plus the axioms given by (1) and (2) form a system denoted by S_θ. The class of models with bases M_α ($\alpha \in \mathrm{A}$) and predicates P_γ ($\gamma \in \Gamma$), $\theta_1, \ldots, \theta_t$ which satisfy S_θ is denoted by $\mathcal{K}_\theta$. Sentences in S_θ contain $\approx$, but in connection only with variables for bases M_α where $\alpha \in \mathrm{A}_0$, since equations involving elements of the remaining bases have been changed to equivalences of various sorts.

Every $\mathcal{K}$-model is trivially converted to a $\mathcal{K}_\theta$-model by defining $x\,\theta_i y$ to be equivalent to $x = y$.

Conversely, let $\mathfrak{M}_\theta$ be a $\mathcal{K}_\theta$-model. According to (1), the relation θ_i is an equivalence on U_i, and so, U_i splits into θ_i-classes $[u]$ for $u \in U_i$. By M_α/θ_i

we denote the collection of those residue classes in U_i/θ_i having representatives in M_α ($\alpha \in \mathrm{A}_i$); for $\alpha \in \mathrm{A}_0$, we let $M_\alpha/\approx$ be M_α, with each element forming its own residue class. For $\gamma \in \Gamma$ we put

$$P_\gamma([u_1], ..., [u_n]) = P_\gamma(u_1, ..., u_n) . \tag{3}$$

The axioms (2) guarantee that the predicates P_γ are well defined on the residue classes by (3). Thus every $\mathcal{K}_\theta$-model $\mathfrak{M}_\theta$ yields a well-defined model $\mathfrak{M} = \mathfrak{M}_\theta/\theta$, which – as in the case of a single base [56] – is easily seen to be a $\mathcal{K}$-model.

We shall need the following observation later on. Suppose for some $\alpha \in \mathrm{A}$ there are no equality signs in S linking any element of M_α with any other element; let $\mathfrak{M}$ be a $\mathcal{K}$-model in which an equivalence relation θ has been defined on M_α by some means, and suppose $\mathfrak{M}$, θ satisfy (1) and (2). It is clear that the sentences of S will be satisfied in the factor model $\mathfrak{M}/\theta$, defined as above (straightforwardly on the bases other than M_α).

Moreover, the factor model $\mathfrak{M}/\theta$ can be viewed as a submodel of $\mathfrak{M}$. We choose a representative from each residue class in M_α/θ and call the set of these representatives M'_α. Now let $\mathfrak{M}'$ be the submodel of $\mathfrak{M}$ with bases M'_α and $M'_\beta = M_\beta$ for $\beta \neq \alpha$. The definition (3) shows that the map $u \to [u]$ is an isomorphism between $\mathfrak{M}'$ and $\mathfrak{M}'/\theta$.

As already mentioned, the study of axiomatizable classes of multibase models naturally reduces to a consideration of classes of single-base models through the process of *unification* of quantifiers, which is now described (cf. [144]).

Let $\mathcal{K}$ be the class of multibase models with bases M_α ($\alpha \in \mathrm{A}$) and predicates P_γ ($\gamma \in \Gamma$). Let $\mathcal{K}^*$ denote the class of models with a single base M and predicates V_α, P^*_γ ($\alpha \in \mathrm{A}, \gamma \in \Gamma$). The predicates V_α are one-place, and P^*_γ has the same rank as P_γ, but its arguments range only over M. From every FOPL formula Φ of the type of $\mathcal{K}$ we construct a corresponding $\mathcal{K}^*$-formula Φ^* by the following recursion:

(a) If Φ is quantifier-free, then we replace $\boldsymbol{P}_\gamma$ in Φ by $\boldsymbol{P}^*_\gamma$ to obtain Φ^*.

(b) If $\Phi = (\exists \boldsymbol{x}^\alpha)\Phi_1$, then $\Phi^* = (\exists \boldsymbol{x})(\boldsymbol{V}_\alpha(\boldsymbol{x})\ \&\ \Phi^*_1)$.

(c) If $\Phi = (\boldsymbol{x}^\alpha)\Phi_1$, then $\Phi^* = (\boldsymbol{x})(\boldsymbol{V}_\alpha(\boldsymbol{x}) \to \Phi^*_1)$.

Given a $\mathcal{K}$-model $\mathfrak{M}$ satisfying a sentence Φ, we put $M = \bigcup_{\alpha \in \mathrm{A}} M_\alpha$, $V_\alpha(u)$ iff $u \in M_\alpha$, and

$$P^*_\gamma(u_1, ..., u_n) = \begin{cases} P_\gamma(u_1, ..., u_n), \text{ if defined} \\ \mathsf{F}, \text{ otherwise} . \end{cases} \tag{4}$$

The result is a $\mathcal{K}^*$-model $\mathfrak{M}^*$ satisfying the sentence Φ^*. Conversely, if $\mathfrak{M}^*$ is a $\mathcal{K}^*$-model satisfying the sentence Φ^*, then letting M_α be the set of all $u \in M$ for which $V_\alpha(u)$ holds, and defining P_γ in accord with (4), we get a $\mathcal{K}$-model $\mathfrak{M}$ satisfying the original sentence Φ. In order that this be a 1–1 correspondence between $\mathcal{K}$- and $\mathcal{K}^*$-models, it is necessary to require $\mathfrak{M}^*$ to have no "extra" elements. If the number of bases is finite, e.g. if $\mathrm{A} = \{1, 2, ..., r\}$, this can be accomplished by restricting $\mathcal{K}^*$ to models satisfying

$$(\boldsymbol{x})(V_1(\boldsymbol{x}) \vee V_2(\boldsymbol{x}) \vee ... \vee V_r(\boldsymbol{x})),$$

which guarantees that $M = \bigcup_{\alpha \in \mathrm{A}} M_\alpha$ in the $\mathcal{K}$-model constructed from the $\mathcal{K}^*$-model $\mathfrak{M}^*$.

The process of *specialization* (or *relativization* (cf. [105], [132])) can be used when one wants to express with general quantifiers that a formula Φ with general quantifiers holds when its bound variables $\boldsymbol{x}_i$ range over subsets M_i defined by formulas $\Psi_i(\boldsymbol{x})$ with one free variable. This we do by first writing Φ as a formula Φ_R with relativized quantifiers, then unifying these to obtain Φ_R^*, which has general quantifiers; substituting $\Psi_i(\boldsymbol{x})$ for $V_i(\boldsymbol{x})$ in Φ_R^* gives the desired formula.

§2.2. *Extrinsic local theorem*

In the single-base case we have the well-known basic local theorem for FOPL:

Compactness theorem [I], [II]: *If every finite subset of an infinite collection* S *of sentences of* FOPL *is consistent, then the whole collection is consistent.* ■

Furthermore, the sentences in S can contain $\approx$, as well as any number (finite or infinite) of different predicate or individual symbols.

Now suppose that the given system S is multibase. It is consistent iff there is some multibase model of appropriate type in which all members of S are valid (it is convenient and unambiguous to call such a model an S-*model*). Applying the unification process, we construct a single-base system S* from S. The consistency of every finite subset of S* follows from the consistency of every finite subset of S; therefore, there is an S*-model $\mathfrak{M}^*$. Constructing a model $\mathfrak{M}$ from $\mathfrak{M}^*$ as in §2.1, we see that S is consistent. Thus, *the compactness theorem holds for systems of multibase axioms.* ■

This immediately implies

Theorem 2 (extrinsic local theorem for projective correspondences): *Let σ be a projective correspondence among (multibase) models from projective classes $\mathcal{K}_1, \ldots, \mathcal{K}_s$, and let $\mathfrak{M}_1, \ldots, \mathfrak{M}_s$ belong to these classes. Suppose for every choice of finite subdiagrams $D_f(\mathfrak{M}_1), \ldots, D_f(\mathfrak{M}_s)$, there are σ-corresponding models $\mathfrak{N}_1 \in \mathcal{K}_1, \ldots, \mathfrak{N}_s \in \mathcal{K}_s$ such that $D_f(\mathfrak{M}_i)$ is realizable in $\mathfrak{N}_i$ $(i = 1, \ldots, s)$. Then there are σ-corresponding models $\mathfrak{N}_1 \in \mathcal{K}_1, \ldots, \mathfrak{N}_s \in \mathcal{K}_s$ in which $\mathfrak{M}_1, \ldots, \mathfrak{M}_s$ are embeddable as submodels.*

Let $S_1, \ldots, S_s, S_0$ be axiom systems characterizing $K_1, \ldots, K_s$, and σ; let $D(\mathfrak{M}_1), \ldots, D(\mathfrak{M}_s)$ be the diagrams of the models in question. We can assume that the auxiliary predicates and the individual constants in different collections of axioms have distinct notations. The collection

$$S = D(\mathfrak{M}_1) \cup \ldots \cup D(\mathfrak{M}_s) \cup S_0 \cup \ldots \cup S_s$$

can be viewed as an axiom system defining a class of multibase models. The intersection of any finite subset of S with any diagram $D(\mathfrak{M}_i)$ is included in some finite subdiagram $D_f(\mathfrak{M}_i)$; it follows that every finite subset of S is consistent. By the compactness theorem, S itself is consistent. If $\mathfrak{N}$ is an S-model, and $\mathfrak{N}_1, \ldots, \mathfrak{N}_s$ its $\mathcal{K}_1$-, ..., $\mathcal{K}_s$-projections, then $\mathfrak{M}_i \subseteq \mathfrak{N}_i$ $(i = 1, \ldots, s)$, for S includes the diagrams of $\mathfrak{M}_1, \ldots, \mathfrak{M}_s$. Moreover, the models $\mathfrak{N}_1, \ldots, \mathfrak{N}_s$ σ-correspond. ■

Setting $s = 1$ in Theorem 2, we get

Corollary 3: *If every finite subdiagram of a model $\mathfrak{M}$ is realizable in some member of a fixed projective class $\mathcal{K}$, then $\mathfrak{M}$ is isomorphic to a submodel of some $\mathcal{K}$-model.* ■

In accord with [VII], we call a class $\mathcal{K}$ of models *pseudoaxiomatizable* iff for every FOPL axiom system **S**, if each finite subset of S is satisfiable in some $\mathcal{K}$-model, then the whole system **S** is satisfiable in some $\mathcal{K}$-model. Repeating the argument used in proving Theorem 2, we obtain the stronger

Remark: *Every projective class of models is pseudoaxiomatizable* ([3]). ■

For an example of an application of Theorem 2 let us look at strong homomorphisms. As mentioned above, the relation "the model $\mathfrak{M}$ is a strong homomorphic image of the model $\mathfrak{N}$" is projective. Therefore, if every finite submodel of a model $\mathfrak{M}$ is the strong homomorphic image of a submodel of a member of some fixed projective class $\mathcal{K}$, then $\mathfrak{M}$ itself is a strong homomorphic image of some $\mathcal{K}$-model.

According to [II] and [M4], the classes of RN-, RI-, and Z-groups are projective (cf. §3.3). It is known that subgroups of groups from these classes belong to the same classes. Therefore, from Corollary 1 we get an extrinsic local theorem:

If every finite subset A of a group $\mathfrak{G}$*, viewed as a partial group, is embeddable in a* T-*group* (T = RN, RI, Z), *then* $\mathfrak{G}$ *is a* T-*group.* ■

We recall that the usual (intrinsic) local theorem for the indicated groups has the form:

If every finitely generated subgroup of a group $\mathfrak{G}$ *is a* T-*group, then* $\mathfrak{G}$ *is a* T-*group* (cf. [II], [M4]). ■

Comparing these two theorems, we see that the former is the stronger; indeed, the structure of the partial group A does not determine the structure of the subgroup $\mathfrak{H}$ generated by A in $\mathfrak{G}$, nor does the embeddability of A in a T-group imply the embeddability of $\mathfrak{H}$.

Lastly, we mention one more corollary:

Corollary 4: *Let the projective class* $\mathcal{K}$ *be a subclass of the axiomatizable class* $\mathcal{K}_0$ *such that* $\mathcal{K}_0$*-submodels of* $\mathcal{K}$*-models are again* $\mathcal{K}$*-models. Then* $\mathcal{K}$ *is universally axiomatizable in* $\mathcal{K}_0$.

To prove it you compare Theorems 1 and 2. ■

From Corollary 2 and the projectiveness of the classes of RN-, RI-, and Z-groups, one deduces the outright axiomatizability of these classes. This fact will be obtained directly in §3.3.

§2.3. *Boundedness and extendability for correspondences*

From the well-known Löwenheim-Skolem theorem and the theorem on extendability of infinite models (cf. [I], §§5, 6) we can easily derive corresponding theorems for multibase models by using the unification process as above. From these we obtain theorems on correspondences and projective classes.

Theorem 3 (boundedness): *For every projective correspondence* σ *on fixed projective single-base classes* $\mathcal{K}_1, \ldots, \mathcal{K}_s$*, there is an infinite cardinal number* $\mathfrak{m} = \mathfrak{m}(\sigma)$ *such that if* $\mathfrak{M}_1 \in \mathcal{K}_1, \ldots, \mathfrak{M}_s \in \mathcal{K}_s$ *are* σ*-corresponding models, and* $D_1, \ldots, D_s$ *are sets of elements of these models and bounded in power by some cardinal* $\mathfrak{n} \geqslant \mathfrak{m}$*, then there are* σ*-corresponding models* $\mathfrak{N}_1 \in \mathcal{K}_1, \ldots, \mathfrak{N}_s \in \mathcal{K}_s$ *such that* $D_i \subseteq \mathfrak{N}_i \stackrel{\cdot}{\subseteq} \mathfrak{M}_i$ *and the power of* $\mathfrak{N}_i$ *does not exceed* $\mathfrak{n}$ $(i = 1, \ldots, s)$.

Let S be a system of FOPL axioms characterizing σ; we shall include in S all of the axioms characterizing the classes $\mathcal{K}_1, \ldots, \mathcal{K}_s$, as well. Let $\mathfrak{M}$ be a multibase S-model which has $\mathfrak{M}_1, \ldots, \mathfrak{M}_s$ as projections. Applying the unification procedure to S, we get an axiom system S*, and from $\mathfrak{M}$ we get an S*-model $\mathfrak{M}^*$. We put

$$\mathfrak{m} = \aleph_0 + \mathfrak{m}_1 ,$$

where $\mathfrak{m}_1$ is the power of S*; note that $\mathfrak{m}$ depends only on σ. The model $\mathfrak{M}^*$ includes the set

$$D = D_1 \cup \ldots \cup D_s$$

whose cardinality does not exceed $\mathfrak{n}$. Since $\mathfrak{m}$ is sufficiently large, the classical Löwenheim-Skolem theorem tells us there is an S*-model $\mathfrak{N}^*$ of power not greater than $\mathfrak{n}$ such that $D \subseteq \mathfrak{N}^* \subseteq \mathfrak{M}^*$. Returning from $\mathfrak{M}^*$ to $\mathfrak{M}$ we get from $\mathfrak{N}^*$ an S*-submodel $\mathfrak{N}$ whose projections $\mathfrak{N}_1, \ldots, \mathfrak{N}_s$ satisfy all the requirements in the theorem. ■

For $s = 1$ we have the

Corollary: *For every projective single-base class $\mathcal{K}$ there is an infinite cardinal number $\mathfrak{m}$ such that if the cardinality of a subset D of some $\mathcal{K}$-model $\mathfrak{M}$ is not greater than $\mathfrak{n} \geqslant \mathfrak{m}$, then $\mathfrak{M}$ has a $\mathcal{K}$-submodel containing D and not exceeding $\mathfrak{n}$ in power.* ■

For axiomatizable classes the number $\mathfrak{m}$, generally speaking, coincides with the number of basic predicates associated with the class. In the case of projective classes the auxiliary predicates, as well as the fundamental, are essential for finding $\mathfrak{m}$. E.g., the class $\mathcal{K}^{\mathfrak{m}}$ of all sets of power not less than $\mathfrak{m} \geqslant \aleph_0$ is projective. According to the proof of the theorem above, the characterization of $\mathcal{K}^{\mathfrak{m}}$ requires not fewer than $\mathfrak{m}$ predicates, whereas $\mathcal{K}^{\mathfrak{m}}$ has no basic predicates.

Theorem 4 (extendability): *Let σ be a projective correspondence on projective single-base classes $\mathcal{K}_1, \ldots, \mathcal{K}_s$; let $\mathfrak{n}$ be a cardinal not less than $\mathfrak{m}$, the number given by the theorem on boundedness. Then for any σ-corresponding infinite models $\mathfrak{M}_1 \in \mathcal{K}_1, \ldots, \mathfrak{M}_s \in \mathcal{K}_s$, there are σ-corresponding models $\mathfrak{N}_1 \in \mathcal{K}_1, \ldots, \mathfrak{N}_s \in \mathcal{K}_s$, each having power $\mathfrak{n}$, such that $\mathfrak{M}_i \subseteq \mathfrak{N}_i$, $\mathfrak{M}_i \neq \mathfrak{N}_i$ if the power of $\mathfrak{M}_i$ does not exceed $\mathfrak{n}$ $(i = 1, \ldots, s)$.*

For any $t \leqslant s$, if for every natural number m there are models $\mathfrak{M}_1^{(m)} \in \mathcal{K}_1, \ldots, \mathfrak{M}_s^{(m)} \in \mathcal{K}_s$ with $\mathfrak{M}_1^{(m)}, \ldots, \mathfrak{M}_t^{(m)}$ having at least m elements, then there are σ-corresponding models $\mathfrak{N}_1 \in \mathcal{K}_1, \ldots, \mathfrak{N}_s \in \mathcal{K}_s$ of which $\mathfrak{N}_1, \ldots, \mathfrak{N}_t$ are infinite.

We argue as in [I], § 6. Let S be an axiom system defining σ and including characterizations of $\mathfrak{K}_1, \ldots, \mathfrak{K}_s$. Let $D(\mathfrak{M}_i)$ be the diagram of $\mathfrak{M}_i$, and let $D_i = \{ \boldsymbol{d}_{i\xi} : \xi \in \Xi \}$ $(i = 1, \ldots, s)$ be sets of individual constants of power $\mathfrak{n}$. We assume these symbols to be distinct from all those found in S and the diagrams. Let R_i be the collection of all possible sentences of the form $\boldsymbol{d}_{i\xi} \not\approx \boldsymbol{d}_{i\zeta}$ or $\boldsymbol{d}_{i\xi} \not\approx \boldsymbol{c}_i$, where $\xi, \zeta \in \Xi$, $\xi \neq \zeta$, $c_i \in \mathfrak{M}_i$ (cf. the diagram construction), and consider the axiom system

$$T = S \cup D(\mathfrak{M}_{i_1}) \cup \ldots \cup D(\mathfrak{M}_{i_p}) \cup R_{i_1} \cup \ldots \cup R_{i_p} ,$$

where $i_1, \ldots, i_p$ are those i for which $\mathfrak{M}_i$ has power $\leqslant \mathfrak{n}$. Any finite subset $T_f \subseteq T$ contains only a finite number of individual constants and, therefore, can be realized in $\mathfrak{M}_1, \ldots, \mathfrak{M}_s$, assigning the $\boldsymbol{d}_{i\xi}$ to distinct elements differing from those whose designations appear explicitly in T_f. By the compactness theorem there is a T-model $\mathfrak{P}$. The projections $\mathfrak{P}_{i_1}, \ldots, \mathfrak{P}_{i_p}$ of $\mathfrak{P}$ include the corresponding sets $D_{i_k} \cup \mathfrak{M}_{i_k}$ $(k = 1, \ldots, p)$, which have power $\mathfrak{n}$. According to Theorem 3, in $\mathfrak{P}_1, \ldots, \mathfrak{P}_s$ are σ-corresponding submodels $\mathfrak{N}_1 \in \mathfrak{K}_1, \ldots, \mathfrak{N}_s \in \mathfrak{K}_s$ bounded in power by $\mathfrak{n}$, with $\mathfrak{N}_{i_1}, \ldots, \mathfrak{N}_{i_p}$ including $D_{i_1} \cup \mathfrak{M}_{i_1}, \ldots, D_{i_p} \cup \mathfrak{M}_{i_p}$ and, therefore, having power $\mathfrak{n}$. Applying the entire argument again guarantees that $\mathfrak{N}_1, \ldots, \mathfrak{N}_s$ all have power $\mathfrak{n}$.

The proof of the second part of the theorem is the same except that for R_i we take the collection of all sentences of the form $\boldsymbol{d}_{ik} \not\approx \boldsymbol{d}_{il}$, for $k \neq l$, $(k, l = 1, 2, \ldots)$. ■

For $s = 1$ we have the

Corollary: *For every infinite model* $\mathfrak{M}$ *in a projective single-base class* $\mathfrak{K}$ *there is a proper* $\mathfrak{K}$*-supermodel* $\mathfrak{N}$ *of any previously prescribed power* $\mathfrak{n}$, *as long as* $\mathfrak{n}$ *is not less than the power of* $\mathfrak{M}$ *and* $\mathfrak{m}$, *where* $\mathfrak{m}$ *is the cardinal for* $\mathfrak{K}$ *given by the theorem on boundedness.*

If for every natural number n, $\mathfrak{K}$ *contains a model with at least n elements, then* $\mathfrak{K}$ *contains an infinite model.* ■

Theorems 3 and 4 and their corollaries are formulated only for single-base classes. Analogous statements for multibase classes are also true if the power of a model is taken to mean the power of the union of its bases (4).

§ 3. Quasiuniversal subclasses

§ 3.1. *Stability*

Let

$$\Phi = (\mathfrak{O}_1 \mathfrak{x}_1) \ldots (\mathfrak{O}_m \mathfrak{x}_m)\, \Psi(\mathfrak{x}_1, \ldots, \mathfrak{x}_m, \mathfrak{x}_{m+1}, \ldots, \mathfrak{x}_n) \tag{5}$$

be a multibase second-order formula in prenex form, where $\mathfrak{x}_1, \ldots, \mathfrak{x}_n$ are predicate or individual variables of fixed sort relative to the bases M_α ($\alpha \in A$) (cf. §1.1). It may happen that in this formula all of the quantifiers over individual variables ranging over the sets M_β ($\beta \in B \subseteq A$) are universal. In such a case we shall say that Φ has B-*universal form*. Moreover, these universal quantifiers need not be located sequentially in Φ, but may be interspersed with existential and universal quantifiers over individual variables of other sorts as well as predicate quantifiers.

Theorem 5 (cf. [III] and [163]): *If a multibase* SOPL *sentence* Φ *has* B-*universal form and is satisfied in some model* $\mathfrak{M}$, *then* Φ *is satisfied in every* B-*submodel* (cf. §1.1) *of* $\mathfrak{M}$.

We need several new notions for the proof of this theorem. Let $\mathcal{K}$ be a class of multibase models with bases M_α ($\alpha \in A$) and basic predicates $P_\gamma = P_\gamma^{i_1, \ldots, i_{k_\gamma}}$ ($\gamma \in \Gamma$, $i_p = i_p(\gamma) \in A$). We say that a *second-order* relation (predicate) $Z(x_1, \ldots, x_m, X_1, \ldots, X_n)$ – where $x_1, \ldots, x_m$ and $X_1, \ldots, X_n$ are individual and first-order predicate variables with sorts $\alpha_1, \ldots, \alpha_m$ and $\rho_1, \ldots, \rho_n$, respectively – is determined on $\mathcal{K}$-models iff for every $\mathcal{K}$-model $\mathfrak{M}$, every sequence $\langle u_1, \ldots, u_m, U_1, \ldots, U_n\rangle$ of elements of $\mathfrak{M}$ and predicates defined on $\mathfrak{M}$ with the sorts $\alpha_1, \ldots, \alpha_m, \rho_1, \ldots, \rho_n$ is associated with one of the truth-values T or F in the name of Z.

The relation Z is called *formular* in $\mathcal{K}$ iff one can find a SOPL formula $\Omega(\boldsymbol{x}_1, \ldots, \boldsymbol{x}_m, \boldsymbol{X}_1, \ldots, \boldsymbol{X}_n)$ with appropriate sorts whose value in every $\mathcal{K}$-model $\mathfrak{M}$ for all $u_1, \ldots, u_m, U_1, \ldots, U_n$ from $\mathfrak{M}$ coincides with the value of $Z(u_1, \ldots, u_m, U_1, \ldots, U_n)$. The formula $\Omega(\boldsymbol{x}_1, \ldots, \boldsymbol{x}_m, \boldsymbol{X}_1, \ldots, \boldsymbol{X}_n)$ may contain – in addition to the free variables $\boldsymbol{x}_i$, $\boldsymbol{X}_j$ – bound individual and predicate variables, as well as the predicate constants associated with the class $\mathcal{K}$.

The relation Z is called B-*stable* ($B \subseteq A$) in $\mathcal{K}$ iff in every $\mathcal{K}$-model $\mathfrak{M}$ and for all $u_1, \ldots, u_m, U_1, \ldots, U_n$ from $\mathfrak{M}$ and any B-submodel $\mathfrak{M}_0$ of $\mathfrak{M}$ containing $u_1, \ldots, u_m$, the truth of $Z(u_1, \ldots, u_m, U_1, \ldots, U_n)$ in $\mathfrak{M}$ implies the definition and truth of $Z(u_1, \ldots, u_m, U_1^0, \ldots, U_n^0)$ in $\mathfrak{M}_0$, where U_j^0 is the restriction of U_j to the bases of $\mathfrak{M}_0$ ($j = 1, \ldots, n$).

From the definition of submodel it immediately follows that the basic predicates of the class $\mathcal{K}$, along with their negations, are stable (i.e., A-stable) in $\mathcal{K}$. Conjunctions and disjunctions of B-stable relations are B-stable. Therefore, every relation Z defined in $\mathcal{K}$ by a quantifier-free formula is stable in $\mathcal{K}$.

It is also easy to verify that *if the relation Z is* B-*stable in* $\mathcal{K}$, *then so are the relations defined in* $\mathcal{K}$-*models by the formulas*

$$(\boldsymbol{X}_i)\boldsymbol{Z}, \quad (\exists \boldsymbol{X}_i)\boldsymbol{Z}, \quad (\boldsymbol{x}_j)\boldsymbol{Z}, \quad (\exists \boldsymbol{x}_k)\boldsymbol{Z},$$

for $i = 1, ..., n$, *and* $j, k = 1, ..., m$, *as long as the sort of* $\boldsymbol{x}_k$ *does not belong* to B.

Indeed, in the $\mathcal{K}$-model $\mathfrak{M}$ let $u_1, ..., u_m$ and $U_1, ..., U_{n-1}$ be elements and predicates such that the relation given by

$$Y(u_1, ..., u_m, U_1, ..., U_{n-1}) = (\boldsymbol{X})Z(u_1, ..., u_m, U_1, ..., U_{n-1}, \boldsymbol{X})$$

is true. That is, for any predicate U on $\mathfrak{M}$,

$$Z(u_1, ..., u_m, U_1, ..., U_{n-1}, U)$$

is true, and so,

$$Z(u_1, ..., u_m, U_1^0, ..., U_{n-1}^0, U^0)$$

is defined and true in the B-submodel $\mathfrak{M}_0$ containing $u_1, ..., u_m$. If U ranges over all predicates on $\mathfrak{M}$ with the same sort as $\boldsymbol{X}$, then U^0 runs over all predicates on $\mathfrak{M}_0$ with this sort. Therefore,

$$Y(u_1, ..., u_m, U_1, ..., U_{n-1})$$

is true on $\mathfrak{M}_0$, and the relation Y is B-stable.

Let us look at one other case. Suppose for $u_1, ..., u_{m-1}, U_1, ..., U_n$ the relation given by

$$Y(u_1, ..., u_{m-1}, U_1, ..., U_n) = (\exists \boldsymbol{x})Z(u_1, ..., u_{m-1}, \boldsymbol{x}, U_1, ..., U_n)$$

is true in $\mathfrak{M}$. This means for some u in $\mathfrak{M}$,

$$Z(u_1, ..., u_{m-1}, u, U_1, ..., U_n)$$

is true. Since the sort of $\boldsymbol{x}$ does not belong to B, u belongs to any B-submodel $\mathfrak{M}_0$, which we assume also contains $u_1, ..., u_{m-1}$. But Z is B-stable, so

$$Z(u_1, ..., u_{m-1}, u, U_1^0, ..., U_n^0)$$

is defined and true in $\mathfrak{M}_0$, and so

$$Y(u_1, ..., u_{m-1}, U_1^0, ..., U_n^0)$$

is true in $\mathfrak{M}_0$, and Y is B-stable in $\mathfrak{K}$. The second and third cases are handled analogously. ▲

The theorem to be proved follows at once from these remarks. Indeed, let the sentence Φ be of the form (5). Since Ψ is quantifier free, it is stable. By assumption, none of the $(\mathfrak{O}_1\, \mathfrak{x}_1), \ldots, (\mathfrak{O}_m\, \mathfrak{x}_m)$ is of the form $(\exists \boldsymbol{x}^\alpha)$ where $\alpha \in$ B. Therefore, the sentence Φ is B-stable, which is equivalent to the assertion of the theorem. ■

§3.2. *The intrinsic local theorem*

Let us consider an arbitrary class $\mathfrak{K}$ of models with basic sets M_α $(\alpha \in \mathrm{A})$ and predicates P_γ $(\gamma \in \Gamma)$. As in the definition of projective subclasses, we introduce auxiliary sets of objects N_β $(\beta \in \mathrm{B})$ and predicates Q_δ $(\delta \in \Delta)$, defined on the M_α, N_β. These predicates are generally of mixed sort; that is, some of the arguments vary over the M_α, and some over the N_β. Let $\boldsymbol{R}_\lambda$ $(\lambda \in \Lambda)$ be a system of predicate symbols; these will be viewed as variables ranging over predicates defined on the M_α.

We consider systems of axioms K, R_λ, S of the following form:

(I) **K** is a collection of sentences of FOPL for bases M_α, N_β $(\alpha \in \mathrm{A}, \beta \in \mathrm{B})$ and fundamental predicates P_γ, Q_δ $(\gamma \in \Gamma, \delta \in \Delta)$. The quantifiers on variables over the M_α are universal; in the case of the N_β the quantifiers may be either universal or existential.

(II) For $\lambda \in \Lambda$, the system R_λ consists of FOPL sentences over bases M_α $(\alpha \in \mathrm{A})$ with predicate symbols $\boldsymbol{P}_\gamma$ $(\gamma \in \Gamma)$ and $\boldsymbol{R}_\lambda$. All quantifiers are universal.

(III) S is a collection of SOPL sentences over bases M_α, N_β $(\alpha \in \mathrm{A}, \beta \in \mathrm{B})$. The predicate constants in S lie among the $\boldsymbol{P}_\gamma$ and $\boldsymbol{Q}_\delta$; the predicate variables – all bound, of course – are all taken from $\{\boldsymbol{R}_\lambda : \lambda \in \Lambda\}$. Individual quantifiers associated with the M_α are assumed to be universal, the others arbitrary.

Quantifiers of the form $(\boldsymbol{R}_\lambda)$, $(\exists \boldsymbol{R}_\lambda)$ in the sentences of S are assumed to be relativized to the set of all predicates on the M_α of the same sort as $\boldsymbol{R}_\lambda$ and satisfying the axioms in R_λ $(\lambda \in \Lambda)$.

In all three types of axioms the equality sign can appear, but in connection only with individuals – not with predicates. The sets of fundamental symbols and axioms in each group may be infinite.

Let **L** denote the axiom system resulting from the combination of **K**, S, and the R_λ. We say that a model $\mathfrak{M}$ with bases M_α $(\alpha \in \mathrm{A})$ and predicates P_γ $(\gamma \in \Gamma)$ *satisfies* the system **L** iff it is possible to find auxiliary sets N_β $(\beta \in \mathrm{B})$ and to define predicates Q_δ $(\delta \in \Delta)$ in $\{M_\alpha : \alpha \in \mathrm{A}\} \cup \{N_\beta : \beta \in \mathrm{B}\}$, such that the extended and enriched model $\langle M_\alpha, N_\beta; P_\gamma, Q_\delta : \alpha \in \mathrm{A}, \beta \in \mathrm{B}, \gamma \in \Gamma, \delta \in \Delta \rangle$ satisfies all the sentences in **L**.

A class $\mathcal{L}$ of models is called *quasiuniversal* iff it is the class of all models satisfying an axiom system **L** of the indicated type; note that all quantifiers in **L** over elements of a model are universal.

Analogously, a subclass $\mathcal{L}$ of a class $\mathcal{K}$ of models is called *quasiuniversal in* $\mathcal{K}$ iff $\mathcal{L}$ is the intersection of $\mathcal{K}$ with some quasiuniversal class $\mathcal{L}_0$, that is, $\mathcal{L}$ consists of all $\mathcal{K}$-models satisfying some quasiuniversal system of axioms.

In order to get a clearer picture of quasiuniversal classes, let us look at a few examples.

We shall prove that *every class of nontrivial models (those with more than one element of each sort) which admits an axiomatization by means of axioms in Skolem form:*

$$(\boldsymbol{x}_1) \dots (\boldsymbol{x}_m)(\exists \boldsymbol{y}_1) \dots (\exists \boldsymbol{y}_n)\, \Psi(\boldsymbol{x}_1, \dots, \boldsymbol{x}_m, \boldsymbol{y}_1, \dots, \boldsymbol{y}_n) \tag{6}$$

is quasiuniversal.

Indeed, we shall show that every sentence (6) is equivalent to the following pair of quasiuniversal axioms (first- and second-order):

$$(\boldsymbol{x}_1) \dots (\boldsymbol{x}_m)(\boldsymbol{y}_1) \dots (\boldsymbol{y}_n)(\Psi(\boldsymbol{x}_1, \dots, \boldsymbol{x}_m, \boldsymbol{y}_1, \dots, \boldsymbol{y}_n) \to$$

$$\to \boldsymbol{R}(\boldsymbol{x}_1, \dots, \boldsymbol{x}_m))\,, \tag{7}$$

$$(\boldsymbol{R})(\boldsymbol{x}_1) \dots (\boldsymbol{x}_m)\, \boldsymbol{R}(\boldsymbol{x}_1, \dots, \boldsymbol{x}_m)\,, \tag{8}$$

where the quantifier $(\boldsymbol{R})$ is specialized as indicated above and means: "for all $\boldsymbol{R}$ satisfying axiom (7)".

For if (6) is true in some nontrivial model $\mathfrak{M}$ and the predicate R on $\mathfrak{M}$ satisfies (7), then clearly

$$(\boldsymbol{x}_1) \dots (\boldsymbol{x}_m)\, \boldsymbol{R}(\boldsymbol{x}_1, \dots, \boldsymbol{x}_m)$$

holds in $\mathfrak{M}$. Suppose (6) is false. Then there are $u_1, \dots, u_m$ in $\mathfrak{M}$ such that for all $v_1, \dots, v_n$ in $\mathfrak{M}$, $\Psi(u_1, \dots, u_m, v_1, \dots, v_n)$ is false. Take the predicate $R(x_1, \dots, x_m)$ to be true on $\langle u_1, \dots, u_m \rangle$ and false on all other sequences (and there are others) ([5]). R satisfies axiom (7), but does not satisfy the condition $(\boldsymbol{x}_1) \dots (\boldsymbol{x}_m)\, \boldsymbol{R}(\boldsymbol{x}_1, \dots, \boldsymbol{x}_m)$. ∎

As a second example we take any class $\mathcal{K}$ of (single-base) algebras with basic operations $f_i(x_1, \dots, x_{n_i})$ $(i = 1, \dots, s)$. Checking the axioms:

$$\boldsymbol{x}R\boldsymbol{x} \;\&\; (\boldsymbol{x}R\boldsymbol{y} \to \boldsymbol{y}R\boldsymbol{x}) \;\&\; (\boldsymbol{x}R\boldsymbol{y} \;\&\; \boldsymbol{y}R\boldsymbol{z} \to \boldsymbol{x}R\boldsymbol{z})\,, \tag{9}$$

$$x_1 R y_1 \;\&\; \ldots \;\&\; x_{n_i} R y_{n_i} \;\&\; u \approx f_i(x_1, \ldots, x_{n_i}) \;\&$$

$$\& \; v \approx f_i(y_1, \ldots, y_{n_i}) \to uRv \qquad (i = 1, \ldots, s)\,, \tag{10}$$

$$(R)(x)(y)(u)(v)(\neg xRy \to (uRv \to u \approx v))\,, \tag{11}$$

we note that (9) and (10) signify that R is a congruence relation, while (11) says that every congruence that does not identify everything coincides with the equality relation. Hence, the axioms (9), (10), and (11) are equivalent to the assertion that an algebra is (homomorphically) simple, and, consequently, *the subclass of (homomorphically) simple algebras in any class $\mathcal{K}$ of algebras is quasiuniversal.* ■

Turning to the axioms:

$$R(x_1) \;\&\; \ldots \;\&\; R(x_{n_i}) \;\&\; y \approx f_i(x_1, \ldots, x_{n_i}) \to R(y)\,,$$

$$(R)(x)(y)(z)(R(x) \;\&\; R(y) \;\&\; x \not\approx y \to R(z))\,,$$

we conclude that *the collection of those algebras with no non-trivial proper subalgebras forms a quasiuniversal subclass in a given class of algebras of any fixed type.* ■

In particular, the subclass of all prime fields and the subclass of all groups of finite prime order are quasiuniversal. Neither is projective, since for every natural number n each contains models with more than n elements, but does not contain uncountable models and, hence, violates the extendability result for projective classes (cf. §2.3).

Inasmuch as simple abelian groups do not form a projective class, the class of all simple groups is not projective.

It becomes clear that quasiuniversal subclasses of axiomatizable classes of models profoundly differ from axiomatizable and projective subclasses. A quasiuniversal subclass can even consist of one infinite model, as in the case of the class of prime fields of characteristic zero. Nevertheless, we have the following fundamental theorem:

Theorem 6 (intrinsic local theorem): *If the model $\mathfrak{M}$ has a local system of submodels belonging to a quasiuniversal class $\mathfrak{L}$ ([6]), then $\mathfrak{M}$ belongs to $\mathfrak{L}$. In particular, the union of an increasing chain of $\mathfrak{L}$-models is an $\mathfrak{L}$-model.*

The proof employs the method of "objectification" of predicates, whereby predicates of various orders, ranks, and sorts are viewed as elements of additional basic sets, and sentences of higher order are rewritten as multibase FOPL sentences.

Let $\mathfrak{M}$ be a model with bases M_α $(\alpha \in \mathrm{A})$ and predicates P_γ $(\gamma \in \Gamma)$, and with a local system $\{\mathfrak{M}_\xi : \xi \in \Xi\}$ of submodels satisfying a quasiuniversal axiom system $\mathbf{L} = \mathbf{K} \cup \bigcup_{\lambda \in \Lambda} \mathrm{R}_\lambda \cup \mathrm{S}$ with the structure described above. With every predicate $R_\lambda^{(\mu)}$ defined on the M_α and satisfying R_λ we associate a new element $r_\lambda^{(\mu)}$ in a 1–1 fashion; for each $\lambda \in \Lambda$, the set of all such $r_\lambda^{(\mu)}$ is called U_λ. In the system $\{M_\alpha : \alpha \in \mathrm{A}\} \cup \{U_\lambda : \lambda \in \Lambda\}$ we define new predicates E_λ $(\lambda \in \Lambda)$ as follows: if

$$R_\lambda = R_\lambda(x_1, \ldots, x_{p_\lambda}) \quad (x_i \in M_{\alpha_i}),$$

then

$$E_\lambda = E_\lambda(r_\lambda, x_1, \ldots, x_{p_\lambda}) \quad (r_\lambda \in U_\lambda,\ x_i \in M_{\alpha_i}),$$

and, by definition,

$$E_\lambda(r_\lambda^{(\mu)}, u_1, \ldots, u_{p_\lambda}) = R_\lambda^{(\mu)}(u_1, \ldots, u_{p_\lambda}) \tag{12}$$

for all $r_\lambda^{(\mu)} \in U_\lambda$ and $u_i \in M_{\alpha_i}$ $(i = 1, \ldots, p_\lambda)$.

Thus, $\mathfrak{M}$ gives rise to the model

$$\mathfrak{M}^* = \langle M_\alpha, U_\lambda; P_\gamma, E_\lambda : \alpha \in \mathrm{A}, \gamma \in \Gamma, \lambda \in \Lambda \rangle$$

with a greater number of bases and predicates. Let R_λ^* denote the system of axioms obtained from those in R_λ by replacing every occurrence of $\boldsymbol{R}_\lambda(\boldsymbol{x}_1, \ldots, \boldsymbol{x}_p)$ in them with $\boldsymbol{E}_\lambda(\boldsymbol{r}_\lambda, \boldsymbol{x}_1, \ldots, \boldsymbol{x}_p)$ and prefixing each resulting formula with $(\boldsymbol{r}_\lambda)$ so that it governs the entire formula. Similarly, let S^* denote the collection of axioms obtained from those of S by substituting $\boldsymbol{E}_\lambda(\boldsymbol{r}_\lambda, \boldsymbol{x}_1, \ldots, \boldsymbol{x}_p)$ for $\boldsymbol{R}_\lambda(\boldsymbol{x}_1, \ldots, \boldsymbol{x}_p)$ and changing the quantifiers $(\boldsymbol{R}_\lambda)$, $(\exists \boldsymbol{R}_\lambda)$ to $(\boldsymbol{r}_\lambda)$ and $(\exists \boldsymbol{r}_\lambda)$. The axioms in the R_λ^* and S^* are seen to be FOPL sentences referring to M_α, U_λ, N_β as basic sets and P_γ, E_λ, Q_δ as fundamental predicates.

From the construction of $\mathfrak{M}^*$ and the formula (12) it is seen that for all $\lambda \in \Lambda$, R_λ^* is automatically satisfied by $\mathfrak{M}^*$. If, in addition, we can construct sets N_β $(\beta \in \mathrm{B})$ and define predicates Q_δ $(\delta \in \Delta)$ in the system $\{M_\alpha, N_\beta : \alpha \in \mathrm{A}, \beta \in \mathrm{B}\}$ such that the resulting multibase model (with the M_α, U_λ, N_β as bases and the P_γ, E_λ, Q_δ as basic predicates) satisfies the rest of the system $\mathbf{L}^* = \mathbf{K} \cup \bigcup_{\lambda \in \Lambda} \mathrm{R}_\lambda^* \cup \mathrm{S}^*$, then $\mathfrak{M}$ satisfies L, i.e., $\mathfrak{M}$ belongs to $\mathcal{L}$.

Let us begin with the axiom system

$$\mathrm{T}^* = \mathrm{D}(\mathfrak{M}^*) \cup \mathrm{L}^*,$$

which we shall now show is consistent. By the compactness theorem it suffices to show that every finite subset T_f^* of T^* is consistent. Since $T = T_f^* \cap D(\mathfrak{M}^*)$ is finite, the sentences in it have reference to only a finite number of elements $a_1, ..., a_m$ in $\mathfrak{M}$ and $r_{\lambda_1}^1, ..., r_{\lambda_n}^n$ in $\bigcup_{\lambda \in \Lambda} U_\lambda$. By assumption, there is a $\tau \in \Xi$ such that $a_1, ..., a_m$ are contained in appropriate bases of the $\mathcal{L}$-submodel $\mathfrak{M}_\tau$. If it happens that $\lambda_j = \lambda_k$, then there are $b_1, ..., b_p$ in $\mathfrak{M}$ such that

$$E_{\lambda_j}(r_{\lambda_j}^j, b_1, ..., b_p) \neq E_{\lambda_k}(r_{\lambda_k}^k, b_1, ..., b_p) , \tag{13}$$

which follows from (12) and the 1–1 correspondence between elements of $\bigcup_{\lambda \in \Lambda} U_\lambda$ and certain predicates on $\mathfrak{M}$. We can assume that in every such case the appropriate $b_1, ..., b_p$ are also contained in $\mathfrak{M}_\tau$ (7).

$\mathfrak{M}_\tau^*$ is constructed from $\mathfrak{M}_\tau$ as above and is generally not a submodel of $\mathfrak{M}^*$. By construction, the predicates $R_{\lambda_1}^1, ..., R_{\lambda_n}^n$ corresponding to the elements $r_{\lambda_1}^1, ..., r_{\lambda_n}^n$ satisfy $\mathrm{R}_{\lambda_1}, ..., \mathrm{R}_{\lambda_n}$ in $\mathfrak{M}$. For $i = 1, ..., n$ we denote the restriction of $R_{\lambda_i}^i$ to $\mathfrak{M}_\tau$ by ${}^0R_{\lambda_i}^i$; since the axioms in R_{λ_i} are universal, the predicate ${}^0R_{\lambda_i}^i$ satisfies R_{λ_i} in $\mathfrak{M}_\tau$; therefore, in the set $U_{\lambda_i}^\tau$ constructed for $\mathfrak{M}_\tau^*$ there is an element ${}^0r_{\lambda_i}^i$ such that for all appropriate $u_1, ..., u_p$ in $\mathfrak{M}_\tau$, we have the predicate equation

$$E_{\lambda_i}^\tau({}^0r_{\lambda_i}^i, u_1, ..., u_p) = E_{\lambda_i}(r_{\lambda_i}^i, u_1, ..., u_p) . \tag{14}$$

From (13), (14), and the construction of $\mathfrak{M}^*$ and $\mathfrak{M}_\tau^*$ it follows that the map $r_{\lambda_i}^i \to {}^0r_{\lambda_i}^i$ $(i = 1, ..., n)$ is a 1–1 correspondence. We let T_τ be the set of all formulas obtained from those in T by replacing the constants $\boldsymbol{r}_{\lambda_1}^1, ..., \boldsymbol{r}_{\lambda_n}^n$ with ${}^0\boldsymbol{r}_{\lambda_1}^1, ..., {}^0\boldsymbol{r}_{\lambda_n}^n$. We now convince ourselves that $T_\tau \subseteq D(\mathfrak{M}_\tau^*)$, that is, every sentence in T becomes true in $\mathfrak{M}_\tau^*$ after the above substitution. For the sentences in T have the form:

$$\boldsymbol{a}_i \approx \boldsymbol{a}_j, \quad \boldsymbol{a}_i \approx \boldsymbol{r}_{\lambda_k}^k, \quad \boldsymbol{r}_{\lambda_k}^k \approx \boldsymbol{r}_{\lambda_l}^l, \quad \boldsymbol{P}_\gamma(\boldsymbol{a}_{i_1}, ..., \boldsymbol{a}_{i_s}),$$

$$\boldsymbol{E}_{\lambda_k}(\boldsymbol{r}_{\lambda_k}^k, \boldsymbol{a}_{j_1}, ..., \boldsymbol{a}_{j_p}) ,$$

or the negations of these formulas. Thus it is clear from the choice of τ, the construction of $\mathfrak{M}_\tau^*$, and the relations (13) and (14) that the sentences of T_τ are true in $\mathfrak{M}_\tau^*$.

Hence $\mathfrak{M}_\tau^*$ satisfies $T \cup L^*$ under the interpretation of $\boldsymbol{r}_{\lambda_i}^i$ as ${}^0r_{\lambda_i}^i$ $(i = 1, ..., n)$ – in the sense that $\mathfrak{M}_\tau^*$ can be enriched to make a full-fledged $T \cup L^*$-model. Since $T_f^* \subseteq T \cup L^*$, we conclude that every finite subset of T^* is consistent, and so, T^* itself has a model

$$\mathfrak{N}^* = \langle M'_\alpha, N_\beta, U'_\lambda; P'_\gamma, Q_\delta, E'_\lambda : \alpha \in \mathrm{A}, \beta \in \mathrm{B}, \gamma \in \Gamma, \delta \in \Delta, \lambda \in \Lambda \rangle .$$

As T^* includes the diagram of $\mathfrak{M}^*$, $M'_\alpha \supseteq M_\alpha$ ($\alpha \in \mathrm{A}$), $U'_\lambda \supseteq U_\lambda$ ($\lambda \in \Lambda$), and $\mathfrak{M}^*$ is a submodel of the projection of $\mathfrak{N}^*$ onto the M'_α, U'_λ.

In all the sentences in L^* the quantifiers on variables ranging over the M'_α are universal. Therefore, by Theorem 5 the submodel $\mathfrak{N}^*_1$ of $\mathfrak{N}^*$, obtained by restricting the predicates to the M_α, N_β, U'_λ, also satisfies T^*.

For every $\lambda \in \Lambda$ we define an equivalence relation θ_λ on U'_λ by putting $r\theta r' = \mathsf{T}$ for $r, r' \in U'_\lambda$ iff

$$E'_\lambda(r, u_1, ..., u_p) = E'_\lambda(r', u_1, ..., u_p)$$

for all appropriate $u_1, ..., u_p$ in $\bigcup_{\alpha \in \mathrm{A}} M_\alpha$. In the sentences of L the equality sign appears only with individual variables, so in L^* the variables $\boldsymbol{r}_\lambda$ occur only as arguments of the corresponding $\boldsymbol{E}_\lambda$; from this fact and the reasoning behind (13) we see that the θ_λ can be used to "relativitize equality" in $\mathfrak{N}^*_1$ as described in §2.1. This yields a T^*-model $\mathfrak{N}^*_1/\theta$ with bases M_α, N_β, $U'_\lambda/\theta_\lambda$ ($\alpha \in \mathrm{A}$, $\beta \in \mathrm{B}$, $\lambda \in \Lambda$).

For every $\lambda \in \Lambda$ we see that for any fixed $r \in U'_\lambda$, the expression $\boldsymbol{E}_\lambda(r, \boldsymbol{x}_1, ..., \boldsymbol{x}_p)$ defines in $\mathfrak{N}^*_1$ a $p(\lambda)$-ary predicate on the M_α that satisfies R_λ in $\mathfrak{M}$, since $\mathfrak{N}^*_1$ satisfies R^*_λ. But all such predicates are represented by elements of U_λ as seen from the definition of this set. This gives a 1–1 correspondence between $U'_\lambda/\theta_\lambda$ and U_λ for each $\lambda \in \Lambda$. These maps in turn induce a map from $\mathfrak{N}^*_1/\theta$ onto the submodel $\mathfrak{N}^*_0$ of $\mathfrak{N}^*_1$ with bases M_α, N_β, U_λ ($\alpha \in \mathrm{A}$, $\beta \in \mathrm{B}$, $\lambda \in \Lambda$) which is seen to be an isomorphism (cf. §? !). Hence $\mathfrak{N}^*_0$ satisfies L^* outright, and its projection onto the M_α, U_λ is $\mathfrak{M}^*$. ∎

Stronger than Theorem 6 in form only is its

Corollary: *If a model $\mathfrak{M}$ in a class $\mathcal{K}$ has a local system of submodels belonging to a quasiuniversal subclass $\mathcal{L}$ of $\mathcal{K}$, then $\mathfrak{M}$ belongs to $\mathcal{L}$, also.*

For by definition $\mathcal{L} = \mathcal{K} \cap \mathcal{L}_0$ for some quasiuniversal class $\mathcal{L}_0$. $\mathfrak{M}$ has a local system of $\mathcal{L}$- and, therefore, $\mathcal{L}_0$-submodels. By Theorem 6, $\mathfrak{M} \in \mathcal{L}_0$; by assumption $\mathfrak{M} \in \mathcal{K}$, so $\mathfrak{M} \in \mathcal{L}$. ∎

Theorem 7: *If the quasiuniversal class $\mathcal{L}$ is characterized by an axiom system $\mathrm{L} = \mathrm{K} \cup \bigcup_{\lambda \in \Lambda} \mathrm{R}_\lambda \cup \mathrm{S}$ such that for every $\lambda \in \Lambda$, R_λ is empty or consists solely of identically true sentences, then $\mathcal{L}$ is universally axiomatizable.*

According to Theorem 5, every submodel of an $\mathcal{L}$-model is an $\mathcal{L}$-model itself. Thus the extrinsic local theorem for $\mathcal{L}$ takes the form: *if every finite subdiagram of a model $\mathfrak{M}$ is realizable in some $\mathcal{L}$-model, then $\mathfrak{M}$ is an $\mathcal{L}$-model.* This result together with Theorem 1 will give Theorem 7.

We can demonstrate this extrinsic local theorem for $\mathcal{L}$ by repeating the argument for Theorem 6, but taking the $\mathfrak{M}_\xi$ ($\xi \in \Xi$) to be all the $\mathcal{L}$-models required for realization of all possible finite subdiagrams of the model $\mathfrak{M}$. This repetition would be literal, except the $\mathfrak{M}_\xi$ are not submodels of $\mathfrak{M}$. This fact, however, is used only to guarantee the existence of a predicate ${}^0R^i_{\lambda i}$ on $\mathfrak{M}_\tau$ taking the same values on $F = \{a_1, ..., a_m, b_1, ..., b_t\}$ as $R^i_{\lambda_i}$ does in $\mathfrak{M}$, and satisfying R_{λ_i} in $\mathfrak{M}_\tau$ ($i = 1, ..., n$). Since these axioms are trivial in the present case, any predicate on $\mathfrak{M}_\tau$ can be chosen for ${}^0R^i_{\lambda_i}$ as long as it has the same values on the "realizing image" of F as $R^i_{\lambda_i}$ takes on F. ∎

The chief special case of Theorem 7 – when $\mathcal{K}$ is empty, and the sentences in S do not refer to the supplementary bases N_β – was known before (cf. [88] and [III]). In essence, Theorem 7 asserts the eliminability of the bound predicate variables $\boldsymbol{R}_\lambda$, and the auxiliary predicates Q_δ and bases N_β from K and S in the sense of Ackermann [1].

§3.3. *Applications*

The concept of solvability of a group greatly ramifies on passage from finite to infinite groups; contrasted with the single class of solvable finite groups are the equally natural classes of RN-, RI-, and Z-groups, plus $\overline{\mathrm{RN}}$-, $\overline{\mathrm{RI}}$-, $\overline{\mathrm{Z}}$-, and $\tilde{\mathrm{N}}$-groups, and even others (cf. [II], [81], and [M4]). Local theorems are known for all the indicated classes. For the "lower" classes of RN-, RI-, and Z-groups, as well as orderable groups, these theorems were first proved by the author [II], [M2], with the aid of the compactness theorem for FOPL. Proofs of the local theorems for $\overline{\mathrm{RI}}$- and $\overline{\mathrm{Z}}$-groups were obtained by the author, and for $\tilde{\mathrm{N}}$-groups by Baer [6], but specific group-theoretic methods were used. We now show that local theorems for all the "higher" classes of $\overline{\mathrm{RN}}$-, $\overline{\mathrm{RI}}$-, $\overline{\mathrm{Z}}$-, and $\tilde{\mathrm{N}}$-groups are special cases of the intrinsic local theorem proved above, while the well-known *intrinsic* local theorems for the lower classes of RN-, RI-, and Z-groups can be replaced with stronger *extrinsic* local theorems.

A system $\mathfrak{S}$ of subsets of a set M linearly ordered by inclusion is called *complete* iff it contains the union and intersection of any collection of its members, as well as the set M ([8]). For every complete, linearly ordered system $\mathfrak{S} = \{M_\alpha : \alpha \in \mathrm{A}\}$ of subsets of M, we define a predicate $R_\mathfrak{S}$ (or R, for short) on M by putting $xRy = \mathrm{T}$ iff some member of $\mathfrak{S}$ contains x, but does not contain y. Clearly, R satisfies the axioms

$$*\mathrm{A}:\ \neg\, \boldsymbol{xRx}$$

$$*\mathrm{B}:\ \boldsymbol{xRy} \,\&\, \boldsymbol{yRz} \rightarrow \boldsymbol{xRz}$$

$$*\mathrm{C}:\ \boldsymbol{xRz} \,\&\, \neg\, \boldsymbol{yRz} \rightarrow \boldsymbol{xRy}$$

For $y \in M$, let R^y be the set of all $x \in M$ for which $xRy = \mathsf{T}$; it is easy to see that

$$R^y = \bigcup_{y \notin M_\alpha} M_\alpha , \qquad \text{and} \qquad M_\alpha = \bigcap_{y \notin M_\alpha} R^y ,$$

i.e., each R^y belongs to $\mathfrak{S}$, and each $M_\alpha \in \mathfrak{S}$ can be represented as the intersection of an appropriate collection of the R^y.

On the other hand, suppose R is an arbitrary binary predicate defined on M satisfying the conditions *A–*C. The collection of the sets R^y ($y \in M$), defined as above, is linearly ordered by inclusion; it is not difficult to see from *A–*C that this collection can be extended to a complete, linearly ordered system $\mathfrak{S}$ by adding M and all possible intersections of subcollections of the R^y; moreover, $R_{\mathfrak{S}} = R$. Thus, *for any set M, the complete, linearly ordered (by inclusion) systems of subsets of M are in* 1–1 *correspondence with the predicates on M satisfying* *A–*C. ■

We remark that the system $\mathfrak{S}_1$ refines $\mathfrak{S}$, i.e. $\mathfrak{S}_1 \supseteq \mathfrak{S}$, iff the corresponding predicates R_1, R satisfy the condition $\boldsymbol{xRy \to xR_1y}$.

In case the set in question is a group $\mathfrak{G}$, and we wish to investigate complete, linearly ordered systems (sequences) of subgroups of $\mathfrak{G}$ containing the identity subgroup, then we add the following axioms to *A–*C:

$$\text{*D:} \quad \boldsymbol{xRz \mathbin{\&} yRz \to xy^{-1}Rz} ,$$

$$\text{*E:} \quad \boldsymbol{x \not\approx xx^{-1} \to xx^{-1}Rx} .$$

The axiom

$$\text{*F:} \quad \boldsymbol{xRy \to y^{-1}xyRy}$$

is obviously equivalent to the demand that for any pair of adjacent subgroups in the system, the smaller be a normal divisor of the larger (*normality* of the system).

The axiom

$$\text{*G:} \quad \boldsymbol{x \not\approx xx^{-1} \mathbin{\&} \neg xRy \mathbin{\&} \neg yRx \to xyx^{-1}y^{-1}Rx}$$

is equivalent to the requirement that the factor group of any two consecutive subgroups exist and be abelian (*solvability* of the system); the axiom

$$\text{*H:} \quad \boldsymbol{xRy \to z^{-1}xzRy}$$

is equivalent to requiring every subgroup in the system to be a normal divisor of $\mathfrak{G}$ (*invariance* of the system); the axiom

$$\text{*I:}\quad xx^{-1} \not\approx x \to xyx^{-1}y^{-1}Rx$$

is equivalent to demanding that all of the subgroups in the system be normal divisors of $\mathfrak{G}$, and that the factor group $\mathfrak{G}_{\alpha'}/\mathfrak{G}_\alpha$ of any two consecutive subgroups lie in the center of $\mathfrak{G}/\mathfrak{G}_\alpha$ (*centrality* of the system).

We have been dealing mainly with predicates and not with operations, so we shall assume in what follows that all the axioms *A–*I are written with predicates. E.g., if $P(x, y, z)$, $Q(x, y)$ are the predicates $xy = z$ and $y = x^{-1}$, then *E becomes

$$Q(x, y) \;\&\; P(x, y, z) \;\&\; x \not\approx z \to zRx\,.$$

It is important to note that *A–*I are universal FOPL sentences, even when written with predicates; the quantifiers have been suppressed above for clarity.

The property of a group being an RN-, RI-, or Z-group can be expressed by the SOPL axiom

$$\text{(RN):}\quad (\exists R)(\text{*A \& *B \& *C \& *D \& *E \& *G})\,,$$

$$\text{(RI):}\quad (\exists R)(\text{*A \& *B \& *C \& *D \& *E \& *G \& *H})\,,$$

or

$$\text{(Z):}\quad (\exists R)(\text{*A \& *B \& *C \& *D \& *E \& *I})\,,$$

respectively, where the quantifier $(\exists R)$ is not specialized.

The form of (RN), (RI), and (Z) satisfies the hypotheses of Theorem 7, which implies

Theorem 8: RN-, RI-, *and* Z-*groups form universal subclasses of the class of groups; hence, the extrinsic local theorem holds for these subclasses.* ■

Passing on to $\overline{\text{RN}}$-, $\overline{\text{RI}}$-, $\overline{\text{Z}}$-, and $\widetilde{\text{N}}$-groups, we let $\Phi_1(R)$, $\Phi_2(R)$, $\Phi_3(R)$ denote the second-order matrices (obtained by dropping $(\exists R)$) of the formulas (RN), (RI), (Z). We put

$$\Phi(R) = \text{*A \& *B \& *C \& *D \& *E \& *F}\,,$$

$$\Psi(R) = \text{*A \& *B \& *C \& *D \& *E \& *G}\,.$$

The property of a group being an $\overline{\mathrm{RN}}$-, $\overline{\mathrm{RI}}$-, or $\overline{\mathrm{Z}}$-group can be expressed by the axiom

$$(\forall_{\Phi} R)(\exists_{\Phi_1} R_1)(u)(v)(uRv \to uR_1 v) ,$$

$$(\forall_{\Psi} R)(\exists_{\Phi_2} R_1)(u)(v)(uRv \to uR_1 v) ,$$

or

$$(\forall_{\Psi} R)(\exists_{\Phi_3} R_1)(u)(v)(uRv \to uR_1 v) ,$$

respectively.

These axioms have the form indicated in the definition of quasiuniversal classes. Thus, $\overline{\mathrm{RN}}$-, $\overline{\mathrm{RI}}$-, *and* $\overline{\mathrm{Z}}$*-groups form quasiuniversal subclasses of the class of groups. By the corollary to Theorem 6 the intrinsic local theorem holds for these subclasses.* ■

To write a definition for $\widetilde{\mathrm{N}}$-groups we must characterize the predicate R corresponding to a subgroup system of the form $\{e\} \subseteq \mathfrak{G}_1 \subseteq \mathfrak{G}$. Clearly, it suffices to supplement *A–*E with the axiom

$$\text{*J: } x \not\approx xx^{-1} \;\&\; xRy \to \neg yRz .$$

If Ω is the conjunction of *A–*E and *J, then a group satisfies

$$(\forall_{\Omega} R)(\exists_{\Phi_1} R_1)(u)(v)(uRv \to uR_1 v)$$

iff it is an $\widetilde{\mathrm{N}}$-group. This axiom, too, has the form required by the definition of quasiuniversal class. Consequently, $\widetilde{\mathrm{N}}$*-groups form a quasiuniversal subclass of the class of groups, and so, the intrinsic local theorem holds for this subclass.* ■

Analogous statements are true for the class of groups admitting linear group-orderings (*orderable* groups), and for the class of those groups in which it is possible to extend any partial group-order to a linear group-order (*freely orderable* groups) (cf. [M2]).

Indeed, let $\Phi(R)$ be the FOPL sentence

$$xRx \;\&\; (xRy \;\&\; yRz \to xRz) \;\&\; (xRy \;\&\; yRx \to x \approx y) \;\&$$

$$\& \; (xRy \to wxzRwyz) ,$$

and let $\Psi(R)$ be the conjunction of $\Phi(R)$ and the sentence

$$xRy \vee yRx .$$

Then the property of a group being orderable is expressed by the SOPL sentence

$$(\exists R)\,\Phi(R)\,;$$

a group is free orderable iff it satisfies the sentence

$$(\forall_\Phi R)(\exists_\Psi R_1)(u)(v)(uRv \rightarrow uR_1v)\,.$$

As above, this implies that *the orderable groups form a universal subclass of the class of groups, while the freely orderable groups form a quasiuniversal subclass.* ■

The first assertion was proved by Łoś [86], who gave explicit universal axioms characterizing orderable groups. From the second assertion we deduce the

Corollary: *The intrinsic local theorem holds for the subclass of freely orderable groups.* ■

This corollary is a new result, apparently. A special case of it was indicated by Łoś [88], who showed that the union of an increasing chain of freely orderable groups, embedded one in the other, is a freely orderable group.

The above results indicate a family resemblance between the classes of RN-, RI-, etc., groups, on one hand, and the classes of orderable and freely orderable groups, on the other. This resemblance is more heavily limned by considering the negation of R in the first case. In fact, letting $P = {\sim}R$, we can rewrite *A–*C in the form

$$xPx, \quad xPy \,\&\, yPz \rightarrow xPz, \quad xPy \vee yPx\,;$$

such a P is said to be a *quasiorder.* RN-, RI-, and Z-groups are, therefore, quasiorderable – with the quasiorder subject to various other conditions.

When we combine the notions of RN-, $\overline{\text{RN}}$-, etc., groups with that of ordered or partially ordered groups, it is natural to demand that the complete systems of subgroups involved in the definitions consist of convex subgroups. The previous arguments show that the classes of groups so obtained are universal or quasiuniversal subclasses, as the case may be. In particular, from the general intrinsic local theorem we can deduce a whole series of new concrete local theorems for groups (cf. [M4]).

NOTES

([1]) This notion of submodel is a bit too general as it neither permits nor prohibits empty M'_α; this leads to a certain uncertainty (more literary than factual) in a few passages, which the reader will have to parse for himself.

([2]) If several of the M_α have elements in common, an ambiguity can arise in the association of constants and the assignment of sorts. It is more convenient to associate a single constant with each element of $\bigcup\{M_\alpha\colon \alpha \in \mathrm{A}\}$ and allow it to have several sorts; it is, however, more consistent with the coming definition of realizability to take the disjoint union here, letting each constant have only one sort. The issue could be avoided by requiring all bases to be disjoint, for this would not hinder the constructions to follow.

([3]) The cited definition includes a boundedness property, which is proved for projective classes in § 2.3. Here, arbitrary individual constants can appear in the members of S.

([4]) The author suggests with validity (and ambiguity) that we can replace "power of a model" with "powers of the bases of a model"; indeed, similar formulations of boundedness and extendability with respect to individual bases are possible – and provable by similar techniques.

([5]) This R should be replaced with its negation; the assumption of nontriviality is seen to be unnecessary.

([6]) In other words, if every finite subset of the disjoint union of the bases of $\mathcal{L}$ is included in some $\mathcal{L}$-submodel of $\mathfrak{M}$.

([7]) This provision has been added in translation and serves to correct a slight shortcoming in the original proof by allowing the diagram to be used unchanged.

([8]) It is important that $\mathfrak{S}$ contain the empty set, too; this can be viewed as the union of the empty collection of members of $\mathfrak{S}$ (and M as its intersection).

CHAPTER 12

REGULAR PRODUCTS OF MODELS

Introduction

In this article regular products are offered as generalizations of direct products of models. Theorems of A. Mostowski [105] *and R. Vaught* [179] *on direct products are special cases of theorems to be proved on regular products. As a preliminary, we study a particular form of model correspondence, a notion defined in* [XI].

In [XI, §1.2] the concept of an axiomatizable correspondence between models of fixed classes was introduced. The first section of this article is dedicated to an examination of those model correspondences which can be prescribed by formulas of first-order predicate logic (FOPL) that contain no auxiliary predicate symbols. This analysis is based on an elementary lemma on the reduction of FOPL formulas with separable variables. Offered as an illustration is S. Feferman's result [179] on formulas true in the direct product of a finite number of models.

In §2 we study the product of a possibly infinite number of models and introduce the new concept of a regular product of models, generalizing the notion of direct product. Relying on the idea of separable variables used in §1 and on the theorem of H. Behmann [2] on the normal form for formulas with unary predicate symbols, for every closed formula (or: sentence, axiom) Φ of FOPL concerning a regular product of models $\mathfrak{M}_\alpha$ $(\alpha \in \mathrm{A})$, we effectively construct an expression Ψ equivalent to Φ that is a propositional combination (i.e., combined by means of &, $\vee$, $\neg$) of a finite number of statements Θ of the form: "among the factors are models $\mathfrak{M}_{\alpha_1}, \ldots, \mathfrak{M}_{\alpha_m}$, with $\alpha_{i_1} \neq \alpha_{j_1}, \ldots, \alpha_{i_p} \neq \alpha_{j_p}$, in which $\Phi_{k_1}, \ldots, \Phi_{k_m}$ are respectively true", where the Φ_{k_i} are effectively constructed FOPL sentences. Thus, in order to be able to judge the truth or falsity of arbitrary FOPL sentences concerning a given regular product, it is sufficient (and, in general, necessary) to be able to decide the matter just for sentences of the indicated form concerning the factors. This is the fundamental result of the present article.

Assuming the number of factors in a proper (in the sense of §2) regular product $\Pi\mathfrak{M}_\alpha$ to be infinite, and all factors to be isomorphic to a single model $\mathfrak{M}_0$, we immediately see that the above statements Θ reduce to ones of the form: "Φ_{k_1} & ... & Φ_{k_m} is true in $\mathfrak{M}_0$". Consequently, in the case of a regular power $\Pi\mathfrak{M}_\alpha = \mathfrak{M}_0^{\aleph_\xi}$ the question of the truth of a FOPL sentence Φ in $\mathfrak{M}_0^{\aleph_\xi}$ reduces (uniformly) to the problem of the truth in $\mathfrak{M}_0$ of an appropriate FOPL sentence effectively constructed from Φ. For direct powers this was proved by A. Mostowski [105].

Let us assume that in the proper regular product $\mathfrak{M} = \Pi\mathfrak{M}_\alpha$ the factors fall into classes of mutually isomorphic models, that every such class contains and infinite numbers of models, and that $\{\mathfrak{M}_\beta : \beta \in \mathrm{B}\}$ is a system of representatives from these classes. Then each Θ mentioned above reduces to the conjunction of statements of the form: "among the models $\mathfrak{M}_\beta$ ($\beta \in \mathrm{B}$) there is a model in which the sentence Φ' is true". The negation of such a statement has the form: "$\neg\Phi'$ is true in every model $\mathfrak{M}_\beta$ ($\beta \in \mathrm{B}$)". Consequently, we obtain an algorithm for distinguishing between those FOPL sentences true in $\mathfrak{M}$ and those false if there is an algorithm for telling whether or not a FOPL sentence is simultaneously true in all the $\mathfrak{M}_\beta$. For direct products this result was formulated by Vaught [179].

Besides the greater generality of the results of §2 compared to those obtained by Mostowski and Vaught, the proofs in §2 seem simpler than Mostowski's, basically thanks to the use of Behmann's transformation. I cannot compare the methods of §2 with Vaught's since at present his results have been published only without proofs.

§1. Splitting correspondences

§1.1. Let $\mathfrak{K}_1, \mathfrak{K}_2$ be classes of models with fundamental predicates denoted generically by P_γ ($\gamma \in \Gamma$) and Q_δ ($\delta \in \Delta$), respectively. The bases of the models in these classes are written uniformly as M and N, respectively. We introduce symbols $\boldsymbol{R}_\lambda$ ($\lambda \in \Lambda$) for new predicates R_λ, generally of a many-sorted character, i.e., part of the arguments of each R_λ range over the set M, and part over N. Let S be a system of FOPL sentences written with the help of the predicate symbols $\boldsymbol{P}_\gamma$, $\boldsymbol{Q}_\delta$, $\boldsymbol{R}_\lambda$. The equality sign $\approx$ can occur in these sentences. All the individual variables and quantifiers are assumed to be specialized. Thus, if the individual variable $\boldsymbol{x}$ is of the first sort, then the quantifier $(\exists\boldsymbol{x})$ means "in M there is an element x such that..." (cf. [XI], §1.1). Models $\mathfrak{M} \in \mathfrak{K}_1$, $\mathfrak{N} \in \mathfrak{K}_2$ are said to S-*correspond* iff it is possible to define predicates R_λ on their bases M and N such that all the axioms in S become true.

Correspondences defined in this fashion are called *axiomatizable.* We analogously define axiomatizable correspondences among models of any number of classes (cf. [XI]).

Although in the definitions above the classes $\mathcal{X}_1, \mathcal{X}_2$ could have been arbitrary, in considering axiomatizable correspondences we shall assume that the $\mathcal{X}_i$ are the classes of all models of the respective given types. If $\mathcal{X}_1, \mathcal{X}_2$ are themselves (first-order) axiomatizable, then we assume S includes the FOPL axiom systems characterizing these classes; the matter thus reduces to the basic case of classes of all models. We further note that individual constant symbols are not permitted in the axioms. Should these be needed, we can introduce new unary predicate symbols instead, inasmuch as the number of predicate symbols is not limited and may be infinite.

Among the simpler correspondences are those admitting axiomatizations by means of axioms in which the auxiliary mixed predicate symbols $\boldsymbol{R}_\lambda$ do not occur, and $\approx$ links only variables of the same sort – axioms which thus refer only to the predicates associated with the classes being considered. These are called *splitting correspondences.* As a sort of justification for this name we have

Theorem 1: *Every splitting correspondence among models in classes* $\mathcal{X}_1, ..., \mathcal{X}_s$ *can be characterized by a system of axioms of the form*

$$\varphi_1^{(\mu)} \vee \varphi_2^{(\mu)} \vee ... \vee \varphi_s^{(\mu)} \qquad (\mu \in \mathrm{M}),$$

where each $\varphi_i^{(\mu)}$ *is a sentence written with the symbols associated with* $\mathcal{X}_i$ $(i = 1, ..., s)$.

This theorem is immediately implied by the following purely combinatorial lemma:

Lemma: *Suppose*

$$\psi = (\mathrm{O}_1 \boldsymbol{x}_1) \dots (\mathrm{O}_m \boldsymbol{x}_m)\, \varphi(\boldsymbol{x}_1, ..., \boldsymbol{x}_m, \boldsymbol{x}_{m+1}, ..., \boldsymbol{x}_q)$$

is a FOPL *formula with the* O_j *being quantifiers and* $\boldsymbol{x}_1, ..., \boldsymbol{x}_q$ *being the only variables in the quantifier-free* φ. *Suppose* $\boldsymbol{x}_1, ..., \boldsymbol{x}_q$ *can be decomposed into disjoint classes* $I_1, ..., I_s$ *such that no atomic formula* $\boldsymbol{P}(\boldsymbol{x}_{i_1}, ..., \boldsymbol{x}_{i_n})$ *or* $\boldsymbol{x}_{i_1} \approx \boldsymbol{x}_{i_2}$ *in* φ *involves variables of different classes. Then by an effective process we can find formulas* ψ_{ij} *such that* ψ *is logically equivalent to*

$$\chi = (\psi_{11} \vee ... \vee \psi_{1k_1}) \,\&\, ... \,\&\, (\psi_{p1} \vee ... \vee \psi_{pk_p});$$

moreover, for each ψ_{ij}, its variables are contained in one of the indicated classes, and its quantifier prefix is a subprefix of that of ψ.

The process of reducing ψ to the form χ basically coincides with a procedure known from the theory of formulas with unary predicates (cf. [2]), and the proof is given here only for the completeness of presentation.

The proof proceeds by induction on the number of quantifiers in ψ. If ψ is quantifier free, then the assertion is obvious since as the ψ_{ij} we can take the atomic formulas $P(x_{i_1}, ..., x_{i_n})$, $x_{i_1} \approx x_{i_2}$ and their negations. We assume the lemma is true for $m-1$ quantifiers. Then for the formula

$$(\eth_2 x_2) \dots (\eth_m x_m)\, \varphi(x_1, ..., x_m, x_{m+1}, ..., x_q)$$

we can find one of the form χ equivalent to it, so we have the equivalence

$$\psi \leftrightarrow (\eth_1 x_1)((\psi_{11} \vee \dots \vee \psi_{1k_1}) \;\&\; \dots \;\&\; (\psi_{p1} \vee \dots \vee \psi_{pk_p}))$$

Assume $\eth_1 = \forall$; then

$$\psi \leftrightarrow (x_1)(\psi_{11} \vee \dots \vee \psi_{1k_1}) \;\&\; \dots \;\&\; (x_1)(\psi_{p1} \vee \dots \vee \psi_{pk_p})\,.$$

Collecting those disjuncts in $\psi_{i1} \vee \dots \vee \psi_{ik_i}$ whose variables are in the same class as x_1 into one member, say ψ_{i1}, we can rewrite ψ in the desired form

$$((x_1)\, \psi_{11} \vee \dots \vee \psi_{1k_1'}) \;\&\; \dots \;\&\; ((x_1)\, \psi_{p1} \vee \dots \vee \psi_{pk_p'})\,.$$

If $\eth_1 = \exists$, then we rewrite the χ given by the induction hypothesis in disjunctive normal form and proceed dually. –

With the Lemma proved we now get a proof of Theorem 1, for in every axiom in the system characterizing the correspondence the individual variables sortwise connected with the class $\mathcal{K}_i$ occur only as arguments of predicates associated with $\mathcal{K}_i$ and do not mix with individual variables connected with other classes. ∎

§1.2. A class $\mathcal{K}$ of models is called *minimal* iff it is axiomatizable and includes no proper axiomatizable subclasses. Obviously, the minimal classes are those classes characterizable by *complete* systems of FOPL axioms ([1]). Every model is contained in one and only one minimal class. As a system of axioms for this class we can take the collection of all FOPL sentences true in the given model.

Let σ be some correspondence between models in classes $\mathcal{K}_1, \mathcal{K}_2$. Then to every subclass $\mathcal{L}_1 \subseteq \mathcal{K}_1$ corresponds a well-defined subclass $\mathcal{L}_1 \subseteq \mathcal{K}_2$ consisting of all those $\mathcal{K}_2$-models which σ-correspond to at least one $\mathcal{L}_1$-model. The class $\mathcal{K}_1$, like any class, splits into minimal axiomatizable subclasses, which have corresponding subclasses in $\mathcal{K}_2$ of a complicated nature, generally speaking.

If σ is a splitting correspondence between models in the classes $\mathcal{K}_1$, $\mathcal{K}_2$, then to every minimal subclass $\mathcal{L}^m \subseteq \mathcal{K}_1$ there corresponds an axiomatizable subclass $\mathcal{L}^m\sigma \subseteq \mathcal{K}_2$; moreover, $\mathcal{L}^m\sigma = \mathfrak{M}\sigma$ for any $\mathfrak{M} \in \mathcal{L}^m$.

Indeed, let σ be prescribed by the split axiom system $\{\varphi_1^\mu \vee \varphi_2^\mu : \mu \in \mathrm{M}\}$, and $\mathcal{L}^m$ by the complete system $\mathbf{X} = \{\chi^\nu : \nu \in \mathrm{N}\}$. In view of the completeness of the system $\mathbf{X}$, for every $\mu \in \mathrm{M}$, $\mathbf{X}$ logically implies either φ_1^μ or $\neg\varphi_2^\mu$ (but not both, by the consistency of $\mathbf{X}$); so either φ_1^μ holds in all $\mathcal{L}^m$-models, or $\neg\varphi_1^\mu$ does. The class $\mathcal{L}^m\sigma$ is characterized in $\mathcal{K}_2$ by the system $\{\varphi_2^\mu : \mu \in \mathrm{M}$ *and* $\mathbf{X}$ *implies* $\neg\varphi_1^\mu\}$. In view of the minimality of $\mathcal{L}^m$, for every $\mathfrak{M} \in \mathcal{L}^m$, an arbitrary sentence ψ is true in $\mathfrak{M}$ iff $\mathbf{X}$ implies ψ, whence $\mathcal{L}^m\sigma = \mathfrak{M}\sigma$. ■

By analogy with the notion of pseudoaxiomatizable classes of models (cf. [VII]), let us agree to say that a model correspondence is *pseudoaxiomatizable* iff for every system T of FOPL sentences written with the aid of predicate symbols associated with the classes considered and auxiliary many-sorted predicate symbols whose arguments are connected with the classes considered, if for every finite subset T_0 of T, there are σ-corresponding models satisfying T_0, then there are σ-corresponding models satisfying all of T.

In an analogous fashion we introduce the notions of pseudoprojective and pseudosplitting correspondences: for the former we permit the axioms in $\mathbf{T}$ to involve auxiliary many-sorted predicate symbols relating arguments ranging over auxiliary sets, as well as over the bases of models in the classes under consideration (cf. [XI], §1.2); for the latter the members of $\mathbf{T}$ must refer only to the fundamental predicates of the classes considered. (2)

From the FOPL compactness theorem (multibase version – cf. [XI], §2.2) it immediately follows that *every projective correspondence is pseudoprojective.* ■

We remark that for correspondences, pseudoprojectiveness implies pseudoaxiomatizability, and pseudoaxiomatizability implies pseudosplitting.

Theorem 2: *A correspondence σ on the classes $\mathcal{K}_1$, $\mathcal{K}_2$ is a splitting correspondence iff it is pseudosplitting and for every minimal subclass $\mathcal{L}^m$ of $\mathcal{K}_1$ and every model $\mathfrak{M}$ in $\mathcal{L}^m$, $\mathcal{L}^m\sigma$ is axiomatizable and equal to $\mathfrak{M}\sigma$.*

Necessity was established above. We now prove sufficiency. We are assuming, in effect, that σ is characterizable by relations of the form

$$\mathbf{X} \to \psi \, ,$$

where $\mathbf{X} = \{\chi^\nu : \nu \in \mathrm{N}\}$ is a complete system of sentences of the type of $\mathcal{K}_1$, and ψ is a sentence of the type of $\mathcal{K}_2$. From the pseudosplitting of σ it follows that each relation of this form can be replaced with a sentence of the form

$$\chi_1 \,\&\, \dots \,\&\, \chi_p \to \psi \, ,$$

where $\{\chi_1, \dots, \chi_p\}$ is an appropriate finite subset of $\mathbf{X}$. Putting $\varphi = \chi_1 \,\&\, \dots \,\&\, \chi_p$, we see that σ is characterized by axioms

$$\neg\varphi \vee \psi$$

having split form. ■

§1.3. As an example we mention the problem studied by Feferman (cf. [179]) on reducing a sentence concerning the direct product of a finite number of models to sentences concerning the factors.

Let $\mathfrak{M}_1, \mathfrak{M}_2$ be similar models with bases M_1, M_2 and fundamental predicates $P_\gamma^{(1)}$ and $P_\gamma^{(2)}$ $(\gamma \in \Gamma)$. The *direct product* of $\mathfrak{M}_1, \mathfrak{M}_2$ is the model $\mathfrak{M}$ whose base is the set $M_1 \times M_2 = \{\langle a^1, a^2 \rangle : a^1 \in M_1, a^2 \in M_2\}$ and whose fundamental predicates P_γ $(\gamma \in \Gamma)$ are defined by the predicate equations

$$P_\gamma(\langle a_1^1, a_1^2 \rangle, \dots, \langle a_{n_\gamma}^1, a_{n_\gamma}^2 \rangle) =$$
$$= P_\gamma^{(1)}(a_1^1, \dots, a_{n_\gamma}^1) \,\&\, P_\gamma^{(2)}(a_1^2, \dots, a_{n_\gamma}^2) \, . \quad (^3) \tag{1}$$

Let

$$(\mathrm{O}_1 \boldsymbol{x}_1) \dots (\mathrm{O}_m \boldsymbol{x}_m) \, \varphi(\boldsymbol{x}_1, \dots, \boldsymbol{x}_m)$$

be a sentence of FOPL concerning $\mathfrak{M}$. Replacing each expression

$$\boldsymbol{P}_\gamma(\boldsymbol{x}_{i_1}, \dots, \boldsymbol{x}_{i_n})$$

occurring in this sentence with

$$\boldsymbol{P}_\gamma^{(1)}(\boldsymbol{x}_{i_1}^1, \dots, \boldsymbol{x}_{i_n}^1) \,\&\, \boldsymbol{P}_\gamma^{(2)}(\boldsymbol{x}_\gamma^2, \dots, \boldsymbol{x}_{i_n}^2),$$

and each quantifier $(\mathrm{Ò}_i x_i)$ with the pair of quantifiers $(\mathrm{Ò}_i x_i^1)(\mathrm{Ò}_i x_i^2)$, we obtain an equivalent sentence ψ with specialized individual variables, part of which concern M_1, the remainder M_2. This decomposition of the set of variables satisfies the conditions of the Lemma in § 1.1, and so, by that procedure ψ is reduced to a sentence of the form

$$(\psi_1^1 \vee \psi_1^2) \,\&\, \ldots \,\&\, (\psi_s^1 \vee \psi_s^2)\,,$$

where the sentences $\psi_1^1, \ldots, \psi_s^1$ concern the first factor, and $\psi_1^2, \ldots, \psi_s^2$ concern the second.

We have only been considering the case when $\mathfrak{M}_1$, $\mathfrak{M}_2$ each have a single base. If these models are multibase (cf. [XI]), then their direct product may be definable in two essentially different ways. Suppose, e.g., that $\mathfrak{M}_1$ and $\mathfrak{M}_2$ have bases M_1, N_1 and M_2, N_2, respectively. Then it is natural to let their direct product be the model with bases $M_1 \times M_2$, $N_1 \times N_2$ on which predicates are defined according to the relation (1). For these products what was said above is valid without any change.

Suppose now that these models $\mathfrak{M}_1$, $\mathfrak{M}_2$ have their first base in common: $M_1 = M_2 = M$. Then the model $\mathfrak{M}$ with bases M, $N_1 \times N_2$ and predicates P_γ, again given by (1), is naturally called the *direct product of* $\mathfrak{M}_1$, $\mathfrak{M}_2$ *over* M. Direct products of this sort are met, e.g., in the study of groups and rings with a fixed domain of operators.

The transformation discussed above is not immediately applicable to direct products over fixed sets. On the other hand, the theorem of A. Horn [59], which gives a sufficient condition for a sentence satisfied by every factor to be true in the direct product, remains valid for both types of direct products. Clearly, for this one must demand that the sentence has no equality signs connecting variables over the fixed base with variables over the remaining bases. In particular, Horn's theorem is valid for prenex sentences of the second-order predicate logic in which the only quantifiers over predicates are existential.

§ 2. Regular products

§ 2.1. Turning to direct products of a possibly infinite number of models, we generalize the notion of direct product itself. For simplicity let us assume that all the models considered are single-base.

Suppose we have some system of models

$$\mathfrak{M}_\alpha = \langle M_\alpha; P_\gamma^{(\alpha)} : \gamma \in \Gamma_\alpha \rangle \qquad (\alpha \in \mathrm{A})\,.$$

We do not demand that the models be similar, nor that $\mathfrak{M}_\alpha$, $\mathfrak{M}_\beta$ be distinct for $\alpha \neq \beta$. Suppose we have an index set Λ, and for every $\alpha \in A$, a system of formulas of FOPL

$$\varphi_\lambda^\alpha(\boldsymbol{x}_1, ..., \boldsymbol{x}_{m_\lambda}) = \tag{2}$$

$$= (\mathrm{O}_1 \boldsymbol{y}_1) ... (\mathrm{O}_{p_\lambda} \boldsymbol{y}_{p_\lambda}) \,\hat{\varphi}_\lambda^\alpha(\boldsymbol{x}_1, ..., \boldsymbol{x}_{m_\lambda}, \boldsymbol{y}_1, ..., \boldsymbol{y}_{p_\lambda}) \quad (\lambda \in \Lambda),$$

where O_i is $\forall$ or $\exists$, and for each $\lambda \in \Lambda$, $\hat{\varphi}_\lambda^\alpha$ is an open formula with free variables $\boldsymbol{x}_1, ..., \boldsymbol{x}_m, \boldsymbol{y}_1, ..., \boldsymbol{y}_{p_\lambda}$ and predicate symbols from among the members of $\{\boldsymbol{P}_\gamma^{(\alpha)}(\boldsymbol{x}_1, ..., \boldsymbol{x}_{n_\gamma})\colon \gamma \in \Gamma_\alpha\}$. The equality sing can also occur freely in the formulas (2).

For each $\lambda \in \Lambda$, we introduce a predicate symbol

$$\boldsymbol{S}_\lambda(\boldsymbol{\alpha}, \boldsymbol{x}_1, ..., \boldsymbol{x}_{m_\lambda}),$$

intending $\boldsymbol{\alpha}$ to range over the index set A and $\boldsymbol{x}_1, ..., \boldsymbol{x}_{m_\lambda}$ to range over the cartesian product $M = \Pi M_\alpha$ of the bases of the given models. In fact, the S_λ are taken to represent predicates defined on the pair $\langle A, M\rangle$ by the predicate equations

$$S_\lambda(\alpha, x_1, ..., x_{m_\lambda}) = \varphi_\lambda^\alpha(x_1^\alpha, ..., x_{m_\lambda}^\alpha) \quad (\lambda \in \Lambda), \tag{3}$$

where x^α denotes the element of M_α that is the αth projection of the element $x \in M$, while the value of the right-hand side is computed in $\mathfrak{M}_\alpha$ according to the formulas (2).

With the help of the symbols $\boldsymbol{S}_\lambda$ we define a new system (indexed by a set Γ) of predicates on M by the equations

$$P_\gamma(x_1, ..., x_{n_\gamma}) = \theta_\gamma(x_1, ..., x_{n_\gamma}) =$$

$$= (\mathrm{O}'_1 \boldsymbol{\alpha}_1) ... (\mathrm{O}'_{q_\gamma} \boldsymbol{\alpha}_{q_\gamma}) \,\hat{\theta}_\gamma(\boldsymbol{\alpha}_1, ..., \boldsymbol{\alpha}_{q_\gamma}, x_1, ..., x_{n_\gamma}) \tag{4}$$

$$(\gamma \in \Gamma,\ x_i \in M);$$

here, $\hat{\theta}_\gamma$ is an open FOPL formula with predicate symbols among the $\boldsymbol{S}_\lambda$ and with, possibly, the equality sign, but linking only variables among $\boldsymbol{\alpha}_1, ..., \boldsymbol{\alpha}_{q_\gamma}$. The quantifiers in (4) are specialized and refer to the index set A.

The model $\mathfrak{M}$ with base M and fundamental predicates P_γ ($\gamma \in \Gamma$), defined on M by (4), is called a *regular product* of the models $\mathfrak{M}_\alpha$ ($\alpha \in A$). The formulas in (3) and (4) determine the *type* of regular product. Clearly, if we have a regular product $\mathfrak{M}$ of models $\mathfrak{M}_\alpha$ $(\alpha \in \mathrm{A})$, then we can speak of the regular product $\mathfrak{M}(\mathrm{B})$ (of the same type) of the models $\mathfrak{M}_\beta$ $(\beta \in \mathrm{B})$, where B is an arbitrary subset of A. ([4])

As a basic variety of regular products we consider that case when all the $\mathfrak{M}_\alpha$ are similar, $\Lambda = \Gamma = \Gamma_\alpha$ $(\alpha \in \mathrm{A})$, $n_\gamma = m_\gamma$ $(\gamma \in \Gamma)$, and the equations (3) reduce to the form

$$S_\lambda(\alpha, x_1, ..., x_{m_\lambda}) = \varphi_\lambda^\alpha(x_1^\alpha, ..., x_{m_\lambda}^\alpha) = P_\lambda^{(\alpha)}(x_1^\alpha, ..., x_{m_\lambda}^\alpha)$$

for all α and λ. Regular products of this kind are called *proper*. Varying the formulas in (4), we obtain different types of proper products. E.g., using

$$\theta_\gamma(\boldsymbol{x}_1, ..., \boldsymbol{x}_{n_\gamma}) = (\boldsymbol{\alpha})\, \boldsymbol{S}_\gamma(\boldsymbol{\alpha}, \boldsymbol{x}_1, ..., \boldsymbol{x}_{n_\gamma})$$

to define P_γ $(\gamma \in \Gamma)$, we get the usual direct product of the $\mathfrak{M}_\alpha$. Putting

$$\theta_\gamma(\boldsymbol{x}_1, ..., \boldsymbol{x}_{n_\gamma}) =$$
$$= (\boldsymbol{\alpha})(\boldsymbol{\beta})(\boldsymbol{\alpha} \not\approx \boldsymbol{\beta} \to \boldsymbol{S}_\gamma(\boldsymbol{\alpha}, \boldsymbol{x}_1, ..., \boldsymbol{x}_{n_\gamma}) \vee \boldsymbol{S}_\gamma(\boldsymbol{\beta}, \boldsymbol{x}_1, ..., \boldsymbol{x}_{n_\gamma}))\,,$$

or

$$\theta_\gamma(\boldsymbol{x}_1, ..., \boldsymbol{x}_{n_\gamma}) =$$
$$= (\boldsymbol{\alpha})(\boldsymbol{\beta})(\boldsymbol{\delta})(\boldsymbol{\alpha} \approx \boldsymbol{\beta} \vee \boldsymbol{\beta} \approx \boldsymbol{\delta} \vee \boldsymbol{\alpha} \approx \boldsymbol{\delta} \vee \boldsymbol{S}_\gamma(\boldsymbol{\alpha}) \vee \boldsymbol{S}_\gamma(\boldsymbol{\beta}) \vee \boldsymbol{S}_\gamma(\boldsymbol{\delta}))\,,$$

we obtain proper products of other types.

From these last examples it is seen that proper products are not obliged to be associative, even when they are commutative. Besides that, the proper product of algebras may not even by an algebra.

Theorem 3: *There is a finitistic process such that a given formula* $\Phi(\boldsymbol{x}_1, ..., \boldsymbol{x}_m)$ *of the form*

$$(\mho_{m+1}\boldsymbol{x}_{m+1}) \dots (\mho_n \boldsymbol{x}_n)\, \Phi^*(\boldsymbol{x}_1, ..., \boldsymbol{x}_m, \boldsymbol{x}_{m+1}, ..., \boldsymbol{x}_n)\,,$$

where $\mho_i$ *is* $\forall$ *or* $\exists$, *and* Φ^* *is an open* FOPL *formula with free variables* $\boldsymbol{x}_1, ..., \boldsymbol{x}_n$ *and predicate symbols among* $\boldsymbol{P}_\gamma(\boldsymbol{x}_1, ..., \boldsymbol{x}_{n_\gamma})$ $(\gamma \in \Gamma)$ ([5]), *and given a system of formulas*

$$(\mho_1''\boldsymbol{\alpha}_1) \dots (\mho_{q_\gamma}''\boldsymbol{\alpha}_{q_\gamma})\, \hat{\theta}_\gamma(\boldsymbol{\alpha}_1, ..., \boldsymbol{\alpha}_{q_\gamma}, \boldsymbol{x}_1, ..., \boldsymbol{x}_{n_\gamma}) \quad (\gamma \in \Gamma)\,, \qquad (5)$$

where each $\hat{\theta}_\gamma$ *is an open* FOPL *formula with predicate symbols among* $\boldsymbol{S}_\lambda(\boldsymbol{\alpha}, \boldsymbol{x}_1, ..., \boldsymbol{x}_{m_\lambda})$ $(\lambda \in \Lambda)$ *and, possibly, equalities among the* α_i, *this process permits one to uniformly construct:* (i) *a formula* $\Psi(\boldsymbol{x}_1, ..., \boldsymbol{x}_m)$ *of the form*

$$(\mho'_1 \boldsymbol{\alpha}_1) \dots (\mho'_r \boldsymbol{\alpha}_r)\, \Psi^*(\boldsymbol{\alpha}_1, ..., \boldsymbol{\alpha}_r, \boldsymbol{x}_1, ..., \boldsymbol{x}_m)\,,$$

where Ψ^* *is an open* FOPL *formula with predicate symbols* $\boldsymbol{T}_\nu(\boldsymbol{\alpha}, \boldsymbol{x}_1, ..., \boldsymbol{x}_{m_\nu})$ $(\nu \in \mathrm{N})$ *and, perhaps, equalities among the* $\boldsymbol{\alpha}_i$; (ii) *for every* $\nu \in \mathrm{N}$, *a formula* $\psi_\nu(\boldsymbol{x}_1, ..., \boldsymbol{x}_{m_\nu})$ *with free variables* $\boldsymbol{x}_1, ..., \boldsymbol{x}_{m_\nu}$ *and predicate symbols* $\boldsymbol{S}^+_\lambda(\boldsymbol{x}_1, ..., \boldsymbol{x}_{m_\lambda})$ $(\lambda \in \Lambda)$. *These new formulas* (i), (ii) *have the following properties. Suppose* A *is an index set, and suppose for every* $\alpha \in \mathrm{A}$, *we have a model* $\mathfrak{M}_\alpha$ *with base* M_α, *and for every* $\lambda \in \Lambda$, *a formula* $\varphi^\alpha_\lambda(\boldsymbol{x}_1, ..., \boldsymbol{x}_{m_\lambda})$ *of the sort described under* (2). *Let* $\mathfrak{M} = \langle M; P_\gamma\colon \gamma \in \Gamma\rangle$ *be the regular product of the models* $\mathfrak{M}_\alpha$ $(\alpha \in \mathrm{A})$ *determined by the formulas* (5) *and the* φ^α_λ *in accord with* (3) *and* (4). *We define predicates* S^+_λ *in each* $\mathfrak{M}_\alpha$ *by the predicate equations*

$$S^+_\lambda(x_1, ..., x_{m_\lambda}) = \varphi^\alpha_\lambda(x_1, ..., x_{m_\lambda}) \quad (\lambda \in \Lambda; x_j \in M_\alpha)\,;$$

we define predicates T_ν *on the pair* $\langle A, M\rangle$ *by the equations*

$$T_\nu(\alpha, x_1, ..., x_{m_\nu}) = \psi_\nu(x^\alpha_1, ..., x^\alpha_{m_\nu}) \quad (\nu \in \mathrm{N})\,, \tag{6}$$

where the right-hand side is computed in $\mathfrak{M}_\alpha$, *and where* $x^\alpha \in M_\alpha$ *is the* αth *projection of* $x \in M$. *Then we find that the formula*

$$\Phi(\boldsymbol{x}_1, ..., \boldsymbol{x}_m) \leftrightarrow \Psi(\boldsymbol{x}_1, ..., \boldsymbol{x}_m)$$

is valid in the multibase model $\langle A, M; P_\gamma, T_\nu\colon \gamma \in \Gamma, \nu \in \mathrm{N}\rangle$ *(interpreting the* $\boldsymbol{\alpha}_i$ *to run over* A, *the* $\boldsymbol{x}_j$ *over* M*).*

The proof is carried out by induction on the number of quantifiers in the formula Φ. If Φ has no quantifiers, then replacing the $\boldsymbol{P}_\gamma$ occurring in it with the corresponding formulas (5) and bringing the quantifiers (all with variables $\boldsymbol{\alpha}_i$) to the front, we obtain the desired formula Ψ (with $\mathrm{N} = \Lambda$, and $\boldsymbol{T}_\nu = \boldsymbol{S}_\nu$ for $\nu \in \mathrm{N}$), and putting

$$\psi_\nu(\boldsymbol{x}_1, ..., \boldsymbol{x}_{m_\nu}) = \boldsymbol{S}^+_\nu(\boldsymbol{x}_1, ..., \boldsymbol{x}_{m_\nu}) \qquad (\nu \in \mathrm{N})\,,$$

we have all that is required.

Now suppose Theorem 3 is valid for all formulas, the number of whose quantifiers is less than that of Φ (assuming $n \geqslant m+1$). Then, in a finitistic fashion, for the formula $(\mathrm{O}_{m+1}\boldsymbol{x}_{m+2}) \dots (\mathrm{O}_n\boldsymbol{x}_n)\,\Phi^*$ we can find Ψ and ψ_ν ($\nu \in \mathrm{N}$) with the structure indicated in (i) and (ii) such that the formula

$$(\mathrm{O}_{m+2}\boldsymbol{x}_{m+2}) \dots (\mathrm{O}_n\boldsymbol{x}_n)\,\Phi^* \leftrightarrow$$
$$\leftrightarrow (\mathrm{O}'_1\boldsymbol{\alpha}_1) \dots (\mathrm{O}'_r\boldsymbol{\alpha}_r)\Psi^*(\boldsymbol{\alpha}_1, \dots, \boldsymbol{\alpha}_r, \boldsymbol{x}_1, \dots, \boldsymbol{x}_m, \boldsymbol{x}_{m+1})$$

is valid in the enriched version of any regular product using the given formulas (5); thus, so is

$$\Phi(\boldsymbol{x}_1, \dots, \boldsymbol{x}_m) \leftrightarrow$$
$$\leftrightarrow (\mathrm{O}_{m+1}\boldsymbol{x}_{m+1})(\mathrm{O}'_1\boldsymbol{\alpha}_1) \dots (\mathrm{O}'_r\boldsymbol{\alpha}_r)\Psi^*(\boldsymbol{\alpha}_1, \dots, \boldsymbol{\alpha}_r, \boldsymbol{x}_1, \dots, \boldsymbol{x}_m, \boldsymbol{x}_{m+1})\,.$$

It only remains for us to transform the right-hand side of this second equivalence. To be definite, let us assume that $\mathrm{O}_{m+1} = \forall$; in case O_{m+1} were $\exists$, it would suffice to consider the dual formulas. According to the inductive hypothesis, Ψ^* is constructed with the aid of $\neg$, &, $\vee$ from $\boldsymbol{\alpha}_i \approx \boldsymbol{\alpha}_j$ and $\boldsymbol{T}_\nu(\boldsymbol{\alpha}_i, \boldsymbol{x}_{j_1}, \boldsymbol{x}_{j_2}, \dots)$. Considering $\boldsymbol{x}_1, \dots, \boldsymbol{x}_{m+1}$ to be parameters, we can view the formula

$$\Psi = (\mathrm{O}'_1\boldsymbol{\alpha}_1) \dots (\mathrm{O}'_r\boldsymbol{\alpha}_r)\,\Psi^*(\boldsymbol{\alpha}_1, \dots, \boldsymbol{\alpha}_r, \boldsymbol{x}_1, \dots, \boldsymbol{x}_{m+1})$$

as a parametrized formula with $\approx$ and unary predicate symbols $\boldsymbol{T}_\xi(\boldsymbol{\alpha})$ ($\xi \in \Xi$); that is, we have replaced each expression $\boldsymbol{T}_\nu(\boldsymbol{\alpha}_i, \boldsymbol{x}_{j_1}, \boldsymbol{x}_{j_2}, \dots)$ with $\boldsymbol{T}_\xi(\boldsymbol{\alpha}_i)$ at every occurrence of the former (ξ depends in a 1–1 fashion on $\nu, j_1, \dots, j_{m_\nu}$, only). Therefore, by means of Behmann's transformations (see [2], p. 44), we can show Ψ is equivalent to a conjunction of formulas of the form

$$(\boldsymbol{\alpha}_1) \dots (\boldsymbol{\alpha}_u)(\boldsymbol{\alpha}_{i_1} \approx \boldsymbol{\alpha}_{j_1} \vee \dots \vee \boldsymbol{\alpha}_{i_p} \approx \boldsymbol{\alpha}_{j_p} \vee$$
$$\vee\, \boldsymbol{T}^{\epsilon(1)}_{\xi_1}(\boldsymbol{\alpha}_1) \vee \dots \vee \boldsymbol{T}^{\epsilon(v)}_{\xi_v}(\boldsymbol{\alpha}_v) \vee (\exists \boldsymbol{\beta}_1)\mathrm{X}_1(\boldsymbol{\beta}_1) \vee \dots \vee (\exists \boldsymbol{\beta}_w)\mathrm{X}_w(\boldsymbol{\beta}_w))\,,$$

where

$$\mathrm{X}_i(\boldsymbol{\beta}_i) = \boldsymbol{T}^{\epsilon(i,1)}_{\xi_{i1}}(\boldsymbol{\beta}_i)\ \&\ \dots\ \&\ \boldsymbol{T}^{\epsilon(i,q)}_{\xi_{iq}}(\boldsymbol{\beta}_i) \qquad (i = 1, \dots, w)\,,$$

and ϵ is a function with values $1, -1$, while $\varphi^1 = \varphi$ and $\varphi^{-1} = \neg\varphi$.

Putting

$$X(\boldsymbol{\beta}) = X_1(\boldsymbol{\beta}) \vee \ldots \vee X_w(\boldsymbol{\beta})$$

and using the permutability of the quantifiers $(\boldsymbol{x}_{m+1})$ and $(\boldsymbol{\alpha}_i)$, we reduce the initial formula Φ to a conjunction of (parametrized) formulas of the form

$$(\boldsymbol{\alpha}_1) \ldots (\boldsymbol{\alpha}_u)[\boldsymbol{\alpha}_{i_1} \approx \boldsymbol{\alpha}_{j_1} \vee \ldots \vee \boldsymbol{\alpha}_{i_p} \approx \boldsymbol{\alpha}_{j_p} \vee$$
$$\vee (\boldsymbol{x}_{m+1})(\boldsymbol{T}_{\xi_1}^{\epsilon(1)}(\boldsymbol{\alpha}_1) \vee \ldots \vee \boldsymbol{T}_{\xi_v}^{\epsilon(v)}(\boldsymbol{\alpha}_v) \vee (\exists \boldsymbol{\beta})\, X(\boldsymbol{\beta}))] \,;$$

we have only to be able to appropriately transform a formula $\Phi_1(\boldsymbol{x}_1, \ldots, \boldsymbol{x}_m)$ of the form

$$(\boldsymbol{x}_{m+1})(\boldsymbol{T}_{\xi_1}^{\epsilon(1)}(\boldsymbol{\alpha}_1) \vee \ldots \vee \boldsymbol{T}_{\xi_v}^{\epsilon(v)}(\boldsymbol{\alpha}_v) \vee (\exists \boldsymbol{\beta})\, X(\boldsymbol{\beta}))\,,$$

which can be rewritten as

$$(\boldsymbol{x}_{m+1})[Y_1(\boldsymbol{\alpha}_1) \vee \ldots \vee Y_v(\boldsymbol{\alpha}_v) \vee$$
$$\vee (\exists \boldsymbol{\beta})\, (\boldsymbol{\beta} \not\approx \boldsymbol{\alpha}_1 \;\&\; \ldots \;\&\; \boldsymbol{\beta} \not\approx \boldsymbol{\alpha}_v \;\&\; X(\boldsymbol{\beta}))]\,,$$

where

$$Y_i(\boldsymbol{\alpha}_i) = \boldsymbol{T}_{\xi_i}^{\epsilon(i)}(\boldsymbol{\alpha}_i) \vee X(\boldsymbol{\alpha}_i) \qquad (i = 1, \ldots, v)\,.$$

By construction, the formulas $Y_i(\boldsymbol{\alpha})$, $X(\boldsymbol{\alpha})$ are built with the aid of $\neg$, &, $\vee$ from expressions of the form $\boldsymbol{T}_\nu(\boldsymbol{\alpha}, \boldsymbol{x}_{j_1}, \boldsymbol{x}_{j_2}, \ldots)$ and thus, by assumption, will have the same values in any regular product formed by using the given (5) as the corresponding propositional combinations of the formulas $\psi_\nu(\boldsymbol{x}_{j_1}^\alpha, \boldsymbol{x}_{j_2}^\alpha, \ldots)$, interpreting the predicate symbols as in the statement of Theorem 3. Let $Y_i^*(\boldsymbol{x}_1, \ldots, \boldsymbol{x}_{m+1})$, $X^*(\boldsymbol{x}_1, \ldots, \boldsymbol{x}_{m+1})$ be those formulas – whose free variables are among those shown – obtained from $Y_i(\boldsymbol{\alpha})$, $X(\boldsymbol{\alpha})$ by substituting the formulas $\psi_\nu(\boldsymbol{x}_{j_1}, \boldsymbol{x}_{j_2}, \ldots)$ for the expressions $\boldsymbol{T}_\nu(\boldsymbol{\alpha}, \boldsymbol{x}_{j_1}, \boldsymbol{x}_{j_2}, \ldots)$. Then in any regular product $\mathfrak{M}$ of models $\mathfrak{M}_\alpha$ $(\alpha \in A)$, specified by the given formulas (5), the formula

$$\Phi_1(\boldsymbol{x}_1, \ldots, \boldsymbol{x}_m) \leftrightarrow$$
$$\leftrightarrow (\boldsymbol{x}_{m+1})[Y_1^*(\boldsymbol{x}_1^{\alpha_1}, \ldots, \boldsymbol{x}_{m+1}^{\alpha_1}) \vee \ldots \vee Y_v^*(\boldsymbol{x}_1^{\alpha_v}, \ldots, \boldsymbol{x}_{m+1}^{\alpha_v}) \vee$$
$$\vee (\exists \boldsymbol{\beta})\, (\boldsymbol{\beta} \not\approx \boldsymbol{\alpha}_1 \;\&\; \ldots \;\&\; \boldsymbol{\beta} \not\approx \boldsymbol{\alpha}_v \;\&\; X^*(\boldsymbol{x}_1^\beta, \ldots, \boldsymbol{x}_{m+1}^\beta))]$$

is valid (with the interpretations given in Theorem 3). Here, the quantifier $(\boldsymbol{x}_{m+1})$ can be replaced with a possibly infinite collection of specialized quantifiers $(\boldsymbol{x}^{\alpha}_{m+1})$ $(\alpha \in \mathrm{A})$, since the requirement that x vary over the cartesian product M is equivalent to the requirement that its projections x^{α} vary independently, each over its own M_{α}. The piece quantified by $(\exists \boldsymbol{\beta})$ can be represented as a disjunction

$$\bigvee_{\beta \in \mathrm{A}_1} \mathrm{X}^*(\boldsymbol{x}^{\beta}_1, \ldots, \boldsymbol{x}^{\beta}_{m+1}) ,$$

taking $\mathrm{A}_1 = \mathrm{A} \sim \{\alpha_1, \ldots, \alpha_v\}$. From these facts it follows that Φ_1 is equivalent in $\mathfrak{M}$ (enriched) to

$$(\boldsymbol{x}_{m+1})(\mathrm{Y}^*_1(\boldsymbol{x}^{\alpha_1}_1, \ldots, \boldsymbol{x}^{\alpha_1}_{m+1}) \vee \ldots \vee \mathrm{Y}^*_v(\boldsymbol{x}^{\alpha_v}_1, \ldots, \boldsymbol{x}^{\alpha_v}_{m+1})) \vee$$

$$\vee (\exists \boldsymbol{\beta})(\boldsymbol{\beta} \not\approx \boldsymbol{\alpha}_1 \,\&\, \ldots \,\&\, \boldsymbol{\beta} \not\approx \boldsymbol{\alpha}_v \,\&\, (\boldsymbol{x}^{\beta}_{m+1})\mathrm{X}^*(\boldsymbol{x}^{\beta}_1, \ldots, \boldsymbol{x}^{\beta}_{m+1})) . \quad (7)$$

Introducing the formula

$$\chi(\boldsymbol{x}_1, \ldots, \boldsymbol{x}_m) = (\boldsymbol{x}_{m+1})\mathrm{X}^*(\boldsymbol{x}_1, \ldots, \boldsymbol{x}_m, \boldsymbol{x}_{m+1})$$

and the corresponding predicate U defined on the pair $\langle A, M\rangle$ by the predicate equation

$$U(\alpha, x_1, \ldots, x_m) = \chi(x^{\alpha}_1, \ldots, x^{\alpha}_m) ,$$

computing the right-hand side in $\mathfrak{M}_{\alpha}$ as in (6), we can rewrite the second disjunct of (7) in its final form:

$$(\exists \boldsymbol{\beta})(\boldsymbol{\beta} \not\approx \boldsymbol{\alpha}_1 \,\&\, \ldots \,\&\, \boldsymbol{\beta} \not\approx \boldsymbol{\alpha}_v \,\&\, \boldsymbol{U}(\boldsymbol{\beta}, \boldsymbol{x}_1, \ldots, \boldsymbol{x}_m)) . \quad (8)$$

In practice the number of variables in $\boldsymbol{U}$ may be less than $m+1$ since it depends on the number of actual free variables in χ. Now we have only to transform the expression

$$(\boldsymbol{x}_{m+1})(\mathrm{Y}^*_1(\boldsymbol{x}^{\alpha_1}_1, \ldots, \boldsymbol{x}^{\alpha_1}_{m+1}) \vee \ldots \vee \mathrm{Y}^*_v(\boldsymbol{x}^{\alpha_v}_1, \ldots, \boldsymbol{x}^{\alpha_v}_{m+1})) . \quad (9)$$

If it were known for elements α_i $(i = 1, \ldots, v)$ of A that $\alpha_i \neq \alpha_j$ for $i \neq j$, then by an earlier remark, (7) would be equivalent in $\mathfrak{M}$ for these indices to the formula

$$(\boldsymbol{x}^{\alpha_1}_{m+1})\, \mathrm{Y}^*_1(\boldsymbol{x}^{\alpha_1}_1, \ldots, \boldsymbol{x}^{\alpha_1}_{m+1}) \vee \ldots \vee (\boldsymbol{x}^{\alpha_v}_{m+1})\, \mathrm{Y}^*_v(\boldsymbol{x}^{\alpha_v}_1, \ldots, \boldsymbol{x}^{\alpha_v}_{m+1}) ,$$

and the reduction would be complete. In order to reduce the matter to this case we do the following.

Let τ be an arbitrary partition of the set $J = \{1, ..., v\}$ into pairwise disjoint non-empty subsets $J_1, ..., J_t$. Let the formula $\tau^{\#}(\boldsymbol{\alpha}_1, ..., \boldsymbol{\alpha}_v)$ be the conjunction of all the formulas $\boldsymbol{\alpha}_i \approx \boldsymbol{\alpha}_j$ for i, j in the same τ-class J_k, and all the formulas $\boldsymbol{\alpha}_i \not\approx \boldsymbol{\alpha}_j$ for i, j in different τ-classes. Since the disjunction of the formulas $\tau^{\#}(\boldsymbol{\alpha}_1, ..., \boldsymbol{\alpha}_v)$ – taken with respect to all possible partitions τ of J – is identically true, (9) is logically equivalent to a disjunction of formulas of the form

$$\tau^{\#}(\boldsymbol{\alpha}_1, ..., \boldsymbol{\alpha}_v) \;\&\; (9) \;. \tag{10}$$

Let the members of the τ-classes be labeled without repetitions: $J_k = \{\tau(1,k), ..., \tau(s_k, k)\}$ $(k = 1, ..., t)$. Putting

$$Z_k^*(\boldsymbol{x}_1, ..., \boldsymbol{x}_{m+1}) = Y_{\tau(1,k)}^* \vee ... \vee Y_{\tau(s_k, k)}^* \;,$$

and letting $\boldsymbol{\beta}_k = \boldsymbol{\alpha}_{\tau(1,k)}$ for $k = 1, ..., t$, we can rewrite (10) equivalently as

$$\tau^{\#}(\boldsymbol{\alpha}_1, ..., \boldsymbol{\alpha}_v) \;\&$$

$$\& \; (\boldsymbol{x}_{m+1})(Z_1^*(\boldsymbol{x}_1^{\beta_1}, ..., \boldsymbol{x}_{m+1}^{\beta_1}) \vee ... \vee Z_t^*(\boldsymbol{x}_1^{\beta_t}, ..., \boldsymbol{x}_{m+1}^{\beta_t})) \;.$$

Whenever some of the indices $\beta_1, ..., \beta_t$ happen to coincide in A, the first conjunct of this formula – and with it the whole formula – is false. Therefore, the transformation of the second conjunct may be carried out under the assumption that $\beta_1, ..., \beta_t$ are all distinct. According to the above remark, in this case the second conjunct is equivalent to the formula

$$(\boldsymbol{x}_{m+1}^{\beta_1}) \, Z_1^*(\boldsymbol{x}_1^{\beta_1}, ..., \boldsymbol{x}_{m+1}^{\beta_1}) \vee ... \vee (\boldsymbol{x}_{m+1}^{\beta_t}) \, Z_t^*(\boldsymbol{x}_1^{\beta_t}, ..., \boldsymbol{x}_{m+1}^{\beta_t}) \;.$$

Introducing for $k = 1, ..., t$ the formulas

$$\chi_k(\boldsymbol{x}_1, ..., \boldsymbol{x}_m) = (\boldsymbol{x}_{m+1}) \, Z_k^*(\boldsymbol{x}_1, ..., \boldsymbol{x}_m, \boldsymbol{x}_{m+1})$$

and the corresponding predicates U_k determined on $\langle A, M \rangle$ by the predicate equations

$$U_k(\alpha, x_1, ..., x_m) = \chi_k(x_1^\alpha, ..., x_m^\alpha) \;,$$

computing the right-hand side in $\mathfrak{M}_\alpha$ as in (6), we can rewrite (10) in its final form:

$$\tau^{\#}(\boldsymbol{\alpha}_1, ..., \boldsymbol{\alpha}_\upsilon) \;\&$$

$$\& \; \boldsymbol{U}_1(\boldsymbol{\alpha}_{\tau(1,1)}, \boldsymbol{x}_1, ..., \boldsymbol{x}_m) \vee ... \vee \boldsymbol{U}_t(\boldsymbol{\alpha}_{\tau(1,t)}, \boldsymbol{x}_1, ..., \boldsymbol{x}_m) \,. \quad (11)$$

Therefore, by performing all of the reductions indicated, from the formula Φ we obtain a FOPL formula Ψ_0 equivalent to it in $\mathfrak{M}$ and having the form

$$(\mathfrak{O}_1^* \boldsymbol{\alpha}_1) \ldots (\mathfrak{O}_s^* \boldsymbol{\alpha}_s) \, \Psi_0^*(\boldsymbol{\alpha}_1, ..., \boldsymbol{\alpha}_s, \boldsymbol{x}_1, ..., \boldsymbol{x}_m) \,,$$

where Ψ_0^* is built with the aid of $\neg$, &, $\vee$ from expressions of the form $\boldsymbol{\alpha}_i \approx \boldsymbol{\alpha}_j$ and $\boldsymbol{U}_\pi(\boldsymbol{\alpha}_i, \boldsymbol{x}_{j_1}, \boldsymbol{x}_{j_2}, ...)$; these new predicate symbols $\boldsymbol{U}_\pi$ $(\pi \in \Pi)$, introduced into Ψ_0 via (8) and (11), are interpreted according to equations

$$U_\pi(\alpha, x_1, ..., x_{m_\pi}) = \chi_\pi(x_1^\alpha, ..., x_{m_\pi}^\alpha) \quad (\pi \in \Pi) \,,$$

where the right-hand side is computed in $\mathfrak{M}_\alpha$ as in (6), and the formulas χ_π have free variables $\boldsymbol{x}_1, ..., \boldsymbol{x}_{m_\pi}$ $(m_\pi \leqslant m)$ and predicate symbols solely among the $\boldsymbol{S}_\lambda^+$ $(\lambda \in \Lambda)$. Moreover, the formulas Ψ_0, χ_π $(\pi \in \Pi)$ are obtained effectively, and their structure is independent of any choices concerning the regular product $\mathfrak{M}$; hence, they satisfy the conditions (i) and (ii) in the statement of the theorem. ∎

§**2.2.** We now consider a series of corollaries implied by Theorem 3. First of all, we examine in detail the case when the formula Φ is closed.

Corollary 1: *Suppose the formula Φ in the hypotheses of Theorem 3 is closed. Then the corresponding formula*

$$\Psi = (\mathfrak{O}'_1 \boldsymbol{\alpha}_1) \ldots (\mathfrak{O}'_r \boldsymbol{\alpha}_r) \, \Psi^*(\boldsymbol{\alpha}_1, ..., \boldsymbol{\alpha}_r)$$

is closed and involves, besides $\approx$, only unary predicate symbols $\boldsymbol{T}_\nu(\boldsymbol{\alpha})$ $(\nu \in \mathrm{N})$. For any models $\mathfrak{M}_\alpha$ $(\alpha \in \mathrm{A})$ and any formulas φ_λ^α $(\alpha \in \mathrm{A}, \lambda \in \Lambda)$ as in (2), *the regular product $\mathfrak{M}$ of the $\mathfrak{M}_\alpha$ of the type specified by the φ_λ^α and* (5) *determines – via* (6) *– predicates $T_\nu(\alpha)$ $(\nu \in \mathrm{N})$ on the set* A. *Let $\mathfrak{N}(\mathrm{A})$ be the model $\langle \mathrm{A}; T_\nu : \nu \in \mathrm{N} \rangle$ so produced. If $\mathrm{B} \subseteq \mathrm{A}$, then the model $\mathfrak{N}(\mathrm{B})$ determined by the* B*-subproduct $\mathfrak{M}(\mathrm{B})$ of $\mathfrak{M}$ is a submodel of $\mathfrak{N}(\mathrm{A})$, while the equivalence $\Phi \leftrightarrow \Psi$ holds when Φ is interpreted in $\mathfrak{M}(\mathrm{B})$, and Ψ in $\mathfrak{N}(\mathrm{B})$.*

Indeed, as in Theorem 3, for each $\alpha \in \mathrm{A}$ the value of $T_\nu(\alpha)$ depends only on $\mathfrak{M}_\alpha$ and φ_λ^α $(\lambda \in \Lambda)$, and not on any $\mathfrak{M}_{\alpha'}$ for $\alpha' \neq \alpha$. This means $\mathfrak{N}(\mathrm{B})$ is a submodel of $\mathfrak{N}(\mathrm{A})$ for every subset $\mathrm{B} \subseteq \mathrm{A}$. The second assertion of the corollary follows immediately from Theorem 3. ■

Let us say that a system of subsets $\{R_\iota : \iota \in \mathrm{I}\}$ of an arbitrary set R is a *local system on R* iff every finite subset of R is included in some member of the system.

Corollary 2: *Let $\mathfrak{M}$ be a regular product of models $\mathfrak{M}_\alpha$ $(\alpha \in \mathrm{A})$, and let $\{\mathrm{A}_\iota : \iota \in \mathrm{I}\}$ be a local system on* A; *suppose Φ is a* FOPL *sentence with predicate symbols appropriate to $\mathfrak{M}$. If Φ is true in $\mathfrak{M}(\mathrm{A}_\iota)$ for every $\iota \in \mathrm{I}$, then Φ is true in $\mathfrak{M}$.*

For any such product $\mathfrak{M}$ and sentence Φ, we construct the corresponding sentence Ψ with unary predicate symbols $\boldsymbol{T}_\nu(\boldsymbol{\alpha})$ $(\nu \in \mathrm{N})$ and the model $\mathfrak{N}(\mathrm{A})$. According to Behmann [2], the sentence Ψ, as a formula with unary predicate symbols, is equivalent to a sentence in Skolem form

$$(\boldsymbol{\alpha}_1) \ldots (\boldsymbol{\alpha}_u)(\exists \boldsymbol{\alpha}_{u+1}) \ldots (\exists \boldsymbol{\alpha}_v)\, \Psi_0(\boldsymbol{\alpha}_1, \ldots, \boldsymbol{\alpha}_v) \; .$$

By assumption this sentence is true in the submodel $\mathfrak{N}(\mathrm{A}_\iota)$ for every $\iota \in \mathrm{I}$; since the bases of the $\mathfrak{N}(\mathrm{A}_\iota)$ form a local system on A, this sentence is true in $\mathfrak{N}(\mathrm{A})$, as well. This means Φ is true in $\mathfrak{M}$. ■

From Corollary 2, by a well-known method, we derive the statement:

Let $\mathcal{K}$ be a (first-order) axiomatizable class of models. Let $\mathfrak{M}$ be a regular product of models $\mathfrak{M}_\alpha$ $(\alpha \in \mathrm{A})$, and suppose $\{\mathrm{A}_\iota : \iota \in \mathrm{I}\}$ is a local system of subsets of A. *If for every $\iota \in \mathrm{I}$, the A_ι-subproduct of $\mathfrak{M}$ belongs to $\mathcal{K}$, then $\mathfrak{M}$ itself belongs to $\mathcal{K}$.* ■

In particular, *If the axiomatizable class $\mathcal{K}$ of models contains the proper regular product of a fixed type of any finite number of $\mathcal{K}$-models, then it contains the proper product of the same type of any infinite system of $\mathcal{K}$-models, as well.* ■

As already mentioned, these corollaries, in the case when the regular product is direct, have been shown by Vaught [179]. They were formulated as problems by J. Łoś [89].

As an illustration we consider an example from the theory of RN-, RI-, and Z-groups (see [81] for the definitions). In [XI, §2.2] it was shown that these classes of groups are axiomatizable. In addition, it is known that the direct product of any two groups in one of these classes is again a group in that class. From the above statement about axiomatizable classes we learn that the *complete direct product of any number of* RN- *or* RI- *or Z-groups is a group of the same sort.* ■

Apparently, this property has not been explicitly noted, even though a group-theoretic proof presents no difficulty.

§**2.3.** Without further calculation, we can use Theorem 3 to extend Mostowski's theorem on the decidability of direct powers of decidable models to regular powers of decidable models and, partially, to regular products.

Suppose we have a FOPL sentence Φ – with predicate symbols among $\boldsymbol{P}_\gamma$ $(\gamma \in \Gamma)$ – whose value in an appropriate regular product $\mathfrak{M}$ of models $\mathfrak{M}_\alpha$ $(\alpha \in \mathrm{A})$ we wish to establish. By the first corollary of Theorem 3, we can find a sentence Ψ with unary predicate symbols $\boldsymbol{T}_\nu$ $(\nu \in \mathrm{N})$ that is equivalent to Φ in a certain sense. On the basis of Behmann's theorem [2], we can reduce Ψ to a disjunction of formulas of the form

$$(\exists \boldsymbol{\alpha}_1) \ldots (\exists \boldsymbol{\alpha}_u)\, (\boldsymbol{\alpha}_{i_1} \not\approx \boldsymbol{\alpha}_{j_1} \,\&\, \ldots \,\&\, \boldsymbol{\alpha}_{i_p} \not\approx \boldsymbol{\alpha}_{j_p} \,\&$$

$$\& \; \boldsymbol{T}_{\nu_1}^{\epsilon(1)}(\boldsymbol{\alpha}_1) \,\&\, \ldots \,\&\, \boldsymbol{T}_{\nu_u}^{\epsilon(u)}(\boldsymbol{\alpha}_u)) \,\&\, (\boldsymbol{\alpha})\, \mathrm{W}(\boldsymbol{\alpha})\,, \tag{12}$$

where $\mathrm{W}(\boldsymbol{\alpha})$ is a conjunction of formulas of the form

$$\boldsymbol{T}_{\mu_1}^{\zeta(1)}(\boldsymbol{\alpha}) \vee \ldots \vee \boldsymbol{T}_{\mu_q}^{\zeta(q)}(\boldsymbol{\alpha})\,,$$

$\nu_i, \mu_j \in \mathrm{N}$, and ϵ, ζ are functions taking the values $1, -1$.

To make things simple let us assume the regular product $\mathfrak{M}$ to be proper. Then for each $\nu \in \mathrm{N}$, we can write a sentence ψ_ν^* with predicate symbols among the $\boldsymbol{P}_\gamma$ such that the value of $T_\nu(\alpha)$ in $\mathfrak{M}$ coincides with the value of ψ_ν^* in $\mathfrak{M}_\alpha$ when we interpret each $\boldsymbol{P}_\gamma$ as $P_\gamma^{(\alpha)}$. The nature of the formulas (12) gives us

Theorem 4: *Suppose we have predicate symbols* $\boldsymbol{P}_\gamma(\boldsymbol{x}_1, \ldots, \boldsymbol{x}_{n_\gamma})$ $(\gamma \in \Gamma)$ *and appropriate formulas of the kind in* (4). *Let* Φ *be a* FOPL *sentence with predicates among the* $\boldsymbol{P}_\gamma$. *Then in a finitistic manner we can construct a finite number of sentences* $\Phi_1, \ldots, \Phi_s$ *with predicate symbols among the* $\boldsymbol{P}_\gamma$ *such that if* $\mathfrak{M}$ *is the proper regular product (of the given type) of models* $\mathfrak{M}_\alpha$ $(\alpha \in \mathrm{A})$, *then the truth of* Φ *in* $\mathfrak{M}$ *is equivalent to the truth of a finitistically constructible sentence built with the aid of* $\neg, \&, \vee$ *from statements of the form: "among the factors are models* $\mathfrak{M}_{\alpha_1}, \ldots, \mathfrak{M}_{\alpha_m}$ *in which the sentences* $\Phi_{k_1}, \ldots, \Phi_{k_m}$ *are respectively true; what's more,* $\alpha_{i_1} \neq \alpha_{j_1}, \ldots, \alpha_{i_p} \neq \alpha_{j_p}$*".*

In fact, according to (12), the truth of Φ in $\mathfrak{M}$ is equivalent to the truth of a finitistically constructible sentence which is a propositional combination of statements of the desired form and statements of the form: "in every factor the sentence Φ' is true"; the negation of the latter is of the desired form. Since negation is permitted, by taking $\Phi_1, \ldots, \Phi_s$ to be appropriate propositional combinations of the ψ_ν^* discussed above, we complete the proof. ■

Theorem 4 implies this sharpening of a corollary of Theorem 3:

Assume a type (4) *for proper products to be given. For every* FOPL *sentence* Φ *with appropriate symbols, we can find a natural number n such that for the proper product* $\mathfrak{M}$ *(of the given type) of any models* $\mathfrak{M}_\alpha$ $(\alpha \in A)$, *if* Φ *is true in all possible subproducts of n of these factors, then* Φ *is true in* $\mathfrak{M}$. ■

A model $\mathfrak{A}$ is called *decidable* iff there is a regular algorithm which enables us to decide for every FOPL sentence Φ the question of whether Φ is true in $\mathfrak{A}$ or false. From Theorem 4 we get the

Corollary: *Every proper regular power of a decidable model is decidable.*

Indeed, if all the factors in a proper product are isomorphic to some fixed model $\mathfrak{M}_0$, then the statements in Theorem 4 reduce to: "the sentences $\Phi_{k_1}, \ldots, \Phi_{k_m}$ are true in $\mathfrak{M}_0$, and there are at least q factors". By assumption, all such statements are decidable. Consequently, the proper power under consideration is decidable. ■

It was shown in the Introduction how Theorem 4 can be used to obtain a result analogous to the situation Vaught considered: in the product each factor is isomorphic to an infinite number of other factors.

NOTES

(1) It is clear that all the classes of models in the above definition should be non-empty. In particular, the complete systems mentioned are also consistent.

(2) In parallel with the apparently necessary change in the definition of pseudo-axiomatizability in [VII], it may be interesting to permit auxiliary *single-base* predicates (in particular, unary predicates representing individual constants) to occur in the members of T and the earlier system S. Apart from Theorem 1, analogous results cannot be expected.

(3) Absolute equality can be treated implicitly as one of the P_γ.

([4]) $\mathfrak{M}(B)$ will be called the B-*subproduct* of $\mathfrak{M}$, although it is not generally a submodel of $\mathfrak{M}$.

([5]) It is assumed throughout that $\approx$ does not occur in Φ; it can occur, however, in the $\varphi^{\alpha}_{\lambda}$. Hence, it is easy to include true equality among the predicates in a regular product.

CHAPTER 13

SMALL MODELS

Let $\mathcal{K}$ be an axiomatizable class of models, i.e., the class of all models of the form $\mathfrak{M} = \langle M; P_i; a_j: i \in I, j \in J\rangle$ satisfying a certain system of axioms (i.e., closed formulas, sentences) of first-order predicate logic (FOPL) involving only $\boldsymbol{P}_i\ (i \in I)$ and $\boldsymbol{a}_j\ (j \in J)$ – symbols designating the fundamental predicates and distinguished elements in every model similar to $\mathfrak{M}$. The total number of symbols $\boldsymbol{P}_i, \boldsymbol{a}_j$ (the sum of the powers of I and J) is called the *order* of $\mathfrak{M}$, while the cardinality of the base M of $\mathfrak{M}$ is its *power.* A model whose power is not less than its order, but still infinite, is called *regular*; all other models are said to be *small.* It is well known (see, e.g., [I]) that every regular model $\mathfrak{M}$ in the axiomatizable class $\mathcal{K}$ can be properly isomorphically embedded in some $\mathcal{K}$-model of any previously prescribed power not less than that of $\mathfrak{M}$. In the present article we examine singularities encountered on extending small models, and consider some related problems.

§1. For instance, small models in the following two classes extend singularly.

Example 1: The basic predicates of models in the first class are unary predicates $P_\alpha(x)$ $(\alpha \in \mathrm{A})$, where the index set A is the set of all possible infinite sequences $\alpha = \langle \alpha_1, \alpha_2, \ldots\rangle$ $(\alpha_i = 0, 1)$. The class $\mathcal{K}_1$ consists of all such models satisfying the axioms

$$(\exists \boldsymbol{x})\boldsymbol{P}_\alpha(\boldsymbol{x}) \quad (\alpha \in \mathrm{A})\,; \tag{1}$$

$$(\exists \boldsymbol{x}_1) \ldots (\exists \boldsymbol{x}_n)(\boldsymbol{x}_1 \not\approx \boldsymbol{x}_2\ \&\ \boldsymbol{x}_1 \not\approx \boldsymbol{x}_3\ \&\ \ldots\ \&\ \boldsymbol{x}_{n-1} \not\approx \boldsymbol{x}_n) \rightarrow$$

$$\rightarrow (\boldsymbol{x})(\neg \boldsymbol{P}_\lambda(\boldsymbol{x}) \vee \neg \boldsymbol{P}_\mu(\boldsymbol{x}))\,, \tag{2}$$

for $n = 2^m$ and $[\lambda]_m = [\mu]_m$, where $\lambda, \mu \in \mathrm{A}$ and $m = 1. 2. \ldots$. Here and elsewhere, $[\lambda]_m$ denotes the initial segment $\langle \lambda_1, \ldots, \lambda_m\rangle$ of the infinite sequence $\lambda = \langle \lambda_1, \lambda_2, \ldots\rangle$.

Let $\mathfrak{M}$ be an arbitrary infinite $\mathfrak{K}_1$-model. By (1) for every $\alpha \in \mathrm{A}$, there is an element x_α in $\mathfrak{M}$ (i.e., in the base of $\mathfrak{M}$) for which $P_\alpha(x_\alpha)$ is true; by (2), x_α and $x_{\alpha'}$ are distinct if $\alpha \neq \alpha'$. Thus, any infinite $\mathfrak{K}_1$-model includes a submodel of the power of the continuum. At the same time, for every $m \geqslant 1$, $\mathfrak{K}_1$ contains a finite model of power 2^m. Indeed, let B_m be the set of all sequences $\beta = \langle \beta_1, ..., \beta_m \rangle$ $(\beta_i = 0, 1)$ of length m. Taking $P_\alpha(\beta)$ to be true iff $\beta = [\alpha]_m$, we obtain the desired $\mathfrak{K}_1$-model $\langle \mathrm{B}_m ; P_\alpha \colon \alpha \in \mathrm{A} \rangle$. Consequently, the class $\mathfrak{K}_1$, while containing finite models of arbitrarily great powers and, hence, infinite models, contains no countable models.

Example 2: The signature of the class $\mathfrak{K}_2$ consists of unary operation symbols $\boldsymbol{f}_\alpha(\boldsymbol{x})$ $(\alpha \in \mathrm{A})$ and individual constants $\boldsymbol{a}_\beta$ $(\beta \in \mathrm{B})$, where A is the same as in Example 1, and B is the set of all finite sequences $\beta = \langle \beta_1, ..., \beta_p \rangle$ $(\beta_i = 0, 1; p = 1, 2, ...)$; we also introduce the subsets $\mathrm{B}_{<m} \subseteq \mathrm{B}$ consisting of all sequences of lengths less than m. The class $\mathfrak{K}_2$ is determined by the FOPL axioms

$$(\boldsymbol{x})\Big(\mathop{\&}_{\beta \in \mathrm{B}_{<m}} \boldsymbol{x} \not\approx \boldsymbol{a}_\beta \rightarrow \boldsymbol{f}_\lambda(\boldsymbol{x}) \not\approx \boldsymbol{f}_\mu(\boldsymbol{x})\Big) \tag{3}$$

for $[\lambda]_m \neq [\mu]_m$ and $m = 1, 2, ...$.

Taking $f_\alpha(\beta) = [\alpha]_m$, where m is the length of β, we obtain a countable $\mathfrak{K}_2$-model $\mathfrak{B} = \langle \mathrm{B}; f_\alpha; \beta \colon \alpha \in \mathrm{A}, \beta \in \mathrm{B} \rangle$. Let $\mathfrak{M}$ be an arbitrary $\mathfrak{K}_2$-model with base M including $\mathfrak{B}$ as a proper submodel, and let $x_0 \in M \sim \mathrm{B}$. In M are elements $f_\alpha(x_0)$ $(\alpha \in \mathrm{A})$, which by (3) are distinct for distinct α. Thus, every proper $\mathfrak{K}_2$-extension of the countable $\mathfrak{K}_2$-model $\mathfrak{B}$ has power not less than that of the continuum.

§2. The examples above show that, at least for countable and finite models, the bounds indicated by the following theorem cannot be lowered.

Theorem 1: *If the axiomatizable class $\mathfrak{K}$ contains an infinite model $\mathfrak{M}$ of power $\mathfrak{m}$, then $\mathfrak{M}$ has a proper $\mathfrak{K}$-extension of power $\mathfrak{m}^{\aleph_0}$. If $\mathfrak{K}$ contains models of powers $\mathfrak{m}_1 < \mathfrak{m}_2 < ...$, then $\mathfrak{K}$ contains a model of power $\mathfrak{n}$ satisfying the condition $\mathfrak{m}_1 + \mathfrak{m}_2 + ... \leqslant \mathfrak{n} \leqslant \mathfrak{m}_1 \cdot \mathfrak{m}_2 \cdot ...$.*

For purposes of the proof we first eliminate the distinguished element symbols $\boldsymbol{a}_j$ $(j \in J)$, if there are any, from the signature of $\mathfrak{K}$ in favor of new unary predicate symbols $\boldsymbol{E}_j(\boldsymbol{x})$ $(j \in J)$, adding the sentences

$$(\exists \boldsymbol{x})\, \boldsymbol{E}_j(\boldsymbol{x})\ \&\ (\boldsymbol{x})(\boldsymbol{y})(\boldsymbol{E}_j(\boldsymbol{y}) \rightarrow \boldsymbol{x} \approx \boldsymbol{y}) \quad (j \in J)$$

to the list of axioms characterizing the class $\mathcal{K}$ after rewriting this list in terms of the new predicate symbols. According to the second ϵ-theorem [57], we can limit our attention in the proof to those cases when the signature of $\mathcal{K}$ consists of operation symbols $f_i(x_1, ..., x_{m_i})$ $(i \in I)$ and predicate symbols $P_j(x_1, ..., x_{n_j})$ $(j \in J)$, and all the axioms defining $\mathcal{K}$ have the form

$$(x_1) \dots (x_{p_\nu})\, \Phi_\nu(x_1, ..., x_{p_\nu}),$$

where Φ_ν is quantifier free, but may contain function symbols. Let $\mathfrak{M}$ be an infinite $\mathcal{K}$-model with power $\mathfrak{m}$ and base M. We introduce a new individual constant $\boldsymbol{a}$ and assign it rank 1. The symbols of rank 0 are individual constants in 1–1 correspondence with the elements of the model $\mathfrak{M}$; in general, the symbols of rank $k+1$ $(k \geqslant 1)$ are sequences of the form $\langle \mathfrak{c}_1, ..., \mathfrak{c}_{m_i}, f_i \rangle$, where $i \in I$, and $\mathfrak{c}_1, ..., \mathfrak{c}_{m_i}$ are symbols of ranks $\leqslant k$, at least one of which has rank k. Let C be the set of symbols of all ranks. Let D be the set of axioms obtained from the following cases (I)–(IV).

(I) For all rank 0 symbols $\boldsymbol{c}$, we take the sentence $\boldsymbol{a} \not\approx \boldsymbol{c}$; we also take all sentences of the form $\boldsymbol{c} \not\approx \boldsymbol{c}'$, $P_j(\boldsymbol{c}_1, ..., \boldsymbol{c}_{n_j})$, $\neg P_j(\boldsymbol{c}_1, ..., \boldsymbol{c}_{n_j})$, $f_i(\boldsymbol{c}_1, ..., \boldsymbol{c}_{m_i}) \approx \boldsymbol{c}$ ($\boldsymbol{c}, \boldsymbol{c}', \boldsymbol{c}_k$ – rank 0 symbols) true in $\mathfrak{M}$ under the given correspondence.

(II) For each symbol $\langle \mathfrak{c}_1, ..., \mathfrak{c}_{m_i}, f_i \rangle$ of rank $\geqslant 1$, we take the equation

$$f_i(\mathfrak{c}_1, ..., \mathfrak{c}_{m_i}) \approx \langle \mathfrak{c}_1, ..., \mathfrak{c}_{m_i}, f_i \rangle .$$

(III) For every ν and all $\mathfrak{c}_1, ..., \mathfrak{c}_{p_\nu} \in C$, we take the sentence $\Phi_\nu(\mathfrak{c}_1, ..., \mathfrak{c}_{p_\nu})$.

(IV) Let $\langle b_1, b_2, ... \rangle$ be a simple infinite sequence of distinct elements of $\mathfrak{M}$. For $n = 1, 2, ...$, we define a map from C onto M as follows: for every rank 0 symbol $\boldsymbol{c}$, which corresponds to the element $c \in M$, we put $\boldsymbol{c}(b_n) = c$; we let $\boldsymbol{a}(b_n) = b_n$; we use the recursion

$$\langle \mathfrak{c}_1, ..., \mathfrak{c}_{m_i}, f_i \rangle (b_n) = f_i(\mathfrak{c}_1(b_n), ..., \mathfrak{c}_{m_i}(b_n))$$

for symbols of all higher ranks. Finally, we take the sentences $\mathfrak{c} \approx \mathfrak{c}'$ for every $\mathfrak{c}$, $\mathfrak{c}' \in C$ such that $\mathfrak{c}(b_n) = \mathfrak{c}'(b_n)$ for every n.

The system D of sentences from (I)–(IV) is consistent. Indeed, let D_0 be any finite subset of D; then only a finite number of the symbols in C participate in the writing of the axioms in D_0, and thus only the elements of a finite subset $M_0 \subseteq M$ are referred to. Suppose $b_{n_0} \in M \sim M_0$. Interpreting each symbol $\mathfrak{c} \in C$ occurring in a member of D_0 as $\mathfrak{c}(b_{n_0}) \in M$, we easily see that all the axioms in D_0 become true in $\mathfrak{M}$; hence, D_0 is consistent. Consequently, D is consistent, since D_0 was an arbitrary finite subset. Thus, we can convert

C into a model $\mathfrak{N}$ (with a base C' and similar to $\mathfrak{M}$) that, by virtue of (I)–(III), is a member of $\mathfrak{K}$ and a proper extension of $\mathfrak{M}$. To calculate the power of $\mathfrak{N}$ it suffices to note that the elements of C' are in 1–1 correspondence with the sequences $\langle \mathfrak{c}(b_1), \mathfrak{c}(b_2), \ldots \rangle$ ($\mathfrak{c} \in C$). Hence, the power of $\mathfrak{N}$ is not greater than $\mathfrak{m}^{\aleph_0}$. (¹)

To prove the second assertion of the theorem we let $\mathfrak{M}_1, \mathfrak{M}_2, \ldots$ be the given sequence of models the powers of whose bases $M_1, M_2, \ldots$ are respectively $\mathfrak{m}_1 < \mathfrak{m}_2 < \ldots$; we take S to be a set of individual constants of power $\mathfrak{m}_1 + \mathfrak{m}_2 + \ldots$. For $n = 1, 2, \ldots$, let τ_n be a 1–1 mapping from M_n into S such that $\tau_n(M_n) \subseteq \tau_{n+1}(M_{n+1})$, while $\bigcup \tau_n(M_n) = S$. The elements of S are called symbols of rank 0; the symbols of rank $k+1$ ($k \geqslant 0$) are the sequences of the form $\langle \mathfrak{c}_1, \ldots, \mathfrak{c}_{m_i}, f_i \rangle$, where $i \in I$, and $\mathfrak{c}_1, \ldots, \mathfrak{c}_{m_i}$ are symbols of ranks $\leqslant k$, at least one having rank k. Let C be the collection of all these symbols. For each n, we define a map from a subset of C onto M_n by putting $\boldsymbol{c}(n) = \tau_n^{-1}(\boldsymbol{c})$ for $\boldsymbol{c} \in \tau_n(M_n)$, and putting

$$\langle \mathfrak{c}_1, \ldots, \mathfrak{c}_{m_i}, f_i \rangle (n) = f_i(\mathfrak{c}_1(n), \ldots, \mathfrak{c}_{m_i}(n))$$

wherever possible. The domains of these maps form an increasing chain of subsets whose union is C. We let D be the set of all sentences:

(I′) $\boldsymbol{c} \not\approx \boldsymbol{c}'$ for distinct $\boldsymbol{c}, \boldsymbol{c}' \in S$;

(II′) $f_i(\mathfrak{c}_1, \ldots, \mathfrak{c}_{m_i}) \approx \langle \mathfrak{c}_1, \ldots, \mathfrak{c}_{m_i}, f_i \rangle$ for $i \in I$;

(III′) $\Phi_\nu(\mathfrak{c}_1, \ldots, \mathfrak{c}_{p_\nu})$ for every ν;

(IV′) $\mathfrak{c} \approx \mathfrak{c}'$ for $\mathfrak{c}, \mathfrak{c}'$ such that $\mathfrak{c}(n) = \mathfrak{c}'(n)$ for all sufficiently large n; here, here $\mathfrak{c}$, $\mathfrak{c}'$, $\mathfrak{c}_k$ are arbitrary elements of C.

As above, we conclude that D is consistent; this leads us directly to a $\mathfrak{K}$-model $\mathfrak{N}$ whose base C' is derived from C by means of the identifications (IV′), and whose power $\mathfrak{n}$ meets the requirements of the theorem. ■

§3. A class $\mathfrak{K}$ of models with signature specifying operations and predicates only, characterizable by means of universal axioms, has the property that the union of an increasing chain of $\mathfrak{K}$-models embedded one in another is again a $\mathfrak{K}$-model. Using Theorem 1 and the generalized continuum hypothesis (GCH), we easily deduce that *every infinite $\mathfrak{K}$-model $\mathfrak{M}$ admits a $\mathfrak{K}$-extension of any given power greater than that of $\mathfrak{M}$.* ■

The second ε-theorem previously cited shows that this assertion holds for arbitrary axiomatizable classes.

§4. Let $\mathfrak{K}$ be a class of models; let $\mathfrak{p} \leqslant \mathfrak{q}$ be cardinal numbers. Then $\mathfrak{K}_\mathfrak{p}$, $\mathfrak{K}^\mathfrak{q}$, $\mathfrak{K}_\mathfrak{p}^\mathfrak{q}$ denote the classes of $\mathfrak{K}$ models $\mathfrak{M}$ whose power $\mathfrak{m}$ satisfies the corresponding inequality: $\mathfrak{p} \leqslant \mathfrak{m}$, $\mathfrak{m} \leqslant \mathfrak{q}$, $\mathfrak{p} \leqslant \mathfrak{m} \leqslant \mathfrak{q}$.

Theorem 2: *Let $\mathfrak{K}$, $\mathfrak{L}$ be axiomatizable classes of similar models; suppose for some infinite cardinal $\mathfrak{p}$, $\mathfrak{K}_{\mathfrak{p}}^{\mathfrak{p}} \subseteq \mathfrak{L}$. Then $\mathfrak{K}_{\aleph_0} \subseteq \mathfrak{L}$ if $\mathfrak{p}$ is not less than the order of $\mathfrak{K}$, and $\mathfrak{K}_{\aleph_0}^{\mathfrak{p}} \subseteq \mathfrak{L}$ in any case.*

This theorem includes Vaught's theorem on completeness [177] and is proved much like the latter. We prove, e.g., the second assertion. Let **T** be a system of axioms characterizing $\mathfrak{K}$, and let Φ be any sentence holding throughout $\mathfrak{L}$. We must show Φ holds in every infinite $\mathfrak{K}$-model of power $\mathfrak{m} \leqslant \mathfrak{p}$. Suppose contrarily that the infinite $\mathfrak{K}$-model $\mathfrak{M}$ of power $\mathfrak{m}$ satisfies $\neg\Phi$. The system $\mathbf{T} \cup \{\neg\Phi\}$, having an infinite model $\mathfrak{M}$ of power $\mathfrak{m}$, must by § 3 have a model $\mathfrak{N}$ of power $\mathfrak{p}$, as well. Then $\mathfrak{N} \in \mathfrak{K}_{\mathfrak{p}}^{\mathfrak{p}}$ and, consequently, $\mathfrak{N} \in \mathfrak{L}$, but this means $\mathfrak{N}$ satisfies $\neg\Phi$, which is impossible. Inasmuch as the result of § 3 depends on the GCH, the second assertion of Theorem 2 has been proved only under the assumption of the GCH. ∎

§5. In [XI, 1.2] the notion of projective class was introduced. For single-base models this is equivalent to the following. Let $\mathfrak{K}$ be an axiomatizable class of models whose signature consists of predicate symbols $\boldsymbol{P}_i$ $(i \in I)$, $\boldsymbol{Q}_j$ $(j \in J)$, and $\boldsymbol{R}(\boldsymbol{x})$. We take a $\mathfrak{K}$-model $\mathfrak{M}$ and look at its submodel $\mathfrak{M}_0$ whose base is composed of those elements a in $\mathfrak{M}$ for which $R(a)$ is true; in addition, we delete the predicates Q_j, so that the signature of $\mathfrak{M}_0$ refers only to the P_i. The class of all models $\mathfrak{M}_0$ obtained from $\mathfrak{K}$-models in this fashion is said to be *projective*. It can be shown that Theorem 1 and its corollary in § 3 hold for projective classes, as well.

NOTE

(1) The technique of § 3 can be used to construct an extension of power $\mathfrak{m}^{\aleph_0}$. Note that if $\mathfrak{m} < \mathfrak{p} < \mathfrak{m}^{\aleph_0}$, then $\mathfrak{p}^{\aleph_0} = \mathfrak{m}^{\aleph_0}$ – without recourse to the GCH.

CHAPTER 14

FREE SOLVABLE GROUPS

Groups isomorphic to the factor group $\mathfrak{F}/\mathfrak{F}^{(n)}$ of a free group $\mathfrak{F}$ with respect to its nth commutator subgroup are called *free (n-step) solvable groups* ($n = 1, 2, \ldots$). Below we derive some properties of these groups, and on the basis of these we then show that the elementary theory of any free solvable non-commutative group is not recursively decidable in the sense of Tarski [166].

§**1**. Auslander and Lyndon [4] have proved that if the quotient $\mathfrak{F}/\mathfrak{N}$ of a free group $\mathfrak{F}$ by a normal subgroup $\mathfrak{N}$ is infinite, then the center of the factor group $\mathfrak{F}/[\mathfrak{N},\mathfrak{N}]$ is trivial. Using this result, we easily establish

Theorem 1: *Let $\mathfrak{N}$ be a normal subgroup of the free group $\mathfrak{F}$ such that $\mathfrak{F}/\mathfrak{N}$ is torsion-free. Then any two commuting elements u, v of the group $\mathfrak{F}_0 = \mathfrak{F}/[\mathfrak{N},\mathfrak{N}]$ are either members of $\mathfrak{N}_0 = \mathfrak{N}/[\mathfrak{N},\mathfrak{N}]$ or powers of one and the same element of $\mathfrak{F}_0$.*

By considering the subgroup of $\mathfrak{F}$ generated by $u, v, \mathfrak{N}$ rather than the whole group, we reduce the matter to the case when the factor group $\mathfrak{F}/\mathfrak{N}$ is abelian with two generators u, v. Now three cases are possible: (1) $\mathfrak{F}/\mathfrak{N} = \{1\}$; (2) $\mathfrak{F}/\mathfrak{N}$ is free abelian with two free generators u, v; (3) $\mathfrak{F}/\mathfrak{N}$ is free cyclic. In the first case, $u, v \in \mathfrak{N}_0$, *finis.* In the second case, let a, b be free generators of an auxiliary free metabelian group $\mathfrak{H}$. ([1]) Since $\mathfrak{F}$ is free, there is a homomorphism τ from $\mathfrak{F}$ into $\mathfrak{H}$ with $u^\tau = a, v^\tau = b$. Since τ maps $\mathfrak{N}$ into $[\mathfrak{H},\mathfrak{H}]$ (= the center of $\mathfrak{H}$), we have $[a, b] = [u, v]^\tau = 1$, which contradicts the non-commutativity of $\mathfrak{H}$. Finally, in the third case, if we take $c \in \mathfrak{F}$ such that $c\mathfrak{N}$ generates $\mathfrak{F}/\mathfrak{N}$, we have in $\mathfrak{F}_0$: $u = c^k a_1,\ v = c^l a_2$ for some $a_1, a_2 \in \mathfrak{N}$. Suppose $kl \neq 0$, and put $s = k/(k, l),\ t = l/(k, l)$. Then $u^t = u^s a$ for some $a \in \mathfrak{N}$ that commutes with u in $\mathfrak{F}_0$. But this means a is a central element in the group $(u, \mathfrak{N})/[\mathfrak{N},\mathfrak{N}]$. Inasmuch as $(u, \mathfrak{N})$ is free and the index of $\mathfrak{N}$ in it is infinite, the Auslander-Lyndon result tells us that $a \in [\mathfrak{N}, \mathfrak{N}]$, and so, $u^t = u^s$ in $\mathfrak{F}_0$. Take integers p, q such that $sp + tq = 1$, and set $w = u^p v^q \in \mathfrak{F}_0$; then $u = w^s, v = w^t$, as was desired. Similar arguments show that the case $kl = 0$ with $u, v \notin \mathfrak{N}_0$ is impossible. ∎

For the free group $\mathfrak{F}$, all the factor groups $\mathfrak{F}/\mathfrak{F}^{(i)}$ are torsion-free. Therefore, by setting $\mathfrak{N} = \mathfrak{F}^{(n-1)}$ in Theorem 1, we find that *any two commuting elements of a free solvable group* $\mathfrak{G}$ *are either members of the highest nontrivial commutator subgroup of* $\mathfrak{G}$ *(starting with* $\mathfrak{G}^{(0)} = \mathfrak{G}$*) or powers of one and the same element.* ■

This in turn immediately implies that if *an element of a free solvable group commutes with all its conjugates, then it lies in the last nontrivial term of the derived series of the group.* ■

Following Kontorovič [78], we say a group is an R-*group* iff for any elements x, y of the group, if $x^m = y^m$ for some nonzero integer m, then $x = y$.

Theorem 2: *If the factor group* $\mathfrak{F}/\mathfrak{N}$ *of a free group* $\mathfrak{F}$ *by a normal subgroup* $\mathfrak{N}$ *is an* R-*group, then* $\mathfrak{F}_0 = \mathfrak{F}/[\mathfrak{N}, \mathfrak{N}]$ *is also an* R-*group.*

Let $x, y \in \mathfrak{F}$, and suppose $x^m \equiv y^m \ (mod\ \mathfrak{N}')$. Then $x^m \equiv y^m \ (mod\ \mathfrak{N})$, which gives $x \equiv y \ (mod\ \mathfrak{N})$, by assumption. Thus, $x = ya$ for some $a \in \mathfrak{N}$, and $(xa)^m \equiv x^m \ (mod\ \mathfrak{N}')$. Since $(xa)^m = x^m$ implies $a^{x^{m-1}} \ldots a^x a = 1$, where a^y is $y^{-1}ay$, and transforming this in turn by x gives $a^{x^m} a^{x^{m-1}} \ldots a^{x^2} a^x = 1$, so that $a^{x^m} = a$, we thus have $x^m a = a x^m$ in $\mathfrak{F}_0$. This means a lies in the center of the group $(x^m, \mathfrak{N})/[\mathfrak{N}, \mathfrak{N}]$. The group $(x^m, \mathfrak{N})$ is free, while $(x^m, \mathfrak{N})/\mathfrak{N}'$ is torsion-free. By the cited Auslander-Lyndon theorem, we know $a \in \mathfrak{N}'$, i.e., $x = y$ in $\mathfrak{F}_0$. ■

Corollary: *Every free solvable group is an* R-*group.*

For free 1-step solvable groups, which are abelian, the assertion is obvious. To continue the proof by induction, we assume every free n-step solvable group is an R-group. Let $\mathfrak{F}$ be a free group; then $\mathfrak{F}^{(n)}$ is a normal subgroup of $\mathfrak{F}$, and the factor group $\mathfrak{F}/\mathfrak{F}^{(n)}$ is an R-group. From Theorem 2 we conclude that the free $(n+1)$-step solvable group $\mathfrak{F}/\mathfrak{F}^{(n+1)}$ is an R-group. ■

§2. Let $\mathfrak{K}$ be a class of models with signature $\Sigma = \langle \boldsymbol{P}_1, \ldots, \boldsymbol{P}_s; \boldsymbol{a}_1, \ldots, \boldsymbol{a}_t \rangle$, where $\boldsymbol{P}_1, \ldots, \boldsymbol{P}_s$ are predicate symbols, and $\boldsymbol{a}_1, \ldots, \boldsymbol{a}_t$ are individual constant symbols. Let us assume that in every $\mathfrak{K}$-model $\mathfrak{M}$ some submodel has been selected arbitrarily and is called the σ-submodel $\sigma(\mathfrak{M})$ of $\mathfrak{M}$. We say that σ-submodels are *elementary in* $\mathfrak{K}$ iff there is a formula $\Phi(\boldsymbol{x})$ of first-order predicate logic (FOPL) whose extralogical symbols are contained in Σ and whose only free variable is $\boldsymbol{x}$ such that for every $\mathfrak{K}$-model $\mathfrak{M}$ and every element u in $\mathfrak{M}$ (i.e., in the base of $\mathfrak{M}$), $\Phi(u)$ is true in $\mathfrak{M}$ iff $u \in \sigma(\mathfrak{M})$.

When considering classes of groups we assume that the signature consists of predicate symbols for multiplication and inversion, although we shall use operation notation for conciseness. When we speak of groups or classes of groups with *fixed elements,* we shall mean that individual constants for the chosen fixed elements also appear in the signatures of the corresponding models.

Lemma: *For each n, the successive commutator subgroups are elementary in the class of all free n-step solvable groups.*

Indeed, if $\mathfrak{G}$ is a free n-step solvable group, then according to a corollary of Theorem 1, we have for all $u \in \mathfrak{G}$:

$$u \in \mathfrak{G}^{(n-1)} \quad \text{iff} \quad (\forall y \in \mathfrak{G})(u \cdot y^{-1}uy = y^{-1}uy \cdot u)\,,$$

$$u \in \mathfrak{G}^{(n-2)} \quad \text{iff} \quad (\forall y \in \mathfrak{G})(uy^{-1}uy \cdot u^{-1}y^{-1}u^{-1}y \in \mathfrak{G}^{(n-1)})\,,$$

$$\dots$$

$$u \in \mathfrak{G}' \quad \text{iff} \quad (\forall y \in \mathfrak{G})(uy^{-1}uy \cdot u^{-1}y^{-1}u^{-1}y \in \mathfrak{G}'')\,. \blacksquare \qquad (1)$$

It is interesting to note that the first derived subgroup is not commonly elementary. In fact, we observe the following:

Remark: *The commutator subgroup is not elementary in any (first-order) axiomatizable class $\mathfrak{X}$ of groups among whose factor groups appears, for arbitrarily large finite p, a free metabelian group with p free generators (rank p).*

For the proof we introduce the FOPL formulas

$$\Psi_m(\boldsymbol{x}) = (\exists y_1 \dots y_m z_1 \dots z_m)(\boldsymbol{x} \approx [y_1, z_1] \dots [y_m, z_m])$$

for $m = 1, 2, \dots$.

Let S be the system of closed FOPL formulas (or: sentences, axioms) characterizing the class $\mathfrak{X}$; suppose the notion of commutator subgroup is defined in $\mathfrak{X}$ by some FOPL formula $\Phi(\boldsymbol{x})$. Then the infinite system $S \cup \{\Phi(\boldsymbol{c})\} \cup \{\neg \Psi_m(\boldsymbol{c})\colon m = 1, 2, \dots\}$, where $\boldsymbol{c}$ is an individual constant, is contradictory and – by the compactness theorem – has a finite contradictory subset. This means that if q is the greatest index among the sentences $\neg \Psi_m(\boldsymbol{c})$ appearing in this subset (and some do), then the sentence

$$(\boldsymbol{x})(\Phi(\boldsymbol{x}) \rightarrow \Psi_q(\boldsymbol{x})) \qquad (2)$$

is valid in every $\mathfrak{X}$-group. In other words, in any $\mathfrak{X}$-group each element of the commutator subgroup must be representable as the product of q commutators.

Let $\mathfrak{H}$ be a free metabelian group of rank q' $(q' \geqslant 4q)$, and let $\mathfrak{G}$ be a $\mathcal{K}$-group having $\mathfrak{H}$ as a factor group. An immediate computation shows us that every element of the commutator subgroup of $\mathfrak{H}$ is the product of $q'-1$ commutators, although $\mathfrak{H}'$ does contain elements not expressible as the product of q commutators. Hence $\mathfrak{G}$ also has elements in its commutator subgroup that are not products of q commutators. Therefore, (2) fails in $\mathfrak{G}$, and the remark is verified. ■

On the other hand, it is easy to indicate an infinite family of axiomatizable classes of groups in which the commutator subgroup is elementary. E.g., such are the classes $\mathcal{K}_m$, where each $\mathcal{K}_m$ is characterized by the axiom

$$(\boldsymbol{x})(\Psi_{m+1}(\boldsymbol{x}) \rightarrow \Psi_m(\boldsymbol{x}))\ .$$

In $\mathcal{K}_m$ the commutator subgroup is defined by the formula $\Psi_m(\boldsymbol{x})$. From what we saw earlier we know that every metabelian group with $m+1$ generators belongs to $\mathcal{K}_m$, while free metabelian groups of sufficiently large ranks do not. Therefore, there are infinitely many distinct classes among the $\mathcal{K}_m$.

Subgroups that are elementary in classes of groups whose signatures have no individual constants are characteristic subgroups, and the search for them appears interesting not only for this or that axiomatizable class, but also for the more important individual groups.

§**3**. Let $\mathcal{K}$ be a class of models with signature Σ. Let $\mathrm{T}(\Sigma)$ be the collection of all FOPL sentences of signature Σ; let $\mathrm{T}(\mathcal{K})$ be the subset of $\mathrm{T}(\Sigma)$ consisting of all those sentences true in every $\mathcal{K}$-model. $\mathrm{T}(\mathcal{K})$ is called the *elementary theory* of the class $\mathcal{K}$, or when $\mathcal{K} = \{\mathfrak{M}\}$, of the model $\mathfrak{M}$. The elementary theory of $\mathcal{K}$ is said to be *(recursively) decidable* iff there is an algorithm enabling one to decide for every sentence in $\mathrm{T}(\Sigma)$ the question of its membership in $\mathrm{T}(\mathcal{K})$.

Theorem 3: *For $n \geqslant 2$, $k \geqslant 2$, the elementary theory of the free n-step solvable group $\mathfrak{G}$ with k free generators, two of which are fixed, is undecidable.*

By "two fixed free generators" we mean – as mentioned above – that the signature of $\mathfrak{G}$ consists of predicate symbols for multiplication and inversion and individual constant symbols $\boldsymbol{a}$, $\boldsymbol{b}$ for any two distinct free generators of $\mathfrak{G}$. We introduce the FOPL formulas

$$\zeta(\boldsymbol{x}) = \boldsymbol{xa} \approx \boldsymbol{ax}$$

$$\theta(\boldsymbol{x},\boldsymbol{y}) = \boldsymbol{xa} \approx \boldsymbol{ax}\ \&\ \boldsymbol{yb} \approx \boldsymbol{by}\ \&\ \boldsymbol{xy}^{-1} \in \mathfrak{G}'\ ,\ (^2)$$

where $xy^{-1} \in \mathfrak{G}'$ stands for the corresponding formula derived from (1). By Theorem 1 we see that in $\mathfrak{G}$, $\zeta(c)$ holds iff there is an m such that $c = a^m$, while $\theta(c, d)$ holds iff there is an m such that $c = a^m$ and $d = b^m$.

Finally, we introduce formulas

$$S(x, y, z) = xy \approx z \,, \tag{3}$$

$$P(x, y, z) = (\exists uv)(uxb \approx xbu \;\&\; \theta(y, v) \;\&\; uv^{-1}z^{-1} \in \mathfrak{G}') \,. \tag{4}$$

These formulas define two binary operations (written $c \oplus d$, $c * d$, respectively) on the subset of $\mathfrak{G}$ consisting of all elements satisfying $\zeta(x)$, i.e., all elements of the form a^m. Moreover, $a^s \oplus a^t = a^{s+t}$ and $a^s * a^t = a^{st}$. The first is immediate from (3). Assuming $P(a^s, a^t, a^r)$ to hold, we find the relations $ca^sb = a^sbc$ and $\theta(a^s, d)$ imply $c = (a^sb)^k$, $d = b^t$; with $cd^{-1}a^{-r} \in \mathfrak{G}'$, this shows $k = t$, $r = st$; thus, $a^s * a^t = a^{st}$. Therefore, the elementary theory T_0 of the arithmetic of the integers (with +, ·) is weakly interpretable in the elementary theory $T(\mathfrak{G}; a, b)$ of the group $\mathfrak{G}$ with fixed elements a, b. Since T_0 is undecidable by Church's theorem, $T(\mathfrak{G}; a, b)$ is undecidable. ■

From the weak interpretability of T_0 in $T(\mathfrak{G}; a, b)$ we derive as usual (cf. [166]) the undecidability of the elementary theory $T(\mathfrak{G})$ of the group $\mathfrak{G}$ without any fixed elements, and for $n = 2, 3, \ldots$, the existence of a finitely axiomatizable class $\mathfrak{S}_n$ of n-step solvable groups whose elementary theory is essentially undecidable. ([3])

NOTES

([1]) A group is *metabelian* iff it satisfies the identity

$$(\forall)\,(xyx^{-1}y^{-1} \cdot z \approx z \cdot xyx^{-1}y^{-1}) \,.$$

([2]) The third conjunct in $\theta(x, y)$ should be replaced with $(\exists w)(wab \approx abw \;\&\; wxy \in \mathfrak{G}')$; cf. [R3].

([3]) In the terminology of [166], this interpretation of T_0 in $T(\mathfrak{G}; a, b)$ is *relative*, not weak. The methods there also show that any subtheory of $T(\mathfrak{G})$ with the same signature is undecidable.

CHAPTER 15

A CORRESPONDENCE BETWEEN RINGS AND GROUPS

Let $\mathfrak{K}_1, \mathfrak{K}_2$ be two classes of models with signatures Σ_1, Σ_2, respectively. As usual, by the elementary theory $T(\mathfrak{K}_i)$ of the class $\mathfrak{K}_i$ we mean the collection of all closed formulas (sentences, axioms) of first-order predicate logic (FOPL) whose predicate symbols belong to Σ_i that are true in all $\mathfrak{K}_i$-models (cf. [166]). We shall say that $\mathfrak{K}_1$ is *syntactically included* in $\mathfrak{K}_2$ iff there is an algorithm enabling one to construct for every FOPL sentence Φ of signature Σ_1, a corresponding sentence Φ^* of signature Σ_2 such that $\Phi \in T(\mathfrak{K}_1)$ iff $\Phi^* \in T(\mathfrak{K}_2)$. The classes $\mathfrak{K}_1, \mathfrak{K}_2$ are *syntactically equivalent* iff each is syntactically included in the other.

In the case of syntactical equivalence every elementary problem (i.e., formulatable in FOPL) concerning one of the classes can be transformed into an equivalent problem concerning the other. A well-known example is the correspondence between associative skewfields and certain projective planes. In the present article we investigate, from the indicated point of view, a correspondence between the class of all (not necessarily associative) rings with identity and a certain class of metabelian groups with two fixed elements. For significant classes of rings this correspondence induces syntactical equivalences. In a comparatively large number of cases the algorithmic undecidability of elementary theories of classes of rings has been established [139]. Our correspondence allows us to obtain from this a whole series of classes of metabelian groups with undecidable theories. Among these are, e.g., the class of all metabelian groups, each free metabelian group of rank > 1, and the class of all metabelian groups satisfying the identity $(x)(x^p \approx 1)$, where p is an odd prime. A simple argument lets us derive from these results the undecidability of the elementary theory of an noncommutative free nilpotent group of an arbitrary given nilpotence class.

The basic results of this article were presented at the Second Colloquium on Algebra held in April, 1959 in Moscow. A short report on them was published as [M7].

§1. The direct mapping

Let $\mathfrak{R}$ be an arbitrary ring, not necessarily associative. On the set $\sigma(\mathfrak{R})$ of triples $\langle a, b, c\rangle$ of elements of $\mathfrak{R}$ we define a binary operation by means of the relation

$$\langle a, b, c\rangle * \langle x, y, z\rangle = \langle a+x,\ b+y,\ bx+c+z\rangle\,. \tag{1}$$

It is easy to verify that the operation $*$ is associative, the triple $\langle 0, 0, 0\rangle$ is an identity element for it, and

$$\langle a, b, c\rangle^{-1} = \langle -a,\ -b,\ ba-c\rangle \tag{2}$$

is a 2-sided inverse. Thus, $\sigma(\mathfrak{R})$ with the operation $*$ is a group. From (1) it follows that every triple of the form $\langle 0, 0, c\rangle$ is a central element in the group $\sigma(\mathfrak{R})$; the equation

$$\langle a, b, c\rangle * \langle x, y, z\rangle * \langle a, b, c\rangle^{-1} * \langle x, y, z\rangle^{-1} = \langle 0, 0, bx-ya\rangle$$

shows that for any $\mathfrak{R}$, $\sigma(\mathfrak{R})$ is metabelian. ([1])

Consider now an arbitrary FOPL sentence

$$\Phi = (\mathrm{O}_1 \boldsymbol{x}_1) \ldots (\mathrm{O}_m \boldsymbol{x}_m)\ \Phi_0(\boldsymbol{x}_1, \ldots, \boldsymbol{x}_m)\,,$$

where $\mathrm{O}_i = \forall, \exists$, and the open Φ_0 involves at most one extralogical symbol, the multiplication sign. The requirement that Φ be true in $\sigma(\mathfrak{R})$ is equivalent to some demand laid on the ring $\mathfrak{R}$. This latter can again be expressed as the validity of a certain FOPL sentence $\sigma(\Phi)$, this time in $\mathfrak{R}$. To construct $\sigma(\Phi)$ it suffices to replace every quantifier $(\mathrm{O}_i \boldsymbol{x}_i)$ in Φ with three quantifiers $(\mathrm{O}_i \boldsymbol{x}_i')(\mathrm{O}_i \boldsymbol{x}_i'')(\mathrm{O}_i \boldsymbol{x}_i''')$, and to replace every expression $\boldsymbol{x}_i \boldsymbol{x}_j \approx \boldsymbol{x}_k$ with the formula

$$\boldsymbol{x}_i' + \boldsymbol{x}_j' \approx \boldsymbol{x}_k' \ \&\ \boldsymbol{x}_i'' + \boldsymbol{x}_j'' \approx \boldsymbol{x}_k'' \ \&\ \boldsymbol{x}_i''' + \boldsymbol{x}_j''' + \boldsymbol{x}_j'' \boldsymbol{x}_j' \approx \boldsymbol{x}_k'''\ .$$

Therefore, the transformation $\Phi \to \sigma(\Phi)$ is a syntactical embedding of $\{\mathfrak{R}\}$ in $\{\sigma(\mathfrak{R})\}$ (or simply, of $\mathfrak{R}$ in $\sigma(\mathfrak{R})$). In particular, if a given class $\mathcal{R}$ of rings has a decidable elementary theory, then so will the corresponding class of groups $\sigma(\mathcal{R})$, for σ does not depend on the choice of the ring $\mathfrak{R}$.

§2. Groups with distinguished elements

In what follows we assume that the every ring $\mathfrak{R}$ has an identity element 1 such that for any element $x \in \mathfrak{R}$, $1 \cdot x = x \cdot 1 = x$. Then in the corresponding group $\sigma(\mathfrak{R})$ it is natural to distinguish the two elements

$$a_1 = \langle 1, 0, 0 \rangle, \qquad a_2 = \langle 0, 1, 0 \rangle; \tag{3}$$

we shall view $\sigma(\mathfrak{R})$ not as a group, but as a more complicated structure: a group with a pair of distinguished elements, to which we assign the individual constant designations $\boldsymbol{a}_1, \boldsymbol{a}_2$. This structure will at times be called an *enriched group,* for short. From the point of view of model theory an enriched group is an algebraic system with signature $\langle \cdot ; \boldsymbol{a}_1, \boldsymbol{a}_2 \rangle$ consisting of a base set, a basic binary operation $\cdot$ (abbreviated by juxtaposition, as usual), and two basic distinguished (but not necessarily distinct) elements a_1, a_2. In accord with the general theory of models (cf. [XI]), enriched groups $\mathfrak{G}$, $\mathfrak{G}'$ are said to be *isomorphic* iff there is an ordinary group isomorphism from $\mathfrak{G}$ onto $\mathfrak{G}'$ that maps the pair $\langle a_1, a_2 \rangle$ of distinguished elements of $\mathfrak{G}$ onto the pair $\langle a_1', a_2' \rangle$ of distinguished elements of $\mathfrak{G}'$. We note that by choosing different pairs of distinguished elements in a given group, we can get nonisomorphic enriched groups.

We define analogously homomorphisms and direct (cartesian) products of enriched groups. Subgroups of an enriched group $\mathfrak{G}$ are ordinary subgroups which contain the distinguished elements a_1, a_2 of $\mathfrak{G}$ as their own distinguished elements.

In what follows, every ring $\mathfrak{R}$ will be viewed as a ring with a distinguished element, its identity. This means that as subrings of $\mathfrak{R}$ we only consider those ordinary subrings containing the identity element of $\mathfrak{R}$, etc.

The transformation σ in §1 can now be viewed as a map associating with every ring $\mathfrak{R}$ with distinguished identity element 1, an enriched group $\sigma(\mathfrak{R})$ with distinguished elements $a_1 = \langle 1, 0, 0 \rangle$, $a_2 = \langle 0, 1, 0 \rangle$. The group $\mathfrak{G} = \sigma(\mathfrak{R})$ and its elements a_1, a_2 possess the following properties:

(A1) The group $\mathfrak{G}$ is metabelian; in other words, any elements $x, y, z \in \mathfrak{G}$ satisfy the relation:

$$xyx^{-1}y^{-1} \cdot z = z \cdot xyx^{-1}y^{-1}.$$

(A2) The subsets of $\mathfrak{G}$ consisting of elements commuting with a_1, a_2 form unenriched abelian subgroups $\mathfrak{G}_1$, $\mathfrak{G}_2$, respectively, of the group $\mathfrak{G}$.

(A3) The intersection of $\mathfrak{G}_1$ and $\mathfrak{G}_2$ is the center $\mathfrak{Z}$ (unenriched) of $\mathfrak{G}$.

(A4) For any elements z_1, z_2 of the center $\mathfrak{Z}$ of $\mathfrak{G}$, there is an element $x \in \mathfrak{G}$ such that

$$a_1 x a_1^{-1} x^{-1} = z_1 , \quad a_2 x a_2^{-1} x^{-1} = z_2 . \tag{4}$$

(A5) There are homomorphisms f_1, f_2 from $\mathfrak{Z}$ into $\mathfrak{G}_1, \mathfrak{G}_2$, respectively, such that $f_1(c) = a_1$, $f_2(c) = a_2^{-1}$, where $c = a_2 a_1 a_2^{-1} a_1^{-1}$, and for every $z \in \mathfrak{Z}$,

$$a_2 f_1(z)\, a_2^{-1} f_1(z)^{-1} = a_1 f_2(z)\, a_1^{-1} f_2(z)^{-1} = z .$$

In fact, a quick calculation shows $\mathfrak{G}_1$ consists of all triples of the form $\langle u, 0, w\rangle$, $\mathfrak{G}_2$ consists of all triples of the form $\langle 0, v, w\rangle$, while $\mathfrak{Z}$ actually consists of all the triples $\langle 0, 0, w\rangle$ $(u, v, w \in \mathfrak{R})$. If $z_1 = \langle 0, 0, w_1\rangle$ and $z_2 = \langle 0, 0, w_2\rangle$, then $x = \langle w_2, -w_1, 0\rangle$ is a solution for the equations (4). As the homomorphisms we seek, we can take $f_1(z) = \langle w, 0, 0\rangle$ and $f_2(z) = \langle 0, -w, 0\rangle$ for $z = \langle 0, 0, w\rangle$.

§3. The inverse mapping

Let $\mathcal{G}_4$ be the class of all enriched groups satisfying the conditions (A1)–(A4). Since these conditions are easily written in the form of FOPL sentences, $\mathcal{G}_4$ is a finitely axiomatizable class. Let $\mathfrak{G}$ be a $\mathcal{G}_4$-group with the group operation $\cdot$ and distinguished elements a_1, a_2. We define two new binary operations $+, \times$ on the center $\mathfrak{Z}$ of $\mathfrak{G}$ by putting

$$z_1 + z_2 = z_1 z_2 , \tag{5}$$

$$z_1 \times z_2 = x_2 x_1 x_2^{-1} x_1^{-1} \tag{6}$$

for all $z_1, z_2 \in \mathfrak{Z}$, where x_1, x_2 are any elements of $\mathfrak{G}$ satisfying the conditions

$$x_1 a_2 = a_2 x_1, \quad x_2 a_1 = a_1 x_2, \quad a_i x_i a_i^{-1} x_i^{-1} = z_i \quad (i = 1, 2). \tag{7}$$

We now show that the base of $\mathfrak{Z}$ with the operations $+, \times$ is a ring $\mathfrak{Z}^*$ with identity element $c = a_2 a_1 a_2^{-1} a_1^{-1}$. Indeed, by (A4) there are elements $x_1, x_2 \in \mathfrak{G}$ satisfying (7). Suppose

$$a_i y_i a_i^{-1} y_i^{-1} = z_i \qquad (i = 1, 2)$$

for some $y_1 \in \mathfrak{G}_2$, $y_2 \in \mathfrak{G}_1$. From $a_i y_i a_i^{-1} y_i^{-1} = a_i x_i a_i^{-1} x_i^{-1}$ it follows that

$x_i^{-1}y_ia_i = a_ix_i^{-1}y_i$, i.e., the element $x_i^{-1}y_i$ commutes with a_i, but it also commutes with a_{3-i} by the other relations assumed; thus by (A3), $x_i^{-1}y_i \in \mathfrak{Z}$ $(i = 1, 2)$. But then we get $y_2y_1y_2^{-1}y_1^{-1} = x_2x_1x_2^{-1}x_1^{-1}$. Therefore, the operation $\times$ is well-defined by (6).

Addition in $\mathfrak{Z}^*$ is the group operation in $\mathfrak{Z}$, and so, $\mathfrak{Z}^*$ is an abelian group with respect to its first operation. To prove the distributivity relations

$$(u+v) \times w = u \times w + v \times w, \quad w \times (u+v) = w \times u + w \times v\,, \qquad (8)$$

we first take $x, y \in \mathfrak{G}_2$ and $t \in \mathfrak{G}_1$ such that $u = a_1xa_1^{-1}x^{-1}$, $v = a_1ya_1^{-1}y^{-1}$, and $w = a_2ta_2^{-1}t^{-1}$. By (A2), x and y commute, so

$$a_1(xy)a_1^{-1}(xy)^{-1} = a_1xa_1^{-1}x^{-1} \circ a_1ya_1^{-1}y^{-1} = u + v \;, (^2)$$

which gives

$$(u+v) \times w = txyt^{-1}y^{-1}x^{-1} = txt^{-1}x^{-1} \cdot tyt^{-1}y^{-1} = u \times w + v \times w\,.$$

We establish the second relation in (8) similarly. Finally, a straightforward claculation shows $c \times z = z \times c = z$ for all $z \in \mathfrak{Z}^*$.

The correspondence associating with every $\mathcal{G}_4$-group $\mathfrak{G}$ the ring $\mathfrak{Z}^*$ constructed above is denoted by τ.

If $\Psi = (\mathrm{O}_1\boldsymbol{x}_1) \ldots (\mathrm{O}_n\boldsymbol{x}_n)\,\Psi_0(\boldsymbol{x}_1, \ldots, \boldsymbol{x}_n)$ is a FOPL sentence with symbols $+, \times, \mathbf{1}$ concerning rings with identity, then $\tau(\Psi)$ will denote the sentence with symbols appropriate to enriched groups obtained from Ψ by replacing each quantifier $(\mathrm{O}_i\boldsymbol{x}_i)$ with the quantifier $(\mathrm{O}_i^{\zeta}\boldsymbol{x}_i)$ specialized with respect to the formula $\zeta(\boldsymbol{x}) = (\boldsymbol{u})\;(\boldsymbol{ux} \approx \boldsymbol{xu})$, and by replacing the symbol $\mathbf{1}$ in Ψ with the expression $\boldsymbol{a}_2\boldsymbol{a}_1\boldsymbol{a}_2^{-1}\boldsymbol{a}_1^{-1}$, every subformula of the form $\boldsymbol{x}_i + \boldsymbol{x}_j \approx \boldsymbol{x}_k$ with $\boldsymbol{x}_i \cdot \boldsymbol{x}_j \approx \boldsymbol{x}_k$, and every subformula of the form $\boldsymbol{x}_i \times \boldsymbol{x}_j \approx \boldsymbol{x}_k$ with the formula

$$(\exists \boldsymbol{uv})\,(\boldsymbol{x}_k \approx \boldsymbol{uvu}^{-1}\boldsymbol{v}^{-1}\ \&\ \boldsymbol{ua}_1 \approx \boldsymbol{a}_1\boldsymbol{u}\ \&\ \boldsymbol{va}_2 \approx \boldsymbol{a}_2\boldsymbol{v}\ \&$$

$$\&\ \boldsymbol{x}_i \approx \boldsymbol{a}_1\boldsymbol{va}_1^{-1}\boldsymbol{v}^{-1}\ \&\ \boldsymbol{x}_j \approx \boldsymbol{a}_2\boldsymbol{ua}_2^{-1}\boldsymbol{u}^{-1})\,;$$

we are using the group notation, e.g. $^{-1}$, as a shorthand for the actual FOPL expressions involving only multiplication.

Clearly, Ψ is true in the ring $\tau(\mathfrak{G})$ iff $\tau(\Psi)$ is true in the group $\mathfrak{G}$; therefore, the elementary theory of the ring $\tau(\mathfrak{G})$ is syntactically included in that of the enriched group $\mathfrak{G}$. In particular, if the first is undecidable, then so is the second.

E.g., let $\mathfrak{G}$ be the free metabelian group with two free generators a_1, a_2, considered to be distinguished elements. Then every element of $\mathfrak{G}$ can be uniquely represented in the form $a_1^k a_2^l c^m$, where k, l, m are integers and $c = a_2 a_1 a_2^{-1} a_1^{-1}$; the multiplication in $\mathfrak{G}$ then takes the form

$$a_1^k a_2^l c^m \circ a_1^p a_2^q c^r = a_1^{k+p} a_2^{l+r} c^{lp+m+r} ,$$

for any integers k, l, etc. From this it follows that (A1)–(A4) hold for $\mathfrak{G}$, and $\tau(\mathfrak{G})$ is isomorphic to the usual ring of integers. By Church's theorem, the ring of integers is undecidable (cf. [166]). Therefore, the elementary theory of a free metabelian group with two free generators is undecidable.

§4. The reciprocity of the correspondences σ and τ

From the foregoing results we see that for any ring $\mathfrak{R}$ with identity,

$$\tau(\sigma(\mathfrak{R})) = \mathfrak{R}.$$

the equation indicating isomorphism. Concerning the corresponding syntactical transformations, we have the following equivalence: for every FOPL sentence Ψ with symbols $+$, $\times$, $\mathbf{1}$, the sentence

$$\sigma(\tau(\Psi)) \leftrightarrow \Psi$$

holds in every ring with identity, the definition of σ being extended to occurrences of $\boldsymbol{a}_1$, $\boldsymbol{a}_2$ via (3).

We now show that if $\mathcal{G}_5$ is the subclass of $\mathcal{G}_4$ consisting of those groups satisfying (A5) as well, then for any $\mathcal{G}_5$-group $\mathfrak{G}$,

$$\sigma(\tau(\mathfrak{G})) = \mathfrak{G},$$

and for every FOPL sentence Φ with symbols $\bullet$, $\boldsymbol{a}_1$, $\boldsymbol{a}_2$, the sentence

$$\tau(\sigma(\Phi)) \leftrightarrow \Phi$$

holds in the enriched group $\mathfrak{G}$.

To prove this we introduce the operations $+$, $\times$ on the center $\mathfrak{Z}$ of $\mathfrak{G}$ via the relations (5), (6), and let $\mathfrak{H} = \sigma(\mathfrak{Z}^*)$ be the enriched group of triples that corresponds to the ring $\mathfrak{Z}^*$ so constructed. We must show $\mathfrak{H}$ is isomorphic to $\mathfrak{G}$.

By assumption there are homomorphisms f_1, f_2 from the center $\mathfrak{Z}$ into the subgroups $\mathfrak{G}_1$, $\mathfrak{G}_2$ of $\mathfrak{G}$ with the properties described in (A5). With every triple $h = \langle h_1, h_2, h_3 \rangle$ in $\mathfrak{H}$ we associate the element $\pi(h) = f_1(h_1) f_2(h_2)^{-1} h_3$ of the group $\mathfrak{G}$.

We now prove that the map π is a homomorphism from $\mathfrak{H}$ into $\mathfrak{G}$. Let $h = \langle h_1, h_2, h_3 \rangle$ and $k = \langle k_1, k_2, k_3 \rangle$ be arbitrary elements of $\mathfrak{H}$. By (1) and (5),

$$\pi(h * k) = f_1(h_1 k_1) f_2(k_2^{-1} h_2^{-1}) (h_2 \times k_1) h_3 k_3 ,$$

whence

$$\pi(h * k) \pi(k)^{-1} = f_1(h_1 k_1) f_2(h_2^{-1}) f_1(k_1^{-1}) (h_2 \times k_1) h_3 . \tag{9}$$

From the relations

$$a_1 f_2(h_2) a_1^{-1} f_2(h_2)^{-1} = h_2, \quad a_2 f_1(k_1) a_2^{-1} f_1(k_1)^{-1} = k_1$$

we obtain

$$h_2 \times k_1 = f_1(k_1) f_2(h_2) f_1(k_1^{-1}) f_2(h_2^{-1}) ,$$

according to (6) and (7). Therefore, we can rewrite (9) as

$$\pi(h * k) \pi(k)^{-1} = f_1(h_1 k_1) f_1(k_1)^{-1} f_2(h_2)^{-1} h_3$$

$$= f_1(h_1) f_2(h_2)^{-1} h_3 ,$$

whence $\pi(h * k) = \pi(h) \cdot \pi(k)$.

In the enriched group $\mathfrak{H}$ the distinguished elements are $a_1^* = \langle c, e, e \rangle$, $a_2^* = \langle e, c, e \rangle$, where e is the identity element in $\mathfrak{G}$, and $c = a_2 a_1 a_2^{-1} a_1^{-1}$. According to (A5),

$$\pi(a_1^*) = f_1(c) f_2(e) e = a_1 , \quad \pi(a_2^*) = f_1(e) f_2(c)^{-1} e = a_2 .$$

Thus, π is a homomorphism of enriched groups.

We now compute the kernel of π. Suppose for some $h = \langle h_1, h_2, h_3 \rangle \in \mathfrak{H}$, $\pi(h) = e$. Then $f_1(h_1) = f_2(h_2) h_3^{-1}$, so by (A3), $f_1(h_1) \in \mathfrak{Z}$; by virtue of (A5),

$$h_1 = a_2 f_1(h_1) a_2^{-1} f_1(h_1)^{-1} = e .$$

We similarly find that $h_2 = e$, and also $h_3 = e$, i.e., h is the identity element of $\mathfrak{H}$. Therefore, the kernel of π is trivial.

It only remains to show that π maps $\mathfrak{H}$ onto $\mathfrak{G}$ to establish it as the desired isomorphism. Let g be an arbitrary element of $\mathfrak{G}$. We set

$$g_i = a_i g a_i^{-1} g^{-1} \quad (i = 1, 2)\ , \qquad h_1 = f_1(g_2)\ ,$$
$$h_2 = f_2(g_1)\ , \qquad g_3 = h_1^{-1} g h_2^{-1}\ . \tag{10}$$

If we can show g_3 is central in $\mathfrak{G}$, then we shall have $g^* = \langle g_2, g_1^{-1}, g_3 \rangle \in \mathfrak{H}$ such that $\pi(g^*) = g$; thus, we shall have proved π is onto.

By (A5) we have $a_2 h_1 a_2^{-1} h_1^{-1} = g_2$; by comparison with (10) we get $h_1^{-1} g a_2 = a_2 h_1^{-1} g$, so $h_1^{-1} g \in \mathfrak{G}_2$; consequently, $g_3 = h_1^{-1} g \cdot h_2^{-1} \in \mathfrak{G}_2$. Since the equation $uvu^{-1} \cdot v^{-1} = v^{-1} \circ uvu^{-1}$ holds for arbitrary elements in a metabelian group, from $a_1 h_2 a_1^{-1} h_2^{-1} = g_1$ and $a_1 g a_1^{-1} g^{-1} = g_1$ we derive $a_1^{-1} h_2^{-1} a_1 h_2 = a_1^{-1} g^{-1} a_1 g$; from this we get $g h_2^{-1} a_1 = a_1 g h_2^{-1}$, so $g h_2^{-1} \in \mathfrak{G}_1$; consequently, $g_3 = h_1^{-1} \circ g h_2^{-1} \in \mathfrak{G}_1$. Thus, $g_3 \in \mathfrak{G}_1 \cap \mathfrak{G}_2$; therefore, by (A3) $g_3 \in \mathfrak{Z}$, as required.

We have proved the following

Theorem 1: *The map σ is a* 1–1 *correspondence between the class $\mathcal{R}$ of all rings with identity and the class $\mathcal{G}_5$ of all enriched metabelian groups satisfying* (A2)–(A5). *Moreover, if a ring $\mathfrak{R} \in \mathcal{R}$ satisfies a* FOPL *sentence Ψ, then the group $\sigma(\mathfrak{R})$ satisfies $\tau(\Psi)$; conversely, if a $\mathcal{G}_5$-group $\mathfrak{G}$ satisfies a* FOPL *sentence Φ, then the ring $\tau(\mathfrak{G})$ satisfies $\sigma(\Phi)$.* ■

Of the conditions (A1)–(A5) characterizing the class $\mathcal{G}_5$, (A5) is more complicated than the rest. In the next section we indicate some smaller classes of groups satisfying analogues of Theorem 1 while admitting simpler characterizations.

§5. Some special cases

If p is a positive integer, we say that a ring $\mathfrak{R}$ has *characteristic p* (or, is a *char p ring*) iff for all $x \in \mathfrak{R}$, $px = 0$, and if $mx = 0$ (m – an integer), then either $x = 0$ or p divides m. A group $\mathfrak{G}$ is called a *p-group* iff for every $x \in \mathfrak{G}$, $x^p = e$. An *rring* is a ring whose additive group is completely divisible and torsion-free.

Theorem 2: *For every odd prime p, the map σ becomes a* 1–1 *correspondence between the class of char p rings with identity and the class of enriched metabelian p-groups satisfying* (A2)–(A4).

Suppose $\mathfrak{R}$ is a char p ring which an identity element, and $\mathfrak{R} = \sigma(\mathfrak{G})$ is the corresponding group. Let $g = \langle g_1, g_2, g_3 \rangle$ be an element of $\mathfrak{G}$; then

$$g^p = \langle pg_1, pg_2, \frac{p(p-1)}{2} g_2 g_1 + pg_3 \rangle = e .$$

Thus $\mathfrak{G}$ is a p-group. For any p-group $\mathfrak{G} \in \mathcal{G}_4$, $\tau(\mathfrak{G})$ is obviously a char p ring. We have only to show that $\mathfrak{G}$ satisfies (A5). The map

$$x \to a_2 x a_2^{-1} x^{-1} \quad (x \in \mathfrak{G}_1) \tag{11}$$

is a homomorphism from $\mathfrak{G}_1$ onto the center $\mathfrak{Z} \subseteq \mathfrak{G}$. But $\mathfrak{Z}$ is an abelian p-group, and thus it is possible to view it as a vector space over the prime field $\mathfrak{P}$ of characteristic p. Choose a basis $\{z_\alpha \colon \alpha \in \mathrm{A}\}$ for $\mathfrak{Z}$ over $\mathfrak{P}$; for each z_α, choose a preimage $u_\alpha \in \mathfrak{G}_1$ with respect to the map (11). We can assume $z_{\alpha_0} = c = a_2 a_1 a_2^{-1} a_1^{-1}$ and $u_{\alpha_0} = a_1$. If $z = \Sigma n_\alpha z_\alpha$ ($n_\alpha \in \mathfrak{P}$, $n_\alpha = 0$ for almost all $\alpha \in \mathrm{A}$) is an arbitrary element of $\mathfrak{Z}$, then we set $f_1(z) = \Sigma n_\alpha u_\alpha$. Since $\mathfrak{G}_1$ is abelian by (A2), f_1 is a well-defined homomorphism, and $f_1(c) = a_1$. As $a_2 u_\alpha a_2^{-1} u_\alpha^{-1} = z_\alpha$ ($\alpha \in \mathrm{A}$), we have (returning to multiplicative notation in $\mathfrak{G}$)

$$a_2 f_1(z) a_2^{-1} f_1(z)^{-1} = a_2 \prod u_\alpha^{n_\alpha} \, a_2^{-1} \prod u_\alpha^{-n_\alpha} = \prod z_\alpha^{n_\alpha} = z .$$

We define f_2 analogously, then prove $f_2(c) = a_2^{-1}$ and $a_1 f_2(z) a_1^{-1} f_2(z)^{-1} = z$ for all $z \in \mathfrak{Z}$. ∎

We can apply similar arguments to the case when $\mathfrak{R}$ is an rring with identity and $\mathfrak{G}$ is a completely divisible torsion-free $\mathcal{G}_4$-group, obtaining thereby

Theorem 3: *The map* σ *becomes a* 1–1 *correspondence between the class of all rrings with identity and the class of all enriched completely divisible torsion-free metabelian groups satisfying* (A2)–(A4). ∎

From the proof of Theorem 2 we see that in the study of the correspondences σ, τ it is natural to consider not simply rings, but also algebras over a given field $\mathfrak{P}$ of characteristic $p \neq 2$. Then in place of groups we must consider metabelian groups over the field $\mathfrak{P}$, i.e., groups having, besides the group multiplication, an operation for raising any element x of the group to the power x^α of any element $\alpha \in \mathfrak{P}$, subject to the conditions:

$$x^\alpha x^\beta = x^{\alpha+\beta}, \quad (x^\alpha)^\beta = x^{\alpha\beta}, \quad x^\alpha y^\beta = y^\beta x^\alpha (x^{-1} y^{-1} x y)^{\alpha\beta} .$$

In constructing the group $\sigma(\mathfrak{R})$ in §1 we just add the definition

$$\langle x, y, z\rangle^{\alpha} = \langle \alpha x,\ \alpha y,\ \frac{\alpha(\alpha-1)}{2}\ yx + az\rangle \tag{12}$$

to (1) and (2); in constructing the ring $\tau(\mathfrak{G})$ of §3 we add the definition $\alpha z = z^{\alpha}$ to (5) and (6). The rest of the reasoning remains in force, and as a result we have

Theorem 4: *The map* σ, *extended by* (12), *becomes a* 1–1 *correspondence between the class of algebras with identity over the fixed field* $\mathfrak{P}$ *of characteristic* $p \neq 2$ *and the class of all enriched metabelian groups over* $\mathfrak{P}$ *satisfying* (A2)–(A4). ■

As final observations we note that the ring $\tau(\mathfrak{G})$, constructed from a $\mathcal{G}_4$-group $\mathfrak{G}$, has no zero divisors if for any elements $x_1 \in \mathfrak{G}_1, x_2 \in \mathfrak{G}_2, x_1x_2 = x_2x_1$ implies $x_1 \in \mathfrak{Z}$ or $x_2 \in \mathfrak{Z}$. If, in addition, for any $g \in \mathfrak{G}_1 \sim \mathfrak{Z}$, $h \in \mathfrak{G}_2 \sim \mathfrak{Z}$ and any $z \in \mathfrak{Z}$, there are $x \in \mathfrak{G}_2, y \in \mathfrak{G}_1$ such that $z = gxg^{-1}x^{-1} = yhy^{-1}h^{-1}$, then $\tau(\mathfrak{G})$ is a skewfield. The conditions indicated are also necessary.

§6. Reductions and interpretations of classes of models

Herein we recall the definitions of several concepts, some of which have already been used in the previous sections. For formal theories they are systematically set forth in [166]. We introduce them in the framework of the theory of classes of models (or of algebraic systems).

Let $\mathfrak{K}$ be a class of models whose signature $\Sigma(\mathfrak{K})$ consists of predicate symbols $\boldsymbol{P}_i$ (of rank n_i) and individual constant symbols $\boldsymbol{a}_j$. We can assume the index i (and j) runs over either the sequence of natural numbers or some finite initial segment of it. With regard to the rank, in the case of an infinite number of predicate symbols we demand that the rank n_i be a general recursive function of the index $\boldsymbol{i}$. Under these assumptions, all FOPL formulas of signature $\Sigma(\mathfrak{K})$ can be enumerated in a natural fashion. We say that the class $\mathfrak{K}$ (or more accurately, its elementary theory $\mathrm{T}(\mathfrak{K})$) is *(recursively) decidable* iff the set S of the numbers of all FOPL sentences true in every $\mathfrak{K}$-model is a recursive set of natural numbers. In the contrary case, $\mathrm{T}(\mathfrak{K})$ is *undecidable.* The class $\mathfrak{K}$ and its elementary theory $\mathrm{T}(\mathfrak{K})$ are said to be *essentially undecidable* iff every nonempty subclass of $\mathfrak{K}$ is undecidable (cf. [166]).

Reflecting a usage in [XI], a class $\mathfrak{K}_2$ of models is called an *enriched subclass* of the class $\mathfrak{K}_1$ iff the signature of $\mathfrak{K}_1$ is included in the signature of $\mathfrak{K}_2$ and every $\mathfrak{K}_2$-model is a $\mathfrak{K}_1$-model (when the extra predicates and distinguished elements, if any, are dropped). An enriched subclass $\mathcal{K}_2$ of the class $\mathcal{K}_1$ is *inessentially enriched* iff its signature results from that of $\mathcal{K}_1$ by the addition of some set of individual constant symbols and every $\mathcal{K}_1$-model becomes a $\mathcal{K}_2$-model for some choice of new distinguished elements.

The class $\mathcal{K}_2$ is a *finitely axiomatizable enriched subclass* of $\mathcal{K}_1$ iff it is an enriched subclass and there is a FOPL sentence Φ of signature $\Sigma(\mathcal{K}_2)$ such that $\mathcal{K}_2$ consists of those enriched $\mathcal{K}_1$-models satisfying Φ.

An *impoverishment* of the class $\mathcal{K}$ is a class with signature included in the signature of $\mathcal{K}$, composed of all those models obtained from $\mathcal{K}$-models by dropping the corresponding extra predicates and distinguished elements.

Suppose in every $\mathcal{K}$-model we have somehow defined a unary predicate $P(x)$, true for at least one element of the model; then the *P-reduction* of is the class of all submodels of $\mathcal{K}$-models whose bases consist of all those elements for which P is true. Below, P is always assumed to be *formular* in $\mathcal{K}$, i.e., defined by a fixed FOPL formula $\rho(\boldsymbol{x})$ in each $\mathcal{K}$-model; in this case we speak of the ρ-reduction of $\mathcal{K}$.

Interpreting (the theory of) the class $\mathcal{K}_2$ of models in (the theory of) the class $\mathcal{K}_1$ consists of doing the following. With every predicate symbol $\boldsymbol{P}_i^{n_i}$ in the signature of $\mathcal{K}_2$ we associate a FOPL formula $\Theta_i(\boldsymbol{x}_1, ..., \boldsymbol{x}_{n_i})$ of signature $\Sigma(\mathcal{K}_1)$, and with every individual constant $\boldsymbol{a}_j$ in $\Sigma(\mathcal{K}_2)$ we associate a defining formula $\varphi_j(\boldsymbol{x})$ of signature $\Sigma(\mathcal{K}_1)$ and choose a formular predicate defined by $\rho(\boldsymbol{x})$ in $\mathcal{K}_1$ so that all sentences

$$(\exists \boldsymbol{x})\,\rho(\boldsymbol{x})\ \&\ (\exists^{\rho}\boldsymbol{x})\,\varphi_i(\boldsymbol{x})\ \&$$

$$\&\ (\forall^{\rho}\boldsymbol{xy})\,(\varphi_i(\boldsymbol{x})\ \&\ \varphi_i(\boldsymbol{y}) \rightarrow \boldsymbol{x} \approx \boldsymbol{y})$$

belong to $\mathrm{T}(\mathcal{K}_1)$, and so that every sentence $\Phi \in \mathrm{T}(\mathcal{K}_2)$ becomes a member Φ^{**} of $\mathrm{T}(\mathcal{K}_1)$ when transformed as follows: each quantifier in Φ is replaced with a similar ρ-specialized quantifier, and each occurrence of $\boldsymbol{P}_i(\boldsymbol{x}_1, ..., \boldsymbol{x}_{n_i})$ is replaced with $\Theta_i(\boldsymbol{x}_1, ..., \boldsymbol{x}_{n_i})$; if the resulting sentence is $\Phi^*(\boldsymbol{a}_{j_1}, ..., \boldsymbol{a}_{jk})$, where $\boldsymbol{a}_1, ..., \boldsymbol{a}_{jk}$ are the individual constants, if any, appearing in Φ, then

$$\Phi^{**} = (\exists \boldsymbol{u}_1 \ldots \boldsymbol{u}_k)\,(\Phi^*(\boldsymbol{u}_1, ..., \boldsymbol{u}_k)\ \&\ \varphi_{j_1}(\boldsymbol{u}_1)\ \&\ \ldots\ \&\ \varphi_{jk}(\boldsymbol{u})) .$$

If for any sentence Φ with appropriate symbols, the sentence Φ^{**} resulting from the transformation above belongs to $\mathrm{T}(\mathcal{K}_1)$ only when the original Φ belongs to $\mathrm{T}(\mathcal{K}_2)$, then the interpretation is called *exact*.

From the point of view of the theory of model classes this means a (first-order) axiomatizable class $\mathcal{K}_2$ is intepretable in an axiomatizable class $\mathcal{K}_1$ iff either $\mathcal{K}_1$ is empty, or $\mathcal{K}_2$ is a nonempty subclass of an impoverished reduction of a formular enrichment of $\mathcal{K}_1$. The interpretation of $\mathcal{K}_2$ in $\mathcal{K}_1$ is exact iff $\mathcal{K}_2$ is actually such an impoverishment; if it is exact, then $\mathcal{K}_2$ is empty iff $\mathcal{K}_1$ is empty.

We see immediately that if the elementary theory of a class $\mathcal{K}_2$ is undecidable and exactly interpretable in the elementary theory of a class $\mathcal{K}_1$, then the elementary theory of $\mathcal{K}_1$ is undecidable, too. If a finitely axiomatizable class $\mathcal{K}_4$ is essentially undecidable and interpretable in a nonempty class $\mathcal{K}_3$, then $\mathcal{K}_3$ is undecidable and includes an essentially undecidable subclass that is finitely axiomatizable with respect to $\mathcal{K}_3$; thus, this holds for every superclass $\mathcal{K}' \supseteq \mathcal{K}_3$, as well (cf. [166]).

§7. The undecidability of sundry classes of metabelian groups

Let $\mathfrak{G}$ be an enriched metabelian group satisfying (A2)–(A4), and let $\mathfrak{R} = \tau(\mathfrak{G})$ be the corresponding ring with identity. Then the formulas from (5) and (6), together with the reducing formula $\rho(\boldsymbol{x}) = (\boldsymbol{u})(\boldsymbol{ux} \approx \boldsymbol{xu})$ and the definition of the identity element, determine an exact interpretation of $\mathfrak{R}$ in $\mathfrak{G}$, where we identify a model with the class consisting of just that model. These same formulas also give an interpretation of any class $\mathcal{R}$ of rings with identity in the corresponding class $\mathcal{G} = \sigma(\mathcal{R})$ of groups. If we take $\mathcal{R}$ to be finitely axiomatizable and essentially undecidable and consider any class $\mathcal{G}' \supseteq \mathcal{G}$ of enriched groups, we find that $\mathcal{G}'$ is undecidable and includes an essentially undecidable subclass, finitely axiomatizable inside $\mathcal{G}'$.

In order to get rid of the distinguished elements, we can avail ourselves of the following

Remark [166] : *If $\mathcal{K}$ is a class of models with distinguished elements $a_1, \ldots, a_m$, and $\mathcal{K}$ is axiomatized by the* FOPL *sentence $\Phi(\boldsymbol{a}_1, \ldots, \boldsymbol{a}_m)$, then its impoverishment $\mathcal{K}^*$ by $\boldsymbol{a}_1, \ldots, \boldsymbol{a}_m$, characterized by the sentence $(\exists \boldsymbol{u}_1 \ldots \boldsymbol{u}_m)\, \Phi(\boldsymbol{u}_1, \ldots, \boldsymbol{u}_m)$, is syntactically equivalent to $\mathcal{K}$.* ([3])

For $\Psi(\boldsymbol{a}_1, \ldots, \boldsymbol{a}_m) \in \mathrm{T}(\mathcal{K})$ iff $(\boldsymbol{u}_1 \ldots \boldsymbol{u}_m)(\Phi \to \Psi) \in \mathrm{T}(\mathcal{K}^*)$; the other way is obvious. ■

Corollary: *The class $\mathcal{M}$ of all metabelian groups includes a finitely axiomatizable (outright), essentially undecidable subclass; thus, $\mathcal{M}$ and every superclass $\mathcal{M}' \supseteq \mathcal{M}$ are undecidable.* ■

R.M. Robinson [139] has given an interpretation I of a finite system of axioms $\boldsymbol{Q}$ with an essentially undecidable theory in the ring of all polynomials in one variable with coefficients from the prime field of any given characteristic p. If $\mathcal{R}_p$ is the class of all char p rings, then viewing I as an interpretation of Q in a part of $\mathcal{R}_p$, we see that $\mathcal{R}_p$ includes a finitely axiomatizable subclass with an essentially undecidable theory.

In § 5 we saw that for $p \neq 2$, $\mathcal{R}_p$ corresponds to a class of metabelian p-groups. Consequently, *for* $p > 2$ *the class* $\mathcal{G}^p$ *of all groups satisfying the identity* $(\boldsymbol{x})(\boldsymbol{x}^p \approx \mathbf{1})$ *includes a finitely axiomatizable and essentially undecidable subclass; in particular, the elementary theory of* $\mathcal{G}^p$ *is undecidable.* ■

The group with generators a_m, b_m, c_m $(m = 1, 2, \ldots)$ and defining relations

$$\boldsymbol{b}_n\boldsymbol{a}_m \approx \boldsymbol{a}_m\boldsymbol{b}_n\boldsymbol{c}_{m+n}\,, \quad \boldsymbol{a}_m^p \approx \boldsymbol{b}_m^p \approx \boldsymbol{c}_m^p \approx \mathbf{1}\,,$$

$$\boldsymbol{a}_m\boldsymbol{c}_n \approx \boldsymbol{c}_n\boldsymbol{a}_m\,, \quad \boldsymbol{b}_m\boldsymbol{c}_n \approx \boldsymbol{c}_n\boldsymbol{b}_m \qquad (m, n = 1, 2, \ldots)$$

corresponds to the ring of polynomials in one variable over the prime field of characteristic p, and so for $p > 2$, this group is undecidable.

§ 8. Nilpotent groups

The undecidability of the elementary theory of the free metabelian group with two free generators was established in § 3. Actually, there is the more general

Theorem 5: *The elementary theory of the free metabelian group* $\mathfrak{G}_n$ *with* n *free generators is undecidable for* $n \geqslant 2$.

Let $a_1, \ldots, a_n$ be free generators of $\mathfrak{G}_n$. We introduce the formula

$$\rho(\boldsymbol{x}) = (\exists \boldsymbol{y})(a_1\boldsymbol{y} \approx \boldsymbol{y}a_1 \;\&\; \boldsymbol{x} \approx a_2\boldsymbol{y}a_2^{-1}\boldsymbol{y}^{-1} \;\&\; \boldsymbol{x}a_1 \approx a_1\boldsymbol{x}).$$

Since $a_1u = ua_1$ implies u is of the form $a_1^m z$, where z is central in $\mathfrak{G}_n$, the predicate defined by $\rho(\boldsymbol{x})$ is true just for elements of the form $(a_2a_1a_2^{-1}a_1^{-1})^m$; i.e., $\rho(\boldsymbol{x})$ defines in $\mathfrak{G}_n$ the center $\mathfrak{Z}$ of the free subgroup with generators a_1, a_2. Introducing the operations of addition and multiplication on $\mathfrak{Z}$ via the relations (5), (6), we obtain an interpretation of the ring of integers in $\mathfrak{G}_n$ (enriched). Therefore, the elementary theory of the group $\mathfrak{G}_n$ is undecidable. ■

Theorem 5 can be extended to free k-step nilpotent groups by using the following observation.

Suppose $\mathcal{K}$ is a class of models, and $\theta(\boldsymbol{x}, \boldsymbol{y})$ is a FOPL formula defining a formular predicate in each $\mathcal{K}$-model which is an equivalence relation – reflexive, symmetric, transitive. For each $\mathcal{K}$-model $\mathfrak{M}$, we form the factor model $\mathfrak{M}/\theta$ and let $\mathcal{K}_\theta$ be the class of all such factor models. Then *if the class* $\mathcal{K}$ *is decidable, so is* $\mathcal{K}_\theta$.

To prove this it suffices to note that every sentence of FOPL concerning $\mathfrak{M}/\theta$ is easily transformed into an equivalent sentence concerning $\mathfrak{M}$. ∎

Theorem 6: *The elementary theory of the free k-step nilpotent group of rank n is undecidable for* $n \geqslant 2, k \geqslant 2$.

For $k = 2$ Theorem 6 coincides with Theorem 5. Suppose the theorem is true for $k = s-1$ $(s > 2)$ and $\mathfrak{G}$ is the free s-step nilpotent group with n free generators $(n \geqslant 2)$. Consider the formula

$$\theta(\boldsymbol{x}, \boldsymbol{y}) = (\exists \boldsymbol{z})((\boldsymbol{u})(\boldsymbol{ux} \approx \boldsymbol{xu}) \ \&\ \boldsymbol{x} \approx \boldsymbol{yz}) .$$

The factor model $\mathfrak{G}/\theta$ is the factor group of $\mathfrak{G}$ by its center, i.e., it is a free $(s-1)$-step nilpotent group of rank n, which is undecidable by supposition. By the remark above this implies $\mathfrak{G}$ itself has an undecidable elementary theory. ∎

NOTES

([1]) A group is *metabelian* iff it is 2-step nilpotent (iff its commutator subgroup is included in its center).

([2]) This is based on (A1), not (A2). In fact, (A2) is not required in the constructions in this section; it is a consequence of (A3) and (A5).

([3]) Therefore, the conclusion of the last sentence of § 6 remains true if $\mathcal{K}_4$ is interpretable in some inessential enrichment of $\mathcal{K}_3$.

CHAPTER 16

THE UNDECIDABILITY OF THE ELEMENTARY THEORIES OF CERTAIN FIELDS

Let $\mathcal{K}$ be a class of models with fundamental predicate symbols $\boldsymbol{P}_\gamma(\boldsymbol{x}_1, ..., \boldsymbol{x}_{n_\gamma})$ $(\gamma \in \Gamma)$ and individual constants $\boldsymbol{a}_\delta$ $(\delta \in \Delta)$. These constitute the *signature* $\Sigma(\mathcal{K})$ of the class $\mathcal{K}$.

A relation $P(x_1, ..., x_n)$, defined in every $\mathcal{K}$-model in some arbitrary fashion, is called *formular* (or *elementary, arithmetic*) iff there is a formula $\Phi(\boldsymbol{x}_1, ..., \boldsymbol{x}_n)$ of first-order predicate logic (FOPL) with free variables $\boldsymbol{x}_1, ..., \boldsymbol{x}_n$ and extralogical symbols only from among the $\boldsymbol{P}_\gamma, \boldsymbol{a}_\delta$ such that in every $\mathcal{K}$-model $\mathfrak{M}$, each sequence $\langle u_1, ..., u_n \rangle$ of elements of (the base set of) $\mathfrak{M}$ satisfies Φ iff $P(u_1, ..., u_n)$ is true.

The *elementary theory* of the class $\mathcal{K}$ is the collection $\mathbf{T}(\mathcal{K})$ of all closed FOPL formulas (sentences) of signature $\Sigma(\mathcal{K})$ that are true in every $\mathcal{K}$-model. $\mathbf{T}(\mathcal{K})$ is said to be *(recursively) decidable* iff there is an algorithm for deciding for every sentence of the above form the question of whether it is true in all $\mathcal{K}$-models.

In case $\mathcal{K}$ is a class of fields, we take the basic predicates to be the relations of being the sum or being the product of two elements of the base of the field. If in the class consisting of an individual field of characteristic zero the property of belonging to the prime subfield is formular, then the elementary theory of this field is undecidable (see [166]).

In 1949 J. Robinson [134] showed that in the field of rational numbers the property of being a natural number is formular. In the more recent paper [136] she has extended this result to all algebraic fields of finite degree over the rationals. By the same token she has proved the undecidability of their elementary theories.

One may conjecture that fields of rational functions in one or several independent variables have undecidable elementary theories, and that so do fields of formal power series, at least over fields with undecidable theories. In the present article the second of these conjectures is proved under a certain additional limitation on the base field. Regarding the first conjecture, here we show only that the field of rational functions in one variable with coefficients in a real closed field has an undecidable theory ([1]). These results are obtained by means of very rudimentary algebraic facts; this may be of methodological interest, inasmuch as the cited results of J. Robinson were proved with the help of rather subtle theorems of the theory of algebraic numbers lying beyond the pale of ordinary college courses in abstract algebra.

§1. The field of rational functions

Let $\mathfrak{R}$ be a real closed field. A relation $x \leqslant y$ of order (unique) is defined in $\mathfrak{R}$ by the formula

$$(\exists \boldsymbol{v})\,(\boldsymbol{v}^2 \approx \boldsymbol{y} - \boldsymbol{x}); \tag{1}$$

therefore, we can view $\mathfrak{R}$ as an ordered field in what follows.

We let $\mathfrak{R}(x)$ be the field of rational functions in the variable x with coefficients in $\mathfrak{R}$, and $\mathfrak{R}[x]$ the ring of polynomials in x with coefficients in $\mathfrak{R}$. We shall need the following well-known

Lemma 1: *If in the ring $\mathfrak{R}[x]$ nonzero polynomials u, v, w have the relation*

$$u^4 + v^4 = w^4 ,$$

then u is a constant. ∎

This immediately implies

Lemma 2: *In the field $\mathfrak{R}(x)$ the property of being a constant is formular: the formula*

$$\mathrm{Con}(\boldsymbol{u}) = (\exists \boldsymbol{v})\,(1 + \boldsymbol{u}^4 \approx \boldsymbol{v}^4) \tag{2}$$

works. ∎

Therefore, in $\mathfrak{R}(x)$ the formula

$$\mathrm{Pos}(\boldsymbol{u}) = \mathrm{Con}(\boldsymbol{u})\ \&\ \boldsymbol{u} \not\approx \mathbf{0}\ \&\ (\exists \boldsymbol{v})\,(\boldsymbol{u} \approx \boldsymbol{v}^2) \tag{3}$$

defines the property of u being a positive constant of $\mathfrak{R}(x)$.

An element $u \in \mathfrak{R}(x)$ is called a *function without real poles* iff it can be represented in the form of a quotient of two polynomials from $\mathfrak{R}[x]$ with the divisor having no roots in $\mathfrak{R}$. We let $\mathfrak{A}$ be the subring of $\mathfrak{R}(x)$ consisting of all the functions in $\mathfrak{R}(x)$ without real poles. The formulas (2), (3), used to define the notion of positive constant, continue to work inside $\mathfrak{A}$. In $\mathfrak{A}$ we introduce the predicate of divisibility via the formula

$$u \mid v = (\exists w)(v \approx uw) .$$

Let $N(u)$ be the unary predicate: the element $u \in \mathfrak{A}$ is a natural number, i.e., is a positive natural multiple of the identity element in $\mathfrak{A}$.

Lemma 3: *The property of being natural is formular in* $\mathfrak{A}$; *indeed,* $N(u)$ *is defined in* $\mathfrak{A}$ *by the* FOPL *formula*

$$\mathrm{Nat}(u) = \mathrm{Pos}(u) \,\&\, (vw)(v+1 \mid w \;\&\; (z)(\mathrm{Pos}(z) \;\&$$

$$\& \; v+z \mid w \to v+z+1 \mid w) \to v+u \mid w) .$$

For suppose u is a natural element in $\mathfrak{A}$. Then $\mathrm{Nat}(u)$ is true by the usual principle of complete induction, which in fact is expressed by $\mathrm{Nat}(u)$ for the predicate of being divisible by $v+u$. It remains to prove that $\mathrm{Nat}(u)$ is false when u is not natural. In other words, we have to show that for every positive nonnatural constant $u \in \mathfrak{A}$, there are elements $v, w \in \mathfrak{A}$ for which

$$v+1 \mid w \;\&\; (\forall z \in \mathfrak{A})(\mathrm{Pos}(z) \;\&\; v+z \mid w \Rightarrow v+z+1 \mid w) \;\&\; v+u \nmid w \qquad (4)$$

is true. We shall show that the elements

$$v = x^2 - u - 1 , \quad w = (x^2 - u)(x^2 - u + 1) \ldots (x^2 - u + [u])$$

satisfy (4). ([2]) That $v+1 \mid w$ and $v+u \nmid w$ is obvious. Let z be a positive constant in $\mathfrak{A}$ such that $v+z \mid w$. Two cases are possible:

(i) $z > u$. The function $v+z+1 = x^2 + (z-u)$ has no real roots (i.e., in $\mathfrak{R}$), so $v+z+1$ divides w in $\mathfrak{A}$.

(ii) $z \leqslant u$. The condition $v+z \mid w$ requires that the roots $\pm\sqrt{u+1-z}$ of the polynomial $v+z$ be roots of w, i.e., that $z = k+1$, where k is a rational integer, $0 \leqslant k \leqslant [u]$. As $z \leqslant u$, so $k < [u]$, and the roots of the polynomial $v+z+1$, namely $\pm\sqrt{u-z}$, are both also roots of w; thus $v+z+1$ is a divisor of w in $\mathfrak{A}$. Therefore, these elements v, w satisfy (4). ■

Theorem 1: *In* $\mathfrak{R}(x)$ *the property of being a natural number is formular.*

In view of Lemma 3 it suffices to show that in $\mathfrak{R}(x)$ the property of not having real poles is formular. For this we need the well-known fact,

Lemma 4: *Let* $f(x) \in \mathfrak{R}(x)$. *Then* $f(a) \geqslant 0$ *for all* $a \in \mathfrak{R}$ *iff* $f(x) = v^2 + w^2$ *for some* $v, w \in \mathfrak{R}(x)$.

In other words, the condition "$f(a) \geqslant 0$ for all $a \in \mathfrak{R}$" is defined in $\mathfrak{R}(x)$ by the formula

$$\mathrm{E}(\boldsymbol{f}) = (\exists \boldsymbol{vw})(\boldsymbol{f} \approx \boldsymbol{v}^2 + \boldsymbol{w}^2). \tag{5}$$

To make the presentation complete we produce a short proof of Lemma 4. The function $f(x)$ can obviously be represented in $\mathfrak{R}(x)$ as $h(x)g(x)/w^2$, where $h(x)$ in $\mathfrak{R}[x]$ splits over $\mathfrak{R}$ into linear factors, while $g(x)$ has no real roots and, therefore, is a product of factors of the form $(x-a)^2 + b^2$ $(a, b \in \mathfrak{R})$. Since for all $c \in \mathfrak{R}$ the value $h(c)$ is nonnegative, every root of $h(x)$ must have even multiplicity; so $h(x) = z^2$ for some $z \in \mathfrak{R}[x]$. On the other hand, by applying the identity

$$(\alpha^2 + \beta^2)(\gamma^2 + \delta^2) = (\alpha\gamma - \beta\delta)^2 + (\alpha\delta + \beta\gamma)^2$$

several times, we can write $g(x)$ in the form $u^2 + v^2$ for some $u, v \in \mathfrak{R}[x]$; thus, $f(x)$ assumes the desired form:

$$f(x) = \left(\frac{uz}{w}\right)^2 + \left(\frac{vz}{w}\right)^2 \quad \blacktriangle$$

We introduce the following formulas ($\mathrm{E}(\boldsymbol{u})$ is defined in (5)):

$$\mathrm{D}(\boldsymbol{u}) = (\exists \boldsymbol{v})(\mathrm{Pos}(\boldsymbol{v}) \mathbin{\&} \mathrm{E}(\boldsymbol{v} - \boldsymbol{u}^2)),$$

$$\mathrm{C}(\boldsymbol{u}) = \mathrm{D}(\boldsymbol{u}) \mathbin{\&} (\boldsymbol{v})(\mathrm{Con}(\boldsymbol{v}) \to \neg \mathrm{D}(\boldsymbol{u}(\boldsymbol{x} - \boldsymbol{v})^{-1})),$$

$$\mathrm{B}(\boldsymbol{u}) = \boldsymbol{u} \not\approx \boldsymbol{0} \mathbin{\&} (\mathrm{C}(\boldsymbol{u}) \vee \mathrm{C}(\boldsymbol{u}^{-1})),$$

$$\mathrm{A}(\boldsymbol{u}) = (\exists \boldsymbol{vw})(\mathrm{D}(\boldsymbol{v}) \mathbin{\&} \mathrm{B}(\boldsymbol{w}) \mathbin{\&} \boldsymbol{u} \approx \boldsymbol{vw}).$$

For $f \in \mathfrak{R}(x)$ the truth of $\mathrm{D}(f)$ is equivalent to the boundedness of f, i.e., to f having no real poles and nonpositive degree. $\mathrm{C}(f)$ holds iff f has no real poles or roots and has nonpositive degree; thus, $\mathrm{B}(f)$ is true iff f has no real poles or roots. Finally, we see that the truth of $\mathrm{A}(f)$ is equivalent to f having no real poles. Therefore, the property of not having real poles is formular in $\mathfrak{R}(x)$. ■

We note that the symbol $\boldsymbol{x}$ explicitly occurs in the formula $\mathrm{C}(u)$. ([3])

Consequently, Theorem 1 has been proved under the assumption that the fundamental concepts of the field $\mathfrak{R}(x)$ (as reflected in its signature) are the operations of addition and multiplication and the distinguished element x, the basic transcendental element in $\mathfrak{R}(x)$.

If the notion of being a natural number is formular in a field or ring, then, as is well known [136], the elementary theory of the ring is undecidable. Furthermore, the presence of individual constants in the signature of the theory is not essential, and they can be dropped. Thus Theorem 1 implies the

Corollary: *The elementary theory of the field $\mathfrak{R}(x)$ with signature $\langle +, \cdot \rangle$ is not recursively decidable.* ■

§2. Fields of formal power series

Suppose $\mathfrak{F}$ is a field, and x an arbitrary object not lying in $\mathfrak{F}$. Then $\mathfrak{F}\{x\}$ denotes the field of formal power series in x of the form $u = 0$ or

$$u = u_k x^k + u_{k+1} x^{k+1} + u_{k+2} x^{k+2} + \dots$$

$$(k = 0, 1, 2, \dots;\ u_i \in \mathfrak{F},\ u_k \neq 0) \tag{6}$$

relative to the usual operations of addition and multiplication. Further we put $\mathfrak{F}_0 = \mathfrak{F}$, $\mathfrak{F}_1 = \mathfrak{F}_0\{x_1\}$, $\mathfrak{F}_2 = \mathfrak{F}_1\{x_2\}$, etc. For each nonzero element $u \in \mathfrak{F}_m = \mathfrak{F}\{x_1, \dots, x_m\}$, the number k appearing in the expression (6) of u with respect to $\mathfrak{F}_{m-1}$ is called the *order* of u (relative to $\mathfrak{F}_{m-1}$). The element 0 is considered to have positive order.

We introduce the formula

$$\mathrm{I}_n(\boldsymbol{u}) = \neg(\exists \boldsymbol{v})(1 + \boldsymbol{u}^n \approx \boldsymbol{v}^n) \qquad (n > 0)\,.$$

For the present we shall assume the characteristic of $\mathfrak{F}$ is zero. If $u \in \mathfrak{F}_m$ has positive order, then for $n > 0$ the common binomial expansion formula

$$(1 + u^n)^{1/n} = 1 + \frac{1}{n} u^n + \frac{1-n}{2!\,n^2} u^{2n} + \dots \tag{7}$$

shows $\mathrm{I}_n(u)$ is false. Similarly, if u has negative order, then by setting

$$1 + u^* = u_k^{-1} x^l u\,, \quad w = u_k^{-1} x^l v\,, \quad 1 + z = (u_k^{-1} x^l)^n + (1 + u^*)^n\,,$$

where in (6) for u we have $k = -l,\ l > 0, x = x_m, u_i \in \mathfrak{F}_{m-1}$ $(i = k, k+1, \ldots)$, we convert the equation $1 + u^n = v^n$ into the equivalent one $1 + z = w^n$. Since the order of z is positive, putting

$$w = 1 + \frac{1}{n} z + \frac{1-n}{2!n^2} z^2 + \ldots \tag{8}$$

gives a solution to the latter equation, and so, $\mathrm{I}_n(u)$ is again false.

Finally, if u has order 0, then

$$u = u_0 + u_1 x_m + u_2 x_m^2 + \ldots \qquad (u_i \in \mathfrak{F}_{m-1},\ u_0 \neq 0)\,.$$

Now if the equation $1 + u^n = v^n$ has a solution

$$v = v_0 + v_1 x_m + v_2 x_m^2 + \ldots \qquad (v_i \in \mathfrak{F}_{m-1}) \tag{9}$$

in $\mathfrak{F}_m$, then $1 + u_0^n = v_0^n$. Conversely, a computation similar to the ones above shows that for $v_0 \in \mathfrak{F}_{m-1}$, if we have $1 + u_0^n = v_0^n$, then there is a solution to $1 + u^n = v^n$ of the form (9), at least in case $v_0 \neq 0$ ([4]). Thus, if $\mathrm{I}_n(u_0)$ is true in $\mathfrak{F}_{m-1}$, then $\mathrm{I}_n(u)$ holds in $\mathfrak{F}_m$.

By applying this reasoning to the fields $\mathfrak{F}_{m-1}, \ldots, \mathfrak{F}_1$ successively, we find that for any $u \in \mathfrak{F}_{m-1}$, if u can be decomposed in the following fashion:

$$\left.\begin{aligned} u &= u_0^{(0)} + u_1^{(0)} x_m + \ldots\,, \\ u_0^{(0)} &= u_0^{(1)} + u_1^{(1)} x_{m-1} + \ldots\,, \\ &\ldots \\ u_0^{(m-2)} &= u_0^{(m-1)} + u_1^{(m-1)} x_1 + \ldots\,, \end{aligned}\right\} \tag{10}$$

where $u_i^{(j)} \in \mathfrak{F}_{m-j-1}$, and if $\mathrm{I}_n(u_0^{(m-1)})$ is true in $\mathfrak{F}$, then $\mathrm{I}_n(u)$ is true in $\mathfrak{F}_m$. We note that the maps $u \to u_0^{(j)}$ are homomorphisms: for any $v, w \in \mathfrak{F}_m$ which have expansions of the form (10), i.e., have nonnegative orders,

$$(v+w)_0^{(j)} = v_0^{(j)} + w_0^{(j)}\,, \quad (vw)_0^{(j)} = v_0^{(j)} w_0^{(j)}$$

$$(j = 0, 1, \ldots, m-1)\,. \tag{11}$$

Now let's suppose the field $\mathfrak{F}$ satisfies the following condition: there are natural numbers r, n such that every $a \in \mathfrak{F}$ can be written as the sum $a = a_1 + \ldots + a_r$ of r elements a_i for which none of the equations $1 + a_i^n = b_i^n$ has a solution in $\mathfrak{F}$ $(i = 1, \ldots, r)$.

With this stipulation we see that the formula

$$\mathrm{H}(\boldsymbol{u}) = \mathrm{H}_{r,n}(\boldsymbol{u})$$

$$= (\exists \boldsymbol{u}_1, \ldots, \boldsymbol{u}_r)(\mathrm{I}_n(\boldsymbol{u}_1) \,\&\, \ldots \,\&\, \mathrm{I}_n(\boldsymbol{u}_r) \,\&\, \boldsymbol{u} \approx \boldsymbol{u}_1 + \ldots + \boldsymbol{u}_r)$$

is true in $\mathfrak{F}_m$ just for those elements having nonnegative orders.

Keeping (11) in mind, we see that the subset of $\mathfrak{F}_m$ consisting of elements satisfying $\mathrm{H}(\boldsymbol{u})$ forms a subring $\mathfrak{H}$. The elements of $\mathfrak{H}$ having inverses in $\mathfrak{H}$ are characterized in $\mathfrak{F}_m$ by the formula

$$\mathrm{G}(\boldsymbol{u}) = \mathrm{H}(\boldsymbol{u}) \,\&\, (\exists \boldsymbol{v})(\mathrm{H}(\boldsymbol{v}) \,\&\, \boldsymbol{uv} \approx \mathbf{1})\,.$$

We find from a quick calculation and (11) that an element $u \in \mathfrak{H}$ is invertible in $\mathfrak{H}$ iff $u_0^{(m-1)} \neq 0$ in (9). Therefore, the formula $\neg\mathrm{G}(\boldsymbol{u}) \,\&\, \mathrm{H}(\boldsymbol{u})$ defines in $\mathfrak{F}_m$ the set elements $u \in \mathfrak{H}$ which have $u_0^{(m-1)} = 0$ in (9). This subset of $\mathfrak{H}$ forms an ideal $\mathfrak{J}$. Two elements $u, v \in \mathfrak{H}$ are congruent *mod* $\mathfrak{J}$ iff they have the same constant term. Accordingly the factor ring $\mathfrak{H}/\mathfrak{J}$ is isomorphic to the field $\mathfrak{F}$.

Till now we have been supposing $\mathfrak{F}$ has characteristic 0. We needed this to make sense of the binomial expansions (7), (8). If, however, the binomial expansion is rewritten in the form

$$(1 + w)^{1/n} = 1 + \frac{1}{n} w + \frac{q_3}{n^{2^2}} w^2 + \frac{q_3}{n^{3^3}} w^3 + \ldots\,,$$

it is easy to see that the coefficients $q_2, q_3, \ldots$ are integers, and so, it is all right for fields of prime characteristic $p > 0$, as long as p and n are relatively prime.

Let $\mathfrak{F}$ have characteristic p, and suppose for some natural numbers r, n the sentence $(\boldsymbol{u})\mathrm{H}_{r,n}(\boldsymbol{u})$ is valid in $\mathfrak{F}$. If $n = p^e n_0$ with $(n_0, p) = 1$, then any relation $1 + a^n = b^n$ is equivalent to $1 + a^{n_0} = b^{n_0}$. So $(\boldsymbol{u})\mathrm{H}_{r,n_0}(\boldsymbol{u})$ is valid in $\mathfrak{F}$, and we can take n_0 in place of n, bringing us to the case when p and n are relatively prime, and all our arguments valid. We immediately obtain

Theorem 2: *Suppose the field $\mathfrak{F}$ has an undecidable elementary theory and for some natural numbers r, n satisfies the sentence $(\boldsymbol{u})\mathrm{H}_{r,n}(\boldsymbol{u})$. Then*

the elementary theory of each field $\mathfrak{F}_m = \mathfrak{F}\{x_1, ..., x_m\}$ $(m = 1, 2, ...)$ *of formal power series is not recursively decidable.*

If, on the contrary, $\mathfrak{F}_m$ were to have a decidable theory, then so would the ring $\mathfrak{H}$, which is defined by a FOPL formula in $\mathfrak{F}_m$. The ideal $\mathfrak{J}$ in $\mathfrak{H}$ is also given by a predicate formular in $\mathfrak{F}_m$. Like $\mathfrak{H}$, the factor ring $\mathfrak{H}/\mathfrak{J}$ would have a decidable theory, which is the same as that of $\mathfrak{F}$, contradicting the hypothesis. ■

The conditions laid on the field $\mathfrak{F}$ in Theorem 2 are seen to hold for the field of rational numbers and the field of rational functions in one variable over an arbitrary real closed field. For both field have undecidable theories. In the first the equation $1 + u^4 = v^4$ is not solvable for $u \neq 0$, while in the second this equation has no solution if u is not constant. Consequently, in both fields every element can be represented as the sum $a_1 + a_2$ of elements a_1, a_2 such that the equations

$$1 + a_1^4 = b_1^4, \quad 1 + a_2^4 = b_2^4,$$

have no solutions. We have proved the

Corollary: *The elementary theory of each field* $\mathfrak{F}\{x_1, ..., x_m\}$ *is undecidable in case* $\mathfrak{F}$ *is either the field of rational numbers or the field of rational functions in one variable over some real closed field.* ■

The supply of fields $\mathfrak{F}\{x_1, ..., x_m\}$ in Theorem 2 can be augmented as follows. We consider a system $\{x_\nu : \nu \in \mathrm{N}\}$ of distinct variables indexed by a linearly ordered set N. The fields of the form

$$\mathfrak{F}\{x_{\nu_1}, x_{\nu_2}, ..., x_{\nu_s}\} \quad (\nu_1 < \nu_2 < ... < \nu_s) \tag{12}$$

compose a local system in the sense that any two of them are included in a third member of the system. Therefore, one may speak of the union of all the fields (12). This union will itself be a field $\mathfrak{F}_\mathrm{N}$. Every element $u \in \mathfrak{F}_\mathrm{N}$ belongs to some subfield $\mathfrak{F}' = \mathfrak{F}\{x_{\nu_1}, ..., x_{\nu_m}\}$, and if the equation $1 + u^n = v^n$ is solvable for this u, then the solution also belongs to $\mathfrak{F}'$. The arguments leading us to Theorem 2 are thus valid for any field of the sort $\mathfrak{F}_\mathrm{N}$. In particular, if the base field $\mathfrak{F}$ is the field of rational numbers or the field of rational functions in one variable over a real closed field, then the elementary theory of the field $\mathfrak{F}_\mathrm{N}$ of power series is not recursively decidable.

NOTES

(1) Undecidability was subsequently established for the elementary theory of any pure transcendental extension of: (1) a real closed field (A. Tarski [165]), and (2) an arbitrary formally real field (R.M. Robinson [140]); he also gives for the first time a proof dating from 1950 for (3) the field of rational numbers. Yu.L. Eršov [30] has proved the undecidability of the theory of the simple transcendental extension of any finite field with characteristic $p \neq 2$. Concerning the second conjecture, he has shown that every field of power series in a finite number of variables has an undecidable theory if the base field does. The reader is also referred to J. Robinson's survey [137] on the decision problem for fields.

(2) The largest integer $[u]$ not greater than u is defined for all positive constants $u \in \mathfrak{R}(x)$ only if the order on $\mathfrak{R}$ is archimedean. In fact, as A. Tarski has pointed out in conversations, his results [162] show that if $\mathfrak{R}$ is nonarchimedean ordered, then the subset Nat($\boldsymbol{u}$) defines in $\mathfrak{A}$ contains every "infinite" positive constant. The formula

$$\mathrm{Nat}^*(\boldsymbol{u}) = \mathrm{Nat}(\boldsymbol{u})\ \&\ (\exists \boldsymbol{v}\boldsymbol{w})(\boldsymbol{z})\ (\mathrm{Pos}(\boldsymbol{z}) \rightarrow (\boldsymbol{v}+\boldsymbol{z} \mid \boldsymbol{w} \leftrightarrow \mathrm{Nat}(\boldsymbol{z})\ \&\ \boldsymbol{z} \leqslant \boldsymbol{u}))$$

can be substituted for Nat($\boldsymbol{u}$) in nonarchimedean cases (here, $\boldsymbol{z} \leqslant \boldsymbol{u}$ is the formula given by (1)). This is a version of Tarski's formula in [R4] adapted to the present framework (cf. [XVII, Note 1]). The error and its correction are also discussed in [137, p. 307].

(3) See the Remark (Chapter 17) following this article.

(4) In the original the author seems to claim, wrongly, that this also holds for $v_0 = 0$. Either way, it has no bearing on the main discussion. Cf. [R4].

CHAPTER 17

A REMARK CONCERNING "THE UNDECIDABILITY OF THE ELEMENTARY THEORIES OF CERTAIN FIELDS" [XVI]

Theorem 1 of the preceding article asserts that in the field $\mathfrak{R}(x)$ of rational functions in one variable over a real closed field $\mathfrak{R}$, the property of being a natural number is formular (elementary) if we take the signature to be $\langle +, \cdot ; \boldsymbol{x} \rangle$. With a few insignificant changes, however, we can get the stronger result that this property is formular in $\mathfrak{R}(x)$ even when the signature consists of $+$ and $\cdot$ only.

We take $\mathfrak{A}$ in Lemma 3 to be the subring of all bounded functions and replace the elements v, w in the proof of that lemma with

$$v = (x^2 + (u+1)^{-1})^{-1} - u - 1 \ ,$$

$$w = (v+1)(v+2) \dots (v+1+[u]) \ .$$

Since the formula $\mathrm{D}(\boldsymbol{u})$ – in which the symbol $\boldsymbol{x}$ does not occur – already defines the new subring $\mathfrak{A}$ in $\mathfrak{R}(x)$, the formulas $\mathrm{C}(\boldsymbol{u})$, $\mathrm{B}(\boldsymbol{u})$, $\mathrm{A}(\boldsymbol{u})$ are unnecessary; thus the stronger statement is readily obtained ([1]). In particular, no longer needed is the remark on the eliminability of distinguished elements made without proof at the end of §1.

NOTE

([1]) As in the preceding chapter (cf. Note 2 there), the formula $\mathrm{Nat}(\boldsymbol{u})$ only works when $\mathfrak{R}$ is archimedean-ordered. On the other hand, by replacing $\boldsymbol{y} \mid \boldsymbol{z}$ with $(\exists \boldsymbol{w})(\mathrm{D}(\boldsymbol{w})\ \&\ \boldsymbol{z} \approx \boldsymbol{wy})$ throughout the formula $\mathrm{Nat}^*(\boldsymbol{u})$, we obtain Tarski's formula [R4] involving only $+$, $\cdot$ and defining the set of natural numbers outright in the simple transcendental extension of any real closed field.

CHAPTER 18

CONSTRUCTIVE ALGEBRAS. I

Contents

page

§1. Algebraic systems
- 1.1. Functions, operations, predicates 151
- 1.2. Generating sets. Terms 153
- 1.3. Primitive and quasiprimitive classes 157
- 1.4. Defining relations 159
- 1.5. Algebras of recursive functions 161

§2. Numbered sets
- 2.1. Mappings of numbered sets 165
- 2.2. Unireducibility for numberings 171
- 2.3. Equivalent numberings 178

§3. Numbered algebraic systems
- 3.1. R-numberings of algebraic systems 187
- 3.2. Subsystems 192
- 3.3. Homomorphisms and congruence relations 196

§4. Finitely generated algebras
- 4.1. General finitely generated algebras 201
- 4.2. Finitely presented algebras 206

Introduction

The concept of an algebraic system is one of the most common central notions of contemporary algebra. One of its more important special cases is the concept of an algebra as a system $\mathfrak{A}$ consisting of an arbitrary set A and a finite sequence of operations $f_i(x_1, \ldots, x_{r_i})$ $(i = 1, \ldots, l)$ defined on A. Therefore, in considering matters of general algebra concerning the idea of constructiveness, it is natural to ask: which algebras are to be regarded as constructive? The answer is more or less clear: an algebra $\mathfrak{A}$ is constructive iff its base set A consists of constructive elements and is itself constructively prescribed, while the basic operations f_i are also constructive. A certain indeterminacy, however, is hidden in this. For obviously the notion of constructiveness must be made precise, and this precision can be gained by widely differing methods, ranging from the classical Gödel–Church–Kleene approach to the newer ones of A.A. Markov [102], A.N. Kolmogorov [77], and a number of others. The notion of constructive algebra admits a corresponding series of refinements.

Constructive algebraic systems are encountered in a number of works, including A. Mostowski [104, 106] and A.V. Kuznecov [83, 84], and especially A. Fröhlich and J.C. Shepherdson [41]. In the first papers mentioned, the base A is a set of natural numbers, and in the remainder, a set of words over some finite alphabet.

A second approach to this problem is laid out by A.N. Kolmogorov, V.A. Uspenskiĭ [173, 174], H.G. Rice [126], and H. Rogers [141]. Well known is the important role partial recursive functions and their standard (Gödel or, as here, Kleene) numbering play in the theory of algorithms. A.N. Kolmogorov was apparently the first to note that the Kleene numbering, which is fundamental to the theory of computable operations, is not at all unique as regards the construction of this theory, and that it would be interesting to study in some manner all the numberings of the collection of partial recursive functions, or all of those with a given property. With varying emphasis this program was realized in the articles cited, and among these quite clearly in Rogers' paper.

In the present survey we attempt to unite the two approaches by introducing the concept of a *numbered algebra.*

An algebra $\mathfrak{A}$ is said to be *numbered* iff it is accompanied by a map α from some set D_α of natural numbers onto the base A of the algebra $\mathfrak{A}$. The map α is called a *numbering* of $\mathfrak{A}$; it may or may not be 1–1. A function $F_i(x_1, ..., x_{r_i})$ defined on the natural numbers *represents* the operation f_i of the algebra $\mathfrak{A}$ relative to the numbering α iff

$$f_i(\alpha x_1, ..., \alpha x_{r_i}) = \alpha F_i(x_1, ..., x_{r_i}) \qquad (x_1, ..., x_{r_i} \in D_\alpha),$$

i.e., iff when we apply F_i to numbers $x_1, ..., x_{r_i}$ of given elements of A, we get a number for the result of performing the operation f_i on the given elements.

The numbering α can be viewed as a sort of coordinatization of the algebra $\mathfrak{A}$. In a certain sense a numbered algebra is the analogue of an analytic manifold, a Lie group, etc. This analogy has influenced to some extent the choice of problems considered in this article. The situation in the theory of numbered algebras is singular, however, for the notion of isomorphism of numbered algebras is supplanted by that of their recursive equivalence as the fundamental relation in this theory; this was noted by Fröhlich and Shepherdson [41] with regard to the theory of fields.

On the basis of the character of the set D_α, the partition of this set into classes corresponding to the map α, and the functions F_i representing the basic operations of $\mathfrak{A}$ in the numbering α, we introduce in §3 several variations on the concept of a constructive numbered algebra. Since constructiveness is defined for numbered algebras, it becomes necessary to prove its invariance under renumberings, homomorphisms, etc.

Part of these questions can be formulated not only for numbered algebras, but also for arbitrary numbered sets. It is thus appropriate as a preliminary to examine general numbered sets; this is done in §2. In particular, we show that a series of theorems proved by J. Myhill [111], A.A. Mučnik [109], and H. Rogers [141], concerning numberings of partial recursive functions and reducibility of sets, hold in fact for numberings of arbitrary sets, as well.

In §4 we investigate constructive numberings of finitely generated algebras and of algebras with a finite number of defining relations. We limit ourselves to establishing only general facts regarding isomorphism or equivalence of the possible constructive numberings of the indicated algebras. Also given here are proofs of reformulated theorems of A.V. Kuznecov and J.C.C. McKinsey, as well as of a new theorem on the existence of constructive numberings for algebras satisfying certain additional conditions.

For terminological unity we include in §1 an exposition of familiar notions from the general theory of algebras.

As already mentioned, the purpose of this part of the survey is to elaborate a system of ideas required for a general theory of constructive algebras; only a very limited number of concrete results on particular problems appear here. The author hopes to correct this shortcoming in a second part to this survey, where results of an algorithmic or constructive character concerning groups, rings, fields and, in a different vein, deductive algebras will be presented as systematically as possible. ([1])

The present article is an expansion of the report read by the author on September 20, 1960 at the All-Union Colloquium on General Algebra in Sverdlovsk. Individual results from this survey were presented at a meeting of the Ivanovo Mathematical Society held in June, 1960, and in lectures given in October at the University of Kazahstan in Alma-Ata.

§1. Algebraic systems

§1.1. *Functions, operations, predicates*

Let A, B be two arbitrary nonempty sets. A rule f that associates with every n-sequence $\langle a_1, ..., a_n\rangle$ of elements of A, a unique well-defined element $f(a_1, ..., a_n)$ of B is called a (*total*) *n-ary* or *n-place function* defined on A and taking values in B. If for such a correspondence we relax the condition that every ordered n-tuple $\langle a_1, ..., a_n\rangle$ $(a_i \in A)$ be associated with some element of B, then the function is said to be *partial* (not necessarily total) on A. The set of all such n-tuples for which the partial function f is defined is called the *domain* (of existence or definition) of the function f. The subset of B consisting of all values actually assumed by f is called the *range* of f. A unary partial function g is often said to *map* a set M into a set N in case M is included in the domain of g and N contains every value taken by g on M.

Two partial functions are said to be *equal* iff they have the same domain of definition, and their values are equal for every n-tuple in their domain.

A partial n-ary function f, defined on a set A, whose values belong to the same set A is called an *n-ary partial operation* on A. A partial operation defined for all arguments from A is called a (*total*) *operation* on A. An n-ary function $P(x_1, ..., x_n)$ defined on the set A and taking values in some fixed two-element set, whose elements will be denoted by T, F, is called an *n-ary predicate* on A. If for elements $a_1, ..., a_n$ in A we find $P(a_1, ..., a_n) = \mathsf{T}$, we say the predicate P is *true* for $\langle a_1, ..., a_n\rangle$, or that $a_1, ..., a_n$ are in the relation P. If, however, $P(a_1, ..., a_n) = \mathsf{F}$, then we say P is *false* for $\langle a_1, ..., a_n\rangle$. It will be convenient to assume $\mathsf{T} = 1$, $\mathsf{F} = 0$; thus a predicate becomes a function with values among the natural numbers. ([2])

For formal reasons it is sometimes desirable to consider 0-ary functions and predicates, as well, with the understanding that these are simply individual elements of the set of possible values.

A set A together with an arbitrarily prescribed finite sequence $\langle a_1, ..., a_l\rangle$ of its own elements and finite sequences $\langle f_1, ..., f_m\rangle$ of total operations, $\langle g_1, ..., g_n\rangle$ of partial operations, and $\langle P_1, ..., P_p\rangle$ of predicates defined on A is called an *algebraic system.*

The sequence

$$\langle 0, ..., 0; r_1, ..., r_m; s_1, ..., s_n; t_1, ..., t_p\rangle ,$$

where there are l zeroes, and r_i is the rank (or arity) of the operation f_i, i.e., the number of its argument places, s_j is the rank of the partial operation g_j, while t_k is the rank of the predicate P_k $(i, j, k = 1, 2, ...)$, is called the *simi-*

larity type of the given algebraic system. The set A is the *base set* of the system, the elements $a_1, \ldots, a_l$ are its *distinguished elements,* and the f_i, g_j, P_k are its *basic (total) operations, partial operations,* and *predicates.*

An algebraic system having no basic predicates is known as a *partial algebra.* If it has neither fundamental predicates nor basic partial operations, it is called an *algebra.* Finally, an algebraic system having no basic operations, partial or total, is called a *model.*

Algebraic systems having the same type are said to be *similar.* A collection of similar systems is called a *class* of systems. The notation

$$\langle A; a_1, \ldots, a_l; f_1, \ldots, f_m; g_1, \ldots, g_n; P_1, \ldots, P_p \rangle \,, \tag{1}$$

represents the algebraic system $\mathfrak{A}$ with base A, distinguished elements $a_1, \ldots, a_l$, basic operations $f_1, \ldots, f_m$, partial operations $g_1, \ldots, g_n$, and predicates $P_1, \ldots, P_p$.

Let

$$\mathfrak{B} = \langle B; b_1, \ldots, b_l; f_1', \ldots, f_m'; g_1', \ldots, g_n'; P_1', \ldots, P_p' \rangle$$

be a second algebraic system similar to $\mathfrak{A}$. A one-valued mapping from the set A into the set B is called a *mapping of the system* $\mathfrak{A}$ *into the system* $\mathfrak{B}$. A mapping φ from $\mathfrak{A}$ into $\mathfrak{B}$ is a *homomorphism* of $\mathfrak{A}$ into $\mathfrak{B}$ iff it satisfies the following conditions:

$$\left.\begin{array}{ll}
\text{(i)} & \varphi a_i = b_i \qquad (i = 1, \ldots, l)\,; \\[1ex]
\text{(ii)} & \varphi f_i(x_1, \ldots, x_{r_i}) = f_i'(\varphi x_1, \ldots, \varphi x_{r_i}) \\
 & \qquad\qquad (i = 1, \ldots, m;\, x_1, \ldots, x_{r_i} \in A)\,; \\[1ex]
\text{(iii)} & x = g_j(x_1, \ldots, x_{s_j}) \Rightarrow \varphi x = g_j'(\varphi x_1, \ldots, \varphi x_{s_j}) \\
 & \qquad\qquad (j = 1, \ldots, n;\, x, x_1, \ldots, x_{s_j} \in A)\,; \\[1ex]
\text{(iv)} & P_k(x_1, \ldots, x_{t_k}) \Rightarrow P_k'(\varphi x_1, \ldots, \varphi x_{t_k}) \\
 & \qquad\qquad (k = 1, \ldots, p;\, x_1, \ldots, x_{t_k} \in A)\,;
\end{array}\right\} \tag{2}$$

here, the symbol $\Rightarrow$ stands for "implies". A 1–1 homomorphism of $\mathfrak{A}$ onto $\mathfrak{B}$ whose inverse map is a homomorphism from $\mathfrak{B}$ onto $\mathfrak{A}$ is called an *isomorphism* of $\mathfrak{A}$ onto $\mathfrak{B}$.

We note that in the definition of a homomorphism between two systems, the basic notions in the first correspond to those in the second occupying the same positions in the standard notation (1) for algebraic systems. E.g., let D be the set of natural numbers 0, 1, 2, ..., and let +, X denote the usual arithmetic operations of addition and multiplication of numbers. Consider the algebras

$$\mathfrak{A} = \langle D; +, \times \rangle, \qquad \mathfrak{B} = \langle D; \times, + \rangle .$$

These algebras are not isomorphic, since if φ were an isomorphism from $\mathfrak{A}$ onto $\mathfrak{B}$, we would have

$$\varphi(x \times y) = \varphi x + \varphi y$$

for all natural numbers x, y, and consequently for any $y \in D$,

$$\varphi 0 = \varphi 0 + \varphi y ,$$

i.e., $\varphi y = 0$, so φ would not be a 1–1 correspondence.

§1.2. *Generating sets. Terms*

Suppose $h(x_1, ..., x_s)$ is a partial operation defined on a set A. A subset $M \subseteq A$ is said to be *closed under h* iff for every sequence $\langle u_1, ..., u_s \rangle$ of elements of M, the existence of $h(u_1, ..., u_s)$ implies $h(u_1, ..., u_s) \in M$. The subset M is *closed under a set of partial operations* iff it is closed under each member of the set.

It follows from these definitions that the intersection of any system M_ζ ($\zeta \in Z$) of subsets closed under given partial operations either is empty or is itself a subset closed under these operations.

Let M be a subset of the set A, and let $g_1, ..., g_n$ be partial operations on A. Let N be the intersection of all subsets of A which are closed under $g_1, ..., g_n$ and include M. This subset N is the smallest closed subset that includes M; it is called the set *generated* by M in A under $g_1, ..., g_n$; the elements of the generating set M are called *generators.*

To give a more detailed description of the structure of the set N we use terms, defined as follows.

We consider some finite set of signs $\mathbf{h}_1, ..., \mathbf{h}_s$, with each of which is associated a natural number. These signs are called *function* or *operation symbols,* and their associated natural numbers are their *arities* or *ranks.* We also assume another set of symbols, distinct from the ones above, is given; these are called *individual symbols.*

Terms are certain finite sequences (strings) of function and individual symbols and the three special symbols) (, . Their structure is determined by the following recursion scheme:

(i) The string consisting of a single individual symbol is a term.

(ii) If $\mathfrak{a}_1, \ldots, \mathfrak{a}_{n_i}$ are terms and $\boldsymbol{h}_i$ is an n_i-ary function symbol, then the string $\boldsymbol{h}_i(\mathfrak{a}_1, \ldots, \mathfrak{a}_{n_i})$ is also a term.

Example 1: Suppose $\boldsymbol{h}$ is a unary function symbol and $\boldsymbol{a}$ is an individual symbol. Then the strings $\boldsymbol{a}, \boldsymbol{h}(\boldsymbol{a}), \boldsymbol{h}(\boldsymbol{h}(\boldsymbol{a})), \ldots$ are terms.

Example 2: Let $\boldsymbol{f}, \boldsymbol{g}$ be binary function symbols, $\boldsymbol{a}, \boldsymbol{b}$ individual symbols. Then the strings $\boldsymbol{f}(\boldsymbol{a}, \boldsymbol{a}), \boldsymbol{g}(\boldsymbol{a}, \boldsymbol{b}), \boldsymbol{f}(\boldsymbol{a}, \boldsymbol{g}(\boldsymbol{a}, \boldsymbol{b})), \ldots$ are terms.

Terms are often written in an abbreviated form employing the so-called operator notation. Namely, instead of $\boldsymbol{g}(\mathfrak{a}, \mathfrak{b})$ and $\boldsymbol{h}(\mathfrak{a}, \mathfrak{b}, \mathfrak{c})$, we write $(\mathfrak{a})\boldsymbol{g}(\mathfrak{b})$ and $(\mathfrak{a})\boldsymbol{h}(\mathfrak{b})\boldsymbol{h}(\mathfrak{c})$, where $\boldsymbol{g}, \boldsymbol{h}$ are function symbols and $\mathfrak{a}, \mathfrak{b}, \mathfrak{c}$ are terms; the parentheses may be dropped if it leads to no ambiguity. In particular, with this notation the terms in Examples 1 and 2 might assume the forms $\boldsymbol{a}, \boldsymbol{ah}, \boldsymbol{ahh}, \ldots$ and $\boldsymbol{afa}, \boldsymbol{agb}, \boldsymbol{af}(\boldsymbol{agb}), \ldots$, respectively.

Now we introduce the idea of the value of a term for given values of the individual and function symbols in and on a given base set A. Giving a value to an individual symbol means, by definition, assigning it a well-defined element (its *value*) in the set A. Giving a value to an n-ary function symbol $\boldsymbol{f}$ means assigning it some concrete n-ary operation f defined on the set A. In addition to function symbols we sometimes consider partial function symbols, as well. The definition of terms with these remains as before, while as values for partial function symbols we admit any partial operations of the appropriate rank defined on A.

If the base and values of individual and function symbols are given, then the value of a term $\mathfrak{a}$ is defined, inductively:

(i′) If the term $\mathfrak{a}$ consists of a single individual symbol, then the value of $\mathfrak{a}$ is the value of the symbol.

(ii′) If the term $\mathfrak{a}$ has the form $\boldsymbol{f}(\mathfrak{a}_1, \ldots, \mathfrak{a}_n)$, where $\boldsymbol{f}$ is an n-ary function symbol whose value is f, and the values of the terms $\mathfrak{a}_1, \ldots, \mathfrak{a}_n$ are defined and equal respectively to $\mathfrak{a}_1^0, \ldots, \mathfrak{a}_n^0$, then the value of the term $\mathfrak{a}$ is the value of the operation f at the point $\langle \mathfrak{a}_1^0, \ldots, \mathfrak{a}_n^0 \rangle$. When $\boldsymbol{f}$ is a partial function symbol and f is not defined at the point $\langle \mathfrak{a}_1^0, \ldots, \mathfrak{a}_n^0 \rangle$, the value of $\mathfrak{a}$ is considered undefined. The value of the term $\boldsymbol{f}(\mathfrak{a}_1, \ldots, \mathfrak{a}_n)$ is also taken to be undefined if the value of any of the terms $\mathfrak{a}_1, \ldots, \mathfrak{a}_n$ is undefined.

Example 3: Let Q be the set of rational numbers, and let $\boldsymbol{+}, \boldsymbol{\times}$ be the binary function symbols, $\boldsymbol{:}$ a binary partial function symbol, whose values we take to be the usual arithmetic operations of addition, multiplication, and division of numbers. Then for $\boldsymbol{x}^0 = 0,\ \boldsymbol{y}^0 = 1,\ \boldsymbol{z}^0 = 2$, the term $\boldsymbol{x} + (\boldsymbol{y} : \boldsymbol{z})$ has value $\frac{1}{2}$, while the values of the terms $(\boldsymbol{x} \times \boldsymbol{y}) : \boldsymbol{x}$, $(\boldsymbol{z} + \boldsymbol{y}) : \boldsymbol{x}$, $\boldsymbol{z} : (\boldsymbol{y} : \boldsymbol{x})$ are undefined.

Now suppose that in some term $\mathfrak{a}$ containing function (or partial function) symbols $\boldsymbol{f}_1, \ldots, \boldsymbol{f}_s$ and individual symbols $\boldsymbol{a}_1, \ldots, \boldsymbol{a}_m, \boldsymbol{x}_1, \ldots, \boldsymbol{x}_n$, values are specified for the function symbols and $\boldsymbol{a}_1, \ldots, \boldsymbol{a}_m$, but values for $\boldsymbol{x}_1, \ldots, \boldsymbol{x}_n$ are not fixed. Then for every sequence of values for $\boldsymbol{x}_1, \ldots, \boldsymbol{x}_n$, $\mathfrak{a}$ will have a definite value (or be undefined); thus we can view the values of $\mathfrak{a}$ as determining an n-place function (or partial function) with respect to the variables $\boldsymbol{x}_1, \ldots, \boldsymbol{x}_n$. The term $\mathfrak{a}$, or better $\mathfrak{a}^0$ is a convenient notation for this function, as long as we know which symbols are the variables. Therefore, from a number of given operations on the set A, we can obtain an even infinite number of new operations, writable in the form of terms. These new operations are sometimes called *termal*, sometimes *polynomial,* operations in the given ones.

We return to the study of sets of generators. Let $g_1, \ldots, g_n$ be partial operations defined on a set A, and let M be a subset of A. By N denote the subset of A generated by M under the operations given. Let us consider terms whose notation contains only function symbols for the partial functions $g_1, \ldots, g_n$ and individual symbols for the elements of M. It is easily verified that the values of all possible terms in these symbols, which have the values specified, belong to N. Furthermore, the set of values of all such terms is closed under the partial operations $g_1, \ldots, g_n$. Consequently, the set N generated in A by M under $g_1, \ldots, g_n$ is the set of values of all terms written with symbols standing for the partial operations $g_1, \ldots, g_n$ and for the individual elements of M.

We now consider an arbitrary algebraic system

$$\mathfrak{A} = \langle A; a_1, \ldots, a_l; f_1, \ldots, f_m; g_1, \ldots, g_n; P_1, \ldots, P_p \rangle .$$

Let A_0 be a subset of the base A closed under the operations $f_1, \ldots, f_m$ and containing all the distinguished elements $a_1, \ldots, a_l$. We let $f_1', \ldots, f_m', P_1', \ldots, P_m'$ be the operations and predicates on A_0 of the same ranks as $f_1, \ldots, f_m$, $P_1, \ldots, P_p$ and with the same values that the latter operations and predicates take for sequences of elements of A_0. Besides these, we introduce a partial operation g_j' on A_0 corresponding to g_j in $\mathfrak{A}$ ($j = 1, \ldots, n$). The value of $g_j'(x_1, \ldots, x_{s_j})$ for $x_1, \ldots, x_{s_j} \in A_0$ is defined and equal to $g_j(x_1, \ldots, x_{s_j})$ if the latter is defined and belongs to A_0. Otherwise, $g_j'(x_1, \ldots, x_{s_j})$ is undefined. Under these conditions the algebraic system

$$\mathfrak{A}_0 = \langle A_0; a_1, \ldots, a_l; f_1', \ldots, f_m'; g_1', \ldots, g_n'; P_1', \ldots, P_p' \rangle$$

is called a *subsystem* of the algebraic system $\mathfrak{A}$.

It is clear that every subset of A closed under the basic operations and containing the distinguished elements is the base of a uniquely defined subsystem of $\mathfrak{A}$; moreover, every subsystem of $\mathfrak{A}$ has the same type as $\mathfrak{A}$. If the set A_0 is closed under $g_1, \ldots, g_n$ as well as $f_1, \ldots, f_m$, then the corresponding subsystem is called a *closed* subsystem of $\mathfrak{A}$.

A subset M of A is a *generating set* for the subsystem $\mathfrak{A}_0$ in the algebraic system $\mathfrak{A}$ iff $M \cup \{a_1, \ldots, a_l\}$ generates A_0 in A under $f_1, \ldots, f_m, g_1, \ldots, g_n$. The system $\mathfrak{A}$ is *finitely generated* iff it itself has a generating set that is finite.

When studying algebraic systems of a fixed type

$$\langle 0, \ldots, 0; r_1, \ldots, r_m; s_1, \ldots, s_n; t_1, \ldots, t_p \rangle,$$

it is customary to introduce corresponding individual, operation, partial operation, and predicate symbols of the given ranks. The sequence of these symbols is called the *signature* of an algebraic system of this type ([3]). Given an arbitrary algebraic system $\mathfrak{A}$ of this type, we take the values of the signature symbols to be the corresponding distinguished elements and basic operations, partial operations, and predicates in $\mathfrak{A}$. Thus, the signature symbols become formal designations for the basic notions in each algebraic system of the given type. The signature symbols are interpreted with fixed values in each algebraic system, but these values may vary on passage from one system to another of this type.

Suppose $\mathfrak{K}$ is a class of algebraic systems with signature Σ. Then a *Σ-term* is a term whose function signs all belong to Σ. If a Σ-term $\mathfrak{a}$ contains individual symbols $\boldsymbol{x}_1, \ldots, \boldsymbol{x}_q$, it is more explicitly written as $\mathfrak{a}(\boldsymbol{x}_1, \ldots, \boldsymbol{x}_q)$. Those individual symbols which occur in the signature Σ have values fixed in each system with signature Σ and are therefore called *individual constants.* The remaining individual symbols are known as *individual variables* and come from some adequately huge supply. Their values may be chosen arbitrarily in each algebraic system.

By induction on the length of terms one easily proves the following proposition: if φ is a homomorphism from an algebraic system $\mathfrak{A}$ with signature Σ into an algebraic system $\mathfrak{B}$ with the same signature, and $\mathfrak{a}(\boldsymbol{x}_1, \ldots, \boldsymbol{x}_q)$ is an arbitrary Σ-term, then the values $\mathfrak{a}^0$, $\mathfrak{a}^1$ of this term in $\mathfrak{A}$ and $\mathfrak{B}$ are related by the equation

$$\varphi\mathfrak{a}^0(u_1, \ldots, u_q) = \mathfrak{a}^1(\varphi u_1, \ldots, \varphi u_q) \tag{3}$$

for all possible values u_i of the symbols $\boldsymbol{x}_i$ $(i = 1, \ldots, q)$ in $\mathfrak{A}$ (i.e., in the base of $\mathfrak{A}$) for which the left-hand side is defined.

Suppose the set M generates the algebraic system $\mathfrak{A}$. Then each element of

$\mathfrak{A}$ is a value of some Σ-term of the form $\mathfrak{a}(\boldsymbol{x}_1, \ldots, \boldsymbol{x}_q)$ when the $\boldsymbol{x}_i$ receive values from among the elements of M or the distinguished elements of $\mathfrak{A}$. By (3) the homomorphic image of every element in $\mathfrak{A}$ is uniquely determined if the images of the given generators are known.

§1.3. *Primitive and quasiprimitive classes*

Let us assume Σ is a fixed signature for algebraic systems. Let $\mathfrak{a}(\boldsymbol{x}_1, \ldots, \boldsymbol{x}_n, \boldsymbol{y}_1, \ldots, \boldsymbol{y}_p)$ and $\mathfrak{b}(\boldsymbol{x}_1, \ldots, \boldsymbol{x}_n, \boldsymbol{z}_1, \ldots, \boldsymbol{z}_q)$ be two Σ-terms with individual variables $\boldsymbol{x}_1, \ldots, \boldsymbol{x}_n, \boldsymbol{y}_1, \ldots, \boldsymbol{y}_p, \boldsymbol{z}_1, \ldots, \boldsymbol{z}_q$, only the $\boldsymbol{x}_i$ occurring in both $\mathfrak{a}$ and $\mathfrak{b}$. The remaining symbols appearing in the notation of these terms are thus assumed to belong to Σ. We say that the *identity* $\mathfrak{a} \approx \mathfrak{b}$ *is valid* (or *holds*) in an algebraic system $\mathfrak{A}$ with signature Σ iff the value of the left-hand side coincides with the value of the right-hand side for any values of the variables $\boldsymbol{x}_i, \boldsymbol{y}_j, \boldsymbol{z}_k$ in $\mathfrak{A}$. In addition, if there are basic partial operations in $\mathfrak{A}$, we only require that whenever the value of one of the terms $\mathfrak{a}$, $\mathfrak{b}$ is defined for given values of the $\boldsymbol{x}_i, \boldsymbol{y}_j, \boldsymbol{z}_k$, the other have a defined and equal value ([4]).

The concept of conditional identity is an immediate generalization of that identity, for a *conditional identity* is an expression of the form

$$\mathfrak{a}_1 \approx \mathfrak{b}_1 \;\&\; \ldots \;\&\; \mathfrak{a}_r \approx \mathfrak{b}_r \rightarrow \mathfrak{a} \approx \mathfrak{b}, \tag{4}$$

where $\mathfrak{a}_1, \mathfrak{b}_1, \ldots, \mathfrak{a}_r, \mathfrak{b}_r, \mathfrak{a}, \mathfrak{b}$ are terms of signature Σ and &, $\rightarrow$ are the usual propositional signs.

We say that the conditional identity (4) is *valid* in an algebraic system $\mathfrak{A}$ with signature Σ iff whenever the equations $\mathfrak{a}_i^0 = \mathfrak{b}_i^0$ $(i = 1, \ldots, r)$ all hold for particular values of their variables in $\mathfrak{A}$, then $\mathfrak{a}^0 = \mathfrak{b}^0$ holds for these values. When partial function signs occur in Σ, these equations – as above – are considered to hold either if both sides are defined and have equal values, or if neither side is defined.

A class $\mathfrak{K}$ of algebraic systems with signature Σ is called *primitive* (or *equational* or a *variety*) iff it consists of all algebraic systems with signature Σ in which the members of a fixed set S of identities of signature Σ are all valid. $\mathfrak{K}$ is *finitely primitive* iff S can be chosen finite.

Analogously, the class of all algebraic systems with fixed signature Σ in which the whole of a given set of conditional identities of signature Σ is valid is called a *quasiprimitive class* (or a *quasivariety*) (cf. [IV]). When this set of conditional identities can be chosen to be finite, the class $\mathfrak{K}$ is said to be *finitely quasiprimitive.*

Example 4: An algebra with one basic binary operation is called a *groupoid.* A *semigroup* is a groupoid in which the identity

$$x \times (y \times z) \approx (x \times y) \times z \tag{5}$$

is valid, interpreting the binary function symbol $\times$ as the basic operation.

A *semigroup with identity element* is an algebra with signature consisting of the individual symbol e and the binary operation symbol $\times$ in which the identities (5),

$$x \times e \approx x, \quad e \times x \approx x$$

are valid.

A *left cancellative semigroup* is a semigroup in which the conditional identity

$$x \times y \approx x \times z \rightarrow y \approx z$$

is valid.

A semigroup in which

$$y \times x \approx z \times x \rightarrow y \approx z$$

is valid is called a *right cancellative semigroup.*

A semigroup with left and right cancellation is called a (*two-sided*) *cancellative semigroup.*

From these definitions it is clear that groupoids, semigroups, and semigroups with identity element compose varieties of algebras, while semigroups with left, right, and two-sided cancellation form quasivarieties of algebras.

Example 5: A *group* is an algebra with two operations, one binary, one unary, in which (5),

$$x^{-1} \times (x \times y) \approx y, \quad (y \times x) \times x^{-1} \approx y \tag{6}$$

are valid, where $\times$ is the binary function symbol, $^{-1}$ the unary.

Consequently, groups form a finitely primitive class.

Abelian groups are characterized by (5), (6) and $x \times y \approx y \times x$. Thus the class of all abelian groups is also finitely primitive.

Example 6: A *ring* is an algebra with two binary operations and one unary operation in which the identities

$$x + (y + z) \approx (x + y) + z, \quad (-x) + (x + y) \approx y,$$

$$x + y \approx y + x, \quad x \times (y + z) \approx (x \times y) + (x \times z)$$

are valid, where $+, \times, -$ are symbols for the first, second, and third operations, respectively. A ring is *associative* iff (5) is valid in it. Hence, rings and associative rings also form finitely primitive classes.

We note that inasmuch as every identity $\mathfrak{a} \approx \mathfrak{b}$ is equivalent with respect to validity to the conditional identity $\boldsymbol{x} \approx \boldsymbol{x} \to \mathfrak{a} \approx \mathfrak{b}$, every primitive class of algebraic systems is a quasiprimitive class at the same time.

It is easy to check that every primitive class contains all homomorphic images of any of its members. A homomorphic image of a left cancellative semigroup, however, need not have left cancellation. Therefore, the class of left cancellative semigroups can serve as an example of a finitely quasiprimitive class that is not primitive.

§1.4. *Defining relations*

Let $\mathcal{K}$ be the class of all algebras whose signature Σ consists of individual symbols $\boldsymbol{a}_\gamma$ $(\gamma \in \Gamma)$ and function symbols $\boldsymbol{f}_\delta$ of rank r_δ $(\delta \in \Delta)$. We fix some set S of conditional identities, all of whose function symbols belong to Σ. Let $\mathcal{K}_0$ be the class of all algebras with signature Σ such that each is generated by the set $\{a_\gamma \colon \gamma \in \Gamma\}$ of its distinguished elements under its operations f_δ $(\delta \in \Delta)$, and in each all members of S are valid. The class $\mathcal{K}_0$ is not a quasivariety, since the requirement that an algebra be generated by its distinguished elements cannot be expressed by means of conditional identities.

Consider now two arbitrary $\mathcal{K}_0$-algebras

$$\mathfrak{A} = \langle A; a_\gamma; f_\delta \colon \gamma \in \Gamma, \delta \in \Delta \rangle, \quad \mathfrak{B} = \langle B; b_\gamma; g_\delta \colon \gamma \in \Gamma, \delta \in \Delta \rangle .$$

If there is a homomorphism from $\mathfrak{A}$ into $\mathfrak{B}$, by §1.1 it must carry a_γ onto b_γ $(\gamma \in \Gamma)$. But the a_γ generate $\mathfrak{A}$, so any homomorphism is determined by their images. In other words, there is at most one homomorphism from a given $\mathcal{K}_0$-algebra into any other given $\mathcal{K}_0$-algebra, and it will be onto if it exists.

It is easily proved [IV] that no matter what S is chosen, there is up to isomorphism a unique $\mathcal{K}_0$-algebra admitting homomorphic mappings onto every other $\mathcal{K}_0$-algebra. This algebra is called the *algebra with generators* $\boldsymbol{a}_\gamma$ $(\gamma \in \Gamma)$ *and defining relations* S.

An algebra $\mathfrak{A}$ is said to be *finitely presented by conditional identities* iff it is isomorphic to an algebra with a finite number of generators and a finite number of defining relations. If these relations can be taken to be simply identities, then $\mathfrak{A}$ is *finitely presented by identities.* In the sequel, finitely presented algebras should be understood as those finitely presented by conditional identities, unless the contrary is explicitly stated.

Defining relations are generally used in the following ways: (1) a quasivariety

$\mathfrak{K}$ of algebras is specified by means of some set S of conditional identities; (2) the signature Σ of $\mathfrak{K}$ is augmented with new individual constants $\boldsymbol{c}_\zeta$ ($\zeta \in Z$), and a set S_1 of conditional identities is chosen such that in its members no individual variables occur – only the symbols a_γ ($\gamma \in \Gamma$) for the distinguished elements in $\mathfrak{K}$-algebras and the added $\boldsymbol{c}_\zeta$ ($\zeta \in Z$). Such conditional identities are called *constant relations.* In this situation we say that an algebra $\mathfrak{A}$ is *presented by generators* $\boldsymbol{c}_\zeta$ ($\zeta \in Z$) *and (constant) defining relations* S_1 *in the quasivariety* $\mathfrak{K}$ iff $\mathfrak{A}$ has generators $\boldsymbol{a}_\gamma, \boldsymbol{c}_\zeta$ and defining relations $S_1 \cup S$.

Suppose in some quasivariety $\mathfrak{K}$ the algebra $\mathfrak{A}$ is presented by generators $\boldsymbol{c}_\zeta$ ($\zeta \in Z$) and defining relations S_1, and the algebra $\mathfrak{B}$ is presented by generators $\boldsymbol{d}_\eta$ ($\eta \in H$) and defining relations S_2, where the $\boldsymbol{c}_\zeta$ are all distinct from the $\boldsymbol{d}_\eta$. Then the algebra $\mathfrak{C}$ presented in in $\mathfrak{K}$ by the generators $\boldsymbol{c}_\zeta, \boldsymbol{d}_\eta$ ($\zeta \in Z$, $\eta \in H$) and constant defining relations $S_1 \cup S_2$ is called the *free composition* (or *free product*) of $\mathfrak{A}$ and $\mathfrak{B}$ in $\mathfrak{K}$. This definition is generalized by isomorphism, i.e., if $\mathfrak{A}$, $\mathfrak{B}$, $\mathfrak{C}$ are respectively isomorphic to $\mathfrak{A}_1$, $\mathfrak{B}_1$, $\mathfrak{C}_1$, then we say $\mathfrak{C}_1$ is the free composition of $\mathfrak{A}_1$ and $\mathfrak{B}_1$ in $\mathfrak{K}$, without reference to to presentations.

We similarly define the $\mathfrak{K}$-free composition of any system of $\mathfrak{K}$-algebras. That free compositions are uniquely determined up to isomorphism is easily proved.

Example 7: Let $\mathfrak{K}$ be the class of all algebras with signature consisting of the operation symbols $\boldsymbol{f}_\delta$ ($\delta \in \Delta$). A $\mathfrak{K}$-algebra $\mathfrak{A}$ with generators $\boldsymbol{c}_\zeta$ ($\zeta \in Z$) and an empty set of defining relations is called an *absolutely free algebra with free generators* $\boldsymbol{c}_\zeta$ ($\zeta \in Z$). We can represent this algebra more concretely as follows. Let A be the set of all terms in the symbols $\boldsymbol{f}_\delta, \boldsymbol{c}_\zeta$ ($\delta \in \Delta, \zeta \in Z$). On A we define operations f_δ for $\delta \in \Delta$ by putting

$$f_\delta(\mathfrak{a}_1, \ldots \mathfrak{a}_{r_\delta}) = \boldsymbol{f}_\delta(\mathfrak{a}_1, \ldots, \mathfrak{a}_{r_\delta}) \qquad (\mathfrak{a}_1, \ldots, \mathfrak{a}_{r_\delta} \in A)\,.$$

Then $\langle A; f_\delta : \delta \in \Delta \rangle$ is the absolutely free $\mathfrak{K}$-algebra with free generators $\boldsymbol{c}_\zeta$. We take the operation f_δ to be the value of the signature symbol $\boldsymbol{f}_\delta$ ($\delta \in \Delta$), and the term $\boldsymbol{c}_\zeta \in A$ to be the value of the generator symbol $\boldsymbol{c}_\zeta$ ($\zeta \in Z$).

An algebra $\mathfrak{A}$ is *locally absolutely free* iff every finite subset of $\mathfrak{A}$ generates an absolutely free subalgebra.

It is easy to show that every absolutely free algebra is locally absolutely free, and that every subalgebra of an absolutely free algebra is absolutely free.

It is also easily proved that *a* $\mathfrak{K}$*-algebra* $\mathfrak{A}$ *is locally absolutely free iff all the formulas* (⁵)

$$f_\delta(x_1, ..., x_{r_\delta}) \not\approx f_\epsilon(y_1, ..., y_{r_\epsilon}) \quad (\delta, \epsilon \in \Delta;\ \delta \neq \epsilon)\,,$$

$$f_\delta(x_1, ..., x_{r_\delta}) \approx f_\delta(y_1, ..., y_{r_\delta}) \to x_1 \approx y_1 \,\&\, ... \,\&\, x_{r_\delta} \approx y_{r_\delta} \quad (\delta \in \Delta)\,,$$

are valid in it for all values of the x_i, y_j. ■

In particular, a groupoid is locally absolutely free iff the formula

$$x \times y \approx u \times v \to x \approx u \,\&\, y \approx v \tag{7}$$

is valid in it. This shows the class $\mathcal{L}$ of all locally (absolutely) free groupoids is a quasivariety.

One can easily construct locally free groupoids that are not free. E.g., such are the "completely divisible" locally free groupoids – those satisfying the added condition that every element factor into the product of two others.

§1.5. *Algebras of recursive functions*

We introduce the following fixed designations:

D = the set of all natural numbers 0, 1, 2, ... ;

$\mathcal{F}$ = the set of all unary total functions defined on D and taking values in D;

$\mathcal{F}_p$ = the set of all unary partial functions defined on D and taking values in D.

We consider these four operations defined on $\mathcal{F}_p$:

(i) *Addition of functions.* Let $f, g \in \mathcal{F}_p$. We define a new partial function h on D by putting $h(x) = f(x) + g(x)$ when both $f(x)$ and $g(x)$ are defined, and taking $h(x)$ to be undefined whenever either of $f(x), g(x)$ is undefined ($x \in D$). The partial function h is called the *sum* of f and g, symbolically: $h = f + g$.

(ii) *Composition of functions.* For $f, g \in \mathcal{F}_p$ we define a new partial function h on D by putting $h(x) = f(g(x))$ if $g(x)$ and $f(g(x))$ are both defined, and taking $h(x)$ to be undefined for all other $x \in D$. The partial function h is called the composition of f and g, symbolically; $h = f * g$.

(iii) *Iteration of a function.* Let $f \in \mathcal{F}_p$. We define a new function h on D by setting $h(0) = 0$ and $h(x+1) = f(h(x))$ if $h(x)$ and $f(h(x))$ are defined ($x \in D$). If for some $n \in D$, $h(n)$ or $f(h(n))$ is undefined, then we take $h(x)$ to be undefined for all $x > n$. The partial function h is called the *iteration* of f and denoted by ιf or f^ι.

(iv) *Inversion of a function.* For every $f \in \mathcal{F}_p$ we define a new function $f^{-1} \in \mathcal{F}_p$, the *inverse* of f, by putting $f^{-1}(a) = b$ if $f(x)$ is defined for $x \leqslant b$

and $f(b) = a$, while $f(x) \neq a$ for $x < b$. In the contrary case $f^{-1}(a)$ is considered as undefined.

On $\mathcal{F}_p$ the four operations $+, *, , ^{-1}$ are total. On $\mathcal{F}$ the operation of inversion will not be total, for the inverse of a total function may not be defined everywhere. Let λ, κ be the elements of $\mathcal{F}$ given by

$$\lambda(x) = x + 1, \quad \kappa(x) = x - [\sqrt{x}]^2$$

for $x \in D$ ($[y]$ is the greatest integer not exceeding y). The algebras

$$\mathfrak{R}^\iota = \langle \mathcal{F}; \lambda, \kappa; +, *, \iota \rangle ,$$

$$\mathfrak{R} = \langle \mathcal{F}_p; \lambda, \kappa; +, *, ^{-1} \rangle ,$$

and the partial algebra

$$\mathfrak{R}^0 = \langle \mathcal{F}; \lambda, \kappa; +, *, ^{-1} \rangle$$

are of fundamental importance in the theory of computable functions.

By $\mathcal{F}_{\mathrm{prim}}$ we denote the set of functions generated in $\mathfrak{R}^\iota$ by $\{\lambda, \kappa\}$ under all the operations, by $\mathcal{F}_{\mathrm{gr}}$ the set generated in $\mathfrak{R}^0$ by $\{\lambda, \kappa\}$ under all the operations $+, *, ^{-1}$, and by $\mathcal{F}_{\mathrm{pr}}$ the set generated by $\{\lambda, \kappa\}$ in $\mathfrak{R}$ under all its operations. Functions in $\mathcal{F}_{\mathrm{prim}}$ are called *primitive recursive*, those in $\mathcal{F}_{\mathrm{gr}}$ *general recursive*, and those in $\mathcal{F}_{\mathrm{pr}}$ *partial recursive*.

According to theorems of R.M. Robinson [138] and J. Robinson [135], these definitions of primitive recursive, general recursive, and partial recursive functions are equivalent to their usual definitions using the process of primitive recursion and the μ-operator, as presented, e.g., in Kolmogorov and Uspenskiĭ [77].

From the same theorem of J. Robinson we see that the operation of iteration is termal in the algebra $\mathfrak{R}^0$. Thus every primitive recursive function is also general recursive. Furthermore, from Kleene's representation for partial recursive functions it follows that $\mathcal{F}_{\mathrm{gr}} = \mathcal{F} \cap \mathcal{F}_{\mathrm{pr}}$; that is, every total partial recursive function is general recursive. Therefore, it should not be ambiguous if general recursive functions are sometimes simply called recursive.

The algebras

$$\mathfrak{R}_{\mathrm{prim}} = \langle \mathcal{F}_{\mathrm{prim}}; +, *, \iota \rangle ,$$

$$\mathfrak{R}_{\mathrm{pr}} = \langle \mathcal{F}_{\mathrm{pr}}; +, *, ^{-1} \rangle ,$$

and the partial algebra

$$\mathcal{R}_{gr} = \langle \mathcal{F}_{gr}; +, *, ^{-1} \rangle$$

are called the algebras of primitive recursive, partial recursive, and general recursive functions, respectively. All three are finitely generated. As will be shown later (in §4.2), neither the first nor the second is finitely presented. The third algebra is partial, and no concept of finite presentation for partial algebras has been introduced.

Let $f \in \mathcal{F}_p$. The subalgebra $\mathfrak{A}_f$ of $\mathfrak{R}$ with the generator f (and the distinguished elements λ, κ) is called the *degree of unsolvability* of f. The functions f, g have the *same* degree of unsolvability iff $\mathfrak{A}_f = \mathfrak{A}_g$. When $\mathfrak{A}_f \subseteq \mathfrak{A}_g$ and $\mathfrak{A}_f \neq \mathfrak{A}_g$, the degree of unsolvability of f is *less* than that of g. When $\mathfrak{A}_f \nsubseteq \mathfrak{A}_g$ and $\mathfrak{A}_g \nsubseteq \mathfrak{A}_f$, f and g have *incomparable* unsolvability degrees. In particular, all partial recursive functions have one and the same degree of unsolvability, namely $\mathfrak{R}_{pr}$, the least among the degrees of unsolvability.

Many definitions and theorems in what follows have identical formulations for primitive recursive, general recursive, and partial recursive functions. Therefore, for brevity's sake we introduce the special symbol **R** meaning "primitive recursive", "general recursive", or "partial recursive" – for short: **R = prim, gr, pr.**

For $n \geqslant 2$, an n-place partial function f defined on D with values in D – a "numerical" function – is called an **R**-*function* iff for any unary **R**-functions $h_1, \ldots, h_n, f(h_1(x), \ldots, h_n(x))$ defines a (unary) **R**-function.

This definition can be given in a more convenient form. We consider the sequence

$$\langle 0, 0\rangle, \langle 0, 1\rangle, \langle 1, 0\rangle, \langle 0, 2\rangle, \langle 1, 1\rangle, \langle 2, 0\rangle, \langle 0, 3\rangle, \langle 1, 2\rangle, \ldots$$

of all ordered pairs of natural numbers. Let $\nu(x, y)$ be the number of the place in this sequence of the pair $\langle x, y\rangle$, with $\langle 0, 0\rangle$ occupying the 0th place; thus, $\nu(0, 0) = 0$, $\nu(0, 1) = 1$, etc. Let $\mathfrak{l}(z)$ and $\mathfrak{r}(z)$ be the corresponding left and right members of the pair with number z. Thus we have

$$\nu(\mathfrak{l}(z), \mathfrak{r}(z)) = z, \quad \mathfrak{l}(\nu(x, y)) = x, \quad \mathfrak{r}(\nu(x, y)) = y$$

for all $x, y, z \in D$. It is easy to deduce [119] the relation

$$\nu(x, y) = \tfrac{1}{2}((x + y)^2 + 3x + y)$$

and similar expressions for the functions $\mathfrak{l}$ and $\mathfrak{r}$. This implies, in particular, that ν, $\mathfrak{l}$, $\mathfrak{r}$ are primitive recursive.

The recursion scheme

$$\nu_2(x_1, x_2) = \nu(x_1, x_2) ,$$

$$\nu_{n+1}(x_1, ..., x_{n+1}) = \nu(x_1, \nu_n(x_2, ..., x_{n+1}))$$

lets us define a series of n-ary functions ν_n ($n = 2, 3, ...$). For every $n \geqslant 2$, ν_n is a primitive recursive 1–1 correspondence between D and D^n, the set of all ordered n-tuples of natural numbers. The number $\nu_n(x_1, ..., x_n)$ is called the *standard number* of the n-tuple $\langle x_1, ..., x_n \rangle$. For $\nu_n(x_1, ..., x_n) = y$, we have

$$x_1 = \mathfrak{l}(y) = \mathfrak{l}_1^n(y) ,$$

$$x_2 = \mathfrak{l}(\mathfrak{r}(y)) = \mathfrak{l}_2^n(y) ,$$

$$\cdots$$

$$x_n = \mathfrak{r}(\mathfrak{r}(... \mathfrak{r}(y))) = \mathfrak{l}_n^n(y) .$$

Consequently, the function $\mathfrak{l}_i^n$, whose value at any point $y \in D$ is the ith term of the n-tuple with standard number y, is primitive recursive ($n \geqslant 2$; $i = 1, ..., n$).

Let $f(x_1, ..., x_n)$ be any n-place function ($n \geqslant 2$). If f is an **R**-function, then the unary function f^π, where

$$f^\pi(x) = f(\mathfrak{l}_1^n(x), ..., \mathfrak{l}_n^n(x)) , \tag{8}$$

is also an **R**-function. Conversely, from (8) we get

$$f(x_1, ..., x_n) = f^\pi(\nu_n(x_1, ..., x_n)) ,$$

so when f^π is an **R**-function, f is too. Therefore, $f \to f^\pi$ gives a 1–1 mapping from $\mathcal{F}_\mathbf{R}$ onto $\mathcal{F}_\mathbf{R}^{(n)}$, the set of all n-ary **R**-functions. Let M be a subset of the set of natural numbers. The function $\chi_M \in \mathcal{F}$ with value 1 at points of M and value 0 off of M is called the *characteristic function* of M. The set M is *primitive recursive* or *(general) recursive* as its characteristic function is primitive recursive or general recursive. The set M is *recursively enumberable* iff it is either the empty set (written $\emptyset$) or the set of all values taken by some unary primitive recursive function. In what follows, recursively enumerable sets will occasionally be called *partial recursive*. Thus **R**-*sets* makes sense in every case. In particular, it is easy to prove [71] that M is an **R**-set iff it is the set of all solutions to the equation $\psi(x) = 0$ for some unary **R**-function ψ.

In the sequel we shall encounter the functions $\dot{-}$, *sg*, $\overline{sg}$, *exp*, defined for all $x, y \in D$ as follows (cf. [71, 119]):

$$x \dot{-} y = \begin{cases} x - y & \text{if } x \geqslant y, \\ 0 & \text{if } x < y; \end{cases}$$

$$\overline{sg}(x) = 1 \dot{-} x;$$

$$sg(x) = 1 - \overline{sg}(x).$$

We let p_n be the $(n+1)$th prime number, so $p_0 = 2, p_1 = 3$, etc. If $x = p_0^{k_0} p_1^{k_1} \ldots p_i^{k_i} \ldots$ is the decomposition of the number $x \geqslant 1$ into prime factors, then we put $exp(i, x) = k_i$ $(i \in D)$. To make *exp* total on D^2 we put $exp(i, 0) = 0$.

All these functions, as well as the ordinary arithmetic functions xy, $[x/y]$ (with $[x/0] = x$), $[\sqrt[x]{y}]$, $|x - y|$, $rem(x, y) = x - y[x/y]$ are primitive recursive.

§2. Numbered sets

§2.1. *Mappings of numbered sets*

Let A be an arbitrary nonempty finite or countably infinite set. A *numbering* α of the set A is a single-valued mapping from some subset D_α of $D = \{0, 1, 2, \ldots\}$, the set of all natural numbers, onto A. The set D_α is called the *number set* of the numbering α. If $n \in D_\alpha$ and $\alpha n = a$, we say n is an α-*number* (or α-*index*) of the element a.

The set A together with one of its numberings is called a *numbered set* and sometimes written as $\langle A, \alpha \rangle$.

Naturally associated with every numbering α of A is an equivalence relation θ_α defined on D_α as follows: two numbers $m, n \in D_\alpha$ are θ_α-*equivalent* iff $\alpha m = \alpha n$, i.e., iff they are α-numbers of one and the same element of A. Under θ_α the set D_α splits into equivalence classes, the collection of which is denoted by D_α/θ_α. The numbering α naturally induces a 1–1 mapping, the *canonical map*, of D_α/θ_α onto A.

When α is 1–1, every $a \in A$ has a unique α-number. If $D_\alpha = D$, α is said to be a *simple* numbering. A nonempty subset $E \subseteq D$ has the *trivial* numbering $\alpha x = x$, with $D_\alpha = E$.

A subset E' of a set $E \subseteq D$ is called an **R**-*subset in* E [77] iff $E' = E \cap M$, where M is an **R**-set in the usual absolute sense (**R** = **prim**, **gr**, **pr**). A subset C

of a numbered set A with numbering α is called an **R**-*subset* (relative to α) iff the set of all α-numbers of elements of C is an **R**-subset in D_α.

For $n \geqslant 2$, the numbering α of the set A induces a numbering $\alpha^{(n)}$ of the set A^n of all n-tuples $\langle a_1, ..., a_n\rangle$ of elements in A. Namely, let $D_\alpha^{(n)}$ be the set of standard numbers of the n-tuples $\langle x_1, ..., x_n\rangle \in D_\alpha^n$. Then for $x_1, ..., x_n \in D_\alpha$, we put

$$\alpha^{(n)} \nu_n(x_1, ..., x_n) = \langle \alpha x_1, ..., \alpha x_n\rangle ;$$

so $\alpha^{(n)}$ is a numbering of A^n with number set $D_\alpha^{(n)}$.

A set M of n-tuples ($n \geqslant 1$) of elements in A is an **R**-*set* (relative to the numbering α of A) iff the set of all $\alpha^{(n)}$-numbers of elements of M is an **R**-subset in $D_\alpha^{(n)}$ (with $\alpha^{(1)} = \alpha$, $D_\alpha^{(1)} = D_\alpha$).

A predicate $P(a_1, ..., a_n)$ defined on the numbered set $\langle A, \alpha\rangle$ is called an **R**-*predicate* (relative to α) iff the set of all n-tuples for which P is true is an **R**-set with respect to α.

Correspondingly, an n-ary predicate P^* defined on a subset E of D is called an **R**-*predicate* iff $E = \emptyset$ or P^* is an **R**-predicate relative to the trivial numbering of E, i.e., iff the standard numbers of the n-tuples for which P^* is true form an **R**-subset in $E^{(n)}$; P^* is an **R**-predicate *absolutely* when $E^{(n)}$ can be replaced with $D^{(n)}$ (or simply D) in the preceding ([6]).

The numbering α of a numbered set A is called *positive* iff the number set D_α and the associated equivalence relation θ_α are partial recursive. The numbering α is *negative* iff D_α and $\sim\theta_\alpha$ (the negation of the predicate θ_α) are partial recursive. The numbering α is *decidable* iff D_α and θ_α are general recursive. These three kinds of numberings are the ones usually encountered in the most important concrete cases.

According to Post [121], if a numerical set and its complement (in D) are both recursively enumerable, then they are both general recursive. This implies: *if a simple numbering is simultaneously positive and negative, then it is decidable.* ∎

A map φ from a set A with numbering α into a set B with numbering β is called an **R**-*map* from $\langle A, \alpha\rangle$ into $\langle B, \beta\rangle$ (or more briefly, from A into B) iff there is a unary numerical **R**-function f mapping D_α into D_β and satisfying

$$\varphi(\alpha x) = \beta f(x) \quad (x \in \mathrm{D}_\alpha) . \tag{9}$$

A 1–1 **R**-map from A onto B is called an **R**-*monomorphism* of A onto B. An **R**-*equivalence* between A and B is an **R**-monomorphism from A onto B whose inverse mapping from B onto A is also an **R**-monomorphism.

The numbered set A is **R**-*monomorphic* to the numbered set B iff there exists an **R**-monomorphism from A onto B. A and B are **R**-*equivalent* iff there exists an **R**-equivalence between A and B.

A 1–1 map φ from the set A with numbering α onto the set B with numbering β is called an **R**-*unimorphism* iff there exists a unary **R**-function f mapping D_α 1–1 onto D_β and satisfying (9). An **R**-unimorphism from A onto B whose inverse is an **R**-unimorphism from B onto A is called an **R**-*isomorphism* of A onto B.

The numbered set A is **R**-*unimorphic* to the numbered set B iff there exists an **R**-unimorphism from A onto B. A and B are **R**-*isomorphic* iff there exists an **R**-isomorphism from A onto B.

Clearly, every **R**-unimorphism is an **R**-monomorphism at the same time, while every **R**-isomorphism is also an **R**-equivalence. In particular, **R**-isomorphism for two numbered sets implies **R**-equivalence. Moreover, the relations of **R**-isomorphism and **R**-equivalence are reflexive, symmetric, and transitive.

The relations of **R**-monomorphism and **R**-unimorphism are reflexive and transitive, but not, in general, symmetric. Therefore, in the general case these relations are weaker than the corresponding relations, **R**-equivalence and **R**-isomorphism. There are, however, some important cases when unimorphism and monomorphism turn out to be equivalent to isomorphism and equivalence. Such cases are indicated by

Theorem 2.1.1: *Let $\langle A, \alpha\rangle$ and $\langle B, \beta\rangle$ be numbered sets. Assume the number set D_α is recursively enumerable. Then every* **pr**-*unimorphism of A onto B is actually a* **pr**-*isomorphism of A onto B; if A and B are in fact* **pr**-*isomorphic, then D_β is also recursively enumerable. If θ_β is absolutely recursively enumerable* (7), *then every* **pr**-*monomorphism from A onto B is a* **pr**-*equivalence.*

To prove the first assertion of the theorem we assume φ is a **pr**-unimorphism from A onto B, and we let f be a unary **pr**-function mapping D_α 1–1 onto D_β and satisfying the condition (9). By definition, D_α is the set of all values assumed by some $h \in \mathcal{F}_{\text{prim}}$ (for $D_\alpha \neq \emptyset$), i.e., D_α is the range of h. Consequently, D_β is the range of the **gr**-function $f * h$ and thus is recursively enumerable. The function g inverse on D_β to f is given by

$$g(x) = h(\mu z(x = f(h(z)))) \quad (x \in D_\beta)$$

(the operator μz is read as "the least $z \in D$, if it exists, such that... "); therefore, g is a **pr**-function, as well as a 1–1 map from D_β onto D_α satisfying condition (9), appropriately rewritten for the inverses. Thus A is a **pr**-isomorphic to B by the map φ.

Passing to the second assertion, we let ρ be a unary **prim**-function having the nonempty set

$$\{\nu(x, y) : x, y \in D_\beta \text{ and } x\, \theta_\beta\, y\}$$

as its range; thus for all $x, y \in D_\beta$, we have $x\, \theta_\beta\, y$ iff there is a $u \in D$ such that $x = \mathfrak{l}(\rho(u))$ and $y = \mathfrak{r}(\rho(u))$; the **prim**-functions ν, $\mathfrak{l}$, $\mathfrak{r}$ were introduced in §1.5.

Let f be a unary **pr**-function realizing a given **pr**-monomorphism φ of A onto B. So for every $x \in D_\beta$ we can find a number $w \in D_\alpha$ such that $x\, \theta_\beta f(w)$. Therefore, the unary **pr**-function g with domain D_β defined by

$$g(x) = h(\mathfrak{r}(\mu z(|x - \mathfrak{l}(\rho(\mathfrak{l}(z)))|$$

$$+ |f(h(\mathfrak{r}(z))) - \mathfrak{r}(\rho(\mathfrak{l}(z)))| = 0))),$$

where h enumerates D_α, is a map from D_β into D_α satisfying (9), rewritten for the inverse situation. Thus φ is a **pr**-equivalence of A onto B. ∎

Theorem 2.1.2: *If a numbered set $\langle A, \alpha\rangle$ has a general recursive number set D_α and is* **pr**-*unimorphic (***pr**-*monomorphic) to a numbered set $\langle B, \beta\rangle$, then A is* **gr**-*unimorphic (***gr**-*monomorphic) to B.*

We assume the existence of a **pr**-unimorphism φ and with it, a unary pr-function f mapping D_α 1–1 onto D_β and satisfying (9). Since D_α is a recursive set, the function $g \in \mathcal{F}$ given by

$$g(x) = \begin{cases} f(x) & \text{if } x \in D_\alpha, \\ 0 & \text{if } x \notin D_\alpha, \end{cases}$$

is general recursive, maps D_α 1–1 onto D_β, and satisfies (9) for φ. Thus A is **gr**-unimorphic to B by the given map φ. The case of a **pr**-monomorphism is proved similarly. ∎

From Theorems 2.1.1 and 2.1.2 we see that the relations of **gr**-unimorphism and **gr**-isomorphism coincide on the class of numbered sets with general recursive number sets.

Theorem 2.1.3: *Let φ be an* **R**-*map from a set A with numbering α into a set B with numbering β, and let $N \subseteq B$ be an* **R**-*subset in B. Then the complete φ-preimage M of N is an* **R**-*subset in A.*

Let f be a unary **R**-function realizing the map φ. By hypothesis N is an **R**-subset in B. This means that for some unary **R**-function g, the set of all solutions to the equation $g(x) = 0$ which lie in D_β is equal to the set of all β-numbers of elements of N. As is easily checked, the set of all solutions of $g(f(x)) = 0$ which lie in D_α coincides with the set of all α-numbers of elements in the preimage M of N. Since $g * f$ is an **R**-function, M is an **R**-subset in A. ■

We now want to extend the definition of **R**-predicates given above to operations and D-valued functions defined on numbered sets.

Let $g(a_1, ..., a_n)$ be a partial function with natural numbers as values defined on a set A with numbering α. We say g is an **R**-*function* (relative to α) iff there exists an n-place ordinary numerical **R**-function G such that

$$g(\alpha x_1, ..., \alpha x_n) = G(x_1, ..., x_n) \quad (x_1, ..., x_n \in D_\alpha), \tag{10}$$

in the usual sense of strong equality: both sides are defined and equal or both are undefined for any particular elements $x_1, ..., x_n \in D_\alpha$.

Comparing this definition with that of **R**-predicate, we see that for **R** = **prim**, **gr**, an n-ary predicate P on A is an **R**-predicate with respect to α iff as a numerically valued function (cf. §1.1), P is an **R**-function relative to α.

For **R** = **pr**, a modified condition holds: P is partial recursive on A iff the n-ary function with value equal to 1 at points where P is true and undefined elsewhere is partial recursive on A.

According to §1.1, a partial function $h(a_1, ..., a_n)$ defined on a set A and taking values in the same set is called a *partial operation* on A. If A has a numbering α, then h is called an **R**-*operation* on A relative to this numbering iff there is an n-ary numerical **R**-function H such that

$$h(\alpha x_1, ..., \alpha x_n) = \alpha H(x_1, ..., x_n)$$

for those n-tuples $\langle x_1, ..., x_n \rangle \in D_\alpha^n$ for which the left-hand side is defined, and the domain of h is an **R**-set relative to α.

Theorem 2.1.4: *Let φ be an* **R**-*monomorphism from a set A with numbering α onto a set B with numbering β. If φ transforms a numerically valued function $g(a_1, ..., a_n)$ (a predicate $P(a_1, ..., a_n)$) defined on A into an* **R**-*function g_1 (an* **R**-*predicate P_1) on B, then g is an* **R**-*function (P is an* **R**-*predicate).*

Of course, that φ transforms g into g_1 means

$$g(a_1, ..., a_n) = g_1(\varphi a_1, ..., \varphi a_n) \quad (a_1, ..., a_n \in A),$$

in the sense of strong equality.

Let G_1 be an n-ary function satisfying

$$g_1(\beta x_1, ..., \beta x_n) = G_1(x_1, ..., x_n) \qquad (x_1, ..., x_n \in D_\beta) ,$$

strongly, and let f be a unary **R**-function realizing the monomorphism φ. Then the n-ary numerical function G defined by

$$G(x_1, ..., x_n) = G_1(f(x_1), ..., f(x_n)) \quad (x_i \in D)$$

will be an **R**-function satisfying (10).

Consideration of a predicate on A reduces to consideration of the set of all sequences of arguments for which the predicate is true; thus, Theorem 2.1.4 for the predicate case reduces to Theorem 2.1.3. ∎

Theorem 2.1.5: *Let φ be an* **R**-*equivalence mapping the numbered set $\langle A, \alpha\rangle$ onto the numbered set $\langle B, \beta\rangle$. Then D-valued* **R**-*functions,* **R**-*predicates, and* **R**-*operations defined on A are transformed by φ into* **R**-*functions,* **R**-*predicates, and* **R**-*operations on B, respectively.*

The validity of this assertion for numerically valued functions and predicates follows from the preceding theorem immediately. Let us turn to the case when we are given an n-ary partial operation h on A and wish to learn about its φ-transform h_1, the operation on B strongly satisfying the equation

$$\varphi h(a_1, ..., a_n) = h_1(\varphi a_1, ..., \varphi a_n) \qquad (a_i \in A) .$$

Suppose the unary **R**-function f realizes the monomorphism φ from A into B, while the unary **R**-function g realizes the inverse monomorphism φ^{-1}. Let H be an n-ary numerical **R**-function representing the operation h on A. For any particular numbers $x_1, ..., x_n \in D_\beta$, we have to find a β-number for the element $h_1(\beta x_1, ..., \beta x_n)$ of B, assuming this element is defined. To begin, we note that $g(x_i) \in D$ is an α-number of the element $\varphi^{-1}(\beta x_i) \in A$ $(i = 1, ..., n)$, therefore, $f(H(g(x_1), ..., g(x_n)))$ is one of the β-numbers of $h_1(\beta x_1, ..., \beta x_n)$. Consequently, the numerical partial function H_1 defined by

$$H_1(x_1, ..., x_n) = f(H(g(x_1), ..., g(x_n))) \quad (x_i \in D)$$

corresponds to the operation h_1 on B. Since it is a composition of **R**-functions, H_1 is itself an **R**-function. Since φ^{-1} is an **R**-map, Theorem 2.1.3 shows the domain of h_1 is an **R**-set relative to β. ∎

Corollary: *Under an* **R**-*isomorphism a positively (negatively) numbered set is carried onto a positively (respectively, negatively) numbered set. If the numbered sets* $\langle A, \alpha\rangle$, $\langle B, \beta\rangle$ *have recursively enumerable number sets, and A is positively (negatively) numbered and* **R**-*equivalent to B, then B is positively (negatively) numbered, too.*

Indeed, the positiveness (negativeness) of a numbering means that the number set is recursively enumerable, while the relation of equality (inequality) of two elements in the numbered set is a binary partial recursive predicate. Therefore, the assertions of this corollary follow immediately from Theorems 2.1.1 and 2.1.5. ■

§2.2. *Unireducibility for numberings*

So far we have been comparing numberings α, β of, generally speaking, distinct sets A, B. Now we turn to a more detailed study of numberings of one and the same set A.

As in the preceding, the symbol **R** denotes any one of the expressions "primitive recursive", "general recursive", or "partial recursive".

Let α, β be numberings of a set A. We say that the unary numerical partial function f *reduces* α to β iff the domain of f includes D_α and for every $x \in D_\alpha$, we have $f(x) \in D_\beta$ and $\alpha x = \beta f(x)$, i.e., iff for every α-number x of any element of A, $f(x)$ is defined and is a β-number of the same element.

The numbering α is **R**-*multireducible* or **R**-*reducible,* for short, to β (in symbols: $\alpha \leqslant_{\mathrm{Rm}} \beta$) iff there exists an **R**-function reducing α to β.

The numbering α is **R**-*unireducible* to β (in symbols: $\alpha \leqslant_{\mathrm{R1}} \beta$) iff there exists **R**-function reducing α to β that is 1–1 on D_α.

Lastly, α is **R**-*unimorphic* to β (in symbols: $\alpha \leqslant_{\mathrm{R}} \beta$) iff there exists an **R**-function reducing α to β and mapping D_α 1–1 onto D_β.

It is clear that the relations $\leqslant_{\mathrm{Rm}}, \leqslant_{\mathrm{R1}}, \leqslant_{\mathrm{R}}$ are reflexive and transitive, and that for each meaning of **R** they are connected by the implications

$$\alpha \leqslant_{\mathrm{R}} \beta \Rightarrow \alpha \leqslant_{\mathrm{R1}} \beta \Rightarrow \alpha \leqslant_{\mathrm{Rm}} \beta\,.$$

If D_α is a recursive set, and α is reduced to β by the function $f \in \mathcal{F}_{\mathrm{pr}}$, then the function g defined by

$$g(x) = \begin{cases} f(x) & \text{if } x \in D_\alpha\,, \\ 0 & \text{if } x \notin D_\alpha\,, \end{cases}$$

is general recursive and reduces α to β. Therefore, for a recursive number set

D_α, the partial recursive (multi-) unireducibility of α to β is equivalent to its general recursive (multi-) unireducibility to β. For arbitrary D_α, D_β, however, all the forms of reducibility introduced above are inequivalent. We give only an example showing that **pr**-multireducibility in general differs from **gr**-multireducibility.

Example 8: As we know, there are partial recursive functions that cannot be extended to general recursive functions [71]. Let $F(x)$ be such a function. Let D' be the domain of F, and C its range. We define two numberings α, β on C by letting $x \in D'$ be an α-number of $F(x) \in C$ (so $\alpha x = F(x)$ and $D_\alpha = D'$), and by letting $y \in C$ be a β-number for itself (so β is trivial and $D_\beta = C$). Suppose the function f reduces α to β. The condition for this can be written as $F(x) = f(x)$ $(x \in D_\alpha)$, signifying that f is an extension of F. Since no general recursive function extends F, α is not **gr**-reducible to β; however, α is reduced to β by the partial recursive function F, while D_α is recursively enumerable, being the domain of a partial recursive function.

Recall that a numbering α of a set A is called *simple* when D_α is the set of all natural numbers. The following obvious theorem shows that under pretty broad assumptions the study of arbitrary numberings can be reduced to that of simple ones.

Theorem 2.2.1: *For every numbering α of the set A with an infinite recursively enumerable set D_α, there exists a simple numbering β of A that is* **gr**-*unimorphic to α.*

As is well known, every infinite recursively enumerable set is the range of some 1–1 general recursive function [71]. Therefore,

$$D_\alpha = \{h(0), h(1), h(2), \dots\},$$

where $h \in \mathcal{F}_{gr}$ is an appropriate 1–1 function. We define β by putting

$$\beta x = \alpha h(x) \quad (x = 0, 1, 2, \dots).$$

This numbering β on A is simple, and h **gr**-unimorphically reduces β to α. ■

We return to the consideration of arbitrary numberings α, β of some non-empty set A. The numbering α is said to be **R**-*multiequivalent* or **R**-*equivalent*, for short, to β (in symbols: $\alpha \equiv_{Rm} \beta$) iff α is **R**-multireducible to β, and β is **R**-multireducible to α. We shall call α **R**-*uniequivalent* to β (in symbols: $\alpha \equiv_{R1} \beta$) iff α is **R**-unireducible to β, and β is **R**-unireducible to α. Finally, α is **R**-*isomorphic* to β (in symbols: $\alpha \equiv_{R} \beta$) iff α and β are **R**-unimorphic to each other.

These definitions can be represented briefly by the single scheme:

$$\alpha \equiv_S \beta \Leftrightarrow \alpha \leqslant_S \beta \quad \text{and} \quad \beta \leqslant_S \alpha \qquad (\mathbf{S} = \mathbf{Rm}, \mathbf{R1}, \mathbf{R})\,.$$

The various relations of multiequivalence, uniequivalence, and isomorphism thus defined are clearly reflexive, symmetric, and transitive.

If we compare the several concepts of unimorphism and isomorphism for numberings of one and the same set with the concepts introduced in §2.1 of unimorphism and isomorphism of different numbered sets, we see immediately that **R**-unimorphism (**R**-isomorphism) of the numberings α, β of the set A means exactly that the identity map on A is an **R**-unimorphism (**R**-isomorphism) of $\langle A, \alpha\rangle$ onto $\langle A, \beta\rangle$. In particular, it follows from the results of §2.1 that if a numbering α of A with recursively enumerable number set D_α is **R**-unimorphic to another numbering β of A, then these numberings are in fact **R**-isomorphic.

Coming back to simple numberings, we now generalize a theorem of Myhill [111] concerning the theory of reducibility of problems.

Theorem 2.2.2: **R**-*uniequivalence of simple numberings* α, β *of a set* A *is equivalent to their* **R**-*isomorphism.*

We have already remarked that **R**-isomorphism of numberings trivially implies their **R**-uniequivalence. For proving the converse let us agree, following Myhill to define a *finite correspondence* as any finite sequence

$$\langle\langle x_0, y_0\rangle, \langle x_1, y_1\rangle, ..., \langle x_k, y_k\rangle\rangle \tag{11}$$

of pairs of natural numbers satisfying the conditions

$$x_i = x_j \Leftrightarrow y_i = y_j\,,$$

$$\alpha x_i = \beta y_i \qquad (i, j = 0, ..., k)\,.$$

The *number* of a finite correspondence of the form (11) is the natural number

$$n = \prod_{i=0}^{k} p_i^{1+\nu(x_i, y_i)}\,, \tag{12}$$

where $p_0 = 2, p_1 = 3, ...$ are the successive prime numbers, and $\nu(x, y)$ is the standard number of the pair $\langle x, y\rangle$, computed according to §1.5.

We assume there is a unary **R**-function f unireducing α to β and a unary

R-function g unireducing β to α. Using f, g we want to construct binary numerical **R**-functions S, T such that if n is the number of a finite correspondence (11), then $S(m, n)$ is the number of a correspondence

$$\langle\langle x_0, y_0\rangle, ..., \langle x_k, y_k\rangle, \langle m, y_{k+1}\rangle\rangle\,, \tag{13}$$

and $T(m, n)$ is the number of

$$\langle\langle x_0, y_0\rangle, ..., \langle x_k, y_k\rangle, \langle x_{k+1}, m\rangle\rangle\,,$$

where m is an arbitrary natural number, and x_{k+1}, y_{k+1} are certain numbers depending on m and n.

We begin by constructing the function S. Suppose we are given a natural number m and a correspondence (11) with number n.

(i) If $m \in \{x_0, ..., x_k\}$ and i is the least index such that $m = x_i$, then we put $y_{k+1} = y_i$.

(ii) If $m \notin \{x_0, ..., x_k\}$ and $f(m) \notin \{y_0, ..., y_k\}$, then we set $y_{k+1} = f(m)$.

(iii) If $m \notin \{x_0, ..., x_k\}$, but $f(m) \in \{y_0, ..., y_k\}$, then we let i_1 be the smallest index for which $f(m) = y_{i_1}$. If in turn $f(x_{i_1}) \in \{y_0, ..., y_k\}$, then let i_2 be the least index such that $f(x_{i_1}) = y_{i_2}$, etc. We show that the numbers $y_{i_1}, y_{i_2}, ...$ are pairwise distinct. Indeed, suppose for $u \geqslant 1$ the numbers $y_{i_1}, ..., y_{i_u}$ are pairwise distinct and $f(x_{i_u}) = y_{i_{u+1}}$. If it were to happen that $y_{i_{u+1}} = y_{i_v}$ for some $1 \leqslant v \leqslant u$, then we would have $f(x_{i_u}) = f(x_{i_{v-1}})$, taking x_{i_0} to be m. But f is 1–1, so this would give $x_{i_u} = x_{i_{v-1}}$, and either $x_{i_u} = m$ or $y_{i_u} = y_{i_{v-1}}$, contradicting the hypotheses in either case. Thus the y_{i_u} are all distinct. Therefore, for some $s \leqslant k+1$ we have

$$f(m) = y_{i_1},\ f(x_{i_1}) = y_{i_2}, ...,\quad f(x_{i_{s-1}}) = y_{i_s},\quad f(x_{i_s}) \notin \{y_0, ..., y_k\}\,. \tag{14}$$

Finally, we let $y_{k+1} = f(x_{i_s})$.

It is easy to see that in all three cases the sequence (13) so obtained will be a finite correspondence.

Moreover, the function S is determined. By interchanging the roles of f, $x_0, ..., x_k$ and $g, y_0, ..., y_k$, we define the function T analogously. It remains to prove that S, T are **R**-functions. For this task we define auxiliary numerical functions φ, ψ, ψ', χ, χ', ω, σ as follows:

$$\varphi(n) = \mu z(\overline{sg}(\mathrm{rem}(n, p_z)) = 0)\,,$$

$$\psi(i, n) = \mathrm{l}(\exp(i, n) \dot{-} 1) + n \cdot \overline{sg}(\exp(i, n)) ,$$

$$\psi'(i, n) = \mathrm{r}(\exp(i, n) \dot{-} 1) + n \cdot \overline{sg}(\exp(i, n)) .$$

Obviously, if n is the number of a correspondence of the form (11), then $\varphi(n) = k + 1$, while

$$\psi(i, n) = \begin{cases} x_i & \text{if } i < k+1 , \\ n & \text{if } i \geqslant k+1 ; \end{cases}$$

$$\psi'(i, n) = \begin{cases} y_i & \text{if } i < k+1 ; \\ n & \text{if } i \leqslant k+1 . \end{cases}$$

Now we set

$$\chi(m,n) = \mu z((\varphi(n) \dot{-} z) \cdot \prod_{i=0}^{z} |m - \psi(i, n)| = 0) ,$$

$$\chi'(m, n) = \mu z((\varphi(n) \dot{-} z) \cdot \prod_{i=0}^{z} |m - \psi'(i, n)| = 0) .$$

Since

$$m \in \{x_0, ..., x_k\} \Leftrightarrow \prod_{i=0}^{k} |m - x_i| = 0 ,$$

we find that

$$\chi(m, n) = \begin{cases} i \quad \text{if } m \notin \{x_0, ..., x_{i-1}\}, \ m = x_i, \ i < k+1 , \\ k+1 \quad \text{otherwise;} \end{cases}$$

$$\chi'(m, n) = \begin{cases} i \quad \text{if } m \notin \{y_0, ..., y_{i-1}\}, \ m = y_i, \ i < k+1 , \\ k+1 \quad \text{otherwise.} \end{cases}$$

Finally, we put

$$\omega(n, y, 0) = \chi'(y, n) ,$$

$$\omega(n, y, t+1) = \chi'(f(\psi(\omega(n, y, t), n)), n) .$$

Now it is easy to see that for case (iii), when (14) holds we have

$$\omega(n, f(m), 0) = i_1, \quad \ldots, \quad \omega(n, f(m), s-1) = i_s,$$

$$\omega(n, f(m), s) = k + 1,$$

and consequently, $s = \rho(m, n)$, where ρ is defined by

$$\rho(m, n) = \sum_{i=0}^{k} \prod_{j=0}^{i} \operatorname{sg} |\omega(n, f(m), j) - (k+1)| .$$

With the aid of the indicated auxiliary functions the definition of the function S can be expressed by the following scheme:

$$S(m, n) = \begin{cases} n \cdot p_{(n)}^{\exp(\chi(m,n),n)} & \text{if } \chi(m, n) < \varphi(n), \\ n \cdot p_{(n)}^{1+\nu(m,f(m))} & \text{if } \chi(m, n) \geqslant \varphi(n) \text{ and } \chi'(m, n) \geqslant \varphi(n), \\ n \cdot p_{(n)}^{1+\nu(m,f(\psi(\omega(n,f(m)), \rho(m,n)-1),n)))} & \text{otherwise}. \end{cases}$$

From the formulas defining φ, ψ, ψ', χ, χ', it is clear that these functions are primitive recursive. As regards the functions ω and ρ, they are primitive recursive or general recursive depending on whether f is a **prim**- or **gr**-function; since α is simple, f is total, so the cases $\mathbf{R} = \mathbf{gr}, \mathbf{pr}$ coincide. Therefore, if f is an $\mathbf{R}$-function, so is S by the above scheme.

We can prove analogously that if g is an $\mathbf{R}$-function, so is T.

The function S permits (by giving the number (12) of the result) any finite correspondence (11) to be extended by a pair $\langle m, y_{k+1} \rangle$ whose left-hand member can be any natural number m. Similarly, the function T allows any finite correspondence (11) to admit a new pair $\langle x_{k+1}, m \rangle$ whose right-hand member is an arbitrary given natural number m. As a starting point we take the correspondence $\langle\langle 0, f(0) \rangle\rangle$, consisting of a single pair; by successively applying S and T to it, we progressively extend it to a map from D onto D.

For a more precise description of this map, we introduce a function η whose value $\eta(r)$ is the number (12) of the rth extension of the initial sequence. This function is given by

$$\eta(0) = p_0^{1+\nu(0,f(0))},$$

$$\eta(r+1) = \begin{cases} T(r/2, \eta(r)) & \text{if } r \text{ is even}, \\ S((r+1)/2, \eta(r)) & \text{if } r \text{ is odd}. \end{cases}$$

Since the function η is generated by primitive recursion from the **R**-function

$$S([\frac{n+1}{2}], z) \cdot (n - 2[\frac{n}{2}]) + T([\frac{n}{2}], z) \cdot |n - 2[\frac{n}{2}] - 1|,$$

it is itself an **R**-function.

Finally, let h be the function such that if x is the left-hand member of a pair in the finite correspondence with number $\eta(r)$ for some r, then $h(x)$ is the right-hand member. That h is a 1–1 partial function is immediate from the definition of finite correspondences. According to the definition of η, the last pair in the finite correspondence with number $\eta(2x)$ has the form $\langle x, y_{2x} \rangle$, so

$$h(x) = y_{2x} = \mathfrak{r}(\exp(2x, \eta(2x)) - 1),$$

showing h is total and, in fact, is an **R**-function. Similarly, the last pair in the finite correspondence with number $\eta(2y+1)$ is $\langle x_{2y+1} \rangle$, hence

$$h^{-1}(y) = x_{2y+1} = \mathfrak{l}(\exp(2y+1, \eta(2y+1)) - 1);$$

thus h maps D onto D, and its inverse is an **R**-function.

By the definition of finite correspondence, h **R**-unimorphically reduces α to β, while its inverse h^{-1} **R**-unimorphically reduces β to α; hence α and β are **R**-isomorphic. ∎

The obvious example below shows that Theorem 2.2.2 fails for nonsimple numberings, generally speaking.

Example 9: Let the set A consist of a single element, and determine two numberings α, β of A by taking D_α to be the set of all even numbers and D_β to be any set not recursively enumerable which contains all the even numbers. The identity map **prim**-unireduces α to β, while multiplication by 2 **prim**-unireduces β to α. However, α cannot possibly be even **pr**-isomorphic to β, for the set of all values taken by a **pr**-function for even arguments must certainly be recursively enumerable.

We can conveniently formulate an extension of Theorem 2.2.2 to more general numberings if we agree to let $\mathbf{R}'$ stand for one of the properties: "general recursive" or "partial recursive" for functions (correspondingly: "recursive" or "recursively enumerable" for sets).

Corollary: *If numberings α, β of a set A are* $\mathbf{R}'$*-uniequivalent and their number sets D_α, D_β are* $\mathbf{R}'$*-sets, then α is* $\mathbf{R}'$*-isomorphic to β.*

If one of the number sets is finite, then so is the other, and the result follows trivially. So let us assume both are infinite. Then by Theorem 2.2.2 there exist simple numberings α', β' of A that are $\mathbf{R}'$-unimorphic to α, β, respectively. Since D_α, D_β are recursively enumberable, the $\mathbf{R}'$-unimorphism of α' to α and β' to β implies their respective $\mathbf{R}'$-isomorphism. Hence,

$$\alpha' \equiv_{\mathrm{R}'} \alpha \equiv_{\mathrm{R}'1} \beta \equiv_{\mathrm{R}'} \beta' ;$$

this means $\alpha' \equiv_{\mathrm{R}'1} \beta'$. By virtue of Theorem 2.2.2 this implies $\alpha' \equiv_{\mathrm{R}'} \beta'$; therefore, $\alpha \equiv_{\mathrm{R}'} \beta$. ■

§2.3. *Equivalent numberings*

According to §2.2 two numberings α, β of a set A are $\mathbf{R}$-equivalent when there is an $\mathbf{R}$-function reducing α to β, as well as an $\mathbf{R}$-function reducing β to α, i.e., in case there is an $\mathbf{R}$-procedure enabling us, given an α-number of an arbitrary element $a \in A$, to find a β-number for a, and there is also an $\mathbf{R}$-procedure permitting the reverse. Clearly, if we are interested not only in the properties of the set A, but also in those of its numberings, we should consider only numberings isomorphic in some sense to be "identical". On the other hand, if the properties of A itself interest us, while its numberings are regarded as just an added tool, then it is natural to consider as "identical" those numberings equivalent with respect to a given sort of function. Therefore, it is normally important to know whether some kind of complicated numbering of A may not be equivalent to a less complicated one, e.g., a simple or even 1–1 simple numbering. In this scene, clearly, special roles should be played on the one hand by numberings equivalent in the desired sense to 1–1 simple ones, and on the other by "stable" numberings, which are isomorphic in the desired sense to any numberings so equivalent to them. We shall now examine these two sorts of numberings. As above, $\mathbf{R}' = \mathbf{gr}, \mathbf{pr}$.

Theorem 2.3.1: *In order that a numbering α of a set A be $\mathbf{R}'$-equivalent to a simple 1–1 numbering of A, it is necessary and sufficient that the following two conditions hold:*

(i) *The number set D_α includes a* **pr**-*set M having a nonempty intersection with every θ_α-class in D_α, while the number of equivalence classes composing D_α/θ_α is infinite.*

(ii) *Viewed as a numerical function, θ_α is an $\mathbf{R}'$-function on D_α relative to the trivial numbering* (cf. §1.1, §2.1), *i.e., there exists a binary numerical $\mathbf{R}'$-function L strongly satisfying*

$$L(x, y) = \theta_\alpha(x, y) \qquad (x, y \in D_\alpha) . \tag{15}$$

To prove this theorem we first assume α is $\mathbf{R}'$-equivalent to a simple 1–1 numbering β. Thus, A must be infinite, and with it D_α/θ_α. Let f, g be $\mathbf{R}'$-functions respectively reducing α to β and β to α. Then letting $M = \{g(0), g(1), \ldots\}$ fulfills condition (i), and putting $L(x, y) = \overline{\text{sg}}\,|f(x) - f(y)|$ satisfies (ii).

Conversely, let M be a subset of D_α containing at least one element of each class belonging to D_α/θ_α, and suppose M, being nonempty, is the range of the function $\varphi \in \mathcal{F}_{\text{prim}}$. We introduce an auxiliary function ψ specified by

$$\psi(0) = 0\,,$$

$$\psi(n+1) = \mu z \sum_{i=0}^{n} L(\varphi(\psi(i)), \varphi(z)) = 0)\,,$$

where L is an $\mathbf{R}'$-function satisfying (15).

We are assuming the set D_α/θ_α is infinite, so ψ is a general recursive function; moreover,

$$L(\varphi(\psi(i)), \varphi(\psi(n+1))) = 0 \quad (i = 0, \ldots, n)\,. \tag{16}$$

We now construct a simple numbering β of A by setting $\beta n = \alpha\varphi(\psi(n))$. From (16) it follows that β is 1–1.

The $\mathbf{R}'$-functions f, g defined by

$$f(x) = \mu z(\psi(z) = \mu y(L(x, \varphi(y)) = 1))\,,$$

$$g(x) = \varphi(\psi(x)) \quad (x \in D)$$

reduce α to β and β to α, respectively. ■

If we consider only numberings of infinite sets with recursively enumerable number sets, then condition (i) holds automatically. In particular, *a simple numbering α of an infinite set A is recursively equivalent to a 1–1 simple numbering of A iff θ_α is recursive.* ■ ([8])

On the other hand, if the numbering is 1–1, we can take $L(x, y) = \overline{\text{sg}}\,|x - y|$ and immediately deduce from Theorem 2.3.1 that a 1–1 numbering α of an infinite set A is $\mathbf{R}'$-equivalent to a simple 1–1 numbering iff the number set D_α is recursively enumerable.

In turn it follows that (i) and (ii) in Theorem 2.3.1 are also necessary and sufficient conditions that a numbering α of an infinite set A be $\mathbf{R}'$-equivalent to a 1–1 numbering of A with recursively enumerable number set.

Now we examine in more detail the conditions under which the $\mathbf{R}'$-multireducibility of a numbering α to a numbering β entails the $\mathbf{R}'$-unireducibility of α to β.

In order that a numbering α of a set A be $\mathbf{R}'$-unireducible to a numbering β of A, it is obviously necessary that the number of different α-numbers of a given arbitrary element in A not exceed the number of its distinct β-numbers. When D_α is finite, this condition is also sufficient. If the property is made effective, it becomes a general sufficient condition for $\mathbf{R}'$-multireducibility to yield $\mathbf{R}'$-unireducibility.

Theorem 2.3.2: *Suppose the number set D_α of the numbering α of the set A is an $\mathbf{R}'$-set, and α is reduced by the $\mathbf{R}'$-function f to the numbering β of A. Also suppose there exists a binary numerical $\mathbf{R}'$-function Q such that $Q(f(x),y)$ is defined and $\beta Q(f(x),y) = \beta f(x)$ for all $x \in D_\alpha$, $y \in D$, and that the number of different values $Q(f(x),y)$ ($y \in D$) for any given $x \in D_\alpha$ is not less than the number of distinct α-numbers of $\alpha x \in A$. Then α is $\mathbf{R}'$-unireducible to β.*

We can assume D_α is infinite. We construct a partial function g unireducing α to β by the following effective procedure. Since D_α is assumed to recursively enumerable, it is the range of some unary 1–1 **gr**-function φ. Introducing the notation $a_n = \varphi(n)$ ($n = 0, 1, \ldots$), we set $g(a_0) = f(a_0)$ and define further values of g by the following recursion. Let $b_i = g(a_i)$ be already defined for $i = 0, \ldots, n$. If $f(a_{n+1})$ is distinct from $b_0, \ldots, b_n$, put $g(a_{n+1}) = f(a_{n+1})$. If $f(a_{n+1}) = b_m$ for some $0 \leqslant m \leqslant n$, find the minimal s such that $Q(b_m, s)$ is distinct from $b_0, \ldots, b_n$, and put $g(a_{n+1}) = Q(b_m, s)$. We can always find such an s because there are at least as many different values $Q(b_m, y)$ ($y \in D$) as different α-numbers for the element αa_{n+1} of A.

The function g so constructed obviously unireduces α to β. Since an algorithm has been specified for calculating the value $g(x)$ for every x belonging to the $\mathbf{R}'$-set D_α, by the Church–Kleene thesis g is a **pr**-function; when $\mathbf{R}' = \mathbf{gr}$, g can be extended to a **gr**-function. So α is $\mathbf{R}'$-unireducible to β. ∎

A numbering α of a set A is said to have *infinite classes* iff every element of A has infinitely many different α-numbers.

We shall say a numbering α has $\mathbf{R}'$*-infinite classes* iff there exists a binary numerical $\mathbf{R}'$-function Q such that $Q(x, y)$ is defined and $\alpha Q(x, y) = \alpha x$ for all $x \in D_\alpha$, $y \in D$, and that the set of values $\{Q(x, y): y \in D\}$ is infinite for every $x \in D_\alpha$.

From Theorem 2.3.2 we immediately obtain

Corollary 1: *Suppose the number set D_α of the numbering α is an $\mathbf{R}'$-set and α is $\mathbf{R}'$-multireducible to the numbering β with $\mathbf{R}'$-infinite classes. Then α is $\mathbf{R}'$-unireducible to β.* ∎

By applying this corollary twice and keeping the corollary to Theorem 2.2.2 in mind, we can derive

Corollary 2: *Suppose the numberings α, β have* **R**$'$*-infinite classes, and* **R**$'$ *number sets. If α and β are* **R**$'$*-equivalent, then they are* **R**$'$*-isomorphic.* ■

Recall that a numbering α is said to be *positive* when its number set D_α and equivalence relation θ_α are both partial recursive.

Remark: *Every positive numbering with infinite classes has* **pr**-*infinite classes.*

In fact, the recursive enumerability of θ_α means there is a partial recursive function L such that $L(x, y)$ is defined for all $x, y \in D_\alpha$ and (15) holds. ([9]) Let us represent D_α as the range of some unary **prim**-function φ. Then the function Q defined by

$$Q(x, y) = \varphi(y) \cdot L(x, \varphi(y)) + x \cdot \overline{\text{sg}}(L(x, \varphi(y))) \quad (x \in D_\alpha, y \in D)$$

has all the properties required in the definition of a numbering with **pr**-infinite classes. ■

This remark and Corollaries 1 and 2 yield

Corollary 3: *If the number set of the numbering α is recursively enumerable, and if α is* **pr**-*reducible to a positive numbering β with infinite classes, then α is* **pr**-*unireducible to β. Furthermore, if positive numberings α, β with infinite classes are* **pr**-*equivalent, they are also* **pr**-*isomorphic.* ■

We have pointed out some cases where the **R**$'$-equivalence of numberings α, β entails their **R**$'$-isomorphism. In the above the conditions laid on the numberings were symmetrical in α and β, on the whole. We now want to strengthen the conditions on α so that the conditions on β can be weakened considerably without losing the desirable property that α and β are isomorphic if they are equivalent, relative to some given class of functions.

We introduce a definition: a numbering α is said to be **R**-*stable* iff D_α is an **R**-set and α is **R**-isomorphic to every **R**-equivalent numbering with an **R** number set.

As **R** ranges over **prim, gr, pr**, we obtain three kinds of stability. The last two are so related: a **pr**-stable numbering with a general recursive number set is **gr**-stable.

The following straightforward theorem shows **R**$'$-stable numberings are always rather complicated.

Theorem 2.3.3: *No θ_α-class of an* $\mathbf{R}'$*-stable numbering* α *can be recursive* **($\mathbf{R}'$ = gr, pr).**

Let D_β be the set of numbers obtained by multiplying the members of D_α by 2, and define a new numbering β (of the same set A as α) by setting $\beta x = \alpha(x/2)$ for $x \in D_\beta$. Let us suppose that, contrary to the assertion of Theorem 2.3.3, the set $\alpha^{-1}(a)$ consisting of all α-numbers of some element $a \in A$ is indeed a **gr**-set. Then the set $\beta^{-1}(a)$ is also recursive. First we consider the case when $\beta^{-1}(a)$ contains more than one element. Let s be some fixed element of $\beta^{-1}(a)$ and let $D_{\beta*} = (D_\beta \sim \beta^{-1}(a)) \cup \{s\}$, which becomes the number set of the numbering β^* defined by putting $\beta^* x = \beta x$ $(x \in D_{\beta*})$. Then there are $\mathbf{R}'$-functions f, g such that

$$f(x) = x \qquad (x \in D)\,,$$

$$g(x) = \begin{cases} x & \text{if } x \in D_\beta \sim \beta^{-1}(a) \\ s & \text{if } x \in \beta^{-1}(a); \end{cases}$$

f reduces β^* to β, g reduces β to β^*. Thus β^* and β are $\mathbf{R}'$-equivalent. The numbering β is $\mathbf{R}'$-stable because it is **prim**-isomorphic to α. Since $D_{\beta*}$ is an $\mathbf{R}'$-set and β^* is $\mathbf{R}'$-equivalent to β, it must be that β^* and β are $\mathbf{R}'$-isomorphic. This, however, is impossible, for the element a has only one β^*-number, but more than one β-number. If we can get another contradiction when $\beta^{-1}(a)$ consists of a single number, we shall have proved the theorem. Assuming $\beta^{-1}(a)$ has only one element and taking $D_{\beta*} = D_\beta \cup \{1\}$, $\beta^* 1 = a$, $\beta^* x = \beta x$ $(x \in D_\beta)$, we can apply an argument similar to the one above to obtain the same contradictory conclusion. ■

In order to formulate conditions sufficient for the stability of a numbering, we introduce the new concept of the completeness of a numbering. Namely, a numbering α of a set A is called *complete* iff it satisfies the following requirements:

(i) We can choose two particular elements e, e' of D_α that are α-numbers of distinct elements of A.

(ii) There exists an effective process whereby, given an arbitrary unary **gr**-function h, we can find a number $m \in D_\alpha$ such that $h(m) \in D_\alpha$ and $\alpha h(m) = \alpha m$.

More precisely, (ii) requires there to be a **gr**-function φ such that $\alpha h(\varphi(n)) = \alpha\varphi(n)$, where n is the number of h in the sense of Kleene [71].

Every complete numbering is simple. For we can take h to be the constant function whose every value is equal to some given number s. Then by (ii) there must be a number m for which $h(m) \in D_\alpha$, whence $s \in D_\alpha$. ■

The typical example of a complete numbering, which in fact motivated introducing the notion of completeness, is the *Kleene numbering* ξ of $\mathcal{F}_{\text{pr}}$, the set of all 1-place numerical partial recursive functions. To see this, we take e, e' to be, e.g., ξ-numbers for the functions f, g such that $f(x) = x$, $g(x) = x + 1$. According to the Recursion Theorem (Kleene [71]), for every ternary partial recursive function F there is a primitive recursive function $\varphi \in \mathcal{F}_{\text{prim}}$ such that $F(a, \varphi(a), y) = U(\varphi(a), y)$ $(a, y \in D)$ strongly, where U is Kleene's universal function: n is a ξ-number of the unary function whose value at $x \in D$ is $U(n, x)$, when defined.

Let n be a Kleene number of an arbitrary given **gr**-function h; thus $h(x) = U(n, x) = x \in D$. We define the function F by strongly setting

$$F(a, x, y) = U(U(a, x), y) \quad (a, x, y \in D)\,.$$

By the Recursion Theorem there is a **prim**-function φ such that $U(U(a, \varphi(a)), y) = U(\varphi(a), y)$ strongly for $a, y \in D$. With $m = \varphi(n)$ we have $U(m, y) = U(U(n, m), y)$ $(y \in D)$, i.e., $\xi m = \xi h(m)$.

By translating arguments of Rogers [141] into the language of numberings we easily prove

Theorem 2.3.4: *Let* α *be an arbitrary complete numbering. Then* (I) α *has* **gr**-*infinite equivalence classes;* (II) *if* α *is* **gr**-*reducible to a numbering* β, *then* α *is* **gr**-*unireducible to* β; (III) α *is* $\mathbf{R}'$-*stable* ($\mathbf{R}'$ = **gr**, **pr**).

Suppose the **gr**-function f reduces α to β. According to Theorem 2.3.2, to prove (II) it suffices to construct a binary function Q with the properties there described. For brevity's sake we shall just sketch an effective procedure for computing the values of Q.

To begin with we construct a binary **gr**-function S satisfying

$$\alpha S(x, y) = \alpha x\,,$$

$$f(S(x, y)) \neq f(S(x, z)) \qquad (x, y, z \in D;\; y \neq z)\,. \tag{17}$$

We start by defining $S(x, 0) = x$ for a given $x \in D$, and proceed further by recursion. Suppose $S(x, 0), \ldots, S(x, r)$ have been defined and satisfy (17) for $y, z = 0, \ldots, r$. Let h_1, h_2 be unary **gr**-functions such that for all $t \in D$,

$$h_1(t) = \begin{cases} x \text{ if } f(t) \notin \{f(S(x,0)), \ldots, f(S(x,r))\}, \\ e \text{ otherwise;} \end{cases}$$

$$h_2(t) = \begin{cases} e' \text{ if } f(t) \in \{f(S(x,0)), \ldots, f(S(x,r))\}, \\ e \text{ otherwise.} \end{cases}$$

From $S(x, 0), \ldots, S(x, r)$ and a Kleene number for f we can compute Kleene numbers n_1, n_2 for the functions h_1, h_2; using φ, the **gr**-function promised us by the completeness of α, we can calculate $m_1 = \varphi(n_1), m_2 = \varphi(n_2)$.

If $f(m_1) \notin \{f(S(x, 0)), \ldots, f(S(x, r))\}$, we set $S(x, r+1) = m_1$. The relation $\alpha h_1(m_1) = \alpha m_1$ gives us $\alpha m_1 = \alpha x$, so $S(x, r+1)$ satisfies (17).

If $f(m_1) \in \{f(S(x, 0)), \ldots, f(S(x, r))\}$, then we set $S(x, r+1) = m_2$. Now $f(m_1) = f(S(x, i)$ for some $0 \leqslant i \leqslant r$, which tells us $\alpha m_1 = \alpha S(x, i)$ since f is a reducing function. Thus $\alpha m_1 = \alpha x$ by (17); because $\alpha m_1 = \alpha h(m_1) = \alpha e$, we finally learn that $\alpha e = \alpha x$. If it were to happen that $f(m_2) = f(S(x, j))$ for some $0 \leqslant j \leqslant r$, we would have $\alpha e' = \alpha m_2 = \alpha x$, contradicting the facts: $\alpha e = \alpha x$ and $\alpha e \neq \alpha e'$. Therefore, $f(m_2) \notin \{f(S(x, 0)), \ldots, f(S(x, r))\}$, and this in turn gives $\alpha m_2 = \alpha e = \alpha x$. Consequently, the value chosen as $S(x, r+1)$ satisfies (17) for $y, z = 0, \ldots, r+1$.

We thus construct S. From the regularity of the construction process it follows that S is general recursive.

Since $\alpha S(x, y) = \alpha x$ and $S(x, y) \neq S(x, z)$ for $y \neq z$ $(x, y, z \in D)$, the numbering α has **gr**-infinite classes. On the other hand, the function Q such that

$$Q(x, y) = f(S(x, y)) \quad (x, y \in D)$$

fulfills the demands of Theorem 2.3.2, so α is **gr**-unireducible to β. Finally, suppose in addition that β is $\mathbf{R}'$-reducible to α and that D_β is an $\mathbf{R}'$-set. Because α has $\mathbf{R}'$-infinite classes, β is $\mathbf{R}'$-unireducible to α; by the corollary to Theorem 2.2.2, this implies α is $\mathbf{R}'$-isomorphic to β. Therefore, α is $\mathbf{R}'$-stable for both meanings of $\mathbf{R}'$. ∎

We saw earlier that the usual Kleene numbering ξ of the set $\mathcal{F}_{\text{pr}}$ is complete. Hence, as a special case of Theorem 2.3.4, we obtain the following theorem of Rogers [141]:

Corollary: *Every numbering of $\mathcal{F}_{\text{pr}}$ which has a recursively enumerable number set and is* **pr**-*equivalent to the Kleene numbering is actually* **pr**-*isomorphic to it.* ∎

Let ψ_n be the element of $\mathcal{F}_{\mathrm{pr}}$ with ξ-number n; this is not a unique notation since ξ is not 1–1. With each ψ_n is associated a recursively enumerable set $W_n = \{\psi_n(x): x \in D\}$. By considering n to be a Post number of the set W_n, we obtain the so-called Post numbering of the set $\mathcal{W}$ of all recursively enumerable subsets of D. From the completeness of the Kleene numbering of $\mathcal{F}_{\mathrm{pr}}$ we can obviously deduce completeness for the Post numbering of $\mathcal{W}$. Therefore, the indicated theorem of Rogers also holds for numberings of the set $\mathcal{W}$.

The next theorem was proved by Rice [126] for the Post numbering of $\mathcal{W}$ and later reproved by Uspenskiĭ [174] for the Kleene numbering of $\mathcal{F}_{\mathrm{pr}}$.

Theorem 2.3.5: *Every set A with a complete numbering α has but two subsets general recursive with respect to α: A itself and the empty set $\emptyset$.*

In other words, a nonempty proper subset M of D must be nonrecursive if whenever it contains an α-number of an arbitrary element $a \in A$, it contains all α-numbers of a.

Let us assume to the contrary that such a set M is recursive. Then its complement $\widetilde{M} = D {\sim} M$ is also recursive. Since complete numberings have infinite classes, M and $\widetilde{M}$ are both infinite. Let $M, \widetilde{M}$ be the respective ranges of appropriate 1–1 functions $g, h \in \mathcal{F}_{\mathrm{gr}}$. Define the functions $\varphi, f \in \mathcal{F}$ as follows:

$$\varphi(x) = \mu z(|x - g(z)| \cdot |x - h(z)| = 0)\,,$$

$$f(x) = g(\varphi(x)) \cdot \overline{\mathrm{sg}}\, |x - h(\varphi(x))| + h(\varphi(x)) \cdot \overline{\mathrm{sg}}\, |x - g(\varphi(x))|\,.$$

The function f is general recursive, but for every $m \in D$ we have $\alpha f(m) \neq \alpha m$, contradicting the completeness of α. ■

Theorem 2.3.5 can obviously be restated as: *every unary* **gr**-*predicate on a numbered set with complete numbering is constant.* ■

In the definition of complete numbering we demand not only the solvability of the equation $\alpha f(m) = \alpha m$ for every $f \in \mathcal{F}_{\mathrm{gr}}$, but also the existence of an algorithm for finding solutions uniformly. If we only require the existence of solutions, then the numbering might be called *formally complete.* In proving Theorem 2.3.5 only this formal completeness was used, so the theorem is valid for arbitrary formally complete numberings.

Let us say (following Post [121]) that a total function $f \in \mathcal{F}$ *reduces* a set of natural numbers M to another such set N iff for every $x \in D$,

$$x \in M \Leftrightarrow f(x) \in N\,.$$

We say M is *(recursively) reducible* to N iff there exists a **gr**-function f reducing

M to N. When f can be chosen 1–1, M is *(recursively) unireducible* to N. When M is reducible to N and N to M, we say M is *recursively equivalent* to N. Finally, M is *recursively isomorphic* to N iff there exists a **gr**-function $f \in \mathcal{F}_{\mathrm{gr}}$ which maps $\boldsymbol{D}$ 1–1 onto $\boldsymbol{D}$ and carries M onto N.

In order to establish the connection between these notions and the corresponding ideas in the theory of numberings, we consider simple numberings of a fixed two-element set A, whose elements are denoted by T, F. With every subset M of D we associate the simple numbering α_M of A determined by

$$\alpha_M n = \begin{cases} \mathsf{T} & \text{if } n \in M, \\ \mathsf{F} & \text{if } n \notin M. \end{cases}$$ (10)

Note that for $\mathsf{T} = 1$, $\mathsf{F} = 0$ the numbering α_M is none other than the characteristic function χ_M of the set M.

It is clear that the concepts of recursive reducibility, unireducibility, equivalence, and isomorphism have the same meaning for sets as the respective general recursive notions have for the corresponding numberings of A. Therefore, every theorem concerning the **gr**-reducibility, etc. of numberings yields as a special case (possibly vacuous) a theorem about the reducibility, etc. of sets of numbers. E.g., from Theorem 2.2.2 we get the following known

Theorem 2.3.6 (Myhill [111]): *If each of the sets, M, N is recursively unireducible to the other, then they are recursively isomorphic.* ∎

Let us provisionally agree that a set $M \subseteq D$ is *effectively infinite* iff M includes an infinite recursively-enumerable subset. Obviously, the numbering α_M has **gr**-infinite classes iff the set M and its complement $\widetilde{M}$ are both effectively infinite (11). From Corollary 2 of Theorem 2.3.2 we learn that *if effectively infinite sets with effectively infinite complements are recursively equivalent, they are recursively isomorphic.* ∎

Following Mučnik [109], for any $k \geqslant 1$ we say that a system $\langle M_1, \ldots, M_k \rangle$ of numerical sets is *reducible* to a system $\langle N_1, \ldots, N_k \rangle$ iff there exists a **gr**-function $f \in \mathcal{F}_{\mathrm{gr}}$ such that

$$x \in M_i \Leftrightarrow f(x) \in N_i \quad (x \in D;\ i = 1, \ldots, k).$$

The notions of unireducibility, recursive equivalence, and recursive isomorphism for systems of subsets of D are defined analogously.

Let $A_k = \{0, 1, \ldots, k\}$. With every system of pairwise disjoint nonempty sets $M_1, \ldots, M_k$ we associate a simple numbering α of A_k defined by

$$\alpha n = \begin{cases} i & \text{if } n \in M_i \quad (i = 1, ..., k), \\ 0 & \text{if } n \notin M_1 \cup ... \cup M_k. \end{cases}$$

Recursive reducibility, etc., of systems of sets are seen to be equivalent to the respective properties of the corresponding numberings, as long as $D \sim \cup M_i \neq \emptyset$. Systems satisfying this last condition are considered *normal.* For unnormal systems it suffices to take $A_k = \{1, ..., k\}$. Theorem 2.2.2, applied to numberings of A_k, gives us

Theorem 2.3.7: *Recursively uniequivalent systems of pairwise disjoint non-empty sets are recursively isomorphic.* ■

This represents a slight strengthening of Theorem 1 in the cited article by Mučnik.

The general formulation of Theorem 2.2.2 (in its Corollary) can clearly be understood as an isomorphism theorem for arbitrary – finite and infinite – systems of sets.

§3. Numbered algebraic systems

§3.1. **R**-*numberings of algebraic systems*

Suppose we have an algebraic system (see §1.1)

$$\mathfrak{A} = \langle A; a_1, ..., a_l; f_1, ..., f_m; g_1, ..., g_n; P_1, ..., P_p \rangle,$$

where $a_1, ..., a_l$ are the distinguished elements, $f_1, ..., f_m$ the operations (all total), $g_1, ..., g_n$ the partial operations, and $P_1, ..., P_p$ the predicates, all defined on the base set A. By supplementing $\mathfrak{A}$ with some numbering α of the set A, we obtain a more complex object $\langle \mathfrak{A}, \alpha \rangle$ which we shall call a *numbered algebraic system.* The ordinary, unnumbered algebraic systems will, for occasional contrast, be called *abstract algebraic systems.* The styles of homomorphism and isomorphism introduced in §1.1 for abstract systems will at times be called *abstract* in order to differentiate them from those defined below in connection with numbered systems.

A numbering α of a system $\mathfrak{A}$ is called an **R**-*numbering* iff all the basic operations (including the partial ones) and predicates of $\mathfrak{A}$ are respectively **R**-operations and **R**-predicates relative to α.

The concepts of **R**-map, **R**-monomorphism, **R**-equivalence, etc., introduced in §2.1 for numbered sets can be extended in the following way to numbered algebraic systems.

Let us consider two arbitrary similar numbered algebraic systems $\mathfrak{A}$, $\mathfrak{B}$, whose bases A, B have the numberings α, β. A homomorphism of $\mathfrak{A}$ into $\mathfrak{B}$ that is an **R**-map of $\langle A, \alpha\rangle$ into $\langle B, \beta\rangle$ in the sense of §2.1 is an **R**-*homomorphism* of $\langle\mathfrak{A}, \alpha\rangle$ into $\langle\mathfrak{B}, \beta\rangle$ (or simply, of $\mathfrak{A}$ into $\mathfrak{B}$).

An abstract isomorphism from $\mathfrak{A}$ onto $\mathfrak{B}$ that is also an **R**-homomorphism is called an **R**-*monomorphism* of $\mathfrak{A}$ onto $\mathfrak{B}$. An abstract isomorphism from $\mathfrak{A}$ onto $\mathfrak{B}$ that is an **R**-*equivalence* between $\langle A,\alpha\rangle$ and $\langle B, \beta\rangle$ is called an **R**-*equivalence* between the numbered systems $\mathfrak{A}$ and $\mathfrak{B}$. Analogously, an **R**-*unimorphism* (**R**-*isomorphism*) of $\mathfrak{A}$ onto $\mathfrak{B}$ is an abstract isomorphism from $\mathfrak{A}$ onto $\mathfrak{B}$ that is an **R**-unimorphism (**R**-isomorphism) from $\langle A, \alpha\rangle$ onto $\langle B, \beta\rangle$, as well.

The numbered algebraic systems $\mathfrak{A}$, $\mathfrak{B}$ are **R**-*equivalent* iff there exists an **R**-equivalence mapping $\mathfrak{A}$ onto $\mathfrak{B}$. $\mathfrak{A}$ and $\mathfrak{B}$ are **R**-*isomorphic* iff there exists **R**-isomorphism from $\mathfrak{A}$ onto $\mathfrak{B}$.

The concepts of **R**-equivalence and **R**-isomorphism of systems are central to the theory of numbered algebraic systems. Of their significance, we can repeat what was said in §2.3 concerning the corresponding notions for numberings of a given set. Namely, if the purely algebraic properties of numbered systems interest us, we should consider equivalent systems (in some sense) as "identical". But if we are interested in both the algebraic properties and those of the numberings, then we should regard as "identical" only those systems isomorphic with respect to their numberings.

The definitions of monomorphism, etc., for numbered algebraic systems have been chosen so that all the theorems in §2.1 concerning numbered sets remain valid for numbered systems. In particular, from Theorem 2.1.5 we immediately get

Theorem 3.1.1: *If an algebraic system* $\mathfrak{A}$ *has an* **R**-*numbering and is* **R**-*equivalent to a numbered algebraic system* $\mathfrak{B}$, *then the numbering of* $\mathfrak{B}$ *is also an* **R**-*numbering.* ■

By analogy with positively and negatively numbered sets, a numbered algebraic system $\langle\mathfrak{A}, \alpha\rangle$ is called *positively (negatively) numbered* iff α is a **pr**-numbering of $\mathfrak{A}$ and positively (negatively) numbers the base A.

The corollary of Theorem 2.1.5 shows that if of two **pr**-equivalent algebraic systems with recursively enumerable number sets, one is positively (negatively) numbered, then so is the other.

Furthermore, if an algebraic system is numbered positively (negatively), then so is every **pr**-isomorphic system.

To enable immediate use of the results of §2.2 and §2.3 in the study of numbered systems, we make yet another definition.

Suppose φ is an abstract isomorphism from the algebraic system $\mathfrak{A}$ with numbering α onto the abstract algebraic system $\mathfrak{B}$. We now define a numbering β of $\mathfrak{B}$ by putting $D_\beta = D_\alpha$ and

$$\beta n = \varphi(\alpha n) \quad (n \in D_\beta) , \tag{18}$$

i.e., by taking $\beta = \varphi * \alpha$. The numbering β is called the *translation* of α from $\mathfrak{A}$ to $\mathfrak{B}$ under φ. By (18) the map is a **prim**-isomorphism from $\langle \mathfrak{A}, \alpha \rangle$ onto $\langle \mathfrak{B}, \beta \rangle$.

To avoid repetition we introduce the symbol **Q**, which can assume any of the meanings: mono-, equi-, uni-, iso-. Thus for numbered sets (for numberings) the expression **Q**-*morphic*, as **Q** varies, will mean: monomorphic (reducible), equivalent (equivalent), unimorphic (unimorphic), isomorphic (isomorphic).

Suppose now we are given some abstract isomorphism φ from an algebraic system $\mathfrak{B}$ with numbering β' onto an algebraic system $\mathfrak{A}$ with numbering α. Translating β' from $\mathfrak{B}$ to $\mathfrak{A}$ by means of φ, we obtain a new numbering $\beta = \varphi * \beta'$ of the system $\mathfrak{A}$. Thus, the abstract system $\mathfrak{A}$ presents two aspects: as the numbered system $\langle \mathfrak{A}, \alpha \rangle$ and as the numbered system $\langle \mathfrak{A}, \beta \rangle$. It is easy to see that the abstract isomorphism φ is an **RQ**-morphism from $\langle \mathfrak{B}, \beta' \rangle$ onto $\langle \mathfrak{A}, \alpha \rangle$ iff the numbering β is **RQ**-morphic to the numbering α.

This has the following immediate consequence. Let α, β be two numberings of the abstract system $\mathfrak{A}$. Then $\langle \mathfrak{A}, \alpha \rangle$ is **RQ**-morphic to $\langle \mathfrak{A}, \beta \rangle$ iff there exists an abstract automorphism φ of $\mathfrak{A}$ such that the numbering $\varphi * \alpha$ is **RQ**-morphic to the numbering α.

This corollary of the definitions can be alternatively formulated as an assertion, which for ease of reference we call a theorem.

Theorem 3.1.2: *Let φ be an abstract automorphism of an algebraic system $\mathfrak{A}$ with numbering α. Translating α by φ, we get a new numbering $\varphi * \alpha$ of the system $\mathfrak{A}$. In order that $\langle \mathfrak{A}, \varphi * \alpha \rangle$ be* **RQ**-*morphic to $\langle \mathfrak{A}, \alpha \rangle$* ([12]), *it is necessary and sufficient that φ be an* **RQ**-*morphism of $\langle \mathfrak{A}, \alpha \rangle$ onto itself.* ■

This theorem and the remark preceding it allow us to make a quick survey of all numbered algebraic systems that can be obtained from a given abstract system by laying various numberings on it.

Let us consider some numbering α of a system $\mathfrak{A}$. How can we find all numberings β of $\mathfrak{A}$ such that $\langle \mathfrak{A}, \beta \rangle$ is **R**-equivalent to $\langle \mathfrak{A}, \alpha \rangle$?

Here's the answer. Let $\mathcal{E}_\alpha$ be the set of all numberings of $\mathfrak{A}$ **R**-equivalent to α; let $\mathcal{N}_\alpha$ be the collection of all numberings of $\mathfrak{A}$ obtainable as translations of those in $\mathcal{E}_\alpha$ by abstract automorphisms of $\mathfrak{A}$. Then $\mathcal{N}_\alpha$ is just the set of all numberings β of $\mathfrak{A}$ for which $\langle \mathfrak{A}, \beta \rangle$ is **R**-equivalent to $\langle \mathfrak{A}, \alpha \rangle$.

The set $\mathcal{N}_\alpha$ breaks up into classes of **R**-equivalent numberings. Theorem

3.1.2 helps us add a few details regarding these classes. Let $\mathfrak{G}$ be the group of all abstract automorphisms of $\mathfrak{A}$, and let $\mathfrak{G}_\alpha$ be the subgroup of $\mathfrak{G}$ consisting of all **R**-equivalences of $\langle \mathfrak{A}, \alpha\rangle$ onto itself. Then $\mathfrak{G}_\alpha$ is invariant in $\mathfrak{G}$ ([13]), and the classes of **R**-equivalent numberings in $\mathfrak{N}_\alpha$ are in 1–1 correspondence with the cosets of $\mathfrak{G}_\alpha$ in $\mathfrak{G}$.

Analogous statements can be made concerning the numberings of an abstract algebraic system $\mathfrak{A}$ that convert $\mathfrak{A}$ into **R**-isomorphic numbered systems.

Example 10: Let $\mathfrak{A}$ be the algebra with base

$$A = \{a_0, a_1, \dots\} \quad (a_i \neq a_j \text{ for } i, j \in D, \ i \neq j)$$

and binary operation $\times$ with

$$a_m \times a_n = a_n \qquad (m, n \in D) .$$

$\mathfrak{A}$ is clearly a semigroup; moreover, every 1–1 map of A onto itself is an automorphism of $\mathfrak{A}$. This implies any two simple 1–1 numberings of A turn $\mathfrak{A}$ into **prim**-isomorphic numbered semigroups. Now let $f \in \mathcal{F} \sim \mathcal{F}_{gr}$ be a nonrecursive 1–1 function mapping D onto itself. We introduce two simple 1–1 numberings α, β of $\mathfrak{A}$ by setting

$$\alpha n = a_n, \quad \beta n = a_{f(n)} \qquad (n \in D) .$$

The numberings α, β are not even **pr**-equivalent, although the numbered algebras $\langle \mathfrak{A}, \alpha\rangle, \langle \mathfrak{A}, \beta\rangle$ are **prim**-isomorphic.

The structure of a numbered algebraic system $\langle \mathfrak{A}, \alpha\rangle$ is completely determined by specifying the number set D_α, the equivalence relation θ_α and functions representing the basic operations and predicates in coordinate form. It is natural to distinguish certain classes of numbered systems depending on whether this numerical set and predicate and some choice of numerical coordinate functions are partial, general, or primitive recursive for each system in the class.

(I) A numbered algebraic system is called *positive (negative)* iff its numbering is partial recursive and positive (negative).

From the remarks made at the end of §2.1 it follows that a numbered algebraic system **pr**-isomorphic to a positive (negative) system is itself positive (negative). Furthermore, if a numbered system with recursively enumerable number set is **pr**-equivalent to a positive (negative) system, it is also positive (negative).

In addition, Theorem 2.2.1 and its proof imply every positive (negative) system with infinite number set is **pr**-isomorphic to a positive (negative) system whose numbering is simple.

(II) A numbered algebraic system $\langle \mathfrak{A}, \alpha \rangle$ is called *general recursive* iff D_α is recursively enumerable and θ_α, the basic operations, and the basic predicates are general recursive.

Theorem 3.1.1 implies that if one of two **gr**-equivalent numbered systems having recursively enumerable number sets is general recursive, then so is the other.

Theorems 3.1.1 and 2.1.1 show that every numbered system **gr**-isomorphic to a general recursive system is itself general recursive.

(III) A numbered system $\langle \mathfrak{A}, \alpha \rangle$ is called *constructive* iff α, D_α and θ_α are general recursive (i.e., iff α is a decidable **gr**-numbering).

From Theorem 3.1.1, again, we learn that if either of two **gr**-equivalent numbered systems with general recursive number sets is constructive, then so is the other. In addition, Theorem 2.1.1 and its proof tell us that a **pr**-isomorphism (**pr**-equivalence) from one constructive system onto another is actually a **gr**-isomorphism (**gr**-equivalence).

Theorem 3.1.3: *Every infinite constructive algebraic system is* **gr**-*isomorphic to a constructive, simply numbered system and* **gr**-*equivalent to a constructive system with a* 1–1 *simple numbering.*

Every infinite constructive system with 1–1 *numbering is* **gr**-*isomorphic to a constructive system with simple* 1–1 *numbering.*

Let $\langle \mathfrak{A}, \alpha \rangle$ be an infinite constructive system. This means D_α is general recursive and D_α/θ_α is infinite. We know, too, that as θ_α is a **gr**-predicate, it is a **gr**-function on D_α. By virtue of Theorem 2.3.1 the numbering α is **gr**-equivalent to some simple 1–1 numbering β on the base of $\mathfrak{A}$. Therefore, $\langle \mathfrak{A}, \alpha \rangle$ is **gr**-equivalent to the constructive system $\langle \mathfrak{A}, \beta \rangle$, as desired.

The first and third assertions remain to be proved. To do this we let $\varphi \in \mathcal{F}_{gr}$ be 1–1 **gr**-function whose range is D_α. We define a new numbering β on $\mathfrak{A}$ by putting

$$\beta x = \alpha\varphi(x) \quad (x \in D)\,.$$

Now by letting

$$\psi(x) = \begin{cases} \mu y(\varphi(y) = x) & \text{if } x \in D_\alpha\,, \\ 0 & \text{if } x \notin D_\alpha\,, \end{cases}$$

we obtain a **gr**-function ψ that is inverse to φ on D_α. Hence, α and β are **gr**-isomorphic; by earlier observations, $\langle \mathfrak{A}, \alpha\rangle$ is **gr**-isomorphic to $\langle \mathfrak{A}, \beta\rangle$, and $\langle \mathfrak{A}, \beta\rangle$ is constructive. If α is $1-1$, then obviously β is $1-1$, too. ■

The theorem just proved implies, in particular, that every infinite constructive algebraic system is **gr**-equivalent to a trivially numbered system with base D in which appropriate numerical **gr**-functions serve as the total operations and predicates, while numerical **pr**-functions with **gr**-domains serve as the partial operations.

(IV) Lastly, we say that an algebraic system is *primitive recursive* iff it has a $1-1$ **prim**-numbering whose number set is primitive recursive.

According to Theorem 3.1.1, if either of two **prim**-equivalent algebraic systems having $1-1$ numberings with primitive recursive number sets is primitive recursive, then so is the other.

From (IV) it is clear that up to **prim**-isomorphism the only primitive recursive systems are those trivially numbered systems in each of which the base is a primitive recursive subset of D, and the total operations and predicates are numerical **prim**-functions restricted in domain to the base, while numerical **prim**-functions restricted in domain to **prim**-subsets of the base serve as partial operations.

§3.2. *Subsystems*

Let $\langle \mathfrak{A}, \alpha\rangle$ be an arbitrary numbered algebraic system with base A; let $\mathfrak{B}$ be an abstract subsystem of $\mathfrak{A}$ with nonempty base B. Denoting by D_β the collection of all α-numbers of elements of B, we introduce a numbering β on B by setting

$$\beta n = \alpha n \quad (n \in D_\beta)\,.$$

The numbered system $\langle \mathfrak{B}, \beta\rangle$ so obtained is called a *numbering subsystem* of $\langle \mathfrak{A}, \alpha\rangle$. Since all functions representing the basic notions of $\mathfrak{A}$ in coordinate form relative to α will simultaneously represent the basic notions of $\mathfrak{B}$ with respect to β, we see that if α is an **R**-numbering of $\mathfrak{A}$, then β is an **R**-numbering of $\mathfrak{B}$ (**R** = **prim**, **gr**, **pr**).

A subsystem $\mathfrak{B}$ of the numbered system $\langle \mathfrak{A}, \alpha\rangle$ is called an **R**-*subsystem* iff the base B of $\mathfrak{B}$ is an **R**-subset of A relative to α.

We note that in the terminology adopted in the present and preceding subsections, a numbering **gr**-subsystem of a numbered system need not be a **gr**-system. E.g., let E be a set of numbers that wholly includes no infinite **pr**-subset of D, but suppose the intersection of E with some **gr**-set C is infinite. We trivially number E and endow it with the multiplication $x \times y = y$ seen earlier; this turns E into a numbered semigroup $\mathfrak{E}$. The numbering subsystem with base $E \cap C$ is then a **gr**-subsystem of $\mathfrak{E}$, but cannot be a **gr**-system in the sense of §3.1 (II) because its number set $E \cap C$ is not recursively enumerable.

It is easy to verify, however, that a numbering **prim**-subsystem of a **prim**-system is primitive recursive, a numbering **gr**-subsystem of a constructive (general recursive) system is constructive (general recursive), and a numbering **pr**-subsystem of a positive (negative) system is positive (negative).

We also note that *every general recursive system* $\langle \mathfrak{A}, \alpha \rangle$ *is* **pr**-*isomorphic to a constructive system.*

For by assumption D_α is recursively enumerable. We can further assume D_α to be infinite. By Theorem 2.2.1 there exists a simple numbering β of the base A of $\mathfrak{A}$ that is **gr**-unimorphic to the numbering α. Thus the identity map is a **gr**-unimorphism from $\langle \mathfrak{A}, \beta \rangle$ onto $\langle \mathfrak{A}, \alpha \rangle$. Since this map transforms the basic operations and predicates of $\langle \mathfrak{A}, \beta \rangle$, as well as the equality relation, into the **gr**-operations and **gr**-predicates of $\langle \mathfrak{A}, \alpha \rangle$, the basic notions of $\mathfrak{A}$ are general recursive relative to β by Theorem 2.1.4. The remark quickly follows. ∎

Theorem 3.2.1: *Let M be an absolutely recursively enumerable subset of the number set D_α of an algebraic system $\mathfrak{A}$ with* **pr**-*numbering α. Then the subsystem $\mathfrak{M}$ generated by $\alpha(M)$ in $\mathfrak{A}$ is the α-image of an appropriate absolutely recursively enumerable subset T of D_α.*

Let $f_1, ..., f_n$ be all the basic operations, total and partial, of the numbered system $\langle \mathfrak{A}, \alpha \rangle$. By hypothesis, there exist numerical **pr**-functions $F_1, ..., F_n$ satisfying the equation

$$F_i(x_1, ..., x_{s_i}) = f_i(\alpha x_1, ..., \alpha x_{s_i}) \tag{19}$$

$$(x_1, ..., x_{s_i} \in D_\alpha;\ i = 1, ..., n)$$

strongly, i.e., one side is defined iff the other is. Let $F_0 \in \mathcal{F}_{gr}$ be a **gr**-function whose range is equal to M. We now consider the set B of all terms constructed from the individual symbols $\overline{F_0(0)}, \overline{F_0(1)}, \overline{F_0(2)}, ...$ and the function symbols $\boldsymbol{F}_1, ..., \boldsymbol{F}_n$, which will designate the corresponding functions.

The *standard* (or Gödel) numbering of the set B is defined by recursion as follows. The standard number $\#\overline{F_0(m)}$ of the term $\overline{F_0(m)}$ is the natural number 3^m $(m = 0, 1, 2, ...)$. We continue the definition by assuming the terms $\mathfrak{a}_1, ..., \mathfrak{a}_{s_i}$ have numbers $\#\mathfrak{a}_1, ..., \#\mathfrak{a}_{s_i}$; then the term $\boldsymbol{F}_i(\mathfrak{a}_1, ..., \mathfrak{a}_{s_i})$ has the number

$$\#\boldsymbol{F}_i(\mathfrak{a}_1, ..., \mathfrak{a}_{s_i}) = 2^{i+1} \cdot p_1^{\#\mathfrak{a}_1} \cdots p_{s_i}^{\#\mathfrak{a}_{s_i}}\ , \tag{20}$$

where $i = 1, ..., n$, and p_j is the jth prime number ($p_1 = 3$).

Since the individual and function symbols have values already determined, every term in B either has a definite value in D or has no defined value. The latter can happen because the functions F_i are not necessarily total. We introduce the function $H \in \mathcal{F}_{\mathrm{p}}$ by putting $H(m)$ equal to the value of the term in B whose standard number is n; this is well defined, for the standard numbering is obviously 1–1.

By assumption, all the functions F_i are partial recursive, meaning there is an effective process for calculating the values of each F_i ($i = 0, ..., n$). There is also an effective method for calculating the value of a term, given its standard number. This means H is partial recursive; therefore, its range is a partial recursive set T. We have only to show the set $\alpha(T)$ is the base of the subsystem of $\mathfrak{M}$ generated by $\alpha(M)$. But this is obvious, since (19) implies the value of each term in B is an α-number of the value of the corresponding term constructed from symbols for the operations f_i and the elements $\alpha F_0(m)$ of the generating set. ■

From this theorem we conclude that the generated subsystem $\mathfrak{M}$ is the α-image of a recursively enumerable subset of the number set D_α. This gives us no right to conclude $\mathfrak{M}$ is a **pr**-subsystem of $\langle \mathfrak{A}, \alpha \rangle$, for the latter requires the set of *all* α-numbers of elements of $\mathfrak{M}$ to be recursively enumerable in D_α, not just the set of α-numbers obtained with the aid of terms. The following simple remark shows, however, that in the majority of important cases, $\mathfrak{M}$ actually is a **pr**-subsystem in this sense.

Remark 1: *If a set A has a positive numbering* α *then the* α*-image of every recursively enumerable subset M of the number set* D_α *is a recursively enumerable subset of A relative to* α.

We have to establish the recursive enumerability of N, the set of all α-numbers of elements of $\alpha(M)$. By assumption, D_α is a **pr**-set, and θ_α is a **pr**-predicate on D_α, and thus absolutely. Therefore, the set of all θ_α-equivalent pairs of numbers from D_α can be represented in the form

$$\{\langle \varphi(t), \psi(t) \rangle : t \in D\}$$

for some $\varphi, \psi \in \mathcal{F}_{\mathrm{prim}}$. Let $\chi \in \mathcal{F}_{\mathrm{prim}}$ have range equal to M. Then N is the set of all those $x \in D$ such that for some $u, v \in D$, we have $x = \varphi(u)$, $\psi(u) = \chi(v)$. Consequently, N is recursively enumerable. ■

Combining Remark 1 with Theorem 3.2.1, we get the

Corollary: *If the numbered algebraic system* $\langle \mathfrak{A}, \alpha \rangle$ *is positive, then the* α*-image of any* **pr**-*subset of the number set* D_α *generates a* **pr**-*subsystem in* $\mathfrak{A}$. ■

From an earlier remark we see that every numbering **pr**-subsystem of a general recursive system is **pr**-isomorphic to a general recursive system. Therefore, *every numbering subsystem generated in a general recursive system by a* **pr**-*subset is* **pr**-*isomorphic to a general recursive system.* ■ ([14])

Example 11: Termal subsystems. Let 𝔐 be an algebraic system with base A and with basic operations (total and partial) $f_1, ..., f_n$. Let T be some set of terms involving function symbols $f_1, ..., f_n$ for the operations and auxiliary individual symbols $x_1, x_2, ...$. Let U be the subset of A consisting of all values taken by terms in T as the values of the variables $x_1, x_2, ...$ range independently over A. The subsystem 𝔐 generated in 𝔄 by U is called the *termal (verbal)* subsystem defined in 𝔄 by the set T of terms.

If the variables range in value not over the whole base, but only over some subset $B \subseteq A$, then the submodel 𝔑 generated by the set V of the so-determined values of terms in T is called the termal subsystem defined in 𝔄 by T restricted to B.

Above we introduced the standard numbering of a certain set of terms. We adapt this notion to the present situation by letting the number $\#x_j$ of the term x_j be the natural number 3^j, continuing the definition by the recursion rule (20).

The set T is called *recursively enumerable* iff the set of standard numbers of the terms in T is recursively enumerable. If we now repeat the previous argument word for word, we get the following generalization of Theorem 3.2.1.

Remark 2: *Let M be an absolutely recursively enumerable subset of the number set D_α of an algebraic system 𝔄 with* **pr**-*numbering α; let T be a recursively enumerable set of terms in the variables $x_1, x_2, ...$ and the function symbols corresponding to 𝔄. Then the termal subsystem 𝔑 generated in 𝔄 by T restricted to $\alpha(M)$ is the α-image of an appropriate absolutely recursively enumerable subset of D_α.* ■

In particular, the termal numbering subsystem defined in a positive system by a **pr**-set of terms is itself positive.

We also mention another special case: *the termal numbering subsystem generated in a general recursive system by a* **pr**-*set of terms is* **pr**-*isomorphic to a general recursive system.* ■ ([15])

The termal subgroup defined in a group 𝔊 by the single term $[x,y] = x^{-1}y^{-1}xy$ is clearly the *commutator subgroup* (or first derived group) of 𝔊; the termal subgroup defined in 𝔊 by the term $[[x_1,x_2],[x_3,x_4]]$ is the second commutator subgroup of 𝔊, and so on. On the other hand, the termal subgroup given by the term $[[x_1,x_2],x_3]$ is the second member of the lower central series of the group 𝔊, the termal subgroup determined by

$[[[x_1, x_2], x_3], x_4]$ is the third member of the lower central series, etc. Analogous notions can be similarly introduced in the theory of rings.

According to the above theorem and related remarks, the successive commutator subgroups and the terms of the lower central series of a general recursive group are also general recursive groups.

§3.3. *Homomorphisms and congruence relations*

For the time being we narrow our attention from general algebraic systems to algebras alone. As we have said already, a well-defined map φ from the base A of an algebra $\mathfrak{A}$ onto the base B of a similar algebra $\mathfrak{B}$ is called a *homomorphism* from $\mathfrak{A}$ onto $\mathfrak{B}$ iff

$$\varphi f_i(u_1, ..., u_{r_i}) = g_i(\varphi u_1, ..., \varphi u_{r_i})$$

$$(u_1, ..., u_{r_i} \in A;\ i = 1, ..., m)\ ,$$

where the f_i and g_i are the basic operations of $\mathfrak{A}$ and $\mathfrak{B}$, respectively. The *congruence relation* associated with the homomorphism φ is the binary predicate σ on A defined by the condition

$$u\sigma v \Leftrightarrow \varphi u = \varphi v \quad (u, v \in A)\ . \tag{21}$$

It is apparent that the relation σ is reflexive, symmetric, and transitive and in addition satisfies:

$$u_1 \sigma v_1 \text{ and ... and } u_{r_i} \sigma v_{r_i} \Rightarrow f_i(u_1, ..., u_{r_i})\ \sigma f_i(v_1, ..., v_{r_i})$$

$$(u_j, v_j \in A; i = 1, ..., m)\ . \tag{22}$$

Abstractly, every binary predicate σ on A that is reflexive, symmetric, and transitive and also satisfies (22) is called a *congruence relation* on $\mathfrak{A}$.

Given a congruence σ on $\mathfrak{A}$, we can decompose A into classes of elements related to each other by σ, and convert the collection A/σ of all these congruence classes into an algebra similar to $\mathfrak{A}$ by putting

$$f_i^*([u_1], ..., [u_{r_i}]) = [f_i(u_1, ..., u_{r_i})]$$

$$(u_j \in A;\ i = 1, ..., m)\ , \tag{23}$$

where $[u]$ denotes the class in A/σ containing the element $u \in A$. From (22) it follows that the f_i^* determined by (23) are well-defined operations on A/σ, while the map φ: $u \to [u]$ is a homomorphism from $\mathfrak{A}$ onto the resulting algebra $\mathfrak{A}/\sigma$ which has σ for its congruence relation as given by (21). This map φ is called the *canonical homomorphism* of $\mathfrak{A}$ onto $\mathfrak{A}/\sigma$.

Now suppose $\mathfrak{A}$ is numbered by α. Defining a numbering α^* of A/σ (the *canonical numbering*) by setting

$$\alpha^* n = [\alpha n] \quad (n \in D_\alpha),$$

we turn the factor algebra $\mathfrak{A}/\sigma$ into a numbered algebra $\langle \mathfrak{A}/\sigma, \alpha^* \rangle$. If $F_i(x_1, ..., x_{r_i})$ is a numerical function representing the basic operation f_i of $\mathfrak{A}$ in coordinate form with respect to α, i.e., if F_i satisfies

$$f_i(\alpha x_1, ..., \alpha x_{r_i}) = \alpha F_i(x_1, ..., x_{r_i}) \quad (x_j \in D_\alpha),$$

then this same function represents the operation f_i^* of $\mathfrak{A}/\sigma$ relative to α^*. Consequently, *if α is an* **R**-*numbering* (**R** = **prim, gr, pr**) *of the algebra* $\mathfrak{A}$, *then α^* is an* **R**-*numbering of the factor algebra* $\mathfrak{A}/\sigma$. ■

The equivalence relation θ_{α^*} corresponding to the numbering α^* of $\mathfrak{A}/\sigma$ characterizes the set of pairs $\langle x, y \rangle$ of numbers in D_α for which $\alpha^* x = \alpha^* y$, i.e., for which $[\alpha x] = [\alpha y]$, but this is the same as αx and αy being σ-congruent. In other words, θ_{α^*} characterizes the set of pairs of α-numbers of elements in A that are in the relation σ. Therefore, θ_{α^*} is an **R**-predicate iff σ is an **R**-predicate relative to α.

We now return to the general situation. Let φ be an **R**-homomorphism of the numbered algebra $\langle \mathfrak{A}, \alpha \rangle$ onto some numbered algebra $\langle \mathfrak{B}, \beta \rangle$. Let σ be the associated congruence on $\mathfrak{A}$. The map φ induces the abstract isomorphism

$$\varphi^*: [u] \to \varphi u \quad (u \in A)$$

of the factor algebra $\mathfrak{A}/\sigma$ onto the algebra $\mathfrak{B}$. Any unary numerical function H representing φ in coordinate form, i.e., satisfying $\varphi(\alpha n) = \beta H(n)$ $(n \in D_\alpha)$, obviously represents φ^* relative to α^*, β. Hence, φ^* is an **R**-monomorphism of $\langle \mathfrak{A}/\sigma, \alpha^* \rangle$ onto $\langle \mathfrak{B}, \beta \rangle$. Thus we have established

Theorem 3.3.1: *The canonical numbering α^* of the factor algebra $\mathfrak{A}/\sigma$ of an algebra $\mathfrak{A}$ with* **R**-*numbering α by a congruence σ is an* **R**-*numbering having the same number set as α. The equivalence relation θ_{α^*} is an* **R**-*predicate iff the congruence relation σ is an* **R**-*predicate relative to α. If φ is an* **R**-*homomorphism from $\langle \mathfrak{A}, \alpha \rangle$ onto a numbered algebra $\langle \mathfrak{B}, \beta \rangle$, then the canonical isomorphism of $\langle \mathfrak{A}/\sigma, \alpha^* \rangle$ onto $\langle \mathfrak{B}, \beta \rangle$ is an* **R**-*monomorphism.* ■

In particular, we observe that the factor algebra (canonically numbered) of a positive (negative) algebra by a congruence that is a **pr**-predicate (the negation of a **pr**-predicate) is positive (negative). Furthermore, a factor algebra of a constructive (general recursive) algebra is constructive (general recursive) iff the corresponding congruence is general recursive.

From the second part of Theorem 2.1.1 we immediately obtain

Theorem 3.3.2: *Suppose the numbered algebra $\langle \mathfrak{A}, \alpha \rangle$ with recursively enumerable number set is* **pr**-*homomorphically mapped onto the numbered algebra $\langle \mathfrak{B}, \beta \rangle$ with positive numbering. Then the canonical monomorphism from the corresponding canonically numbered factor algebra of $\mathfrak{A}$ onto $\langle \mathfrak{B}, \beta \rangle$ is a* **pr**-*equivalence.* ∎

This theorem and the remarks above yield the important

Corollary: *Every positive algebra that is a* **pr**-*homomorphic image of a positive algebra $\mathfrak{A}$ is* **pr**-*equivalent to the factor algebra of $\mathfrak{A}$ by some* **pr**-*congruence.* ∎

In view of Theorems 2.1.1. and 3.1.1, a similar statement holds for constructive algebras, too. Namely, if a constructive algebra $\mathfrak{A}$ is mapped **gr**-homomorphically onto a constructive algebra $\mathfrak{B}$, then the canonical monomorphism of the corresponding factor algebra $\mathfrak{A}/\sigma$ is a **gr**-equivalence, while σ is general recursive on $\mathfrak{A}$.

This shows that up to **gr**-equivalence the only **gr**-homomorphic images of a given constructive algebra are its canonically numbered factor algebras by its various **gr**-congruences.

Up to now we have been studying how homomorphisms can be specified by means of congruences. A congruence on an algebra $\mathfrak{A}$ can be viewed as the collection of pairs of elements of $\mathfrak{A}$ congruent to one another. In the general theory of algebras, however, one not uncommonly investigates the possibility of determining homomorphisms by means of sets of single elements of the algebra rather than sets of pairs. Below we indicate a general sort of algebra class in which such a specification can be realized in the simplest way possible.

Let $\mathfrak{K}$ be an arbitrary class of similar (abstract) algebraic systems of the form (1) (cf. §1.1). We consider an arbitrary term $\mathfrak{a}(\boldsymbol{x}_1, \ldots, \boldsymbol{x}_s)$ constructed from individual constants $\boldsymbol{a}_1, \ldots, \boldsymbol{a}_l$, individual variables $\boldsymbol{x}_1, \ldots, \boldsymbol{x}_s$, and function symbols $\boldsymbol{f}_1, \ldots, \boldsymbol{f}_m, \boldsymbol{g}_1, \ldots, \boldsymbol{g}_n$. Let $\mathfrak{A}$ be a fixed system in $\mathfrak{K}$. Then the values of the symbols $\boldsymbol{a}_1, \ldots, \boldsymbol{a}_l, \boldsymbol{f}_1, \ldots, \boldsymbol{f}_m, \boldsymbol{g}_1, \ldots, \boldsymbol{g}_n$ in the term $\mathfrak{a}$ are thus fixed; in turn they determine the value (when defined) of $\mathfrak{a}$ in $\mathfrak{A}$ for each arbitrary choice of values among the elements of $\mathfrak{A}$ for the variables $\boldsymbol{x}_i$. The term $\mathfrak{a}$ so defines an s-ary partial operation on the base of every algebraic system with the signature of $\mathcal{K}$. As mentioned before, these partial operations are called *termal* or *polynomial* operations.

Along with termal operations we must consider defined operations of a slightly more general form, the so-called quasitermal operations.

An *atomic formula* is an expression of the form $\boldsymbol{P}_i(\mathfrak{a}_1, ..., \mathfrak{a}_{t_i})$ or of the form $\mathfrak{a} \approx \mathfrak{b}$, where $\boldsymbol{P}_i$ is a predicate symbol from the signature of $\mathcal{K}$, and $\mathfrak{a}, \mathfrak{b}, \mathfrak{a}_1, ..., \mathfrak{a}_{t_i}$ are terms built from the signature symbols $\boldsymbol{a}_1, ..., \boldsymbol{a}_l, \boldsymbol{f}_1, ..., \boldsymbol{f}_m, \boldsymbol{g}_1, ..., \boldsymbol{g}_n$, plus some individual variables $\boldsymbol{x}_1, ..., \boldsymbol{x}_r$.

Expressions constructed from atomic formulas with conjunction, disjunction and negation signs by the usual rules are called *open formulas,* while those expressions consisting of atomic formulas combined with conjunction and disjunction signs alone are called *positive* open formulas. Lastly, an $\exists$-*formula* is a formula of first-order predicate logic (FOPL) of the form

$$(\exists \boldsymbol{y}_1) ... (\exists \boldsymbol{y}_t)\, \Phi(\boldsymbol{x}_1, ..., \boldsymbol{x}_s, \boldsymbol{y}_1, ..., \boldsymbol{y}_t)\,, \tag{24}$$

where Φ is an open formula with the individual variables $\boldsymbol{x}_1, ..., \boldsymbol{x}_s, \boldsymbol{y}_1, ..., \boldsymbol{y}_t$. If Φ is a positive open formula in (24) is called a *positive* $\exists$-formula.

For every choice of values in the base of the algebraic system $\mathfrak{A} \in K$ for the individual variables $\boldsymbol{x}_1, ..., \boldsymbol{x}_s$, the formula (24) has one of the values: true, false, undefined (possible when some of the basic operations of $\mathfrak{A}$ are not total). Therefore, every $\exists$-formula represents a certain predicate on $\mathfrak{A}$, possibly not totally defined (naturally, such predicates are said to be partial). These predicates are called $\exists$-predicates.

Let $P(x_1, ..., x_s)$ be the predicate characterized in $\mathfrak{A}$ by a given formula of the form (24). Suppose that for every $x_1, ..., x_{s-1}$ in $\mathfrak{A}$, there is at most one element x_s in $\mathfrak{A}$ such that $P(x_1, ..., x_s)$ is true. If we let $F(x_1, ..., x_{s-1})$ be this element (when it exists), we determine a new operation on $\mathfrak{A}$. This is a partial operation, but it may turn out to be total.

An operation $G(x_1, ..., x_{s-1})$ specified on each member of $\mathcal{K}$ by any means whatever is called an *open, positive,* or $\exists$-*operation* iff there is an open, positive $\exists$-formula, respectively, that defines G in all $\mathcal{K}$-systems as described in the preceding paragraph. In particular, an open and positive operation is called *quasitermal.* A characterization of these can be found in [IX].

Positive $\exists$-predicates and $\exists$-operations have the following property of *persistence* with respect to homomorphisms: *for every homomorphism φ of a $\mathcal{K}$-system $\mathfrak{A}$ onto another $\mathcal{K}$-system $\mathfrak{B}$ and for every positive $\exists$-predicate $P(x_1, ..., x_r)$ and every positive $\exists$-operation $F(x_1, ..., x_s)$ on $\mathcal{K}$-systems, we have*

$$P(x_1, ..., x_r) \text{ true in } \mathfrak{A} \Rightarrow P(\varphi x_1, ..., \varphi x_r) \text{ true in } \mathfrak{B}\,(x_j \in \mathfrak{A}),$$

$$\varphi F(x_1, ..., x_s) = F(\varphi x_1, ..., \varphi x_s) \quad (x_j \in \mathfrak{A})\,, \tag{25}$$

where the right side of the equation in (25) is defined whenever the left side is.

The proof is carried out easily by induction on the length of the formula Φ in (24). ■

In the general theory of algebraic systems an important role is played by predicates and operations characterizable by FOPL formulas not only of the form (24), but also of the general form

$$(\eth_1 y_1) \ldots (\eth_t y_t)\, \Phi(x_1, ..., x_s, y_1, ..., y_t)\,,$$

where $\eth_1, ..., \eth_t$ are universal or existential quantifiers in any sort of order. The next theorem fixes the special role played by $\exists$-predicates and $\exists$-operations in the theory of effective numbered systems. As usual, **R** = **prim, gr, pr.**

Theorem 3.3.3: *Suppose the algebraic system* $\mathfrak{A}$ *has an* **R**-*numbering with recursively enumerable number set. Then:* (i) *all* ([16]) *open predicates and all termal operations on* $\mathfrak{A}$ *are* **R**-*predicates and* **R**-*operations, respectively;* (ii) *in case* **R** = **gr**, *every total* $\exists$*-operation on* $\mathfrak{A}$ *is general recursive;* (iii) *all* ([16]) $\exists$*-predicates and all* ([16]) *partial* $\exists$*-operations on* $\mathfrak{A}$ *are partial recursive.*

All the assertions of this theorem are immediately deduced from the generally known properties of **R**-functions (cf. [71]), and we omit the proofs. ■

Theorem 3.3.4: *Suppose in the class* $\mathcal{K}$ *of algebras there exists a binary positive* $\exists$*-operation* $\circ$ *that is totally defined for each K-algebra* $\mathfrak{A}$ *and satisfies*

$$x \circ x = y \circ y\,, \tag{26}$$

$$x \circ y = x \circ x \Rightarrow x = y \tag{27}$$

for all x, y in $\mathfrak{A}$. *Then:* (i) *every congruence relation* σ *on* $\mathfrak{A}$ *is uniquely determined by the* σ*-class containing the element* $e = x \circ x$; (ii) *assuming* $\mathfrak{A}$ *has an* $\mathbf{R}'$*-numbering* α *with recursively enumerable number set, the congruence* σ *on* $\mathfrak{A}$ *is an* $\mathbf{R}'$*-predicate relative to* α *iff the* σ*-class containing e is an* $\mathbf{R}'$*-set relative to* α*; moreover,* (iii) *if* $\circ$ *is a* **prim**-*operation (with respect to* α*), then the primitive recursiveness of* σ *is equivalent to the primitive recursiveness of the* σ*-class containing e* ($\mathbf{R}'$ = **gr, pr**).

Let φ be the canonical homomorphism of $\mathfrak{A}$ onto $\mathfrak{A}/\sigma$. Since the operation $\circ$ is positive, it is homomorphically persistent, i.e., the $\exists$-formula defining $\circ$ in $\mathcal{K}$ defines a binary partial operation (call it $\circ$, too) in any homomorphic image of a $\mathcal{K}$-algebra and (25) holds. In particular, in $\mathfrak{A}/\sigma$ we have $\varphi(a \circ b) = \varphi a \circ \varphi b$ for all $a, b \in \mathfrak{A}$. But $a\sigma b$ iff $\varphi a = \varphi b$, so in view of (26), (27) we find

$$a\sigma b \Rightarrow \varphi a \circ \varphi b = \varphi a \circ \varphi a \Rightarrow \varphi(a \circ b) = \varphi(a \circ a) \Rightarrow (a \circ b)\sigma e \ ,$$

and conversely,

$$(a \circ b)\sigma e \Rightarrow \varphi a \circ \varphi b = \varphi a \circ \varphi a \Rightarrow \varphi a = \varphi b \Rightarrow a\sigma b \ .$$

Thus if we let $[e]$ be the σ-class containing e, we get

$$a\sigma b \Leftrightarrow a \circ b \in [e] \ ,$$

proving the first assertion.

To prove (ii) we let ρ be a binary numerical $\mathbf{R}'$-function representing $\circ$ relative to α. Such a function is known to exist by virtue of Theorem 3.3.3. Let E be any unary numerical partial function representing the set $[e]$ relative to α, i.e., $\alpha x \in [e] \Leftrightarrow E(x) = 1$ for $x \in D_\alpha$. Then $E(\rho(x, y))$ gives a binary function representing the relation σ with respect to α in the obvious sense. Therefore, if the class $[e]$ is an $\mathbf{R}'$-set, σ is an $\mathbf{R}'$-predicate. Conversely, if S is a binary numerical function representing σ relative to α, and n is any α-number for e, then $S(x, n)$ gives a unary function representing $[e]$; thus if σ is an $\mathbf{R}'$-predicate, $[e]$ is an $\mathbf{R}'$-set. This same argument works for proving (iii). ■

As a special case of Theorem 3.3.4 we have: a congruence on a group (ring) with an $\mathbf{R}$-numbering is an $\mathbf{R}$-congruence iff the corresponding normal divisor (ideal) is an $\mathbf{R}$-set.

§4. Finitely generated algebras

§4.1. *General finitely generated algebras*

Let $\mathfrak{A}$ be an algebra with fundamental operations $f_i(u_1, ..., u_{r_i})$ $(i = 1, ..., m)$ and no distinguished elements – the latter merely for the sake of convenience. In accord with §1.2 we say the algebra $\mathfrak{A}$ is *finitely generated* iff there exists a finite number of elements of $\mathfrak{A}$, denoted by $a_1, ..., a_l$ with repetitions possible, that together generate the whole algebra. If we consider $a_1, ..., a_l$ to be distinguished elements, adding individual symbols $\boldsymbol{a}_1, ..., \boldsymbol{a}_l$ to the signature of $\mathfrak{A}$, we convert $\mathfrak{A}$ into an algebra, also known as $\mathfrak{A}$, of similarity type

$\tau = \langle 0, ..., 0; r_1, ..., r_m \rangle$ (cf. §1.1). That $\mathfrak{A}$ is finitely generated by $\{a_1, ..., a_l\}$ now means exactly that $\mathfrak{A}$ has no subalgebra of type τ other than itself, i.e., that $\mathfrak{A}$ is a minimal algebra of type τ.

Let T be the collection of all possible terms constructed from the individual constants $\boldsymbol{a}_1, ..., \boldsymbol{a}_l$ and the function symbols $\boldsymbol{f}_1, ..., \boldsymbol{f}_m$. For $1 \leqslant i \leqslant m$, let $\mathfrak{a}_1, ..., \mathfrak{a}_{r_i}$ be any terms in T: by taking the term $\boldsymbol{f}_i(\mathfrak{a}_1, ..., \mathfrak{a}_{r_i})$ to be the result of applying $\boldsymbol{f}_i$ to $\mathfrak{a}_1, ..., \mathfrak{a}_{r_i}$, we obtain an r_i-ary operation on T. In this manner we turn T into an algebra $\mathfrak{T}$ of type τ. According to §1.4, $\mathfrak{T}$ is a free algebra with free generators $\boldsymbol{a}_1, ..., \boldsymbol{a}_l$ in the class of all algebras of type τ.

We introduce a numbering of T by taking 3^j to be the number $\#\boldsymbol{a}_j$ of $\boldsymbol{a}_j$ ($j = 1, ..., l$) and proceeding further via the recursion conditions ($i = 1, ..., m$):

$$\#\boldsymbol{f}_i(\mathfrak{a}_1, ..., \mathfrak{a}_{r_i}) = 2^i \cdot p_1^{\#\mathfrak{a}_1} \dots p_{r_i}^{\#\mathfrak{a}_{r_i}} \tag{28}$$

when we know the number $\#\mathfrak{a}_k$ of the term $\mathfrak{a}_k \in T$ ($k = 1, ..., r_i$).

The numbering of T so defined is called the *standard numbering* of the algebra $\mathfrak{T}$ and is denoted by γ^*. The number set D_γ of this numbering is not equal to D, but it is primitive recursive, as is easily checked.

From (28) we see that the ith basic operation in $\mathfrak{T}$ (the application of $\boldsymbol{f}_i$ to r_i terms) is represented relative to γ^* by the function F_i, where

$$F_i(x_1, ..., x_{r_i}) = 2^i \cdot p_1^{x_1} \dots p_{r_i}^{x_{r_i}} \quad (x_k \in D) . \tag{29}$$

Since the functions $F_1, ..., F_m$ are primitive recursive, the numbering γ^* is primitive recursive. Besides that, γ^* is a 1–1 numbering, and so, the algebra $\mathfrak{T}$ is primitive recursive in the sense of §3.2.

Let φ be the map of $\mathfrak{T}$ onto $\mathfrak{A}$ under which each term in T is sent to its obvious value in $\mathfrak{A}$. With the aid of φ we can translate the numbering γ^* to a numbering γ with number set D_γ by setting

$$\gamma n = \varphi(\gamma^* n) \quad (n \in D_\gamma) .$$

The map φ is a homomorphism of $\mathfrak{T}$ onto $\mathfrak{A}$, so the functions F_i representing the fundamental operations of $\mathfrak{T}$ relative to γ^* will also represent the basic operations of $\mathfrak{A}$ with respect to γ. Thus, γ is a primitive recursive numbering of $\mathfrak{A}$. This will be called the *standard numbering* of the finitely generated algebra $\mathfrak{A}$ (relative to the selected generators $a_1, ..., a_l$).

Let σ be the congruence on $\mathfrak{T}$ corresponding to the homomorphism φ. According to the definitions above, the canonical isomorphism of $\mathfrak{T}/\sigma$ onto $\mathfrak{A}$ has the property that corresponding elements of $\mathfrak{T}/\sigma$ and $\mathfrak{A}$ have identical standard numbers.

In addition to the standard numbering it is sometimes convenient to use another special numbering of $\mathfrak{A}$ constructed as follows. We supplement the basic operations of $\mathfrak{A}$ by adding the new operation f_{m+1}, where $f_{m+1}(u) = u$ for all $u \in \mathfrak{A}$, and denote this enrichment of $\mathfrak{A}$ by $\mathfrak{A}_0$. The algebra $\mathfrak{A}_0$ is a minimal algebra of type $\langle 0, ..., 0; r_1, ..., r_m, 1\rangle$. Since the bases of $\mathfrak{A}_0$ and $\mathfrak{A}$ are one and the same, the standard numbering δ of $\mathfrak{A}_0$ is also a numbering of $\mathfrak{A}$, called the *extended* standard numbering (relative to $a_1, ..., a_l$).

A standard numbering can be 1–1 in case $\mathfrak{A}$ is a free algebra. In contrast, an extended standard numbering always has infinite classes of numbers. Indeed, if n_0 is a δ-number of some element $a \in \mathfrak{A}$, then $2^{m+1} \cdot p_1^{n_0}$ is a δ-number of the element $f_{m+1}(a)$, i.e., of the very element a. The function $2^{m+1} \cdot p_1^x$ is primitive recursive; hence, the binary numerical function Q defined by the scheme

$$Q(x, 0) = x\ ,$$

$$Q(x, n+1) = 2^{m+1} \circ p_1^{Q(x,n)}$$

is primitive recursive.

By what was just said, whenever n_0 is a δ-number of an element $a \in \mathfrak{A}$, the natural numbers $Q(n_0, 0)$, $Q(n_0, 1)$, ... are distinct δ-numbers of a. Therefore, the numbering δ has **prim**-infinite classes in the sense of §2.2.

The functions F_i representing the basic operations of $\mathfrak{A}$ relative to γ obviously represent them relative to δ, as well. Hence, the extended standard numbering δ of $\mathfrak{A}$ is primitive recursive. It is not hard to convince one's self that, in general, the standard numberings and extended standard numberings of $\mathfrak{A}$ are all **prim**-equivalent. This immediately follows from the next theorem.

Theorem 4.1.1: *The standard and extended standard numberings of a finitely generated algebra* $\mathfrak{A}$ *relative to any given finite sequence of generators are* **R**-*reducible to any* **R**-*numbering* α *of* $\mathfrak{A}$ (**R** = **prim, gr, pr**).

Suppose the numerical **R**-functions $G_i(x_1, ..., x_{r_i})$ $(i = 1, ..., m)$ represent the basic operations of the algebra $\mathfrak{A} = \langle A; f_1, ..., f_m\rangle$, and suppose $n_1, ..., n_l$ are α-numbers of the generators $a_1, ..., a_l$ of $\mathfrak{A}$. We define the function $f \in \mathcal{F}$ according to the scheme:

$$f(x) = \begin{cases} G_i(f(\exp(1,x)), ..., f(\exp(r_i, x))) & \text{if} \\ & \exp(0,x) = i \text{ for some } 1 \leqslant i \leqslant m, \\ n_j \quad \text{if } x = 3^j & \text{for some } 1 \leqslant j \leqslant l\,, \\ 0 \quad \text{in all other cases.} & \end{cases} \qquad (30)$$

Since this is a scheme of regressive recursion, reducible in the usual way (cf. [119]) to a primitive recursion, and since the given functions G_i and *exp* are **R**-functions, f will also be an **R**-function. We want to show f reduces γ and δ to α, that is,

$$\gamma x = \alpha f(x) \qquad (x \in D_\gamma)\ ,$$

$$\delta x = \alpha f(x) \qquad (x \in D_\delta)\ . \tag{31}$$

Because every γ-number of an element in $\mathfrak{A}$ is at the same time one of its δ-numbers, it suffices to prove (31) alone. The smallest member of D_δ is 3, which is a δ-number for the element a_1. According to (30), $f(3) = n_1$, and thus $a_1 = \delta 3 = \alpha f(3)$.

Suppose $y > 3$ belongs to D_δ and for all $x \in D_\delta$, (31) holds for x if $x < y$. If $y = 3^j$, then by (30) we find $f(y) = n_j$, and so, $a_j = \delta y = \alpha f(y)$. If $exp(0,y) = i$ for some $1 \leqslant i \leqslant m$, then $exp(1,y), \ldots, exp(r_i, y)$ are members of D_δ less than y; hence, by virtue of the inductive hypothesis,

$$\delta \exp(k,y) = \alpha f(\exp(k,y)) \qquad (k = 1, \ldots, r_i)\ .$$

On the other hand, from the appropriate case in (30) we see that

$$f(y) = G_i(f(\exp(1,y)), \ldots, f(\exp(r_i, y)))\ ;$$

consequently,

$$\begin{aligned} \alpha f(y) &= f_i(\alpha f(\exp(1,y)), \ldots, \alpha f(\exp(r_i, y))) \\ &= f_i(\delta \exp(1,y), \ldots, \delta \exp(r_i, y)) \\ &= \delta F_i(\exp(1,y), \ldots, \exp(r_i, y)) = \delta y\ , \end{aligned}$$

where the F_i, given by (29), represent the f_i relative to δ. ∎

This theorem motivates the following definition: a numbering α of an arbitrary algebraic system $\mathfrak{A}$ is called a *Gödel* ***R**-numbering* iff α is an **R**-numbering of $\mathfrak{A}$ with **R** number sets and is **R**-reducible to any other **R**-numbering of $\mathfrak{A}$.

Theorem 4.1.1 asserts that standard and extended standard numberings of a finitely generated algebra $\mathfrak{A}$ are Gödel **R**-numberings for each meaning of **R**.

The above definition implies all Gödel **R**-numberings of an algebraic system are **R**-equivalent.

From Theorem 4.1.1 and Corollary 2 of Theorem 2.3.2 we deduce the immediate

Corollary: *Every Gödel* **R**′*-infinite classes of a finitely generated algebra is* **R**′*-isomorphic to any extended standard numbering of this algebra* (**R**′ = **gr**, **pr**). ■

The most important characteristics of any numbering α are the nature of its number set D_α and the nature of the equivalence relation θ_α engendered by α on D_α. The number sets D_γ, D_δ corresponding to standard numberings of a finitely generated algebra have a simple structure: they are primitive recursive. As regards the corresponding equivalence relations θ_γ, θ_δ, they can be recursive of all three types – **prim**, **gr**, and **pr** – but they might not even be recursively enumerable. Which of these occurs depends on the internal structure of the algebra. E.g., we have

Theorem 4.1.2: *If the finitely generated algebra* $\mathfrak{A}$ *admits a positive numbering, i.e., a* **pr**-*numbering* α *whose* D_α *and* θ_α *are partial recursive, then the set of Gödel* **pr**-*numberings of* $\mathfrak{A}$ *coincides with the set of all the positive numberings of* $\mathfrak{A}$. *If* $\mathfrak{A}$ *admits at least one constructive numbering, i.e., a* **gr**-*numbering* α *whose* D_α *and* θ_α *are general recursive, then the set of all such numberings of* $\mathfrak{A}$ *coincides with the set of all Gödel* **gr**-*numberings of this algebra.*

For let α be a **pr**-numbering of $\mathfrak{A}$ with partial recursive D_α and θ_α. Since any δ is **pr**-reducible to α, Theorem 2.1.1 shows δ is **pr**-equivalent to α; in particular, by Theorem 2.1.5 the equivalence relation θ_δ is partial recursive, thus proving the first claim.

If α now has general recursive D_α and θ_α, then by using Theorem 2.1.2 we deduce the **gr**-equivalence of α and δ from their **pr**-equivalence. Theorem 2.1.5 in turn implies θ_δ is general recursive, which proves the second claim. ■

Theorem 4.1.2 shows that all constructive numberings of a finitely generated algebra $\mathfrak{A}$ are **gr**-equivalent, and if any such numberings of $\mathfrak{A}$ exist, the standard numberings must themselves be constructive.

Furthermore, keeping the corollary of Theorem 4.1.1 in mind, we see that if the finitely generated algebra $\mathfrak{A}$ has at least one constructive numbering, then it has a constructive numbering with **gr**-infinite classes; in addition, all numberings like the latter are **gr**-isomorphic to the extended standard numberings of $\mathfrak{A}$.

When a finite number of generators of $\mathfrak{A}$ are fixed, each element of $\mathfrak{A}$ is representable as the value of a term in T, the set previously described. $\mathfrak{A}$ is said to have a *recursively solvable word problem* iff there exists an algorithm enabling one to tell from the notations of any two terms in T whether or not their values in $\mathfrak{A}$ are equal. Since from the notation of any term in T we can effectively find its standard number, and from a standard number we can effectively recover the notation of the term to which it corresponds, we see the word problem for $\mathfrak{A}$ is effectively equivalent to deciding for arbitrary natural numbers x, y whether or not they are γ-numbers of the same element in $\mathfrak{A}$, i.e., whether or not x and y are θ_γ-equivalent. But existence of an algorithm for deciding the question of θ_γ-equivalence is tantamount to the general recursiveness of θ_γ.

Therefore, *among finitely generated algebras, the algebras with recursively solvable word problems are just those admitting constructive numberings.* ■

An algebra admitting a constructive numbering could be called "constructivizable" by analogy with topologizable or orderable groups. But to be brief we shall call constructivizable finitely generated algebras simply *constructive,* since all constructive numberings of such an algebra are **gr**-equivalent.

An example of a nonconstructive finitely generated algebra is the algebra $\mathfrak{R}_{\mathrm{pr}}$ of of all unary numerical partial recursive functions defined in §1.5. For it is easy to see that the standard numberings of $\mathfrak{R}_{\mathrm{pr}}$ are **prim**-equivalent to the Kleene numbering. But in the words of §2.3, the Kleene numbering is complete and stable, and the equivalence relation associated with a stable numbering is never partial recursive. So $\mathfrak{R}_{\mathrm{pr}}$ is not constructive.

The properties of stable numberings imply that all Gödel **gr**-numberings of $\mathfrak{R}_{\mathrm{pr}}$ are not only **gr**-equivalent, but even **gr**-isomorphic.

§4.2. *Finitely presented algebras*

In §1.4 we introduced the notion of an algebra defined by a given system of conditional identities and a given sequence of generators. An algebra $\mathfrak{A}$ definable by means of finite systems of conditional identities and generators is called a finitely presented algebra. In practice we shall assume some arbitrary finite presentation is fixed for $\mathfrak{A}$.

Theorem 4.2.1: *The equivalence relation θ_γ corresponding to the standard numbering γ of an arbitrary finitely presented algebra $\mathfrak{A}$ is partial recursive. Hence, all Gödel* **pr**-*numberings of a finitely presented algebra are positive.*

Let $\boldsymbol{a}_1, \ldots, \boldsymbol{a}_l$ be individual constant symbols for the fixed generators of $\mathfrak{A}$ and let

$$\mathrm{I}_i = \mathfrak{c}_1 \approx \mathfrak{c}_2 \; \& \ldots \& \; \mathfrak{c}_{2r-1} \approx \mathfrak{c}_{2r} \to \mathfrak{c}_{2r+1} \approx \mathfrak{c}_{2r+2} \tag{32}$$

be a typical member of the fixed finite system $\{\mathrm{I}_1, \ldots, \mathrm{I}_p\}$ of conditional identities defining $\mathfrak{A}$. Every term $\mathfrak{c}_j$ occurring in (32) involves only symbols $f_1, ..., f_m$ for the basic operations of $\mathfrak{A}$, the constants $\boldsymbol{a}_1, ..., \boldsymbol{a}_l$, and individual variables $\boldsymbol{x}_1, ..., \boldsymbol{x}_n$; it will be convenient to write $\mathfrak{c}_j$ as $\mathfrak{c}_j(\boldsymbol{x}_1, ..., \boldsymbol{x}_n)$, even when fewer variables actually appear in $\mathfrak{c}_j$. Let E be the subset of D^2 consisting of all pairs of θ_γ-equivalent numbers, where γ is the standard numbering of $\mathfrak{A}$ relative to the given generators. Recall that γ is induced by the standard numbering γ^* of the set T of all terms involving only $\boldsymbol{a}_1, ..., \boldsymbol{a}_l$ and $f_1, ..., f_m$; γ and γ^* have the same infinite primitive recursive number set D_γ.

Let us now take any term $\mathfrak{c}_j(\boldsymbol{x}_1, ..., \boldsymbol{x}_n)$ from among those occurring in (32), and replace the variables $\boldsymbol{x}_1, ..., \boldsymbol{x}_n$ with arbitrary terms $\mathfrak{a}_1, ..., \mathfrak{a}_n$ from T. We get a term $\mathfrak{c}_j(\mathfrak{a}_1, ..., \mathfrak{a}_n)$ in T as the result. The standard number of the term $\mathfrak{c}_j(\mathfrak{a}_1, ..., \mathfrak{a}_n)$ depends functionally on the standard numbers of the terms $\mathfrak{a}_1, ..., \mathfrak{a}_n$. This correspondence can be realized by an n-ary numerical **prim**-function h_j, as is clear from formula (28) in § 4.1.

In terms of the set E, the validity of (32) in $\mathfrak{A}$ is equivalent to the validity of the following statement:

(A_i) for all $x_1, ..., x_n \in D_\gamma$, if the pairs

$$\langle h_{2k-1}(x_1, ..., x_n), h_{2k}(x_1, ..., x_n)\rangle \quad (k = 1, ..., r)$$

belong to E, then so does the pair

$$\langle h_{2r+1}(x_1, ..., x_n), h_{2r+2}(x_1, ..., x_n)\rangle .$$

Let $g \in \mathcal{F}_{\mathrm{gr}}$ be a 1–1 function enumerating D_γ; we can assume $g(0) = 3$. The symmetry, transitivity, and reflexivity of θ_γ are then expressed by the statements:

(B) for all $x, y \in D_\gamma$, if $\langle x, y\rangle \in E$, then $\langle y, x\rangle \in E$;

(C) for all $x, y, z \in D_\gamma$, if $\langle x, y\rangle \in E$ and $\langle y, z\rangle \in E$, then $\langle x, z\rangle \in E$;

(D) for all $x \in D$, if $\langle g(x), g(x)\rangle \in E$, then $\langle g(x+1), g(x+1)\rangle \in E$;

(E) $\langle 3,3\rangle \in E$.

In fact, E is the smallest subset of D^2 for which (A_1), ..., (A_p), (B), (C), (D), (E) are satisfied. Let $E_0 \subseteq D$ be the set of standard numbers of the pairs composing E.

We introduce numerical **gr**-functions G_i, H_1, H_2, H_3 corresponding to (A_i), (B), (C), (D) by setting

$$G_i(u_1, \dots, u_n, v_1, \dots, v_r) = \begin{cases} \nu(h_{2r+1}(\mathfrak{l}(u_1), \dots, \mathfrak{l}(u_n)), h_{2r+1}(\mathfrak{l}(u_1), \dots, \mathfrak{l}(u_n))) \\ \qquad \text{if } v_k = \nu(h_{2k-1}(\mathfrak{l}(u_1), \dots, \mathfrak{l}(u_n)), \\ \qquad h_{2k}(\mathfrak{l}(u_1), \dots, \mathfrak{l}(u_n))) \quad (k = 1, \dots, r), \\ 24 \qquad \text{otherwise;} \end{cases}$$

$$H_1(u) = \nu(\mathfrak{r}(u), \mathfrak{l}(u)) ;$$

$$H_2(u, v) = \begin{cases} \nu(\mathfrak{l}(u), \mathfrak{r}(v)) & \text{if } \mathfrak{r}(u) = \mathfrak{l}(v), \\ 24 & \text{otherwise;} \end{cases}$$

$$H_3(u) = \begin{cases} \nu(g(x+1), g(x+1)) & \text{if } \mathfrak{l}(u) = \mathfrak{r}(u) = g(x) \\ & \qquad \text{for some } x \in D, \\ 24 & \text{otherwise.} \end{cases}$$

The functions G_i, H_1, H_2, H_3 are so defined that if numbers of arbitrary pairs in E are taken as arguments, then the value of each function is the standard number either of the pair $\langle 3, 3\rangle \in E$ or of the pair in E obtained from the given pairs by an application of the corresponding general rule (A$_i$), (B), (C), (D) ($i = 1, \dots, p$). Let us take a quick look at the algebra $\mathfrak{C}$ with base D and basic operations $G_1, \dots, G_p, H_1, H_2, H_3$. E_0 is the smallest subset of D containing $24 = \nu(g(0), g(0))$ and closed under the operations of $\mathfrak{C}$, i.e., it is the base of the subalgebra generated in $\mathfrak{C}$ by the set $\{24\}$. The algebra $\mathfrak{C}$ is constructively numbered by the trivial numbering. On the basis of Theorem 3.2.1 we conclude that E_0 is recursively enumerable, and with it E. This means θ_γ is partial recursive. ■

This theorem is easily generalized to the case of algebras presented by certain infinite systems of conditional identities. Namely, suppose we have a **gr**-function $s \in \mathcal{F}_{\mathrm{gr}}$ and three simply infinite sequences of symbols: one of individual constants $\boldsymbol{a}_1, \boldsymbol{a}_2, \dots$, one of individual variables $\boldsymbol{x}_1, \boldsymbol{x}_2, \dots$, and one function symbols $\boldsymbol{f}_1, \boldsymbol{f}_2, \dots$ of rank $s(1), s(2), \dots$, respectively. We number the set U of all terms constructed from these symbols by letting $\boldsymbol{a}_i$ and $\boldsymbol{x}_i$ have the numbers 3^i and 5^i and putting

$$\#\boldsymbol{f}_i(\mathfrak{b}_1, \dots, \mathfrak{b}_{s(i)}) = 2^i \cdot p_1^{\#\mathfrak{b}_1} \dots p_{s(i)}^{\#\mathfrak{b}_{s(i)}}$$

for longer terms. Using this numbering we can number the set of all possible conditional identities involving members of U by taking the number of the conditional identity (32) to be

$$\prod_{j=1}^{2r+2} p_j^{\#c_j(x_1,\dots,x_n)} .$$

By repeating the proof of Theorem 4.2.1 almost word for word, we now find that *the standard numbering of an algebra presented by a finite or countable number of generators and a recursively enumerable set of conditional identities has a partial recursive equivalence relation.* ∎

By inversion we immediately get the

Corollary: *If the standard numbering γ of a finitely generated algebra $\mathfrak{A}$ does not have a partial recursive equivalence relation θ_γ, then $\mathfrak{A}$ cannot be finitely or even effectively presented.* ∎

In particular, this tells us the algebra $\mathfrak{R}_{pr}$ of unary numerical partial recursive functions is not finitely (or even effectively) presented.

On the other hand, Novikov's group [116], whose word problem is not recursively solvable, is an example of a finitely presented nonconstructive algebra.

No general algorithm exists for deciding from the form of a finite presentation whether the corresponding finitely presented algebra is constructive or not (cf. [116]). A precise formulation and proof of this will be given later in this survey. But here and now we submit two different simple conditions that assure a finitely presented algebra is constructive.

Among the congruence relations on an arbitrary algebra $\mathfrak{A}$ are always these two trivial ones: the null congruence, under which no distinct elements of $\mathfrak{A}$ are congruent, and the unit congruence, under which all elements of $\mathfrak{A}$ are congruent. The algebra $\mathfrak{A}$ is said to be *simple* iff it has no nontrivial congruences.

Theorem 4.2.2 (Kuznecov [83]): *Every simple finitely presented algebra is constructive.*

If a given finitely presented simple algebra $\mathfrak{A}$ has only one element, there is nothing to prove. So suppose $\mathfrak{A}$ has more than one element. Let $a_1, \dots, a_l$ be generators of this algebra. According to the proof of Theorem 4.2.1, we can effectively find $g, h \in \mathcal{F}_{\text{prim}}$ such that

$$\langle g(0), h(0)\rangle, \langle g(1), h(1)\rangle, \dots \tag{33}$$

is a listing of all pairs of θ_γ-equivalent natural numbers.

We now take any two terms $\mathfrak{a}, \mathfrak{b} \in T$ with γ^* numbers m, n and ask ourselves, "Are the values of these terms in $\mathfrak{A}$ equal or not, i.e., does the pair $\langle m, n\rangle$ appear in (33) or not?" To decide this question we adjoin $\mathfrak{a} \approx \mathfrak{b}$ to the relations defining $\mathfrak{A}$ and let $\mathfrak{A}_1$ be the algebra presented by this new system of defining relations. We select functions $g_1, h_1 \in \mathcal{F}_{\text{prim}}$ such that

$$\langle g_1(0), h_1(0)\rangle, \langle g_1(1), h_1(1)\rangle, \dots \tag{34}$$

is a listing of θ_{γ_1}-equivalent pairs, where γ_1 is the standard numbering of $\mathfrak{A}_1$ relative to the generators $a_1, \dots, a_l$.

If $\mathfrak{a} \approx \mathfrak{b}$ is valid in $\mathfrak{A}$, then $\langle m, n\rangle$ appears in the sequence (33). But if $\mathfrak{a} \not\approx \mathfrak{b}$ is valid in $\mathfrak{A}$, then the algebra $\mathfrak{A}_1$ must have but one element, for it is a homomorphic image of $\mathfrak{A}$, which is simple. Hence, in (34) we must encounter somewhere the pairs $\langle 3, 3^2\rangle, \langle 3, 3^3\rangle, \dots, \langle 3, 3^l\rangle$. On examining pairs from the first and second lists alternately, after a finite number of steps either we find $\langle m, n\rangle$ in the first sequence and learn that $\mathfrak{a} \approx \mathfrak{b}$ holds in $\mathfrak{A}$, or we find all the pairs $\langle 3, 3^2\rangle, \dots, \langle 3, 3^l\rangle$ in the second sequence and know that $\mathfrak{a} \approx \mathfrak{b}$ is not valid in $\mathfrak{A}$. Thus we recursively solve the word problem for $\mathfrak{A}$; this means $\mathfrak{A}$ is constructive. ∎

This argument clearly works also for algebras presented by recursively enumerable systems of defining relations.

Now suppose $\langle \mathfrak{A}, \alpha\rangle$ is a positive numbered algebra in the sense of §3.1. In particular, this means D_α can be represented as the range of some appropriate $h \in \mathcal{F}_{\text{prim}}$. Besides $\mathfrak{A}$ we also consider the absolutely free algebra $\mathfrak{B}$ with individual symbols $\boldsymbol{a}_0, \boldsymbol{a}_1, \dots, \boldsymbol{a}_n, \dots$ as generators and with term-forming operations involving function symbols $\boldsymbol{f}_i$ corresponding to the basic operations of $\mathfrak{A}$. We assign certain natural numbers to the terms composing the base of $\mathfrak{B}$ by taking the number $\%\boldsymbol{a}_n$ of the term $\boldsymbol{a}_n$ to be $h(n)$, and putting

$$\%\boldsymbol{f}_i(\mathfrak{a}_1, \dots, \mathfrak{a}_{r_i}) = F_i(\%\mathfrak{a}_1, \dots, \%\mathfrak{a}_{r_i})$$

for longer terms; here, F_i is a fixed r_i-ary numerical **pr**-function representing the operation f_i of $\mathfrak{A}$ relative to α.

In some effective fashion we list all pairs of terms from $\mathfrak{B}$ in the sequence

$$\langle \mathfrak{b}_0, \mathfrak{c}_0\rangle, \langle \mathfrak{b}_1, \mathfrak{c}_1\rangle, \dots, \langle \mathfrak{b}_n, \mathfrak{c}_n\rangle, \dots$$

Let

$$\langle b_0, c_0\rangle, \langle b_1, c_1\rangle, \dots, \langle b_n, c_n\rangle, \dots$$

be the corresponding listing of the numbers of the above terms, so that $b_n = \%\mathfrak{b}_n, c_n = \%\mathfrak{c}_n$. By hypothesis, θ_α is a partial recursive predicate, so all pairs of θ_α-equivalent natural numbers can be effectively listed in a sequence

$$\langle d_0, e_0\rangle, \langle d_1, e_1\rangle, \ldots, \langle d_n, e_n\rangle, \ldots$$

Now we construct a sequence of identities $\mathfrak{a}_n \approx \mathfrak{a}'_n$ in the following manner.

Step 1: Compare the pairs $\langle b_0, c_0\rangle$ and $\langle d_0, e_0\rangle$. If they coincide, then we put $\mathfrak{a}_0 = \mathfrak{b}_0$ and $\mathfrak{a}'_0 = \mathfrak{c}_0$; if they are not identical, we take $\mathfrak{a}_0 = \mathfrak{b}_0, \mathfrak{a}'_0 = \mathfrak{b}_0$.

Step 2: Compare the sequence of pairs $\langle b_0, c_0\rangle, \langle b_1, c_1\rangle$ with the sequence of pairs $\langle d_0, e_0\rangle, \langle d_1, e_1\rangle$. If $\langle b_0, c_0\rangle$ occurs in the second sequence, we take $\mathfrak{a}_1 = \mathfrak{b}_0, \mathfrak{a}'_1 = \mathfrak{c}_0$; if not, we set $\mathfrak{a}_1 = \mathfrak{b}_0, \mathfrak{a}'_1 = \mathfrak{b}_0$. Next we look at the pair $\langle b_1, c_1\rangle$. If it appears in the second sequence, we take $\mathfrak{a}_2 = \mathfrak{b}_1, \mathfrak{a}'_2 = \mathfrak{c}_1$.

Step 3: Now we take the first three members of each sequence: $\langle b_0, c_0\rangle$, $\langle b_1, c_1\rangle, \langle b_2, c_2\rangle$ and $\langle d_0, e_0\rangle, \langle d_1, e_1\rangle, \langle d_2, e_2\rangle$. If the pair $\langle b_0, c_0\rangle$ appears in the second initial segment, then we put $\mathfrak{a}_3 = \mathfrak{b}_0, \mathfrak{a}'_3 = \mathfrak{c}_0$; if not, we set $\mathfrak{a}_3 = \mathfrak{b}_0, \mathfrak{a}'_3 = \mathfrak{b}_0$. Next we do the same thing with $\langle b_1, c_1\rangle$, etc.

The result is an effectively constructed sequence of formal equations

$$\mathfrak{a}_0 \approx \mathfrak{a}'_0, \quad \mathfrak{a}_1 \approx \mathfrak{a}'_1, \ldots, \mathfrak{a}_n \approx \mathfrak{a}'_n, \ldots . \tag{35}$$

Let $\mathfrak{C}$ be the algebra with generators $\boldsymbol{a}_0, \boldsymbol{a}_1, \ldots, \boldsymbol{a}_n, \ldots$ and defining relations (35). Let φ be the map from $\mathfrak{B}$ onto $\mathfrak{A}$ such that $\varphi(\mathfrak{a}) = \alpha(\%\mathfrak{a})$ $(\mathfrak{a} \in \mathfrak{B})$. This map is an abstract homomorphism of $\mathfrak{B}$ onto $\mathfrak{A}$; moreover, in the obvious sense the pairs $\langle \mathfrak{a}_n, \mathfrak{a}'_n\rangle$ determine the congruence σ on $\mathfrak{B}$ corresponding to φ. Clearly, $\mathfrak{C}$ is naturally isomorphic to the factor algebra $\mathfrak{B}/\sigma$. It is easy to see that the canonical isomorphism of $\mathfrak{B}/\sigma$ onto $\mathfrak{A}$ induces a **pr**-monomorphism of $\langle \mathfrak{C}, \gamma\rangle$ onto $\langle \mathfrak{A}, \alpha\rangle$; here, γ is the standard numbering of the effectively presented algebra $\mathfrak{C}$. We have thus proved

Theorem 4.2.3: *Every positive algebra* $\langle \mathfrak{A}, \alpha\rangle$ *is a* **pr**-*monomorphic image of an effectively presented algebra with standard numbering. If* $\mathfrak{A}$ *is also simple, the translated standard numbering is a constructive numbering of* $\mathfrak{A}$. ■

For any algebra $\mathfrak{A}$, the number of elements in the factor algebra $\mathfrak{A}/\sigma$ is called the *index* of the arbitrary congruence σ on $\mathfrak{A}$. Somewhat akin to the simple algebras are those infinite algebras, all of whose non-unit congruences have finite index.

Theorem 4.2.4: *Let* $\langle \mathfrak{A}, \alpha\rangle$ *be a finitely generated positive infinite algebra, all of whose congruences other than the unit congruence have finite index. Then any standard numbering of* $\mathfrak{A}$ *is constructive.*

Let $a_1, \ldots, a_l$ be fixed generators of $\mathfrak{A}$. If in the proof of the preceding theorem we take the algebra $\mathfrak{T}$ of §4.1 to be the free algebra and the number $\%\boldsymbol{a}_j$ of the term $\boldsymbol{a}_j \in T$ to be any α-number of the generator a_j in $\mathfrak{A}$, then we can view $\mathfrak{A}$ as presented by the corresponding recursively enumerated sequence of defining relations (35).

We consider arbitrary terms $\mathfrak{a}, \mathfrak{b} \in T$. By adjoining $\mathfrak{a} \approx \mathfrak{b}$ to (35) we obtain a system of relations that defines a homomorphic image $\mathfrak{B}$ of $\mathfrak{A}$ whose standard numbering we effectively verify as positive, using the facts following the proof of Theorem 4.2.1. Hence we can uniformly and effectively list all identities involving terms from T that are valid in $\mathfrak{B}$ in a sequence

$$\mathfrak{b}_0 \approx \mathfrak{b}'_0, \quad \mathfrak{b}_1 \approx \mathfrak{b}'_1, \ldots, \mathfrak{b}_n \approx \mathfrak{b}'_n, \ldots \tag{36}$$

analogous to (35). If $\mathfrak{a} \approx \mathfrak{b}$ is valid in $\mathfrak{A}$, then it must appear in the sequence (35). If $\mathfrak{a} \not\approx \mathfrak{b}$ is valid in $\mathfrak{A}$, then the algebra $\mathfrak{B}$, presented by (36), must be finite.

We define the *height* of a term in T as usual: the height of $\boldsymbol{a}_i$ is 0; if the heights of terms $\mathfrak{p}_1, \ldots, \mathfrak{p}_{r_i}$ have been defined and s is their maximum, then the height of $f_i(\mathfrak{p}_1, \ldots, \mathfrak{p}_{r_i})$ is $s+1$.

We let $\mathbf{L}_n$ be the set of all possible identities of the form $\mathfrak{p} \approx \mathfrak{q}$, where $\mathfrak{p}$ has height n, while the height of $\mathfrak{q}$ is less. A subset of $\mathbf{L}_n$ is called *complete* iff every term of height n appears on the left-hand side of at least one member of the subset. Let $\mathbf{L}_n^1, \ldots, \mathbf{L}_n^{t_n}$ be all the complete subsets of $\mathbf{L}_n$.

It is clear that if in any algebra $\mathfrak{C}$ all the identities in some complete subset $\mathbf{L}_n^k$ are valid, then $\mathfrak{C}$ is finite. Conversely, if $\mathfrak{C}$ is a finite algebra, then there exists an $n \in D$ such that all members of an appropriate complete subset $\mathbf{L}_n^k$ are valid in $\mathfrak{C}$.

Returning now to the terms $\mathfrak{a}, \mathfrak{b}$, we develop an effective procedure $\mathbf{P}$ as follows. The nth step of $\mathbf{P}$ consists of first looking for $\mathfrak{a} \approx \mathfrak{b}$ among the initial n members of the sequence (35). If it appears there, we terminate the procedure, knowing $\mathfrak{a} \approx \mathfrak{b}$ holds in $\mathfrak{C}$. If this identity is not found, we look at the initial segment of (36) of length n to see whether or not it includes the complete set $\mathbf{L}_j^k$ for some $1 \leqslant j \leqslant n$ and some $1 \leqslant k \leqslant t_j$. If it does, we know $\mathfrak{D}$ is finite and, therefore, $\mathfrak{a} \not\approx \mathfrak{b}$ holds in $\mathfrak{A}$. If none of these complete sets is included in this segment, then we perform the $(n+1)$th step of $\mathbf{P}$.

The demands on $\mathfrak{A}$ are such that $\mathbf{P}$ must terminate at some step, giving an answer to the question: "Are the values of $\mathfrak{a}, \mathfrak{b}$ in $\mathfrak{A}$ equal or not?" Since (36) – and with it $\mathbf{P}$ – is uniform with regard to the choice of $\mathfrak{a}, \mathfrak{b}$, this amounts to a recursive solution to the word problem for the algebra $\mathfrak{A}$. ∎

We shall consider one more test for the constructiveness of an algebra. Let

$\mathcal{K}$ be some class of algebras. An algebra $\mathfrak{A}$ – not necessarily a member of $\mathcal{K}$ – is said to be *approximated by* $\mathcal{K}$*-algebras* iff for every two distinct elements a, b of $\mathfrak{A}$, there exists a homomorphism φ of $\mathfrak{A}$ into some $\mathcal{K}$-algebra such that $\varphi(a) \neq \varphi(b)$.

Theorem 4.2.5 (cf. McKinsey [98]): *Suppose the algebra* $\mathfrak{A}$ *is presented by the finite number of generators* $\boldsymbol{a}_1, ..., \boldsymbol{a}_l$ *and the finite system* S *of conditional identities, and suppose* $\mathfrak{A}$ *is approximated by finite algebras in which all the members of* S *are valid. Then* $\mathfrak{A}$ *is constructive.*

In order to learn for arbitrary terms $\mathfrak{a}, \mathfrak{b} \in T$ whether or not the identity $\mathfrak{a} \approx \mathfrak{b}$ is valid in $\mathfrak{A}$, it is sufficient to apply to this identity the effective procedure whose nth step acts as follows. As a preliminary we list all complete sets L_j^k in the order

$$\mathrm{L}_1^1, ..., \mathrm{L}_1^{t_1}, \mathrm{L}_2^1, ..., \mathrm{L}_2^{t_2}, ... \qquad (37)$$

Now we let S_n be the union of S and the nth member of the sequence (37). Let $\mathfrak{B}_n$ be an algebra presented by the generators $\boldsymbol{a}_1, ..., \boldsymbol{a}_l$ and the defining relations S_n. As we mentioned above, $\mathfrak{B}_n$ must be finite; moreover, the number of its elements does not exceed a bound effectively computable from the number of basic operations, their ranks, and the numbers n and l. In such a situation we can obviously decide whether or not $\mathfrak{a} \approx \mathfrak{b}$ holds in $\mathfrak{B}_n$. If it is not valid in $\mathfrak{B}_n$, then it fails in $\mathfrak{A}$ *a fortiori*, and we terminate the procedure with a negative answer in hand.

If $\mathfrak{a} \approx \mathfrak{b}$ holds in $\mathfrak{B}_n$, then we consider the identity $\mathfrak{a}_n \approx \mathfrak{a}_n'$ from the sequence (35) corresponding to the algebra $\mathfrak{A}$. If it coincides with $\mathfrak{a} \approx \mathfrak{b}$, we learn the latter is valid in $\mathfrak{A}$, so we terminate the procedure. If $\mathfrak{a}_n \approx \mathfrak{a}_n'$ differs from $\mathfrak{a} \approx \mathfrak{b}$, we perform the $(n+1)$th step of the procedure.

By assumption, for any terms $\mathfrak{a}, \mathfrak{b} \in T$, either $\mathfrak{a} \approx \mathfrak{b}$ holds in $\mathfrak{A}$, or $\mathfrak{a} \not\approx \mathfrak{b}$ holds in at least one of the finite algebras $\mathfrak{B}_n$. That means this uniform procedure must terminate after a finite number of steps in every case; the word problem for $\mathfrak{A}$ is thus recursively solved. ∎

NOTES

(1) This project is continued in [XXII], [XXIV], and special numbered sets are investigated in [XXV], [XXVII], and [XXVIII].

(2) This common convention and the simplications it permits were added in translation.

([3]) In practice, having the same signature becomes equivalent to having the same type. The type above is finite, and this is usually sufficient, but in § 1.4 and § 4.2 we must admit infinite types.

([4]) At this point in the original the author, citing [71], refers to the common use of a sign other than = in formal equations, a practice observed throughout this translation.

([5]) According to [R5], the author suggested the formulas $\mathfrak{a}(\boldsymbol{x}, \boldsymbol{x}_1, \ldots, \boldsymbol{x}_n) \approx \boldsymbol{x} \rightarrow \boldsymbol{u} \approx \boldsymbol{v}$ ($\mathfrak{a}$ is any term other than $\boldsymbol{x}$ in which $\boldsymbol{x}$ occurs) be added to these and to (7). In a revision of [M8] and in [XXIII], however, Mal'cev adds correct formulas: $\mathfrak{a}(\boldsymbol{x}, \boldsymbol{x}_1, \ldots, \boldsymbol{x}_n) \not\approx \boldsymbol{x}$ ($\mathfrak{a} \neq \boldsymbol{x}$). Thus, while the class $\mathcal{L}$ below is not quasiprimitive, it is got from a quasivariety by removing the one-element algebras.

([6]) While Mal'cev does not define these two notions, he uses them often, apparently with the meanings given them in this paragraph added in translation.

([7]) This is tantamount to assuming β is positive.

([8]) The author (*via* [R5]) has supplied the word "infinite", missing from this and the following two paragraphs in the original.

([9]) With D_α assumed to be a **pr**-set, θ_α is a **pr**-function iff α is both positive and negative. The remark as stated, however, is not difficult to prove.

([10]) This is a numbering of A iff $M \neq \emptyset, D$. Hence, some adjustments must be made in the discourse following, although the results are valid.

([11]) This is faulty. Judging from [R5], it is the author's wish that the whole paragraph be disregarded.

([12]) Should we not ask the identity map on $\mathfrak{A}$ to be the morphism here? Equivalently we could have "In order that $\varphi * \alpha$ be **RQ**-morphic to α, ... ", as suggested in [R5]. The theorem then becomes true, trivial, useful and distinct from the preceding remark. Everything results from different readings of (9) in § 2.1.

([13]) $\mathfrak{G}_\alpha$ is not necessarily a normal subgroup of $\mathfrak{G}$. We do have the following easily checked relations for every $\varphi \in \mathfrak{G}$ (cf. [R5]): $\varphi \mathfrak{G}_\alpha \varphi^{-1} = \mathfrak{G}_{\varphi * \alpha}$, $\varphi * \mathcal{E}_\alpha = \mathcal{E}_{\varphi * \alpha}$. These suggest an interesting natural map from $\mathfrak{N}_\alpha / \equiv_{\mathbf{Rm}}$ onto the set of left cosets of the normalizer of $\mathfrak{G}_\alpha$ in $\mathfrak{G}$.

([14]) Actually, a numbering **pr**-subsystem of a **gr**-system *is* a **gr**-system. The whole paragraph remains valid, moreover, when we replace "general recursive" with "constructive".

([15]) The preceding note applies here, too.

([16]) When **R** = **pr**, the conclusions (i), (iii) hold only if "positive" is inserted in these three places.

CHAPTER 19

THE UNDECIDABILITY OF THE ELEMENTARY THEORY OF FINITE GROUPS

In the study of elementary theories of classes of algebras, the question of the algorithmic decidability of the elementary theory of the class of all finite groups naturally arises. This question was mentioned as an open problem in the well-known book [166] of A. Tarski, A. Mostowski, and R.M. Robinson. In the present article the elementary theory of the class of all finite groups is shown to be undecidable. In the course of the proof we establish, as corollaries, the undecidability of the elementary theories of several other classes of finite groups, as well as various classes of rings and semigroups.

§1. Let $\mathfrak{K}$ be an arbitrary fixed field. By $\mathcal{R}$ we denote the class of all rings – not necessarily associative – that are algebras of finite (linear) dimension over $\mathfrak{K}$. By a *formula* we mean any formula of first-order predicate logic (FOPL), whose only extralogical symbol, in addition to the equality sign $\approx$, is a binary operation symbol for multiplication; the usual abbreviations, such as juxtaposition for multiplication, are employed in the notation here. A *sentence* is a closed formula, which may contain unquantified individual constant symbols. The formula

$$\boldsymbol{ax} \approx \boldsymbol{x} \,\&\, \boldsymbol{x}^2 \approx \boldsymbol{x} \,\&\, \boldsymbol{x} \not\approx \boldsymbol{0}$$

is abbreviated by $\langle \boldsymbol{a}, \boldsymbol{x} \rangle$. The individual constants $\boldsymbol{a}$, $\boldsymbol{b}$ will play a special role in what follows. A sentence $\Phi(\boldsymbol{a}, \boldsymbol{b})$ with individual constants $\boldsymbol{a}$, $\boldsymbol{b}$ is called *normal* iff Φ is the conjunction of a sentence Φ^* with the same individual constants, the sentences

$$(1^a)\colon \quad (x)(y)(x \not\approx y \,\&\, \langle a,x\rangle \,\&\, \langle a,y\rangle \to (xy \approx x \vee yx \approx y) \,\&$$

$$\& \,(xy \approx 0 \vee yx \approx 0)),$$

(2^a): $(x)(y)(z)(\langle a,x\rangle \,\&\, \langle a,y\rangle \,\&\, \langle a,z\rangle \,\&\, xy \not\approx 0 \,\&\, yz \not\approx 0 \rightarrow xz \not\approx 0)$,

(3^a): $a^2 \approx 0 \,\&\, (x)(\langle a,x\rangle \rightarrow xa \approx 0)$,

(4^{ab}): $ab \approx 0 \,\&\, ba \approx a \,\&\, (x)(y)(\langle a,x\rangle \,\&\, \langle b,y\rangle \rightarrow xy \approx 0)$,

and the sentences (1^b), (2^b), (3^b) obtained from the above by substituting $\boldsymbol{b}$ for $\boldsymbol{a}$. A ring $\mathfrak{R} \in \mathcal{R}$ is called a $\Phi(a, b)$*-ring* iff one can distinguish elements a, b in it so that $\Phi(\boldsymbol{a},\boldsymbol{b})$ is true in $\langle \mathfrak{R}, a, b\rangle$ (from another point of view: iff the sentence $(\exists \boldsymbol{a})(\exists \boldsymbol{b})\Phi(\boldsymbol{a},\boldsymbol{b})$ is valid in $\mathfrak{R}$). The notation can be shortened to "Φ-ring" if no confusion is likely. An element u of a Φ-ring $\mathfrak{R}$ is an *a-element* iff $\langle a, u\rangle$ is true in $\mathfrak{R}$.

Modifying a basic definition from [168], we say that the formula Φ *represents* the unary numerical function f in the class $\mathcal{R}$ iff these three conditions are fulfilled: (i) Φ is a normal sentence; (ii) in any Φ-ring in $\mathcal{R}$ containing exactly m a-elements, there are exactly $f(m)$ b-elements ($m = 0, 1, 2, \ldots$); (iii) for every $m = 0, 1, 2, \ldots$, there is a Φ-ring in $\mathcal{R}$ that contains exactly m a-elements.

Theorem 1: *For every normally specified general recursive function f, we can effectively construct a sentence Φ representing f in the class $\mathcal{R}$.*

By a theorem of J. Robinson [135], Theorem 1 will be proved if we succeed in constructing formulas representing the functions κ, λ, where $\kappa(x) = x+1$, $\lambda(x) = x - [\sqrt{x}]^2$, and we can demonstrate an effective method for constructing formulas representing the functions given by $g(x) + h(x)$, $g(h(x))$, and $g^{-1}(x)$ (i.e., $\mu y(g(y) = x)$) when we already have formulas representing g and h in $\mathcal{R}$. Here, we shall construct formulas only for κ and g^{-1}; the remaining formulas are similarly constructed.

We denote by $\Psi_1(\boldsymbol{a}, \boldsymbol{b}, \boldsymbol{c})$ the conjunction of the following formulas

$$(x)(\langle a,x\rangle \rightarrow \langle b, cx\rangle), \tag{1}$$

$$(x)(y)(\langle a,b\rangle \,\&\, \langle a,y\rangle \,\&\, cx \approx cy \rightarrow x \approx y), \tag{2}$$

$$(\exists z)\,[\langle b,z\rangle \,\&\, (x)(\langle a,x\rangle \rightarrow cx \not\approx z) \,\&\, \\ \&\, (y)(\langle b,y\rangle \,\&\, y \not\approx z) \rightarrow (\exists x)(\langle a,x\rangle \,\&\, y \approx cx))]\,. \tag{3}$$

Let $\Psi(\boldsymbol{a},\boldsymbol{b})$ be the conjunction of $(\exists \boldsymbol{c})\Psi_1(\boldsymbol{a}, \boldsymbol{b}, \boldsymbol{c})$ and (1^a)–(3^a), (1^b)–(3^b), (4^{ab}). Then Ψ represents the function κ in $\mathcal{R}$. Indeed, suppose in the ring

$\mathfrak{R} \in \mathcal{R}$ there exist a, b, c with the properties described in (1), (2), (3). According to (3), there is an element z such that the set of b-elements in $\mathfrak{R}$ consists of z and all the a-elements multiplied by c on the left. Since by (2) multiplication of different a-elements by c gives distinct b-elements, the number of b-elements in $\mathfrak{R}$ is one greater than the number of a-elements.

In order to construct a Ψ-ring containing a specified number m of a-elements, we take a linear space over $\mathfrak{K}$ with basis elements $a, b, c, x_1, \ldots, x_m, y_1, \ldots, y_{m+1}$ and introduce multiplication on the basis by setting $ba = a$, $ax_i = x_i = x_i x_j$, $cx_i = y_i$, $by_i = y_i = y_i y_j$ $(i \leqslant j; i = 1, \ldots, m; j = 1, \ldots, m+1)$, and by requiring all other products of pairs of basis elements to be equal to 0. The algebra so obtained is clearly a Ψ-ring in which $x_1, \ldots, x_m$ are the a-elements.

§2. Suppose the function g is represented by the formula $\Phi(a, b)$, and the equation $g(x) = m$ is solvable for every $m = 0, 1, 2, \ldots$ By definition, $g^{-1}(m)$ is the least solution to the corresponding equation. We introduce a formula $\Phi_1(a, b, c)$ that expresses in every ring $\mathfrak{R} \in \mathcal{R}$ the following properties: (I) the set $\mathfrak{R}_c$ of elements $x \in \mathfrak{R}$ for which $cx = x$ forms a subring in $\mathfrak{R}$; (II) the elements a, b belong to $\mathfrak{R}_c$; (III) $\Phi(a, b)$ is true in $\mathfrak{R}_c$. We can take $\Phi_1(a, b, c)$ to be the conjunction of the formula

$$(x)(y)(cx \approx x \mathbin{\&} cy \approx y \rightarrow c \cdot xy \approx xy) \mathbin{\&} ca \approx a \mathbin{\&} cb \approx b$$

and the relativization of $\Phi(a, b)$ with respect to the formula $\rho(x) = cx \approx x$.

By $\Psi_1(a, b, a', b', c)$ we denote the conjunction of the formula $\Phi_1(a', b', c)$ and the formula ([1])

$$(x)(\langle a, x\rangle \leftrightarrow cx \approx x \mathbin{\&} \langle b', x\rangle) \mathbin{\&} (x)(\langle b, x\rangle \leftrightarrow cx \approx x \mathbin{\&} \langle a', x\rangle) \mathbin{\&} \tag{4}$$

$$\mathbin{\&} (x)\{cx \approx x \mathbin{\&} \langle a', x\rangle \rightarrow (\exists uvde)[\Phi_1(u, v, d) \mathbin{\&} \tag{5}$$

$$\mathbin{\&} (y)(cy \approx y \mathbin{\&} \langle a', y\rangle \mathbin{\&} yx \approx y \mathbin{\&} y \not\approx x \rightarrow d \cdot ey \approx ey \mathbin{\&} \langle u, ey\rangle) \mathbin{\&} \tag{6}$$

$$\mathbin{\&} (y)(z)(cy \approx y \mathbin{\&} \langle a', y\rangle \mathbin{\&} cz \approx z \mathbin{\&} \langle a', z\rangle \mathbin{\&} y \not\approx z \rightarrow ey \not\approx ez) \mathbin{\&} \tag{7}$$

$$\mathbin{\&} (y)(dy \approx y \mathbin{\&} \langle u, y\rangle \rightarrow (\exists z)(cz \approx z \mathbin{\&} \langle a', z\rangle \mathbin{\&} xz \approx 0 \mathbin{\&} y \approx ez)) \mathbin{\&} \tag{8}$$

$$\& ((\exists f)(y)(y')(cy \approx y \,\&\, \langle b', y\rangle \,\&\, cy' \approx y' \,\&\, \langle b', y'\rangle \,\&\, fy \approx fy' \rightarrow$$

$$\rightarrow y \approx y' \,\&\, d\cdot fy \approx fy \,\&\, \langle v, fy\rangle \,\&\, (\exists z)(dz \approx z \,\&\, \langle v, z\rangle \,\&\, z\cdot fy \approx 0)) \vee$$

$$\vee (\exists f)(y)(y')(dy \approx y \,\&\, \langle v, y\rangle \,\&\, dy' \approx y' \,\&\, \langle v, y'\rangle \,\&\, fy \approx fy' \rightarrow$$

$$\rightarrow y \approx y' \,\&\, c\cdot fy \approx fy \,\&\, \langle b', fy\rangle \,\&\, (\exists z)(cz \approx z \,\&\, \langle b', z\rangle \,\&\, z\cdot fy \approx 0)))] \}. \tag{9}$$

Let $\Psi(a,b)$ be the conjunction of $(\exists a'b'c)\ \Psi_1(a, b, a', b', c)$ and (1^a)–(3^a), (1^b)–(3^b), (4^{ab}). Then Ψ represents g^{-1}. Indeed, suppose in the ring $\mathfrak{R} \in \mathcal{R}$ there are elements a, b, a', b', c for which Ψ_1 is true, and there are exactly m a-elements. Then $\mathfrak{R}_c$ is a subring containing a', b' in which $\Phi(a', b')$ is true. By (4) the set of b'-elements of $\mathfrak{R}_c$ coincides with the set of a-elements of $\mathfrak{R}$, while the set of a'-elements of $\mathfrak{R}_c$ is just the set of b-elements of $\mathfrak{R}$; thus $m = g(n)$, where n is the number of b-elements in $\mathfrak{R}$. By $(1^{a'})$–$(3^{a'})$ the a'-elements of $\mathfrak{R}$ can be uniquely denoted as $x_1, \ldots, x_n$ so that the relations $x_l x_i = x_l x_l = x_l$, $x_i x_l = 0$ $(l < i)$ hold. According to (5), for every $0 \leqslant j < n$, there exist elements u, v, d in $\mathfrak{R}$ such that $\mathfrak{R}_d$ is a $\Phi(u, v)$-ring. The conditions (6)–(8) guarantee that the u-elements of $\mathfrak{R}_d$ are in 1–1 correspondence with $x_1, \ldots, x_j$, while (9) shows that the number of v-elements in $\mathfrak{R}_d$ differs from m, hence $g(j) \neq m$ (some vacuous cases occur here, of course).

Therefore, if in the Ψ-ring $\mathfrak{R}$ the number of a-elements is m and the number of b-elements is n, then $g(n) = m$, while $g(j) \neq m$ $(0 \leqslant j < n)$; i.e., $n = g^{-1}(m)$.

It remains to verify (iii). Let m be given; put $n = g^{-1}(m)$. By hypothesis, for every $i = 0, 1, \ldots, n$, there exists a $\Phi(a_i, b_i)$-ring $\mathfrak{R}_i$ containing i a_i-elements $x_{i1}, \ldots, x_{ii}$ and $g(i)$ b_i-elements $y_{i1}, \ldots, y_{ig(i)}$ with the relations $ax_{iq} = x_{iq}x_{ir} = x_{iq}x_{iq} = x_{iq}$, $x_{ir}x_{iq} = 0$ $(q < r)$, $b_i y_{is} = y_{is}y_{it} = y_{is}y_{is} = y_{is}$, $y_{it}y_{is} = 0$ $(s < t)$, as well as $a_i b_i = a_i^2 = b_i^2 = 0$, $b_i a_i = a_i$; some cases may be vacuous. From these relations it follows that the a_i-elements, the b_i-elements, a_i, and b_i together form a linearly independent subset of $\mathfrak{R}_i$; this set can be extended to a basis for $\mathfrak{R}_i$ by the addition of suitable elements $w_{i1}, \ldots, w_{ik_i}$. Assuming the algebras $\mathfrak{R}_i$ have no elements in common, we formally construct a new algebra $\mathfrak{R}$ as follows. The basis of $\mathfrak{R}$ consists of the chosen basis elements of all the $\mathfrak{R}_i$ plus the new elements a, b, c, d_j, e_j, f_j $(j = 0, \ldots, n-1)$. The multiplication of basis elements that come from $\mathfrak{R}_i$ is performed as in $\mathfrak{R}_i$ $(i = 0, \ldots, n)$; we further put $ba = a$, $bx_{nr} = x_{nr}$ $(r = 1, \ldots, n)$, $ay_{ns} = y_{ns}$ $(s = 1, \ldots, m)$, $cx = x$ $(x \in \mathfrak{R}_n)$. For every $0 \leqslant j < n$, we set $d_j x = x$ $(x \in \mathfrak{R}_j)$, $e_j x_{nl} = x_{jl}$ $(l < j)$, and $f_j y_{ns} = y_{js}$ $(s = 1, \ldots, m)$ if $g(j) > m$, or $f_j y_{jt} = y_{nt}$ $(t = 1, \ldots, g(j))$

if $g(j) < m$. The remaining products of pairs of basis elements we set equal to 0. The resulting algebra $\mathfrak{R}$ satisfies (1^a)–(3^a), (1^b)–(3^b), (4^{ab}), and $\Psi_1(a, b, a_n, b_n, c)$ is true in $\mathfrak{R}$. Thus Ψ represents the function g^{-1} in $\mathcal{R}$. This completes the discussion of Theorem 1. ■

§3. From Theorem 1 we easily obtain the

Corollary: *For every normally specified general recursive function f, we can effectively construct a sentence $\Upsilon(\boldsymbol{a}, \boldsymbol{b})$ representing this function in the class $\mathcal{R}^1$ of rings with identity that are algebras of finite dimension over $\mathfrak{K}$.*

Let $\Phi(\boldsymbol{a}, \boldsymbol{b})$ be a formula representing f in $\mathcal{R}$. By $\Upsilon_1(\boldsymbol{a}, \boldsymbol{b}, \boldsymbol{a}', \boldsymbol{b}', \boldsymbol{c})$ we denote the conjunction of the formula $\Phi_1(\boldsymbol{a}', \boldsymbol{b}', \boldsymbol{c})$ and the formula

$$(x)(\langle \boldsymbol{a}, x\rangle \leftrightarrow \boldsymbol{c}x \approx x \;\&\; \langle \boldsymbol{a}', x\rangle) \;\&\; (x)(\langle \boldsymbol{b}, x\rangle \leftrightarrow \boldsymbol{c}x \approx x \;\&\; \langle \boldsymbol{b}', x\rangle)\,.$$

Let $\Upsilon(\boldsymbol{a}, \boldsymbol{b})$ be the conjunction of (1^a)–(3^a), (1^b)–(3^b), (4^{ab}) and $(\exists \boldsymbol{a}'\boldsymbol{b}'\boldsymbol{c})\ \Upsilon_1(\boldsymbol{a}, \boldsymbol{b}, \boldsymbol{a}', \boldsymbol{b}', \boldsymbol{c})$. Then Υ represents f in $\mathcal{R}^1$. Indeed, the properties (i), (ii) the sentence Υ clearly has. As to (iii), suppose $\mathfrak{R} \in \mathcal{R}$ is a $\Phi(a', b')$-ring containing exactly m a'-elements. We adjoin new elements a, b, c, e to a basis for $\mathfrak{R}$ and set $eu = ue = cu = u$, $uc = 0$ for $u \in \mathfrak{R}$, and $ba = a$, $a^2 = b^2 = 0$. By further putting $ax = x$, $by = y$ for each a'-element x and each b'-element y in $\mathfrak{R}$, we obtain an Υ-ring having an identity element e and containing m a-elements. ■

§4. From Theorem 1 and its corollary we obtain in the usual way:

Theorem 2: *The class $\mathcal{R}$ of all rings that are finite dimensional algebras over a fixed field $\mathfrak{K}$ and the class $\mathcal{R}^1$ of all $\mathcal{R}$-rings with identity have undecidable elementary theories.*

To prove this it suffices to take a general recursive function f with a non-recursive range of values and to construct a formula $\Phi(\boldsymbol{a}, \boldsymbol{b})$ representing f in $\mathcal{R}^1$. For then the falsity of the sentence

$$\Phi(\boldsymbol{a}, \boldsymbol{b}) \;\&\; (\exists x_1 \ldots x_n)\,\Big(\mathop{\&}_{i \neq j} (x_i \not\approx x_j \;\&\; \langle \boldsymbol{b}, x_i\rangle) \;\&\; \;\&\; (y)(\langle \boldsymbol{b}, y\rangle \rightarrow \bigvee_i y \approx x_i)\Big)$$

in every $\mathcal{R}^1$-ring (for any choices of distinguished elements a, b) is equivalent to the unsolvability of the equation $f(x) = n$. ■

Corollary: *The elementary theory of the class* $\mathcal{R}_\pi^F$ *of all finite rings satisfying the identity* $(x)(\pi x \approx 0)$, *where* π *is any fixed prime number, as well as of the class* $\mathcal{R}^F$ *of all finite rings, is undecidable.*

The first part follows from Theorem 2 on taking $\mathfrak{K}$ to be the prime field of characteristic π, while the second is implied by the first, because $\mathcal{R}_\pi^F$ is a finitely axiomatizable subclass relative to $\mathcal{R}^F$. ■

§5. A correspondence between the class of rings with identity and a certain class of metabelian groups was established and studied in [XV]. Under this correspondence, the class of all rings of odd prime characteristic π with identity is mapped onto a certain finitely axiomatizable class of metabelian groups satisfying the identity $(x)(x^\pi \approx 1)$. An effective method was indicated whereby for every FOPL sentence concerning rings with identity we can construct a sentence appropriate to groups such that if the first sentence holds in a ring with identity iff the second is true in the corresponding group. Since finite rings correspond to finite groups, from the corollary of Theorem 2 we immediately deduce

Theorem 3: *For every odd prime* π, *the elementary theory of the class of all finite metabelian groups in which* $(x)(x^\pi \approx 1)$ *holds is undecidable.* ■

For each π, the class indicated is a finitely axiomatizable subclass in the class of all finite groups, the class of all finite metabelian groups, the class of all finite semigroups, etc. Therefore, the elementary theories of all these classes are undecidable.

Let $\mathfrak{L}$ be a metabelian Lie ring of odd prime characteristic π. We can make a metabelian π-group out of $\mathfrak{L}$ by defining a new operation of multiplication: we set $xy = x + y + \frac{1}{2}[x, y]$. We then have the relations $[x, y] = xyx^{-1}y^{-1}$, $x+y = xy[y, x]^{1/2}$. Conversely, by using the latter relations to define the sum and bracket of any two elements in a given metabelian group, we obtain a metabelian Lie ring of characteristic π. Hence the class of all finite metabelian Lie rings of odd prime characteristic π is syntactically equivalent (cf. [XV]) to the class of all finite metabelian π-groups and, along with the latter, has an undecidable elementary theory.

A metabelian Lie ring is associative, whence it follows that the elementary theories of the class of finite two-step nilpotent rings and the class of finite associative rings are undecidable.

NOTE

(¹) This formula has been rearranged in translation to make it clearer and more accurate.

CHAPTER 20

ELEMENTARY PROPERTIES OF LINEAR GROUPS

Introduction

Elementary properties of a group, field, or in general any algebraic system $\mathfrak{A}$ are those properties of $\mathfrak{A}$ that we can express in the language of first-order predicate logic (FOPL), taking as primitives the basic operations and predicates of the system $\mathfrak{A}$. Consequently, in the realm of the "elementary" theory of groups we can pose, among others, the following questions:

(a) Which of the group-theoretic notions ordinarily defined by unrestricted logical means (e.g., by using predicate logics of higher orders) admit definitions in the FOPL language?

(b) What algorithmic structure does this or that set of FOPL sentences have?

(c) Under what conditions do nonisomorphic groups nonetheless have identical elementary properties?

In the present article questions of this sort are considered for the following matrix groups ($n \geqslant 2$):

$GL(n, \mathfrak{K})$ – the multiplicative group of all nonsingular matrices of order n over the field $\mathfrak{K}$ (the *general linear group*);

$SL(n, \mathfrak{K})$ – the multiplicative group of all $n \times n$ matrices with determinant 1 (the *special linear group*);

$PG(n, \mathfrak{K})$ – the factor group of $GL(n, \mathfrak{K})$ by its center (the *projective group*);

$PS(n, \mathfrak{K})$ – the factor group of $SL(n, \mathfrak{K})$ by its center (the *special projective group*).

Throughout the article we shall assume $\mathfrak{K}$ is a field of characteristic zero, although with a few reservations the basic results can be extended to fields of prime characteristic. The notation $A_1 \dotplus \ldots \dotplus A_s$ will indicate the direct (Kronecker) sum of the matrices $A_1, \ldots, A_s$, so

$$A \dotplus B = \begin{pmatrix} A & \\ & B \end{pmatrix}$$

schematically. I_r denotes the identity matrix of order r.

I denotes an identity matrix whose order is provided by the context; O denotes such a zero matrix.

A matrix C of square cells of the form

$$C = \begin{pmatrix} A & I & & \\ & A & I & \\ & & \ddots & \\ & & & A \end{pmatrix},$$

where the characteristic polynomial of the cell A is irreducible over $\mathfrak{K}$, is called a *generalized Jordan cell.* The number of copies of A is the *degree* of C. Thus the order of a Jordan cell is the product of its index and degree. The characteristic polynomial of A is called the *root polynomial* of the cell C.

As is well known (cf., e.g., [M3], p. 131]), every matrix M of order n over $\mathfrak{K}$ can be represented as $U^{-1}AU$, where $U \in GL(n, \mathfrak{K})$ and A has the form

$$(A_1^{(1)} \dotplus \ldots \dotplus A_{k_1}^{(1)}) \dotplus \ldots \dotplus (A_1^{(s)} \dotplus \ldots \dotplus A_{k_s}^{(s)}) \,. \tag{1}$$

Here, the matrices $A_1^{(i)}, \ldots, A_{k_i}^{(i)}$ are Jordan cells constructed from the same matrix $A^{(i)}$; what's more, the characteristic polynomials of $A^{(1)}, \ldots, A^{(s)}$ are distinct and irreducible over $\mathfrak{K}$. Let m_j^i be the index of $A_j^{(i)}$, and n_i the order of the fundamental matrix $A^{(i)}$. The configuration

$$\chi = \{[(m_1^1, n_1), \ldots (m_{k_1}^1, n_1)] \ \ldots \ [(m_1^s, n_s) \ldots (m_{k_s}^s, n_s)]\ \}$$

of these numbers is called the *Segre characteristic* (over $\mathfrak{K}$) of the matrix $M = U^{-1}AU$. The Segre characteristic of M over the given field $\mathfrak{K}$ is uniquely determined apart from the order of distribution of the pairs inside each pair of brackets and the order of the bracketed systems themselves (cf. [M3], p. 131).

We shall use $\mathfrak{K}^*$ throughout to denote the algebraic closure of the field $\mathfrak{K}$. Every matrix M over $\mathfrak{K}$ is simultaneously a matrix with entries in $\mathfrak{K}^*$. But the Segre characteristics of M over $\mathfrak{K}$ and over $\mathfrak{K}^*$ are in general different. Since only polynomials of degree 1 are irreducible over $\mathfrak{K}^*$, the Segre characteristic over $\mathfrak{K}^*$ always consists of pairs of the form $(m, 1)$; we shall abbreviate the notation by using, instead of the pair $(m, 1)$, its first member only. The Segre characteristic over $\mathfrak{K}^*$ is called the *absolute* Segre characteristic of a given matrix. In order to compute the absolute characteristic of a matrix from its Segre characteristic over $\mathfrak{K}$, we first replace each pair by its first member, then copy the ith bracketed system n_i times, where n_i is the second member just discarded from each of its pairs. E.g., if $\chi = \{[(2, 2)]\ [(1,1)]\}$, then $\chi^* = \{[2]\ [2]\ [1]\}$.

The matrix A in the above is called the *Jordan form* of the matrix M over $\mathfrak{K}$. In a certain sense, the Segre characteristic takes into account only the arithmetical structure of the Jordan form.

The *basic elementary group predicate* is the relation $P(x, y, z)$, equivalent to $xy = z$. The *basic ring predicates* are the relations $S(x, y, z)$ and $P(x, y, z)$, respectively equivalent to $x + y = z$ and $xy = z$.

A formula Φ of FOPL with equality $\approx$ (cf. [117]) in which no predicate symbol other than $\boldsymbol{P}$ occurs is called a *group formula.* The operation notation $\boldsymbol{x} \cdot \boldsymbol{y} \approx \boldsymbol{z}$, etc., will lead to convenient abbreviations in writing group formulas. We similarly define *ring formulas.*

A group formula with free individual variables, specifies a *formular predicate* defined on each group. If Φ is a closed FOPL group formula (i.e., it contains no free variables), then we call Φ a *group sentence.*

In §1 a method is indicated whereby for each of the groups $GL(n, \mathfrak{K})$, $SL(n, \mathfrak{K})$ and for every Segre characteristic χ, we can construct a group formula $\Phi(\boldsymbol{x})$ that is true in the given group for exactly those matrices with Segre characteristic χ. Thus in each of these groups the Segre characteristic of an element can be defined elementarily within the group itself.

With the help of the results of §1 we establish in §2 that the groups $G(m, \mathfrak{K})$ and $G(n, \mathfrak{L})$ $(n \geqslant 3, G = GL, SL, PG, PS)$ are indistinguishable by their elementary properties (in other terms, they have the same elementary, or arithmetic, type) iff $m = n$ and the fields $\mathfrak{K}$, $\mathfrak{L}$ are elementarily indistinguishable (i.e., elementarily equivalent). In conclusion it is proved that for each of these groups the set of FOPL group sentences true in the given group is recursively equivalent to the set of all ring sentences true in the corresponding base field.

§1. The elementary nature of the Segre characteristic

§1.1. *The elementariness of diagonalizability*

The centers of both $GL(n, \mathfrak{K})$ and $SL(n, \mathfrak{K})$ consist of scalar matrices (cf. [183]). Hence the group formula

$$(\boldsymbol{X})(\boldsymbol{XM} \approx \boldsymbol{MX})$$

gives an elementary characterization of the scalar matrices in the indicated groups, i.e., this formula is true for just those matrices M with Segre characteristic $\{[(1,1) \ldots (1,1)]\}$.

Somewhat more complicated to prove is

Lemma 1: *In each of the groups $GL(n, \mathfrak{K})$, $SL(n, \mathfrak{K})$ the formula*

$$\mathrm{Cm}(M) = (X)(Y)(XM \approx MX \,\&\, YM \approx MY \rightarrow XY \approx YX)$$

is true just for those matrices whose Segre characteristics have the form $\{[(m_1, n_1)], \ldots, [(m_s, n_s)]\}$. *In other words,* $\mathrm{Cm}(M)$ *holds in these groups iff M has a Jordan form with only one Jordan cell for each root polynomial, i.e., for each irreducible factor of the characteristic polynomial of M.*

Indeed, let the matrix M have a characteristic of the form indicated in the lemma, and let X, Y be matrices commuting with M. We reduce M to Jordan normal form A over $\mathfrak{K}^*$ by means of the matrix U: $A = U^{-1}MU$. Let $X_0 = U^{-1}XU$, $Y_0 = U^{-1}YU$. Because X_0 and Y_0 commute with A, we can immediately conclude (cf. [M3], p. 146) that $X_0Y_0 = Y_0X_0$; hence $XY = YX$.

Conversely, suppose the characteristic of the matrix M does not have the form indicated. We bring M into generalized Jordan form (1) over $\mathfrak{K}$. By hypothesis, among the cells $A_j^{(i)}$ occur at least two with the same root polynomial. Let these be $A_1^{(1)}$ and $A_2^{(1)}$. Consider the matrices

$$X = X_1 \dot{+} I_t, \qquad Y = Y_1 \dot{+} I_t,$$

where X_1, Y_1 are matrices with variables as elements having the same order as the matrix

$$B = A_1^{(1)} \dot{+} A_2^{(1)},$$

and t is chosen so that X will have the same order as A. Dividing X_1, Y_1 into cells of the same size as in B, and rewriting the relations

$$BX_1 = X_1B, \quad BY_1 = Y_1B$$

in the form of linear relations among the cells in the matrices X_1, Y_1, we easily find values for these cells equal to either I or O such that the substituted matrices $\hat{X}_1$, $\hat{Y}_1$ have determinant 1 and commute with B, but for every positive number m, the matrices $\hat{X}_1^m$, $\hat{Y}_1^m$ fail to commute. These calculations are completely similar to those performed on p. 146 of [M3], and we shall omit them here. As a quick example, we note that if

$$B = A^{(1)} \dot{+} A^{(1)},$$

then we can take

$$\hat{X}_1 = \begin{pmatrix} I_l & I_l \\ & I_l \end{pmatrix}, \qquad \hat{Y}_1 = \begin{pmatrix} I_l & \\ I_l & I_l \end{pmatrix},$$

where l is the order of $A^{(1)}$. ■

We now exhibit a formula describing those matrices similar over $\mathfrak{K}^*$ to diagonal matrices with pairwise distinct diagonal elements. Over $\mathfrak{K}$ these are the matrices whose Segre characteristics have the form $\{[(1, n_1)] \ \ldots \ [(1, n_s)]\}$.

Lemma 2: *In each of the groups $GL(n, \mathfrak{K})$, $SL(n, \mathfrak{K})$ the formula*

$$\mathrm{Dd}_r(M) = \mathrm{Cm}(M) \ \& \ (X)(X^{-1}MX{\cdot}M \approx M{\cdot}X^{-1}MX \to MX^{r!} \approx X^{r!}M)$$

is true when $r = n$ for any matrix M similar over $\mathfrak{K}^*$ to a diagonal matrix with pairwise distinct diagonal elements. If $\mathrm{Dd}_r(M)$ is true for some positive number r and matrix M, then M is similar over $\mathfrak{K}^*$ to a diagonal matrix with distinct diagonal elements.

Let us assume M has a characteristic of the form mentioned above. Put M in diagonal form over $\mathfrak{K}^*$ and suppose $X^{-1}MX \cdot M = M \cdot X^{-1}MX$. Then $X^{-1}MX$ is again diagonal, and its diagonal elements are those of M, but possibly in a different order. Hence, in each row and in each column of X there is but one nonzero element, i.e., X is a *monomial* matrix. Therefore, $X^{n!}$ is diagonal and $X^{n!}M = MX^{n!}$.

Let M be any matrix in $GL(n, \mathfrak{K})$ not diagonalizable over $\mathfrak{K}^*$, but satisfying $\mathrm{Cm}(M)$. To prove the second assertion in the lemma, it is sufficient to construct for every such M, a matrix $X \in SL(n, \mathfrak{K})$ such that no positive power of X commutes with M, yet $X^{-1}MX \cdot M = M \cdot X^{-1}MX$.

First put M in Jordan form (1) over $\mathfrak{K}$. Since M is not diagonalizable over $\mathfrak{K}^*$, there is a cell among the $A_j^{(i)}$ with index greater than 1. Let this be

$$A_1^{(1)} = \begin{pmatrix} A^{(1)} & I_m & \\ & \ddots & \\ & & A^{(1)} \end{pmatrix},$$

and let i be its index. Setting

$$X_1 = \alpha^{1-i}(\alpha^{2i-2} I_m \dot{+} \alpha^{2i-4} I_m \dot{+} \ldots \dot{+} I_m),$$

we find for any nonzero element $\alpha \in \mathfrak{K}$:

$$X_1^t A_1^{(1)} X_1^{-t} = \begin{pmatrix} A^{(1)} & & \alpha^{2t} I_m \\ & \ddots & \\ & & A^{(1)} \end{pmatrix} \qquad (t = 1, 2, \dots) ;$$

consequently,

$$X^{-1}AX \cdot A = A \cdot X^{-1}AX , \qquad X^t A \neq A X^t$$

$$(t = 1, 2, \dots) , \ |X| = 1 ,$$

where $X = X_1 \dot{+} I_q$ for some identity matrix I_q of appropriate order. ∎

Lemma 3: *In each group $GL(n, \mathfrak{K})$, $SL(n, \mathfrak{K})$ the formula*

$$\mathrm{D}_n(M) = (\exists X)(\mathrm{Dd}_n(X) \,\&\, MX \approx XM)$$

is true for those and only those matrices that are reducible to diagonal form over $\mathfrak{K}^*$.

Suppose $\mathrm{D}_n(M)$ is true, and let X be a matrix with the asserted properties. On reducing X to diagonal form over $\mathfrak{K}^*$, we simultaneously reduce M to diagonal form, for matrices commuting with a diagonal matrix with distinct diagonal elements are themselves diagonal.

Conversely, suppose M is diagonalizable over $\mathfrak{K}^*$. Then the Jordan form for M over $\mathfrak{K}$ will look like

$$A = (A^{(1)} \dot{+} \dots \dot{+} A^{(1)}) \dot{+} \dots \dot{+} (A^{(s)} \dot{+} \dots \dot{+} A^{(s)}) ,$$

where $A^{(1)}, \dots, A^{(s)}$ are matrices with distinct characteristic polynomials irreducible over $\mathfrak{K}$. The matrix A certainly commutes with the matrix

$$X = (\alpha_1^{(1)} A^{(1)} \dot{+} \dots \dot{+} \alpha_{k_1}^{(1)} A^{(1)}) \dot{+} \dots$$
$$\dot{+} (\alpha_1^{(s)} A^{(s)} \dot{+} \dots \dot{+} \alpha_{k_s}^{(s)} A^{(s)}) .$$

Clearly, elements $\alpha_j^{(i)}$ can be chosen from $\mathfrak{K}$ so that the determinant of X is equal to 1, and the characteristic roots of X are simple. ∎

§1.2. *The elementariness of the characteristics of diagonalizable matrices*
We consider arbitrary diagonal matrices

$$A = \alpha_1 E_{11} + \ldots + \alpha_n E_{nn}, \qquad B = \beta_1 E_{11} + \ldots + \beta_n E_{nn},$$

with coefficients in $\mathfrak{K}^*$. Let us agree to write $A \,\eta\, B$ iff for all $i, j = 1, \ldots, n$, from $\alpha_i \neq \alpha_j$ we conclude $\beta_i \neq \beta_j$, i.e., iff the matrix A is in a certain sense smoother than B.

Lemma 4: *In each group* $GL(n, \mathfrak{K})$, $SL(n, \mathfrak{K})$ *the formula*

$$M \leqslant N = \mathrm{D}_n(M) \,\&\, \mathrm{D}_n(N) \,\&\, (X)(XN \approx NX \rightarrow XM \approx MX) \qquad (2)$$

is true for those and only those matrices that are simultaneously similar over $\mathfrak{K}^*$ *to diagonal matrices* A, B *such that* $A \,\eta\, B$.

If M, N are jointly similar to matrices A, B such that $A \,\eta\, B$, then clearly $M \leqslant N$ (*per*(2)). Conversely, suppose $M \leqslant N$ for matrices M, N. From $\mathrm{D}_n(N)$ holding, we learn that N can be reduced over $\mathfrak{K}$ to the form

$$B = (B^{(1)} \dot{+} \ldots \dot{+} B^{(1)}) \dot{+} \ldots \dot{+} (B^{(s)} \dot{+} \ldots \dot{+} B^{(s)}),$$

where the matrices $B^{(1)}, \ldots, B^{(s)}$ have distinct characteristic polynomials that are irreducible over $\mathfrak{K}$. From $MN = NM$ it follows that by the same similarity transformation M is brought into the form

$$A = A_1 \dot{+} \ldots \dot{+} A_s,$$

where A_i is a matrix with cell structure paralleling that of

$$B_i = B^{(i)} \dot{+} \ldots \dot{+} B^{(i)}$$

for $i = 1, \ldots, s$. Since $A \leqslant B$, any matrix X_i that commutes with B_i must also commute with A_i. By taking X_i to be various celled matrices constructed from identity and null cells, which thus commute with B_i, one easily concludes that A_i must have the form

$$A^{(i)} \dot{+} \ldots \dot{+} A^{(i)}, \qquad (A^{(i)} B^{(i)} = B^{(i)} A^{(i)}).$$

As we diagonalize $B^{(1)}, \ldots, B^{(s)}$ over $\mathfrak{K}^*$, we simultaneously diagonalize $A^{(1)}, \ldots, A^{(s)}$. Since the characteristic roots of the $B^{(i)}$ are all different, we see that after the indicated transformation the matrices A, B are in the relation η. ■

From the proof of Lemma 4 we see that, in particular, if matrices M, N can be transformed into diagonal matrices A, B such that $A\ \eta\ B$, then any reduction of M and N to diagonal form yields matrices in the relation η.

We let $\mathcal{D}$ be the set of all diagonal matrices in the group $GL(n, \mathfrak{K})$ or $SL(n, \mathfrak{K})$, as the case may be. The relation η restricted to $\mathcal{D}$ is reflexive and transitive. We introduce an equivalence re ation θ on $\mathcal{D}$ by putting

$$A\ \theta\ B \Leftrightarrow A\ \eta\ B \text{ and } B\ \eta\ A\,.$$

The factor set $\mathcal{D}/\theta$ is the set of all equivalence classes with respect to θ; the relation η induces a lattice order $\leqslant$ on $\mathcal{D}/\theta$. The class of matrices equivalent to a given matrix $A = \alpha_1 E_{11} + \ldots + \alpha_n E_{nn}$ is completely determined by the partition of $P = \{1, 2, \ldots, n\}$ into subsets consisting of indices of equal diagonal elements in A. E.g., if $A = E_{11} + 2E_{22} + E_{33} + 3E_{44}$, then the associated partition π is $\{\{1,3\}, \{2\}, \{4\}\}$. Moreover, if $A, B \in \mathcal{D}$ and $A\ \eta\ B$, and π_1, π_2 are the corresponding partitions of P, then π_2 is a refinement of π_1 (in symbols: $\pi_1 \leqslant \pi_2$). In other words, the lattice $\mathfrak{D} = \langle \mathcal{D}/\theta, \leqslant \rangle$ is naturally isomorphic to the lattice of all partitions of P, with which it will be identified in the sequel.

The connection between the partition π associated with a diagonal matrix A and the Segre characteristic of A is seen from the following example. Let

$$A = E_{11} + 7E_{22} + E_{33} + E_{44} + 7E_{55}\,,$$

then $\pi = \{\{1,3\}, \{2,5\}, \{4\}\}$. If we replace the inner braces with brackets and each number with 1, we obtain the expression $\chi = \{[1,1]\ [1,1]\ [1]\}$, which is just the Segre characteristic of A.

The operation of recovering a partition from a given Segre characteristic is not well defined. E.g., the partitions $\pi_1 = \{\{1,3\}, \{2,5\}, \{4\}\}$, $\pi_2 = \{\{1,2\}, \{3,4\}, \{5\}\}$ match the same characteristic $\chi = \{[1,1]\,[1,1]\,[1]\}$. It is clear, however, *that any two partitions leading to the same characteristic are mapped onto each other by appropriate automorphisms of the lattice* $\mathfrak{D}$. ■

It is also easy to prove the converse: *any two partitions conjugate under automorphisms of* $\mathfrak{D}$ *yield the same Segre characteristic.* ■

We enumerate the partitions corresponding to elements of $\mathcal{D}/\theta$ as $\pi_1, \ldots, \pi_r$. Let $\Omega^{(n)}(\boldsymbol{x}_1, \ldots, \boldsymbol{x}_r)$ be a FOPL formula involving the symbol $\leqslant$ such that $\Omega^{(n)}(\pi_1, \ldots, \pi_r)$ is the conjunction of all relations of the form $\pi_i \leqslant \pi_j$, $\pi_i \nleqslant \pi_j$ true in the lattice $\mathfrak{D}$. In other words, $\Omega^{(n)}$ is the *diagram* of $\mathfrak{D}$.

Thus, if for certain $i_1, ..., i_r$ the relation $\Omega^{(n)}(\pi_{i_1}, ..., \pi_{i_r})$ is true in $\mathfrak{D}$, then the mapping $\pi_k \to \pi_{i_k}$ $(k = 1, ..., r)$ is an automorphism of $\mathfrak{D}$, and π_k and π_{i_k} yield the same Segre characteristic. Hence if π_i has characteristic χ, then the formula

$$\Omega_\chi(x_i) = (\exists x_1 \ldots x_{i-1} x_{i+1} \ldots x_r)\, \Omega^{(n)}(x_1, ..., x_r)$$

is true in $\mathfrak{D}$ for those and only those elements of $\mathcal{D}/\theta$ with characteristic χ. Using these conventions, we can state

Theorem 1: *In each group $GL(n, \mathfrak{K})$, $SL(n, \mathfrak{K})$ with $n \geqslant 3$, the formula*

$$\Phi_\chi(X_i) = (\exists X_1 \ldots X_{i-1} X_{i+1} \ldots X)\, (\Omega^{(n)}(X_1, ..., X_r)\ \&$$

$$\&\ \underset{j}{\&} D_n(X_j)\ \&\ \underset{j,k}{\&}\ X_j X_k \approx X_k X_j)\,,$$

where $\leqslant$ is replaced with the formula from (2), is true for just those matrices that are diagonalizable over $\mathfrak{K}$ and have characteristic χ.

Suppose $X_i = U^{-1} A U$, where A belongs to $\mathcal{D}$ and has characteristic χ. For each other diagonal Segre characteristic, we choose one matrix in $\mathcal{D}$ with this characteristic; we thus obtain a sequence of diagonal matrices $A_1, ..., A_r$ with $A_i = A$ such that $\Omega^{(n)}(A_1, ..., A_r)$ holds. But then the matrices

$$X_j = U^{-1} A_j U \quad (j = 1, ..., r)$$

satisfy

$$\Omega^{(n)}(X_1, ..., X_r)\ \&\ \underset{j}{\&}\ D_n(X_j)\ \&\ \underset{j,k}{\&}\ X_j X_k = X_k X_j\,;$$

this means $\Phi_\chi(X_i)$ is true.

Conversely, suppose $\Phi_\chi(X_i)$ is true. Then X_i is among matrices $X_1, ..., X_r$ that commute with each other, are diagonalizable over $\mathfrak{K}^*$, and satisfy $\Omega^{(n)}(X_1, ..., X_r)$ in the lattice constructed from the set of all diagonal matrices over $\mathfrak{K}^*$. But this lattice and the lattice $\mathfrak{D}$ are naturally isomorphic. Therefore, for every partition π of the index set P, there is a matrix among $X_1, ..., X_r$ corresponding to π. Let us consider the partitions $\{\{1\}, \{2, ..., n\}\}, ..., \{\{n\}, \{1, ..., n-1\}\}$; let $X_{i_1}, ..., X_{i_n}$ be matrices whose equivalence classes correspond to these partitions, and which thus have only two characteristic roots. One of these is a simple root of the characteristic polynomial, while the other root must have a higher multiplicity, for n exceeds 2. Since the elements of each matrix X_{i_k} and the coefficients of its characteristic polynomial belong to $\mathfrak{K}$, both roots of this polynomial lie in $\mathfrak{K}$. Consequently, the matrices $X_{i_1}, \ldots, X_{i_n}$ are simultaneously diagonalizable not only over $\mathfrak{K}^*$, but also over $\mathfrak{K}$. But when $X_{i_1}, \ldots, X_{i_n}$ have diagonal form, any matrix commuting with them is itself diagonal; i.e., the matrices $X_1, \ldots, X_r$ are simultaneously reducible over $\mathfrak{K}$ to diagonal form. Among these we find X_i. By the remarks made preceding the theorem, the characteristic of X_i is none other than χ. ■

Earlier we indicated a formula $\mathrm{D}_n(M)$ characterizing those matrices diagonalizable over $\mathfrak{K}^*$. We have now constructed for each diagonal Segre characteristic χ, a formula $\Phi_\chi(M)$ that is true for just those matrices similar over $\mathfrak{K}$ to diagonal matrices with characteristic χ. By taking the disjunction of the formulas $\Phi_\chi(M)$ over all possible diagonal χ, we obtain a formula $\Phi(M)$ true for exactly the matrices diagonalizable over the field $\mathfrak{K}$.

These arguments are not applicable to $GL(2, \mathfrak{K})$ or $SL(2, \mathfrak{K})$; they will be treated separately in the next subsection.

The formulas used in the construction of the $\Phi_\chi(M)$ lead to a proof of

Theorem 2: *For each $n \geqslant 2$, there exists a group sentence Ψ_n true in the groups $GL(n, \mathfrak{K})$, $SL(n, \mathfrak{K})$ and false in the groups $GL(m, \mathfrak{K})$, $SL(m, \mathfrak{K})$ for $m \neq n$.*

Indeed, consider the group sentence

$$\Psi_n^* = (\exists \boldsymbol{x}_1 \ldots \boldsymbol{x}_r)(\mathop{\&}_j \mathrm{D}_n(\boldsymbol{x}_j) \mathbin{\&} \mathop{\&}_{j,k} \boldsymbol{x}_j\boldsymbol{x}_k \approx \boldsymbol{x}_k\boldsymbol{x}_j \mathbin{\&} \Omega^{(n)}(\boldsymbol{x}_1, \ldots, \boldsymbol{x}_r)).$$

This sentence is true in $GL(n, \mathfrak{K})$ and $SL(n, \mathfrak{K})$ as we have already seen. In $GL(m, \mathfrak{K})$, $SL(m, \mathfrak{K})$ for $m < n$, the sentence Ψ_n^* cannot be true, for the lattice of partitions of $\{1, \ldots, m\}$ has fewer elements than the lattice of partitions of $\{1, \ldots, n\}$, while $\mathrm{D}_n(X)$ characterizes the matrices diagonalizable over $\mathfrak{K}^*$ not only in $GL(n, \mathfrak{K})$, $SL(n, \mathfrak{K})$, but also in $GL(m, \mathfrak{K})$, $SL(m, \mathfrak{K})$ for all $m < n$. Therefore, the sentence

$$\Psi_n = \Psi_n^* \mathbin{\&} \neg\Psi_{n+1}^*$$

has the required property. ■

§1.3. *The elementariness of the Segre characteristic in the general case*

For now we assume $n \geqslant 3$. We consider the formula

$$\Upsilon_1(B, A) = \Phi(B) \mathbin{\&} BA \approx AB \mathbin{\&} (X)(XA \approx AX \to XB \approx BX) \mathbin{\&}$$

$$\mathbin{\&} (Y)(\Phi(Y) \mathbin{\&} YA \approx AY \mathbin{\&} (X)(XA \approx AX \to XY \approx YX) \to Y \leqslant B).$$

In order to clarify the structure of a matrix B that finds itself in the relation Υ_1 to an arbitrary given matrix A, we reduce A to Jordan form (1) over $\mathfrak{K}$ and put

$$A_i = A_1^{(i)} \dotplus \ldots \dotplus A_{k_i}^{(i)} \qquad (i = 1, \ldots, s)\,,$$

so $A = A_1 \dotplus \ldots \dotplus A_s$. Let Y be an arbitrary matrix diagonalizable over $\mathfrak{K}$ that commutes with all matrices commuting with A. By a well-known theorem (cf [M3], p. 147), Y can be represented as $\varphi(A)$ for some polynomial φ over $\mathfrak{K}$. Consequently,

$$Y = (Y_1^{(1)} \dotplus \ldots \dotplus Y_k^{(1)}) \dotplus \ldots \dotplus (Y_1^{(s)} \dotplus \ldots \dotplus Y_{k_s}^{(s)}) \quad (Y_j^{(i)} = \varphi(A_j^{(i)}))$$

By hypothesis, $Y_j^{(i)}$ is diagonalizable over $\mathfrak{K}$ and commutes with $A_j^{(i)}$. Thus, $Y_j^{(i)}$ is a scalar matrix, for otherwise $A_j^{(i)}$ would break diagonally over $\mathfrak{K}$ into finer cells, which is impossible. Hence, $\varphi(A_j^{(i)}) = \beta_{ij} I^{(ij)}$, and so, $\varphi(A^{(i)}) = \beta_{ij} I^{(i)}$, where $A^{(i)}$ is the cell with characteristic polynomial irreducible over $\mathfrak{K}$ from which the Jordan cells $A_1^{(i)}, \ldots, A_{k_i}^{(i)}$ are constructed. Thus, $\beta_{i1} = \ldots = \beta_{ik_i}$. This tells us Y has the form

$$\beta_1 I^{(1)} \dotplus \ldots \dotplus \beta_s I^{(s)}\,. \tag{3}$$

The matrices B for which $\Upsilon_1(B, A)$ holds are the smoothest among such matrices Y, i.e., those Y that in (3) have $\beta_1, \ldots, \beta_s$ all different.

Now let us examine the formula

$$\Upsilon_2(C, A) = \Phi(C) \;\&\; CA \approx AC \;\&$$

$$\&\; (X)(\Phi(X) \;\&\; XA \approx AX \;\&\; XC \approx CX \rightarrow X \leqslant C)\,.$$

Let C, A be matrices for which this formula is true. In particular, C is diagonalizable over $\mathfrak{K}$, so let us transform C over $\mathfrak{K}$ into

$$C = \gamma_1 I^{(1)} \dotplus \ldots \dotplus \gamma_t I^{(t)}\,,$$

where $\gamma_1, \ldots, \gamma_t$ are distinct elements of $\mathfrak{K}$. Because A and C commute, A must reduce to a parallel cellular form:

$$A = A_1 \dotplus \ldots \dotplus A_t\,.$$

We note that $A_1, \ldots, A_t$ cannot be broken diagonally into finer cells over $\mathfrak{K}$. Indeed, if A_1 e.g. could equal $A' \dot{+} A''$, then the matrix

$$X = \gamma' I' \dot{+} \gamma'' I'' \dot{+} \gamma_2 I^{(2)} \dot{+} \ldots \dot{+} \gamma_t I^{(t)}$$

would commute with A, C, but would be smoother than C, contradicting the assertion of Υ_2. Therefore, for any matrices A, B, C such that $\Upsilon_1(B, A)$, $\Upsilon_2(C, A)$ hold, there exists a similarity transformation over $\mathfrak{K}$ simultaneously reducing A to Jordan form (1), B to the form (3), and C to the form

$$(\gamma_{11} I^{(11)} \dot{+} \ldots \dot{+} \gamma_{1k_1} I^{(1k_1)} \dot{+} \ldots \dot{+} (\gamma_{s1} I^{(s1)} \dot{+} \ldots \dot{+} \gamma_{sk_s} I^{(sk_s)}) . \qquad (4)$$

We consider one last formula:

$$\Upsilon_3(D, A) = \mathrm{D}_n(D) \,\&\, DA \approx AD \,\&$$

$$\& \,(X)(\mathrm{D}_n(X) \,\&\, XA \approx AX \,\&\, XD \approx DX \rightarrow X \lessdot D) ,$$

which differs from Υ_2 only in that diagonalizability over $\mathfrak{K}$ has been weakened to diagonalizability over $\mathfrak{K}^*$.

Let A be an arbitrary member of $GL(n, \mathfrak{K})$ or $SL(n, \mathfrak{K})$, and let B, C, D be other members of the same group such that $CD = DC$ and $\Upsilon_1(B, A)$, $\Upsilon_2(C, A)$, $\Upsilon_3(D, A)$ hold. We transform D over $\mathfrak{K}$ into

$$(D^{(1)} \dot{+} \ldots \dot{+} D^{(1)}) \dot{+} \ldots \dot{+} (D^{(s)} \dot{+} \ldots \dot{+} D^{(s)}) ,$$

where $D^{(1)}, \ldots, D^{(s)}$ are matrices with distinct characteristic polynomials, all irreducible over $\mathfrak{K}$. From the commutativity of C, D and the other assumptions, it follows that C is simultaneously reduced to the form (4), while B and A assume the respective forms (3) and (1).

In § 1.2 we constructed formulas $\Phi_\chi(M)$ for all Segre characteristics χ of matrices diagonalizable over $\mathfrak{K}$. Now we shall see how to construct, for any Segre characteristic χ with order n, a formula $\Phi_\chi(M)$ true in $GL(n, \mathfrak{K})$ or $SL(n, \mathfrak{K})$ for just those matrices with characteristic χ over $\mathfrak{K}$. Suppose A is a matrix with characteristic

$$\chi = \{[(m_1^1, n_1) \ldots (m_{k_1}^1, n_1)] \ \ldots \ [(m_1^s, n_s) \ldots (m_{k_s}^s, n_s)]\} ,$$

and suppose B, C, D are matrices that commute with each other and satisfy $\Upsilon_1(B, A)$, $\Upsilon_2(C, A)$, $\Upsilon_3(D, A)$. From the above discussion it follows that B, C

can be jointly transformed over $\mathfrak{K}^*$ into diagonal matrices of the forms (3), (4), respectively, while at the same time D assumes the form

$$D = \sum_{i=1}^{s} \sum_{k=1}^{p_i} \sum_{l=1}^{n_i} \zeta_{il} I_1 ,$$

where

$$p_i = \sum_{j=1}^{k_i} m_j^i \qquad (i = 1, \ldots, s) ;$$

the coefficients β_i are all different in (3), as are the γ_{ij} in (4) and the ζ_{il} here. The now diagonalized matrices B, C, D determine partitions π_1, π_2, π_3 of the set $P = \{1, 2, \ldots, n\}$ ([1]). Let $\pi_4, \ldots, \pi_r$ be the remaining partitions of P. As before, we let $\Omega^{(n)}(\boldsymbol{x}_1, \ldots, \boldsymbol{x}_r)$ be the diagram of the lattice of all these partitions. Then as the desired formula we can take

$$\Phi_\chi(A) = (\exists X_1 \ldots X_r)(\Upsilon_1(X_1, A) \,\&\, \Upsilon_2(X_2, A) \,\&\, \Upsilon_3(X_3, A) \,\&$$

$$\& \mathop{\&}_{i} \mathrm{D}_n(X_i) \,\&\, \mathop{\&}_{i,j} X_i X_j \approx X_j X_i \,\&\, \Omega^{(n)}(X_1, \ldots, X_r)) .$$

Now we turn to the case $n = 2$. As already mentioned, the formulas $\Phi_\chi(M)$ for $GL(2, \mathfrak{K})$, $SL(2, \mathfrak{K})$ have to be specially constructed. The only possible characteristics are

$$\chi_1 = \{[(1,1)(1,1)]\}, \ \chi_2 = \{[(1,2)]\}, \quad \chi_3 = \{[(1,1)]\,[(1,1)]\},$$

$$\chi_4 = \{[(2,1)]\} .$$

χ_1 characterizes the scalar matrices, so we can take

$$\Phi_{\chi_1}(M) = (X)(MX \approx XM) .$$

The matrices with characteristic χ_2 or χ_3 are those reducible over $\mathfrak{K}^*$ to diagonal matrices with distinct diagonal elements. According to Lemma 2, these matrices and only these satisfy $\mathrm{Dd}_2(M)$. Consequently, for the remaining characteristic χ_4 we can take

$$\Phi_{\chi_4}(M) = \neg\Phi_{\chi_1}(M) \,\&\, \neg\,\mathrm{Dd}_2(M) .$$

We now show that the matrices with characteristic χ_3 are described by the formula

$$\Phi_{\chi_3}(M) = \mathrm{Dd}_2(M) \,\&\, (\exists X)(\Phi_{\chi_4}(X) \,\&\, MXM^{-1} \cdot X \approx X \cdot MXM^{-1});$$

then, of course, we can take

$$\Phi_{\chi_2}(M) = \mathrm{Dd}_2(M) \,\&\, \neg\Phi_{\chi_3}(M) .$$

Indeed, suppose $\Phi_{\chi_3}(M)$ is true and X satisfies $\Phi_{\chi_4}(X)$, while $MXM^{-1} \cdot X = X \cdot MXM^{-1}$. Reduce X over $\mathfrak{K}$ to the form $\begin{pmatrix}\xi & 1\\ & \xi\end{pmatrix}$. Then the latter condition on X implies that M becomes a triangular matrix and is, therefore, already diagonalizable over $\mathfrak{K}$, as we wanted to prove. The rest is easy.

§1.4. *Projective groups*

The projective groups $PG(n, \mathfrak{K})$ and $PS(n, \mathfrak{K})$ are the factor groups of the linear groups $GL(n, \mathfrak{K})$ and $SL(n, \mathfrak{K})$ by their centers. These centers consist respectively of all scalar matrices αI $(\alpha \neq 0, \alpha \in \mathfrak{K})$ and of all scalar matrices αI with $\alpha^n = 1$. Since congruence classes of elements of the corresponding linear groups serve as elements of the projective groups, elements of the linear groups can be viewed as representatives of members of the projective groups. Thus the matrices A, B represent the same element of a projective group iff they satisfy the formula

$$(\exists X)(A \approx BX \,\&\, (Y)(XY \approx YX)) \tag{5}$$

in the corresponding linear group. The formula (5) can be rewritten as $(\exists \boldsymbol{\alpha})(A \approx \boldsymbol{\alpha} B)$, where the quantifier $(\exists \boldsymbol{\alpha})$ is interpreted as ranging over the set of nonzero elements of the field $\mathfrak{K}$ if we are considering $GL(n, \mathfrak{K})$, and over the set of roots in $\mathfrak{K}$ of the polynomial $\lambda^n - 1$ in the case of $SL(n, \mathfrak{K})$.

Let $\Psi(X, Y)$ be any group formula. By replacing every occurrence in Ψ of each formula of the form $A \approx B$ with the formula (5), we obtain a new group formula denoted by $\Psi^{\mathrm{p}}(X, Y)$. For an arbitrary matrix $X \in GL(n, \mathfrak{K})$, we let $[X]$ be the collection of all matrices of the form αX $(\alpha \in \mathfrak{K}, \alpha \neq 0)$. The coset $[X]$ is an element of the projective group $PG(n, \mathfrak{K})$. Clearly, for all $X, Y \in GL(n, \mathfrak{K})$, $\Psi([X], [Y])$ is true in $PG(n, \mathfrak{K})$ iff $\Psi^{\mathrm{p}}(X, Y)$ is true in $GL(n, \mathfrak{K})$. The situation $PS(n, \mathfrak{K})$, $SL(n, \mathfrak{K})$ is completely analogous.

Since elements of a projective group can be viewed as a class of matrices differing one from another by a certain sort of nonzero scalar factor, and since the matrices X and αX $(\alpha \neq 0)$ have the same Segre characteristic, it makes sense to speak of the Segre characteristics of members of projective groups, as well.

We now want to show that the Segre characteristic is also an elementary concept for projective groups. To do this we need the following two observations.

Remark 1: *Let A, B be square matrices over $\mathfrak{K}$ with respective orders m, n and let X be a rectangular matrix with m rows and n columns. If the characteristic polynomials of A, B are relatively prime and $AX = XB$, then $X = O$.*

Indeed, $AX = BX$ implies $A^iX = XB^i$ $(i = 1, 2, \dots)$; so for any polynomial $\varphi(\lambda)$ over $\mathfrak{K}$, we have $\varphi(A)X = X\varphi(B)$.

For the relatively prime characteristic polynomials $f(\lambda)$, $g(\lambda)$ of the matrices A, B, we choose polynomials $p(\lambda)$, $q(\lambda)$ over $\mathfrak{K}$ such that $pf + qg = 1$. Since $f(A) = O$, $g(B) = O$, we have

$$X = (p(A)f(A) + q(A)g(A))\,X = q(A)\,Xg(B) = O. \ \blacksquare$$

A celled matrix is called *cell-monomial* iff in each row and in each column there is but one nonzero cell.

Remark 2: *Suppose $A = A_1 \dot{+} \dots \dot{+} A_s$ is a cell-diagonal matrix, and the characteristic polynomials $f_i(\lambda)$ of the diagonal-cells A_i are relatively prime. Suppose further that for the matrix X, the transformed matrix $B = X^{-1}AX$ has a parallel cell-diagonal form $B_1 \dot{+} \dots \dot{+} B_s$ such that B_j and A_{i_j} have the same characteristic polynomial $(j = 1, \dots, s)$. Then X is cell-monomial, and $X^{s!}$ is cell-diagonal.*

Let $X = \| X_{ij} \|_{i,j=1,\dots,s}$. The condition $AX = XB$ gives

$$A_i X_{ij} = X_{ij} B_j \qquad (i, j = 1, \dots, s)\,. \tag{6}$$

Consider the permutations

$$\sigma = \begin{pmatrix} 1 & \dots & s \\ i_1 & \dots & i_s \end{pmatrix}, \qquad \sigma^{-1} = \begin{pmatrix} 1 & \dots & s \\ j_1 & \dots & j_s \end{pmatrix}.$$

By virtue of Remark 1, from (6) it follows that $X_{ij} = O$ if either $j \neq j_i$ or $i \neq i_j$, i.e., X is cell-monomial. But then for any cell diagonal matrix $C = C_1 \dot{+} \dots \dot{+} C_s$ of the same design as A, we have $X^{-1}CX = C_1' \dot{+} \dots \dot{+} C_s'$, where if C_i has the characteristic polynomial $h_i(\lambda)$ $(i = 1, \dots, s)$, then C_j' has the characteristic polynomial $h_{i_j}(\lambda)$ $(j = 1, \ldots, s)$. This means that transforming the cell-diagonal matrix C' by X applies σ to the characteristic polynomials of the diagonal-cells. So X^m will produce the permutation σ^m. Since $\sigma^{s!} = 1$, $X^{s!}$ will shift no characteristic polynomials; it is therefore simply a cell-diagonal matrix. ■

Now we can easily prove the following analog of Lemma 1.

Lemma 1a: *In each group $PG(n, \mathfrak{K})$, $PS(n, \mathfrak{K})$ the formula*

$$\mathrm{PCm}_n(a) = (x)(y)(ax \approx xa \;\&\; ay \approx ya \to x^{n!}y^{n!} \approx y^{n!}x^{n!})$$

is true for just those a with Segre characteristic over $\mathfrak{K}$ of the form $\{[(m_1, n_1)] \ldots [(m_s, n_s)]\}$.

Indeed, suppose a is represented by a matrix A with the indicated sort of characteristic. Then $\mathrm{PCm}_n(a)$ holds in the given projective group iff

$$(X)(Y)(\alpha)(\beta)(\exists\gamma)(AX \approx \alpha XA \;\&$$

$$\& \; AY \approx \beta YA \to X^{n!}Y^{n!} \approx \gamma Y^{n!}X^{n!})$$

holds for A in the corresponding linear group. Let us assume that for appropriate matrices X, Y and scalars α, β, we have $AX = \alpha XA$, $AY = \beta YA$. Reduce A to Jordan form over $\mathfrak{K}$;

$$A = A_1 \dotplus \ldots \dotplus A_s .$$

By Remark 2 the matrices $X^{s!}$, $Y^{s!}$ have the forms

$$X^{s!} = X_1 \dotplus \ldots \dotplus X_s , \qquad Y^{s!} = Y_1 \dotplus \ldots \dotplus Y_s ;$$

moreover,

$$A_i X_i = \alpha^{s!} X_i A_i , \qquad A_i Y_i = \beta^{s!} Y_i A_i .$$

Comparing the determinants of the left- and right-hand sides of these equations, we find that $\alpha^{p_i s!} = \beta^{p_i s!} = 1$, where p_i is the order of A_i. From the relations $X_i^{-1} A_i X_i = \alpha^{s!} A_i$, etc., we now get

$$X_i^{-p_i} A_i X_i^{p_i} = \alpha^{s! p_i} A_i = A_i , \qquad Y_i^{-p_i} A_i Y_i^{p_i} = A_i .$$

Therefore, $X_i^{p_i}$ and $Y_i^{p_i}$ commute with the Jordan cell A_i; thus they commute with each other. Consequently, $X^{n!}Y^{n!} = Y^{n!}X^{n!}$.

Now suppose that when A is reduced to Jordan form it does not have a characteristic of the form indicated in this lemma, i.e., it contains two cells corresponding to the same root polynomial. For such a matrix A we constructed in § 1.1 matrices $Y, X \in SL(n, \mathfrak{K})$ commuting with A, but not with each other. It is easy to verify that not only do X, Y fail to commute, but even $X^t Y^t \neq \gamma Y^t X^t$ for all positive numbers t and any $\gamma \in \mathfrak{K}$. ■

Lemma 2a: *In each group $PG(n, \mathfrak{K})$, $PS(n, \mathfrak{K})$ the formula*

$$\mathrm{PDd}_n(\boldsymbol{a}) = \mathrm{PCm}_n(\boldsymbol{a}^n) \,\&\, (\boldsymbol{x})(\boldsymbol{x}^{-1}\boldsymbol{a}\boldsymbol{x}\cdot\boldsymbol{a} \approx \boldsymbol{a}\cdot\boldsymbol{x}^{-1} \rightarrow \boldsymbol{a}\boldsymbol{x}^{n!} \approx \boldsymbol{x}^{n!}\boldsymbol{a})$$

is true for those and only those elements a for which a^n has a Segre characteristic over $\mathfrak{K}$ of the form $\{[(1,n_1)] \ldots [(1,n_s)]\}$, i.e., reduces over $\mathfrak{K}^*$ to diagonal form with distinct diagonal elements.

Necessity: Suppose $a^n = [A^n]$ has a Segre characteristic of the form indicated; so A reduces over $\mathfrak{K}^*$ to $A = \alpha_1 E_{11} + \ldots + \alpha_n E_{nn}$ with $\alpha_i^n \neq \alpha_j^n$ for $i \neq j$. Now let X be any nonsingular matrix such that $X^{-1}AX\cdot A = \alpha A \cdot X^{-1}AX$. By taking the determinants of both sides, we learn $\alpha^n = 1$. Since $(X^{-1}AX)^n A = \alpha^n A(X^{-1}AX)^n$, we have $X^{-1}AX\cdot A = A\cdot X^{-1}AX$. Consequently, the matrix $X^{-1}AX$ is diagonal with distinct diagonal elements. By Remark 2, this implies $X^{n!}$ is diagonal, so $X^{n!}A = AX^{n!}$.

Sufficiency: If $\mathrm{PCm}_n(A^n)$ is true, but the characteristic of A is not of the indicated sort, then A reduces over $\mathfrak{K}$ to the form $A = A_1 \dot{+} A_1'$, where A_1 is a Jordan cell with index greater than 1. It is easy to see that the matrix X constructed in the proof of Lemma 2 satisfies the conditions $X^{-1}AX\cdot A = A\cdot X^{-1}AX$, $AX^t \neq X^t A$ $(t > 0)$. Therefore, for such a matrix A, the relation $\mathrm{PDd}_n([A])$ is false. ■

Lemma 3a: *In each of the groups $PG(n, \mathfrak{K})$, $PS(n, \mathfrak{K})$ the formula*

$$\mathrm{PD}_n(\boldsymbol{a}) = (\exists \boldsymbol{x})(\mathrm{PDd}_n(\boldsymbol{x}) \,\&\, \boldsymbol{a}\boldsymbol{x} \approx \boldsymbol{x}\boldsymbol{a})$$

is true for those and only those elements diagonalizable over $\mathfrak{K}^*$.

The truth of $\mathrm{PD}_n([A])$ for a matrix A diagonalizable over $\mathfrak{K}^*$ follows immediately from the proof of Lemma 3. Conversely, suppose $\mathrm{PD}_n([A])$ is true for some matrix A; i.e., suppose there exists a matrix X such that X^n reduces over $\mathfrak{K}^*$ to a diagonal matrix with distinct diagonal elements and $AX = \alpha XA$ for some scalar α. From this we conclude $\alpha^n = 1$, $AX^n = \alpha^n X^n A = X^n A$; hence A is diagonalizable over $\mathfrak{K}^*$. ■

The reasoning in § 1.2 and § 1.3 can be carried over to the case of projective groups almost word for word, yielding as a result

Theorem 3: *For each of the groups $GL(n, \mathfrak{K})$, $SL(n, \mathfrak{K})$, $PG(n, \mathfrak{K})$, $PS(n, \mathfrak{K})$ and for every Segre characteristic χ appropriate to order n, there exists a group formula $\Phi_\chi(\boldsymbol{a})$ true in the corresponding group for just those elements a possessing the characteristic χ over $\mathfrak{K}$. The form of $\Phi_\chi(\boldsymbol{a})$ does not depend on the properties of the char 0 base field $\mathfrak{K}$.*

For G = GL, SL, PG, PS and for every n, there exists a group sentence Ψ_n true in $G(n, \mathfrak{K})$, but false in $G(m, \mathfrak{K})$ for $m \neq n$. The form of Ψ_n is independent of the nature of the base field. ∎

Next we take up the question of exactly which linear and projective groups have the same elementary theories.

§ 2. Elementary (arithmetic) types of linear and projective groups

We say that two classes of groups have the same *elementary* (or *arithmetic*) *type* iff every group sentence true in all members of one class is also true in all members of the other. Two individual groups of the same elementary type are said to be *elementarily equivalent.*

A subgroup $\mathfrak{H}$ of a group $\mathfrak{G}$ is called an *elementary subgroup* iff there is a group formula $\Psi(\boldsymbol{x})$ such that $\mathfrak{H}$ consists of exactly the elements of $\mathfrak{G}$ for which Ψ is true in $\mathfrak{G}$. A subgroup $\mathfrak{H}$ (or even just a subset) of the group $\mathfrak{G}$ is *elementary relative to* $a_1, \ldots, a_m \in \mathfrak{G}$ iff there exists a group formula $\Psi(\boldsymbol{x}, \boldsymbol{a}_1, \ldots, \boldsymbol{a}_m)$ involving individual constants standing for these elements of $\mathfrak{G}$ such that Ψ is true in $\mathfrak{G}$ for the members $x \in \mathfrak{H}$ alone.

Using the results of § 1, we shall specify conditions under which linear and projective groups have the same elementary type. Along the way we shall establish the relative elementariness of several subgroups of the groups under study.

§ 2.1. *The relative elementariness of certain subgroups*

Lemma 5: *The commutator subgroup of $GL(n, \mathfrak{K})$ is $SL(n, \mathfrak{K})$; this is an elementary subgroup of $GL(n, \mathfrak{K})$.*

(Here, as elsewhere in this article, we are assuming $\mathfrak{K}$ is a field of characteristic 0. Lemma 5 is true for fields of arbitrary characteristic, as long as they have more than two elements [183].)

The coincidence of the commutator subgroup of $GL(n, \mathfrak{K})$ with $SL(n, \mathfrak{K})$ is well known (cf. [183]). From the usual proof of this result it is easy to discern a stronger property of the derived group of $GL(n, \mathfrak{K})$: each of its elements, i.e., each matrix in $SL(n, \mathfrak{K})$, can be represented as a product of a fixed number t (depending on n) of commutators from $GL(n, \mathfrak{K})$. Therefore, the truth of the formula

$$(\exists X_1 Y_1 \dots X_t Y_t)(A \approx X_1^{-1} Y_1^{-1} X_1 Y_1 \cdot \dots \cdot X_t^{-1} Y_t^{-1} X_t Y_t)$$

for a matrix A in the group $GL(n, \mathfrak{K})$ is equivalent to the membership of A in $SL(n, \mathfrak{K})$. $SL(n, \mathfrak{K})$ is thus an elementary subgroup of $GL(n, \mathfrak{K})$. ■

In each of the linear groups $GL(n, \mathfrak{K})$, $SL(n+1, \mathfrak{K})$ with $n \geqslant 2$, we shall now unearth relatively elementary subgroups isomorphic to $GL(2, \mathfrak{K})$.

Let $\mathfrak{G}$ be one of the groups $GL(n, \mathfrak{K})$ $(n \geqslant 3)$ or $SL(n, \mathfrak{K})$ $(n \geqslant 4)$. Consider the group formula $\Phi_{\chi_0}(A)$ characterizing the matrices $A \in \mathfrak{G}$ with Segre characteristic $\chi_0 = \{[1, 1, 1]\ [1]\ \dots\ [1]\}$ over $\mathfrak{K}$, i.e., those matrices reducible over $\mathfrak{K}$ to diagonal form with one triple characteristic root, the rest (if any) simple. So suppose $A \in \mathfrak{G}$ has the form

$$\alpha_1(E_{11} + E_{22} + E_{33}) + \alpha_4 E_{44} + \dots + \alpha_n E_{nn}\,, \tag{7}$$

where $\alpha_i \neq \alpha_j$ for $i \neq j$. The collection of all members of $\mathfrak{G}$ commuting with A forms a subgroup $\mathfrak{H}$ consisting of all matrices of the form $M \dotplus D$, where $M \in GL(3, \mathfrak{K})$ and D is a diagonal matrix, arbitrary if $\mathfrak{G} = GL(n, \mathfrak{K})$, but satisfying $|D| = |M|^{-1}$ when $\mathfrak{G} = SL(n, \mathfrak{K})$. Hence, in either case, the commutator subgroup of $\mathfrak{H}$ consists of the matrices of the form

$$M \dotplus (E_{44} + \dots + E_{nn}) \qquad (M \in SL(3, \mathfrak{K}))\,;$$

therefore, this subgroup is isomorphic to $SL(3, \mathfrak{K})$. We have proved

Lemma 6: *For each of the groups $GL(n, \mathfrak{K})$, $SL(n, \mathfrak{K})$ with $n \geqslant 3$, there is a formula $\Psi(X, A)$ with the property that for every matrix A satisfying $\Phi_{\chi_0}(A)$ – χ_0 as above – the collection of matrices X such that $\Psi(X, A)$ holds in the corresponding group forms a subgroup isomorphic to $SL(3, \mathfrak{K})$.*

Indeed, for the groups $GL(n, \mathfrak{K})$ $(n \geqslant 3)$ and $SL(n, \mathfrak{K})$ $(n \geqslant 4)$, we can take

$$\Psi(X, A) = (\exists X_1 Y_1 \dots X_t Y_t)\big(\mathop{\&}_i (AX_i \approx X_i A \;\&\; AY_i \approx Y_i A) \;\&$$
$$\& \; X \approx \prod_i X_i^{-1} Y_i^{-1} X_i Y_i\big)\,,$$

where t depends on n. And $\Psi(X, A) = X \approx X$ works for $SL(3, \mathfrak{K})$. ■

Now consider a matrix of the form

$$B = \alpha(E_{11} + E_{22}) + \beta E_{33} \qquad (\alpha \neq \beta,\ \alpha^2\beta = 1) \tag{8}$$

in the group $SL(3, \mathfrak{K})$. The matrices $T \in SL(3, \mathfrak{K})$ commuting with B have the form

$$T = T_1 \dotplus \gamma I_1 \quad (T_1 \in GL(2, \mathfrak{K}),\ \gamma = |T_1|^{-1}) . \tag{9}$$

Since the scalar γ is uniquely determined by the matrix T_1, which can be any member of $GL(2, \mathfrak{K})$, the collection of all such T is a subgroup isomorphic to $GL(2, \mathfrak{K})$. By combining this result with Lemma 6, we easily prove

Lemma 7: *For each of the groups $GL(n, \mathfrak{K})$ $(n \geqslant 2)$, $SL(n, \mathfrak{K})$ $(n \geqslant 3)$ there exist formulas $\Theta(\boldsymbol{A}, \boldsymbol{B})$, $\Xi(\boldsymbol{X}, \boldsymbol{A}, \boldsymbol{B})$ with the following property: whenever elements A, B of the group at hand satisfy Θ, the elements X in the relation Ξ to A, B form a subgroup of this group isomorphic to $GL(2, \mathfrak{K})$.*

For in the case of $GL(2, \mathfrak{K})$, any identically true formulas work. When we turn to one of the groups $GL(n, \mathfrak{K})$, $SL(n, \mathfrak{K})$ with $n \geqslant 3$, we can take $\Theta(\boldsymbol{A}, \boldsymbol{B})$ to be $\Phi_{\chi_0}(\boldsymbol{A})$ & $\Psi(\boldsymbol{B}, \boldsymbol{A})$. In order to construct Ξ, we let $\Phi_{\chi_1}(\boldsymbol{B})$ be the formula true in $SL(3, \mathfrak{K})$ for just the matrices with Segre characteristic $\chi_1 = \{[1,1]\,[1]\}$ over $\mathfrak{K}$. Then the relativization of the formula $\Phi_{\chi_1}(\boldsymbol{B})$ & $\boldsymbol{XB} \approx \boldsymbol{BX}$ to the set of elements Y characterized by $\Psi(\boldsymbol{Y}, \boldsymbol{A})$ will work as $\Xi(\boldsymbol{X}, \boldsymbol{A}, \boldsymbol{B})$. ∎

Lemma 8: *For each of the projective groups $PG(n, \mathfrak{K})$, $PS(n, \mathfrak{K})$* with $n \geqslant 3$, *there exist formulas $\Theta(\boldsymbol{a}, \boldsymbol{b})$, $\Xi(\boldsymbol{x}, \boldsymbol{a}, \boldsymbol{b})$ with the following property: whenever elements a, b of the given group satisfy Θ, the elements x satisfying $\Xi(x, a, b)$ form a subgroup isomorphic to the quotient group $GL(2, \mathfrak{K})/\mathfrak{Z}$, where $\mathfrak{Z}$ consists of all matrices of the form αI_2 with $\alpha^3 = 1$.*

Arguing as above, we start with the formula $\Phi_{\chi_0}(\boldsymbol{a})$ true for those and only those elements of the given projective group $\mathfrak{G}$ that can be represented by matrices diagonalizable over $\mathfrak{K}$ to the form (7). Let A be a matrix of this form in the corresponding linear group $\mathfrak{G}_0$, and let $\mathfrak{H}$ be the centralizer in $\mathfrak{G}$ of the element $[A]$. Let $\mathfrak{H}_0$ be the subgroup of $\mathfrak{G}_0$ consisting of all matrices M such that for some scalar $\gamma \in \mathfrak{K}$,

$$AM = \gamma MA . \tag{10}$$

The group $\mathfrak{H}$ is the factor group of the group $\mathfrak{H}_0$ by the subgroup $\mathfrak{Z}_0$ consisting of all scalar matrices in $\mathfrak{G}_0$. In (10) we see that A and γA must have the same characteristic roots. Thus, when we multiply the set of characteristic values of A by γ, we have to get the same set of scalars back again. But only one of the roots has multiplicity greater than 1; under multiplication by γ, it must go onto itself, so $\gamma = 1$. Hence, $\mathfrak{H}_0$ is none other than the centralizer of the matrix A in $\mathfrak{G}_0$. If $\mathfrak{H}'$ ($\mathfrak{H}'_0$) is the commutator subgroup of $\mathfrak{H}$ ($\mathfrak{H}_0$), then we have found out that

$$\mathfrak{H}' \cong \mathfrak{H}'_0/(\mathfrak{H}'_0 \cap \mathfrak{Z}_0) \cong SL(3, \mathfrak{K})/\mathfrak{Z}_1 = PS(3, \mathfrak{K}),$$

where $\mathfrak{Z}_1$ is the subgroup of $SL(3, \mathfrak{K})$ consisting of all scalar matrices in this group.

Now construct a formula $\Phi^*(\boldsymbol{b})$ such that $\Phi^*([B])$ is true in $PS(3, \mathfrak{K})$ iff B reduces over $\mathfrak{K}$ to the form (8). Let B have this form, and let $\mathfrak{N}$ be the centralizer in $PS(3, \mathfrak{K})$ of the element $[B]$. Let $\mathfrak{N}_0$ be the subgroup of $SL(3, \mathfrak{K})$ consisting of matrices T of the form (9). Then $\mathfrak{N}$ is the factor group of $\mathfrak{N}_0$ by the intersection of $\mathfrak{N}_0$ with the center $\mathfrak{Z}_1$ of $SL(3, \mathfrak{K})$; this means $\mathfrak{N}$ is isomorphic to $GL(2, \mathfrak{K})/\mathfrak{Z}$, where $\mathfrak{Z}$ consists of all matrices of the form αI_2 for $\alpha \in \mathfrak{K}, \alpha^3 = 1$. The proof of Lemma 8 is completed by constructing the required formulas exactly as above. ■

§2.2. *The group* $GL(2, \mathfrak{K})/\mathfrak{Z}$

Lemmas 7 and 8 show that in each of the groups $GL(n, \mathfrak{K})$, $SL(n+1, \mathfrak{K})$, $PG(n, \mathfrak{K})$, $PS(n+1, \mathfrak{K})$ for $n \geqslant 2$, there is a relatively elementary subgroup isomorphic to $GL(2, \mathfrak{K})/\mathfrak{Z}$, where $\mathfrak{Z} = \{I_2\}$ in the linear cases, $\mathfrak{Z} = \{\alpha I_2 : 0 \neq \alpha \in \mathfrak{K}\}$ in the case of $PG(2, \mathfrak{K})$, and $\mathfrak{Z} = \{\alpha I_2 : \alpha \in \mathfrak{K} \text{ and } \alpha^3 = 1\}$ in the remaining projective cases. In order to avoid considering all these cases separately, we shall let $\mathfrak{Z}$ vary over arbitrary central subgroups of $GL(2, \mathfrak{K})$.

Lemma 9: *There are group formulas* $\Gamma(\boldsymbol{a})$, $\Delta(\boldsymbol{x}, \boldsymbol{a})$, $\Sigma(\boldsymbol{x}, \boldsymbol{y}, \boldsymbol{z})$, $\Pi(\boldsymbol{x}, \boldsymbol{y}, \boldsymbol{z}, \boldsymbol{a})$ *possessing the following property: for any central subgroup* $\mathfrak{Z}$ *of* $GL(2, \mathfrak{K})$ *and for any element a of the group* $\mathfrak{G} = GL(2, \mathfrak{K})/\mathfrak{Z}$ *satisfying* $\Gamma(a)$ *in* $\mathfrak{G}$, *if the set* $\hat{K}$ *consists of all elements* $x \in \mathfrak{G}$ *such that* $\Delta(x, a)$ *holds in* $\mathfrak{G}$, *and if operations* $\oplus$, $\otimes$ *are defined on* $\hat{K}$ *by the rules*

$$x \oplus y = z \Leftrightarrow \Sigma(x, y, z) \quad \text{holds in } \mathfrak{G}, \tag{11}$$

$$x \otimes y = z \Leftrightarrow \Pi(x, y, z, a) \quad \text{holds in } \mathfrak{G}, \tag{12}$$

then the algebraic system $\langle \hat{K}; \oplus, \otimes \rangle$ *is a field isomorphic to* $\mathfrak{K}$.

To begin with, we show that in the group $\mathfrak{G}$ the formula

$$\Lambda(a) = (\exists x)(x^{-1}ax\cdot a \approx a\cdot x^{-1}ax \;\&\; a^2x^2 \not\approx x^2a^2)$$

is true for those and only those elements representable by matrices that can be reduced over $\mathfrak{K}$ to the form $\begin{pmatrix}\alpha & \alpha\\ & \alpha\end{pmatrix}$, $\alpha \neq 0$.

Well, $[A]$ satisfies Λ in $\mathfrak{G}$ iff the matrix A satisfies the FOPL formula (2)

$$\Lambda^{\mathrm{p}}(A) = (\exists X)(\exists \lambda \in \mathfrak{Z})(\forall \mu \in \mathfrak{Z})$$

$$(X^{-1}AX\cdot A \approx \lambda A\cdot X^{-1}AX \;\&\; A^2X^2 \not\approx \mu X^2A^2)\,.$$

If A reduces to the form $\begin{pmatrix}\alpha & \alpha\\ & \alpha\end{pmatrix}$, then $\Lambda^{\mathrm{p}}(A)$ is clearly true: as X we can take $2E_{11} + E_{22}$.

Conversely, suppose $\Lambda^{\mathrm{p}}(A)$ holds, but A is not similar to a matrix of the indicated form. Then either A is scalar, which is impossible since $A^2X^2 \neq X^2A^2$ for some X; or A reduces over $\mathfrak{K}^*$ to the form $A = \alpha E_{11} + \beta E_{22}$ with $\alpha \neq \beta$. The latter is also impossible, for $X^{-1}AX\cdot A = A\cdot X^{-1}AX\cdot \lambda I_2$ tells us – on taking determinants – that $\lambda = \pm 1$. So X is either diagonal or has the form $\gamma E_{12} + \delta E_{21}$; in both cases, $X^2A^2 = A^2X^2$.

Now we prove the subtler assertion: in the group $\mathfrak{G}$ the formula

$$\Lambda_1(a) = (\exists xyz)(\Lambda(x) \;\&\; y^{-1}xy\cdot x \approx x\cdot y^{-1}xy \;\&$$

$$\&\; z^{-1}xz\cdot x \approx x\cdot z^{-1}xz \;\&\; a \approx y^{-1}z^{-1}yz)$$

is true for those and only those elements a representable by matrices that can be reduced over $\mathfrak{K}$ to the form $\begin{pmatrix}\alpha & \beta\\ & \alpha\end{pmatrix}$, where $\alpha I_2 \in \mathfrak{Z}$, $\beta \in \mathfrak{K}$.

Put

$$\Lambda_1^{\mathrm{p}}(A) = (\exists XYZ)(\exists \lambda\mu\nu \in \mathfrak{Z})(\Lambda^{\mathrm{p}}(X) \;\&\; Y^{-1}XY\cdot X \approx \lambda X\cdot Y^{-1}XY \;\&$$

$$\&\; Z^{-1}XZ\cdot X \approx \mu X\cdot Z^{-1}XZ \;\&\; A \approx \nu Y^{-1}Z^{-1}YZ)\,.$$

Then for every matrix $A \in GL(2, \mathfrak{K})$, the truth of $\Lambda_1([A])$ in $\mathfrak{G}$ is equivalent to the truth of $\Lambda_1^{\mathrm{p}}(A)$ in $GL(2, \mathfrak{K})$.

Suppose for some $A \in GL(2, \mathfrak{K})$ that $\Lambda_1^{\mathrm{p}}(A)$ is true; hence, there exist matrices X, Y, Z with the properties asserted by Λ_1^{p}. Since $\Lambda^{\mathrm{p}}(X)$ holds, we can view X as having the form $\begin{pmatrix}\xi & \xi\\ & \xi\end{pmatrix}$. The relation $Y^{-1}XY\cdot X = \lambda X\cdot Y^{-1}XY$ shows $Y = \begin{pmatrix}\eta_1 & \eta_2\\ & \eta_3\end{pmatrix}$. Similarly, Z must have the form $\begin{pmatrix}\zeta_1 & \zeta_2\\ & \zeta_3\end{pmatrix}$. As

$A = Y^{-1}Z^{-1}YZ \cdot \nu I_2$ for some scalar matrix $\nu I_2 \in \mathfrak{Z}$, we find

$$A = \begin{pmatrix} \nu & \beta \\ & \nu \end{pmatrix} \tag{13}$$

for some scalar $\beta \in \mathfrak{K}$.

The converse – that a matrix A of the form (13) satisfies Λ_1^{p} in $GL(2, \mathfrak{K})$ – is easily verified.

At last we introduce the formulas

$$\Sigma(x, y, z) = xy \approx z ,$$

$$\Pi(x, y, z, a) = ((x \approx 1 \vee y \approx 1) \,\&\, z \approx 1) \vee (x \not\approx 1 \,\&\, y \not\approx 1 \,\&$$

$$\& \, (\exists uv)(x \approx u^{-1}au \,\&\, y \approx v^{-1}av \,\&\, z \approx v^{-1}u^{-1}auv))$$

and set

$$\Gamma(a) = \Lambda(a) \,\&\, \Lambda_1(a) ,$$

$$\Delta(x, a) = \Lambda_1(x) \,\&\, ax \approx xa .$$

Suppose for some $a \in \mathfrak{G}$ that $\Gamma(a)$ is true. We shall assume $a = [A]$, where A has been reduced to the form $\begin{pmatrix} \alpha & \alpha \\ & \alpha \end{pmatrix}$ with $\alpha I_2 \in \mathfrak{Z}$. Every matrix X such that $\Delta([X], [A])$ is true has the form $\lambda\begin{pmatrix} 1 & \xi \\ & 1 \end{pmatrix}$ with $\lambda I_2 \in \mathfrak{Z}$. By associating with every scalar $\xi \in \mathfrak{K}$ the $\mathfrak{Z}$-coset

$$x(\xi) = \left[\begin{pmatrix} 1 & \xi \\ & 1 \end{pmatrix}\right] ,$$

we obtain a 1–1 mapping from $\mathfrak{K}$ onto the set $\hat{K}$ of elements x satisfying $\Delta(x, a)$ in $\mathfrak{G}$.

The definition of Σ immediately shows that

$$x(\xi) \oplus x(\eta) = x(\xi + \eta) .$$

Furthermore, if $x(\xi), x(\eta) \neq I_2$, then $\xi, \eta \neq 0$, and from the relations

$$U^{-1}AU = \lambda_1 \begin{pmatrix} 1 & \xi \\ & 1 \end{pmatrix},$$

$$V^{-1}AV = \lambda_2 \begin{pmatrix} 1 & \eta \\ & 1 \end{pmatrix},$$

$$Z = \lambda V^{-1}U^{-1}AUV ,$$

we learn
$$U=\mu_1\begin{pmatrix}1&\gamma\\&\xi\end{pmatrix},\quad V=\mu_2\begin{pmatrix}1&\delta\\&\eta\end{pmatrix},\quad Z=\mu\begin{pmatrix}1&\xi\eta\\&1\end{pmatrix};$$
thus $\otimes$ is a well-defined operation on $\hat{K}$ and $x(\xi)\otimes x(\eta)=x(\xi\cdot\eta)$. Consequently, $\langle\hat{K};\oplus,\otimes\rangle$ is an isomorphic image of the field $\mathfrak{K}$. ■

§2.3. *The fundamental theorems*

With the results so far established we can easily prove the basic theorems formulated in the Introduction. We begin by combining the results of Lemmas 7, 8 and 9.

Lemma 10: *For each of the groups* $GL(n,\mathfrak{K})$, $PG(n,\mathfrak{K})$, $SL(n+1,\mathfrak{K})$, $PS(n+1,\mathfrak{K})$ *with* $n\geqslant 2$, *there exist group formulas* $\Omega(\boldsymbol{a},\boldsymbol{b},\boldsymbol{c})$, $\Delta^*(\boldsymbol{x},\boldsymbol{a},\boldsymbol{b},\boldsymbol{c})$, $\Sigma^*(\boldsymbol{x},\boldsymbol{y},\boldsymbol{z},\boldsymbol{a},\boldsymbol{b})$, $\Pi^*(\boldsymbol{x},\boldsymbol{y},\boldsymbol{z},\boldsymbol{a},\boldsymbol{b},\boldsymbol{c})$ *with the following property: for any elements a, b, c of the given group* $\mathfrak{G}$ *satisfying* $\Omega(a,b,c)$ *in* $\mathfrak{G}$, *if* $\hat{K}$ *is the set of all* $x\in\mathfrak{G}$ *such that* $\Delta^*(x,a,b,c)$ *holds, then* Σ^*, Π^* *determine operations* $\oplus$, $\otimes$ *on* $\hat{K}$ *via definitions analogous to* (11), (12); *the algebra* $\langle\hat{K};\oplus,\otimes\rangle$ *is a field isomorphic to* $\mathfrak{K}$.

As $\Omega(\boldsymbol{a},\boldsymbol{b},\boldsymbol{c})$ we take the conjunction of $\Theta(\boldsymbol{a},\boldsymbol{b})$ (from Lemma 7 or 8) and the relativization of $\Gamma(\boldsymbol{c})$ (from Lemma 9) to the formula $\rho(\boldsymbol{w})=\Xi(\boldsymbol{w},\boldsymbol{a},\boldsymbol{b})$ (from Lemma 7 or 8, as the case may be). As $\Delta^*(\boldsymbol{x},\boldsymbol{a},\boldsymbol{b},\boldsymbol{c})$, $\Sigma^*(\boldsymbol{x},\boldsymbol{y},\boldsymbol{z},\boldsymbol{a},\boldsymbol{b})$, $\Pi^*(\boldsymbol{x},\boldsymbol{y},\boldsymbol{z},\boldsymbol{a},\boldsymbol{b},\boldsymbol{c})$ we take the relativizations of $\Delta(\boldsymbol{x},\boldsymbol{c})$, $\Sigma(\boldsymbol{x},\boldsymbol{y},\boldsymbol{z})$, $\Pi(\boldsymbol{x},\boldsymbol{y},\boldsymbol{z},\boldsymbol{c})$ – all from Lemma 9 – to the formula $\Xi(\boldsymbol{w},\boldsymbol{a},\boldsymbol{b})$. ■

Theorem 4: *For* $n\geqslant 2$, *each of the groups* $GL(n,\mathfrak{K})$, $PG(n,\mathfrak{K})$, $SL(n+1,\mathfrak{K})$, $PS(n+1,\mathfrak{K})$ *is syntactically equivalent to the field* $\mathfrak{K}$.

This asserts [XV] the existence of two algorithms, one for transforming each FOPL group sentence into a ring sentence such that the former is true in the given group iff the latter is true in the field $\mathfrak{K}$; a reverse algorithm enables us to construct for each ring sentence a group sentence that is true in the given group iff the original sentence is true in $\mathfrak{K}$.

To begin the proof of Theorem 4, we consider an arbitrary group sentence
$$\Phi=(\mho_1\boldsymbol{x}_1)\dots(\mho_r\boldsymbol{x}_r)\Psi(\boldsymbol{x}_1,\dots,\boldsymbol{x}_r)\quad(\mho_i=\forall\text{ or }\exists).$$

The truth of Φ in $GL(n,\mathfrak{K})$ is obviously equivalent to the validity in $\mathfrak{K}$ of the ring sentence $\Phi^{\mathfrak{K}}$ obtained from Φ by the following well-known procedure:

(I) From the formula $\Psi_{r+1}=\Psi(\boldsymbol{x}_1,\dots,\boldsymbol{x}_r)$ we get a new formula $\Psi_{r+1}^{\mathfrak{K}}$ by replacing all the subformulas of Ψ_{r+1} of the form $\boldsymbol{x}_i\approx\boldsymbol{x}_j$ and $\boldsymbol{x}_i\approx\boldsymbol{x}_j\cdot\boldsymbol{x}_k$ are replaced with the corresponding formulas

$$\underset{l,m}{\&}(x_i^{(lm)} \approx x_j^{(lm)}), \quad \underset{l,m}{\&}(x_i^{(lm)} \approx x_j^{(l1)} x_k^{(1m)} + \ldots + x_j^{(ln)} x_k^{(nm)});$$

(II) If the formula $\Psi_{i+1} = (\eth_{i+1} x_{i+1}) \ldots (\eth_r x_r)\Psi$ has been transformed into $\Psi_{i+1}^{\mathfrak{K}}$, then the formula $\Psi_i = (x_i)\Psi_{i+1}$ corresponds to

$$\Psi_i^{\mathfrak{K}} = (x_i^{(11)})(x_i^{(12)}) \ldots (x_i^{(nn)})(\det \| x_i^{(lm)} \| \not\approx 0 \rightarrow \Psi_{i+1}^{\mathfrak{K}}), \tag{14}$$

while the case $\Psi_i = (\exists x_i)\Psi_{i+1}$ yields

$$\Psi_i^{\mathfrak{K}} = (\exists x_i^{(11)})(\exists x_i^{(22)}) \ldots (\exists x_i^{(nn)})$$
$$(\det \| x_i^{(lm)} \| \not\approx 0 \,\&\, \Psi_{i+1}^{\mathfrak{K}}); \tag{15}$$

(III) Finally, we have $\Phi^{\mathfrak{K}} = \Psi_1^{\mathfrak{K}}$.

In the case of $SL(n, \mathfrak{K})$ the appropriate $\Phi^{\mathfrak{K}}$ is obtained by changing $\det \| x_i^{(lm)} \| \not\approx 0$ to $\det \| x_i^{(lm)} \| \approx 1$. For the projective groups $PG(n, \mathfrak{K})$, $PS(n, \mathfrak{K})$, we already know how to transform the group sentence Φ into a group sentence Φ^{p} whose truth in the corresponding linear group $GL(n, \mathfrak{K})$, $SL(n, \mathfrak{K})$ is equivalent to the truth of Φ in the given projective group. We can now apply the above procedures to obtain a ring sentence $(\Phi^{\mathrm{p}})^{\mathfrak{K}}$ with the desired property.

The passage from a group sentence to a ring sentence concerning the field $\mathfrak{K}$ is perfectly straightforward and does not depend on our previous findings. The reverse passage is not so easily achieved, but the earlier constructions bear the burden.

Suppose $\mathfrak{G}$ is one of the groups mentioned in the theorem we are proving, and suppose

$$\Upsilon = (\eth_1 x_1) \ldots (\eth_t x_t)\, \Upsilon_0(x_1, \ldots, x_t)$$

is an arbitrary ring sentence. We want to find out when Υ is true in the base field $\mathfrak{K}$ of the group $\mathfrak{G}$. Recall the formulas Ω, Δ^*, Σ^*, Π^* constructed in Lemma 10 for the group $\mathfrak{G}$. By restricting the quantifiers in Υ to the set of w characterized by $\Delta^*(w, a, b, c)$ and replacing the basic predicate symbols $S(x, y, z)$, $P(x, y, z)$ in Υ_0 with the formulas $\Sigma^*(x, y, z, a, b)$, $\Pi^*(x, y, z, a, b, c)$, we obtain a group formula $\Upsilon^{\#}(a, b, c)$. If $a, b, c \in \mathfrak{G}$ are chosen so that $\Omega(a, b, c)$ is true in $\mathfrak{G}$, and $\hat{K}$, $\oplus$, $\otimes$ are defined as in Lemma 10, then $\Upsilon^{\#}$ asserts the same thing in the field $\langle \hat{K}; \oplus, \otimes \rangle$ as Υ does in the isomorphic field $\mathfrak{K}$. Therefore, the truth of Υ in $\mathfrak{K}$ is equivalent to the truth of the sentence

$$\Upsilon^{\mathfrak{G}} = (\exists \boldsymbol{abc})(\Omega(\boldsymbol{a}, \boldsymbol{b}, \boldsymbol{c}) \,\&\, \Upsilon^{\#}(\boldsymbol{a}, \boldsymbol{b}, \boldsymbol{c}))$$

in the group $\mathfrak{G}$. ∎

Corollary: *For each of the groups $GL(n, \mathfrak{K})$, $PG(n, \mathfrak{K})$, $SL(n+1, \mathfrak{K})$, $PS(n+1, \mathfrak{K})$ with $n \geqslant 2$, the set of all group sentences true in this group is not a recursive set if the corresponding set for the base field $\mathfrak{K}$ is not recursive.* ∎

In other words, if the elementary theory of the field $\mathfrak{K}$ is not recursively decidable, then the elementary theories of all these groups are also undecidable. In particular, all the indicated linear and projective groups over the field of rational numbers have undecidable elementary theories.

Theorem 5: *In order that the groups $G(m, \mathfrak{K}_1)$, $G(n, \mathfrak{K}_2)$ ($G = GL, PG, SL, PS$; $n \geqslant 3$) be of the same elementary type (i.e., be elementarily equivalent), it is necessary and sufficient that $m = n$ and the base fields $\mathfrak{K}_1$, $\mathfrak{K}_2$ themselves be elementarily equivalent.*

The sufficiency of these conditions is obvious. For we transform the arbitrary group sentence Φ concerning the groups $G(n, \mathfrak{K}_1)$, $G(n, \mathfrak{K}_2)$ into the ring sentence $\Phi^{\mathfrak{K}}$ whose form does not depend on the nature of the base fields. The truth of $\Phi^{\mathfrak{K}}$ in $\mathfrak{K}_1$ is equivalent to its truth in $\mathfrak{K}_2$ since $\mathfrak{K}_1$ and $\mathfrak{K}_2$ are elementarily equivalent. Consequently, Φ is true in $G(n, \mathfrak{K}_1)$ iff it is true in $G(n, \mathfrak{K}_2)$. Hence, these groups are elementarily equivalent.

The necessity of the condition $m = n$ follows from Theorem 3. It is also necessary that the elementary types of $\mathfrak{K}_1$ and $\mathfrak{K}_2$ coincide. For let Υ be a ring sentence; the group sentence $\Upsilon^{\mathfrak{G}}$ is the same for $\mathfrak{G} = G(n, \mathfrak{K}_1), G(n, \mathfrak{K}_2)$ since its construction does not depend on the structure of the base field. As $\Upsilon^{\mathfrak{G}}$ is true in $G(n, \mathfrak{K}_1)$ iff it is true in $G(n, \mathfrak{K}_2)$, Υ is valid in $\mathfrak{K}_1$ iff it is valid in $\mathfrak{K}_2$. ∎

Theorem 5 lets us compare the elementary types of the groups within each of the four series *GL, SL, PG, PS*. Concerning groups belonging to different series, it is clear that the *GL*-groups are not elementarily equivalent to the groups in the remaining three series. The coincidence or divergence of the elementary types of groups in the last three series depends, in general, on properties of the base field.

§2.4. *Concluding remarks*

In the preceding, we have examined the series of general linear and projective groups from order $n = 2$ on, as well as the two series of special groups, starting with order $n = 3$. We have been neglecting $SL(2, \mathfrak{K})$ and $PS(2, \mathfrak{K})$.

Whether Theorem 4 is valid for these excepted groups for arbitrary base fields of characteristic 0 remains unclear to the author. Theorem 4, however, is certainly true for these two groups when the field $\mathfrak{K}$ is *almost euclidean,* i.e., when there exists a natural number t such that for every $\alpha \in \mathfrak{K}$, either α or $-\alpha$ is the sum of t squares in $\mathfrak{K}$.

Among the almost euclidean fields we find, e.g., the field of rational numbers, finite algebraic extensions of this field, the fields of complex and real numbers, and others. If $\mathfrak{K}$ is one of these fields, then the elementary theories of $SL(2, \mathfrak{K})$, $PS(2, \mathfrak{K})$, and $\mathfrak{K}$ are recursively syntactically equivalent.

Instead of a field $\mathfrak{K}$ we can consider a ring $\mathfrak{R}$ with identity element and take $GL(n, \mathfrak{R})$ to be the group of all $n \times n$ matrices over $\mathfrak{R}$ whose determinants are invertible in $\mathfrak{R}$; groups $SL(n, \mathfrak{R})$, $PG(n, \mathfrak{R})$, $PS(n, \mathfrak{R})$ can be defined analogously. Under natural restrictions Theorems 4 and 5 can easily be extended to such groups over rings, as long as $n \geqslant 3$. The groups $SL(2, \mathfrak{R})$, $PS(2, \mathfrak{R})$ present special interest when $\mathfrak{R}$ is the ring of rational integers.

In §1 it was shown that a number of important subgroups and subsets of matrix groups are relatively elementary. If we turn to the class of all compact simple Lie groups, we see that many of their subgroups also have this property, e.g., their simple subgroups. It is natural to pose similar problems for simple Lie algebras, as well.

NOTES

(1) These partitions may not be distinct, even for nondiagonal χ.

(2) This is not in general a group formula, for we do not assume this group $\mathfrak{Z}$ of scalar matrices is an elementary subgroup of $GL(2, \mathfrak{K})$.

CHAPTER 21

THE EFFECTIVE INSEPARABILITY OF THE SET OF VALID SENTENCES FROM THE SET OF FINITELY REFUTABLE SENTENCES IN SEVERAL ELEMENTARY THEORIES

Let $\mathfrak{K}$ be some class of models with signature Σ. Formulas of first-order predicate logic (FOPL) whose extralogical constants are contained in Σ are called *$\mathfrak{K}$-formulas.* A closed $\mathfrak{K}$-formula (*$\mathfrak{K}$-sentence*) Φ is said to be (*identically*) *valid in* $\mathfrak{K}$ iff it is true in all models belonging to $\mathfrak{K}$. Φ is *finitely refutable in* $\mathfrak{K}$ iff Φ is false in some finite $\mathfrak{K}$-model. When Φ is true in all finite $\mathfrak{K}$-models, we say it is *finitely valid in* $\mathfrak{K}$. By $\mathbf{T}(\mathfrak{K})$ we denote the set of all $\mathfrak{K}$-sentences valid in $\mathfrak{K}$ (the *elementary theory* of $\mathfrak{K}$), and by $\mathbf{FR}(\mathfrak{K})$ the set of all $\mathfrak{K}$-sentences finitely refutable in $\mathfrak{K}$. In [XIX] it was shown that $\mathbf{FR}(\mathfrak{K})$ is not a recursive set when $\mathfrak{K}$ is the class of all groups or of all associative rings, Lie rings, etc. Using the results of that article, we shall now prove the stronger proposition that $\mathbf{T}(\mathfrak{K})$ and $\mathbf{FR}(\mathfrak{K})$ are effectively inseparable if $\mathfrak{K}$ is one of the classes mentioned. From this we can immediately derive, in particular, the theorem of B.A. Trahtenbrot [169] on the recursive inseparability of the set of logical validities from the set of finitely refutable sentences of FOPL.

§1. Let $\mathcal{L}$ be the class of all rings – not necessarily associative – that are algebras over a fixed prime field $\mathfrak{R}$ of prime characteristic π. Echoing [169], we first indicate an effective procedure whereby for each $\mathcal{L}$-sentence Φ, one can construct a new $\mathcal{L}$-sentence $\Phi^{(m)}$ whose validity in $\mathcal{L}$ is equivalent to the truth of Φ in all $\mathcal{L}$-rings with identity element containing fewer than m elements (cf. Lemma 1).

By $\langle \boldsymbol{q}, \boldsymbol{x}\rangle$ we denote the $\mathcal{L}$-formula

$$\boldsymbol{q}\boldsymbol{x} \approx \boldsymbol{x} \,\&\, \boldsymbol{x}^2 \approx \boldsymbol{x} \,\&\, \boldsymbol{x} \not\approx \boldsymbol{0}\,,$$

and we abbreviate

$$cx \approx x \,\&\, xc \approx x \,\&\, x^2 \approx 0$$

by $[c, x]$. In a given $\mathfrak{L}$-ring $\mathfrak{R}$, the elements x for which $\langle q, x\rangle$ holds in $\mathfrak{R}$ are called *q-elements*; the subset $\mathfrak{R}_c^*$ of $\mathfrak{R}$ consisting of all x such that $[c, x]$ holds in $\mathfrak{R}$ is called the *space belonging to c*; the set of all x for which $px = x$ is denoted by $\mathfrak{R}_p$ $(q, c, p \in \mathfrak{R})$. We let $\Gamma(a)$ be the conjunction of the $\mathfrak{L}$-formulas (1^a) and (2^a) from [XIX], §1. We take $\mathbf{U}(p, q)$ to be the conjunction of the formulas

$$(xy)(\langle q, x\rangle \,\&\, [x, y] \rightarrow py \approx y) \,\&$$

$$\&\, (xy)(px \approx x \,\&\, py \approx y \rightarrow xy \approx 0)\,,$$

$$(xy)(\langle q, x\rangle \,\&\, \langle q, y\rangle \,\&\, x \not\approx y \rightarrow$$

$$\rightarrow (\exists u)(([x, u] \vee [y, u]) \,\&\, (\neg[x, u] \vee \neg[y, u])))\,,$$

$$(x)(\exists y)(px \approx x \rightarrow \langle q, y\rangle \,\&$$

$$\&\, (u)([y, u] \rightarrow u \approx x \vee u \approx 2x \vee \ldots \vee u \approx \pi x))\,,$$

$$(xy)(\exists z)(\langle q, x\rangle \,\&\, \langle q, y\rangle \rightarrow \langle q, z\rangle \,\&$$

$$\&\, (u)([z, u] \rightarrow (\exists vw)([x, v] \,\&\, [y, w] \,\&\, u \approx v + w)))\,.$$

For any $p, q \in \mathfrak{R}$, the truth of $\mathbf{U}(p, q)$ in $\mathfrak{R}$ means: (I) the space belonging to any q-element is included in $\mathfrak{R}_p$, and the product of any two elements of $\mathfrak{R}_p$ is equal to 0; (II) the spaces belonging to different q-elements are distinct; (III) every 0-dimensional or 1-dimensional linear subspace of the linear space $\mathfrak{R}_p$ is the space belonging to some q-element; (IV) the complex sum of the spaces belonging to any two q-elements is itself a space belonging to some q-element. Thus when $\mathbf{U}(p, q)$ holds, every space belonging to a q-element is, according to (I), a linear subspace of $\mathfrak{R}_p$, and by (III), (IV) every finite-dimensional linear subspace of $\mathfrak{R}_p$ belongs to some q-element.

§2. It is easy to calculate that the number of different algebras constructible from a given n-dimensional linear space over $\mathfrak{K}$ by adjoining a multiplication operation is equal to π^{n^3}. Let $t = F(n)$ be the number of different linear subspaces of such a space; by setting $G(t) = n$ we define an inverse function G, which becomes totally defined when we agree to take 0 as its value for those natural numbers t that do not belong to the range of F. From the explicit formula for F we would immediately see that F and G are primitive recursive functions. Therefore, $D(t) = \pi^{G(t)^3}$ defines a primitive recursive function. By the method employed in [XIX] we can construct an $\mathcal{L}$-formula $\Delta(\boldsymbol{a}, \boldsymbol{b})$ with the following properties: (i) if in any $\mathcal{L}$-algebra $\mathfrak{R}$ there are elements a, b satisfying $\Delta(a, b)$, and if the number of a-elements in $\mathfrak{R}$ is equal to t, then the number of b-elements in $\mathfrak{R}$ is equal to $D(t)$; (ii) for every number t, there exists a finite algebra $\mathfrak{R}$ with identity containing elements a, b such that $\Delta(a, b)$ holds, while the number of a-elements in it equals t.

§3. Let $\mathbf{V}(\boldsymbol{q}, \boldsymbol{c}, \boldsymbol{g}, \boldsymbol{a}, \boldsymbol{b})$ denote the formula

$$\Delta^{\#}(\boldsymbol{g}, \boldsymbol{a}, \boldsymbol{b}) \,\&\, (\boldsymbol{xy})(\boldsymbol{gx} \approx \boldsymbol{x} \,\&\, \boldsymbol{gy} \approx \boldsymbol{y} \rightarrow \boldsymbol{g}\cdot\boldsymbol{xy} \approx \boldsymbol{xy}) \,\&$$

$$\& \,(\exists \boldsymbol{z})[(\boldsymbol{y})(\langle \boldsymbol{q}, \boldsymbol{y}\rangle \,\&\, \boldsymbol{y} \subseteq \boldsymbol{c} \rightarrow \boldsymbol{g}\cdot\boldsymbol{zy} \approx \boldsymbol{zy} \,\&\, \langle \boldsymbol{a}, \boldsymbol{zy}\rangle) \,\&$$

$$\& \,(\boldsymbol{x})(\boldsymbol{gx} \approx \boldsymbol{x} \,\&\, \langle \boldsymbol{a}, \boldsymbol{x}\rangle \rightarrow (\exists \boldsymbol{y})(\langle \boldsymbol{q}, \boldsymbol{y}\rangle \,\&\, \boldsymbol{y} \subseteq \boldsymbol{c} \,\&\, \boldsymbol{x} \approx \boldsymbol{zy})) \,\&$$

$$\& \,(\boldsymbol{xy})(\langle \boldsymbol{q}, \boldsymbol{x}\rangle \,\&\, \boldsymbol{x} \subseteq \boldsymbol{c} \,\&\, \langle \boldsymbol{q}, \boldsymbol{y}\rangle \,\&\, \boldsymbol{y} \subseteq \boldsymbol{c} \,\&\, \boldsymbol{zx} \approx \boldsymbol{zy} \rightarrow \boldsymbol{x} \approx \boldsymbol{y})] \;;$$

here, $\Delta^{\#}(\boldsymbol{g}, \boldsymbol{a}, \boldsymbol{b})$ is the relativization of $\Delta(\boldsymbol{a}, \boldsymbol{b})$ to the set of x described by $\boldsymbol{gx} \approx \boldsymbol{x}$, i.e., to the subspace $\mathfrak{R}_g$ in any particular $\mathcal{L}$-algebra $\mathfrak{R}$, while $\boldsymbol{y} \subseteq \boldsymbol{c}$ is an abbreviation for the formula $(\boldsymbol{u})([\boldsymbol{y}, \boldsymbol{u}] \rightarrow [\boldsymbol{c}, \boldsymbol{u}])$, which asserts $\mathfrak{R}_y^* \subseteq \mathfrak{R}_c^*$ for $y, c \in \mathfrak{R}$.

Suppose we have selected elements p, q, c, g, a, b from the $\mathcal{L}$-algebra such that $\mathbf{U}(p, q) \,\&\, \langle q, c\rangle \,\&\, \mathbf{V}(q, c, g, a, b)$ is satisfied. Then the space $\mathfrak{R}_g$ is a subalgebra containing the elements a, b, which satisfy $\Delta(a, b)$ inside $\mathfrak{R}_g$. Furthermore, the number of a-elements in $\mathfrak{R}_g$ equals the number of q-elements y such that $\mathfrak{R}_y^*$ is included in $\mathfrak{R}_c^*$; hence, the number of b-elements in $\mathfrak{R}_g$ is equal to π^{r^3}, where r is the dimension of $\mathfrak{R}_c^*$.

Let $\mathbf{W}(\boldsymbol{c}, \boldsymbol{g}, \boldsymbol{b})$ be the conjunction of the following formulas:

$$(\boldsymbol{xyz})(\boldsymbol{gx} \approx \boldsymbol{x} \,\&\, \langle \boldsymbol{b}, \boldsymbol{x}\rangle \,\&\, [\boldsymbol{c}, \boldsymbol{y}] \,\&\, [\boldsymbol{c}, \boldsymbol{z}] \rightarrow$$

$$\rightarrow (\boldsymbol{xy} \approx \boldsymbol{xz} \rightarrow \boldsymbol{y} \approx \boldsymbol{z}) \,\&\, (\exists \boldsymbol{u})([\boldsymbol{c}, \boldsymbol{u}] \,\&\, \boldsymbol{xy} \cdot \boldsymbol{xz} \approx \boldsymbol{xu}))\,,$$

$$(xy)(gx \approx x \mathbin{\&} \langle b, x\rangle \mathbin{\&} gy \approx y \mathbin{\&} \langle b, y\rangle \mathbin{\&} x \not\approx y \to$$

$$\to (\exists uvww')([c,u] \mathbin{\&} [c,v] \mathbin{\&} [c,w] \mathbin{\&} [c,w'] \mathbin{\&}$$

$$\mathbin{\&} xu\cdot xv \approx xw \mathbin{\&} yu\cdot yv \approx yw' \mathbin{\&} w \not\approx w')) .$$

The truth of $\mathbf{U}(p,q) \mathbin{\&} \langle q, c\rangle \mathbin{\&} \mathbf{W}\langle c, g, b)$ in the algebra $\mathfrak{R}$ for particular elements p, q, c, g, b tells us that for any fixed b-element x of the subalgebra $\mathfrak{R}_g$, the elements of the form xu ($u \in \mathfrak{R}_c^*$) compose a subalgebra $\mathfrak{R}_c^x$ of $\mathfrak{R}_g$ isomorphic to the algebra consisting of the linear space $\mathfrak{R}_c^*$ supplied with the multiplication operation $\otimes$ determined by the condition: $u \otimes v = w \Leftrightarrow xu\cdot xv = xw$. To different b-elements $x, y \in \mathfrak{R}_g$ correspond distinct algebras based on $\mathfrak{R}_c^*$. ([1])

Suppose for certain $p, q, c, g, a, b \in \mathfrak{R}$,

$$\mathbf{U}(p,q) \mathbin{\&} \langle q, c\rangle \mathbin{\&} \mathbf{V}(q, c, g, a, b) \mathbin{\&} \mathbf{W}(c, g, b)$$

is satisfied in $\mathfrak{R}$, and let r be the dimension of $\mathfrak{R}_c^*$. Then the number of b-elements in $\mathfrak{R}_g$ is equal to π^{r^3}, i.e., to the number of distinct multiplication operations that turn $\mathfrak{R}_c^*$ into an $\mathcal{L}$-ring. Therefore, every $\mathcal{L}$-ring having linear dimension r as an algebra is isomorphic to one of the subalgebras $\mathfrak{R}_c^x$ of $\mathfrak{R}_g$.

§4. Let Φ be any FOPL sentence concerning rings. By $\Phi^{\#}(c,x)$ we denote the restriction of Φ to $\mathfrak{R}_c^x$, or more precisely, the formal relativization of Φ to the set of z characterized by the formula

$$\mathbf{S}(z, c, x) = (\exists u)([c,u] \mathbin{\&} z \approx xu) .$$

We also put

$$\mathbf{E}(c,x) = (\exists e)(\mathbf{S}(e,c,x) \mathbin{\&} (z)(\mathbf{S}(z,c,x) \to ze \approx z \mathbin{\&} ez \approx z)) ,$$

$$\mathbf{X}_n(p) = (\exists x_1 \ldots x_n)(px_1 \approx x_1 \mathbin{\&} \ldots \mathbin{\&} px_n \approx x_n \mathbin{\&}$$

$$\mathbin{\&} \textstyle\&\, \alpha_1 x_1 + \ldots + \alpha_n x_n \not\approx 0) ,$$

where the conjunction in $\mathbf{X}_n$ is taken over all possible nonzero sequences $\langle \alpha_1, \ldots, \alpha_n\rangle$ of numbers from the set $\{0, 1, \ldots, \pi-1\}$. For every $p \in \mathfrak{R}$, $\mathbf{X}_n(p)$ is true in $\mathfrak{R}$ iff the dimension of $\mathfrak{R}_p$ is not less than n.

Finally, we introduce the formulas

$$\mathbf{Y}(q, c, g, a, b) = \mathbf{V}(q, c, g, a, b) \mathbin{\&} \mathbf{W}(c, g, b)\,,$$

$$\mathbf{Z}(p, q) = \mathbf{U}(p, q) \mathbin{\&} \Gamma(q) \mathbin{\&}$$

$$\mathbin{\&} (c)(\langle q, c\rangle \rightarrow (\exists gab)\, \mathbf{Y}(q, c, g, a, b))\,,$$

$$\hat{\Phi}(q) = (cgabx)(\langle q, c\rangle \mathbin{\&} \mathbf{Y}(q, c, g, a, b) \mathbin{\&}$$

$$\mathbin{\&} gx \approx x \mathbin{\&} \langle b, x\rangle \mathbin{\&} \mathbf{E}(c, x) \rightarrow \Phi^{\#}(c, x))\,.$$

Lemma 1: *If the sentence Φ is valid in all $\mathfrak{L}$-rings with identity of dimension less than n, then the sentence*

$$(pq)(\neg \mathbf{X}_n(p) \mathbin{\&} \mathbf{Z}(p, q) \rightarrow \hat{\Phi}(q))$$

is identically valid in $\mathfrak{L}$.

For suppose in the $\mathfrak{L}$-ring $\mathfrak{R}$ there are elements p, q, c, g, a, b, x such that $\neg \mathbf{X}_n(p)$, $\mathbf{Z}(p, q)$, $\langle q, c\rangle$, $\mathbf{Y}(q, c, g, a, b)$, $gx = x$, $\langle b, x\rangle$, $\mathbf{E}(c, x)$ all hold. Then the algebras $\mathfrak{R}_c^x$ and $\mathfrak{R}_c^*$ have the same dimension, which must be less than n. In addition, $\mathfrak{R}_c^x$ has an identity element. Hence, $\Phi^{\#}(c, x)$ holds in $\mathfrak{R}$. ∎

Lemma 2: *If Φ is false in some n-dimensional $\mathfrak{L}$-algebra with identity, then the sentence*

$$(pq)(\mathbf{X}_n(p) \mathbin{\&} \mathbf{Z}(p, q) \rightarrow \neg \hat{\Phi}(q))$$

is valid in $\mathfrak{L}$.

Suppose $\mathfrak{R} \in \mathfrak{L}$, and suppose to $\mathfrak{R}$ belong elements p, q satisfying $\mathbf{X}_n(p)$ and $\mathbf{Z}(p, q)$ in $\mathfrak{R}$. Then in $\mathfrak{R}_p$ is an n-dimensional linear subspace belonging to some q-element c. Therefore, there are elements $g, a, b \in \mathfrak{R}$ for which $\mathfrak{R}_g$ is a subalgebra containing a and b and exactly π^{n^3} b-elements x. Translating $\mathfrak{R}_c^*$ on the left by these b-elements, we obtain subalgebras $\mathfrak{R}_c^x$ isomorphic to the π^{n^3} different enrichments of $\mathfrak{R}_c^*$ with multiplication operations. Hence, among the $\mathfrak{R}_c^x$ is an algebra with identity in which Φ is false. It follows that $\neg \hat{\Phi}(q)$ is true in $\mathfrak{R}$. ∎

Lemma 3: *If relative to the class of all $\mathfrak{L}$-rings with identity the $\mathfrak{L}$-sentence Φ is finitely refutable, while the $\mathfrak{L}$-sentence Ψ is finitely valid, then the sen-*

tence

$$(pq)(\mathbf{Z}(p,q) \rightarrow (\hat{\Phi}(q) \rightarrow \hat{\Psi}(q))) \tag{1}$$

is valid in every $\mathcal{L}$-ring.

Suppose Φ is false in some n-dimensional algebra with identity. Then by Lemmas 1 and 2, the sentences

$$(pq)(\neg \mathbf{X}_n(p) \,\&\, \mathbf{Z}(p,q) \rightarrow \hat{\Psi}(q)),$$

$$(pq)(\mathbf{X}_n(p) \,\&\, \mathbf{Z}(p,q) \rightarrow \neg \hat{\Phi}(q))$$

are valid in $\mathcal{L}$; hence, (1) is valid in $\mathcal{L}$. ■

Lemma 4: *If the $\mathcal{L}$-sentence Ψ is false in some finite $\mathcal{L}$-ring with identity, while the $\mathcal{L}$-sentence Φ is true in all finite $\mathcal{L}$-rings with identity, then* (1) *is false in some finite $\mathcal{L}$-ring with identity.*

We shall have proved this lemma if we show that whenever we have a finite-dimensional $\mathcal{L}$-algebra with identity in which Ψ fails, we can construct a finite-dimensional $\mathcal{L}$-algebra containing elements p, q for which $\mathbf{Z}(p, q)$ and $\hat{\Phi}(q)$ are true, but $\hat{\Psi}(q)$ is false. The construction of such an algebra is analogous to the constructions made in [XIX]. We omit it here because of its length. ■

§5. Suppose M^1, M^2 are arbitrary disjoint, recursively enumerable sets of natural numbers. An effective method is suggested in [XIX], §4, whereby from the Post-Kleene numbers of the sets M^1, M^2 we can construct two sequences of $\mathcal{L}$-sentences Φ^i_m ($i = 1, 2; m = 0, 1, 2, \dots$) such that $m \in M^i$ iff Φ^i_m is false in some finite $\mathcal{L}$-ring with identity. Consider the sequence of sentences

$$\Upsilon_m = (pq)(\mathbf{Z}(p,q) \rightarrow (\hat{\Phi}^1_m(q) \rightarrow \hat{\Phi}^2_m(q))).$$

According to Lemmas 3 and 4, if $m \in M^1$, then Υ_m is identically valid in $\mathcal{L}$, but if $m \in M^2$, then Υ_m is finitely refutable in the class of $\mathcal{L}$-rings with identity. In other words, any pair $\langle M^1, M^2\rangle$ of disjoint recursively enumerable sets of natural numbers is recursively reducible to the pair $\langle \mathrm{T}(\mathcal{L}), \mathrm{FR}(\mathcal{L})\rangle$. By taking $\langle M^1, M^2\rangle$ to be an effectively inseparable pair, or by using a theorem of Mučnik [109], we immediately conclude that $\mathrm{T}(\mathcal{L})$ and $\mathrm{FR}(\mathcal{L})$ are effectively inseparable. Thus we have proved

Theorem 1: *The set of identically valid sentences and the set of finitely refutable sentences for the class of all rings with identity element having a given prime characteristic are effectively inseparable.* ∎

By using – as in [XIX], §5 – the correspondence established between rings and groups in [XV], we find that together with Theorem 1 we have proved

Theorem 2: *For every odd prime π, the set of sentences identically valid and the set of sentences finitely refutable in the class of all metabelian π-groups are effectively inseparable; the same holds for the class of all char π rings satisfying the identity*

$$(xyz)(xy \approx -yx \;\&\; x \cdot yz \approx xy \cdot z \approx 0)\,. \blacksquare$$

From Theorem 2 it follows that the indicated sets of sentences are effectively inseparable also for the classes of all groups, all associative rings, all Lie rings, etc.

NOTES

(^{1}) This does not preclude that $\mathfrak{R}^x_c$, $\mathfrak{R}^y_c$ may coincide.

CHAPTER 22

CLOSELY RELATED MODELS AND RECURSIVELY PERFECT ALGEBRAS

This article was inspired by a problem of A. Mostowski which he formulated in [166], p. 84. We consider the arithmetic $\mathfrak{S} = \langle\{0, 1, 2, ...\}; +, \times\rangle$ and ask whether or not there exist a binary operation $*$ on the set of natural numbers $D = \{0, 1, 2, ...\}$ and natural numbers $a_1, ..., a_p$ such that: (i) $\langle D, *\rangle$ is a group; (ii) the relation $x * y = z$ is definable in $\mathfrak{S}$ by a formula of first-order predicate logic (FOPL); (iii) the relations $x + y = z, x \times y = z$ are definable by FOPL formulas in the group $\langle D; *; a_1, ..., a_p\rangle$ with distinguished elements $a_1, ..., a_p$. Below we solve a general problem related to this problem of Mostowski. As a corollary of this solution we obtain a positive answer to Mostowski's problem (cf. Theorems 2 and 4 below).

§1. Closely related models

Suppose $\mathcal{K}_1, \mathcal{K}_2$ are arbitrary classes of models with respective signatures

$$\Sigma_1 = \{\boldsymbol{P}_1, ..., \boldsymbol{P}_s; \boldsymbol{a}_1, ..., \boldsymbol{a}_p\} \quad \text{and} \quad \Sigma_2 = \{\boldsymbol{Q}_1, ..., \boldsymbol{Q}_t; \boldsymbol{b}_1, ..., \boldsymbol{b}_t\}.$$

For $j = 1, 2$, let $\mathbf{F}_j$ be the set of all FOPL formulas whose predicate symbols belong to Σ_j; let $\mathrm{T}(\mathcal{K}_j)$ be the subset of $\mathbf{F}_j$ consisting of all closed formulas (sentences) true in every $\mathcal{K}_j$-model. Let $\rho(\boldsymbol{x})$ be a formula in $\mathbf{F}_2$ in which only one free individual variable $\boldsymbol{x}$ occurs. To prescribe a *homomorphism* φ of $\mathbf{F}_1$ into $\mathbf{F}_2$, we associate with every predicate symbol $\boldsymbol{P}_i(\boldsymbol{x}_i, ..., \boldsymbol{x}_{m_i})$ a specific formula $\Pi_i(\boldsymbol{x}_1, ..., \boldsymbol{x}_{m_i})$ in $\mathbf{F}_2$, and with every individual constant symbol $\boldsymbol{a}_k$ some formula $\mathrm{A}_k(\boldsymbol{x})$ in $\mathbf{F}_2$ with one free variable. If $\Phi \in \mathbf{F}_1$, then Φ^φ is the formula in $\mathbf{F}_2$ obtained by transforming Φ as follows: (I) we replace each occurrence of $\boldsymbol{P}_i(\boldsymbol{y}_1, ..., \boldsymbol{y}_{m_i})$ in Φ with $\Pi_i(\boldsymbol{y}_1, ..., \boldsymbol{y}_{m_i})$; (II) the original quantifiers in Φ are relativized to the predicate $\rho(\boldsymbol{x})$, or more graphically put, they are restricted to the set of x for which $\rho(x)$ is true ([1]); (III) if the result of performing (I), (II) is $\Phi_1(\boldsymbol{a}_{k_1}, ..., \boldsymbol{a}_{k_l})$, where $\boldsymbol{a}_{k_1}, ..., \boldsymbol{a}_{k_l}$ are the individual constants (if any) occurring in Φ, then we put

$$\Phi^{\varphi} = (\exists u_1, \ldots, u_l)(\Phi_1(u_1, \ldots, u_l) \;\&\; \mathrm{A}_{k_1}(u_1) \;\&\; \ldots \;\&\; \mathrm{A}_{k_l}(u_l)) .$$

A homomorphism φ from $\mathbf{F}_1$ into $\mathbf{F}_2$ is called a *relative ρ-interpretation of* $\mathcal{K}_1$ *in* $\mathcal{K}_2$ iff $\mathrm{T}(\mathcal{K}_1)^{\varphi} \subseteq \mathrm{T}(\mathcal{K}_2)$. A relative ρ-interpretation is called simply an *interpretation* when $(\boldsymbol{x})\rho(\boldsymbol{x})$ is valid throughout $\mathcal{K}_2$.

Classes $\mathcal{K}_1, \mathcal{K}_2$ of models are said to be *related* iff there exists an interpretation φ of $\mathcal{K}_1$ in $\mathcal{K}_2$ and an interpretation ψ of $\mathcal{K}_2$ in $\mathcal{K}_1$ such that

$$(\forall)(\Phi^{\varphi\psi} \leftrightarrow \Phi) \in \mathrm{T}(\mathcal{K}_1),$$

$$(\forall)(\Psi^{\psi\varphi} \leftrightarrow \Psi) \in \mathrm{T}(\mathcal{K}_2), \qquad (\Phi \in \mathrm{F}_1, \Psi \in \mathrm{F}_2) ; \tag{1}$$

$$(\boldsymbol{x})(\boldsymbol{x} \approx \boldsymbol{a}_k \leftrightarrow \mathrm{A}_k(\boldsymbol{x})^{\psi}) \in \mathrm{T}(\mathcal{K}_1),$$

$$(\boldsymbol{x})(\boldsymbol{x} \approx \boldsymbol{b}_l \leftrightarrow \mathrm{B}_l(\boldsymbol{x})^{\varphi}) \in \mathrm{T}(\mathcal{K}_2) , \quad (k=1, \ldots, p; l=1, \ldots, q) . \tag{2}$$

Suppose φ is a ρ-interpretation of $\mathcal{K}_1$ in $\mathcal{K}_2$, and $\mathfrak{N} = \langle N; Q_1, \ldots, Q_t; b_1, \ldots, b_q \rangle$ is a $\mathcal{K}_2$-model. We let $\mathfrak{N}^{\varphi}$ denote the Σ_1-model $\langle R; P_1^{\varphi}, \ldots, P_s^{\varphi}; a_1^{\varphi}, \ldots, a_p^{\varphi} \rangle$, where R is the subset of N defined by ρ, P_i^{φ} is the predicate on R defined in $\mathfrak{N}$ by the formula Π_i, and a_k^{φ} is the unique element of R satisfying A_k in $\mathfrak{N}$. The ρ-interpretation φ is said to be *isomorphic* iff for every model $\mathfrak{M} \in \mathcal{K}_1$, there exists a model $\mathfrak{N} \in \mathcal{K}_2$ such that $\mathfrak{N}^{\varphi}$ is isomorphic to $\mathfrak{M}$. The classes $\mathcal{K}_1$ and $\mathcal{K}_2$ are said to be *closely related* iff there exist isomorphic interpretations φ of $\mathcal{K}_1$ in $\mathcal{K}_2$ and ψ of $\mathcal{K}_2$ in $\mathcal{K}_1$ that satisfy the conditions (1) and (2).

By taking classes consisting of single models, we adapt these notions to apply to individual models, as well.

§2. Recursively perfect algebras

We first recall a notion from [XVIII]. A 1–1 map α from a set D_α of natural numbers onto the base M of the model $\mathfrak{M} = \langle M; P_1, \ldots, P_s; a_1, \ldots, a_p \rangle$ is called a *(1–1) numbering* of $\mathfrak{M}$. The numbering α is said to be *constructive* iff D_α is a recursive set, while the predicates $P_1, \ldots, P_s$ become recursive predicates on D_α under the influence of α. The constructively numbered model $\langle \mathfrak{M}, \alpha \rangle$ is said to be *(recursively) steadfast* iff every constructive numbering of $\mathfrak{M}$ is recursively equivalent to α. In [XVIII], §4.1 it was shown that every constructively numbered, finitely generated algebra is steadfast. There are, however, steadfast constructive algebras with no finite sets of generators.

Theorem 1: *Every finite algebraic extension $\mathfrak{K}$ of the field of rational numbers, every special linear group $SL(n, \mathfrak{K})$ and its subgroup $RSL(n, \mathfrak{K})$ of triangular matrices over such a field $\mathfrak{K}$ for $n \geqslant 2$, and every torsion-free, completely divisible nilpotent group of finite rank is a constructively numberable and steadfast algebra.*

For the fields and the nilpotent groups, the proof is carried out easily and directly. The basic steps of the proof for the groups $SL(n, \mathfrak{K})$, $RSL(n, \mathfrak{K})$ are sketched in §3 below. ■

A model or algebra is called (*recursively*) *perfect* iff it is infinite and admits a constructive numbering, and every recursive predicate defined on the model is formular, i.e., represented by some FOPL formula. Gödel's theorem [166] shows that the arithmetic $\mathfrak{S}$ is a perfect algebra. The definitions immediately imply

Theorem 2: *Any two perfect models are closely related to each other.* ■

A relative ρ-interpretation φ of a model $\mathfrak{M} = \langle M; P_1, ..., P_s; a_1, ..., a_p \rangle$ in a model $\mathfrak{N} = \langle N; Q_1, ..., Q_t; b_1, ..., b_q \rangle$ with constructive numbering β is called *recursive* iff the set R defined in $\mathfrak{N}$ by ρ and the formular predicates P_i^φ on R defined by Π_i $(i = 1, ..., s)$ are recursive relative to β. Hence, $\mathfrak{N}^\varphi$ inherits a constructive numbering β^φ from $\mathfrak{N}$ when φ is recursive.

Theorem 3: *Suppose there exists a recursive and isomorphic relative interpretation φ of the perfect model $\mathfrak{M} = \langle M; P_1, ..., P_s \rangle$ in the model $\mathfrak{N} = \langle N; Q_1, ..., Q_t; b_1, ..., b_q \rangle$ with constructive numbering β, and suppose there exists a* FOPL *formula $\Upsilon(\boldsymbol{x}_1, ..., \boldsymbol{x}_r; \boldsymbol{x})$ defining a* 1–1 *recursive map from some subset W of the set R^r of all sequences $\langle x_1, ..., x_r \rangle$ $(x_k \in R)$ onto the whole set N. Then $\mathfrak{N}$ is closely related to $\mathfrak{M}$. If, in addition, $\mathfrak{N}$ is steadfast, then it is perfect.*

Let R be defined by ρ, P_i^φ by Π_i $(i = 1, ..., s)$. The model $\mathfrak{N}^\varphi = \langle R; P_1^\varphi, ..., P_s^\varphi \rangle$ is constructive and abstractly isomorphic to the perfect model $\mathfrak{M}$; therefore, $\mathfrak{N}^\varphi$ is also perfect. (²) The set W is recursively enumerable (relative to β^φ), so there is a recursive predicate $w(x; x_1, ..., x_r)$ on R that maps R 1–1 into W. Since $\mathfrak{N}^\varphi$ is perfect, this relation is represented by a FOPL formula Ω. Now the formula

$$\Xi(\boldsymbol{x}, \boldsymbol{y}) = (\exists \boldsymbol{x}_1 \dots \boldsymbol{x}_r)(\mathop{\&}_j \rho(\boldsymbol{x}_j) \,\&\, \Omega(\boldsymbol{x}; \boldsymbol{x}_1, ..., \boldsymbol{x}_r) \,\&\, \Upsilon(\boldsymbol{x}_1, ..., \boldsymbol{x}_r; \boldsymbol{y}))$$

defines a recursive 1–1 mapping from R onto N. By using formulas

$$\Pi_i'(x_1, \ldots, x_{m_i}) = (\exists u_1 \ldots u_{m_i})(\Xi(u_1, x_1) \,\&\, \ldots \,\&$$

$$\&\, \Xi(u_{m_i}, x_{m_i}) \,\&\, \Pi_i(u_1, \ldots, u_{m_i}))$$

to interpret the fundamental predicates P_i, we obtain an isomorphic and recursive interpretation χ of $\mathfrak{M}$ in $\mathfrak{N}$. Hence, the model $\mathfrak{N}^\chi = \langle N; P_1^\chi, \ldots, P_s^\chi\rangle$ is isomorphic to $\mathfrak{M}$, perfect, and recursively numbered by β. The predicates Q_j, $x = b$ are defined on N and recursive relative to β. By the perfectness of $\mathfrak{N}^\chi$, these predicates are expressed in it by FOPL formulas involving predicate symbols only from among $P_1, \ldots, P_s$.

Thus, $\mathfrak{M}$ and $\mathfrak{N}$ are closely related; moreover, every predicate on N that is recursive relative to β is formular in $\mathfrak{N}$. Hence, if $\mathfrak{N}$ is steadfast, it is perfect. ■

Corollary: *The ring of rational integers, as well as every finite algebraic extension of the field of rational numbers that has no nontrivial automorphisms, is a recursively perfect algebra.*

For if $\mathfrak{K}$ is a finite extension of the field $\mathfrak{Q}$ of rational numbers, and $\alpha \in \mathfrak{K}$ is a primitive element, then according to J. Robinson [136] there exists a formula $\varphi(x)$ defining the set Q of rational numbers in $\mathfrak{K}$, while in $\mathfrak{K}$ the formula

$$\Upsilon(x_1, \ldots, x_n; x) = (\exists a)(x \approx x_1 + x_2 a + \ldots + x_n a^{n-1} \,\&\, f(a) \approx 0)$$

gives a recursive 1–1 mapping of Q^n onto $\mathfrak{K}$; here, f is the irreducible monic polynomial over $\mathfrak{Q}$ having α as a root, and $n = deg(f)$. The rest of the proof is straightforward. ■

§3. Linear groups

According to a remark by Mostowski [166], the automorphism groups of closely related models are isomorphic. A recursively perfect algebras has no nontrivial automorphisms, but every infinite group does. Indeed, a group can be perfect only if it has two distinguished elements at the very least. As the following theorem shows, such perfect algebras actually exist.

Theorem 4: *Let the field $\mathfrak{K}$ be a recursively perfect algebra. For every $n \geqslant 3$, there are $n \times n$ matrices $A, B \in RSL(n, \mathfrak{K})$ and $A' \in SL(n, \mathfrak{K})$ such that $\langle SL(n, \mathfrak{K}); A, A'\rangle$ and $\langle RSL(n, \mathfrak{K}); A, B\rangle$ – groups enriched with two distinguished elements – are recursively perfect algebras.*

Here, $SL(n, \mathfrak{K})$ is the multiplicative group of all $n \times n$ matrices over $\mathfrak{K}$ with determinant 1, while $RSL(n, \mathfrak{K})$ is the subgroup consisting of all (upper) triangular matrices in $SL(n, \mathfrak{K})$. As the special matrix A we take the Jordan cell with 1's on the diagonal, as A' we take its transpose, and as B we choose a certain general sort of diagonal matrix in $SL(n, \mathfrak{K})$.

We shall indicate the general course of the proof for $n = 3$. Similar arguments apply to higher-order matrices. A given constructive numbering of the field $\mathfrak{K}$ naturally induces constructive numberings of the groups $SL(3, \mathfrak{K})$, $RSL(3, \mathfrak{K})$; these will give a reference for recursiveness in what follows.

We start with the group $RSL(3, \mathfrak{K})$ enriched with the distinguished fixed matrices

$$A = E_{11} + E_{22} + E_{33} + E_{12} + E_{23} ,$$

$$B = b_1 E_{11} + b_2 E_{22} + b_3 E_{33} ,$$

where $b_1 b_2^{-1} \neq b_2 b_3^{-1}$, although b_1, b_2, b_3 are otherwise arbitrary elements of $\mathfrak{K}$.

Letting $I = E_{11} + E_{22} + E_{33}$, and putting

$$\rho(X) = (\exists Y)(AY \approx YA \;\&\; Y \cdot BAB^{-1} \approx BAB^{-1} \cdot Y \;\&\; Y^3 \approx X) ,$$

$$\mathrm{Twm}(Y) = YA \approx AY \;\&\; Y^2 \approx 1 \;\&\; Y \not\approx 1 ,$$

$$\mathrm{Id}(W) = (\exists Y)(\mathrm{Twm}(Y) \;\&\; W \approx A^{-1} Y A^{-1} Y) \; (^3) ,$$

we see that ρ defines the set R of all matrices of the form $I + aE_{13}$ $(a \in \mathfrak{K})$, while Id is true for the matrix $I + E_{13}$ and only it. Now we define predicates for binary operations $\oplus, \otimes$ on R by using the following formulas inside $RSL(3, \mathfrak{K})$ (cf. [XX], §2.2):

$$\Sigma(X, Y, Z) = XY \approx Z ,$$

$$\Pi(X, Y, Z) = ((X \approx 1 \vee Y \approx 1) \;\&\; Z \approx 1) \vee (X \not\approx 1 \;\&\; Y \not\approx 1 \;\&$$

$$\& \; (\exists UVW)(UB \approx BU \;\&\; VB \approx BV \;\&\; \mathrm{Id}(W) \;\&$$

$$\& \; UW \approx XU \;\&\; VW \approx YV \;\&\; UVW \approx ZUV)) \; (^4) .$$

We thus obtain a recursive ρ-interpretation of $\mathfrak{K}$ in $\langle RSL(3, \mathfrak{K}); A, B\rangle$; moreover, $a \to I + aE_{13}$ gives an isomorphism from $\mathfrak{K}$ onto the field $\langle R; \oplus, \otimes \rangle$.

If we use some of the special properties of triangular matrices, we can easily construct a formula

$$\Upsilon(X_{11}, X_{22}, X_{33}, X_{12}, X_{13}, X_{23}; X)$$

that is true in $RSL(3, \mathfrak{K})$ only for matrices of the form

$$X_{ij} = I + a_{ij} E_{13} \qquad (1 \leqslant i \leqslant j \leqslant 3, \ \prod_i a_{ii} = 1),$$

$$X = \sum_{i \leqslant j} a_{ij} E_{ij} ;$$

Υ thus represents a recursive 1–1 mapping from a subset of R^6 onto $RSL(3, \mathfrak{K})$.

If we demonstrate the group $RSL(3, \mathfrak{K})$ is steadfast, we can then apply Theorem 3 to conclude the perfectness of $\langle RSL(3, \mathfrak{K}); A, B \rangle$. As already mentioned, $RSL(3, \mathfrak{K})$ naturally inherits a constructive numbering α from the given constructive numbering of the base field $\mathfrak{K}$. Suppose β is another constructive numbering of this group. From the form of ρ and the formulas defining $\oplus$ and $\otimes$ it is easy to see that the set R and the operations $\oplus, \otimes$ are recursive relative to β. Since $\langle R; \oplus, \otimes \rangle$ is a recursively perfect algebra, the restrictions of the numberings α, β to R are recursively equivalent, i.e., there exists an algorithm A whereby from the number of any element of R in either of the numberings α, β one can find its number in the other numbering. From the explicit construction of the formula Υ one could extract an algorithm B for finding the number of a matrix X from the numbers of its "coordinate" matrices X_{ij} $(i \leqslant j)$ – and vice-versa – in any constructive numbering of $RSL(3, \mathfrak{K})$. Knowing the α-number of a matrix X in this group, we find the α-numbers of the matrices X_{ij} with the aid of B. Using A, we find the β-numbers of the X_{ij}, and using B, we find the β-number of X. We now conclude $\langle RSL(3, \mathfrak{K}); A, B \rangle$ is perfect.

We turn to the group $\langle SL(3, \mathfrak{K}); A, A' \rangle$ with distinguished elements

$$A = I + E_{12} + E_{23}, \quad A' = I + E_{21} + E_{32}.$$

The formulas

$$\sigma(X) = XAX^{-1} \cdot A \approx A \cdot XAX^{-1},$$

$$\sigma'(X) = XA'X^{-1} \cdot A' \approx A' \cdot XA'X^{-1}$$

serve to define in this enriched group the subgroups $\mathfrak{G}_1$ of upper triangular matrices and $\mathfrak{G}_2$ of lower triangular matrices with the property that their center diagonal elements cubed are equal to 1. The formula

$$\tau(X) = \sigma(X) \,\&\, \sigma'(X) \,\&\, \boldsymbol{X^6 A \not\approx A X^6}$$

defines in $SL(3, \mathfrak{K})$ the set of diagonal matrices of the form $aE_{11} + abE_{22} + ab^2E_{33}$, where $a^3b^3 = 1$, but $a^6 \neq 1$. Let C satisfy τ, and let B be any matrix commuting with C such that $\neg\sigma(B)$ holds. Then we can use B to define $R \subseteq \mathfrak{G}_1$ and the analogous set $R' \subseteq \mathfrak{G}_2$ ([5]), and to construct formulas

$$\Upsilon_1(X_{11}, X_{22}, X_{33}, X_{12}, X_{13}, X_{23}; X)\,,$$

$$\Upsilon_2(X_{11}, X_{22}, X_{33}, X_{21}, X_{31}, X_{32}; X)$$

"coordinatizing" $\mathfrak{G}_1$, $\mathfrak{G}_2$ by members of R, R'. Every element of $SL(3, \mathfrak{K})$ can be represented in the form $Y_1Y_2Y_3Y_4Y_5$ with Y_1, Y_3, $Y_5 \in \mathfrak{G}_1$ and Y_2, $Y_4 \in \mathfrak{G}_2$; using this fact and the formulas Υ_1, Υ_2, we can easily construct a coordinatizing formula for the whole group $SL(3, \mathfrak{K})$. Arguments similar to those given earlier can be made to establish the steadfastness and perfectness of $\mathfrak{G}_1$ and $\mathfrak{G}_2$ (properly enriched) and of $\langle SL(3, \mathfrak{K}); A, A'\rangle$.

NOTES

([1]) The free variables and individual constants appearing in Φ should be required to have the property ρ, too; cf. [XV], §6.

([2]) It appears we must interpret the definition of perfect model absolutely: for any constructive numbering of the abstract model, every predicate recursive with respect to it has to be formular. Thus, the recursiveness of φ is needed not to show $\mathfrak{N}^{\varphi}$ is perfect, but to conclude β constructively numbers $\mathfrak{N}^{\chi}$ below.

([3]) These two formulas will not define $I + E_{13}$; the formulas

$$\mathrm{Twm}(\boldsymbol{Y}) = \boldsymbol{YB \approx BY} \,\&\, \boldsymbol{Y^2 \approx 1} \,\&\, \boldsymbol{Y \not\approx 1}\,,$$

$$\mathrm{Id}(\boldsymbol{W}) = \rho(\boldsymbol{W}) \,\&\, (\exists \boldsymbol{Y})(\mathrm{Twm}(\boldsymbol{Y}) \,\&\, \boldsymbol{W \approx A^{-1} Y A^{-1} Y})$$

will, as long as B is assumed to have distinct diagonal elements. Note that ρ requires *char* $(\mathfrak{K}) \neq 3$.

([4]) This formula has been corrected to allow for multiplication by I, the zero element of the field being constructed. The commutativity conditions for U, V are unnecessary in any case.

([5]) Since C is diagonal with distinct diagonal elements, it can be used to define $I + E_{13}$ and $I + E_{31}$.

CHAPTER 23

AXIOMATIZABLE CLASSES OF LOCALLY FREE ALGEBRAS OF VARIOUS TYPES

In [M8] the author formulated a theorem on the algorithmic decidability of the elementary theory of every finitely axiomatizable subclass in the class of locally absolutely free algebras with a given signature; a decision procedure was sketched there. This theorem is extended below to the classes of locally free algebras with symmetric basic operations, a detailed description of a corresponding decision algorithm simpler than the one in [M8] is given, and some new properties of these algebras are discovered.

§1. Locally absolutely free algebras

Let $\Sigma = \{f_1, ..., f_s\}$ be a set of symbols, distinct as labelled, to each of which corresponds some natural number n_i, the *rank* of the particular symbol $\boldsymbol{f}_i$. To prescribe an *algebra with signature* Σ we choose a nonempty set A for the base set of the algebra, and with every symbol $\boldsymbol{f}_i$ we associate a concrete n_i-ary operation f_i defined on A and taking values in A. The algebra $\mathfrak{A}$ with base A and basic operations $f_1, ..., f_s$ is denoted by $\langle A; f_1, ..., f_s\rangle$.

A formula $\Phi(\boldsymbol{x}_1, ..., \boldsymbol{x}_n)$ of first-order predicate logic (FOPL) with equality is called a *formula of signature* Σ iff it contains no symbols other than &, $\vee$, $\neg$, $\approx$, $\exists$, $\forall$,), (, comma, individual variables, and operation symbols from Σ. If Φ is a closed formula (*sentence, axiom*), then in every algebra with signature Σ it is either true or false. When Φ contains free individual variables $\boldsymbol{x}_1, ..., \boldsymbol{x}_n$, it has a definite truth-value in an algebra $\mathfrak{A}$ with signature Σ for each selection of values for the $\boldsymbol{x}_i$ in $\mathfrak{A}$ (i.e., in its base); $\Phi(\boldsymbol{x}_1, ..., \boldsymbol{x}_n)$ thus determines an n-ary predicate on $\mathfrak{A}$.

A collection of algebras with one and the same signature is called a *class* of algebras. A class $\mathfrak{K}$ of algebras with signature Σ is said to be (*first-order*) *axiomatizable* iff there exists a system S of FOPL sentences of signature Σ such that an algebra with signature Σ belongs to $\mathfrak{K}$ iff all the sentences in S are true in it. The class $\mathfrak{K}$ is *finitely axiomatizable* iff there exists a finite set S of sentences that has this property.

Suppose the class $\mathfrak{K}$ of algebras has the following properties: (i) $\mathfrak{K}$ contains all isomorphic images of $\mathfrak{K}$-algebras ($\mathfrak{K}$ is *abstract*); (ii) $\mathfrak{K}$ contains all subalgebras of every $\mathfrak{K}$-algebra ($\mathfrak{K}$ is *universal*); (iii) $\mathfrak{K}$ contains the direct (cartesian) product of any system of its members ($\mathfrak{K}$ is *multiplicatively closed*). Then in $\mathfrak{K}$ one can naturally define (cf., e.g., [11], [IV]) algebras prescribed by given systems of generators and defining relations and free algebras with given systems of free generators. Algebras that belong to $\mathfrak{K}$ and are free with respect to $\mathfrak{K}$ are called *$\mathfrak{K}$-free* algebras.

A $\mathfrak{K}$-algebra is called *locally $\mathfrak{K}$-free* iff every finitely generated subalgebra is $\mathfrak{K}$-free. Clearly, locally $\mathfrak{K}$-free algebras exist when all finitely generated subalgebras of $\mathfrak{K}$-free algebras are $\mathfrak{K}$-free, for then all $\mathfrak{K}$-free algebras are themselves locally $\mathfrak{K}$-free.

When $\mathfrak{K}$ is the class of all algebras with signature Σ, $\mathfrak{K}$-free algebras are called *absolutely free.* The absolutely free algebra with a given (finite or infinite) number of free generators is commonly presented in the following form.

We start with the set of distinct symbols $\boldsymbol{a}_1, \boldsymbol{a}_2, \ldots$, which we call terms of length l. Terms of greater length will be certain sequences (or strings) of the symbols $\boldsymbol{f}_i, \boldsymbol{a}_\alpha$,), (and are defined by recursion. Namely, suppose that $\mathfrak{a}_1, \ldots, \mathfrak{a}_{n_i}$ are terms of respective length $l_1, \ldots, l_{n_i}$, and that $\boldsymbol{f}_i$ is an n_i-ary operation symbol from Σ; then the string $\boldsymbol{f}_i(\mathfrak{a}_1, \ldots, \mathfrak{a}_{n_i})$ is considered to be a term of length $l_1 + \ldots + l_{n_i} + 1$. These are said to be *terms of signature* Σ (*Σ-terms*) in the symbols $\boldsymbol{a}_\alpha$ – sometimes thought of as variables, sometimes constants – that are used in their construction. $\varphi(\boldsymbol{a}_{\alpha_1}, \ldots, \boldsymbol{a}_{\alpha_r})$ will denote a term, all of whose variables occur among those indicated.

Let A be the set of all Σ-terms in the variables $\boldsymbol{a}_\alpha$. We define an algebra $\mathfrak{A}$ with signature Σ, base A, and basic operations f_i by putting

$$f_i(\mathfrak{a}_1, \ldots, \mathfrak{a}_{n_i}) = \boldsymbol{f}_i(\mathfrak{a}_1, \ldots, \mathfrak{a}_{n_i}) \quad (\mathfrak{a}_1, \ldots, \mathfrak{a}_{n_i} \in A)$$

for $i = 1, \ldots, s$. The algebra $\mathfrak{A} = \langle A; f_1, \ldots, f_s \rangle$ is the unique (up to isomorphism) absolutely free algebra with signature Σ and generating set $\{\boldsymbol{a}_1, \boldsymbol{a}_2, \ldots\}$. All algebras isomorphic to $\mathfrak{A}$ are also absolutely free.

Algebras whose signatures consist of a single binary operation symbol are called *groupoids*. As the sign for the groupoid operation we shall use the dot, but write ab instead of $a \cdot b$. In particular, the free groupoid with free generators $\boldsymbol{a}, \boldsymbol{b}$ can be represented as the set of all strings $\boldsymbol{a}, \boldsymbol{b}, \boldsymbol{ab}, (\boldsymbol{ab})\boldsymbol{a}, \boldsymbol{a}(\boldsymbol{ba}), \ldots$, with multiplication given by the rules $\boldsymbol{a} \cdot \boldsymbol{ba} = \boldsymbol{a}(\boldsymbol{ba}), \boldsymbol{ab} \cdot \boldsymbol{a} = (\boldsymbol{ab})\boldsymbol{a}$, etc.

From the above construction it is clear that in absolutely free algebras with signature $\Sigma = \{\boldsymbol{f}_1, \ldots, \boldsymbol{f}_s\}$ the following axioms are valid (here and elsewhere, initial universal quantifiers applying to the whole formula have been suppressed for clarity; (C) was omitted from [M8] by oversight):

(A) $$f_i(x_1, \dots, x_{n_i}) \approx f_i(y_i, \dots, y_{n_i}) \to x_1 \approx y_1 \,\&\, \dots \,\&\, x_{n_i} \approx y_{n_i}$$

$$(i = 1, \dots, s),$$

(B) $$f_i(x_1, \dots, x_{n_i}) \not\approx f_j(y_1, \dots, y_{n_j}) \qquad (i \neq j;\ i, j = 1, \dots, s),$$

(C) $\varphi(x, x_1, \dots, x_m) \not\approx x$, φ any term distinct from x in which x actually occurs.

It is clear that these axioms are true in every locally absolutely free algebra. Moreover, the converse holds.

Theorem 1: *In order that an algebra with signature* $\Sigma = \{f_1, \dots, f_s\}$ *be locally absolutely free, it is necessary and sufficient that it satisfy the axioms* (A), (B), (C).

The necessity of these conditions was noted to be obvious, so we only prove their sufficiency. Let $\mathfrak{A}$ be a Σ-algebra in which all the sentences (A), (B), (C) hold. We take an arbitrary finite subset $\{a_1, \dots, a_n\}$ of the base of $\mathfrak{A}$ and consider the subalgebra $\mathfrak{B}$ generated in $\mathfrak{A}$ by this set. From the set $\{a_1, \dots, a_n\}$ we successively delete those elements which are expressible in $\mathfrak{B}$ as the values of terms in constants designating the remaining elements.

Let $\{b_1, \dots, b_r\}$ be the resulting refined set of generators for $\mathfrak{B}$. We want to show that for any terms φ, ψ in $\boldsymbol{b}_1, \dots, \boldsymbol{b}_r$, the values φ^0, ψ^0 in $\mathfrak{B}$ of these terms are equal only if the terms themselves are equal (as strings). We proceed by induction on the minimum l of the lengths of the terms φ, ψ. For $l = 1$ the equation $\varphi^0 = \psi^0$ in question will have the form

$$\chi^0(b_1, \dots, b_r) = \boldsymbol{b}_j^0 = b_j\,.$$

If this equation holds, $\boldsymbol{b}_j$ cannot fail to appear in χ since $b_1, \dots, b_r$ have been chosen to be irredundant; hence, as a consequence of (C), χ must coincide with the term $\boldsymbol{b}_j$. For $l > 1$, the equation $\varphi^0 = \psi^0$ reduces by virtue of (A), (B) to several equations involving terms whose lengths are less than l. ∎

Theorem 1 implies that *the class of all locally absolutely free algebras is axiomatizable.* ∎

We shall prove this class cannot be finitely axiomatized. Let's consider, e.g., the groupoid $\mathfrak{G}$ with formal generators $\boldsymbol{a}, \boldsymbol{b}$ and the single defining relation

$$\boldsymbol{a} \approx \underbrace{((\boldsymbol{aa}) \ldots) \boldsymbol{a}}_{l+1 \text{ times}}$$

in the class of all groupoids. In $\mathfrak{G}$ the axiom (A) is clearly valid, while (B) is inapplicable. In addition, $\mathfrak{G}$ satisfies all the axioms (C) in which the term φ has length less than $2l$, but the sentence $\boldsymbol{x} \not\approx ((\boldsymbol{x}\boldsymbol{x}_1) \ldots) \boldsymbol{x}_l$ is not valid in $\mathfrak{G}$. It follows that the system of axioms (A), (B), (C) is not equivalent to any finite system of FOPL sentences. This argument works for any other non-empty signature, as well.

§2. Ordered groupoids

The groupoid $\mathfrak{G}$ is said to be *ordered* (*partially ordered*) iff there is prescribed a linear (a partial) ordering $\leqslant$ of its elements such that the axiom

(D) $\qquad \boldsymbol{x} < \boldsymbol{y} \rightarrow \boldsymbol{ux} < \boldsymbol{uy} \;\&\; \boldsymbol{xu} < \boldsymbol{yu}$

holds in $\mathfrak{G}$.

When the axiom

(E) $\qquad \boldsymbol{x} < \boldsymbol{xy} \;\&\; \boldsymbol{x} < \boldsymbol{yx}$

is valid in an ordered groupoid $\mathfrak{G}$, it is said to be *tightly ordered.*

A groupoid $\mathfrak{G}$ is *tightly orderable* iff it is possible to define a tight order on it.

Theorem 2: *Locally absolutely free groupoids are just those tightly orderable groupoids satisfying the axiom*

(F) $\qquad \boldsymbol{xy} \approx \boldsymbol{uv} \rightarrow \boldsymbol{x} \approx \boldsymbol{u} \;\&\; \boldsymbol{y} \approx \boldsymbol{v}\,.$

We begin the proof by showing every absolutely free groupoid $\mathfrak{G}$ can be tightly ordered. Let $a_1, a_2, \ldots$ be free generators of $\mathfrak{G}$. We linearly order this set arbitrarily. All elements of $\mathfrak{G}$ are uniquely representable as values of terms in the constants $\boldsymbol{a}_\alpha$. If the term φ is shorter than the term ψ, we set $\varphi^0 < \psi^0$. If the terms φ, ψ have the same length, while the values of all shorter terms have been ordered, and $\varphi^0 = \mathfrak{a}^0 \mathfrak{b}^0$, $\psi^0 = \mathfrak{c}^0 \mathfrak{d}^0$, then we put $\varphi^0 < \psi^0$ if either $\mathfrak{a}^0 < \mathfrak{c}^0$ or $\mathfrak{a}^0 = \mathfrak{c}^0$, $\mathfrak{b}^0 < \mathfrak{d}^0$. This linear ordering certainly satisfies (D), (E).

The property of tight orderability is quasiuniversal in the sense of [XI], §3.2. We just proved that locally absolutely free groupoids are locally tightly orderable. By the intrinsic local theorem (Theorem 6 of [XI]) this implies every locally absolutely free groupoid can be tightly ordered.

We have thus established the necessity of the condition in Theorem 2. The sufficiency is obvious, since for every term $\varphi(\boldsymbol{x}, \boldsymbol{x}_1, \ldots, \boldsymbol{x}_m)$ distinct from $\boldsymbol{x}$ but involving $\boldsymbol{x}$, and for all elements $a, a_1, \ldots, a_m$ of a tightly ordered groupoid $\mathfrak{G}$, we have $a < \varphi^0(a, a_1, \ldots, a_m)$ in $\mathfrak{G}$; thus (C) evidently holds in such groupoids. ■

The class of tightly ordered groupoids is finitely axiomatizable. Theorem 2 shows that by adding the one axiom (F), we obtain a finitely axiomatizable class of ordered groupoids that has as a projection the infinitely axiomatizable class of locally absolutely free groupoids.

In an obvious way Theorem 2 can be extended to arbitrary signatures if appropriate notions of strong orderability are defined.

§3. $\mathfrak{S}$-algebras

By analogy with commutative groupoids, in which the identity $\boldsymbol{xy} \approx \boldsymbol{yx}$ is valid, we can introduce the notion of a Σ-algebra with symmetry conditions on its basic operations for any signature $\Sigma = \{f_1, \ldots, f_s\}$. Suppose $\mathfrak{S}_1, \ldots, \mathfrak{S}_s$ are subgroups of the permutation groups of the corresponding sets $\{1, \ldots, n_1\}$, ..., $\{1, \ldots, n_s\}$. We say that a Σ-algebra $\mathfrak{A}$ is an *algebra with symmetry conditions* $\mathfrak{S} = \langle \mathfrak{S}_1, \ldots, \mathfrak{S}_s \rangle$ – or more briefly an $\mathfrak{S}$*-algebra* – iff the identities

$$f_i(\boldsymbol{x}_1, \ldots, \boldsymbol{x}_{n_i}) \approx f_i(\boldsymbol{x}_{1\pi}, \ldots, \boldsymbol{x}_{n_i\pi}) \quad (\pi \in \mathfrak{S}_i;\ i = 1, \ldots, s) \tag{1}$$

are valid in $\mathfrak{A}$.

The class of all $\mathfrak{S}$-algebras is determined by identities, so it contains free algebras, which we shall call $\mathfrak{S}$*-free* algebras. An algebra $\mathfrak{A}$, all of whose finitely generated subalgebras are $\mathfrak{S}$-free (so $\mathfrak{A}$ is an $\mathfrak{S}$-algebra), will be called *locally* $\mathfrak{S}$*-free.* An $\mathfrak{S}$-free algebra with an arbitrary number of free generators can be explicitly constructed as follows. Let A be the set of all terms of signature Σ in the desired number of individual symbols $\boldsymbol{a}_1, \boldsymbol{a}_2, \ldots$ Let $f_i(\mathfrak{a}_1, \ldots, \mathfrak{a}_{n_i})$ be a specific subterm of a term $\mathfrak{b} \in A$, and let $\pi \in \mathfrak{S}_i$; if we replace $f_i(\mathfrak{a}_1, \ldots, \mathfrak{a}_{n_i})$ inside $\mathfrak{b}$ with the term $f_i(\mathfrak{a}_{1\pi}, \ldots, \mathfrak{a}_{n_i\pi})$, we obtain a term said to result from $\mathfrak{b}$ by an elementary transformation. Two terms are said to be equivalent iff one can be obtained from the other by a finite chain of elementary transformations. Let $[\mathfrak{b}]$ denote the class of terms equivalent to $\mathfrak{b}$, and put

$$f_i([\mathfrak{b}_1], \ldots, [\mathfrak{b}_{n_i}]) = [f_i(\mathfrak{b}_1, \ldots, \mathfrak{b}_{n_i})] \quad (\mathfrak{b}_j \in A;\ i = 1, \ldots, s); \tag{2}$$

let $\mathfrak{A}$ be the $\mathfrak{S}$-algebra whose elements are the equivalence classes of terms in A, and whose basic operations f_i are defined by (2). Then $\mathfrak{A}$ is $\mathfrak{S}$-free with generators $[a_1]$, $[a_2]$, ... In other words, the elements of an $\mathfrak{S}$-free algebra with free generators $a_1, a_2, \ldots$ can be represented by all possible Σ-terms in the $\boldsymbol{a}_\alpha$; the operations f_i work the same on these terms as in the absolutely free case, but terms connected by elementary transformations represent the same element of the algebra.

From this it is clear that the axioms (B), (C), given for locally absolutely free algebras, also hold in any locally $\mathfrak{S}$-free algebra. The axioms (A) are not in general true in locally $\mathfrak{S}$-free algebras, but the following axioms are:

$$(\mathrm{A}^{\mathfrak{S}}) \qquad f_i(x_1, \ldots, x_{n_i}) \approx f_i(y_1, \ldots, y_{n_i}) \to$$
$$\to \bigvee_{\pi \in \mathfrak{S}_i} (x_1 \approx y_{1\pi} \;\&\; \ldots \;\&\; x_{n_i} \approx y_{n_i\pi}) \quad (i = 1, \ldots, s)\,.$$

Theorem 1 immediately generalizes to

Theorem 1$^{\mathfrak{S}}$: *In order that an algebra $\mathfrak{A}$ with signature Σ be locally $\mathfrak{S}$-free, it is necessary and sufficient that $\mathfrak{A}$ satisfy the axioms* ($\mathrm{A}^{\mathfrak{S}}$), (1), (B), (C). ∎

We can similarly extend Theorem 2 to locally $\mathfrak{S}$-free algebras. It suffices to replace the condition (F) with the conditions ($\mathrm{A}^{\mathfrak{S}}$) and (1) in its formulation.

From now on we shall have in mind a fixed signature $\Sigma = \{f_1, \ldots, f_s\}$ and a fixed system of symmetry conditions $\mathfrak{S} = \langle \mathfrak{S}_1, \ldots, \mathfrak{S}_s \rangle$ for Σ, unless we specifically state otherwise. In connection with this, we shall understand "locally free" to mean "locally $\mathfrak{S}$-free".

§4. Special formulas

We use these fixed abbreviations:

$$\mathrm{N}_i(y) = (\forall x_1 \ldots x_{n_i})(y \not\approx f_i(x_1, \ldots, x_{n_i})) \qquad (i = 1, \ldots, s)\,,$$

$$\mathrm{N}_{\mathfrak{p}}(y) = \mathrm{N}_{i_1}(y) \;\&\; \ldots \;\&\; \mathrm{N}_{i_k}(y)$$
$$(\mathfrak{p} = \{i_1, \ldots, i_k\};\; 1 \leqslant i_1 < \ldots < i_k \leqslant s)\,,$$

$$\mathrm{E}_{\mathfrak{p}}^{m} = (\exists y_1 \ldots y_m)(\mathrm{N}_{\mathfrak{p}}(y_1) \;\&\; \ldots \;\&\; \mathrm{N}_{\mathfrak{p}}(y_m) \;\&\; \mathop{\&}_{i \neq j} y_i \not\approx y_j)$$
$$(m = 2, 3, \ldots)\,,$$

$$\mathbf{E}_{\mathfrak{p}}^1 = (\exists y)\,\mathbf{N}_{\mathfrak{p}}(y)\,,$$

$$\mathbf{D}_{\mathfrak{p}}^m = \mathbf{E}_{\mathfrak{p}}^m \quad (m = 1, 2, \dots).$$

The signs &, V are used for the conjunction and disjunction of several formulas.

Elements a in an algebra $\mathfrak{A}$ for which $\mathbf{N}_{\mathfrak{p}}(a)$ is true are called $\mathfrak{p}$-*indecomposable.* The truth in $\mathfrak{A}$ of the sentence $\mathbf{E}_{\mathfrak{p}}^m$ means $\mathfrak{A}$ contains at least m different $\mathfrak{p}$-indecomposable elements, while the truth of $\mathbf{D}_{\mathfrak{p}}^m$ means $\mathfrak{A}$ contains fewer than m $\mathfrak{p}$-indecomposable elements. In particular, when $\mathbf{D}_{\mathfrak{p}}^1$ is true in $\mathfrak{A}$, every element is $\mathfrak{p}$-decomposable.

Let $\mathfrak{A}$ be a locally free algebra, and suppose $\mathfrak{p} \neq \{1, 2, \dots, s\}$, $i \notin \mathfrak{p}$, $1 \leqslant i \leqslant s$. The values in $\mathfrak{A}$ of each Σ-term of the form $f_i(\mathfrak{a}_1, \dots, \mathfrak{a}_{n_i})$ in the FOPL variables are all $\mathfrak{p}$-indecomposable elements. The number of elements in $\mathfrak{A}$ that are the values of even one such term is infinite. We conclude that for all values of $\mathfrak{p}$ except $\{1, \dots, s\}$, all the sentences $\mathbf{E}_{\mathfrak{p}}^m$ are true in $\mathfrak{A}$, but every sentence $\mathbf{D}_{\mathfrak{p}}^m$ is false.

Similarly, in the locally free algebra $\mathfrak{A}$ a sentence of the form $(\forall)\mathbf{N}_{\mathfrak{p}}(f_i(\mathfrak{a}_1, \dots, \mathfrak{a}_{n_i}))$ – where $\mathfrak{a}_1, \dots, \mathfrak{a}_{n_i}$ are Σ-terms in the FOPL variables – is true if $i \notin \mathfrak{p}$, but is false if $i \in \mathfrak{p}$.

For the sentences $\mathbf{D}_{\mathfrak{p}}^m$, $\mathbf{E}_{\mathfrak{p}}^m$ with $\mathfrak{p} = \{1, \dots, s\}$ we introduce the abbreviations $\mathbf{D}^m$, $\mathbf{E}^m$; we let T, F stand for the sentences $\mathbf{D}^1 \vee \mathbf{E}^1$, $\mathbf{D}^1$ & $\mathbf{E}^1$.

We now define *special formulas* as formulas of the form

$$\Phi(x_1, \dots, x_n) = (\exists y_1 \dots y_m)\Big(\mathop{\&}_{i=1}^{p} x_{\alpha_i} \approx \varphi_i \;\&\; \mathop{\&}_{j} x_{\beta_j} \not\approx \psi_j \;\&$$

$$\& \mathop{\&}_{k} y_{\gamma_k} \not\approx \chi_k \;\&\; \mathop{\&}_{l} \mathbf{N}_{\delta_l}(y_{\epsilon_l})\Big)\,, \tag{3}$$

where φ_i, ψ_j, χ_k are Σ-terms in the variables $x_1, \dots, x_n, y_1, \dots, y_m$, the variables $x_{\alpha_1}, \dots, x_{\alpha_p}$ do not occur in any of the terms φ_i, ψ_j, χ_k and are distinct from the variables $x_{\beta_1}, x_{\beta_2}, \dots$, and the indices have the possible ranges: $1 \leqslant \alpha_i, \beta_j \leqslant n$; $1 \leqslant \gamma_k, \epsilon_l \leqslant m$; $1 \leqslant \delta_l \leqslant s$.

Along with special formulas (3) that actually contain free individual variables $x_1, \dots, x_n$, we shall consider special closed formulas of the form (3), i.e., sentences of the form

$$(\exists y_1 \dots y_m)\Big(\mathop{\&}_{k} y_{\gamma_k} \not\approx \chi_k \;\&\; \mathop{\&}_{l} \mathbf{N}_{\delta_l}(y_{\epsilon_l})\Big)\,. \tag{4}$$

We shall be interested in the interpretation of certain FOPL formulas in an algebra $\mathfrak{A}$ when the quantifiers are limited in range to some subset of the base of $\mathfrak{A}$. To provide for this formally we introduce the symbols $(\forall y \in T)$, $(\exists y \in T)$ for writing restricted quantifiers; these are read as "for every element y in T" and "there exists an element y in T such that". If in the algebra $\mathfrak{A}$ the set T is defined by a FOPL formula $\tau(x)$, and $\Psi(y)$ is any FOPL formula, then the equivalences

$$(\forall y \in T)\, \Psi(y) \leftrightarrow (y)(\tau(y) \rightarrow \Psi(y))\,,$$

$$(\exists y \in T)\, \Psi(y) \leftrightarrow (\exists y)(\tau(y)\, \&\, \Psi(y))$$

are valid in $\mathfrak{A}$.

We formulate as lemmas several observations that will be helpful in the sequel.

Lemma 1: *Let $\varphi(x_1, ..., x_p)$ be a term of length d in which x_1 actually occurs. Then in every locally free algebra $\mathfrak{A}$ and for every $a \in \mathfrak{A}$, the equation $a = \varphi^0(x_1, ..., x_p)$ is solvable for at most n^d different elements x_1, where $n = max(n_1, ..., n_s)$.*

For suppose $\mathfrak{A}$ is locally $\mathfrak{S}$-free and the term φ has the form $f_i(x_1, ..., x_{n_i})$. If $\langle x_1^*, ..., x_{n_i}^* \rangle$ is a solution to $a = \varphi^0$, then by $(\mathrm{A}^{\mathfrak{S}})$ every other solution has the form $\langle x_{1\pi}^*, ..., x_{n_i\pi}^* \rangle$ $(\pi \in \mathfrak{S}_i)$. So $a = \varphi^0$ is solvable only when x_1 is given one of the values $x_1^*, ..., x_{n_i}^*$, which do not exceed n in number. Now suppose φ has the form $f_i(\mathfrak{a}_1, ..., \mathfrak{a}_{n_i})$; as part of an inductive hypothesis we can assume the lemma applies to the shorter terms $\mathfrak{a}_1, ..., \mathfrak{a}_{n_i}$. Suppose x_1 occurs in $\mathfrak{a}_j$. By the first argument, the equation $a = \varphi^0$ is solvable for at most n different values of $\mathfrak{a}_j$, say $a_1, ..., a_n$ to be generous. Each of the equations $a_k = \mathfrak{a}_j^0$ has solutions for at most n^{d-1} elements x_1. Therefore, $a = \varphi^0$ is solvable for not more than n^d different elements x_1. ∎

Lemma 2: *Let $\varphi_1, ..., \varphi_r$ be terms in $y_1, ..., y_m, y_{m+1}, ..., y_t$, let d be the maximum of the lengths of these terms, and let $n = max(n_1, ..., n_s)$. In the locally free algebra $\mathfrak{A}$ we choose subsets $T_1, ..., T_m$ of the base and consider a formula*

$$(\forall y_1 \in T_1) \ldots (\forall y_m \in T_m)(y_{\alpha_1} \approx \varphi_1 \vee \ldots \vee y_{\alpha_r} \approx \varphi_r)$$

$$(1 \leqslant \alpha_j \leqslant t) \qquad (5)$$

in which no term φ_i coincides with the variable y_{α_i}, and each subformula $y_{\alpha_i} \approx \varphi$ contains at least one occurrence of one of the bound variables $y_1, ..., y_m$. If each of the sets $T_1, ..., T_m$ contains more than rn^d elements, then (5) is equivalent in $\mathfrak{A}$ to F. *In particular, the formula (5) is identically false in $\mathfrak{A}$ when the quantifiers are unrestricted.*

We prove this by induction on the number of quantifiers in (5). Let $m = 1$; by reordering term indices, we can put the original formula into the form

$$(\forall y_1 \in T_1)(y_{\alpha_1} \approx \varphi_1 \vee ... \vee y_{\alpha_p} \approx \varphi_p \vee y_1 \approx \varphi_{p+1} \vee ... \vee y_1 \approx \varphi_r)$$

$$(2 \leqslant \alpha_i \leqslant r), \qquad (6)$$

where the terms $\varphi_1, ..., \varphi_p$ explicitly involve y_1. We fix values for $y_2, ..., y_t$ in $\mathfrak{A}$. According to Lemma 1, each of the equations $y^0_{\alpha_i} = \varphi^0_i$ can then have no more than n^d solutions for y^0_1; each equation $y^0_1 = \varphi^0_{p+j}$ has by (C) at most one solution for y^0_1. Hence, the disjunction $y_{\alpha_1} \approx \varphi_1 \vee ... \vee y_1 \approx \varphi_r$ will be true in $\mathfrak{A}$ for at most rn^d values of y_1. By assumption the set T_1 contains more than rn^d elements. It follows that (6) is false in $\mathfrak{A}$ for any choice of values for the free variables.

Now we consider a formula of the form (5) with $m \geqslant 2$, assuming meanwhile that Lemma 3 holds for all similar formulas with fewer quantifiers. If every subformula $y_{\alpha_i} \approx \varphi_i$ in (5) contains an occurrence of one of the variables $y_2, ..., y_m$, then by the inductive hypothesis the formula

$$(\forall y_2 \in T_2) ... (\forall y_m \in T_m)(y_{\alpha_1} \approx \varphi_1 \vee ... \vee y_{\alpha_r} \approx \varphi_r)$$

– and with it, the original formula – will be identically false in $\mathfrak{A}$. In the contrary case we put (5) into the form

$$(\forall y_1 \in T_1)(y_{\beta_1} \approx \varphi_{\gamma_1} \vee ... \vee y_{\beta_u} \approx \varphi_{\gamma_u} \vee$$

$$\vee (\forall y_2 \in T_2) ... (\forall y_m \in T_m)(y_{\delta_1} \approx \varphi_{\epsilon_1} \vee ... \vee y_{\delta_v} \approx \varphi_{\epsilon_v})),$$

where $y_2, ..., y_m$ do not occur in the subformulas $y_{\beta_i} \approx \varphi_{\gamma_i}$, but the $y_{\delta_j} \approx \varphi_{\epsilon_j}$ – the remaining disjuncts in (5) – each have an occurrence of at least one of the variables $y_2, ..., y_m$. The subformula

$$(\forall y_2 \in T_2) ... (\forall y_m \in T_m)(y_{\delta_1} \approx \varphi_{\epsilon_1} \vee ... \vee y_{\delta_v} \approx \varphi_{\epsilon_v}) \qquad (7)$$

satisfies the conditions of the lemma and has but $m-1$ quantifiers. Therefore, (7) is equivalent to F in $\mathfrak{A}$, so (5) is equivalent to

$$(\forall y_1 \in T_1)(y_{\beta_1} \approx \varphi_{\gamma_1} \vee \dots \vee y_{\beta_u} \approx \varphi_{\gamma_u}),$$

which is also identically false in $\mathfrak{A}$. ∎

§5. **Standard formulas**

Formulas constructed from expressions of the form $\mathbf{D}^m$, $\mathbf{E}^m$, $\mathbf{N}_i(x)$, and the special formulas (3) with the aid of the connectives & and ∨ only are called *standard formulas.* In particular, standard sentences are built up with the aid of &, ∨ from sentences of the form (4), $\mathbf{D}^m$, $\mathbf{E}^m$.

Theorem 3: *There exists an algorithm* A *whereby for every* FOPL *formula* $\Phi(x_1, \dots, x_n)$ *of signature* Σ, *one can construct a standard formula* $\Phi^{\mathsf{A}}(x_1, \dots, x_n)$ *that is equivalent to* $\Phi(x_1, \dots, x_n)$ *in the class of all locally* $\mathfrak{S}$*-free algebras.*

The algorithm A can also be termed a procedure for reducing a formula to standard form. We start by describing a procedure for the reduction to standard form of formulas of a certain sort, the so-called E-formulas; these include the usual existential FOPL formulas. Then we delineate the reduction of negations of standard formulas to standard form. Alternating these procedures enables us to reduce any Σ-formula to standard form.

E-*formulas* are those formulas constructed with the aid of &, ∨, ∃ alone from expressions of the form $\mathbf{N}_i(x)$, $\mathbf{D}^m$, $\mathbf{E}^m$, $\varphi \approx \psi$, $\varphi \not\approx \psi$, where φ, ψ are Σ-terms in the FOPL variables. ([1]) In particular, all existential prenex formulas built up from terms and all standard formulas are E-formulas.

The reduction to standard form of an arbitrary E-formula $\Phi(x_1, \dots, x_n)$ is carried out as follows. By extracting all quantifiers in Φ, we obtain a formula

$$(\exists y_1 \cdots y_m)\, \Phi_0(x_1, \dots, x_n, y_1, \dots, y_m),$$

where Φ_0 is a positive propositional combination of expressions of the form $\varphi \approx \psi$, $\varphi \not\approx \psi$, $\mathbf{N}_i(x_\alpha)$, $\mathbf{N}_i(y_\beta)$, $\mathbf{D}^k$, $\mathbf{E}^k$. Rewriting Φ_0 in disjunctive form and distributing the quantifiers by means of the logical equivalence

$$(\exists y)(\Gamma \vee \Delta) \leftrightarrow (\exists y)\,\Gamma \vee (\exists y)\,\Delta,$$

we see that we have only to handle formulas of the type $(\exists y_1 \dots y_m)\Psi$, where Ψ is a conjunction of atoms of the form $\varphi \approx \psi$, $\varphi \not\approx \psi$, $\mathbf{N}_i(x_\alpha)$, $\mathbf{N}_i(y_\beta)$, $\mathbf{D}^k$, $\mathbf{E}^k$. In some of these conjuncts the bound variables $y_1, \dots, y_m$ may not appear. Let the conjunction of these be Ψ_0, the conjunction of the remainder Ψ_1. Replacing $(\exists y_1 \dots y_m)\Psi$ with $\Psi_0 \,\&\, (\exists y_1 \dots y_m)\Psi_1$, we proceed to the transformation of the expression $(\exists y_1 \dots y_m)\Psi_1$.

If among the conjuncts in Ψ_1 is an expression of the form $\varphi \approx \psi$, and the lengths of these terms are greater than 1, then the conjunct has the form

$$f_i(\mathfrak{a}_1, \dots, \mathfrak{a}_{n_i}) \approx f_j(\mathfrak{b}_1, \dots, \mathfrak{b}_{n_j}) . \tag{8}$$

If $i \neq j$, (8) is equivalent to F in all locally free algebras by (B). And if $i = j$, the formula (8) is equivalent by $(\mathrm{A}^{\mathfrak{S}})$, (1) to

$$\bigvee_{\pi \in \mathfrak{S}_i} (\mathfrak{a}_1 \approx \mathfrak{b}_{1\pi} \,\&\, \dots \,\&\, \mathfrak{a}_{n_i} \approx \mathfrak{b}_{n_i\pi}) ,$$

which comprises shorter terms.

Suppose now we meet a conjunct $x_\alpha \approx \varphi$ in Ψ_1. If $\varphi = x_\alpha$, the conjunct is logically valid and can be dropped. If the term φ contains x_α but is distinct from it, then by (C) this conjunct – and with it the whole formula $(\exists y_1 \dots y_m)\Psi_1$ – is equivalent to F in the class under consideration. But if x_α does not occur in φ, we substitute φ for x_α at each occurrence in Ψ_1 except in the conjunct $x_\alpha \approx \varphi$ we are studying.

Should in Ψ_1 we come to a conjunct $y_\beta \approx \varphi$ with y_β occurring in φ, then we either drop it or replace it with F, judging by whether φ is y_β or not.

Now assume $y_\beta \approx \varphi$ appears in Ψ_1, and y_β does not occur in φ. By replacing y_β with φ at all other occurrences and renumbering the bound variables, we can put $(\exists y_1 \dots y_m)\Psi_1$ into the form

$$(\exists y_1 \dots y_{m-1})(\exists y_m)(y_m \approx \varphi \,\&\, \Psi_2) ;$$

this in turn will be equivalent to $(\exists y_1 \dots y_{m-1})\Psi_2$, from which y_m is wholly absent. If $\mathbf{N}_i(y_\beta)$ occurs in Ψ_1, then the transformation would convert this conjunct into a more complicated formula $\mathbf{N}_i(\varphi)$, should the length of φ be greater than 1. But in view of what we observed in §4, this member can be replaced with T or F, as determined by the structure of φ.

Finally, suppose we are confronted in Ψ_1 with the conjunct

$$f_i(\mathfrak{a}_1, \dots, \mathfrak{a}_{n_i}) \not\approx f_j(\mathfrak{b}_1, \dots, \mathfrak{b}_{n_j}) .$$

For $i \neq j$ it yields its place to T by virtue of (B). For $i = j$ this conjunct can be replaced with

$$\underset{\pi \in \mathfrak{S}_i}{\&} (\mathfrak{a}_1 \not\approx \mathfrak{b}_{1\pi} \vee \ldots \vee \mathfrak{a}_{n_i} \not\approx \mathfrak{b}_{n_i\pi}),$$

which contains shorter terms.

We can view the above as one "pass" through the general procedure; after putting the new formula in disjunctive form with distributed quantifiers, we apply these "elementary" transformations again. Since the effect of each elementary transformation is to produce formulas with either shorter or fewer terms, after a finite number of steps the procedure must terminate with the entire original formula reduced to standard form.

§6. The reduction of negations of standard formulas

Standard formulas are positive propositional combinations of sentences of the form $\mathbf{D}^m$, $\mathbf{E}^m$ and special formulas of the form (3). Since the negations of the sentences $\mathbf{D}^m$, $\mathbf{E}^m$ are equivalent to the sentences $\mathbf{E}^m$, $\mathbf{D}^m$, the negation of a standard formula is quickly converted into a positive propositional combination of expressions $\mathbf{D}^m$, $\mathbf{E}^m$ and negations of special formulas (3). Therefore, the general problem of reducing the negation of a standard formula to standard form simplifies to the narrower task of reducing to standard form the negation

$$(\forall \boldsymbol{y}_1 \ldots \boldsymbol{y}_m)(\bigvee_i \boldsymbol{x}_{\alpha_i} \not\approx \varphi_i \vee \bigvee_j \boldsymbol{x}_{\beta_j} \approx \psi_j \vee \bigvee_k \boldsymbol{y}_{\gamma_k} \approx \chi_k \vee \bigvee_l \neg \mathrm{N}_{\delta_l}(\boldsymbol{y}_{\epsilon_l})) \tag{9}$$

of an arbitrary special formula (3).

Since we already know from §5 how to reduce any E-formula to standard form, it suffices to find an E-formula equivalent to (9) in every locally $\mathfrak{S}$-free algebra.

Suppose among the terms $\varphi_1, \varphi_2, \ldots$ in (9) the term φ_i has maximal length. If this length is greater than 1, then φ_i has the form $f_q(\mathfrak{a}_1, \ldots, \mathfrak{a}_{n_q})$. Let Υ be the disjunction under quantification in (9) apart from the subformula $\boldsymbol{x}_{\alpha_i} \not\approx \varphi_i$. Then we can represent (9) in the equivalent form

$$\mathrm{N}_q(\boldsymbol{x}_{\alpha_i}) \vee (\exists \mathbf{z}_1 \ldots \mathbf{z}_{n_q})(\boldsymbol{x}_{\alpha_i} \approx f_q(\mathbf{z}_1, \ldots, \mathbf{z}_{n_q}) \,\&$$
$$\& \, (\forall \boldsymbol{y}_1 \ldots \boldsymbol{y}_m)(f_q(\mathbf{z}_1, \ldots, \mathbf{z}_{n_q}) \not\approx f_q(\mathfrak{a}_1, \ldots, \mathfrak{a}_{n_q}) \vee \Upsilon)).$$

Thus the matter reduces to transforming the subformula

$$(\forall y_1 \cdots y_m)(f_q(\mathfrak{z}_1, \ldots, \mathfrak{z}_{n_q}) \not\approx f_q(\mathfrak{a}_1, \ldots, \mathfrak{a}_{n_q}) \vee \Upsilon)$$

into E-form. By $(A^{\mathfrak{S}})$ and (1) this formula is equivalent to the conjunction of all the formulas

$$(\forall y_1 \cdots y_m)(\mathfrak{z}_{1\pi} \not\approx \mathfrak{a}_1 \vee \ldots \vee \mathfrak{z}_{n_q\pi} \not\approx \mathfrak{a}_{n_q} \vee \Upsilon) \quad (\pi \in \mathfrak{S}_q)\,. \tag{10}$$

The new formulas (10) result from (9) on replacing $x_{\alpha_i} \not\approx \varphi_i$ with certain atoms of the form $\mathfrak{z} \not\approx \mathfrak{a}$ in which the term $\mathfrak{a}$ is shorter than φ_i. Therefore, by applying this transformation a finite number of times, we can further simplify the general problem to reducing formulas of the form (9) in which there are either no inequations or only inequations of the type $x_\alpha \not\approx y_\beta$. With appropriate labeling of the variables, such a formula looks like

$$(\forall y_1 \cdots y_m)(x_{11} \not\approx y_1 \vee \ldots \vee x_{1r_1} \not\approx y_1 \vee \ldots \vee$$
$$\vee\, x_{p1} \not\approx y_p \vee \ldots \vee x_{pr_p} \not\approx y_p \vee \Omega(x; y_1, \ldots, y_m)\,,$$

which is obviously equivalent to

$$x_{11} \not\approx x_{12} \vee \ldots \vee x_{11} \not\approx x_{1r_1} \vee \ldots \vee x_{p1} \not\approx x_{p2} \vee \ldots \vee$$
$$\vee\, x_{p1} \not\approx x_{pr_p} \vee (\forall y_{p+1} \cdots y_m)\, \Omega(x; x_{11}, \ldots, x_{p1}, y_{p+1}, \ldots, y_m)$$

Consequently, we are home free if we can reduce to E-form an arbitrary formula of the sort

$$(\forall y_1 \cdots y_m)(\bigvee_j x_{\beta_j} \approx \psi_j \vee \bigvee_k y_{\gamma_k} \approx \chi_k \vee \bigvee_l \neg \mathrm{N}_{\delta_l}(y_{\epsilon_l}))\,, \tag{11}$$

where the ψ_j, χ_k are terms in the variables $x_1, \ldots, x_q, y_1, \ldots, y_m$, and the indices have the possible ranges: $1 \leqslant \beta_j \leqslant q$; $1 \leqslant \gamma_k, \epsilon_l \leqslant m$; $1 \leqslant \delta_l \leqslant s$. We shall abbreviate (11) as $(\forall y_1 \ldots y_m)\Omega$.

If Ω contains disjuncts of the form $x_\beta \approx x_\beta$ or $y_\gamma \approx y_\gamma$, then (11) is equivalent to T; we shall assume Ω contains none of these. Furthermore, all the subformulas not containing any occurrences of the bound variables can be brought beyond the scope of the quantifiers, so that we can assume there are no such subformulas in (11). If after this purging there are also no disjuncts of the form $\neg \mathrm{N}_\delta(y_\epsilon)$ in Ω, then by Lemma 2 the formula (11) is identically false in all locally free algebras. On the other hand, if Ω has no disjuncts of the form $\boldsymbol{x}_\beta \approx \psi$, nor of the form $\boldsymbol{y}_\gamma \approx \chi$, then (11) reduces to a disjunction of formulas of the form $(\boldsymbol{y}) \neg \mathrm{N}_{\mathfrak{p}}(\boldsymbol{y})$ with $\mathfrak{p} \subseteq \{1, ..., s\}$, i.e., either to F or to $\mathbf{D}^1$ (cf. §4).

Thus we can suppose that Ω contains disjuncts of the form $\neg \mathrm{N}_\delta(\boldsymbol{y}_\epsilon)$, as well as of the other form, and that none of the latter resemble $\boldsymbol{x}_\beta \approx \boldsymbol{x}_\beta$ or $\boldsymbol{y}_\gamma \approx \boldsymbol{y}_\gamma$, but each of them contains an occurrence of at least one of the variables $\boldsymbol{y}_1, ..., \boldsymbol{y}_m$. If we let $\boldsymbol{y}_1, ..., \boldsymbol{y}_t$ be the variables among $\boldsymbol{y}_1, ..., \boldsymbol{y}_m$ that actually occur in disjuncts $\neg \mathrm{N}_\delta(\boldsymbol{y}_\epsilon)$ in Ω and replace the free variables $\boldsymbol{x}_1, ..., \boldsymbol{x}_q$ with $\boldsymbol{y}_{m+1}, ..., \boldsymbol{y}_{m+q}$, we can convert (11) into a formula of the form

$$(\forall \boldsymbol{y}_1 \dots \boldsymbol{y}_t)(\mathop{\&}_l \mathrm{N}_{\delta_l}(\boldsymbol{y}_{\epsilon_l}) \rightarrow (\forall \boldsymbol{y}_{t+1} \dots \boldsymbol{y}_m)(\bigvee_k \boldsymbol{y}_{\gamma_k} \approx \chi_k)) , \tag{12}$$

where the χ_k are terms in $\boldsymbol{y}_1, ..., \boldsymbol{y}_{m+q}$, the γ_k belong to $\{1, ..., t\}$.

The subformula $(\forall \boldsymbol{y}_{t+1} \dots \boldsymbol{y}_m)(\bigvee_k \boldsymbol{y}_{\gamma_k} \approx \chi_k)$ in (12) can be written equivalently as

$$\bigvee_u (\boldsymbol{y}_{\gamma_u} \approx \chi_u) \vee (\forall \boldsymbol{y}_{t+1} \dots \boldsymbol{y}_m)(\bigvee_v \boldsymbol{y}_{\gamma_v} \approx \chi_v) ,$$

where the $\boldsymbol{y}_{\gamma_u} \approx \chi_u$ are the disjuncts in which $\boldsymbol{y}_{t+1}, ..., \boldsymbol{y}_m$ do not occur. According to Lemma 2, the expression $(\forall \boldsymbol{y}_{t+1} \dots \boldsymbol{y}_m)(\bigvee_v \boldsymbol{y}_{\gamma_v} \approx \chi_v)$ is identically false in every locally free algebra, so that (12) is equivalent to the expression

$$(\forall \boldsymbol{y}_1 \dots \boldsymbol{y}_t)(\bigvee_l \mathrm{N}_{\delta_l}(\boldsymbol{y}_{\epsilon_l}) \rightarrow \bigvee_u \boldsymbol{y}_{\gamma_u} \approx \chi_u) . \tag{13}$$

By collecting the subformulas $\mathrm{N}_{\delta_l}(\boldsymbol{y}_{\epsilon_l})$ in (13) that have the same variable $\boldsymbol{y}_{\epsilon_l}$, we can rewrite (13) in the form

$$(\forall \boldsymbol{y}_1, ..., \boldsymbol{y}_t)(\mathrm{N}_{\mathfrak{p}_1}(\boldsymbol{y}_1) \,\&\, ... \,\&\, \mathrm{N}_{\mathfrak{p}_t}(\boldsymbol{y}_t) \rightarrow \bigvee_u \boldsymbol{y}_{\gamma_u} \approx \chi_u) . \tag{14}$$

Let T_i generically denote the set of elements in an algebra for which the formula $\mathrm{N}_{\mathfrak{p}_i}(\boldsymbol{y})$ is true $(i = 1, ..., t)$. Using these abbreviations, we can change indices a bit and write (14) as

$$\Theta(\boldsymbol{y}_{m+1}, ..., \boldsymbol{y}_{m+q}) = (\forall \boldsymbol{y}_1 \in T_1) \dots (\forall \boldsymbol{y}_t \in T_t)(\boldsymbol{y}_{\gamma_1} \approx \chi_1 \vee \dots \vee$$

$$\vee\, \boldsymbol{y}_{\gamma_r} \approx \chi_r) .$$

By Lemma 2 the formula Θ is equivalent to F in the locally free algebra $\mathfrak{A}$ whenever the subsets $T_1, \ldots, T_t$ of the base of $\mathfrak{A}$ each contain more than rn^d elements; here, n is the maximum of the ranks of the signature symbols, and d is the maximum of the lengths of the terms $\chi_1, \ldots, \chi_r$. Therefore, for $p = rn^d$ the equivalence

$$\Theta \leftrightarrow \urcorner(\mathbf{D}_{\mathfrak{p}_1}^{p+1} \;\&\; \Theta) \vee \ldots \vee (\mathbf{D}_{\mathfrak{p}_t}^{p+1} \;\&\; \Theta)$$

is valid in every locally free algebra (²).

If we replace every expression of the form $\mathbf{D}_{\mathfrak{p}}^{p+1}$ with the equivalent sentence

$$\mathbf{D}_{\mathfrak{p}}^{1} \vee (\mathbf{D}_{\mathfrak{p}}^{2} \;\&\; \mathbf{E}_{\mathfrak{p}}^{1}) \vee \ldots \vee (\mathbf{D}_{\mathfrak{p}}^{p+1} \;\&\; \mathbf{E}_{\mathfrak{p}}^{p})$$

and note that for each i, $\mathbf{D}_{\mathfrak{p}_i}^{1}$ is true iff T_i is empty, we see that Θ is equivalent to the disjunction of all the formulas

$$\mathbf{D}_{\mathfrak{p}_i}^{j+1} \;\&\; \mathbf{E}_{\mathfrak{p}_i}^{j} \;\&\; \Theta \qquad (i = 1, \ldots, t; j = 1, \ldots, rn^d)\,. \tag{15}$$

Let us scrutinize one of the formulas (15), say $\mathbf{D}_{\mathfrak{p}_1}^{j+1} \;\&\; \mathbf{E}_{\mathfrak{p}_1}^{j} \;\&\; \Theta$. For any choice of values for the free variables in a locally free algebra $\mathfrak{A}$, this formula is true in $\mathfrak{A}$ iff T_1 consists of exactly j elements, each of these satisfying

$$\Xi(y_1) = (\forall y_2 \in T_2) \ldots (\forall y_t \in T_t)(y_{\gamma_1} \approx \chi_1 \vee \ldots \vee y_{\gamma_r} \approx \chi_r)\,.$$

Therefore, our representative of (15) is equivalent in every locally free algebra to the formula

$$\mathbf{D}_{\mathfrak{p}_1}^{j+1} \;\&\; (\exists w_1 \ldots w_j)(\mathbf{N}_{\mathfrak{p}_1}(w_1) \;\&\; \ldots \;\&\; \mathbf{N}_{\mathfrak{p}_1}(w_j) \;\&$$

$$\& \mathop{\&}_{\zeta \neq \eta} w_\zeta \not\approx w_\eta \;\&\; \Xi(w_1) \;\&\; \ldots \;\&\; \Xi(w_j))\,.$$

The subformulas $\Xi(w_\zeta)$ have the same structure as Θ, but have fewer quantifiers. Repeated applications of this last transformation will eventually lead us to an E-formula. ■

§7. The reduction of closed formulas

The algorithm A described in §§5 and 6 enables us to construct for every formula Φ of signature Σ, a formula Φ^{A} equivalent to Φ in every locally $\mathfrak{S}$-free algebra and fabricated with the aid of &, ∨ from sentences $\mathbf{D}^m$, $\mathbf{E}^m$ and special formulas (3), which have a more complex structure. For formulas Φ with no free variables there is the stronger

Theorem 4: *There exists an algorithm* B *whereby for every sentence* Φ *of signature* Σ, *one can construct a sentence* Φ^{B}, *using only* &, ∨ *and the sentences* $\mathbf{D}^m$, $\mathbf{E}^m$ ($m = 1, 2, \dots$), *such that* Φ *and* Φ^{B} *are equivalent in the class of all locally* $\mathfrak{S}$*-free algebras.*

First we apply A to Φ; this results in a positive propositional combination Φ^{A} of the $\mathbf{D}^m$, $\mathbf{E}^m$, and special formulas (3) that is equivalent to Φ in every locally $\mathfrak{S}$-free algebra. Since Φ and Φ^{A} have no free variables, the special formulas occurring in Φ^{A} must have the form

$$(\exists y_1 \dots y_m)(\underset{\alpha,\beta}{\&}\, y_\alpha \not\approx \chi_{\alpha\beta} \,\&\, \mathbf{N}_{\mathfrak{p}_1}(y_1) \,\&\, \dots \,\&\, \mathbf{N}_{\mathfrak{p}_r}(y_r))\,, \tag{16}$$

where the $\chi_{\alpha\beta}$ are terms in $y_1, \dots, y_m$, and $r \leqslant m$. We just have to show (16) can be expressed in terms of the $\mathbf{D}^m$ and $\mathbf{E}^m$.

Suppose one of the sets $\mathfrak{p}_1, \dots, \mathfrak{p}_r$ – e.g. $\mathfrak{p}_1$ – is not equal to $\{1, \dots, s\}$; indeed, suppose $i \in \{1, \dots, s\} \sim \mathfrak{p}_1$. We shall see that under these conditions, (16) is equivalent to the shorter sentence

$$(\exists y_2 \dots y_m)(\underset{\gamma,\delta}{\&}\, y_\gamma \not\approx \chi_{\gamma\delta} \,\&\, \mathbf{N}_{\mathfrak{p}_2}(y_2) \,\&\, \dots \,\&\, \mathbf{N}_{\mathfrak{p}_r}(y_r)) \tag{17}$$

produced by removing the quantifier $(\exists y_1)$ from (16) and then deleting all subformulas in which y_1 occurs. First we note that (17) is a logical consequence of (16). On the other hand, suppose the sentence (17) is true in the locally free algebra $\mathfrak{A}$ and let $\langle y_2^0, \dots, y_m^0\rangle$ be a sequence of elements of $\mathfrak{A}$ that satisfies the quantifier-free matrix of (17) in $\mathfrak{A}$. We define terms $\mathfrak{c}_n$ ($n = 1, 2, \dots$) in the variable y_1 by setting

$$\mathfrak{c}_1 = y_1$$

$$\mathfrak{c}_{n+1} = f_i(\mathfrak{c}_n, \dots, \mathfrak{c}_n) \quad (n = 1, 2, \dots)\,.$$

By an observation made in §4, we know the sentence $(y_1)\mathbf{N}_{\mathfrak{p}_1}(\mathfrak{c}_n)$ is true in $\mathfrak{A}$ for all $n \geqslant 2$. Consider the inequations $y_\xi \not\approx \chi_{\xi\eta}$ discarded from (16) in forming (17). Since each of these contains at least one occurrence of y_1, Lemma 2 tells us there are only finitely many $y_1^0 \in \mathfrak{A}$ such that $\langle y_1^0, y_2^0, \dots, y_m^0\rangle$ satisfies the disjunction $\bigvee_{\xi,\eta} y_\xi \approx \chi_{\xi\eta}$. Consequently, for any fixed $a \in \mathfrak{A}$ and for sufficiently large n, the sequence $\langle \mathfrak{c}_n^0(a), y_2^0, \dots, y_m^0\rangle$ satisfies the conjunction $\underset{\xi,\eta}{\&} y_\xi \not\approx \chi_{\xi\eta} \,\&\, \mathrm{N}_{\mathfrak{p}_1}(y_1)$; this proves (16) is true in $\mathfrak{A}$.

If in (17) we find a subformula $\mathrm{N}_{\mathfrak{q}}(y)$ with $\mathfrak{q} \neq \{1, ..., s\}$, it, too, can be eliminated, etc. This means we can assume (16) has the form

$$(\exists y_1 \dots y_m)(\underset{\alpha,\beta}{\&} y_\alpha \not\approx \chi_{\alpha\beta} \,\&\, \mathrm{N}_{\mathfrak{p}}(y_1) \,\&\, \dots \,\&\, \mathrm{N}_{\mathfrak{p}}(y_r))\,, \tag{18}$$

where $\mathfrak{p} = \{1, ..., s\}$. But

$$\mathrm{N}_{\mathfrak{p}}(y) \to y \not\approx f_i(w_1, ..., w_{n_i})$$

is valid for every i in every locally free algebra. Hence, in (18) we can cross off all inequations whose right-hand sides have length greater than 1. The resulting formula looks like

$$(\exists y_1 \dots y_m)(\underset{\gamma,\delta}{\&} y_\gamma \not\approx y_\delta \,\&\, \mathrm{N}_{\mathfrak{p}}(y_1) \,\&\, \dots \,\&\, \mathrm{N}_{\mathfrak{p}}(y_r))\,;$$

for some $t \leqslant r$, this is equivalent in every infinite Σ-algebra to $\mathbf{E}^t$ – or to T or F in extreme cases. ■

Two algebras with signature Σ are said to be *elementarily* (or *arithmetically*) *equivalent* iff every FOPL sentence of signature Σ that is true in one of them is also true in the other. To know which of the sentences $\mathbf{D}^m$, $\mathbf{E}^m$ are true in an algebra and which are false, we have only to know how many elements of the algebra are indecomposable. Therefore, from Theorem 4 we immediately obtain the

Corollary: *In order that two locally $\mathfrak{S}$-free algebras be elementarily equivalent, it is necessary and sufficient that either each algebra have an infinite number of indecomposable elements, or both algebras have the same finite number* (*possibly* 0) *of indecomposables.* ■

As an example of nonisomorphic elementarily equivalent algebras we point out the groupoid $\mathfrak{A}$ with generators $a_1, a_2, ...$ and defining relations

$$a_n \approx a_{n+1} a_{n+1} \qquad (n = 1, 2, ...)$$

and the groupoid $\mathfrak{B}$ with generators $b_1, b_2, ...$ and defining relations

$$b_n \approx (b_{n+1} b_{n+1})\, b_{n+1} \qquad (n = 1, 2, ...)\,.$$

Both groupoids are locally absolutely free and have no indecomposable elements; this makes them elementarily equivalent. But they can't be isomorphic, because in $\mathfrak{A}$ there is an element a_1 whose successive "square roots" can be extracted *ad infinitum,* whereas there is no such element in $\mathfrak{B}$.

Suppose $\mathfrak{C}$ is an absolutely free groupoid with a countably infinite number of generators, and $\mathfrak{D}$ is an arbitrary locally absolutely free groupoid. The axioms (A), (B), (C) characterizing locally absolutely free algebras are Horn sentences. Therefore, the direct product of any system of locally absolutely free algebras is locally absolutely free. In particular, the direct product $\mathfrak{C} \times \mathfrak{D}$ is a locally absolutely free groupoid. Since $\mathfrak{C} \times \mathfrak{D}$, like $\mathfrak{C}$, has an infinite number of indecomposable elements, $\mathfrak{C} \times \mathfrak{D}$ and $\mathfrak{C}$ are elementarily equivalent.

Let $\mathcal{K}$ be an arbitrary class of algebras with signature Σ. The *elementary theory* $\mathrm{T}(\mathcal{K})$ of the class $\mathcal{K}$ is the collection of all those Σ-sentences that are true in every $\mathcal{K}$-algebra. We make no great distinction between the elementary theory $\mathbf{T}(\mathcal{K})$ and the set $\#\mathrm{T}(\mathcal{K})$ of Gödel numbers of the sentences in $\mathbf{T}(\mathcal{K})$. Thus we say an elementary theory $\mathbf{T}$ is *recursive* (*recursively decidable*) or *primitive recursive* as the set $\#\mathrm{T}$ is recursive or primitive recursive.

A positive natural number n is *spectral* for the class $\mathcal{K}$ iff there is an algebra in $\mathcal{K}$ with exactly $n-1$ indecomposables; the number 0 is *spectral* for $\mathcal{K}$ when some $\mathcal{K}$-algebra has infinitely many indecomposables. The set $\sigma(\mathcal{K})$ consisting of all the numbers spectral for $\mathcal{K}$ is called the *spectrum* of $\mathcal{K}$.

We introduce the abbreviations

$$\mathbf{C}^0 = \mathbf{D}^1, \quad \mathbf{C}^n = \mathbf{E}^n \,\&\, \mathbf{D}^{n+1} \qquad (n = 1, 2, \ldots).$$

The sentence $\mathbf{C}^n$ is true in an algebra $\mathfrak{A}$ iff $\mathfrak{A}$ contains exactly n indecomposable elements. This implies that for $n \geqslant 0$,

$$n + 1 \in \sigma(\mathcal{K}) \Leftrightarrow \neg\, \mathbf{C}^n \notin \mathrm{T}(\mathcal{K}). \tag{19}$$

By Theorem 4 we can algorithmically construct from every sentence Φ, a sentence Φ^* equivalent to Φ in every locally $\mathfrak{S}$-free algebra and formed by taking the disjunction of conjunctions of various sentences of the form $\mathbf{D}^m$, $\mathbf{E}^m$. But $\mathbf{E}^m \,\&\, \mathbf{D}^{m+n+1}$ is logically equivalent to

$$\mathbf{C}^m \vee \mathbf{C}^{m+1} \vee \ldots \vee \mathbf{C}^{m+n};$$

hence, each sentence Φ reduces over the indicated class to one of the forms

$$\mathbf{C}^{m_1} \vee \mathbf{C}^{m_2} \vee \ldots \vee \mathbf{C}^{m_r} \vee \mathbf{E}^{m_r},\ \mathbf{C}^{m_1} \vee \mathbf{C}^{m_2} \vee \ldots \vee \mathbf{C}^{m_r},\ \mathsf{T}, \mathsf{F}, \qquad (20)$$

where the m_i are natural numbers satisfying $0 \leqslant m_1 < \ldots < m_r$.

From this it follows that every finitely axiomatizable subclass in the class of locally $\mathfrak{S}$-free algebras ([3]) has a spectrum similar to one of

$$\emptyset,\ \{n_1, \ldots, n_k\},\ \{0, n_1, \ldots, n_k, n_k+1, \ldots\} \quad (1 \leqslant n_1 < \ldots < n_k)\,;$$

the spectrum of an (infinitely) axiomatizable subclass can be an arbitrary finite set of natural numbers or an arbitrary infinite set of natural numbers that contains 0.

Theorem 5: *Let $\mathfrak{K}$ be an arbitrary class of locally $\mathfrak{S}$-free algebras. Its elementary theory* $\mathrm{T}(\mathfrak{K})$ *is recursive (primitive recursive) iff its spectrum* $\sigma(\mathfrak{K})$ *is recursive (primitive recursive). In particular, the elementary theory of any finitely axiomatizable subclass of locally $\mathfrak{S}$-free algebras, as well as of any single locally $\mathfrak{S}$-free algebra, is primitive recursive.*

The assertions of Theorem 5 regarding recursiveness follow immediately from Theorem 4. And if $\mathrm{T}(\mathfrak{K})$ is primitive recursive, then (19) shows $\sigma(\mathfrak{K})$ is primitive recursive, too. So let us assume $\sigma(\mathfrak{K})$ is primitive recursive. The algorithm reducing an arbitrary Σ-sentence to the canonical form (20) works by applying a series of elementary transformations repeatedly to successive results, so that the sentence with number n is transformed into the sentence with number $\kappa(n)$ after one pass. From the description of the elementary steps in the algorithm we see that κ is a primitive recursive function. Furthermore, it is not hard to produce a primitive recursive function λ such that no more than $\lambda(n)$ passes are required to reduce the sentence with number n to canonical form (20). Let ξ be the characteristic function of the set of sentences in canonical form; thus $\xi(q) = 1$ when q is the number of a sentence (20), and $\xi(q) = 0$ for all other q. Now we define a binary numerical function ρ by requiring

$$\rho(0, n) = n\,,$$

$$\rho(j+1, n) = \kappa(\rho(j, n))\,.$$

The functions ξ and ρ are clearly primitive recursive. This means the function ω defined by

$$\omega(n) = \rho(\min_x (\chi(\rho(x, n)) = 1 \quad \text{and} \quad x \leqslant \lambda(n)), n)$$

is also primitive recursive; $\omega(n)$ is just the number of the canonical sentence to which the Σ-sentence with number n is reduced.

The numbers $r, m_1, \ldots, m_r$ in (20) are computable by primitive recursive functions from the number q of a canonical sentence. The canonical sentence Ψ belongs to $T(\mathcal{K})$ iff $\sigma(\mathcal{K})$ is included in the spectrum of the class of all algebras satisfying Ψ. This inclusion can be expressed by conditions of the form

$$x \in \sigma(\mathcal{K}) \text{ and } x \leqslant m_r + 1 \Rightarrow x = m_1 + 1 \text{ or} \ldots \text{or } x = m_r + 1,$$

etc. Therefore, the characteristic function ξ_0 of the set of all canonical sentences true in every $\mathcal{K}$-algebra is primitive recursive. The characteristic function $\xi_0\omega$ of $T(\mathcal{K})$ is primitive recursive, too. ■

We can easily establish similarly that the elementary theory of a first-order axiomatizable class $\mathcal{K}$ of locally $\mathfrak{S}$-free algebras is recursively enumerable (and $\mathcal{K}$ is recursively axiomatizable) iff the complement of $\sigma(\mathcal{K})$ is recursively enumerable.

As a final observation we note that although there are recursively axiomatizable classes of locally free algebras with undecidable elementary theories, no class of locally free algebras can have an essentially undecidable theory, because every class has a subclass with a finite spectrum, which is consequently decidable.

NOTES

([1]) These formulas are considered to be atomic from now on.

([2]) By §4 we can assume $\mathfrak{p}_i = \{1, \ldots, s\}$ $(i = 1, \ldots, t)$.

([3]) As the author notes, this usage is reserved for classes $\mathcal{K} \subseteq \mathcal{L}$ such that $\mathcal{K}$ consists of all $\mathcal{L}$-algebras satisfying a certain finite system of FOPL axioms; this does not mean that $\mathcal{K}$, as an independent class, is finitely axiomatizable.

CHAPTER 24

RECURSIVE ABELIAN GROUPS

Adopting the terminology of [XVIII], we call a group $\mathfrak{G}$ *constructive* iff it is accompanied by a single-valued mapping α from some recursive set D_α of natural numbers onto $\mathfrak{G}$ such that there are binary recursive general functions θ, f with the properties

$$\theta(x,y) = 1 \Leftrightarrow \alpha x = \alpha y\ , \quad \alpha f(x,y) = \alpha x \cdot \alpha y \quad (x, y \in D_\alpha)\ .$$

A map α with the properties mentioned is called a *constructive numbering* of the group $\mathfrak{G}$. Groups for which constructive numberings exist are called *constructivizable* or *computable.* General problems naturally arise: determining what constructive numberings are admitted by given abstract groups, which subgroups of a given constructive group are recursive or recursively enumerable relative to the given numbering, etc. Below we indicate several initial results in this direction for abelian groups.

§1. From the observations in [XVIII], §3.1 we know that for proving the general affirmative assertions of the theorems below, we can assume the elements of an infinite constructive group are the natural numbers, a general recursive two-place function serves as the group operation, and the numbering is trivial. Examples of constructive abelian groups will be given in the form of groups with defining relations. Moreover, we assume such a group is numbered in the following standard fashion: if in specifying the group the symbols $\boldsymbol{x}_1$, $\boldsymbol{x}_2$, ... are taken for generators, then the element a has the number n iff a can be written in the form $\boldsymbol{x}_1^{n_1} \cdot \boldsymbol{x}_2^{n_2} \cdot \ldots \cdot \boldsymbol{x}_k^{n_k}$, where n is the standard number in the style of [XVIII], §4 of the sequence $\langle n_1, n_2, \ldots, n_k \rangle$ of integers.

Theorem 1: *The periodic part of a constructive group is recursively enumerable. There exist constructive abelian groups whose periodic parts are not recursive.*

An algorithm for enumerating the periodic part is constructed in an obvious fashion. As a constructive abelian group with nonrecursive periodic part we can offer the abelian group with generators $\boldsymbol{x}_1, \boldsymbol{x}_2, \ldots$ and defining relations

$$\boldsymbol{x}^n_{\lambda(n)} \approx \boldsymbol{1} \quad (n = 1, 2, \ldots),$$

where λ is a 1–1 general recursive function whose set of values is not recursive. ■

If $\mathfrak{G}$ is a group, we denote by $\Pi(\mathfrak{G})$ the set of all primes p for which there is an element in $\mathfrak{G}$ of order p. The *p-primary component* $\mathfrak{G}_{(p)}$ of $\mathfrak{G}$ is the set consisting of every element of $\mathfrak{G}$, including the identity element, whose order has the form p^s for some $s = 0, 1, 2, \ldots$.

Theorem 2: *All primary components of a constructive periodic group* $\mathfrak{G}$ *are recursive; the set* $\Pi(\mathfrak{G})$ *is recursively enumerable. For every recursively enumerable set* Π_0 *of prime numbers, there exists a constructive abelian group* $\mathfrak{G}_0$ *such that* $\Pi_0 = \Pi(\mathfrak{G}_0)$. ■

Let $\mathfrak{G}_1, \mathfrak{G}_2, \ldots$ form a sequence of numbered groups, and let α_n be the numbering of $\mathfrak{G}_n$ with number set D_n. We obtain the *standard numbering* of the direct sum $\mathfrak{G} = \mathfrak{G}_1 \oplus \mathfrak{G}_2 \oplus \ldots$ by taking n to be a number of the element $\alpha_1 n_1 \cdot \alpha_2 n_2 \cdot \ldots \cdot \alpha_k n$ of $\mathfrak{G}$ iff n is the standard number of the sequence $\langle n_1, \ldots, n_k \rangle$.

If the characteristic function $\chi_n(x)$ of the set D_n and the functions $\theta_n(x, y)$, $f_n(x, y)$ for $\mathfrak{G}_n$ are recursive in x, y, n, then the direct sum $\mathfrak{G}_1 \oplus \mathfrak{G}_2 \oplus \ldots$ with its standard numbering is a constructive group. In particular, if $\mathfrak{G}$ is a constructive periodic abelian group and $\Pi(\mathfrak{G}) = \{p_{\nu(1)}, p_{\nu(2)}, \ldots\}$, where ν is a 1–1 recursive function, then $\mathfrak{G}$ is recursively equivalent to the direct sum of its primary subgroups $\mathfrak{G}_{(p_{\nu(1)})}, \mathfrak{G}_{(p_{\nu(2)})}, \ldots$ taken with the standard numbering.

§2. Numberings α and β of an algebra $\mathfrak{A}$ are said to be *autoequivalent* iff there exists an abstract automorphism σ of the algebra $\mathfrak{A}$ such that $\sigma\alpha$ and β are recursively equivalent. A computable algebra is *autostable* iff all its constructive numberings are autoequivalent. A computable algebra, all of whose constructive numberings are recursively equivalent, was called *recursively steadfast* in [XXII]. Steadfastness certainly implies autostability, but not conversely. There are no more than a countable number of recursively equivalent constructive numberings of a given abstract algebra. Automorphisms translate constructive numberings into constructive numberings. Therefore, *an algebra with an uncountable group of automorphisms cannot be steadfast.* ■

The abelian group $\mathfrak{A}_{p,n}$ with generators $\boldsymbol{a}_{i,j}$ and defining relations $\boldsymbol{a}^p_{i,j+1} \approx \boldsymbol{a}_{i,j}, \boldsymbol{a}^p_{i,1} \approx \boldsymbol{1}$ (p prime; $i = 1, \ldots, n$; $j = 1, 2, \ldots$) is called the p-divisible,

p-primary abelian group of rank n. Its standard numbering is constructive. The automorphism group of $\mathfrak{A}_{p,n}$ is uncountable, so $\mathfrak{A}_{p,n}$ cannot be steadfast. At the same time it is easy to prove *for every prime p and every $n = 1, 2, \ldots$, the group $\mathfrak{A}_{p,n}$ is autostable.* ■

It is known that completely divisible, torsion-free abelian groups are computable and steadfast ([XXII], §2). The completely divisible and torsion-free abelian group of countable rank $\mathfrak{R}_\infty$ is the direct sum of rank 1 groups and thus admits a constructive numbering. The automorphism group of $\mathfrak{R}_\infty$ is uncountable, so $\mathfrak{R}_\infty$ is not steadfast.

A constructive abelian group $\mathfrak{G}$ is said to have *algorithmic linear dependence* iff there exists an algorithm whereby for every choice of natural numbers $n_1, \ldots, n_k$, one can tell whether or not the elements of $\mathfrak{G}$ with numbers $n_1, \ldots, n_k$ are linearly dependent.

Theorem 3: *In order for a constructive torsion-free abelian group of countable rank to have algorithmic linear dependence, it is necessary and sufficient that it have a recursively enumerable basis. Suppose the group $\mathfrak{R}_\infty$ has a recursively enumerable basis relative to the constructive numbering α. A constructive numbering β of $\mathfrak{R}_\infty$ is autoequivalent to α iff $\mathfrak{R}_\infty$ has a recursively enumerable basis under β. There are constructive numberings of $\mathfrak{R}_\infty$ for which there are no recursively enumerable bases.* ■

To present an example of a completely divisible and torsion-free constructive abelian group in which linear dependence is not algorithmic, we represent $\mathfrak{R}_\infty$ as the additive group of linear forms in the variables $x_1, x_2, \ldots$ with rational coefficients and take the standard numbering of these forms. Let ν be a 1–1 recursive unary function with nonrecursive range. The subspace $\mathfrak{H}$ generated in $\mathfrak{R}_\infty$ by the forms $ix_{2\nu(i)} - x_{2\nu(i)+1}$ $(i = 0, 1, \ldots)$ is easily seen to be recursive. Therefore, the numbering of $\mathfrak{R}_\infty$ induces a constructive numbering of the factor group $\mathfrak{R}_\infty/\mathfrak{H}$. There is no algorithm for linear dependence in $\mathfrak{R}_\infty/\mathfrak{H}$, for the question of the linear dependence of $[x_{2n}]$, $[x_{2n+1}]$ in $\mathfrak{R}_\infty/\mathfrak{H}$ is equivalent to the question of the membership of n in the set of values of ν, which was assumed not to be algorithmically decidable.

Theorem 3 implies, in particular, that *the completely divisible and torsion-free abelian group of countable rank, although computable, is not autostable.* ■

The indicated numbering of $\mathfrak{R}_\infty/\mathfrak{H}$ also gives a negative answer to the question of whether there exists an algorithm whereby in every constructive torsion-free abelian group of rank 2 we can find a pair of basis elements. More precisely: let $U(n, x, y)$ be Kleene's universal partial recursive function, which for different values of n gives all possible binary partial recursive functions. We ask: are there general recursive functions $\varphi(n)$, $\psi(n)$ such that if for a number n,

$U(n, x, y)$ defines a group operation on the set of natural numbers turning it into a completely divisible and torsion-free abelian group of rank 2, then $\varphi(n)$, $\psi(n)$ are linearly independent elements of this group? By Theorem 3 the answer is negative. ([1])

§3. Let $\mathfrak{G}$ be a constructive torsion-free abelian group of rank r with given basis elements $g_1, ..., g_r$. For each $g \in \mathfrak{G}$ let us find integers $m, m_1, ..., m_r$ such that $mg = m_1 g_1 + ... + m_r g_r$, and let us put the element g into correspondence with the form

$$\frac{m_1}{m} + \frac{m_2}{m} x_2 + ... + \frac{m_r}{m} x_r$$

in the independent variables $x_2, ..., x_r$, thereby obtaining a recursive mapping of $\mathfrak{G}$ into the numbered group $\mathfrak{R}_r$ of all these rational linear r-forms. Therefore, *every constructive torsion-free abelian group of rank r is recursively equivalent to an appropriate recursively enumerable subgroup of* $\mathfrak{R}_r$ *containing* $1, x_2, ..., x_r$. ■

A similar result with respect to $\mathfrak{R}_\infty$, naturally numbered, holds for all constructive torsion-free abelian groups with recursively enumerable bases.

We now want to examine in greater detail the subgroups of the group $\mathfrak{R}_1$ of rational numbers under addition. Let $\mathfrak{G}$ be a subgroup of $\mathfrak{R}_1$ that contains the number 1. Let $\Delta(\mathfrak{G})$ be the set of all pairs $\langle i, n \rangle$ of natural numbers for which $p_i^{-n} \in \mathfrak{G}$, where p_i is the ith prime.

Lemma: *A subgroup* $1 \in \mathfrak{G} \subseteq \mathfrak{R}_1$ *is recursive (recursively enumerable) iff the set* $\Delta(\mathfrak{G})$ *is recursive (recursively enumerable).* ■

In the theory of groups one ordinarily describes a subgroup $\mathfrak{G}$ of $\mathfrak{R}_1$ containing 1 by means of its characteristic, a certain sequence of the form $\delta(\mathfrak{G}) = \langle \delta_0, \delta_1, \delta_2 \rangle$, where each δ_i is either a natural number or the symbol ∞. The passage from $\Delta(\mathfrak{G})$ to $\delta(\mathfrak{G})$ is accomplished by the rule: $\delta_i = n$ if $p_i^{-n} \in \mathfrak{G}$, but $p_i^{-n-1} \notin \mathfrak{G}$, and $\delta_i = \infty$ when $p_i^{-n} \in \mathfrak{G}$ for $n = 1, 2, ...$.

We introduce a numerical partial function d by taking $d(i)$ to undefined if $\delta_i = \infty$, and putting $d(i) = \delta_i$ for other values of δ_i. The function d, like the sequence δ, is called the *characteristic* of the subgroup $\mathfrak{G}$. The connection between $\Delta(\mathfrak{G})$ and the corresponding characteristic d is expressed by the relation

$$d(i) = \mu x(\langle i, x \rangle \notin \Delta(\mathfrak{G})) - 1 ,$$

from which directly follows

Theorem 4: *The subgroup $\mathfrak{G}$ is recursive iff its characteristic is representable in the form*

$$d(i) = \mu x(f(i, x) = 0)$$

for some general recursive function f.

The subgroup $\mathfrak{G}$ is recursively enumerable iff its characteristic can be represented in the form

$$d(i) = \mu x(f(i, x) \quad \text{is undefined})$$

for an appropriate partial recursive function f. ■

The number n is called an *ordinary point* of the subgroup $\mathfrak{G}$ iff $\delta_n \neq \infty$.

The set of ordinary points of a recursive subgroup $1 \in \mathfrak{G} \subseteq \mathfrak{R}_1$ is recursively enumerable. Every recursively enumerable set of natural numbers is the set of ordinary points of some appropriate recursive subgroup of $\mathfrak{R}_1$ containing 1.

For if $f(x)$ is a recursive function, then the subgroup $\mathfrak{G}$ with the characteristic

$$d(i) = \mu x(f(x) = i)$$

is, by Theorem 4, recursive and has the range of f as its set of ordinary points. ■

Remark: *Let d be the characteristic of a subgroup $1 \in \mathfrak{G} \subseteq \mathfrak{R}_1$. The set $E^{(n)}$ of those i such that d(i) is either undefined or not less than n is recursive (recursively enumerable) when $\mathfrak{G}$ is recursive (recursively enumerable).* ■

As an example we note the subgroup $\mathfrak{G}'$ with characteristic

$$d(i) = \mu x(U(i, x) \quad \text{is undefined}),$$

where $U(i, x)$ is the Kleene universal function. The set $E^{(n)}$ is just the set of numbers of those functions defined at the points $0, 1, \ldots, n-1$. By a theorem of Rice the set $E^{(n)}$ cannot be recursive. On the basis of Theorem 4 and the remark, we see that $\mathfrak{G}$ is a nonrecursive, recursively enumerable subgroup of $\mathfrak{R}_1$.

NOTE

([1]) The example certainly shows the set of n described is not recursively enumerable.

CHAPTER 25

SETS WITH COMPLETE NUMBERINGS

To Andreĭ Nikolaevič Kolmogorov
on the occasion of
his sixtieth birthday

The fundamental fact of the theory of algorithms is the existence of a two-place partial recursive function (**pr**-function) $U(n, x)$ from which for the various fixed values of the variable n we obtain all unary **pr**-functions. By taking n to be the number of the function U_n defined by $U_n(x) = U(n, x)$, we get the numbering of the set of all unary **pr**-functions that is induced by the universal function U. Modifying the function U in a special way results in certain Kleene and Post numberings (cf. [XVIII]) of functions and sets that have a series of important special properties.

This sort of numbering can be constructed not only for the set of all unary **pr**-functions and the set of all recursively enumerable sets (or **re**-sets), but also for various subsets of these two. To begin with, we introduce those numberings such that for every natural number n, there exists an algorithm for computing the values of the function with number n. A.N. Kolmogorov advanced a program (cf. [174]) for systematically studying the properties of these computable numberings of families of functions and sets. The essential part of this program has been accomplished by V.A. Uspenskiĭ [174–176].

Nevertheless, the study of constructive algebras, begun in roughly the same years, convincingly shows that not only numberings of sets of functions are of interest but numberings of arbitrary sets of objects, as well. Starting from this more general point of view, the author presented the basis for a general theory of numbered sets in the first part [XVIII] of this continuing survey. Among other concepts, the notion of a complete numbering of an arbitrary set was introduced; to a certain degree it generalized the Kleene numbering. In the present article we give a new definition of a complete numbering that differs from the earlier, although a completely numbered set in the new sense will also be completely numbered in the old sense. The theory of complete numberings in the new sense, however, will prove to be richer. In particular, the new theory enables us to unite many important results of Myhill [111], Rogers [141], Mučnik [109], and Smullyan [154] that previously seemed unrelated. The exposition of this theory occupies §§1–5 of this article.

In §6 we describe the structure of families of **pr**-functions whose Kleene numbers form **re**-sets. This part has its source in Rice [126], in which a similar problem was first considered and essentially solved in full for the Post numbering of the collection of **re**-sets. The specific details for the solution of this problem for the Kleene numbering of the **pr**-functions were given by Uspenskiĭ [174]. Our presentation parallels that of Rice [126].

As a supplement, in §§ 7–9 we analyze examples of numbered families of **pr**-functions.

Although the present material is related to the contents of [XVIII], its presentation below does not assume the reader's familiarity with that survey.

§1. Complete numberings

A single-valued map α from a nonempty subset N_α of the set N of natural numbers 0, 1, ... onto a set A of nondescript objects is called a *numbering* of A. If $\alpha n = a$, then n is called an *α-number* of the element a ($n \in N_\alpha$, $a \in A$). The set A together with some numbering α of it is called a *numbered set* and can be written as $\langle A, \alpha \rangle$. The numbering α is *simple* iff $N_\alpha = N$. The numberings we consider below will almost always be simple.

Basic Definition: *A simple numbering α of a set A is* complete *iff there is a special element $e \in A$ such that for every unary partial recursive function f, there exists a unary general recursive function g satisfying the condition*

$$\alpha g(x) = \begin{cases} \alpha f(x) & \text{if } f(x) \text{ is defined,} \\ e & \text{otherwise.} \end{cases} \tag{1}$$

The fundamental example of a complete numbering is the Kleene numbering of the set of all unary **pr**-functions. It is obtained in the following manner. The number

$$\mathfrak{c}(m, n) = \tfrac{1}{2}((m+n)^2 + 3m + n)$$

is called the *Cantor number* of the pair $\langle m, n \rangle$ of natural numbers. By $\mathfrak{l}(x)$, $\mathfrak{r}(x)$ we denote the numbers satisfying the relation $\mathfrak{c}(\mathfrak{l}(x), \mathfrak{r}(x)) = x$.

It is easy to see that

$$[m, n] = \mathfrak{c}(\mathfrak{l}(m), \mathfrak{c}(\mathfrak{r}(m), n))$$

also defines a 1–1 numbering of the set of all pairs of natural numbers.

Let

$$[x_1, x_2, ..., x_s] = [[x_1, x_2], x_3, ..., x_s] \quad (s = 3, 4, ...) .$$

If

$$[x_1, ..., x_s] = n ,$$

then we set

$$[n]_{si} = x_i \quad (i = 1, ..., s; \quad s = 2, 3, ...) .$$

Suppose $U(n, x)$ is a **pr**-function universal for all unary **pr**-functions in x. We put

$$K^{(1)}(n, x) = U(\mathfrak{l}(n), \mathfrak{c}(\mathfrak{r}(n), x) ,$$

$$K^{(s+1)}(x_0, x_1, ..., x_{s+1}) = K^{(s)}([x_0, x_1], x_2, ..., x_{s+1}) \quad (s = 1, 2, ...).$$

Easy reasoning shows $K^{(s)}(n, x_1, ..., x_s)$ is a universal function with parameter n for all s-ary partial recursive functions.

The functions $K^{(s)}$ are called the *Kleene universal functions.*

By letting the number n correspond to the s-ary function $K_n^{(s)} = K^{(s)}(n, x_1, ..., x_s)$, we obtain the *Kleene numbering* $\kappa^{(s)}$ of the set $\mathcal{F}_{\mathrm{pr}}^{(s)}$ of all s-ary **pr**-functions. We write $\langle \mathcal{F}_{\mathrm{pr}}^{(1)}, \kappa^{(1)} \rangle$ more briefly as $\langle \mathcal{F}_{\mathrm{pr}}, \kappa \rangle$ and take the function defined nowhere as the special element e.

Let W_n be the set of all values of the function $K_n^{(1)}$. The map $\omega: n \to W_n$ is called the *Post numbering* of the set $\mathcal{W}$ of all recursively enumerable sets. For ω we take e to be the empty set $\emptyset$.

Let $\langle A, \alpha \rangle$ be a numbered set. If M is a set of natural numbers, αM will denote the α-image of M, the subset of A consisting of all elements, at least one of whose α-numbers belongs to M. If $B \subseteq A$, then $\alpha^{-1}B$ is the set of all α-numbers of the elements of B. A tilde denotes the complement (in N or in A, as the case may be). E.g., $\omega^{-1}e$ (i.e., $\omega^{-1}\{e\}$) is the set of all Post numbers of the empty set, and $\kappa^{-1}\tilde{e}$ is the set of all Kleene numbers of all functions with nonempty domains of definition.

Theorem 1.1: *The Kleene numberings* $\kappa^{(s)}$ $(s = 1, 2, \dots)$ *and the Post numbering* ω *are complete numberings, and the sets* $\kappa^{(s)-1}\tilde{e}$ *and* $\omega^{-1}\tilde{e}$ *are all recursively enumerable.*

We prove completeness for κ alone, since the completeness of the other numberings is an immediate consequence of the completeness of κ.

Let f be a unary **pr**-function. Find a number m such that for all $x, t \in N$,

$$K^{(1)}(f(x), t) = K^{(2)}(m, x, t) = K^{(1)}([m, x], t) .$$

This means the general recursive function (**gr**-function) g defined by $g(x) = [t, x]$ satisfies (1). That $\kappa^{-1}\tilde{e}$, $\omega^{-1}\tilde{e}$ are recursively enumerable is obvious. ■

Theorem 1.2: *For every complete numbering* α *and every* $n \geqslant 1$, *there exists an n-ary* **gr**-*function* p_n *such that for every n-ary* **pr**-*function h we can find a natural number m for which*

$$\alpha p_n(m, x_2, \dots, x_n) = \begin{cases} \alpha h(m, x_2, \dots, x_n) & \text{if } h(m, x_2, \dots, x_n) \text{ is defined,} \\ e & \text{otherwise} \qquad (x_2, \dots, x_n \in N) . \end{cases} \tag{2}$$

We consider the partial function f defined by

$$f(x) = K([x]_{n1}, \dots, [x]_{nn}) \qquad (x \in N) ,$$

where $[x]_{ni}$ is the function indicated above, and we write K instead of $K^{(n)}$, there being no confusion about the arity of the function. By definition, there exists a **gr**-function g such that f, g satisfy (1). Consequently, for any $x_1, \dots, x_n \in N$ we have

$$\alpha g([x_1, \dots, x_n]) = \begin{cases} \alpha K(x_1, \dots, x_n) & \text{if } K(x_1, \dots, x_n) \text{ is defined ,} \\ e & \text{otherwise .} \end{cases}$$

Since $K^{(n)}$ is a universal function, there must be some $t \in N$ such that for all $x_1, \dots, x_n \in N$,

$$h([x_1, x_1], x_2, \dots, x_n) = K(t, x_1, \dots, x_n)$$
$$= K([t, x_1], x_2, \dots, x_n) .$$

Taking $p_n(x_1, \dots, x_n) = g([x_1, \dots, x_n])$ and $m = [t, t]$, we obtain (2). ■

Theorem 1.3: *Suppose α is a complete numbering. Then for every* **pr**-*function $H(x_1, ..., x_n)$ we can find a* **gr**-*function $g(x_2, ..., x_n)$ such that for all $x_1, ..., x_n \in N$,*

$$\alpha g(x_2, ..., x_n) = \begin{cases} \alpha H(g(x_2, ..., x_n), x_2, ..., x_n) & \textit{when this is defined}, \\ e & \textit{otherwise.} \end{cases} \tag{3}$$

Let

$$h(x_1, ..., x_n) = H(p_n(x_1, ..., x_n), x_2, ..., x_n),$$

where p_n is the function obtained in Theorem 1.2. According to that theorem, we can find a number m such that

$$\alpha p_n(m, x_2, ..., x_n) = \begin{cases} \alpha h(m, x_2, ..., x_n) & \text{when this is defined}, \\ e & \text{otherwise.} \end{cases}$$

Therefore, we can take $g(x_2, ..., x_n)$ to be $p_n(m, x_2, ..., x_n)$ and satisfy (3). ∎

From Theorem 1.3 it follows that *every complete numbering α of a set A containing at least two elements is a complete numbering in the sense of* [XVIII], §2.3.

Indeed, the numbering α is complete in the old sense if there exists a **gr**-function g such that

$$\alpha K(n, g(n)) = \alpha g(n)$$

for all values of n such that $K_n^{(1)}$ is a totally defined function. On applying Theorem 1.3 to the function $H(x_1, x_2) = K(x_2, x_1)$, we obtain just such a function $g(x_2)$. ∎

§2. Isomorphism. Factor numberings

In comparing two numberings with each other, we have to distinguish two cases: (a) they number different sets; (b) they number the same set (cf. [XVIII]). We consider these cases separately.

Let α_1, α_2 be simple numberings of the same set A. The numbering α_1 is said to be **gr**-*reducible* to the numbering α_2 iff there exists a unary **gr**-function f such that

$$\alpha_1 n = \alpha_2 f(n) \quad (n \in N). \tag{4}$$

If α_1 is **gr**-reducible to α_2, and α_2 is **gr**-reducible to α_1, then we say α_1 and α_2 are **gr**-*equivalent* (or *multiequivalent*, **m**-*equivalent*). The simple numberings α_1, α_2 are *recursively isomorphic* iff there exists a **gr**-function that maps N 1–1 onto itself and satisfies (4).

Since every complete numbering is complete in the sense of [XVIII], Theorem 2.3.4 of that article implies

Theorem 2.1: *If a simple numbering α_1 of a set A is* **gr**-*equivalent to a complete numbering α_2 of A, then α_1 and α_2 are recursively isomorphic.* ∎

Now suppose α_1 and α_2 are simple numberings of sets A_1 and A_2. The numberings α_1, α_2 are said to be *freely isomorphic* iff there exists a **gr**-function $f(x)$ that maps N 1–1 onto itself and satisfies

$$\alpha_1 m = \alpha_1 n \Leftrightarrow \alpha_2 f(m) = \alpha_2 f(n)$$

for all natural numbers m, n.

Similarly, α_2 is a *homomorphic image* of α_1 iff there is a **gr**-function $f(x)$ mapping N 1–1 onto itself and satisfying,

$$\alpha_1 m = \alpha_1 n \Rightarrow \alpha_2 f(m) = \alpha_2 f(n) \quad (m, n \in N).$$

Let $\langle A, \alpha \rangle$ be an arbitrary simply numbered set, and let σ be an equivalence relation on A. By $[a]_\sigma$ we denote the set of all elements σ-equivalent to $a \in A$. By setting

$$\alpha_0 n = [\alpha n]_\sigma \quad (n \in N),$$

we obtain a numbering α_0 of the collection A/σ of all residue classes. The numbering α_0 is called the *factor numbering* of α relative to σ; α_0 is a homomorphic image of α.

Theorem 2.2: *A homomorphic image of a complete numbering is itself complete.*

For suppose the numbering α_0 of the set A_0 is a homomorphic image of the complete numbering α of the set A, and let f be a recursive permutation effecting the homomorphism. If $e \in A$ is the special element for α, and t is an α-number for e, then we take $e_0 = \alpha_0 f(t)$ as the special element for α_0. We see that the **gr**-function $p_2(x_1, x_2)$ satisfying the condition (2) of Theorem 1.2 for α, e also satisfies (2) for α_0, e_0. This implies α_0 is complete. ∎

For an example we look to the Kleene, Post numberings κ and ω defined above. Since $K_m^{(1)} = K_n^{(1)} \Rightarrow W_m = W_n$, the identity map on N specifies a homomorphism from κ onto ω.

§3. Enumerable families of elements

Let $\langle A, \alpha\rangle$ be a simply numbered set. We shall call nonempty subsets of A *families* (of objects in A). Families consisting of a single element will often be identified with that element. The notion of recursive enumerability can be defined for families in the following three ways:

(i) A family $B \subseteq A$ is *weakly enumerable* iff B is the α-image of some recursively enumerable subset $M \subseteq N$;

(ii) The family B is *strongly enumerable* iff B is the 1–1 α-image of some recursively enumerable set M of numbers, i.e., $\alpha M = B$ and for all $m, n \in M$, $m \neq n \Rightarrow \alpha m \neq \alpha n$;

(iii) The family B is *totally enumerable* (*totally creative, totally recursive*, etc.) iff the set $\alpha^{-1}B$ of all α-numbers of elements of B is recursively enumerable (creative, recursive, etc.).

Theorem 3.1 (cf. [XVIII], §2.3): *The completely numbered set $\langle A, \alpha\rangle$ has no totally recursive families other than the trivial ones – $\emptyset$ and A.*

Suppose to the contrary that $B \neq \emptyset, A$ is a totally recursive family, that $\alpha^{-1}B$ is recursive. Then its complement $\alpha^{-1}\widetilde{B}$ is also recursive. Let $s \in \alpha^{-1}B$, $t \in \alpha^{-1}\widetilde{B}$. We construct a recursive function f by putting

$$f(x,y) = \begin{cases} t & \text{if } x \in \alpha^{-1}B, \\ s & \text{if } x \in \alpha^{-1}\widetilde{B}. \end{cases}$$

According to Theorem 1.3, there is a **gr**-function $g(x)$ such that

$$\alpha g(x) = \alpha f(g(x), x) \quad (x \in N). \tag{5}$$

If $g(0) \in \alpha^{-1}B$, then $f(g(0), 0) = t$, so $\alpha \in B$, $\alpha f(g(0), 0) = \alpha t \in \widetilde{B}$, but this contradicts (5). The assumption that $g(0) \in \alpha^{-1}\widetilde{B}$ leads to a similar contradiction. ∎

Theorem 3.2. *If the totally enumerable family B of elements of the completely numbered set $\langle A, \alpha\rangle$ contains the special element e, then $B = A$.*

Suppose on the contrary that $B \neq A$, and let $t \in \alpha^{-1}\widetilde{B}$. We introduce a function f by setting

$$f(x) = \begin{cases} t & \text{if } x \in \alpha^{-1}B, \\ \text{undefined} & \text{otherwise}. \end{cases}$$

Then f is a **pr**-function. Since α is complete, we can find a **gr**-function $g(x)$ such that

$$\alpha g(x) = \begin{cases} \alpha t & \text{if } x \in \alpha^{-1}B \\ e & \text{if } x \in \alpha^{-1}\widetilde{B}. \end{cases}$$

From this it follows that

$$x \in \alpha^{-1}\widetilde{B} \Leftrightarrow g(x) \in \alpha^{-1}B,$$

i.e., the set $\alpha^{-1}\widetilde{B}$ is multireducible (or **m**-reducible) to the recursively enumerable set $\alpha^{-1}B$; thus the set $\alpha^{-1}\widetilde{B}$ is recursively enumerable. Since complementary **re**-sets are recursive, B and $\widetilde{B}$ are totally recursive; this contradicts Theorem 3.1. ■

Theorem 3.3: *Every nontrivial totally enumerable family B of elements of the completely numbered set $\langle A, \alpha \rangle$ is totally creative.*

Let $\alpha^{-1}B$ be the range of the **gr**-function $F(x)$, and let M be some creative set of natural numbers. We introduce an auxiliary function:

$$f(x) = \begin{cases} F(x) & \text{if } x \in M, \\ \text{undefined} & \text{if } x \in \widetilde{M}. \end{cases}$$

The function f is partial recursive, so we can find a unary **gr**-function g such that

$$\alpha g(x) = \begin{cases} \alpha F(x) & \text{if } x \in M, \\ e & \text{if } x \notin M. \end{cases} \tag{6}$$

By Theorem 3.2, e does not belong to B. Therefore, from (6) we learn that

$$x \in M \Leftrightarrow g(x) \in \alpha^{-1}B,$$

i.e., that the creative set M is **m**-reducible to the **re**-set $\alpha^{-1}B$; thus $\alpha^{-1}B$ is also creative. ■

Theorem 3.4: *Suppose in the completely numbered set* $\langle A, \alpha \rangle$ *the family* $\tilde{e}$ *of all nonspecial elements is totally enumerable. Then every nontrivial family* B *containing* e *is totally productive.*

We define a partial recursive function f by taking some $t \in \alpha^{-1}\widetilde{B}$ and setting

$$f(x) = \begin{cases} t & \text{if } x \in \alpha^{-1}\tilde{e}\,, \\ \text{undefined} & \text{otherwise}\,. \end{cases}$$

We can find a **gr**-function $g(x)$ such that

$$\alpha g(x) = \begin{cases} \alpha t & \text{if } x \in \alpha^{-1}\tilde{e}\,, \\ e & \text{if } x \in \alpha^{-1} e\,; \end{cases}$$

this means

$$x \in \alpha^{-1}\tilde{e} \Leftrightarrow g(x) \in \alpha^{-1} B\,. \tag{7}$$

By Theorem 3.3 the set $\alpha^{-1}e$ is productive. Therefore, by (7) the set $\alpha^{-1}B$ is productive. ■

§4. Completely numbered sets whose every family of nonspecial elements is totally enumerable

Let $\langle A, \alpha \rangle$ be a completely numbered set with the property given in the section heading. Since the collection of all **re**-sets is countable, but the collection of all subsets of an infinite set is uncountable, the set A must be finite, and the set of α-numbers of each of its elements other than e must be recursively enumerable.

Suppose $A = \{e, a_0, a_1, \ldots, a_s\}$. We put $M_i = \alpha^{-1}a_i$ $(i = 0, \ldots, s)$. The sequence $\langle M_0, \ldots, M_s \rangle$ is called the *system associated with the numbering* α.

Conversely, suppose we have a sequence $\langle M_0, \ldots, M_s \rangle$ of nonempty, pairwise disjoint sets of natural numbers whose union does not exhaust N. We define (cf. [XVIII], §2.3) a simple numbering α, *associated with* $\langle M_0, \ldots, M_s \rangle$, on the auxiliary set $A = \{e, a_0, \ldots, a_s\}$ by requiring for all $n \in N$,

$$\alpha n = a_i \Leftrightarrow n \in M_i \qquad (i = 0, \ldots, s)\,,$$

and

$$\alpha n = e \Leftrightarrow n \in N \sim M_0 \sim \ldots \sim M_s.$$

We recall these well-known definitions [109]: the system $\langle M_0, ..., M_s\rangle$ of subsets of N is **m**-*reducible* to the system $\langle P_0, ..., P_s\rangle$ iff there exists a **gr**-function $g(x)$ such that

$$x \in M_i \Leftrightarrow g(x) \in P_i \quad (x \in N; \quad i = 0, ..., s)\,.$$

A system $\langle V_0, ..., V_s\rangle$ of nonempty, pairwise disjoint, and recursively enumerable numerical sets is called **m**-*universal* iff every other sequence of $s+1$ nonempty and pairwise disjoint **re**-sets is **m**-reducible to it. Clearly, $\bigcup V_i \neq N$.

The systems $\langle M_0, ..., M_s\rangle, \langle P_0, ..., P_s\rangle$ are **m**-*equivalent* iff each is **m**-reducible to the other. These systems are *recursively isomorphic* iff one is **m**-reducible to the other by means of a **gr**-function g that maps N 1–1 onto itself.

Theorem 4.1: *If α is a complete numbering of a finite set $A = \{e, a_0, ..., a_s\}$ such that every nonspecial element is totally enumerable, then the associated system of sets $M_i = \alpha^{-1}a_i$ $(i = 0, ..., s)$, is* **m**-*universal. Conversely, if $\langle M_0, ..., M_s\rangle$ is an arbitrary* **m**-*universal system, then the associated numbering is complete with e as special element, and all the other elements are totally enumerable.*

Indeed, suppose α is a complete numbering of $A = \{e, a_0, ..., a_s\}$, and a_i is totally enumerable $(i = 0, ..., s)$. Then the sets $M_0, ..., M_s$ are recursively enumerable, nonempty, and pairwise disjoint. Suppose $\langle P_0, ..., P_s\rangle$ is any other system of nonempty and pairwise disjoint **re**-sets. We consider the function

$$f(x) = \begin{cases} m_i & \text{if } x \in P_i \quad (i = 0, ..., s)\,, \\ \text{not defined} & \text{if } x \in N \sim \bigcup P_i\,, \end{cases}$$

where $m_0, ..., m_s$ are fixed elements of the respective sets $M_0, ..., M_s$. Since f is a **pr**-function, we can find a **gr**-function g such that

$$\alpha g(x) = \begin{cases} \alpha m_i & \text{if } x \in P_i \quad (i = 0, ..., s)\,, \\ e & \text{otherwise.} \end{cases}$$

Consequently, the function g **m**-reduces $\langle P_0, ..., P_s\rangle$ to $\langle M_0, ..., M_s\rangle$. Since the former system was arbitrary within bounds, the latter is **m**-universal.

Conversely, suppose $\langle M_0, ..., M_s\rangle$ is an **m**-universal system, and let α be the associated numbering of the auxiliary set $A = \{e, a_0, ..., a_s\}$. Let f be an arbitrary unary **pr**-function. We put

$$P_i = \{x : f(x) \in M_i\} \quad (i = 0, ..., s) .$$

The sets $P_0, ..., P_s$ are nonempty, pairwise disjoint, and recursively enumerable. Therefore, there exists a **gr**-function $g(x)$ **m**-reducing $\langle P_0, ..., P_s\rangle$ to $\langle M_0, ..., M_s\rangle$ and thus satisfying (1) in the definition of completeness. ∎

A comparison of the definitions of the notions of **m**-equivalence for systems of sets and for numberings shows that two systems of $s+1$ nonempty, pairwise disjoint, nonexhaustive sets of numbers are **m**-equivalent (recursively isomorphic) iff their associated numberings of the set $A = \{e, a_0, ..., a_s\}$ are **m**-equivalent (recursively isomorphic) (cf. [XVIII]). Now by comparing Theorem 2.1 with Theorem 4.1 and observing that all **m**-universal systems of a fixed number $s+1$ of sets are **m**-equivalent, we conclude the important proposition:

Corollary 4.1: *For each $s \geqslant 0$, all* **m**-*universal systems consisting of $s+1$ sets are recursively isomorphic to each other* (Myhill [111], Mučnik [109]). ∎

We inspect some examples of complete numberings of finite sets.

Example 1: In the Post numbering ω let us take all nonempty **re**-sets to be σ-equivalent. Then the factor numbering ω/σ is a complete numbering of the two-element set $\{e, \tilde{e}\}$, where e is the empty set, $\tilde{e}$ the family of all nonempty **re**-sets. In this numbering the element $\tilde{e}$ is totally enumerable; hence, the set $\alpha^{-1}\tilde{e}$ of all Post numbers of nonempty recursively enumerable sets is **m**-universal.

Example 2: Let $\mathcal{W}$ be the collection of all **re**-sets. We decompose $\mathcal{W}$ into three classes: the class consisting of the empty set alone, both of which we write as e; the class a_0 of all **re**-sets containing 1; the class a_1 of remaining **re**-sets. Let σ be the equivalence relation on $\mathcal{W}$ corresponding to this partition. The numbering ω/σ of $\mathcal{W}/\sigma = \{e, a_0, a_1\}$ is complete. The family $\tilde{e}$ is totally enumerable, but the nonspecial elements a_0, a_1 are not.

Theorem 4.1 yields the following generalization of Theorem 3.3:

Theorem 4.2: *Let $\langle A, \alpha\rangle$ be a completely numbered set, and let $B_0, ..., B_s$ be pairwise disjoint, nonempty, totally enumerable families of elements of A. Then the system $\langle \alpha^{-1}B_0, ..., \alpha^{-1}B_s\rangle$ is* **m**-*universal.*

To see this, let B_e be the family of all elements in A not belonging to $B_0 \cup ... \cup B_s$ ([1]). Let σ be the equivalence relation with equivalence classes

$B_e, B_0, ..., B_s$, and consider the numbering $\alpha_0 = \alpha/\sigma$ of the finite set $A_0 = \{B_e, B_0, ..., B_s\}$. This numbering is complete with B_e as its special element. Since $\alpha_0^{-1}B_i = \alpha^{-1}B_i$ $(i = 0, ..., s)$, the nonspecial elements in A_0 are totally enumerable. By Theorem 4.1 the system $\langle\alpha^{-1}B_0, ..., \alpha^{-1}B_s\rangle$ is **m**-universal. ∎

Theorem 4.2 makes it possible to construct **m**-universal systems very easily. For an example we turn to the Kleene numbering κ. Let B_i be the family of all (unary) **pr**-functions that take the value i at the point 0 $(i = 0, ..., s)$. The families B_i are totally enumerable relative to κ (cf. § 6) and satisfy the other requirements of the theorem above. Therefore, $\langle \kappa^{-1}B_0, ..., \kappa^{-1}B_s\rangle$ is an **m**-universal system.

It is known that an **m**-universal pair of sets are effectively inseparable [154]. So, the sets $\kappa^{-1}B_0, \kappa^{-1}B_1$ are effectively inseparable.

§5. Universal series of sets

The results of §4 concerning **m**-universal systems of finite numbers of sets can be extended in an obvious manner to infinite systems of sets. Namely, with every infinite sequence of sets

$$M_0, M_1, ..., M_n, ... \tag{8}$$

is connected a predicate $M(i, x)$ defined by

$$M(i, x) \Leftrightarrow x \in M_i \quad (i, x \in N).$$

A sequence of sets (8) will often be identified with this predicate.

A sequence of sets (8) will be called a *series* iff the predicate $M(i, x)$ is recursively enumerable, and no two sets in (8) have elements in common.

A sequence $M(i, x)$ is **m**-*reducible* to a sequence $P(i, x)$ iff there exists a **gr**-function $g(x)$ such that

$$M(i, x) \Leftrightarrow P(i, g(x)) \quad (i, x \in N).$$

Sequences $M(i, x), P(i, x)$ **m**-reducible to each other are **m**-*equivalent.* A sequence $M(i, x)$ reducible to a sequence $P(i, x)$ by means of a **gr**-function g mapping N 1–1 onto itself is *recursively isomorphic* to $P(i, x)$.

A series $V(i, x)$ is **m**-*universal* iff every series $M(i, x)$ **m**-reduces to it.

(*N.B.:* the notions of universal and creative sequences were introduced by Cleave [21] as a basis for extending the fundamental results of Mučnik and

Smullyan on creative pairs of sets. In particular, Cleave obtained Theorem 5.2 below, but by a different method.)

A complete numbering α of a set A is called *serially complete* iff all the nonspecial elements of A can be arranged in a simple infinite sequence $a_0, a_1, \dots$ $(i \neq j \Rightarrow a_i \neq a_j)$ so that the sequence of sets

$$\alpha^{-1}a_0, \alpha^{-1}a_1, \dots, \alpha^{-1}a_r, \dots \tag{9}$$

is connected with a serial predicate $M(i, x)$ (i.e., so that (9) forms a series). The following analog of Theorem 4.1 holds:

Theorem 5.1: *The series $M(i, x)$ connected with a serially complete numbering α is* **m**-*universal. Conversely, suppose an* **m**-*universal series $M(i, x)$ is given. Then the numbering α of the auxiliary set $A = \{e, a_0, a_1, \dots\}$ defined by the conditions*

$$\alpha n = a_i \Leftrightarrow M(i, n), \qquad \alpha n = e \Leftrightarrow \textit{for all } i \in N, \textit{ not } M(i, n) \tag{10}$$

is a complete and serial numbering.

We prove the first assertion as an illustration. Let $P(i, x)$ be an arbitrary series. Let $M(i, x)$ be represented in parametric form by **gr**-functions $\varphi_M(v)$, $\psi_M(v)$, i.e.,

$$M(i, x) \Leftrightarrow \text{ for some } v \in N,\ i = \varphi_M(v) \text{ and } x = \psi_M(v)\,.$$

We set

$$i(x) = \varphi_P(\min_v(\psi_P(v) = x))\,,$$

$$f(x) = \psi_M(\min_v(\varphi_M(v) = i(x)))$$

and select a function g satisfying (1) for f. From (1) and (10) it follows that

$$P(i, x) \Leftrightarrow M(i, g(x)) \quad (i, x \in N)\,,$$

i.e., that g **m**-reduces the series $P(i, x)$ to the series $M(i, x)$. The second assertion of the theorem is proved in much the same spirit. ■

Theorem 5.2: *All* **m**-*universal series are recursively isomorphic to each other.*

Let $M^1(i, x)$ and $M^2(i, x)$ be **m**-universal series. We take an auxiliary set $A = \{e, a_0, a_1, \ldots\}$ and construct – as in (10) – two numberings α_1, α_2 of A associated with the series M^1, M^2, respectively. From the mutual **m**-reducibility of the series M^1, M^2 follows the **m**-equivalence of the numberings α_1, α_2. Since these numberings are complete, Theorem 2.1 tells us they are recursively isomorphic; hence, the series M^1, M^2 are also isomorphic. ∎

The analog of Theorem 4.2 is

Theorem 5.3: *Let $\langle A, \alpha \rangle$ be a completely numbered set, and suppose there is a sequence $\langle B_0, B_1, \ldots \rangle$ of nonempty, pairwise disjoint families of elements of A such that the predicate $P(i, x)$ defined by*

$$P(i, x) \Leftrightarrow \alpha x \in B_i \quad (i, x \in N)$$

is recursively enumerable. Then $P(i, x)$ is an **m**-*universal series.*

This theorem is deduced from Theorem 5.1 in the same way that Theorem 4.2 was derived from Theorem 4.1. ∎

We again appeal to the Kleene numbering κ for an example. The families B_i of all partial recursive functions taking the value i for the argument 0 $(i = 0, 1, 2, \ldots)$ form a sequence satisfying the hypotheses of Theorem 5.3, as is easily checked. Therefore, $\langle \kappa^{-1}B_0, \kappa^{-1}B_1, \ldots \rangle$ is an **m**-universal series.

§6. Totally enumerable families of partial recursive functions

We shall construct below a system of complete numberings – not at all bizarre – that we shall prove are not mutually isomorphic. To do this we need to know the structure of totally enumerable families of (unary) partial recursive functions relative to the Kleene numbering. Families of **re**-sets totally enumerable in the Post numbering were covered by Rice [126]. Analogous results for **pr**-functions were obtained by Uspenskiĭ [174]. In Uspenskiĭ's work the hypotheses of theorems and the proofs were expressed in topological terms. The results we need concerning totally enumerable sets of functions we shall present here, relying on Rice's reasoning, as several improvements along the way will be of interest. Most functions considered are numerical and unary.

A partial function $f(x)$ is called an *extension* (or continuation) of a partial function $h(x)$ iff for every $n \in N$, if $h(n)$ is defined, then $f(n)$ is defined and equal to $h(n)$. A partial function h is a *finite restriction* of a partial function f iff f is an extension of h, and h has a finite domain of definition. In particular, the nowhere defined function e is a finite restriction of every function f.

The *Gödel number* of a function h defined on a finite set $\{k_0, k_1, \ldots, k_n\}$ $(k_0 < k_1 < \ldots < k_n)$ is the natural number $p_0^{k_0} p_1^{h(k_0)} \ldots p_{2n}^{k_n} p_{2n+1}^{h(k_n)}$, where p_i

is the ith prime. The Gödel number of e is 0 by convention. Functions with finite domains are said to be *finitely defined.*

By repeating the arguments of Rice [126], we immediately get

Theorem 6.1: *Let B be a family of finitely defined functions whose Gödel numbers form a recursively enumerable set. Then the set of all Kleene numbers of all* **pr**-*extensions of the functions in B is also recursively enumerable.* ∎

Similarly, we immediately obtain

Theorem 6.2: *The set M of Gödel numbers of all finitely defined functions contained in a given totally κ-enumerable family B of* **pr**-*functions is recursively enumerable.*

For let $f_n(x)$ be the finitely defined function with Gödel number n. Since $f_n(x)$ is partial recursive in the variables n, x, there is a fixed number t such that

$$f_n(x) = K^{(2)}(t, n, x) = K^{(1)}([t, n], x) .$$

Therefore,

$$M = [[t, N] \cap \kappa^{-1}B]_{22} ,$$

which shows M is recursively enumerable. ∎

Theorem 6.3: *Let h(x) be a* **pr**-*function with recursive domain, and suppose f(x) is some proper* **pr**-*extension of h. Then every subset $M \subseteq N$ that contains all κ-numbers of h, but not one of f, is productive.*

By definition the set $\kappa^{-1}e$ is the collection of all κ-numbers of the function nowhere defined. Let $\chi(x)$ be the function taking value 0 off $\kappa^{-1}e$ and undefined on $\kappa^{-1}e$. Since $\kappa^{-1}\bar{\bar{e}}$ is recursively enumerable, χ is partial recursive. With no loss of generality we can assume that 0 belongs $\kappa^{-1}\bar{\bar{e}}$. Let $\varphi(x)$ be the function whose value is equal to 0 on the domain of h and equal to 1 else here. We select a number t such that for all $n, x \in N$,

$$\chi(n \cdot \varphi(x)) + f(x) = K(t, n, x) = K([t, n], x) .$$

If $n \in \kappa^{-1}e$, then $\chi(n \cdot \varphi(x))$ is defined just where $h(x)$ is defined, so $K([t, n], x) = h(x)$. Therefore, for all $n \in N$,

$$n \in \kappa^{-1}e \Rightarrow [t, n] \in \kappa^{-1}h \subseteq M. \tag{11}$$

If $n \notin \kappa^{-1}e$, then $\chi(n \cdot \varphi(x))$ is defined everywhere, so $K([t, n], x) = f(x)$. Therefore, for all $n \in N$,

$$n \notin \kappa^{-1}e \Rightarrow [t, n] \notin M. \tag{12}$$

The set $\kappa^{-1}e$ is productive, and, according to (11) and (12), is **m**-reducible to M. Hence, M is productive. ■

It requires a bit more patience to prove

Theorem 6.4: *Let $f(x)$ be a* **pr**-*function. Then every set M of natural numbers that contains all κ-numbers of f, but not one κ-number of any finite restriction of f, is productive.*

First, we introduce a new numbering α of the set $\mathcal{F}_{\text{pr}}$ of all unary **pr**-functions by taking n to be an α-number of the function

$$(\alpha n)(x) = K(n, x, x) + \sum_{v=0}^{x} 0 \cdot K(n, v, x) = (\Re K_n^{(2)})(x).\ (^2) \tag{13}$$

From (13) we see that the expression defining αn is partial recursive in n and x. Let r be a number such that

$$(\alpha n)(x) = K([r, n], x).$$

Consequently, $[r, n]$ is a κ-number of the same function that has n as an α-number. This means α is **m**-reducible to κ. Conversely, we choose a number s so that for all $m, x, y \in N$,

$$K^{(3)}(s, m, y, x) = K^{(1)}(m, x). \tag{14}$$

Then we conclude from (13) that if m is a κ-number of a function f, then $[s, m]$ is an α-number for it. Thus the numberings α and κ are **gr**-equivalent. By Theorem 2.1, they are recursively isomorphic. Under a recursive permutation of N, productive sets are transformed into productive sets. Instead of proving Theorem 6.4 for κ, it suffices to prove it for α.

We note that an α-number n of a function $f \in \mathcal{F}_{\text{pr}}$ is just a $\kappa^{(2)}$-number of a binary function $k(x,y) = K_n^{(2)}$ from which f is obtained by means of the operator $\Re$ as defined in (13). Therefore, if we want to find a new α-number m for the function f with the α-number n, we can simply construct a binary **pr**-function $h(x, y)$ such that $h \neq K_n^{(2)}$ and $\Re h = f$, and take m to be any $\kappa^{(2)}$-number for h. We now use this trick to prove the productiveness of the set M indicated in the statement of Theorem 6.4 (for the numbering α).

Let p be some α-number of the given **pr**-function f; thus $f = \Re K_p^{(2)}$. Now suppose we are given a number z for which $W_z \subseteq M$, where

$$W_z = \{u_0, u_1, \dots\}, \quad u_i = K(z, i).$$

We want to describe an effective process, uniform in z, for constructing a binary **pr**-function F distinct from the $K_{u_i}^{(2)}$ such that $\Re F = f$; then we shall have a **pr**-function g with the property that $g(z)$ is a $\kappa^{(2)}$-number of F. This means

$$g(z) \in M \sim W_z ,$$

showing M is productive.

Let $a(x)$, $b(x)$ be **gr**-functions such that the set of the pairs

$$\langle a_0, b_0\rangle, \langle a_1, b_1\rangle, \dots, \langle a_n, b_n\rangle, \dots \qquad (a_n = a(n),\ b_n = b(n))$$

is just the domain of $K_p^{(2)}$. In the process of constructing F, which will have this same domain, we shall also construct a sequence of pairs $\langle c_0, d_0\rangle$, $\langle c_1, d_1\rangle$, ... such that for every $n \in N$, $K^{(2)}(u_n, c_n, d_n)$ is defined. The construction procedure breaks up into steps $\mathbf{A}_0, \mathbf{B}_0, \mathbf{A}_1, \mathbf{B}_1, \dots$; at step $\mathbf{A}_n$ we define $F(a_n, b_n)$, and at step $\mathbf{B}_n$ we define $\langle c_n, d_n\rangle$.

Step $\mathbf{A}_0$: We put

$$F(a_0, b_0) = \begin{cases} K(p, a_0, b_0) & \text{if } a_0 = b_0 , \\ K(p, a_0, b_0) + 1 & \text{if } a_0 \neq b_0 . \end{cases}$$

Step $\mathbf{A}_{n+1}$: Let us assume that the value $F(a_i, b_i)$ and the pair $\langle c_i, d_i\rangle$ at which $K_{u_i}^{(2)}$ is defined have already been constructed for $0 \leqslant i \leqslant n$. We consider three cases:

(i) $\langle a_{n+1}, b_{n+1}\rangle = \langle a_i, b_i\rangle$ for some $i \leqslant n$. Then $F(a_{n+1}, b_{n+1})$ has been defined already.

(ii) The pair $\langle a_{n+1}, b_{n+1}\rangle$ is new (i.e., not among $\langle a_0, b_0\rangle, \dots, \langle a_n, b_n\rangle$), and $a_{n+1} = b_{n+1}$. Then we put $F(a_{n+1}, b_{n+1}) = K(p, a_{n+1}, b_{n+1})$.

(iii) The pair $\langle a_{n+1}, b_{n+1}\rangle$ is new, and $a_{n+1} \neq b_{n+1}$. Then we take $F(a_{n+1}, b_{n+1})$ to be the least natural number distinct from all the values $K(u_i, c_i, d_i)$ for those i such that $0 \leqslant i \leqslant n$ and $\langle c_i, d_i\rangle = \langle a_{n+1}, b_{n+1}\rangle$. If no such i exist, we set $F(a_{n+1}, b_{n+1}) = 0$.

Step $\mathbf{B}_n$: We want to select a pair $\langle c_n, d_n\rangle$ that belongs to the domain of $K_{u_n}^{(2)}$ and satisfies one of the following conditions:

(I) $c_n \neq d_n$ and $\langle c_n, d_n\rangle \notin \{\langle a_0, b_0\rangle, ..., \langle a_n, b_n\rangle\}$;

(II) $c_n = d_n$, $K(p, c_n, d_n)$ is defined, and $K(p, c_n, d_n) \neq K(u_n, c_n, d_n)$.

A pair $\langle c_n, d_n\rangle$ satisfying these conditions actually exists. For the contrary would mean that the function $K_{u_n}^{(2)}$ is defined for only a finite number of non-diagonal pairs, all contained in the set $\{\langle a_0, b_0\rangle, ..., \langle a_n, b_n\rangle\}$, and that the value of $K_{u_n}^{(2)}$ for any diagonal pair $\langle x, x\rangle$ coincides with $K(p, x, x)$, if the latter be defined. But this would imply that the unary partial function $\mathfrak{R}K_{u_n}^{(2)}$ is a finite restriction of f. ([3]) Since u_n is an α-number for $\mathfrak{R}K_{u_n}^{(2)}$ and belongs to M, this would contradict an assumption concerning M.

Therefore, at least one pair $\langle c_n, d_n\rangle$ with the desired properties exist. We can undertake simultaneous computations of the values of $K_p^{(2)}$ and $K_{u_n}^{(2)}$, confident that this effort will lead to a desirable pair in the domain of $K_{u_n}^{(2)}$.

The function F has been constructed. It coincides with no $K_{u_n}^{(2)}$. Indeed, if $\langle c_n, d_n\rangle$ is not in the domain of $K_p^{(2)}$, then $F(c_n, d_n)$ is not defined; since $K(u_n, c_n, d_n)$ is defined, F and $K_{u_n}^{(2)}$ cannot be the same function. Now suppose $K(p, c_n, d_n)$ is defined. If $c_n = d_n$, then by (ii) and (II)

$$F(c_n, d_n) = K(p, c_n, d_n) \neq K(u_n, c_n, d_n).$$

If $c_n \neq d_n$, then by (I) $F(c_n, d_n)$ won't be defined until step $\mathbf{A}_{n+k}$ for some $k > 1$; (iii) then guarantees that $F(c_n, d_n) \neq K(u_n, c_n, d_n)$. Furthermore, $\mathfrak{R}F$ is identical with $\mathfrak{R}K_p^{(2)} = f$, since F and $K_p^{(2)}$ have the same domain and agree on diagonal elements. ■

From the four preceding theorems we immediately deduce the fundamental

Corollary 6.1: *A family of* **pr**-*functions is totally* κ-*enumerable iff it consists of all possible* **pr**-*extensions of finitely defined functions whose set of Gödel numbers is recursively enumerable.* ■

As already mentioned, Corollary 6.1 is included in more general results of V.A. Uspenskiĭ, who kindly communicated the following simple proof of this corollary. In topological language, Corollary 6.1 expresses that a family of **pr**-functions is totally enumerable iff it is effectively open, i.e., iff it is the union of a recursively enumerable system of generalized Baire intervals (see [174] and [175], §10). In this form the corollary immediately follows from Theorem 2 of [174] (since the Kleene numbering is potentially computable) and Theorem 5 of the same article (since $\mathcal{F}_{\mathrm{pr}}$ is ω-separable, and its Kleene numbering is a covering).

§7. Projective families of functions. Computable numberings

If C is a family of unary partial recursive functions, it is natural to distinguish among the numberings of C the so-called computable numberings (Kolmogorov-Uspenskiĭ). Namely, we say the simple numbering γ of the set C is *computable* iff the value of the function in C with γ-number n at the point x is a partial recursive function of n, x, i.e., iff there exists a universal **pr**-function $T(n, x)$ for $\langle C, \gamma\rangle$. Not every family included in $\mathcal{F}_{\text{pr}}$ admits a computable numbering (cf. §9). A family of **pr**-functions that does admit a computable numbering is itself called *computable.*

Let C be a computable family of **pr**-functions and let $T(n, x)$ realize a computable numbering of it. Choose a number t such that for all $n, x \in N$,

$$T(n, x) = K^{(2)}(t, n, x) = K^{(1)}([t, n], x) .$$

From this relation we see that the family C is the κ-image of the recursive set $[t, N]$. We conclude that every computable family of **pr**-functions is weakly κ-enumerable in the sense of §3. The converse is also obvious.

A computable numbering γ of a family C is said to be *principal* iff every other computable numbering of C is **gr**-reducible to γ. Since according to this definition all principal numberings of C are **gr**-equivalent to each other, Theorem 2.1 directly implies

Theorem 7.1: *If a family of* **pr**-*functions has at least one complete principal numbering, then all its principal numberings are complete and recursively isomorphic.* ∎

Without going into further details, we now want to indicate a number of concrete families of functions that admit complete principal numberings.

Let $\mathfrak{P}$ be a computable operator mapping the set $\mathcal{F}_{\text{pr}}$ of all unary **pr**-functions onto some subset $C \subseteq \mathcal{F}_{\text{pr}}$. The operator $\mathfrak{P}$ will be called *projective* when $\mathfrak{P}^2 = \mathfrak{P}$.

With a projective operator $\mathfrak{P}$ we associate a simple numbering π of its range C by setting

$$n = \mathfrak{P}K_n^{(1)} \quad (n \in N) .$$

Since the operator $\mathfrak{P}$ is computable, for some number p we have $\mathfrak{P}K_n^{(1)} = K_{[p,n]}^{(1)}$ for all $n \in N$. Consequently, the function T defined by

$$T(n, x) = K([p, n], x) \quad (n, x \in N)$$

is universal for π; hence, π is a computable numbering of C.

Let us call **pr**-functions $f, g \in \mathcal{F}_{pr}$ *$\mathfrak{P}$-equivalent* iff $\mathfrak{P}f = \mathfrak{P}g$. From $\mathfrak{P}^2 = \mathfrak{P}$ it follows that $f \equiv_{\mathfrak{P}} \mathfrak{P}f$. Therefore, in each $\mathfrak{P}$-equivalence class $[f]_{\mathfrak{P}}$ there is precisely one function from C. This gives a 1–1 correspondence between C and the factor set $\mathcal{F}_{pr}/\mathfrak{P}$ that shows the identity map on N is a free isomorphism of the respective numberings π, $\kappa/\mathfrak{P}$ of these sets. But the factor numbering of a complete numbering is complete (cf. § 2), so π is complete.

Let $T_1(n, x)$ be the universal **pr**-function for another computable numbering γ of C. Take t to be a number such that for all $n, x \in N$,

$$T_1(n, x) = K([t, n], x) .$$

If n is a γ-number for a function $f \in C$, then from

$$f = \mathfrak{P}f = \mathfrak{P}K^{(1)}_{[t,n]} ,$$

it follows that $[t, n]$ is a π-number for f. This means π is principal, since γ was chosen arbitrarily.

A family of **pr**-functions is said to be *projective* iff it is the range of a projective operator. From the above remarks we draw the following conclusions:

Theorem 7.2: *Every projective family of* **pr**-*functions admits complete principal numberings. If the projective family C is included in the projective family C_1, then a principal numbering π of C is isomorphic to some factor numbering of any principal numbering π_1 of C_1.* ■

The notions of computable and principal numberings of families included in $\mathcal{F}_{pr}$ can be naturally extended to arbitrary simply numbered sets in the following way. Let $\langle A, \alpha \rangle$ be a simply numbered set. A simple numbering α_1 of A is *computable* relative to α (or *α-subordinate*) iff α_1 is **gr**-reducible to α. An α-computable numbering α_1 of A is *principal* (relative to α) iff every α-computable numbering is **gr**-reducible to α_1. In other words, the principal numberings of $\langle A, \alpha \rangle$ are just the simple numberings of A that are **gr**-equivalent to α. Therefore, *if the basic numbering α is complete, then all the α-principal numberings are recursively isomorphic to each other.* ■

It is easy, however, to find an example of a completely numbered set with but two elements which has complete computable numberings not isomorphic to the basic numbering. Namely, we partition the set $\mathcal{F}_{pr}$ into two classes: the class $\mathcal{F}_0$ of all **pr**-functions with nonempty domains and ranges equal to $\{0\}$, and the class $\tilde{\mathcal{F}}_0$ of all the remaining functions. The corresponding equivalence relation is denoted by σ. In addition, let τ correspond to the classes e consisting of the function with empty domain and $\tilde{e}$, its complement. The numberings

κ/σ, κ/τ of the two element set are complete, and κ/τ is **gr**-reducible to κ/σ. At the same time, κ/σ and κ/τ are not isomorphic, because neither $\mathcal{F}_0$ nor $\widetilde{\mathcal{F}}_0$ is totally enumerable, but $\tilde{e}$ is.

Nevertheless, the following observation is easy to verify:

Theorem 7.3: *Every complete numbering of a finite set for which all the nonspecial elements are totally enumerable, as well as every serially complete numbering, has the following property: every complete numbering of the same set that is* **gr**-*reducible to the given numbering is principal.*

The first assertion of this theorem is a consequence of Theorem 4.1 and Corollary 4.1; the second follows analogously from Theorems 5.1 and 5.2. E.g., suppose $M(i, x)$ generates a complete numbering α of the set $A = \{e, a_0, a_1, \dots\}$, and suppose the **gr**-function g reduces the complete numbering α_1 of A to α. Then α_1 is serial, being connected with the series $M(i, g(x))$. By Theorem 5.1 the series $M(i, g(x))$ is **m**-universal. Theorem 5.2 now shows α and α_1 are recursively isomorphic. ∎

Returning to projective operators on $\mathcal{F}_{\text{pr}}$, we introduce the following definition: a projective operator $\mathfrak{P}$ is said to be *directedly projective* (or *directive*) iff whenever f is an extension of g, $\mathfrak{P}f$ is an extension of $\mathfrak{P}g$. The range of a directive operator is said to be a *directive family.*

Suppose $C \subseteq \mathcal{F}_{\text{pr}}$ is the family defined by the directive operator $\mathfrak{P}$ which generates the principal numbering π of C. ([4]) Since every κ-number of a function $f \in C$ is also a π-number for f, from the directedness of $\mathfrak{P}$ we can conclude that Theorems 6.3 and 6.4 on the productiveness of certain sets of κ-numbers hold also for the numbering π and – by Theorem 7.1 – for any other principal numbering of C, as well.

Theorem 6.1 obviously applies to any computable numbering of any computable family; Theorem 6.2 immediately extends to principal numberings of arbitrary projective families.

From these observations, as in §6, we deduce a basic corollary:

Corollary 7.1: *Let π be a principal numbering of a directive family C. A subfamily $B \subseteq C$ is totally π-enumerable iff B consists of all possible C-extensions of finitely defined C-functions whose Gödel numbers form a recursively enumerable set.* ∎

We now consider series of simple examples of directive families.

Let M be a nonempty, recursively enumerable set of numbers. Let C_M be the family of all **pr**-functions whose ranges are included in M. For brevity we put

$$C_n = C_{\{0, \dots, n\}} \quad (n \in N).$$

We introduce functions φ_n be setting

$$\varphi_n(x) = \begin{cases} x & \text{if} \quad 0 \leqslant x \leqslant n \,, \\ n & \text{if} \quad x \geqslant n \,. \end{cases}$$

The operators $\mathfrak{P}_n$ defined by

$$(\mathfrak{P}_n f)(x) = \varphi_n(f(x)) \quad (f \in \mathcal{F}_{\mathrm{pr}};\ n, x \in N)$$

are directive and $\mathfrak{P}_n \mathcal{F}_{\mathrm{pr}} = C_n$. Let κ_n denote the principal numbering of C_n induced by $\mathfrak{P}_n$ as described earlier. We can also take $n = \infty$, obtaining $\varphi_\infty(x) = x$, $C_\infty = \mathcal{F}_{\mathrm{pr}}$, and $\kappa_\infty = \kappa$.

The following free isomorphisms are obvious:

(I) If M is a finite set with n elements, then every principal numbering of C_M is isomorphic to the numbering κ_n of C_n.

(II) The numbering κ_0 of C_0 is isomorphic to the Post numbering ω of th set of all **re**-sets.

(III) If M is an infinite recursive set, then every principal numbering of C_M s isomorphic to the Kleene numbering $\kappa = \kappa_\infty$.

(IV) The numbering κ_m is isomorphic to a factor numbering of κ_n whenver $m < n$.

Corollary 7.1 describes the structure of all totally κ_n-enumerable subfamilies of C_n. Using this structure, in §8 we show that the numberings κ, κ_0, κ_1, ... are not isomorphic.

§8. Quasiordered families

Let $\langle A, \alpha \rangle$ be a comp.etely ordered set. Let $\mathcal{E}$ denote the collection of all totally enumerable families of elements of A. The study of this collection's structure can be conducted either in topological language by using $\mathcal{E}$ to introduce an appropriate topology on A along the lines of Uspenskiĭ's development [174, 175] or in the language of partially ordered sets by introducing a suitable order on A. The latter method we now employ.

An element $a \in A$ is *subordinate* to an element $b \in A$ (we write $a \leqslant b$) iff every totally enumerable family in $\mathcal{E}$ that contains a also contains b. It is easy to see that for every $a, b, c \in A$, we have $a \leqslant a$, and if $a \leqslant b$ and $b \leqslant c$, then $a \leqslant c$; this means $\leqslant$ is a partial quasiorder with least element e.

The definition of $\leqslant$ also implies that if the numbered sets $\langle A, \alpha \rangle$ and $\langle B, \beta \rangle$ are isomorphic, then so are the models $\langle A, \leqslant \rangle$ and $\langle B, \leqslant \rangle$.

For every numbered family $\langle C_s, \kappa_s \rangle$ of all unary **pr**-functions whose ranges are included in $\{0, ..., s\}$, the relation $f \leqslant g$ holds iff g is an extension of f (by virtue of Corollary 7.1). Therefore, for κ_s the relation $\leqslant$ is a partial order. Moreover, if $f, g \in C_s$, then the function whose graph is the intersection of the graphs of f and g is the "largest" member of C_s subordinate to both f and g. This means $\langle C_s, \leqslant \rangle$ is a lower semilattice.

As we already noted, the special element e is the smallest element of the semilattice $\langle C_s, \leqslant \rangle$. Corollary 7.1 tells us that the partial functions defined at only one point are the atoms, and the general recursive C_s-functions are the maximal elements in $\langle C_s, \leqslant \rangle$.

In a lower semilattice $\langle A, \leqslant \rangle$, two atoms a_0, b_0 are called *isotopic* iff either $a_0 = b_0$ or a_0, b_0 have no common upper bound in A. Elements $a, b \in A$ are called *isotopic* iff for every atom $a_0 \leqslant a$, there is an isotopic atom $b_0 \leqslant b$, and vice-versa. The nature of $\leqslant$ on C_s implies that the isotopic elements in $\langle C_s, \leqslant \rangle$ are the functions that have the same domains. Therefore, in $\langle C_s, \leqslant \rangle$ the atoms form isotope classes that each contain $s + 1$ functions. With proper interpretation the above remarks also hold for $s = \infty$. Because the isomorphism of the numbered sets under consideration implies the isomorphism of the corresponding semilattices and because in isomorphic semilattices the isotope classes of atoms must correspond, no two numberings among κ, κ_0, κ_1, ... can be isomorphic.

§9. Intrinsically productive families

In §3 we introduced the abstract concept of a weakly enumerable family of elements in a simply numbered set $\langle A, \alpha \rangle$. In §7 the notion of a computable family of unary **pr**-functions was introduced. As we noted, a comparison of these two notions reveals that the computable families of **pr**-functions are just the families weakly enumerable relative to the Kleene numbering κ.

It is clear that every totally enumerable family is weakly enumerable. Rice [127] has found conditions for the complement of a totally enumerable family to be weakly enumerable.

Theorem 9.1: *In each of the numberings κ, κ_0, κ_1, ..., the complement $\widetilde{B}$ of a totally enumerable family B of functions is weakly enumerable iff B consists of all extensions of finitely defined functions whose Gödel numbe*r *form a recursive set.*

Rice proved this theorem for the Post numbering ω. His reasoning, however, remains valid for the other numberings mentioned in the theorem. Since these arguments are comparatively simple, we shall not reproduce them here. ■

According to the definition, a family B of elements of a numbered set

$\langle A, \alpha\rangle$ is weakly enumerable iff there is a nonempty **re**-set M such that $\alpha M = B$. Thus it is natural to call a family B *weakly recursive* iff there is a nonempty recursive set P such that $\alpha P = B$. For complete numberings these notions coincide:

Theorem 9.2: *In a completely numbered set every weakly enumerable family is weakly recursive.*

In the proof of Theorem 2.3.4 of [XVIII] it is shown that for every completely numbered set $\langle A, \alpha\rangle$, there exists a **gr**-function $S(x, y)$ such that

$$\alpha S(x, y) = \alpha x, \quad S(x, y) \neq S(x, z) \quad (y \neq z;\ x, y, z \in N)\,.$$

Suppose the family $B \subseteq A$ is equal to αM, where M is the range of the **gr**-function $f(x)$. We define a new function $g(x)$ by the scheme:

$$g(0) = f(0)\,,$$

$$g(n+1) = S(f(n+1), \min_v(S(f(n+1), v) > g(n)))\,.$$

Since $\alpha g(n) = \alpha f(n) = \alpha f(n)$, we know $B = \alpha P$, where P is the range of g. But g is general recursive and monotonically increasing, so the set P is recursive. ■

We make one more definition: a family B in a simply numbered set $\langle A, \alpha\rangle$ is called *intrinsically productive* iff there exists a **gr**-function $g(x)$ such that for every $n \in N$,

$$W_n \neq \emptyset \text{ and } \alpha W_n \subseteq B \Rightarrow \alpha g(n) \in B \text{ and } \alpha g(n) \notin \alpha W_n\,.$$

As corollaries to this definition we have:

(i) An intrinsically productive family cannot be weakly enumerable;

(ii) Every intrinsically productive family includes a strongly enumerable (cf. § 3) subfamily.

A classical Cantor diagonal argument reveals the productiveness of a host of simple families of functions in the numberings κ_s ($s = \infty, 0, 1, \ldots$).

Theorem 9.3: *In each of the numberings κ_s, a family B is intrinsically productive if it consists of functions with infinite domains and contains the partial characteristic function of every infinite recursive set.*

Recall that the *partial characteristic function* of a set M is the function whose value is equal to 0 on M and is not defined outside of M.

To prove Theorem 9.3 we let $h(x)$ be a **pr**-function possessing these two properties:

(a) If $W_n \neq \emptyset$, then $h(n)$ is defined and $\kappa_s W_n = \kappa_s W_{h(n)}$;

(b) If $h(n)$ is defined, then $K(h(n), i)$ is defined for all $i \in N$.

Suppose for some $n \in N$ that $W_n \neq \emptyset$ and $\kappa_s W_n \subseteq B$. We define functions F_i^n by putting

$$F_i^n(x) = K(K(h(n), i), x) ;$$

all these **pr**-functions have infinite domains, for they belong to B. We successively choose points r_0, s_0 in the domain of F^n subject to the condition $r_0 < s_0$, points r_1, s_1 in the domain of F_1^n such that $s_0 < r_1 < s_1$, points r_2, s_2 in the domain of F_2^n such that $s_1 < r_2 < s_2$, etc. So the set $R^n = \{r_0, r_1, r_2, ...\}$ is infinite and recursive. Let $G^n(x)$ have the value 0 for $x \in R^n$ and be undefined for $x \notin R^n$.

Since $G^n(x)$ is a partial recursive function in n, x, there is a number $t \in N$ such that $G^n(x) = K(t, n, x)$. The function g defined by $g(n) = [t, n]$ shows B is intrinsically productive.

For suppose again that $W_n \neq \emptyset$ and $\kappa_s W_n \subseteq B$. Then by (a) we have

$$\kappa_s W_n = \kappa_s W_{h(n)} = \{F_0^n, F_1^n, ... \} .$$

Now $\kappa_s g(n) = G^n \neq F_i^n$ because F_i^n is defined at s_i, but G^n is not. At the same time, G^n is the partial characteristic function of the infinite recursive set R^n. Consequently, $\kappa_s g(n) \in B \sim \kappa_s W_n$. ∎

The next theorem is similar and has a similar proof:

Theorem 9.4: *In each of the numberings κ_s ($s = \infty$, 1, 2, ...), a family B is intrinsically productive if it consists of* **pr**-*functions with infinite domains and contains all* **gr**-*functions whose ranges are included in* $\{0, 1\}$ *(any two other fixed numbers will work, too).*

Let h, F_i^n, r_i, s_i be defined as in the preceding proof. We define G^n in a different fashion; namely, we put

$$G^n(0) = ... = G^n(r_0) = \overline{sg}\,(F_0^n(r_0)) ,$$

$$G^n(r_0+1) = ... = G^n(r_1) = \overline{sg}(F_1^n(r_1)) ,$$

$$\quad ... \qquad\qquad ... ,$$

where

$$\overline{sg}(x) = \begin{cases} 1 & \text{if } x = 0\,, \\ 0 & \text{if } x > 0\,. \end{cases}$$

Now suppose $W_n \neq \emptyset$ and $\kappa_s W_n \subseteq B$. Then the functions G^n, F_i^n differ in value at the point r_i; hence, $G^n \notin \kappa_s W_n$. But $G^n \in B$, for it is general recursive and has 0, 1 as its only possible values. ∎

The concept of intrinsic productiveness for families of recursively enumerable sets was introduced in Dekker and Myhill [25], where many important families of **re**-sets are shown to be intrinsically productive. A series of results connected with the Kleene numbering and related to the problems considered in §§6 and 9 are contained in Shapiro [149].

NOTES

(¹) We should assume $B_0 \neq A$ to ensure that $B_e \neq \emptyset$ when $s = 0$.

(²) In the original, the limit on the sum was incorrectly set at n. That $\mathfrak{R}$ is a well-defined map from $\mathcal{F}^{(2)}_{\mathrm{pr}}$ onto $\mathcal{F}_{\mathrm{pr}}$ is easily seen.

(³) Not necessarily. This function is finitely defined and becomes a restriction of f when restricted to the domain of f. A remedy is to require p to have the form $[s, m]$ given by (14) and to raise the upper limit of the sum in (13) from x to $x + 1$.

(⁴) Although C is, in general, the range of many projective operators, C also has many principal numberings that are induced by no such projections, as long as it contains at least two functions.

CHAPTER 26

PROBLEMS IN THE THEORY OF CLASSES OF MODELS

Introduction

The theory of models, also often called the theory of algebraic systems, is a discipline on the boundary between abstract algebra and mathematical logic. In the course of the five years since the previous All-Union Mathematical Congress many important papers on the theory of models have appeared, so that today, thanks to these, we can speak of model theory as a developed, independent discipline. In the present report we shall try to outline the contemporary problems in several areas of the theory of classes of models and to survey the basic results in these areas published chiefly in the last few years. A summary of earlier results is included in the well-known survey [107] of A. Mostowski.

§1. Fundamental concepts

§1.1. *Algebras and models*

An *n-place* (or *n-ary*) *operation* on a set M is a function of n arguments defined on M and taking values in this same set. An *n-ary predicate* on M is an n-ary function defined on M and taking values in a particular two-element set. The elements of the latter are called *truth* and *falsity* and written as T and F. If P is an n-ary predicate on M and $a_1, \ldots, a_n \in M$ and $P(a_1, \ldots, a_n) = \mathsf{T}$, then we say the elements $a_1, \ldots, a_n$ are *in the relation P.*

The relations "a divides b", "a is relatively prime to b" can serve as examples of binary predicates defined on the set of natural numbers. The relation "the point A lies between the points B and C" can serve as an example of a ternary predicate on the set of points on a straight line.

Let $f(x_1, \ldots, x_n)$ be an operation on the set M. We introduce a predicate $F(x_1, \ldots, x_n, x)$ by setting

$$F(x_1, \ldots, x_n, x) = \begin{cases} \mathsf{T} \text{ if } f(x_1, \ldots, x_n) = x\,, \\ \mathsf{F} \text{ if } f(x_1, \ldots, x_n) \neq x\,. \end{cases} \tag{1}$$

The predicate F *corresponds* to the operation f. Knowing F, we know f; therefore, the study of operations can be viewed as the study of predicates of a certain special form.

A nonempty set M together with a system of operations $f_1, \ldots, f_s$ arranged in a specific order is called an *algebra*. A nonempty set M supplied with a finite sequence of predicates $P_1, \ldots, P_s$ is called a *model*. The set M, the operations f_i, and the predicates P_i are called the *base set*, the *basic operations*, and the *basic predicates* of the given algebra or model. The indicated algebra and model are denoted more explicitly by

$$\langle M; f_1, \ldots, f_s \rangle\,, \quad \langle M; P_1, \ldots, P_s \rangle\,. \tag{2}$$

Thus, e.g., the algebra whose base is the set of natural numbers and whose single basic operation is the addition of numbers can be denoted by $\langle \{0, 1, 2, \ldots\}, + \rangle$. Similarly, $\langle \{0, 1, 2, \ldots\}; \leqslant, | \rangle$ denotes the model with the same base and with the order relation $\leqslant$ and the divisibility relation $|$ as basic predicates.

If n_i is the rank (arity) of the operation f_i or the predicate P_i in (2), then the sequence $\langle n_1, \ldots, n_s \rangle$ is called the *similarity type* of the corresponding algebra or model. E.g., the type of a groupoid is $\langle 2 \rangle$, the type of a lattice is $\langle 2, 2 \rangle$, and the type of a boolean algebra is $\langle 2, 2, 1 \rangle$.

Let $\mathfrak{A} = \langle M; f_1, \ldots, f_s \rangle$ be an algebra. Let $F_i(x_1, \ldots, x_{n_i}, x)$ be the predicate corresponding to the operation f_i, as defined by the scheme (1). The model $\mathfrak{M} = \langle M; F_1, \ldots, F_s \rangle$ is called the model *corresponding* to the algebra $\mathfrak{A}$ or *representing* this algebra. Since the model $\mathfrak{M}$ completely determines the algebra it represents, the theory of algebras can be viewed as the theory of models of a special form. E.g., the ring $\langle \{0, \pm 1, \pm 2, \ldots\}; +, \times \rangle$ can be viewed as an algebra of type $\langle 2, 2 \rangle$ and as a model $\langle \{0, \pm 1, \pm 2, \ldots\}; S, P \rangle$ of type $\langle 3, 3 \rangle$, where S and P are the predicates representing addition and multiplication.

A *homomorphism* of a model $\mathfrak{M} = \langle M; P_1, \ldots, P_s \rangle$ into a similar model $\mathfrak{M}' = \langle M'; P_1', \ldots, P_s' \rangle$ is a map φ from M into M' such that for any elements $x_1, \ldots, x_{n_i}$ of M, the truth of $P_i(x_1, \ldots, x_{n_i})$ implies the truth of $P_i'(x_1\varphi, \ldots, x_{n_i}\varphi)$ ($i = 1, \ldots, s$). If in addition the map φ is 1–1 and onto, and the inverse map φ^{-1} is also a homomorphism, then φ is called an *isomorphism* from $\mathfrak{M}$ onto $\mathfrak{M}'$. The models $\mathfrak{M}$ and $\mathfrak{M}'$ are said to be *isomorphic* iff there exists an isomorphism from $\mathfrak{M}$ onto $\mathfrak{M}'$.

A homomorphism φ of a model $\mathfrak{M}$ into a model $\mathfrak{M}'$ is said to be *strong* iff the truth of $P_i'(x_1\varphi, ..., x_{n_i}\varphi)$ implies the existence of $y_1, ..., y_{n_i}$ in M such that $y_1\varphi = x_1\varphi, ..., y_{n_i}\varphi = x_{n_i}\varphi$ and $P_i(y_1, ..., y_{n_i})$ is true ($i = 1, ..., s$).

For algebras the concepts of homomorphism and isomorphism are defined as usual; ordinary and strong homomorphisms coincide for algebras. It is easy to see that a map from an algebra into a similar algebra is a homomorphism iff the map is a homomorphism relative to the corresponding models.

§1.2. *Classes of models*

An arbitrary system of similar models is called a *class of models*. The *type* of the class is the type of the models that compose it. Among the classes of models of a given type is the largest, the class of all models of this type. The remaining classes are thus subclasses of this class.

(*N.B.:* If the system of all models of a given type is considered to be a set, then certain antimonies of the set-of-all-sets type can arise. Therefore, in studying classes of models we either limit these classes in some way, or confine ourselves to some axiomatization of set theory, etc.)

A class of models is called *abstract* iff it contains all models isomorphic to any of its members. In what follows, the classes of models we consider will almost always be abstract.

In studying general properties of models in a fixed class $\mathfrak{K}$, it is customary to introduce special symbols designating the first, second, etc. basic predicates in an arbitrary $\mathfrak{K}$-model. The sequence of these symbols with their ranks indicated is called the *signature* of $\mathfrak{K}$. Thus, to specify a model with a given signature, we must indicate a set M and associate with each signature symbol a predicate on M of appropriate rank. This concrete predicate on M is called the *value* of the corresponding predicate symbol in the model with base M thus obtained.

The concepts of algebras and models can be generalized in various directions. Above all, it is often convenient to consider models and algebras in which, besides basic predicates or operations, certain distinguished elements play an essential role. Symbolic designations of these elements are included in the signature of the respective model or algebra, and in the notation of the type their presence is denoted by zeros. E.g., to prescribe an algebra of type $\langle 2; 0, 0\rangle$ means to specify a base M, to define a binary operation on it, and to fix a pair of elements.

Sometimes we encounter a set on which operations, predicates, and distinguished elements are given simultaneously. We adopt the term *algebraic system* for such a structure. Models and algebras are special cases of algebraic systems. On the other hand, by replacing the operations in an algebraic system with their corresponding predicates, we obtain a model whose study yields the properties of the original algebraic system.

Till now we have considered only those systems endowed with a finite number of operations, predicates, and distinguished elements; moreover, each operation and predicate has had only a finite number of arguments. We can depart from these conditions of finiteness in two ways:

(1) We can admit infinite sequences of basic predicates, operations, and distinguished elements, but require the first two to have finite ranks;

(2) We can admit operations and predicates with infinite numbers of arguments.

Although there are papers concerning the second course of action, we shall focus on systems, all of whose operations and predicates depend on a finite number of variables. Since systems with an infinite number of predicates and distinguished elements are seen so frequently, we shall commonly apply the term "model" to any system with arbitrary cardinal numbers of predicates and distinguished elements. The power of the set of signature symbols is called the *order* of the model, and the power of the base of the model is called simply the *power* of the model.

In the development of the general theory of algebraic systems to the present time the following four directions have emerged:

(A) *The general theory of algebras.* In spirit and problems this discipline is closest to classical algebra, e.g., to the abstract theory of groups. Homomorphisms and their kernels, defining relations, direct and free decompositions were and still are the main objects of concern here.

(B) *The theory of classes of models.* The characteristic feature of this theory is that the study of properties of classes of models is carried out in connection with the logical language in which the classes under investigation are defined. Therefore, in the theory of classes of models approximately equivalent roles are played by abstract algebra, in whose language are formulated general properties of classes of models, and mathematical logic with its formal languages, by means of which classes of models are specified.

(C) *Elementary theories.* Suppose with the aid of arbitrary logical means there is given some class $\mathfrak{X}$ of models, which may consist of a single model. Starting with the basic predicates in the class $\mathfrak{X}$, we can use these and other logical means to define a host of new predicates on $\mathfrak{X}$-models. We ask: which of these can be defined in the language of first-order predicate logic (FOPL), the basic classical logic? Other questions: what sort of properties of $\mathfrak{X}$ are describable in the FOPL language? which models are indistinguishable from one another in that language? does there exist an algorithm for recognizing from the notation of their FOPL descriptions those properties held by all

$\mathcal{K}$-models? These and similar problems are basic to the so-called "elementary" theory of the class concerned. A general account of some of the problems mentioned is found in the book [166] of A. Tarski, A. Mostowski, and R.M. Robinson. The elementary theory of rings is analyzed in greater detail by R.M. Robinson [139], the elementary theory of fields by J. Robinson [134], [136] and the author [XVI], the elementary theory of groups by W. Szmielew [155] and the author [XIV], [XV], [XIX]–[XXI], etc.

(D) *Constructive algebras and models.* This direction is closely related to the theory of algorithms and recursive functions. The above definitions of model and algebra deal with an arbitrary base set, arbitrary predicates and operations. If these are qualified with the word "constructive" throughout the definitions, and the notion of constructiveness is appropriately made precise, then definitions of constructive models and algebras are obtained. The theory of such structures is still in its formative stages. An initial summary of the basic concepts and results is included in the author's survey [XVIII], where the relevant literature is indicated.

In what follows we shall restrict ourselves almost exclusively to questions concerning the second of these four sectors, i.e., the theory of classes of models.

§1.3. *The first-order language*

As already mentioned, the fundamental formal language of the theory of models is that of first-order predicate logic with equality (FOPL). This language is explicitly described in every manual of mathematical logic (e.g., D. Hilbert and W. Ackermann [56], P.S. Novikov [117]). Its alphabet consists of: (1) individual variables $\boldsymbol{x}, \boldsymbol{y}, \boldsymbol{z}, \boldsymbol{a}, \boldsymbol{b}, \boldsymbol{x}_1, \boldsymbol{x}_2, \ldots$, the elements of some base set serving as their values; (2) predicate symbols $\boldsymbol{P}, \boldsymbol{R}, \boldsymbol{S}, \boldsymbol{T}, \boldsymbol{P}_1, \boldsymbol{P}_2, \ldots$ of various ranks, predicates of corresponding ranks on the base set serving as their values; (3) logical symbols $\&, \vee, \neg, \rightarrow, \approx, \forall, \exists$ with the respective semantic interpretations *and, or, not, if ... then, equals, for every element, there exists an element ... such that*; (4) the auxiliary punctuation symbols) (, . FOPL *formulas* are finite sequences of these alphabet symbols constructed according to the usual rules.

In the theory of classes of models FOPL formulas are interpreted in the following manner. Let $\mathcal{K}$ be a class of models whose signature Σ consists of the predicate symbols $\boldsymbol{P}_1, \boldsymbol{P}_2, \ldots$ and individual symbols $\boldsymbol{a}_1, \boldsymbol{a}_2, \ldots$ Let Φ be a FOPL formula of signature Σ, i.e., a formula such that Σ contains all its predicate symbols (other than $\approx$), but none of its bound individual variables. A portion of the free variables in Φ can appear in Σ. Let these be $\boldsymbol{a}_{i_1}, \ldots, \boldsymbol{a}_{i_k}$. Suppose the remaining free variables are $\boldsymbol{x}_1, \ldots, \boldsymbol{x}_n$. Finally, suppose the predicate symbols occurring in Φ are $\boldsymbol{P}_{j_1}, \ldots, \boldsymbol{P}_{j_l}$. To indicate all of this, we can write

$$\Phi = \Phi(\boldsymbol{x}_1, \ldots, \boldsymbol{x}_n) = \Phi(\boldsymbol{x}_1, \ldots, \boldsymbol{x}_n; \boldsymbol{a}_{i_1}, \ldots, \boldsymbol{a}_{i_k}; \boldsymbol{P}_{j_1}, \ldots, \boldsymbol{P}_{j_l}) .$$

Now take a model

$$\mathfrak{M} = \langle M; P_1, P_2, \ldots; \; a_1, a_2, \ldots \rangle$$

with signature Σ. In this model the symbols $\boldsymbol{a}_{i_1}, \ldots, \boldsymbol{a}_{i_k}$ and $\boldsymbol{P}_{j_1}, \ldots, \boldsymbol{P}_{j_l}$ designate certain well-defined elements and predicates. If, in addition, values in M are chosen for $\boldsymbol{x}_1, \ldots, \boldsymbol{x}_n$, then the formula Φ is transformed under this semantic interpretation into a statement about the model $\mathfrak{M}$ and the chosen elements $x_1, \ldots, x_n$ that is either true or false. Thus to every choice of n elements (not necessarily distinct) from the model $\mathfrak{M}$ corresponds one of the values T, F. This means we have an n-ary predicate on $\mathfrak{M}$, called the *predicate defined by the formula* $\Phi(\boldsymbol{x}_1, \ldots, \boldsymbol{x}_n)$. A predicate on $\mathfrak{M}$ (on each $\mathfrak{K}$-model) definable there by a FOPL formula of signature Σ is said to be *formular* on $\mathfrak{M}$ (in $\mathfrak{K}$).

A formula with no free occurrences of individual variables is said to be a *closed formula* (or *sentence*). Such a formula is either true or false in every model with whose signature it is compatible, and so can be viewed as an assertion about the properties of the model. Assertions of this form will be called *axioms* ([1]).

The FOPL language is invariant with respect to isomorphisms of models and, therefore, is eminently suited for expressing abstract properties of models.

More explicitly, this means the following. Let $\mathfrak{M}, \mathfrak{M}'$ be two models with the same signature Σ, let φ be an isomorphism from $\mathfrak{M}$ onto $\mathfrak{M}'$, and let Φ be a FOPL formula of signature Σ in which $\boldsymbol{x}_1, \ldots, \boldsymbol{x}_n$ occur as the only free variables. Then for every choice of values $\boldsymbol{x}_1, \ldots, \boldsymbol{x}_n$ in $\mathfrak{M}$, $\Phi(\boldsymbol{x}_1, \ldots, \boldsymbol{x}_n)$ and $\Phi(\boldsymbol{x}_1\varphi, \ldots, \boldsymbol{x}_n\varphi)$ are either both true or both false in the respective models. In particular, if a FOPL sentence is true in some model $\mathfrak{M}$, then it is true in all models isomorphic to $\mathfrak{M}$.

Finally, we mention that a formula of the form

$$(\boldsymbol{x}_1) \ldots (\boldsymbol{x}_m) \Phi(\boldsymbol{x}_1, \ldots, \boldsymbol{x}_m) ,$$

where Φ is a FOPL formula with no quantifiers, is called a *universal formula* or a $\forall$-*formula*. One of the form

$$(\boldsymbol{x}_1) \ldots (\boldsymbol{x}_m) (\exists \boldsymbol{y}_1) \ldots (\exists \boldsymbol{y}_n) \Phi(\boldsymbol{x}_1, \ldots, \boldsymbol{x}_m, \boldsymbol{y}_1, \ldots, \boldsymbol{y}_n)$$

is called a *Skolem formula* or a $\forall\exists$-*formula.* A formula

$$(x_1)\dots(x_m)(\exists y_1)\dots(\exists y_n)(z_1)\dots(z_p)\Phi(x_1,\dots,x_m,y_1,\dots,y_n,z_1,\dots,z_p)$$

is a $\forall\exists\forall$-*formula,* and so on.

§2. Axiomatizable classes of models

§2.1. *General properties*

A class $\mathcal{K}$ of models with arbitrary (possibly infinite) signature Σ is said to be *(first-order) axiomatizable* – or *arithmetic*, as in Tarski [163] – iff the characteristic properties of its members can be described in the FOPL language, i.e., iff there exists a system S, generally infinite, of closed FOPL formulas of signature Σ such that $\mathcal{K}$ contains those and only those models with signature Σ in which all the sentences in S are true. We say that $\mathcal{K}$ is *determined by the axiom system* S.

A class of models is called *finitely axiomatizable* iff it has a finite signature and can be specified by a finite system of axioms. A class of models with finite signature is *recursively axiomatizable* iff it can be characterized by a recursive set of axioms.

As mentioned earlier, every class of algebras can be viewed as a class of models by replacing the basic operations with the corresponding predicates. With this in mind it is easy to see that many important classes of algebras are axiomatizable. E.g., finitely axiomatizable are the classes of all groups, all rings, all lattices, all fields, etc. The classes of all torsion-free groups, all fields of characteristic zero, all algebraically closed fields are recursively but not finitely axiomatizable, as are many others.

Suppose we have a model $\mathfrak{M} = \langle M; P_1, P_2, \dots \rangle$, whose signature may be infinite, and let M' be an arbitrary nonempty subset of M. Let $P_1', P_2', \dots$ be the predicates on M' whose values coincide with the values of $P_1, P_2, \dots$, respectively. The model $\mathfrak{M}' = \langle M'; P_1', P_2', \dots \rangle$ is called a *submodel* of $\mathfrak{M}$ (we write $\mathfrak{M}' \subseteq \mathfrak{M}$). Since the submodel $\mathfrak{M}'$ is completely determined in by the subset M', instead of "the submodel $\langle M'; P_1', P_2', \dots \rangle$" we can simply say "the submodel M'".

If $\mathcal{K}$ is a class of models with signature Σ, and $\mathfrak{M}$ is a model with signature Σ, then the submodels of $\mathfrak{M}$ that belong to $\mathcal{K}$ are called $\mathcal{K}$-*submodels* of $\mathfrak{M}$. E.g., let $\mathcal{K}$ be the class of all groups. Then submodels of a group $\mathfrak{G}$ will not in general be subgroups of $\mathfrak{G}$. The $\mathcal{K}$-submodels of $\mathfrak{G}$ are exactly its subgroups.

Historically, the first general theorem on axiomatizable classes of models was the classical

Theorem of Löwenheim–Skolem: *Let $\mathfrak{K}$ be an axiomatizable class of models whose signature has power $\mathfrak{p}$, and let $\mathfrak{M}$ be a $\mathfrak{K}$-model. Then each set of elements in $\mathfrak{M}$ of power $\mathfrak{m}$ is included in some $\mathfrak{K}$-submodel of $\mathfrak{M}$ of power not greater than* $\mathfrak{m} + \mathfrak{p} + \aleph_0$. ■

Special cases of this theorem: (1) every finite or countable subset of a model in an axiomatizable class $\mathfrak{K}$ with finite signature is included in a finite or countable $\mathfrak{K}$-submodel of this model; (2) every model in an axiomatizable class $\mathfrak{K}$ includes a finite or countable $\mathfrak{K}$-submodel.

The case (2) was first proved by L. Löwenheim [92]. The theorem in its general form was proved by T. Skolem [151].

By the Löwenheim–Skolem theorem, in every nonempty axiomatizable class of models whose signature has infinite power $\mathfrak{p}$, there is a model of power not greater than $\mathfrak{p}$, and in every nonempty axiomatizable class of models with finite signature there is a finite or countable model. The question arises: in an axiomatizable class is there a model of greatest power? In the general case a negative answer is given by the

Theorem on extending models (A.I. Mal'cev [I], §6): *If $\mathfrak{M}$ is an infinite model in an axiomatizable class $\mathfrak{K}$, an $\mathfrak{n}$ is an arbitrary cardinal number, then $\mathfrak{K}$ contains a model of power greater than $\mathfrak{n}$ that includes $\mathfrak{M}$ as a submodel. If an axiomatizable class $\mathfrak{K}$ contains models of arbitrarily large finite powers, then $\mathfrak{K}$ contains an infinite model.* ■

Although a weaker formulation of this theorem appears in [I], the proof given there yields the extension theorem in its full generality. That proof is based on the following property of axiomatizable classes:

Local theorem for FOPL (compactness theorem) (K. Gödel [46], A.I. Mal'cev [I], [II]): *Suppose there is given an infinite system of axiomatizable classes of models of a fixed signature. If the intersection of every finite subsystem of classes in this system is nonempty, then the intersection of the whole system is also nonempty.* ■

For a finite signature this theorem is an immediate consequence of Gödel's completeness theorem. In the general case the proof relies on the axiom of choice. J. Łoś [90] has shown the converse: from the compactness theorem the axiom of choice ([2]) can be derived.

(*N.B.:* Besides the original proof [I] of the compactness theorem, based on the so-called diagrams of models and the reduction of axioms to Skolem form, many other proofs based on essentially different ideas have been published. Cf. e.g., [103], [123], [54]).

§2.2. *Small models*

In the extension theorem the question of for just what powers there are extensions of a given infinite model is left open. In order to answer it let us call a model $\mathfrak{M}$ *regular* when it is infinite and its order (the power of its signature) does not exceed its power. In the contrary case we call $\mathfrak{M}$ *small* [XIII].

From the extension and Löwenheim–Skolem theorems it immediately follows that every regular model $\mathfrak{M}$ in an axiomatizable class $\mathcal{K}$ admits a proper isomorphic embedding in a $\mathcal{K}$-model of any previously selected power not less than the power of $\mathfrak{M}$.

In extending small models we may encounter signularities. E.g., in [XIII], §1 are constructed two classes $\mathcal{K}_1, \mathcal{K}_2$ (each with $2^{\aleph_0}$ signature symbols) that have the following properties: (a) the class $\mathcal{K}_1$ contains models of arbitrarily large finite powers, but every infinite $\mathcal{K}_1$-model has power greater than or equal to the power of the continuum; (b) in the class $\mathcal{K}_2$ there is a countable model, every proper $\mathcal{K}_2$-extension of which has power not less than that of the continuum. Nevertheless, we have the

Theorem ([XIII], §2): *If the axiomatizable class* $\mathcal{K}$ *contains an infinite model* $\mathfrak{M}$ *of power* $\mathfrak{m}$, *then* $\mathfrak{M}$ *has a proper* $\mathcal{K}$*-extension of power* $\mathfrak{m}^{\aleph_0}$. *If* $\mathcal{K}$ *contains models of powers* $\mathfrak{m}_1 < \mathfrak{m}_2 < \ldots$, *then* $\mathcal{K}$ *contains a model whose power lies between* $\mathfrak{m}_1 + \mathfrak{m}_2 + \ldots$ *and* $\mathfrak{m}_1 \cdot \mathfrak{m}_2 \cdot \ldots$. ∎

If we use the generalized continuum hypothesis (GCH), we can prove ([XIII], §3) that *every infinite model* $\mathfrak{M}$ *in an axiomatizable class* $\mathcal{K}$ *admits a* $\mathcal{K}$*-extension of any previously specified power greater than the power of* $\mathfrak{M}$. ∎

These same problems were considered independently by M.O. Rabin [122]. He calls an algebra $\mathfrak{A} = \langle A; f_1, f_2, \ldots \rangle$ *complete* iff for every finitary operation f on A, there is among the basic operations f_s of $\mathfrak{A}$ an operation equal to f. The notion of a complete model is defined analogously. By using the techniques of ultraproducts (see § 4 below), Rabin proves assertions analogous to some of the results formulated above for small models; he also proves the following basic proposition:

Theorem (Rabin [122]): *Assume the* GCH *holds. If the power* $\mathfrak{m}$ *of a complete model* $\mathfrak{M}$ *is less than the first weakly inaccessible cardinal, and if* $\mathfrak{m}^{\aleph_0} > \mathfrak{m}$, *then* $\mathfrak{M}$ *has no proper extensions of power* $\mathfrak{m}$ *that are elementarily equivalent to it.* ∎

Which of these assertions are equivalent to the GCH – and which are provable without its aid – apparently remain open questions.

§2.3. *Completeness and categoricity*

A nonempty class $\mathfrak{K}$ of models is called *categorical* iff all its members are isomorphic, i.e., iff up to isomorphism $\mathfrak{K}$ consists of only one model. A nonempty class $\mathfrak{K}$ with signature Σ is called *complete* iff its members cannot be differentiated from one another in the FOPL language. The latter means that every FOPL sentence of signature Σ that is true in one $\mathfrak{K}$-model is also true in every other $\mathfrak{K}$-model.

From a set-theoretical point of view, the complete, axiomatizable classes with signature Σ are minimal among the axiomatizable classes with the given signature, and each complete but nonaxiomatizable class is a subclass of a complete, axiomatizable class. E.g., let $\mathfrak{M}$ be a model with signature Σ. Let S denote the collection of all FOPL sentences of signature Σ that are true in $\mathfrak{M}$. Then the class $\mathfrak{K}$ of all models with signature Σ in which all the axioms in S are true is the complete and axiomatizable class containing $\mathfrak{M}$.

A system of axioms of signature Σ is called *complete* (*categorical*) iff the class of models with signature Σ that satisfy the given axiom system is complete (respectively, categorical).

The theorem on extending models shows that no categorical class containing infinite models can be axiomatizable. Therefore, the concept of categoricity in a given power, introduced by J. Łoś [87], presents greater interest. According to Łoś, a nonempty class $\mathfrak{K}$ of models is *categorical in power* $\mathfrak{m}$ iff all $\mathfrak{K}$-models of power $\mathfrak{m}$ are isomorphic to one another.

For example, it is known that:

(a) all countable, linearly ordered, dense-in-themselves sets without end points are isomorphic;

(b) all algebraically closed fields of given characteristic that have the same uncountable power are isomorphic.

Hence, the class of densely ordered sets without end points is categorical in $\aleph_0$ (but not in any uncountable power), while the class of all algebraically closed fields of fixed characteristic is categorical in every uncountable power (but not in $\aleph_0$).

Let $\mathfrak{K}$ be a class of models, and let $\mathfrak{p}$, $\mathfrak{q}$ be cardinal numbers. By $\mathfrak{K}_\mathfrak{p}$, $\mathfrak{K}^\mathfrak{q}$, $\mathfrak{K}_\mathfrak{p}^\mathfrak{q}$ we denote the classes of $\mathfrak{K}$-models whose power $\mathfrak{m}$ satisfies the relations $\mathfrak{p} \leqslant \mathfrak{m} \leqslant \mathfrak{q}$, respectively. Then we have

Theorem 1 (A.I. Mal'cev [XIII], §4)): *Suppose the classes $\mathfrak{K}$ and $\mathfrak{L}$ are axiomatizable and similar; suppose $\mathfrak{p}$ is an infinite cardinal number such that $\mathfrak{K}_\mathfrak{p}^\mathfrak{p} \subseteq \mathfrak{L}$. Then $\mathfrak{K}_{\aleph_0} \subseteq \mathfrak{L}$ if $\mathfrak{p}$ is not less than the order of $\mathfrak{K}$, and $\mathfrak{K}_{\aleph_0}^\mathfrak{p} \subseteq \mathfrak{L}$ in any case.* ∎

The second assertion is proved with the aid of the GCH.

Now let us assume the axiomatizable class $\mathcal{K}$ is categorical in an infinite power $\mathfrak{p}$ greater than or equal to the power of the signature of $\mathcal{K}$. Let $\mathfrak{M}$ be a $\mathcal{K}$-model of power $\mathfrak{p}$, and let $\mathcal{L}$ be the axiomatizable and complete class similar to $\mathcal{K}$ and containing $\mathfrak{M}$. Suppose also that all $\mathcal{K}$-models are infinite. Then by Theorem 1, $\mathcal{K}$ is included in $\mathcal{L}$; since $\mathcal{L}$ is minimal, these classes coincide. Thus we have derived

Theorem 2 (R.L. Vaught [132]): *If all models in an axiomatizable class $\mathcal{K}$ are infinite, and $\mathcal{K}$ is categorical in an infinite power not less than its order, then $\mathcal{K}$ is complete.* ■

By analogy with the concept of categoricity in power, we can introduce the notion of a class being complete in a given power. From the above proof of Vaught's theorem it is seen that in the formulation of this theorem, the condition of categoricity in an infinite power can be replaced with the weaker assumption of completeness in an infinite power.

A class $\mathcal{K}$ of models with finite signature Σ is called *recursively decidable* iff there exists an algorithm enabling one to tell for every FOPL sentence of signature Σ whether or not this sentence is true in all $\mathcal{K}$-models. An important connection between completeness and recursive decidability is given by the obvious

Remark: *Every recursively axiomatizable and complete class of models is recursively decidable.* ■

From this observation and Theorem 2 it follows that, in particular, the classes discussed in connection with (a), (b) above are recursively decidable.

A. Ehrenfeucht and A. Mostowski [27] showed that in every axiomatizable class containing infinite models, there are models with arbitrarily large automorphism groups. Using the results of this article, Ehrenfeucht [26] found a series of properties of axiomatizable classes categorical in a power of the form $2^{\mathfrak{m}}$ ($\mathfrak{m} \geqslant \aleph_0$). In particular, he proved that in such classes no linear order relation can be formular.

§2.4. *A set-theoretical characterization of axiomatizable classes*

In connection with the general properties of axiomatizable classes of models stated in §2.1, the problem naturally arises of finding necessary and sufficient conditions characterizing the axiomatizable classes. Since the language in which these tests are to be formulated is not specified, the problem can have a variety of solutions. Thus, J. Łoś [88] characterized the axiomatizable classes in the language of boolean algebras, and J. Mycielski [110] gave a characterization in the language of functions. These works, however, have not yet been published in full, and we shall not present their contents.

In the important article [17], appearing in 1959, C.C. Chang gave a lucid characterization in the language of mappings for those classes describable by Skolem axioms. Suitably generalizing Chang's definitions, A.D. Taĭmanov [158] solved the wider problem of finding set-theoretical characterizations for general axiomatizable classes and for finitely axiomatizable and other sorts of axiomatizable classes of models. Moreover, as simple corollaries he obtained the well-known theorems of Łoś–Tarski, Łoś–Suszko, and A. Robinson, characterizing ∀- and ∀∃-classes. Taĭmanov's results are briefly stated below; other characterizations are given in §4.

Let $\mathfrak{M}_1$ be a submodel of the model $\mathfrak{M}$. An *n-extension* of $\mathfrak{M}_1$ in $\mathfrak{M}$ is a submodel $\mathfrak{M}_2$ of $\mathfrak{M}$ obtained by adjoining to $\mathfrak{M}_1$ not more than n new elements of $\mathfrak{M}$. We also say that $\mathfrak{M}_1$ is an n_1*-submodel* of $\mathfrak{M}$ iff $\mathfrak{M}_1$ contains not more than n_1 elements (or is empty when $n_1 = 0$).

We consider similar models $\mathfrak{M}$, $\mathfrak{N}$ and let $\mathfrak{M}_1$ be an n_1-submodel of $\mathfrak{M}$. An isomorphic mapping φ_1 of $\mathfrak{M}_1$ into $\mathfrak{N}$ is called a (n_1, n_2)*-map* iff every n_2-extension $\mathfrak{N}_2$ of the submodel $\mathfrak{N}_1 = \mathfrak{M}_1^{\varphi_1}$ in $\mathfrak{N}$ can be mapped into under an isomorphism φ_2 that coincides with φ_1^{-1} on $\mathfrak{N}_1$.

Similarly, the map φ_1 is called a (n_1, n_2, n_3)*-map* of $\mathfrak{M}_1$ into $\mathfrak{N}$ iff for every n_2-extension $\mathfrak{N}_2$ of the submodel $\mathfrak{N}_1 = \mathfrak{M}_1^{\varphi_1}$ in $\mathfrak{N}$, there exists an isomorphism φ_2 of $\mathfrak{N}_2$ into $\mathfrak{M}$ that coincides with φ_1^{-1} on $\mathfrak{N}_1$ and such that for every n_3-extension $\mathfrak{M}_3$ of the submodel $\mathfrak{M}_2 = \mathfrak{N}_2^{\varphi_2}$ in $\mathfrak{M}$, there exists an isomorphism φ_3 of $\mathfrak{M}_3$ into $\mathfrak{N}$ coinciding with φ_2^{-2} on $\mathfrak{M}_2$. In a similar fashion we define the notion of an $(n_1, n_2, ..., n_l)$-mapping of an n_1-submodel $\mathfrak{M}_1$ of the model $\mathfrak{M}$ into the model $\mathfrak{N}$.

If for a given submodel $\mathfrak{M}_1$ of a model $\mathfrak{M}$ there exists an $(n_1, ..., n_l)$-map of $\mathfrak{M}_1$ into a model $\mathfrak{N}$, then we write

$$\mathfrak{M} \leqslant (\mathfrak{M}_1; n_1, ..., n_l)\, \mathfrak{N} \tag{3}$$

and say that $\mathfrak{M}_1$ is $(n_1, ..., n_l)$*-mappable* from $\mathfrak{M}$ into $\mathfrak{N}$.

If the relation (3) holds for every n_1-submodel $\mathfrak{M}_1$ of the model $\mathfrak{M}$, then we write

$$\mathfrak{M} \leqslant (n_1, ..., n_l)\, \mathfrak{N} \tag{4}$$

and say that $\mathfrak{M}$ is $(n_1, ..., n_l)$*-embeddable* in $\mathfrak{N}$.

In particular, from these definitions it follows that for $n_1 = 0$ the relation (4) is equivalent to the relation $\mathfrak{N} \leqslant (n_2, ..., n_l)\mathfrak{M}$.

Models $\mathfrak{M}$, $\mathfrak{N}$ are called *$(n_1, \ldots, n_l)$-equivalent* iff each is $(n_1, \ldots, n_l)$-embeddable in the other.

We recall that models $\mathfrak{M}$ and $\mathfrak{N}$ are called *elementarily equivalent* (or *arithmetically equivalent,* or FOPL*-indistinguishable*) iff every FOPL sentence of the signature of $\mathfrak{M}$ that is true in one of the models $\mathfrak{M}$, $\mathfrak{N}$ is also true in the other.

Theorem 3 (Taĭmanov [159]): *In order for similar models $\mathfrak{M}$, $\mathfrak{N}$ to be elementarily equivalent, it is necessary and sufficient that they be $(n_1, \ldots, n_l)$-equivalent for every sequence $\langle n_1, \ldots, n_l \rangle$ of length $l = 1, 2, \ldots$ of natural numbers.* ■

To formulate Taĭmanov's conditions for a class of models to be axiomatizable, we agree to call axioms of the form

$$(\forall \boldsymbol{x}_{11} \ldots \boldsymbol{x}_{1n_1})(\exists \boldsymbol{x}_{21} \ldots \boldsymbol{x}_{2n_2}) \ldots (\mho \boldsymbol{x}_{l1} \ldots \boldsymbol{x}_{ln_l})\, \Phi(\boldsymbol{x}_{11}, \ldots, \boldsymbol{x}_{ln_l}),$$

$$(\exists \boldsymbol{x}_{21} \ldots \boldsymbol{x}_{2n_2}) \ldots (\mho \boldsymbol{x}_{l1} \ldots \boldsymbol{x}_{ln_l})\, \Phi(\boldsymbol{x}_{11}, \ldots, \boldsymbol{x}_{ln_l})$$

$(n_1, n_2, \ldots, n_l)$-axioms and *$(0, n_2, \ldots, n_l)$-axioms,* respectively (here, $\mho = \forall, \exists$ and Φ is a quantifier-free FOPL formula).

Theorem 4 (Taĭmanov [159]): *An abstract class $\mathcal{K}$ of models with finite signature can be determined by $\forall\exists\forall \ldots \mho$-axioms ($\exists\forall\exists \ldots \mho$-axioms) iff whenever an arbitrary model $\mathfrak{M}$ with the signature of $\mathcal{K}$ is $(n_1, \ldots, n_l)$-embeddable ($(0, n_1, \ldots, n_l)$-embeddable) in a $\mathcal{K}$-model for some sequence $\langle n_1, \ldots, n_l \rangle$ of $\mathfrak{M}$ length l, then $\mathcal{K}$ belongs to $\mathcal{K}$.* ■

Theorem 5 (Taĭmanov [158], pt. I): *An abstract class $\mathcal{K}$ of models is axiomatizable iff whenever an arbitrary model $\mathfrak{M}$ of the type of $\mathcal{K}$ is $(n_1, \ldots, n_l)$-embeddable in some $\mathcal{K}$-model for some sequence $\langle n_1, \ldots, n_l \rangle$, then $\mathfrak{M}$ is itself a $\mathcal{K}$-model.* ■

Theorem 6 (Taĭmanov [159]): *Let $\langle n_1, \ldots, n_l \rangle$ be a sequence of natural numbers. Then an abstract class $\mathcal{K}$ of models with finite signature is axiomatizable by an $(n_1, \ldots, n_l)$-axiom (a $(0, n_1, \ldots, n_l)$-axiom) iff every model with this signature that is $(n_1, \ldots, n_l)$-embeddable (respectively, $(0, n_1, \ldots, n_l)$-embeddable) in a $\mathcal{K}$-model is itself a member of $\mathcal{K}$.* ■

For sequences of length 2, the definitions and theorems above were obtained by Chang [17]; for sequences of length 1, Theorems 4 and 6 reduce to theorems of Tarski [163] and Vaught [178], which respectively characterize those classes determined by $\forall$-axioms and by a single $\forall$-axiom.

The conditions for axiomatizability of Chang and Taĭmanov have the form of closure conditions on the class $\mathfrak{K}$: the "kinship" of a model $\mathfrak{M}$ to models in $\mathfrak{K}$ implies the membership of $\mathfrak{M}$ in $\mathfrak{K}$. Developing this old idea, S.R. Kogalovskiĭ [74] put Taĭmanov's conditions into "topological" form and in this way found series of other, still subtler topological properties of axiomatizable classes, including necessary and sufficient conditions.

§2.5. *Categories of models*

In §1.1 we defined the notion of a homomorphic mapping of a model into an arbitrary model of the same type. Every class $\mathfrak{K}$ of models of a given type can be viewed more generally as a system **O** of sets (the bases of the $\mathfrak{K}$-models) together with a system **H** of maps from some of the sets in **O** into others (the homomorphisms between $\mathfrak{K}$-models). The system **H** obviously satisfies the following demands:

(a) The identity map from each set in **O** onto itself belongs to **H**;

(b) For every φ, $\psi \in \mathbf{H}$, if φ is a map of M_1 into M_2 and ψ is a map of M_2 into M_3, then $\varphi\psi$ is a map from M_1 into M_3 and belongs to **H**.

A pair $\langle \mathbf{O}, \mathbf{H} \rangle$ consisting of a system **O** of sets and a system **H** of maps satisfying the conditions (a), (b) is called a *category of sets* or a *concrete category*. Many of the usual algebraic concepts, such as isomorphism, free and direct product, etc., can be defined in arbitrary categories. Therefore, it is natural to ask which properties of classes of models can be expressed in purely categoric terms, if categories of models are taken in the above sense. Thus, in [VIII] the question is posed: in which classes of models can one define the notion of defining relations so that it has the usual properties? As it turns out, it is expedient to give the answer in just the form of conditions laid on the category of models in the class.

Those properties of individual classes of models that are expressible in categoric language are studied in [VIII]–[X]. In many cases it proves possible to give a full categoric characterization of classes of models and of algebras. E.g., in [IX] the categories corresponding to quasiprimitive classes of algebras and to universally axiomatizable classes of models are so characterized.

The investigation of categoric properties of classes of models begun in [VIII]–[X] was continued by S.R. Kogalovskiĭ [73], [74] and J.R. Isbell [62]–[64]. In particular, Kogalovskiĭ [74] managed to find a categoric characterization of universally axiomatizable classes of algebras. By changing the very notion of the category of a class of models, he also found categoric characterizations for general axiomatizable classes of models.

In [62]–[64] Isbell considered general categories of algebras and models, and for these new categories he solved problems analogous to those described above.

Since the research in categories depends on a host of specialized concepts that are not required in the rest of this report, we shall not formulate even the basic results on categories of models, but be content with remarking that this branch of the theory of model classes exists.

§3. Some special axiomatizable classes

Let **V** be some property of classes of models. As above, we shall say that a sentence Φ of signature Φ of signature Σ has the property **V** when the class of models with signature Σ described by Φ has this property. Sometimes, an axiom with a given simple internal structure **W** has the property **V** trivially. It is then natural to ask: (i) is each axiom with property **V** equivalent to an axiom with structure **W**? (ii) does every axiomatizable class of models with property **V** admit an axiomatization by sentences with structure **W**? (iii) what requirements of a structural sort must a class $\mathcal{K}$ with property **V** meet in order to be axiomatizable – and that by axioms with structure **W**?

For an example we take as **V** the property: "all submodels of $\mathcal{K}$-models belong to $\mathcal{K}$". Then the theorems of Tarksi and Vaught mentioned in §2.4 at once answer the questions (i)–(iii) applied to the present case. We shall now state the solutions to these problems for several other properties.

§3.1. *Homomorphically closed classes*

A class $\mathcal{K}$ of models is called *homomorphically closed* iff together with each member $\mathfrak{M}$ it contains every model that is a homomorphic image of the model $\mathfrak{M}$. Similarly, if $\mathcal{L}$ is a subclass of a class $\mathcal{K}$ of models, then $\mathcal{L}$ is *homomorphically closed in* $\mathcal{K}$ (or *relative to* $\mathcal{K}$) iff every $\mathcal{K}$-model that is a homomorphic image of an $\mathcal{L}$-model is also an $\mathcal{L}$-model.

Let Φ be a FOPL formula in prenex conjunctive form. We say that the predicate symbol $\boldsymbol{P}$ has a *positive* (*negative*) *occurrence* in Φ iff in Φ there is an expression of the form $\boldsymbol{P}(\boldsymbol{x}_1, ..., \boldsymbol{x}_k)$ (respectively, $\neg \boldsymbol{P}(\boldsymbol{x}_1, ..., \boldsymbol{x}_k)$), where $\boldsymbol{x}_1, ..., \boldsymbol{x}_k$ are individual symbols.

The formula Φ is said to be *positive in the symbols* $\boldsymbol{P}_1, ..., \boldsymbol{P}_l$ iff none of these symbols has a negative occurrence in Φ. It is *positive* iff no symbol (including $\approx$) occurs negatively in it.

It was observed long ago (E. Marczewski [101]) that all positive axioms are homomorphically closed, and, consequently, if a subclass $\mathcal{L}$ is distinguished in a class $\mathcal{K}$ by positive axioms, then $\mathcal{L}$ is homomorphically closed in $\mathcal{K}$.

(*N.B.:* We say $\mathcal{L}$ is distinguished in $\mathcal{K}$ by certain axioms when $\mathcal{L}$ consists of just those $\mathcal{K}$-models in which the axioms are true.) In 1955 Łoś [88] reported without proof that every axiomatizable, homomorphically closed subclass of a class of algebras is axiomatizable by positive axioms. The author [V] independently announced in 1957 that every axiomatizable and homomorphically closed class of models can be described by positive axioms.

In 1959 R.C. Lyndon published the following general theorem with a detailed proof.

Theorem 7 (Lyndon [96]): *Every axiomatizable and homomorphically closed subclass of an axiomatizable class of models can be distinguished in this class by means of positive axioms.* ■

A basic tool for proving this theorem is the important

Interpolation theorem for FOPL (Lyndon [95]): *Let* Φ *and* Ψ *be* FOPL *sentences; these may contain operation symbols as well as predicate symbols. Suppose* $\Phi \to \Psi$ *is an identically valid* FOPL *formula. Then there exists a sentence* Ω *such that: (a) the sentences* $\Phi \to \Omega$ *and* $\Omega \to \Psi$ *are logically valid; (b) every predicate symbol occurring positively in* Ω *also occurs positively in both* Φ *and* Ψ*; (c) every predicate symbol occurring negatively in* Ω *also occurs negatively in both* Φ *and* Ψ *(the three formulas* Φ, Ψ, Ω *are assumed to have prenex conjunctive form).* ■

If in the formulation of Theorem 7 we strike out the words "positive" and "negative", we obtain the theorem of W. Craig [23] of which this interpolation theorem is a refinement.

Besides the notion of homomorphism, we introduced in §1.1 the notion of strong homomorphism; we recall that the two coincide for algebras. It would be interesting to solve the following open problems connected with this:

Problem 1: How can one describe the structure of those axioms determining classes of models in which all homomorphisms are strong?

Problem 2: In Theorem 1 the condition of the axiomatizability of the subclass $\mathcal{L}$ in the class $\mathcal{K}$ is overly strict. How can it be weakened so that the theorem still remains valid – even when restricted to strong homomorphisms?

§3.2. *Universal and Skolem subclasses*

In §2.4 we cited a theorem of A. Tarski characterizing those classes admitting axiomatizations by universal sentences. Passage from classes to subclasses yields the more general

Theorem 8 (Łoś [89], Tarski [163]): *A subclass $\mathcal{L}$ of a class $\mathcal{K}$ of models with finite signature can be distinguished in $\mathcal{K}$ by universal axioms iff these two conditions are satisfied: (a) every $\mathcal{K}$-submodel of an $\mathcal{L}$-model is an $\mathcal{L}$-model; (b) for every $\mathcal{K}$-model $\mathfrak{M}$, if every finite submodel of $\mathfrak{M}$ is isomorphically embeddable in an $\mathcal{L}$-model, then $\mathfrak{M}$ itself can be embedded in some $\mathcal{L}$-model.* ■

Łoś [89] has pointed out an interesting application of Theorem 8 to this important algebraic problem: given two similar classes $\mathcal{K}_1, \mathcal{K}_2$ of algebras, we want to characterize the class $\mathcal{L}$ of those $\mathcal{K}_1$-algebras that are isomorphically embeddable in $\mathcal{K}_2$-algebras. From Theorem 8 it immediately follows that when the classes $\mathcal{K}_1, \mathcal{K}_2$ are axiomatizable, the class $\mathcal{L}$ is universally axiomatizable (in $\mathcal{K}_1$). E.g., to this time no explicit conditions have been established for the embeddability of associative rings in skewfields; nevertheless, such conditions must be expressible in the form of a system of universal FOPL sentences.

J. Łoś and R. Suszko ([91], pt. II) considered the somewhat more general problem of compatible embeddability. Suppose $\mathfrak{A}_1$, $\mathfrak{A}_2$ are algebras belonging to an axiomatizable class $\mathcal{K}$. Under what conditions is there a $\mathcal{K}$-algebra $\mathfrak{A}$ that includes subalgebras isomorphic to $\mathfrak{A}_1$ and $\mathfrak{A}_2$? It turns out that such a compatible extension $\mathfrak{A}$ exists iff whenever an axiom of the form

$$(x_1) \dots (x_m)(y_1) \dots (y_n)(\Phi(x_1, \dots, x_m) \vee \Psi(y_1, \dots, y_n))$$

is valid throughout $\mathcal{K}$, then at least one of the sentences

$$(x_1) \dots (x_m)\Phi(x_1, \dots, x_m), \quad (y_1) \dots (y_n)\Psi(y_1, \dots, y_n)$$

is valid in both $\mathfrak{A}_1$ and $\mathfrak{A}_2$.

These authors found like conditions for the existence of a compatible extension for any system of algebras in a given axiomatizable class.

Obviously, the union of an increasing chain of groups (ordered by $\subseteq$) is also a group, the union of an increasing chain of rings is a ring, and so on. We ask: for what axiomatizable classes $\mathcal{K}$ of models is it true that the union of an increasing chain of $\mathcal{K}$-models is a $\mathcal{K}$-model? An analogous problem for the FOPL language can be so formulated: for what axioms Φ is it the case that whenever Φ is true in every model in an increasing chain $\mathfrak{M}_1 \subseteq \mathfrak{M}_2 \subseteq \dots$, then Φ is also true in the model $\bigcup \mathfrak{M}_i$? The answers to these questions are given by

Theorem 9 (Łoś and Suszko [91], pt. IV): *An axiomatizable class $\mathcal{K}$ of models can be described by Skolem axioms (i.e., $\forall\exists$-axioms) iff the union of every increasing chain of $\mathcal{K}$-models is a $\mathcal{K}$-model.* ■

This theorem was carried over to subclasses of axiomatizable classes of models by A. Robinson:

Theorem 10 (A. Robinson [133]): *A subclass $\mathcal{L}$ of an axiomatizable class $\mathcal{K}$ of models can be described in $\mathcal{K}$ by Skolem axioms iff $\mathcal{L}$ is axiomatizable and whenever the union of an increasing chain of $\mathcal{L}$-models is a $\mathcal{K}$-model, it is an $\mathcal{L}$-model, as well.* ■

Theorem 10 has been generalized to axioms of arbitrary type $\forall\exists\forall \dots \eth$ ($\eth = \forall, \exists$) by D.A. Zaharov and H.J. Keisler. In presenting these results we use Zaharov's formulation [186].

Let $\mathfrak{M}$ be a submodel of a model $\mathfrak{N}$ and let $\mathfrak{M}_1$, in turn, be a submodel of $\mathfrak{M}$. We write

$$\mathfrak{M} \leqslant (\mathfrak{M}_1; n_1, \dots, n_l)\, \mathfrak{N} \tag{5}$$

just when the identity map on $\mathfrak{M}_1$ is an $(n_1, \dots, n_l)$-map from $\mathfrak{M}_1 \subseteq \mathfrak{M}$ into $\mathfrak{N}$ (as defined in §2.4). If (5) holds for every n_1-submodel $\mathfrak{M}_1$ of $\mathfrak{M}$, we write

$$\mathfrak{M} \leqslant (n_1, \dots, n_l)\, \mathfrak{N}. \tag{6}$$

Finally, we say that $\mathfrak{M}$ is an *l-wise submodel* of $\mathfrak{N}$ (and write $\mathfrak{M} \leqslant_l \mathfrak{N}$) iff (6) holds for every sequence $\langle n_1, \dots, n_l \rangle$ of length l.

Theorem 11 (Zaharov [186]): *An axiomatizable subclass $\mathcal{L}$ of an axiomatizable class $\mathcal{K}$ of models can be distinguished in $\mathcal{K}$ by axioms of the type $\forall\exists\forall \dots \eth$ ($l+1$ symbols, $\eth = \forall, \exists$) iff for every increasing $\leqslant_l$-chain $\mathfrak{M}_1 \leqslant_l \mathfrak{M}_2 \leqslant_l \dots$ of $\mathcal{L}_2$-models, if the union $\bigcup \mathfrak{M}_i$ of this chain belongs to $\mathcal{K}$, then it also belongs to $\mathcal{L}$.* ■

The relation $\mathfrak{M}_1 \leqslant_1 \mathfrak{M}_2$ is equivalent to $\mathfrak{M}_1$ being a submodel of $\mathfrak{M}_2$ in the usual sense. Therefore, on setting $l = 1$ in Zaharov's theorem, we get Theorem 10 as a corollary.

Even broader generalizations of Theorem 10 in various directions have been obtained by A.D. Taĭmanov ([158], pt. II).

§3.3. *Direct products of models*

Suppose with each element α of a set A is associated some model $\mathfrak{M}_\alpha = \langle M_\alpha; P_1^\alpha, P_2^\alpha, \ldots \rangle$ with the fixed signature $\Sigma = \langle \boldsymbol{P}_1, \boldsymbol{P}_2, \ldots \rangle$. Let M be the direct (cartesian) product of the bases M_α, i.e., the set of all functions f defined on A such that $f(\alpha) \in M_\alpha$ for every $\alpha \in$ A. We define predicates P_s on M by making $P_s(f_1, \ldots, f_{n_s})$ true exactly when $P_i^\alpha(f_1(\alpha), \ldots, f_{n_s}(\alpha))$ is true in $\mathfrak{M}_\alpha$ for every $\alpha \in$ A. The model $\mathfrak{M} = \langle M; P_1, P_2, \ldots \rangle$ is called the *cartesian product* of the system of models $\mathfrak{M}_\alpha$ indexed by the set A. Any model isomorphic to $\mathfrak{M}$ is called the *direct product* of the models $\mathfrak{M}_\alpha$.

If the models $\mathfrak{M}_\alpha$ are algebras, then $\mathfrak{M}$ is also an algebra; it is isomorphic to the usual complete direct product of the algebras $\mathfrak{M}_\alpha$.

Thanks to the important role played by direct products in the theory of groups and other algebraic disciplines, in the last ten years quite a few papers devoted to the research of direct products have appeared. In 1951 A. Horn [59] noted that if a sentence of the form

$$(\mho_1 \boldsymbol{x}_1) \ldots (\mho_m \boldsymbol{x}_m)(\mathop{\&}_i (\neg \psi_{i1} \vee \ldots \vee \neg \psi_{i\lambda_i} \vee \psi_i)$$

$$\& \mathop{\&}_j (\neg \chi_{j1} \vee \ldots \vee \neg \chi_{j\mu_j})) \tag{7}$$

(where $\psi_{i\lambda}, \psi_i, \chi_{j\mu}$ are atomic formulas of the form $\boldsymbol{x}_k \approx \boldsymbol{x}_l$ or $\boldsymbol{P}_s(\boldsymbol{x}_{k_1}, \ldots, \boldsymbol{x}_{k_n})$, and $\mho_p = \forall, \exists$) is true in models $\mathfrak{M}_\alpha$ $(\alpha \in \mathrm{A})$ with signature Σ, then this sentence is true in the direct product of the $\mathfrak{M}_\alpha$. It is now accepted practice to call a formula of the form (7) a *Horn sentence.*

We shall call an arbitrary sentence *multiplicatively closed* iff it has the above property, i.e., iff whenever it is true in similar models $\mathfrak{M}_\alpha$ $(\alpha \in \mathrm{A})$, it is also true in their direct product.

In Horn's paper it is proved that a multiplicatively closed universal sentence is equivalent to a Horn sentence. It is natural to wonder whether every multiplicatively closed sentence might not be equivalent to a Horn sentence.

That this is the case for positive sentences and sentences of various other sorts was proved by K. Bing [8]. Recently, this result was obtained by K.I. Appel [3] for sentences involving no predicate symbols other than the equality sign. But A. Morel and C.C. Chang [18] found a necessary condition for an arbitrary sentence to be equivalent to one in Horn form, and with its aid they constructed a multiplicatively closed sentence equivalent to no Horn sentence. In the same article Morel and Chang posed the problem of finding necessary and sufficient conditions that a sentence be reducible to Horn form. Around the same time, R.C. Lyndon [94] conjectured that every multiplicatively closed Skolem sentence can be put in Horn form. In this paper he asserted that multiplicatively closed ∃-sentences are equivalent to Horn sentences. In 1959–60, A.D. Taĭmanov [157] found the desired necessary and sufficient conditions, completing the task set by Morel and Chang. Using these results, Taĭmanov showed that the sentence

$$(\exists x_1 x_2 x_3)(P_1(x_1) \,\&\, P_2(x_2) \,\&$$
$$\& \neg P_3(x_3) \,\&\, (P_1(x_3) \vee P_2(x_3))) \,, \tag{8}$$

although multiplicatively closed, cannot be reduced to Horn form. Since (8) is an ∃-axiom, it serves as a counterexample to Lyndon's conjecture ([3]).

We now turn to results on multiplicatively closed classes, but in passing we mention the paper [118] of A. Oberschelp concerning multiplicatively closed sentences.

A special trend in the theory of multiplicatively closed classes of models was initiated by A. Mostowski [105]; among other things he showed that any direct power of a recursively decidable model is also decidable. Mostowski's results also imply that if an axiomatizable class of models contains all finite direct powers of its members, then it contains arbitrary infinite direct powers of its members. This inspired J. Łoś [88] to make the following conjecture:

Every axiomatizable class of models that contains all finite direct products of its members also contains arbitrary infinite direct products of its members.

This conjecture was proved by R.L. Vaught [179]. In a joint paper [36] S. Feferman and R.L. Vaught gave a detailed proof of the Łoś conjecture and extended it and Mostowski's theorem to a kind of product of models more general than the direct.

Without knowledge of the Feferman–Vaught results, the author published an article [XII] in which he generalized direct products in a different direction and gave a canonical form for expressing sentences concerning a product in terms of sentences concerning the factors. Special cases of these results are those of Mostowski and Vaught cited above.

§3.4. *Subdirect products*

Consider a system of models $\mathfrak{M}_\alpha$ ($\alpha \in \mathrm{A}$) with fixed signature $\Sigma = \langle P_1, P_2, \ldots \rangle$, and let $\mathfrak{M}$ be the cartesian product of the models $\mathfrak{M}_\alpha$. Suppose from the model $\mathfrak{N}$ with signature Σ there are maps φ_α onto the models $\mathfrak{M}_\alpha$ ($\alpha \in \mathrm{A}$). The maps φ_α naturally induce a map φ from $\mathfrak{N}$ into $\mathfrak{M}$. If φ is an isomorphism of

$\mathfrak{N}$ onto some submodel $\mathfrak{M}^*$ of $\mathfrak{M}$, then we say that $\mathfrak{N}$ is decomposed into the subdirect product $\mathfrak{M}^*$ of the models $\mathfrak{M}_\alpha$ with projections φ_α.

Let $\mathcal{K}$ be an abstract class of models. A model $\mathfrak{N}$ is called *subdirectly indecomposable in* $\mathcal{K}$ iff in any decomposition of $\mathfrak{N}$ into a subdirect product of $\mathcal{K}$-models, at least one of the projections is an isomorphism.

G. Birkhoff [10] proved that every algebra with an arbitrary signature Σ is decomposable into a subdirect product of algebras subdirectly indecomposable in the class of all algebras with signature Σ. It is clear that this theorem remains valid when relativized to any homomorphically closed class of algebras. In [V] we find necessary and sufficient conditions for a class $\mathcal{K}$ of models to have the following property: every $\mathcal{K}$-model is the subdirect product of $\mathcal{K}$-models subdirectly indecomposable in $\mathcal{K}$. These conditions yield the corollary: *if a class $\mathcal{K}$ of models is determined by a system of universal axioms and positive axioms, then every $\mathcal{K}$-model is a subdirect product of $\mathcal{K}$-models subdirectly indecomposable in $\mathcal{K}$.* ∎

By analogy with multiplicatively closed classes, we call a class $\mathcal{K}$ of models *subdirectly closed* iff all subdirect products of systems of $\mathcal{K}$-models belong to $\mathcal{K}$. We naturally inquire: by what special sort of axioms are subdirectly closed, axiomatizable classes of models describable? In contrast to the situation for multiplicatively closed classes, the results here are much simpler. The full solution is given by the

Theorem (Lyndon [97]): *An axiomatizable class of models is subdirectly closed iff it can be described a system of by axioms of the form*

$$(x_1) \dots (x_n)(\Phi(x_1, \dots, x_n) \to P_s(x_{j_1}, \dots, x_{j_n})),$$

where Φ is a positive FOPL *formula and P_s is a signature symbol (or $\approx$).* ∎

Now we turn to subclasses. A subclass $\mathcal{L}$ of a class $\mathcal{K}$ of models is called *subdirectly closed* in $\mathcal{K}$ iff every $\mathcal{K}$-model that is decomposable into a subdirect product of $\mathcal{L}$-models is in fact an $\mathcal{L}$-model. The above theorem characterizing the axiomatizable subdirectly closed classes applies, *mutatis mutandis,* to subclasses subdirectly closed in an axiomatizable class (Lyndon [97]).

§3.5. *Convex and quasiaxiomatizable classes*

The intersection of any system of subgroups of an arbitrary group is again a subgroup. Consequently, if $\mathcal{G}$ is the class of groups, then the intersection of any system of $\mathcal{G}$-submodels of an arbitrary $\mathcal{G}$-model is itself a $\mathcal{G}$-model. The same would be true, were $\mathcal{G}$ the class of all rings, all fields, etc. Abstracting from these examples, A. Robinson [128] suggested calling a class $\mathcal{K}$ of models *convex* iff the intersection of any system of $\mathcal{K}$-submodels of an arbitrary $\mathcal{K}$-model either is empty or is a $\mathcal{K}$-model itself. In the paper cited Robinson showed that every axiomatizable convex class of models can be axiomatized by means of Skolem sentences. C.C. Chang strengthened this result by proving

Theorem 12 (Chang [17]): *If in every member of an axiomatizable class $\mathcal{K}$ of models, the intersection of any two $\mathcal{K}$-submodels either is empty or is a $\mathcal{K}$-submodel, then the class $\mathcal{K}$ is describable by Skolem axioms.* ■

Robinson's result is strengthened in another way in [VII] where the hypothesis of the axiomatizability of the class $\mathcal{K}$ is replaced with the assumption of its quasiaxiomatizability, a property defined below.

Let us call a class $\mathcal{K}$ of models *compact* iff for every system **S** of subclasses axiomatizable in $\mathcal{K}$, if the intersection of every finite subsystem of **S** is nonempty, then the intersection of the whole system **S** is nonempty.

A class $\mathcal{K}$ of models is called *locally bounded* (cf. §2.2) iff for every cardinal $\mathfrak{m}$ there is a cardinal $\mathfrak{n}$ (depending only on $\mathfrak{m}$) with this property: in every $\mathcal{K}$-model each subset of the base of power not greater than $\mathfrak{m}$ is included in a $\mathcal{K}$-submodel of power not exceeding $\mathfrak{n}$.

A class $\mathcal{K}$ is called *quasiaxiomatizable* (or *pseudoaxiomatizable*) iff it is both compact and locally bounded.

The compactness and Löwenheim–Skolem theorems from §2.1.show that every axiomatizable class of models is automatically compact and locally bounded. On the other hand, it is easy to construct quasiaxiomatizable classes that are not axiomatizable in the usual sense. Thus the notion of quasiaxiomatizability generalizes the notion of axiomatizability.

Theorem 13 (Mal'cev [VII], §2): *Every abstract, convex, quasiaxiomatizable class is axiomatizable by means of Skolem sentences.* ([4]) ■

Recently, S.R. Kogalovskiĭ observed that the condition of compactness for classes of models can be used in the problem of characterizing those classes that are axiomatizable. We shall say that a class of models is *elementarily closed* iff together with each member it also contains all models elementarily equivalent to it (cf. §2.3). The theorems of A.D. Taĭmanov in §2.4 readily lead to

Theorem 14 (Kogalovskiĭ [75]): *Every compact, elementarily closed class of models is axiomatizable.* ■

It is easy to put this theorem in a more general form. We say the subclass $\mathcal{L}$ of a class $\mathcal{K}$ of models is *elementarily closed in* $\mathcal{K}$ iff every $\mathcal{K}$-model elementarily equivalent to an $\mathcal{L}$-model belongs to $\mathcal{L}$. From Theorem 14 it immediately follows that a subclass $\mathcal{L}$ is axiomatizable in a compact class $\mathcal{K}$ iff $\mathcal{L}$ is compact and elementarily closed in $\mathcal{K}$. The condition that $\mathcal{K}$ be compact can be changed here to the appropriately defined condition that $\mathcal{L}$ be compact in $\mathcal{K}$.

§4. Ultraproducts

In the last few years, the so-called ultraproduct construction, a generalization of the direct product of models, has proved most fruitful for model theory. The seed was sown by J. Łoś [88]; it developed and flourished under the care of A. Tarski and other workers, especially S. Kochen [72] and H.J. Keisler [70]. The techniques of ultraproducts have enabled them to gather under a single style of proof almost all the results presented so far, and to prove a host of powerful new theorems.

§4.1. *Basic definitions*

A nonempty system **D** of nonempty subsets of a set I is called a *filter over* I iff these two conditions are satisfied:

(a) The intersection of any two subsets in **D** also belongs to **D**;

(b) If $A \in \mathbf{D}$ and $A \subseteq B \subseteq I$, then $B \in \mathbf{D}$.

A filter **D** over I is called *principal* iff there exists a set $A \in \mathbf{D}$ such that $\mathbf{D} = \{X\colon A \subseteq X \subseteq I\}$.

Every filter over a finite set is principal. On the other hand, if the intersection of all members of some filter is empty then the filter is automatically nonprincipal.

A filter **D** over I is called an *ultrafilter* iff for every $A \subseteq I$, either A or $I \sim A$ belongs to **D**.

With the aid of the axiom of choice it is easy to prove that every system **F** of subsets of an arbitrary set I can be extended to an ultrafilter over I, as long as the intersection of any finite number of sets in **F** is nonempty.

Now suppose to each element ι of the set I there corresponds some model $\mathfrak{M}_\iota = \langle M_\iota; P_1^\iota, P_2^\iota, \ldots\rangle$ of a fixed type; let **D** be a filter over I.

On the cartesian product $M = \Pi_{\iota\in I} M_\iota$ of the sets M_ι we introduce an equivalence relation $\equiv_{\mathbf{D}}$ by setting

$$x \equiv_{\mathbf{D}} y \Leftrightarrow \{\iota : x^\iota = y^\iota\} \in \mathbf{D} \quad (x, y \in M)\,.$$

Let $M/\mathbf{D}$ denote the factor set, and define a new model

$$\mathfrak{M} = \langle M/\mathbf{D}; P_1, P_2, \ldots\rangle = \prod_{\iota \in \mathrm{I}} \mathfrak{M}_\iota/\mathbf{D}$$

by putting

$$P_s(x_1, \ldots, x_{n_s}) = \mathsf{T} \Leftrightarrow \{\iota : P_s^\iota(x_1^\iota, \ldots, x_{n_s}^\iota) = \mathsf{T}\} \in \mathbf{D}\,.$$

The model $\mathfrak{M}$ is called the *reduced product* of the system $\{\mathfrak{M}_\iota : \iota \in \mathrm{I}\}$ with respect to the filter **D**.

This definition implies that if the filter **D** is principal and generated by the set A, then the reduced product $\Pi_{\iota\in\mathrm{I}}\, \mathfrak{M}_\iota/\mathbf{D}$ is isomorphic to the cartesian product $\Pi_{\iota\in\mathrm{I}}\, \mathfrak{M}_\iota$. The novelty comes when **D** is a nonprincipal ultrafilter.

A reduced product of models relative to an ultrafilter is called an *ultraproduct.* The significance of ultraproducts is determined by their following basic property:

Theorem 15 (cf. [72]): *Suppose we have (i) a* FOPL *formula* $\Phi(\boldsymbol{x}_1, \ldots, \boldsymbol{x}_n)$ *of signature* $\Sigma = \langle \boldsymbol{P}_1, \boldsymbol{P}_2, \ldots\rangle$ *with free individual variables* $\boldsymbol{x}_1, \ldots, \boldsymbol{x}_n$;

(ii) an ultrafilter **D** *over a set* I;

(iii) a system of models $\mathfrak{M}_\iota = \langle M_\iota; P_1^\iota, P_2^\iota, \ldots\rangle$ $(\iota \in \mathrm{I})$ *with signature* Σ, *and elements* $a_1, \ldots, a_n$ *of the cartesian product* $\Pi_{\iota\in\mathrm{I}} M_\iota$.

Let $a_1^\iota, \ldots, a_n^\iota$ *be the projections of* $a_1, \ldots, a_n$ *onto the set* M_ι $(\iota \in \mathrm{I})$. *The relation*

$$\Phi(a_1/\mathbf{D}, \ldots, a_n/\mathbf{D})$$

is true in the ultraproduct $\Pi_{\iota\in\mathrm{I}}\, \mathfrak{M}_\iota/\mathbf{D}$ *iff the set of indices of those models* $\mathfrak{M}_\iota$ *in which* $\Phi(a_1^\iota, \ldots, a_n^\iota)$ *is true belongs to* **D**.

The proof of this theorem is carried out by induction on the number of quantifiers in the formula Φ, assumed to be in prenex form (cf., e.g., [72]). ■

From Theorem 15 it follows that *every axiomatizable class* $\mathfrak{K}$ *of models is closed under ultraproducts, and*

$$\Pi\, \mathfrak{M}_\iota/\mathbf{D} \in \mathfrak{K} \Leftrightarrow \{\iota : \mathfrak{M}_\iota \in \mathfrak{K}\} \in \mathbf{D}\,. \blacksquare$$

If in a reduced product $\Pi\, \mathfrak{M}_\iota/\mathbf{D}$, all the models coincide with a single model $\mathfrak{M}$, then this reduced product is called the *reduced power* of $\mathfrak{M}$ relative to **D** and written $\mathfrak{M}^{\mathrm{I}}/\mathbf{D}$. Reduced powers relative to ultrafilters are called *ultrapowers.*

If we let each element $a \in \mathfrak{M}$ correspond to the element $b \in \mathfrak{M}^{\mathbf{I}}$, all of whose projections coincide with a, then we get a canonical embedding of the model $\mathfrak{M}$ in the reduced power $\mathfrak{M}^{\mathbf{I}}/\mathbf{D}$.

A map φ from a model $\mathfrak{M}$ into a similar model $\mathfrak{N}$ is called an *elementary embedding* of $\mathfrak{M}$ in $\mathfrak{N}$ iff for every formula $\Phi(\boldsymbol{x}_1, ..., \boldsymbol{x}_n)$ of the signature of $\mathfrak{M}$ and with free variables $\boldsymbol{x}_1, ..., \boldsymbol{x}_n$, and for all elements $a_1, ..., a_n \in \mathfrak{M}$, if $\Phi(a_1, ..., a_n)$ is true in $\mathfrak{M}$, then $\Phi(a_1\varphi, ..., a_n\varphi)$ is true in $\mathfrak{N}$.

Theorem 15 shows that the canonical map from a model $\mathfrak{M}$ into an ultrapower $\mathfrak{M}^{\mathbf{I}}/\mathbf{D}$ is an elementary embedding. Therefore, each ultrapower of $\mathfrak{M}$ can be viewed as an elementary extension of $\mathfrak{M}$ lying in every axiomatizable class containing $\mathfrak{M}$.

If the ultrafilter $\mathbf{D}$ is principal, then it is generated by some one-element set $\{\iota\}$. In this case the canonical embedding of $\mathfrak{M}$ in the ultrapower $\mathfrak{M}^{\mathbf{I}}/\mathbf{D}$ is simply an isomorphism of $\mathfrak{M}$ onto $\mathfrak{M}^{\mathbf{I}}/\mathbf{D}$. If the filter $\mathbf{D}$ is not principal and $\mathfrak{M}$ is infinite, then this embedding does not map $\mathfrak{M}$ onto $\mathfrak{M}^{\mathbf{I}}/\mathbf{D}$. Therefore, by taking $\mathbf{I}$ to be an infinite set and $\mathbf{D}$ to be any nonprincipal ultrafilter over $\mathbf{I}$, we get a proper extension of $\mathfrak{M}$ in any axiomatizable class to which $\mathfrak{M}$ belongs. This is a new proof of the extension theorem in §2.1.

With the help of ultrapowers it is possible to forge an algebraic link between the notions of isomorphism and isomorphic embedding and the notions of elementary equivalence and elementary embedding. Let $\mathfrak{M}$, $\mathfrak{M}_1$ be similar models. A map i: $\mathfrak{M} \to \mathfrak{M}_1$ is called a *power embedding* of $\mathfrak{M}$ in $\mathfrak{M}_1$ iff there exist ultrafilters $\langle \mathbf{I}, \mathbf{D} \rangle, \langle \mathbf{I}_1, \mathbf{D}_1 \rangle$ and an isomorphism φ: $\mathfrak{M}^{\mathbf{I}}/\mathbf{D} \to \mathfrak{M}_1^{\mathbf{I}_1}/\mathbf{D}_1$ such that the diagram

$$\begin{array}{ccc} \mathfrak{M}^{\mathbf{I}}/\mathbf{D} & \overset{\varphi}{\to} & \mathfrak{M}_1^{\mathbf{I}_1}/\mathbf{D}_1 \\ j \uparrow & & \uparrow k \\ \mathfrak{M} & \overset{i}{\to} & \mathfrak{M}_1 \end{array}$$

is commutative; here, j and k are the canonical embeddings.

From Theorem 15 it follows that *every power embedding is elementary.* ■ Keisler [70] has proved the converse with the aid of the generalized continuum hypothesis (GCH).

The following slightly more general result is established similarly:

Theorem 16 (Keisler [70]): *If some ultrapowers of the respective models* $\mathfrak{M}$, $\mathfrak{N}$ *are isomorphic, then* $\mathfrak{M}$ *and* $\mathfrak{N}$ *are elementarily equivalent. If the* GCH *holds, then the converse is also true: from the elementary equivalence of models* $\mathfrak{M}$, $\mathfrak{N}$ *follows the isomorphism of appropriate ultrapowers of these models (over the same index set).* ■

The assertion above concerning power embeddings follows from Theorem 16 on enriching the signature with symbols for all the elements of $\mathfrak{M}$.

Without recourse to the GCH we can readily establish

Theorem 17 (T.E. Frayne [39]): *In order that models $\mathfrak{M}, \mathfrak{N}$ be elementarily equivalent, it is necessary and sufficient that there exist an ultrafilter $\langle \mathrm{I}, \mathbf{D}\rangle$ such that $\mathfrak{N}$ is elementarily embeddable in $\mathfrak{M}^{\mathrm{I}}\mathbf{D}$.*

Sufficiency is implied by Theorem 15. To prove necessity (cf. [72]) we arrange all the elements of $\mathfrak{N}$ in a transfinite sequence $b = \langle b_0, b_1, \dots \rangle$. With every every b_α we associate an individual symbol $\boldsymbol{x}_\alpha$; we let T be the set of all FOPL formulas, all of whose free variables are found among the $\boldsymbol{x}_\alpha$. For $\rho \in \mathrm{T}$ the notation $\mathfrak{N} \vdash \rho(b)$ will signify that the formula ρ becomes true in $\mathfrak{N}$ when the free variables are replaced with the corresponding elements of $\mathfrak{N}$. We set $\mathrm{I} = \{\rho : \rho \in \mathrm{T}$ and $\mathfrak{N} \vdash \rho(b)\}$. If $\boldsymbol{x}_{\alpha_1}, \dots, \boldsymbol{x}_{\alpha_n}$ are the free variables occurring in a member ρ of I, then the sentence $(\exists \boldsymbol{x}_{\alpha_1} \dots \boldsymbol{x}_{\alpha_n})\rho$ is true in $\mathfrak{N}$ – and in $\mathfrak{M}$, too. Therefore, there is a sequence a^ρ of elements a^ρ_α in $\mathfrak{M}$ such that $\mathfrak{M} \vdash \rho(a^\rho)$.

We set

$$\Delta^\rho = \{\rho' : \rho' \in \mathrm{I} \text{ and } \mathfrak{M} \vdash \rho'(a^\rho) \qquad (\rho \in \mathrm{I})$$

and put $\mathbf{F} = \{\Delta^\rho : \rho \in \mathrm{I}\}$. Since the intersection of any finite number of sets from the system $\mathbf{F}$ is nonempty, there is an ultrafilter $\mathbf{D}$ over I including $\mathbf{F}$. The map $\varphi \colon \mathfrak{N} \to \mathfrak{M}^{\mathrm{I}}\mathbf{D}$ is defined by $\varphi(b_\alpha) = f_\alpha/\mathbf{D}$ for all α, where $f_\alpha(\rho) = a^\rho_\alpha$ for all $\rho \in \mathrm{I}$. It is easy to see that φ is the desired elementary embedding. ∎

§4.2. *Direct limits and ultralimits*

A *directed system* $\{\mathfrak{M}, j\}$ of models over a directed set I is formed by associating with each element $\alpha \in \mathrm{I}$ a model $\mathfrak{M}_\alpha$ of a fixed type τ, and by associating with each pair $\alpha \leqslant \beta$ of elements in I a homomorphism $j^\beta_\alpha \colon \mathfrak{M}_\alpha \to \mathfrak{M}_\beta$ such that j^α_α is the identity map on $\mathfrak{M}_\alpha$ $(\alpha \in \mathrm{I})$, and for $\alpha < \beta < \gamma$ we have $j^\beta_\alpha j^\gamma_\beta = j^\gamma_\alpha$.

The *direct limit* of the system $\{\mathfrak{M}, j\}$ is the model $\mathfrak{M}_\infty$ of type τ defined as follows. We let M be the set of all pairs of the form $\langle \alpha, a_\alpha\rangle$ for $a_\alpha \in \mathfrak{M}_\alpha$, $\alpha \in \mathrm{I}$. We say the pairs $\langle \alpha, a_\alpha\rangle, \langle \beta, b_\beta\rangle$ in M are equivalent ($\equiv$) when there exists an element $\gamma \in \mathrm{I}$ greater than α, β for which $a_\alpha j^\gamma_\alpha = b_\beta j^\gamma_\beta$. The classes of equivalent pairs form the base set of the model $\mathfrak{M}_\infty$ under construction. We take the basic predicate $P_s(a, b, \dots, c)$ to be true iff there exists an $\alpha \in \mathrm{I}$ such that $a \equiv \langle \alpha, a_\alpha\rangle, \dots, c \equiv \langle \alpha, c_\alpha\rangle$ and $P^\alpha_s(a_\alpha, b_\alpha, \dots, c_\alpha)$ is true in $\mathfrak{M}_\alpha$.

The map j_α: $\mathfrak{M}_\alpha \to \mathfrak{M}_\infty$ defined by setting $a_\alpha j_\alpha$ equal to the equivalence class in $\mathfrak{M}_\infty$ containing the pair $\langle \alpha, a_\alpha \rangle$ is called the *injection* of $\mathfrak{M}_\alpha$ into $\mathfrak{M}_\infty$.

If all the maps j_α^β are monomorphisms, then the injections j_α are obviously isomorphic embeddings, too. But there is also a subtler result:

Theorem 18 [167] : *If in a directed system* $\{\mathfrak{M}, j\}$ *every homomorphism* j_α^β *is an elementary embedding, then each injection* j_α *elementarily embeds* $\mathfrak{M}_\alpha$ *in the limit model* $\mathfrak{M}_\infty$. ■

Suppose we are given a sequence θ (of type ω) of ultrafilters $\langle \mathbf{I}_0, \mathbf{D}_0 \rangle$, $\langle \mathbf{I}_1, \mathbf{D}_1 \rangle$, ... and a model $\mathfrak{M}$. As pointed out above, there is a sequence

$$\mathfrak{M} \to \mathfrak{M}^{\mathbf{I}_0}/\mathbf{D}_0 = \mathfrak{M}_1 \to \mathfrak{M}_0^{\mathbf{I}_1}/\mathbf{D}_1 = \mathfrak{M}_2 \to \mathfrak{M}_2^{\mathbf{I}_2}/\mathbf{D}_2 = \mathfrak{M}_3 \to \ldots$$

of natural embeddings. The direct limit $\mathfrak{M}_\theta$ of this sequence will be a model similar to $\mathfrak{M}$; it is called an *ultralimit.* According to Theorem 18, the injection $\mathfrak{M} \to \mathfrak{M}_\theta$ embeds $\mathfrak{M}$ elementarily in $\mathfrak{M}_\theta$.

Keisler's algebraic characterization(Theorem 16) of elementarily equivalent models in terms of ultrapowers rests on the GCH. Without its support, the analogous theorem for ultralimits can be proved:

Theorem 19 (Kochen [72]): *In order that the models* $\mathfrak{M}$ *and* $\mathfrak{N}$ *be elementarily equivalent, it is necessary and sufficient that there exist a sequence* θ *of ultrafilters such that the ultralimits* $\mathfrak{M}_\theta$, $\mathfrak{N}_\theta$ *are isomorphic.* ■

A class $\mathfrak{K}$ of models is called *model complete* (A. Robinson [132]) iff whenever a $\mathfrak{K}$-model $\mathfrak{M}$ is a submodel of a $\mathfrak{K}$-model $\mathfrak{N}$, then $\mathfrak{M}$ is elementarily embedded in $\mathfrak{N}$.

Theorem 20 (Kochen [72]): *Suppose the class* $\mathfrak{K}$ *of models is closed under isomorphisms and ultrapowers.* $\mathfrak{K}$ *is model complete iff for every pair* $\mathfrak{M} \subseteq \mathfrak{N}$ *of* $\mathfrak{K}$*-models, there exists an ultrafilter* $\langle \mathbf{I}, \mathbf{D} \rangle$ *and an embedding* $\mathfrak{N} \to \mathfrak{M}^{\mathbf{I}}/\mathbf{D}$ *that induces the canonical embedding* $\mathfrak{M} \to \mathfrak{M}^{\mathbf{I}}/\mathbf{D}$. ■

§4.3. *Conditions for the axiomatizability of classes. Elementary relations*

Necessary and sufficient conditions for a class of models to be axiomatizable were given earlier in terms of sequential mappings. The conditions for axiomatizability obtained by Kochen and Keisler are expressed in terms of ultraproducts and have a more algebraic nature. We formulate just a few of the chief results.

Without the GCH:

Theorem 21 (Kochen [72]): *For a class $\mathcal{K}$ of models to be axiomatizable, it is necessary and sufficient that following all hold:*

(a) $\mathcal{K}$ is closed under ultraproducts;

(b) $\mathcal{K}$ and its complement $\widetilde{\mathcal{K}}$ (relative to the class of all models of the type of $\mathcal{K}$) are closed under ultralimits;

(c) $\mathcal{K}$ is closed with respect to isomorphisms, i.e., $\mathcal{K}$ is abstract. ∎

With the GCH one can prove the stronger

Theorem 22 (Keisler [70]): *The axiomatizability of the class $\mathcal{K}$ is equivalent to each of following two assertions:*

(i) $\mathcal{K} = Prod(\mathcal{K})$ and $\widetilde{\mathcal{K}} = Pow(\widetilde{\mathcal{K}})$;

(ii) for some $\mathcal{L} \subseteq \mathcal{K}$, we have

$$\mathcal{K} = \{\mathfrak{M}: Pow(\{\mathfrak{M}\}) \cap Prod(\mathcal{L}) \neq \emptyset\}. \blacksquare$$

Here, *Prod* ($\mathcal{K}$) denotes the class of models, each of which is isomorphic to some ultraproduct of $\mathcal{K}$-models. When *Prod* ($\mathcal{K}$) = $\mathcal{K}$, we say $\mathcal{K}$ is *ultraclosed.* Similarly, *Pow*($\mathcal{K}$) denotes the class of all isomorphs of all possible ultrapowers of $\mathcal{K}$-models.

Conditions for the finite axiomatizability of $\mathcal{K}$ are even simpler to formulate:

Theorem 23 (Kochen [72]): *The class $\mathcal{K}$ is finitely axiomatizable iff both $\mathcal{K}$ and $\widetilde{\mathcal{K}}$ are closed under ultraproducts, ultralimits, and isomorphisms.* ∎

Under the assumption of the GCH there is the stronger

Theorem 24 (Keisler [70]): *A class $\mathcal{K}$ of models is finitely axiomatizable iff both $\mathcal{K}$ and $\widetilde{\mathcal{K}}$ are ultraclosed.* ∎

With the aid of the GCH Keisler also proved the following two tests of separability:

Theorem 25: *If $\mathcal{K} = Prod(\mathcal{K})$, $\mathcal{L} = Pow(\mathcal{L})$, and $\mathcal{K} \cap \mathcal{L} = \emptyset$, then there is an axiomatizable class $\mathcal{K}_1 \supseteq \mathcal{K}$ such that $\mathcal{K} \cap \mathcal{L} = \emptyset$.*

If $\mathcal{K} = Prod(\mathcal{K})$, $\mathcal{L} = Prod(\mathcal{L})$, and $\mathcal{K} \cap \mathcal{L} = \emptyset$, then there is a finitely axiomatizable class $\mathcal{K}_1 \supseteq \mathcal{K}$ such that $\mathcal{K}_1 \cap \mathcal{L} = \emptyset$. ∎

Craig's interpolation theorem (cf. §3.1) is an immediate corollary of the second part of Theorem 25.

We now turn to the problem of elementary relations (formular predicates) on models. Suppose $\mathcal{K}$ is a class of models with signature $\Sigma = \langle \boldsymbol{P}_1, \boldsymbol{P}_2, ...\rangle$, and suppose $\Psi(\boldsymbol{x}_1, ..., \boldsymbol{x}_n)$ is a FOPL formula of signature Σ with free variables $\boldsymbol{x}_1, ..., \boldsymbol{x}_n$. Then in each $\mathcal{K}$-model $\mathfrak{M}$ the formula Ψ defines an n-ary relation $\Psi_{\mathfrak{M}}(x_1, ..., x_n)$, which is called an *elementary* (or *formular*) relation in $\mathcal{K}$. Suppose we are given on each $\mathcal{K}$-model $\mathfrak{M}$ an n-ary predicate $Q = Q_{\mathfrak{M}}(x_1, ..., x_n)$ (e.g., by means of a formula of second-order logic). The problem: what conditions of an algebraic character must the relation Q satisfy in order that there exist a FOPL formula $\Psi(\boldsymbol{x}_1, ..., \boldsymbol{x}_n)$ such that $Q_{\mathfrak{M}} = \Psi_{\mathfrak{M}}$ for every $\mathcal{K}$-model $\mathfrak{M}$?

The answer to this question is given by

Theorem 26 (E.W. Beth [7]): *Let* $\mathcal{K}^*$ *be an axiomatizable class of models with signature* $\Sigma^* = \langle \boldsymbol{Q}, \boldsymbol{P}_1, \boldsymbol{P}_2, ...\rangle$, *where* $\boldsymbol{Q}$ *is an n-ary predicate symbol. In order that there be a* FOPL *formula* $\Psi(\boldsymbol{x}_1, ..., \boldsymbol{x}_n)$ *of signature* $\Sigma = \langle \boldsymbol{P}_1, \boldsymbol{P}_2, ...\rangle$ *such that the sentence*

$$(\forall \boldsymbol{x}_1 ... \boldsymbol{x}_n)(\boldsymbol{Q}(\boldsymbol{x}_1, ..., \boldsymbol{x}_n) \leftrightarrow \Psi(\boldsymbol{x}_1, ..., \boldsymbol{x}_n))$$

is true in every $\mathcal{K}^*$*-model, it is necessary and sufficient that on each model* $\mathfrak{M} = \langle M; P_1, P_2, ...\rangle$ *it be possible to define in no more than one way a new predicate* $Q(x_1, ..., x_n)$ *so that the enriched model* $\mathfrak{M}^* = \langle M; Q, P_1, P_2, ...\rangle$ *is a* $\mathcal{K}^*$*-model.* ■

Using the technique of ultraproducts, Kochen generalized Beth's theorem to the following result:

Theorem 27 (Kochen [72]): *Suppose a relation* $Q(x_1, ..., x_n)$ *is defined in some manner on each model in an axiomatizable class* $\mathcal{K}$ *whose signature does not contain the symbol* $\boldsymbol{Q}$. *Then* Q *is elementary in* $\mathcal{K}$ *iff the following three conditions are all satisfied:*

(a) The $\boldsymbol{Q}$*-enrichment of any ultraproduct of* $\mathcal{K}$*-models is equal to the ultraproduct of the enriched factors;*

(b) The enrichment of any ultralimit of $\mathcal{K}$*-models is equal to the ultralimit of the enriched models;*

(c) If $\mathcal{K}$*-models are isomorphic, then their enrichments are isomorphic.* ■

We mention also that Keisler [70] obtained conditions for Horn axiomatizability (see §3.3) in terms of ultralimits and reduced products, as well as theorems on the existence of universal and homogeneous models.

§5. A few second-order classes of models

§5.1. *Axioms of the second order*

The problems considered in the previous sections were closely tied to the first-order predicate logic and its formal language. The next most important formal language is that of second-order predicate logic (SOPL). Formulas in this language are composed of the same logical symbols &, $\vee$, $\neg$, $\rightarrow$, $\forall$, $\exists$, $\approx$, individual symbols, and predicate symbols as in FOPL. The only difference lies in the use of the quantifiers: in second-order formulas the quantification symbols $\forall$, $\exists$ can bind not only individual variables, but predicate variables as well. The value (true–false) of a SOPL formula in a model whose signature contains all the free predicate and individual variables occurring in the notation of this formula is defined much as for FOPL formulas (cf. [56]). E.g., the formula

$$(\boldsymbol{R})((\exists \boldsymbol{x})\boldsymbol{R}(\boldsymbol{x}) \rightarrow (\exists \boldsymbol{y})(\boldsymbol{R}(\boldsymbol{y})\ \&\ (\boldsymbol{z})(\boldsymbol{R}(\boldsymbol{z}) \rightarrow \boldsymbol{y} \leqslant \boldsymbol{z})))$$

is true in a linearly ordered set iff this set is well ordered (every nonempty subset contains a least element),

The class of all models in which a fixed SOPL formula (axiom) is true is, naturally enough, called a *second-order class.* If a SOPL formula contains neither free predicate, nor free individual variables, then its truth in any model depends solely on the power of the base of that model. We shall call such a closed formula a *sentence*([1]). A second-order class may thus consist of abstract sets of various powers, and the collection of these cardinalities, called the *spectrum* of the class and sentence, completely determines such a class.

A.A. Zykov [187] obtained simple results on the spectra of second-order classes some years ago. Namely, he demonstrated the existence of algorithms whereby:

(i) for every SOPL axiom one can construct an equivalent formula of the form

$$(\mho_1 \boldsymbol{T}_1) \dots (\mho_m \boldsymbol{T}_m)(\mho_{m+1}\boldsymbol{x}_1) \dots (\mho_{m+n}\boldsymbol{x}_n)\, \Phi(\boldsymbol{x}_1, \dots, \boldsymbol{x}_n; \boldsymbol{T}_1, \dots, \boldsymbol{T}_m; \boldsymbol{y}_1, \dots, \boldsymbol{y}_s; \boldsymbol{P}_1, \dots, \boldsymbol{P}_t)\,,$$

where the $\boldsymbol{T}_i, \boldsymbol{P}_j$ are predicate symbols, the $\boldsymbol{x}_k, \boldsymbol{y}_l$ are individual symbols, $\langle \boldsymbol{P}_1, \dots, \boldsymbol{P}_t; \boldsymbol{y}_1, \dots, \boldsymbol{y}_s \rangle$ is the signature of the class of models under consideration, and $\mho_p = \forall, \exists$;

(ii) for every SOPL sentence Φ, one can construct a SOPL sentence Φ^* of of the form

$$(\exists \boldsymbol{S})(\boldsymbol{R})(\mho_1 \boldsymbol{x}_1) \dots (\mho_n \boldsymbol{x}_n)\, \Psi(\boldsymbol{x}_1, \dots, \boldsymbol{x}_n; \boldsymbol{S}, \boldsymbol{R}) \tag{9}$$

$$(\mho_i = \forall, \exists)$$

such that the spectrum σ of the formula Φ is mapped 1–1 onto the spectrum σ^* of Φ^* by the relation

$$\mathfrak{n} = \mathfrak{m} + \mathfrak{m}^l + 2^{\mathfrak{m}^l} \qquad (\mathfrak{m} \in \sigma, \mathfrak{n} \in \sigma^*)$$

($\boldsymbol{S}$ is a binary and $\boldsymbol{R}$ is a unary predicate symbol, and l is the largest of the ranks of the predicate symbols appearing in Φ).

The question, however, of which collections of cardinal numbers can be realized as the spectra of sentences (9) is still open. Clearly, this problem is closely connected with the axiomatics of set theory, but to this day has not been studied in depth.

The SOPL language is exceedingly strong. Many of the important mathematical concepts can be immediately described in it. A detailed elaboration of the theory of the SOPL language thus represents one of the central problems in model theory. At the present time, the only systematic study, more or less, has been of special forms of SOPL obtained by restricting the quantifiers in one way or another. Some of the results awarded these endeavors are described below.

§5.2. *Projective classes*

By definition, a class $\mathcal{K}$ of models with finite signature $\langle \boldsymbol{P}_1, \dots, \boldsymbol{P}_s \rangle$ is called *finitely projective* iff it consists of all models satisfying some fixed SOPL axiom of the form

$$(\exists \boldsymbol{T}_1) \dots (\exists \boldsymbol{T}_m)(\mho_1 \boldsymbol{x}_1) \dots (\mho_n \boldsymbol{x}_n)\, \Phi(\boldsymbol{x}_1, \dots, \boldsymbol{x}_n; \boldsymbol{T}_1, \dots, \boldsymbol{T}_m; \boldsymbol{P}_1, \dots, \boldsymbol{P}_s), \tag{10}$$

where the $\boldsymbol{P}_i$, $\boldsymbol{T}_j$ are predicate symbols, the $\boldsymbol{x}_k$ are individual symbols, and $\mho_l = \forall, \exists$.

We consider the auxiliary class $\mathcal{K}_0$ of models with signature $\langle \boldsymbol{P}_1, \dots, \boldsymbol{P}_s, \boldsymbol{T}_1, \dots, \boldsymbol{T}_m \rangle$ determined by the FOPL axiom

$$(\mho_1 \boldsymbol{x}_1) \dots (\mho_n \boldsymbol{x}_n)\, \Phi(\boldsymbol{x}_1, \dots, \boldsymbol{x}_n; \boldsymbol{T}_1, \dots, \boldsymbol{T}_m; \boldsymbol{P}_1, \dots, \boldsymbol{P}_s)\,. \tag{11}$$

Comparing the axioms (10) and (11), we see that an arbitrary model $\langle M; P_1, ..., P_s \rangle$ belongs to the class $\mathcal{K}$ iff it is possible to define auxiliary predicates $T_1, ..., T_m$ on M so that the enriched model $\langle M; P_1, ..., P_s, T_1, ..., T_m \rangle$ belongs to $\mathcal{K}_0$. In other words, a finitely projective class is the projection of a finitely axiomatizable class. The projection of an arbitrary axiomatizable class is called a *(general) projective* class. A. Tarski [163] proposed that the families of finitely projective and general projective classes be denoted by PAC and PAC_{Δ}, respectively.

As an example, consider the class $\mathcal{K}$ of (associative) rings, with signature consisting of symbols for the predicate $S(x, y, z)$ of addition and the predicate $P(x, y, z)$ of multiplication; these predicates are respectively equivalent to the equations $x + y = z$, $x \cdot y = z$. Let $\mathcal{G}$ be the $\boldsymbol{P}$-projection of $\mathcal{K}$. The finitely projective class $\mathcal{G}$ consists of all those semigroups isomorphic to multiplicative semigroups of associative rings. S.R. Kogalovskiĭ [75] recently showed that the class $\mathcal{G}$ is not axiomatizable.

A significant number of the properties of axiomatizable classes of models are inherited by projective classes. We shall not formulate these properties here, since they are also exhibited by reduced classes, to whose description we now turn.

§5.3. *Reductive classes*

All models that we have considered previously have had but one base set; it is, however, at times convenient to admit models with two or more bases. E.g., Hilbert's axiom system for euclidean geometry has three bases; the sets of points, line, and planes. In axiomatic set theory one often has a set of "elements" and a set of "sets".

We now pass to the general case, but for simplicity we assume the models considered have just two bases, the first and the second. Formulas of the corresponding two-sorted predicate logic have the same form as those of the ordinary, single-sorted FOPL; the difference lies in the division of the individual variables into two "sorts"; in each two-base model, elements of the first base serve as values for variables of the first sort, and elements of the second for variables of the second sort. The basic predicates of a two-sorted model are defined on the pair of bases and will be of different kinds as determined by the sorts of their arguments. Predicates (and predicate symbols) whose every argument is of the first sort are said to be of the first kind. Predicates of the second kind are defined analogously. The remaining predicates are said to be of mixed kind. Thus, the predicates of the first and second kinds are ordinary one-base predicates, and only the mixed predicates are really multibase.

Quantifiers in many-sorted FOPL formulas are interpreted as bounded quantifiers, their ranges restricted to the corresponding bases.

Finally, we can construct and interpret formulas for many-sorted SOPL in a quite analogous fashion.

A class of two-based models with (finite) signature Σ is called *(finitely) axiomatizable* iff it consists of all models with signature Σ satisfying some fixed (finite) system of two-sorted FOPL axioms of signature Σ.

Now let $\mathcal{K}^*$ be an axiomatizable or finitely axiomatizable class of two-sorted models, and suppose the signature of $\mathcal{K}^*$ consists of the predicate symbols $\boldsymbol{P}_1, \ldots, \boldsymbol{P}_s$ of the first kind, the symbols $\boldsymbol{R}_1, \ldots, \boldsymbol{R}_t$ of the second kind, and the mixed predicate symbols $\boldsymbol{S}_1, \ldots, \boldsymbol{S}_q$. Each $\mathcal{K}^*$-model $\mathfrak{M}$ has two bases M_1, M_2. If in $\mathfrak{M}$ we extract the set M_1 and preserve only the predicates $P_1, \ldots, P_s$ defined on it, we get a single-base model $\langle M_1; P_1, \ldots, P_s\rangle$, called the *reduct* of $\mathfrak{M}$.

Definition: *A class $\mathcal{K}$ of ordinary, one-base models is called* reductive (finitely reductive) *iff $\mathcal{K}$ consists of the reducts of all the members of some two-sorted (finitely) axiomatizable class* [XI]. (5)

This defines reductive classes of one-base models in terms of classes of two-base models. To give an appropriate definition involving one-base models only, we consider an arbitrary model $\mathfrak{M} = \langle M; P_1, \ldots, P_t, R\rangle$ with a single base M and a distinguished unary predicate R. We let M_R be the set of elements in M for which R is true. This determines a submodel $\langle M_R; P'_1, \ldots, P'_t, R'\rangle$ of $\mathfrak{M}$ called the ***R**-reduction* of the model $\mathfrak{M}$. We now have the easy

Theorem 28: *A class $\mathcal{K}$ of one-base models with signature $\langle \boldsymbol{P}_1, \ldots, \boldsymbol{P}_s\rangle$ is (finitely) reductive iff $\mathcal{K}$ is the $(\boldsymbol{P}_1, \ldots, \boldsymbol{P}_s)$-projection of the **R**-reduction of some (finitely) axiomatizable class whose signature contains $\boldsymbol{P}_1, \ldots, \boldsymbol{P}_s, \boldsymbol{R}$.* ∎

This theorem gives the desured characterization of reductive classes purely in the theory of one-base models.

In the earlier definition of multibase models had only two bases, but it is easy to see that the family of reductive one-sorted classes is not extended by using n-sorted classes ($n = 3, 4, \ldots$) in the definition.

From Theorem 28 it quickly follows that all (finitely) projective classes are also (finitely) reductive. The author does not know whether the converse is true. The family of reductive classes, however, possesses properties whose validity for the family of projective classes is in doubt. In particular, it is easy to prove ([XI], §§1, 2) that:

(a) the intersection of any (finite) system of (finitely) reductive classes is a (finitely) reductive class;

(b) the union of a finite number of (finitely) reductive classes is a (finitely) reductive class;

(c) a reductive class is quasiaxiomatizable in the sense of § 3.5;

(d) for every reductive class $\mathfrak{K}$, there is an infinite cardinal number $\mathfrak{n}$ such that every subset (of power $\mathfrak{m}$, say) of the base of an arbitrary $\mathfrak{K}$-model $\mathfrak{M}$ is included in a $\mathfrak{K}$-submodel of $\mathfrak{M}$ of power not greater than $\mathfrak{m}+\mathfrak{n}$ (the Löwenheim–Skolem theorem for reductive classes);

(e) for every reductive class $\mathfrak{K}$, there exists a cardinal $\mathfrak{n}$ such that every infinite $\mathfrak{K}$-model of power $\mathfrak{m}$ can be extended to $\mathfrak{K}$-model of any previously chosen power not less than $\mathfrak{m}+\mathfrak{n}$;

(f) the collection of all possible direct products of models in a reductive class is a reductive class;

(g) the collection of all homomorphic images and the collection of all strong homomorphic images of models in a (finitely) reductive class are (finitely) reductive classes;

(h) the collection of all models admitting homomorphic mappings onto members of a given (finitely) reductive class is a (finitely) reductive class, and the same goes for strongly homomorphic mappings.

The properties (a)–(e) all hold for projective classes, too. As regards (g), its status for projective classes is apparently still unknown.

Since the question of the coincidence of projective and reductive classes remains open, it is of interest to note

Theorem 29: *If all models in a finitely reductive class $\mathfrak{K}$ are infinite, then $\mathfrak{K}$ is finitely projective.*

For completeness we present a proof of this theorem. Let $\mathfrak{K}$ be the $(\boldsymbol{P}_1, ..., \boldsymbol{P}_s)$-projection of the $\boldsymbol{R}$-reduction of the class $\mathcal{L}$ of models with signature $\langle \boldsymbol{P}_1, ..., \boldsymbol{P}_s, \boldsymbol{P}_{s+1}, ..., \boldsymbol{P}_t, \boldsymbol{R}\rangle$ satisfying the FOPL axiom $\Phi(\boldsymbol{P}_1, ..., \boldsymbol{P}_t, \boldsymbol{R})$. Let $\mathfrak{M} = \langle M; P_1, ..., P_s\rangle$ be a $\mathfrak{K}$-model, and let $\mathfrak{N}$ be an $\mathcal{L}$-model whose projected $\boldsymbol{R}$-reduction is $\mathfrak{M}$. By the Löwenheim–Skolem theorem there is an $\mathcal{L}$-model $\mathfrak{N}_1 = \langle N_1; P_1, ..., P_t, R\rangle$ of the same power as $\mathfrak{M}$ that satisfies $M \subseteq \mathfrak{N}_1 \subseteq \mathfrak{N}$ (therefore, every $\mathfrak{K}$-model is the projected $\boldsymbol{R}$-reduction of an $\mathcal{L}$-model of the same power). Consequently, there is a predicate $S(x, y)$ defined on N_1 and realizing a 1–1 correspondence between N_1 and M. It thus satisfies the axioms

$$(\boldsymbol{x})(\boldsymbol{y})(\boldsymbol{z})((\boldsymbol{S}(\boldsymbol{x}, \boldsymbol{z}) \,\&\, \boldsymbol{S}(\boldsymbol{y}, \boldsymbol{z}) \rightarrow \boldsymbol{x} \approx \boldsymbol{y}) \,\&$$

$$\& \,(\boldsymbol{S}(\boldsymbol{x}, \boldsymbol{y}) \,\&\, \boldsymbol{S}(\boldsymbol{x}, \boldsymbol{z}) \rightarrow \boldsymbol{y} \approx \boldsymbol{z} \,\&\, \boldsymbol{R}(\boldsymbol{y}))) \,,$$

$$(x)(\exists y)(S(x,y)\ \&\ R(y))\ \&\ (y)(\exists x)(R(y) \to S(x,y)) .$$

Let $\Psi(R, S)$ be the conjunction of these two formulas.

By translating the predicates $P_1, ..., P_t, R, S$ defined on N_1 to the subset M by means of the mapping S, we obtain a model $\mathfrak{M}^* = \langle M; P_1^*, ..., P_t^*, R^*, S^* \rangle$ isomorphic to $\mathfrak{N}_1$ and thus satisfying the axioms

$$\Phi^* = \Phi(P_1^*, ..., P_t^*, R^*) , \quad \Psi^* = \Psi(R^*, S^*) .$$

In addition, the axioms

$$(\forall x_1, ..., x_{n_i})(P_i(x_1, ..., x_{n_i}) \leftrightarrow$$

$$\leftrightarrow (\exists y_1, ..., y_{n_i})(S^*(x_1, y_1)\ \&\ ...\ \&\ S^*(x_{n_i}, y_{n_i})\ \&$$

$$\&\ P_i^*(x_1, ..., x_{n_i})) \qquad (i = 1, ..., s) \tag{12}$$

hold in the model obtained by combining $\mathfrak{M}$ and $\mathfrak{M}^*$.

Let Ω be the conjunction of Φ^*, Ψ^*, and the axioms (12), and let $\mathfrak{K}^*$ be the class of models with signature

$$\langle P_1, ..., P_s, P_1^*, ..., P_t^*, R^*, S^* \rangle$$

that satisfy the axiom Ω. The discussion above shows that on each $\mathfrak{K}$-model can be defined predicates $P_1^*, ..., P_t^*, R^*, S^*$ so that the enriched model satisfies Ω. In other words, every $\mathfrak{K}$-model is the projection of a $\mathfrak{K}^*$-model. Conversely, suppose $\mathfrak{M}$ is the $(P_1, ..., P_s)$-projection of the arbitrary $\mathfrak{K}^*$-model $\mathfrak{M}_1$. Let the $(P_1^*, ..., P_t^*, R^*)$-projection of $\mathfrak{M}_1$ be $\mathfrak{M}_2$; it is an $\mathfrak{L}$-model. The $(P_1^*, ..., P_s^*)$-projected R^*-reduction of $\mathfrak{M}_2$ is isomorphic to $\mathfrak{M}$ by virtue of the truth of Ψ^* and (12) in $\mathfrak{M}_1$. Therefore, the $(P_1, ..., P_s)$-projection of every $\mathfrak{K}^*$-model is a $\mathfrak{K}$-model. Consequently, the class $\mathfrak{K}$, as the projection of the finitely axiomatizable class $\mathfrak{K}^*$, is finitely projective. ■ ([6])

§5.4. *Quasiuniversal classes*

Suppose we have a unary predicate $\mathbf{A}(\mathfrak{M})$ partially defined on a class $\mathfrak{K}$ of models. We say that $\mathbf{A}$ is *local* iff it has the following property: let $\mathfrak{M}$ be a $\mathfrak{K}$-model and let $\{\mathfrak{M}_\gamma : \gamma \in \Gamma\}$ be a system of its submodels such that each finite subset of $\mathfrak{M}$ is included in some $\mathfrak{M}_\gamma$; then if $\mathbf{A}$ is defined at $\mathfrak{M}$ and is defined and true at every $\mathfrak{M}_\gamma$ $(\gamma \in \Gamma)$, then $\mathbf{A}(\mathfrak{M})$ is true.

E.g., let $\mathfrak{K}$ be the class of all groups and **A** be the property of a group being abelian, n-step solvable, or n-step nilpotent (for a fixed natural number n). It is easy to prove in each of these cases that **A** has the localness property, although it is usually formulated in the weaker form: if every finite subset of a group $\mathfrak{G}$ generates a subgroup in $\mathfrak{G}$ with the property **A**, then $\mathfrak{G}$ itself has the property **A**.

The problem of whether a certain property of groups is local (in the class of all groups) frequently pops up in the theory of infinite groups, and its solution is sometimes quite difficult. A general method was indicated in [II] for reducing the problem of localness to the local thoerem for FOPL; with its aid the property of a group having a central system of subgroups, e.g., was shown to be local. This was apparently the first case of applying the theory of models to solve concrete problems in a domain of mathematics not immediately connected with predicate logic.

One immediate consequence of the compactness theorem is the

Theorem (L. Henkin [54]): *If every finite submodel of a model* $\mathfrak{M}$ *is isomorphically embeddable in a member of axiomatizable class* $\mathfrak{K}$*, then* $\mathfrak{M}$ *is itself embeddable in somc* $\mathfrak{K}$*-model.* ∎

This theorem was generalized by Taĭmanov ([158], pt. II) for n-embeddings (cf. §2.4).

The majority of interesting concrete local theorems – e.g. in group theory – involve properties immediately expressible, not in first-order, but in second-order predicate logic. Thus it is of some importance to describe as broad a class as possible of SOPL axioms expressing local properties. The only requirements are that the description be formal (syntactical) and that the formulas be general enough to offer the possibility of proving interesting local theorems in concrete theories. One such class of formulas, called the quasiuniversal formulas, was constructed by the author in [XI]. Their description follows.

Let $\langle \boldsymbol{P}_1, ..., \boldsymbol{P}_s \rangle$ be the signature of the models under study, and let $\boldsymbol{R}_1, ..., \boldsymbol{R}_t$ be auxiliary predicate symbols of arbitrary ranks. Consider a system S consisting of FOPL formulas

$$(\forall \boldsymbol{x}_1 \dots \boldsymbol{x}_{m_i}) \Phi_i(\boldsymbol{x}_1, ..., \boldsymbol{x}_{m_i}; \boldsymbol{P}_1, ..., \boldsymbol{P}_s, \boldsymbol{R}_i) \quad (i = 1, ..., t) \tag{13}$$

and a SOPL formula

$$(\eth_1 \boldsymbol{R}_1)(\forall \boldsymbol{x}_1 \dots \boldsymbol{x}_{n_1}) \dots (\eth_t \boldsymbol{R}_t)(\forall \boldsymbol{x}_{n_{t-1}+1} \dots \boldsymbol{x}_{n_t}) \Psi(\boldsymbol{x}_1, ..., \boldsymbol{x}_{n_t}; \boldsymbol{P}_1, ..., \boldsymbol{P}_s, \boldsymbol{R}_1, ..., \boldsymbol{R}_t), \tag{14}$$

where $\eth_i = \forall, \exists$ and the Φ_i and Ψ are quantifier-free FOPL formulas involving only the symbols indicated. The equality sign can also occur in the formulas (13) and (14), but must connect only individual symbols, not predicate symbols.

We say that the *quasiuniversal* axiom system S is *valid* in the model $\mathfrak{M} = \langle M; P_1, ..., P_s \rangle$ iff the formula (14) is true in $\mathfrak{M}$ under the condition that its quantifiers on $\boldsymbol{R}_1, ..., \boldsymbol{R}_t$ are relativized to the corresponding sets of all the predicates on M satisfying the respective axioms (13).

When the conditions (13) are tautological or completely absent, the validity of S in $\mathfrak{M}$ is equivalent to the truth in $\mathfrak{M}$ of (14) as an ordinary SOPL axiom. In the general case, the validity of S in $\mathfrak{M}$ is equivalent to the truth in $\mathfrak{M}$ of a SOPL prenex axiom with some existential individual quantifiers. Thus, e.g., the quasiuniversal system

$$(x)\Phi(x; P, \boldsymbol{R})$$

$$(\boldsymbol{R})(y)\Psi(y; P, \boldsymbol{R})$$

is equivalent to the formula

$$(\boldsymbol{R})(\exists x)(y)(\Phi \rightarrow \Psi) .$$

So quasiuniversal systems can be viewed as SOPL formulas with a special structure.

Intrinsic local theorem (for quasiuniversal classes) ([XI] §3.2): *Every property of models that can be expressed by a quasiuniversal system of axioms is local.* ∎

This theorem is also true for the more general case of infinite systems of many-sorted formulas of the form (13) and (14).

For a concrete example, we consider the class $\mathcal{K}$ of groups. A binary predicate R defined on groups and satisfying the axiom

$$(\forall xyzuv)((xRy \,\&\, yRx \rightarrow x \approx y) \,\&\, (xRy \,\&\, yRz \rightarrow xRz \,\&\, uxvRuyv))$$

is called a *partial group-ordering,* and a predicate S satisfying the axiom

$$(\forall xyzuv)((xSy \,\&\, ySx \rightarrow x \approx y) \,\&\, (xSy \vee ySx) \,\&$$

$$\& \, (xSy \,\&\, ySz \rightarrow xSz \,\&\, uxvSuyv)) ,$$

in every group is called a *linear* group-order. A group $\mathfrak{G}$ will be called *freely orderable* iff it satisfies the axiom

$$(R)(\exists S)(x)(y)(xRy \rightarrow xSy),$$

where the quantifiers $(\boldsymbol{R})$, $(\exists \boldsymbol{S})$ range over partial and linear group-orderings of $\mathfrak{G}$, respectively. Since the property of a group being freely orderable is quasiuniversal, it is local.

A series of other group-theoretical theorems, all immediate applications of the intrinsic local theorem, are indicated in [XI], §3.3. A local theorem of D.H. McLain [99] is also almost immediately deducible from the intrinsic local theorem.

A class $\mathcal{K}$ of models is naturally called *quasiuniversal* iff it consists of all members of an axiomatizable class $\mathcal{K}_0$ that have a fixed quasiuniversal property. In [XI], §3.2 are a number of examples showing that the behavior of quasiuniversal classes differs essentially from that of axiomatizable or reductive classes. E.g., it turns out that a quasiuniversal class can consist of a single infinite model; also, the class of all prime groups, all prime rings, etc., are quasiuniversal. A more detailed study of quasiuniversal classes has not been performed, but would be of interest.

§5.5. *Formulas with quantifiers on unary predicates*

As already mentioned, little is known about classes of models determined by general SOPL formulas. One might hope to simplify the problem by restricting consideration to those classes defined by SOPL axioms whose only quantified predicate symbols are unary (let us call formulas of this kind *monadic*). But Zykov's theorem in §4.1 shows that the spectral problem for the class of models with one binary predicate S that satisfy a monadic axiom of the form

$$(R)(\eth_1 x_1) \dots (\eth_n x_n)\, \Phi(x_1, \dots, x_n; R, S) \quad (\eth_i = \forall, \exists)$$

is just as general the spectral problem for an arbitrary SOPL sentence.

An essentially simpler case are the *monadic* classes – classes of models with only unary basic predicates that can be axiomatized by monadic axioms. A classical theorem of H. Behmann [2] shows that every monadic class either is empty, i.e., the corresponding axiom is contradictory, or contains a finite model, the number of whose elements does not exceed a bound effectively computable from the length of the defining axiom.

In connection with the theory of automata, the monadic theory of the model

$$\mathfrak{H} = \langle \{0, 1, 2, \dots \}, S \rangle \ ,$$

where S is the relation of immediate successor (i.e., $S(x, y) \Leftrightarrow x + 1 = y$), has lately aroused great interest. Recently, J.R. Büchi [13] succeeded in proving that the set of all monadic formulas true in $\mathfrak{H}$ is algorithmically decidable. The relationship between the monadic theory of $\mathfrak{H}$ and automata theory is examined in detail in the work of S. Kleene, A. Church [20], B.A. Trahtenbrot [170]–[172], et al.

Also of interest is the language of weak monadic formulas. To describe these, let us call a unary predicate defined on a set M *finite* on M iff it is true for only a finite number of elements of M. A monadic formula (or more reasonably, its interpretation) is *weak* iff the predicate quantifiers are restricted to finite predicates. E.g., the formula

$$(R)((\exists x) R(x) \rightarrow (\exists y)(R(y) \ \& \ (z)(R(z) \rightarrow y \leqslant z)))$$

in the weak monadic language is semantically interpreted as the proposition: every nonempty finite subset of the base set has a "least" element.

We look at a second example. Let S be a system of FOPL axioms describing the class of linearly ordered groups with the group operation and the order relation $\leqslant$. Then S and the weak monadic axiom

$$(x)(y)(1 < y < x \rightarrow (\exists R)(\exists z)(R(z) \ \& \ R(1) \ \& \ x \leqslant z \ \& \ \Psi(y, R)),$$

where $\Psi(y, R)$ denotes the formula

$$(u)(v)(R(u) \ \& \ R(v) \ \& \ (w)(R(w) \rightarrow w \leqslant u \lor v \leqslant w) \ \&$$

$$\& \ u < v \rightarrow v \approx uy) \ ,$$

together describe the class of archimedean ordered groups.

As we know, in the class of archimedean ordered groups there is a maximal member, the additive group of real numbers, that is included as an ordered subgroup in no larger archimedean ordered group.

For a final example we turn to the class of linearly ordered sets satisfying the weak monadic axiom

$$(\exists w)(x)(w \leqslant x) \ \& \ (x)(\exists y)(x < y) \ \& \ (x)(\exists R)(y)(y \leqslant x \rightarrow R(y)).$$

It is clear that this class consists of those ordered sets isomorphic to the set of natural numbers.

A class of models (with arbitrary signature) is called *weak monadic* iff it can be described in the weak monadic language. The second and third examples above show it is possible for a weak monadic class to contain a model of power $2^{\aleph_0}$ that is maximal, or even to be categorical. Nevertheless, the following analog of the Löwenheim–Skolem theorem holds:

Theorem (A. Tarski [164]): *Let $\mathfrak{K}$ be a weak monadic class. Then every uncountable $\mathfrak{K}$-model has a $\mathfrak{K}$-submodel of a lesser infinite power.* ■

In particular, this means every categorical weak monadic class consists of a countable model.

Apparently, the properties of weak monadic classes have not yet been studied in great detail. In Büchi [12] are indicated a number of concrete models whose weak monadic theories were investigated in connection with automata problems, as well as for their purely mathematical interest.

NOTES

(^{1}) The free variables (viewed as individual constants) that designate distinguished elements should be allowed to appear in sentences. In practice no distinction is made between sentences and axioms until §5, where for second-order logic, only axioms of the empty signature are called sentences.

(^{2}) Not the full axiom of choice, but the axiom of choice for finite sets (cf. [R6]).

(^{3}) Tarski has pointed out (cf. [R7]) that sentence (2) is equivalent to the Horn sentence

$$(\exists xyz)(P_1(x) \,\&\, P_2(y) \,\&\, \neg P_3(z) \,\&\, (P_2(z) \vee \neg P_3(x))).$$

Lyndon's argument for $\exists$-sentences is sound, and his conjecture for $\forall\exists$-sentences has been confirmed by A. Weinstein (*Notices Amer. Math. Soc.* 11 (1964), 391).

(^{4}) If a convex class $\mathfrak{K}$ of order $\mathfrak{q}$ is axiomatizable, then $\mathfrak{K}^{\mathfrak{q}+\aleph_0}$ (see §2.3) is quasi-axiomatizable and convex, but not axiomatizable if it contains an infinite model. An adequate definition: a class $\mathfrak{K}$ is **quasiaxiomatizable* iff it is locally bounded and every $\mathfrak{n}$-enrichment is compact. For a cardinal $\mathfrak{n}$, the $\mathfrak{n}$-*enrichment* of $\mathfrak{K}$ is the class of the type of $\mathfrak{K}$ augmented by $\mathfrak{n}$ zeros consisting of all possible models created from $\mathfrak{K}$-models by choosing $\mathfrak{n}$ new distinguished elements.

(^{5}) In [XI] such a class $\mathfrak{K}$ is called *projective*.

(^{6}) This proof is easily adapted to show that a general reductive class $\mathfrak{K}$ is projective as long as all the $\mathfrak{K}$-models have infinite powers not less than the order of $\mathcal{L}$.

CHAPTER 27

TOWARD A THEORY OF COMPUTABLE FAMILIES OF OBJECTS

In the course of the last ten years, various authors have introduced the notions of principal numbering (V.A. Uspenskiĭ [174]), complete numbering (A.I. Mal'cev [XVIII], [XXV]), standard and special numberings (A.H. Lachlan [85]), and numerous others. The purpose of this article is to clarify the basic relations among these concepts. In doing this, it is helpful to introduce some new kinds of numberings: normal and subnormal numberings, subspecial numberings, effectively principal numberings, etc.

Typical of the results are the following: a computable family of sets has a computable complete numbering iff it contains a smallest set; principal numberings of such families are isomorphic; effectively principal numberings coincide with subnormal numberings; a family of sets is subnormal iff it is ω-dense and τ-closed in the sense of Uspenskiĭ [174].

In contrast to [174] and [XVIII] the present article considers only simple numberings: by a *numbering* α of a set A of indeterminate objects we understand a well-defined mapping $\alpha\colon N \to A$ of the set N of all natural numbers onto the set A. An element $n \in N$ is called an α-number of the object αn. Subsets of A are called *families* of objects. ([1])

Although the chief interest is aroused when A is the collection of all recursively enumerable sets (**re**-sets) of natural numbers and is numbered by the Post numbering π (cf. [XXV], §1), we shall formulate definitions and theorems for an arbitrary numbered set $\langle A, \alpha\rangle$ whenever possible. Such an approach proves particularly convenient for studying subfamilies of some special family of **re**-sets. E.g., the set $\mathcal{F}_{\mathrm{pr}}$ of all unary partial recursive functions (**pr**-functions) and its subset $\mathcal{F}_{\mathrm{gr}}$ of general recursive functions (**gr**-functions) can be regarded as families of **re**-sets.

§1. Principal and complete numberings

Let α be some numbering of an arbitrary set A. A numbering β of a family

$B \subseteq A$ is **gr**-*multireducible* (or simply, *reducible*) to the numbering α iff there is a **gr**-function $f(x)$ such that

$$\beta n = \alpha f(n) \qquad (n = 0, 1, \dots) .$$

Numberings of the family B that are reducible to α are called *α-computable.* The family B is *α-computable* iff it admits at least one α-computable numbering. If a numbered set $\langle A, \alpha\rangle$ is fixed beforehand, then we shall abbreviate "α-computable" as "computable". In particular, the collection $\mathcal{W}$ of all **re**-sets is always taken to be numbered by the Post numbering π (see [XXV], §1, where the Post numbering is denoted by ω).

Numberings α, α' of one and the same set A are *equivalent* iff each is reducible to the other. They are *isomorphic* iff each is reducible to the other by means of a **gr**-function that maps N 1–1 onto itself.

A numbering β of a family B of objects in the numbered set $\langle A, \alpha\rangle$ is called *α-principal* iff β is α-computable and every α-computable numbering of B is reducible to β.

Theorem 1: *Let the sets $C \subseteq B \subseteq A$ have the respective numberings γ, β, α. Suppose β is α-principal and γ is α-computable; then γ is β-computable.*

Let f, g be **gr**-functions respectively reducing β, γ to α; thus $B = \alpha f(N)$, $C = \beta g(N)$. We introduce a **gr**-function $\varphi(x)$, setting

$$\varphi(2m) = f(m) , \quad \varphi(2m+1) = g(m) \qquad (m \in N) .$$

Since C is a subset of B, we have $B = \alpha\varphi(N)$. Therefore, the map $\delta : N \to B$ defined by

$$\delta n = \alpha\varphi(n) \qquad (n \in N)$$

is an α-computable numbering of the family B. The numbering δ must be reduced to the α-principal numbering β by some **gr**-function $h(x)$. Thus $\alpha\varphi(n) = \beta h(n)$ for all $n \in N$, whence

$$\gamma n = \alpha f(n) = \beta h(2n) \qquad (n \in N) ;$$

this means γ is β-computable. ■

If in the hypotheses of Theorem 1 the numbering β is required merely to be α-computable rather than α-principal, then the conclusion of this theorem may not hold.

As in [XXV], a numbering α of a set A is called a *complete numbering with special element* $e \in A$ iff for every partial recursive function $f(n)$, there exists a **gr**-function $g(n)$ such that for all $n \in N$,

$$\alpha g(n) = \begin{cases} \alpha f(n) & \text{if } f(n) \text{ is defined,} \\ e & \text{if } f(n) \text{ is undefined .} \end{cases}$$

Theorem 2: *Suppose the basic numbering α of the set A is complete with special element e. Then each computable numbering β of an arbitrary computable family B containing e is reducible to a computable numbering β_κ of B that is complete with respect to e.*

We shall show that the desired properties are possessed by the numbering β_κ defined by

$$\beta_\kappa n = \begin{cases} \beta K(n, 0) \text{ if } K(n, 0) \text{ is defined ,} \\ e \quad \text{otherwise ,} \end{cases}$$

where $K(n, x)$ is the Kleene function described in [XXV], §1.

(I) β_κ *is α-computable.* By assumption, there exists a **gr**-function $f(x)$ such that $\beta n = \alpha f(n)$ for all $n \in N$. Since α is complete, there must be a **gr**-function $g(x)$ satisfying

$$\alpha g(n) = \begin{cases} \alpha f(K(n, 0)) \\ e \end{cases} = \begin{cases} \beta K(n, 0) \\ e \end{cases} = \beta_\kappa n \ ,$$

and, consequently, reducing β_κ to α.

(II) β *is reducible to* β_κ. We choose (see [XXV], §1) a number s such that for all $n, x \in N$,

$$n = K^{(2)}(s, n, x) = K^{(1)}([s, n], x) \ .$$

We now have

$$\beta n = \beta K([s, n], 0) = \beta_\kappa [s, n] \qquad (n \in N) \ ;$$

in other words, the **gr**-function $[s, y]$ reduces β to β_κ.

(III) β_κ *is complete with special element e.* By κ we denote the Kleene numbering of the set $\mathcal{F}_{\mathrm{pr}}$ of all **pr**-functions: $\kappa n = K(n, x)$. This numbering is complete with respect to the function Λ defined nowhere. We have for all $m, n \in N$,

$$\kappa m = \kappa n \Rightarrow K(m, 0) = K(n, 0) \Rightarrow \beta_\kappa m = \beta_\kappa n\,,$$

$$\kappa n = \Lambda \Rightarrow K(n, 0) \text{ undefined } \Rightarrow \beta_\kappa n = e\,.$$

In other words, the numbering β_κ is a *homomorphic image* of the complete numbering κ, and thus β_κ is complete ([XXV], Theorem 2.2) with e as its special object. ■

Theorem 2 immediately implies the fundamental

Corollary: *Let α be a complete numbering of a set A with special element e. Then all α-principal numberings of an arbitrary family $B \subseteq A$ containing e are complete at e and are isomorphic to each other.*

Let β be a principal numbering of the family B. According to Theorem 2, β is reducible to a certain computable numbering β_κ. From the computability of β_κ it follows that β_κ is reducible to β; hence, the numbering β is equivalent to the complete numbering β_κ. By the generalizations ([XVIII], Theorem 2.3.4; [XXV], Theorem 2.1) of a theorem of H. Rogers, every numbering equivalent to a complete numbering is isomorphic to it and, therefore, complete itself. The numbering β is thus complete (with special object e). From the definition of principal numbering we see that all principal numberings of the same family are equivalent. Since the ones under discussion are also complete, Rogers' theorem generalized shows they are isomorphic. ■

§2. The α-order and α-topology

In order to characterize the families admitting computable complete numberings, we impose certain restrictions on the structure of the set of totally enumerable families.

According to [XXV], §3, a family B of objects in a numbered set $\langle A, \alpha\rangle$ is called *totally α-enumerable* (α-**te**) iff the set $\alpha^{-1}B$ of all α-numbers of elements of B is recursively enumerable. Thus, every nonempty α-**te** family is α-computable, and the intersection and union of any finite number of α-**te** families are α-**te** families. We also have the obvious

Lemma 1: *The intersection of an α-***te** *family C and an arbitrary family B with α-computable numbering β produces a β-***te** *subfamily of B.*

Indeed, for some **gr**-function f, we have $\beta = \alpha f$. So for every $n \in N$,

$$n \in \beta^{-1}(C \cap B) \Leftrightarrow f(n) \in \alpha^{-1} C;$$

consequently, the set $\beta^{-1}(C \cap B)$ is recursively enumerable. ■

Corollary: *Let $\langle A, \alpha\rangle$ be a numbered set. If a numbering α' of A is reducible to the numbering α, then every* α-**te** *family is also totally α'-enumerable. In particular, equivalent numberings have one and the same set of totally enumerable families.* ■

The invariance of the collection of all totally enumerable families under passage to equivalent numberings allows us to view the structure of this collection as a characteristic of the given numbered set $\langle A, \alpha\rangle$. Using this collection we can introduce a natural topology [173] and partial quasiorder ([XXV], §8) on $\langle A, \alpha\rangle$.

By definition, the *α-open families* of objects in the numbered set $\langle A, \alpha\rangle$ are arbitrary unions of α-**te** families.

An object $a \in A$ is an *α-subobject* of an object $b \in A$ (written $a \leqslant_\alpha b$) iff for every α-**te** family C

$$a \in C \Rightarrow b \in C.$$

The relation $\leqslant_\alpha$ is transitive and reflexive, but not antisymmetric, in general. It is clear that the condition of antisymmetry

$$a \leqslant_\alpha b \quad \text{and} \quad b \leqslant_\alpha a \Rightarrow a = b \qquad (a, b \in A)$$

is equivalent to the demand that A be a T_0-space in the α-topology. On fulfilling this demand, A becomes a partially ordered set. Lemma 1 yields the

Corollary: *Suppose the family $B \subseteq A$ has an α-computable numbering β. Then for all objects a, b in the family B,*

$$a \leqslant_\beta b \Rightarrow a \leqslant_\alpha b.$$

In particular, if A is a T_0-space in the α-topology, then every family $B \subseteq A$ with an α-computable numbering β is a T_0-space under its β-topology. ■

E.g., the set $\mathcal{W}$ of all recursively enumerable sets is a T_0-space relative to the Post numbering π. Therefore, every family of **re**-sets admitting a π-computable numbering forms a T_0-space relative to it.

Lemma 2: *Let $\langle A, \alpha\rangle$ be a numbered set. If a family $B \subseteq A$ has an α-computable complete numbering β with special element $e \in B$, then e is an α-smallest object in B (i.e., e is an α-subobject of every object in B).*

A completely numbered set has no **te**-families containing the special element other than the whole set ([XXV], Theorem 3.2). Thus the special object

e is a β-smallest object in B. By (1) it is also an α-smallest object. ■

A numbering α of a set A is an *upper numbering* iff for every $a \in A$ there exists a **gr**-function $g(x)$ such that for $n \in N$,

$$\alpha g(n) = q \vee \alpha n\,, \tag{2}$$

where $a \vee \alpha n$ is the smallest of the common α-superobjects of the objects a and αn. ([2])

Lemma 3: *Suppose the set A has an upper and complete numbering α with special object e. Then the family B consisting of all α-superobjects of a fixed object $a \in A$ admits an α-principal numbering complete at a.*

We introduce a numbering β of the family B by setting

$$\beta n = a \vee \alpha n \qquad (n \in N)\,.$$

By (2) this numbering is α-computable. On the other hand, for all natural numbers m and n,

$$\alpha m = \alpha n \Rightarrow \beta m = \beta n\,, \qquad \alpha m = e \Rightarrow \beta m = a\,;$$

thus β is a homomorphic image of α. Since α is complete at e, β is complete at a.

It remains to show that the numbering β is α-principal. Suppose B has a numbering γ that is reduced to α by the **gr**-function $f(x)$, i.e., $\gamma n = \alpha f(n)$ for all $n \in N$. Since $a \leqslant_\alpha \gamma n$,

$$\gamma n = a \vee \gamma n = a \vee \alpha f(n) = \beta f(n) \qquad (n \in N)\,;$$

hence, f reduces γ to β. ■

Lemmas 2 and 3 quickly give

Theorem 3: *Suppose the set A is supplied with a complete and upper numbering α. Let b be an object in an arbitrary family $B \subseteq A$. In order that B admit an α-computable numbering complete at b, it is necessary and sufficient that b be the α-smallest element in B.*

Lemma 2 shows the necessity of the condition. To see the sufficiency we assume b is the α-smallest object in B. According to Lemma 3, the family B_0 consisting of all α-superobjects of b has an α-principal numbering β complete at b. By Theorem 1 the family B, an α-computable subfamily of B_0, is also

β-computable. But B contains b; by Theorem 2, B has a β-computable numbering complete at b. ■

As an example we consider the numbered set $\langle \mathcal{W}, \pi \rangle$ of all **re**-sets. The Post numbering π is upper and complete relative to the empty set $\emptyset$. If c, d are **re**-sets, then $c \leqslant_\pi d$ iff $c \subseteq d$. According to Theorem 3, we have: *a family of* **re**-*sets possesses a computable complete numbering iff it contains a smallest set.* ■

We note that Lemma 3 loses its validity when the demand that α be upper is dropped from its hypotheses. For consider the numbered set $\langle \mathcal{F}_{\text{pr}}, \kappa \rangle$ of all **pr**-functions. The relation $c \leqslant_\kappa d$ means exactly that d is an extension of the partial function c ($c, d \in \mathcal{F}_{\text{pr}}$). Let f be a **pr**-function that cannot be extended to a **gr**-function, and let B be the family of all **pr**-functions that are extensions of f. The family B is not computable. For suppose to the contrary that β is a κ-computable numbering of B. Let $V(n, x)$ be the function with β-number n; it is clear that V is partial recursive in the variables n, x. The domain of definition of V is recursively enumerable and can be arranged in a recursive sequence

$$\langle n_0, x_0 \rangle, \langle n_1, x_1 \rangle, \ldots \,. \tag{3}$$

Let x be an arbitrary natural number. Since B certainly contains a function defined at the point x, in the sequence (3) there must be a first pair $\langle n_i, x_i \rangle$ with $x_i = x$. We define a unary function φ by putting $\varphi(x) = V(n_i, x)$ for each $x \in N$. The function φ is general recursive and coincides with f on the domain of f, contradicting that f cannot be so extended.

§3. Normal and subnormal numberings

A numbering β of some family B in a numbered set $\langle A, \alpha \rangle$ is called α-*subnormal* iff β is α-computable and there exists a **pr**-function $g(x)$ such that for all $n \in N$,

$$\alpha n \in B \Rightarrow g(n) \text{ defined and } \alpha n = \beta g(n). \tag{4}$$

The numbering β is called α-*normal* iff β is α-computable and there exists a **gr**-function g satisfying (4) for all $n \in N$.

Correspondingly, a family B is α-*subnormal* (α-*normal*) iff there exists an α-subnormal (α-normal) numbering of B.

Theorem 4: *For α-subnormal (α-normal) families in the numbered set* $\langle A, \alpha\rangle$, *the α-principal numberings coincide with the α-subnormal (α-normal) numberings.*

It is convenient to break the proof into two obvious lemmas.

Lemma 4: *Every α-subnormal numbering* β *of a family* $B \subseteq A$ *is α-principal.*

By assumption, β is α-computable and satisfies (4) along with some function $g \in \mathcal{F}_{\mathrm{pr}}$. Let γ be an α-computable numbering of B reduced on α by $h \in \mathcal{F}_{\mathrm{gr}}$:

$$\gamma n = \alpha h(n) \quad (n \in N)\,. \tag{5}$$

From (5) and (4) we learn that for $n \in N$, $g(h(n))$ is defined and

$$\gamma n = \beta g(h(n))\,;$$

hence, γ is reducible to β. ▲

Lemma 5: *Every numbering* γ *of a family* B *that is equivalent to an α-subnormal (α-normal) numbering* β *of this family is itself α-subnormal (α-normal).*

Since γ is reducible to β, the former is also α-computable. By assumption there is a **pr**-function (**gr**-function) g satisfying (4) and a **gr**-function j reducing β to γ: $\beta x = \gamma j(x)$. This means

$$\alpha n \in B \Rightarrow j(g(n)) \text{ is defined and } \alpha n = \beta g(n) = \gamma j(g(n))\,.$$

Consequently, γ is α-subnormal (α-normal). ▲

We now prove Theorem 4. According to Lemma 4, every α-subnormal numbering is α-principal. Conversely, suppose γ is an α-principal numbering of an α-subnormal (α-normal) numbering of B. By Lemma 4, β is α-principal and, therefore, equivalent to γ. By Lemma 5, γ is α-subnormal (α-normal). ■

Theorem 5: *Suppose* $C \subseteq B \subseteq A$ *are sets with the respective numberings* γ, β, α. *Then*

(i) If γ *and* β *are both α-subnormal (α-normal), then* γ *is β-subnormal (β-normal);*

(ii) If γ *is β-subnormal (β-normal) and* β *is α-subnormal (α-normal), then* γ *is α-subnormal (α-normal).*

We shall verify (i) for subnormal numberings. By the conditions of the theorem, β is reduced to α by some $f \in \mathcal{F}_{\mathrm{gr}}$:

$$\beta n = \alpha f(n) \quad (n \in N)\,, \tag{6}$$

and γ is reduced (5) to α by some $h \in \mathcal{F}_{gr}$. In addition, there is a $g \in \mathcal{F}_{pr}$ satisfying (4) and a $k \in \mathcal{F}_{pr}$ satisfying

$$\alpha n \in C \Rightarrow k(n) \text{ defined and } \alpha n = \gamma k(n)\,. \tag{7}$$

By Theorem 4, β is α-principal; so by Theorem 1, γ is β-computable. In addition, from (6) and (7) we find

$$\beta n \in C \Rightarrow k(f(n)) \text{ defined and } \beta n = \alpha f(n) = \gamma k(f(n))\,.$$

Thus γ is β-subnormal. The remainder of the theorem is proved similarly. ■

The numbered collection $\langle \mathcal{W}, \pi \rangle$ of **re**-sets satisfies two conditions: π is complete at $e = \emptyset$, the empty set; the family of nonempty **re**-sets is totally π-enumerable. We shall see that under such conditions normal and subnormal families differ only in the matter of containing the special object e. ([3])

Theorem 6: *Let $\langle A, \alpha \rangle$ be a completely numbered set with special element e. Then*

(i) Every α-subnormal family B that contains e is α-normal;

(ii) If the family $A \sim \{e\}$ of all nonspecial objects is totally α-enumerable, then adding the special object e to any α-subnormal family $B \subseteq A$ produces an α-normal family $B_1 = B \cup \{e\}$.

First we prove (i). Let β be an α-subnormal numbering of the family B containing e, and let $f \in \mathcal{F}_{gr}$ and $g \in \mathcal{F}_{pr}$ satisfy (6) and (4). We define a new numbering γ of B by setting

$$\gamma n = \begin{cases} \beta g(n) & \text{if } g(n) \text{ is defined}\,, \\ e & \text{otherwise}\,. \end{cases} \tag{8}$$

Thus we have

$$\gamma n = \begin{cases} \alpha f(g(n)) & \text{if } f(g(n)) \text{ is defined}\,, \\ e & \text{otherwise.} \end{cases}$$

Since α is complete at e, there exists $h \in \mathcal{F}_{pr}$ for which

$$\alpha h(n) = \begin{cases} \alpha f(g(n)) & \text{if } f(g(n)) \text{ is defined}\,, \\ e & \text{otherwise}\,. \end{cases}$$

Therefore, $\alpha h(n) = \gamma n$ for every $n \in N$; so γ is α-computable. From (8) and (6) we see

$$\alpha n \in B \Rightarrow \alpha n = \gamma n \ ;$$

hence γ is α-normal. Thus B is α-normal.

Turning to the proof of (ii), we let M be the set of all α-numbers of all nonspecial objects in A. Let B be a family with α-subnormal numbering β that is reduced (5) to α by a **gr**-function f and satisfies (4) for some **pr**-function g. The set M is recursively enumerable. This implies that the function g_1 specified by

$$g_1(n) = \begin{cases} g(n) & \text{if} \quad g(n) \quad \text{defined and} \quad n \in M, \\ \text{undefined otherwise} \end{cases}$$

is partial (but not general) recursive. We define a numbering γ of $B_1 = B \cup \{e\}$ by setting

$$\gamma n = \begin{cases} \beta g_1(n) & \text{if} \quad g_1(n) \quad \text{is defined,} \\ e & \text{otherwise.} \end{cases}$$

As above, we easily verify that γ is α-computable. In addition, from (4) and the definition of g_1 it follows that

$$e \neq \alpha n \in B \Rightarrow \alpha n = \beta g(n) \quad \text{and} \quad n \in M \Rightarrow \alpha n = \gamma n \ ,$$

$$\alpha n = e \Rightarrow n \notin M \Rightarrow g_1(n) \quad \text{undefined} \quad \Rightarrow \gamma n = e \ ;$$

this means that for $n \in N$,

$$\alpha n \in B_1 \Rightarrow \alpha n = \gamma n \ .$$

Thus γ is α-normal. ■

Theorem 7: *Suppose the numbered set $\langle A, \alpha \rangle$ contains an object a such that $A \sim \{a\}$ is totally α-enumerable. If B is an α-normal family containing a as well as other objects, then the family $B_0 = B \sim \{a\}$ is α-subnormal.*

Let β be an α-normal numbering of B, and let $f, g \in \mathcal{F}_{gr}$ satisfy

$$\beta n = \alpha f(n) \ , \quad \alpha n \in B \Rightarrow \alpha n = \beta g(n)$$

for every $n \in N$. The set $M = \alpha^{-1}(A \sim \{a\})$ is nonempty and recursively enumerable. So is the set $M_0 = f^{-1}(M)$; hence, it is the range of values of some **gr**-function $\varphi(x)$. The map γ defined by

$$\gamma n = \alpha f(\varphi(n)) \qquad (n \in N)$$

is an α-computable numbering of B_0. We shall prove that γ is α-subnormal. Introducing the **pr**-function

$$g_1(n) = \min_z (\varphi(z) = g(n)) \, ,$$

we see that

$$\alpha n \in B_0 \Rightarrow \alpha n = \alpha f(g(n)) \quad \text{and} \quad f(g(n)) \in M \, ,$$

$$f(g(n)) \in M \Rightarrow g(n) \in M_0 \Rightarrow g_1(n) \ \text{defined and} \ \ g(n) = \varphi(g_1(n)) \, .$$

Consequently,

$$\alpha n \in B_0 \Rightarrow \alpha n = \alpha f(\varphi(g_1(n))) = \gamma g_1(n) \, . \ \blacksquare$$

We note this application of Theorem 7: if a normal family B of **re**-sets contains a smallest set a, and a is recursive, then the family $B \sim \{a\}$ is subnormal if nonempty.

§4. Effectively principal numberings

Let B be a family included in a numbered set $\langle A, \alpha \rangle$. A natural number n is called a *Post α-number* of the family B iff $B = \alpha(\pi n)$. A natural number n is a *Kleene α-number* of a partial numbering γ of B iff for every $x \in N$, $\gamma x = \alpha K(n, x)$. Of course, π and K are the Post numbering of **re**-sets and the Kleene universal function, respectively. If $K(n, x)$ is defined for all $x \in N$, then the corresponding γ is an α-computable numbering of the family B.

An α-computable numbering β of a family B in $\langle A, \alpha \rangle$ is called *effectively α-principal* iff there is a **pr**-function $w(x)$ such that if n is a Kleene α-number of an α-computable numbering γ of B, then $w(n)$ is a κ-number of a **gr**-function reducing γ to β (in other words, iff there exists $w \in \mathcal{F}_{\text{pr}}$ such that if $\alpha(\pi n) = B$, and $K(n, x)$ is defined for every $x \in N$, then $\alpha K(n, z) = \beta K(w(n), z) \quad (z \in N)$.

Theorem 8: *The α-subnormal numberings of any family B coincide with its effectively α-principal numberings.*

Suppose β is an effectively α-principal numbering of B, and let w be a **pr**-function such that if n is a Kleene α-number of computable numbering γ of B, then $\kappa w(n)$ is a **gr**-function reducing γ to β. Suppose $f \in \mathcal{F}_{gr}$ reduces β to α: $\beta x = \alpha f(x)$. Let n be an arbitrary natural number, and let $C_n = B \cup \{\alpha n\}$. We construct a numbering γ_n of the new family C_n by putting $\gamma_n z = \alpha F(n, z)$, where

$$F(n, 0) = n\ ,$$

$$F(n, x+1) = f(x) \quad (x \in N)\ .$$

The function $F(n, x)$ is general recursive in the variables n, x; hence for some $r \in N$,

$$F(n, x) = K([r, n], x) \quad (n, x \in N)\ .$$

Suppose $n \in N$ and $\alpha n \in B$. Then $C_n = B$, and $[r, n]$ is a Kleene α-number for the computable numbering $\gamma_n x = \alpha K([r, n], x)$ of the family B. Therefore, $\alpha K([r, n], x) = \beta K(w([r, n]), x)$ and

$$\alpha n = \alpha K([r, n], 0) = \beta K(w([r, n]), 0)\ .$$

Introducing the partial recursive function $g(y) = K(w([r, y]), 0)$, we find that for all $n \in N$,

$$\alpha n \in B \Rightarrow g(n) \text{ defined and } \alpha n = \beta g(n)\ ; \tag{9}$$

thus β is α-subnormal.

Conversely, suppose we have a **pr**-function g and a **gr**-function f satisfying (9) and $\beta x = \alpha f(x)$, respectively. Let γ be a computable numbering of B, say $\gamma x = \alpha K(n, x)$ for some fixed $n \in N$. Since $K(n, x) \in B$ for every $x \in N$, we learn from (9) that

$$\alpha K(n, x) = \beta g(K(n, x)) = \beta K([s, n], x) \quad (x \in N)\ ,$$

where s is an appropriate fixed natural number. In other words, a suitable **gr**-function w can be defined by taking $w(y) = [s, y]$. ■

§5. Standard families and precomplete numberings

A different notion of complete numbering was introduced in [XVIII], §2.3: a numbering α is called *complete* iff there exists a **gr**-function $\varphi(x)$ such that

$$\alpha K(n, \varphi(n)) = \alpha\varphi(n) \tag{10}$$

for all those for which the function $K(n, x)$ is defined for every x ([4]), i.e., iff there exists an algorithm whereby from the Kleene number n of a **gr**-function $g(x) = K(n, x)$ one can find a solution x (an α-fixed point for the transformation g) of the equation

$$\alpha g(x) = \alpha x.$$

To avoid confusion with the notion of completeness introduced earlier, numberings complete in the sense of [XVIII] will be called *precomplete numberings* from now on.

Every complete numbering is also precomplete [XXV]. It is not at present known under what conditions the converse is true. In [XVIII], §2.3 it is proved that *equivalent precomplete* numberings are isomorphic (Rogers' theorem generalized). ∎

Theorem 9: *Suppose α precompletely numbers the set A. Then every α-normal numbering β of an arbitrary family $B \subseteq A$ is precomplete, and all α-normal numberings of B are isomorphic to one another.*

In particular, all principal numberings of a normal family of **re**-sets are precomplete and isomorphic.

Let $f, g, \varphi \in \mathcal{F}_{\text{gr}}$ reduce β to α and satisfy (9) and (10), respectively. We choose a natural number s such that

$$f(K(n, g(x))) = K(s, n, x) = K([s, n], x) \quad (n, x \in N).$$

Let n be a number for which $K(n, x)$ is a total function. From (9) and (10) we obtain

$$\beta K(n, g(\varphi([s, n]))) = \alpha f(K(n, g(\varphi([s, n])))) =$$

$$= \alpha K([s, n], \varphi([s, n])) = \alpha\varphi([s, n]) = \beta g(\varphi([s, n])).$$

Thus $g(\varphi([s, n]))$ is a β-fixed point for $K(n, x)$. Hence, β is a precomplete numbering.

According to Theorem 4, all α-normal numberings of B are α-principal and therefore equivalent to one another. By virtue of the generalization of Rogers' theorem, equivalent precomplete numberings are isomorphic; hence, all α-normal numberings of B are isomorphic. ■

In accord with a definition of A.H. Lachlan [85], a numbering β of a family $B \subseteq A$ is called *α-standard* iff β is α-computable and

$$\alpha n \in B \Rightarrow \beta n = \alpha n\,.$$

A family admitting an α-standard numbering is called *α-standard.*

Since every standard numbering is clearly normal, *every α-standard family is α-normal.* ■ The converse also holds:

Theorem 10: *Every α-normal numbering β of an α-normal family B is equivalent to an α-standard numbering of B; the α-normal families thus coincide with the α-standard families.*

Suppose f, g are **gr**-functions such that for all $n \in N$,

$$\beta n = \alpha f(n)\,, \quad \alpha n \in B \Rightarrow \alpha n = \beta g(n)\,.$$

We introduce a new numbering γ of B by putting $\gamma n = \beta g(n)$ $(n \in N)$. From

$$\gamma n = \beta g(n) = \alpha f(g(n))$$

it follows that the numbering γ is α-computable. In addition,

$$\alpha n \in B \Rightarrow \alpha n = \beta g(n) = \gamma n\,.$$

Thus γ is α-standard. The normal numberings β, γ are equivalent by Theorem 4. ■

Corollary: *If the basic numbering α is precomplete, then every α-normal numbering is isomorphic to an α-standard numbering.*

By Theorem 10, each α-normal numbering is equivalent to an appropriate α-standard numbering. According to Theorem 9, these are isomorphic. ■

We observe that a numbering isomorphic to an α-standard numbering is not necessarily α-standard, but it will be α_1-standard for a suitable numbering α_1 isomorphic to α.

Theorem 11: *Let β be a precomplete principal numbering of a family B of*

re-*sets, and let* $C \subseteq B$ *be a* β-*subnormal family. For any* **gr**-*function* $h(x)$, *if*

$$\beta h(0) \subseteq \beta h(1) \subseteq \ldots, \quad \bigcup_i \beta h(i) \in B, \quad \beta h(i) \in C \quad (i \in N),$$

then $\bigcup_i \beta h(i) \in C$.

Lachlan proved this theorem for the particular case when B contains all **re**-sets and C is a π-normal family. His proof can be applied to Theorem 11 almost without change.

We introduce the notation $T_i = \beta h(i)$, $T = \bigcup_i T_i$. Let γ be a β-subnormal numbering of C, and let $g \in \mathcal{F}_{\mathrm{pr}}$ satisfy

$$\beta n \in C \Rightarrow g(n) \text{ defined} \quad \text{and } \beta n = \gamma g(n).$$

Following Lachlan, we choose two strongly enumerated sequences ([5]) of finite sets $T_{i,j}$, $S_{i,j}$ such that

$$T_{i,j} \subseteq T_{i,j+1}, \quad \bigcup_j T_{i,j} = T_i, \quad S_{i,j} \subseteq S_{i,j+1}, \quad \bigcup_j S_{i,j} = \gamma g(i),$$

and $S_{i,j} = \emptyset$ if $g(i)$ is undefined. We put

$$U_x = \bigcup_{i \leqslant x} T_{i,x};$$

hence, $\bigcup_x U_x = T$. The set

$$R_i = T_0 \cup \bigcup \{T_x : x > 0 \text{ and for some } y,$$

$$U_y \supseteq S_{i,x} \text{ and } U_{x-1} \subseteq S_{i,y}\}, \tag{11}$$

is recursively enumerable, and $\{R_i : i \in N\}$ is a computable family (i.e., $i \to R_i$ is π-computable) included in B. Since β is principal, by Theorem 1 there exists a **gr**-function $\psi(n)$ such that $\beta\psi(i) = R_i$. The numbering β is also precomplete. Therefore, we can find a number r satisfying the equation

$$R_r = \beta\psi(r) = \beta r.$$

We can assume that $T_0 \neq \emptyset$, $T_{i,j} \neq \emptyset$. If $g(r)$ is not defined, then $S_{r,j} = \emptyset$; so there are no terms T_x in (11) for R_r, and $R_r = T_0$. But then it follows from $\beta r = T_0 \in C$ that $g(r)$ is defined. Thus $g(r)$ is in fact defined.

Consider the set $\gamma g(r)$. If $\gamma g(r) = T$, then $T \in C$ and we are finished. If

$\gamma g(r) \neq T$, then we can find a number p such that $p \in \gamma g(r) \sim T$ or $p \in T \sim \gamma g(r)$. In the first case, for all sufficiently large x, no y satisfies $U_y \supseteq S_{r,x}$; in the second, for all sufficiently large x, no y satisfies $U_{x-1} \subseteq S_{r,y}$. Therefore, in (11) for R_r there are only a finite number of terms T_x; consequently, $R_r = T_n$ for some n. If for every i, $T_{n+i} = T_{n+i+1}$, then $T = T_n \in C$, proving the theorem. In the contrary case we can choose n so that $R_r = T_n$ and $T_n \neq T_{n+1}$. From $\beta r = T_n \in C$ it follows that $T_n = \gamma g(r)$. Therefore, for all sufficiently large y, we have

$$U_y \supseteq T_{n,y} \supseteq S_{r,n+1}, \quad U_n \subseteq S_{r,y}.$$

Consequently, the set T_{n+1} appears in the representation (11) of R_r; hence, $R_r \supseteq T_{n+1}$, but this contradicts $R_r = T_n \neq T_{n+1}$. ■

Theorem 12: *If a set A with precomplete numbering α is the union of a finite number of α-***te** *families $A_1, \ldots, A_s$, then at least one of these families coincides with A.*

It clearly suffices to prove this for $s = 2$. We can assume that A_1 and A_2 are nonempty and that

$$\alpha^{-1}A_1 = \{\varphi_1(0), \varphi_1(1), \ldots, \varphi_1(n), \ldots\},$$

$$\alpha^{-1}A_2 = \{\varphi_2(0), \varphi_2(1), \ldots, \varphi_2(n), \ldots\},$$

where φ_1, φ_2 are appropriate **gr**-functions. Let $M_1 \subseteq N$ be the set of all those numbers that first appear in the first line before they appear in the second; let M_2 be the set of all numbers first occurring in the second line not later than they occur the first. Since $A_1 \cup A_2 = A$, we know $M_1 \cup M_2 = N$. The sets M_1, M_2 are recursively enumerable and their intersection is empty; hence, these sets are recursive. Suppose $A_1, A_2 \neq A$. Then we can find numbers p, q such that $\alpha p \in A_1 \sim A_2$, $\alpha q \in A_2 \sim A_1$. Consider the **gr**-function

$$g(x) = \begin{cases} q & \text{if } x \in M_1, \\ p & \text{if } x \in M_2. \end{cases}$$

Since the numbering α is precomplete, there must be a fixed point for which $\alpha g(r) = \alpha r$. If $r \in M_1 \subseteq \alpha^{-1}A_1$, then $\alpha r \in A_1$ and $g(r) = q$. From $\alpha g(r) = \alpha r$ we learn that $\alpha q \in A_1$, contradicting $\alpha q \notin A_1$. An analogous contradiction can be obtained under the assumption that $r \in M_2$. ■

Corollary: *If a family B of* **re**-*sets admits a computable precomplete numbering β and contains a minimal set a that is recursive, then a is the smallest set in B.*

Indeed, in the contrary case there is a set $b \in B$ of which a is not a subset. The family B_1 of all those sets in B that contain at least one point of a, and the family B_2 of those that contain at least one point of $N \sim a$ are β-**te** families, each different from B. But their union is B, contradicting Theorem 12.([6]) ■

Lachlan [85] proved Theorem 12 and its corollary in a slightly different form for standard numberings of families of **re**-sets. This corollary is of interest, since it is not known whether every standard family of **re**-sets contains a smallest set, or even whether each family of **re**-sets with a computable precomplete numbering does.

As a simple example of a normal family of **re**-sets, we consider the family B_M of all recursively enumerable supersets of a given **re**-set M. Let r be a π-number for M, and let $S(x, y)$ be a **gr**-function such that $\pi S(m, n) = \pi m \cup \pi n$ ($m, n \in N$). Then the map β determined by $\beta n = \pi S(r, n)$ is a computable numbering of B_M. This numbering is standard inasmuch as

$$\pi n \in B_M \Rightarrow \beta n = \pi r \cup \pi n = \pi n \, .$$

A more complicated example is the family C consisting of every set of the form $\pi^{-1}B$ for an arbitrary **te**-family B of **re**-sets. The normality of C is readily established with the help of Theorem 20 on the structure of the π-**te** families (see the end of §7).

§6. Special and subspecial numberings

A numbering β of a family B in some numbered set $\langle A, \alpha \rangle$ is called *α-subspecial* iff it is α-computable and there exists a **pr**-function $g(x)$ satisfying the conditions

$$\alpha n \in B \Rightarrow g(n) \text{ defined} \quad \text{and } \alpha n = \beta g(n) \, , \tag{12}$$

$$g(n) \text{ defined} \Rightarrow \beta g(n) \leqslant_\alpha \alpha n \tag{13}$$

for all $n \in N$. From (12) we see that a subspecial numbering is subnormal.

When we can find a **gr**-function g satisfying (12) and (13), we say β is *α-special.* Thus a special numbering is normal.

A family B in $\langle A, \alpha \rangle$ is called *α-subspecial* (*α-special*) iff there exists an α-subspecial (α-special) numbering of B.

Assume β is an α-special numbering of a family B in $\langle A, \alpha\rangle$. Suppose the set A contains an α-minimal object a, and let r be an α-number of a. Then from (13) it follows that $\beta g(r) = a$, and so $a \in B$. In particular, *every special family of* **re**-*sets contains the empty set.* ■

A numbering β of a family B is called *specially α-standard* iff β is α-standard and $\beta n \leqslant_\alpha \alpha n$ for all $n \in N$.

Remark: *If the basic numbering α is complete, then each special numbering β of an arbitrary special family B is isomorphic to a specially standard numbering γ of B.*

As in the proof of Theorem 10 we introduce the new numbering γ, where $\gamma n = \beta g(n)$. The conditions (12) and (13) respectively become

$$\alpha n \in B \Rightarrow \alpha n = \gamma n, \qquad \gamma n \leqslant_\alpha \alpha n ,$$

showing that γ is a specially standard numbering of B. By Theorem 9 the normal numberings β, γ are isomorphic. ■

According to Lachlan [85], a family of **re**-sets is called *special* iff it admits a specially standard numbering. The remark above shows that our special families of **re**-sets coincide with Lachlan's.

The next three theorems are almost literal analogs of the corresponding Theorems 5, 6, and 7 concerning normality.

Theorem 13: *Suppose $A \supseteq C \supseteq B$ are sets of arbitrary objects with the respective numberings α, γ, β. Then*

(i) If γ is α-(sub)special, and β is γ-(sub)special, then β is α-(sub)special;

(ii) If β is γ-computable and α-(sub)special, γ is α-computable, and for any objects $a, b \in C$

$$a \leqslant_\alpha b \Rightarrow a \leqslant_\gamma b ,$$

then β is γ-(sub)special.

We prove (ii) for β subspecial. By assumption there exists a **pr**-function $g(x)$ satisfying (12) and (13) and a **gr**-function $f(x)$ satisfying $\beta n = \alpha f(n)$. From this we get

$$\gamma n \in B \Rightarrow \alpha f(n) \in B \Rightarrow g(f(n)) \text{ defined and } \alpha f(n) = \beta g(f(n)) ,$$

that is,

$$\gamma n \in B \Rightarrow g(f(n)) \text{ defined and } \gamma n = \beta g(f(n)) .$$

In addition we have

$$g(f(n)) \text{ defined} \Rightarrow \beta g(f(n)) \leqslant_\alpha \alpha f(n) = \gamma n\ ,$$

whence, by hypothesis, it follows that

$$g(f(n)) \text{ defined} \Rightarrow \beta g(f(n)) \leqslant_\gamma \gamma n\ .$$

Thus β is γ-subspecial. ■

Theorem 14: *If the basic numbering α of the set A is complete at $e \in A$, then adding e to an arbitrary α-subspecial family B produces an α-special family $B_1 = B \cup \{e\}$. In particular, under the assumption that α is complete, α-subspecial families are α-special iff they contain e.* ([7])

Let β be an α-subspecial numbering of B. From the completeness of α it follows that the numbering γ of B_1 defined by

$$\gamma n = \begin{cases} \beta g(n) & \text{if } g(n) \text{ defined}\ , \\ e & \text{otherwise} \end{cases}$$

is α-computable. From (12) and (13) we see that γ is a specially α-standard numbering. ■

Theorem 15: *Suppose in the numbered set $\langle A, \alpha\rangle$ there is an element a such that the family $A \sim \{a\}$ is totally α-enumerable. If an α-subspecial family $B \subseteq A$ contains a and other objects, then the family $B_0 = B \sim \{a\}$ is also α-subspecial.*

Let β be an α-subspecial numbering of B, and let $g \in \mathcal{F}_{pr}$ satisfy (12) and (13) and $f \in \mathcal{F}_{gr}$ satisfy $\beta n = \alpha f(n)$. Let φ, g_1 be the functions and γ the α-subnormal numbering of B_0 constructed in the course of proving Theorem 7. We also have

$$g_1(n) \text{ defined} \Rightarrow \gamma g_1(n) = \alpha f(\varphi(g_1(n))) = \beta g(n) \leqslant_\alpha \alpha n\ ;$$

consequently, γ is α-subspecial. ■

Since the Post numbering π of the set $\mathcal{W}$ of all **re**-sets is complete at the empty set, and the family of all nonempty **re**-sets is totally π-enumerable, the preceding theorems yield this

Corollary: *Every special family of* **re**-*sets contains the empty set. Every subspecial family in* $\mathcal{W}$ *that is not special is obtained from a corresponding special family by deleting the empty set.* ■

We now wish to present Lachlan's basic theorem on the construction of special classes of **re**-sets and to indicate some of its connections with the results of V.A. Uspenskiĭ.

A numbering γ of a family C consisting of finite sets of natural numbers, is called *strongly computable* iff γ is π-computable and the function

$$p(n) = \text{ power of } \gamma n$$

is general recursive. A family of finite sets admitting a strongly computable numbering is called *strongly enumerable.*

Theorem 16: *In order for a family of* **re**-*sets to be subspecial, it is necessary and sufficient that it consist of all limits of monotonic sequences of sets from a fixed strongly enumerable family of finite sets.*

Lachlan [85] proved that a family **re**-sets is special iff it coincides with the collection of all **re**-sets that are unions of monotonic sequences of members of members of some strongly enumerable family of finite sets, one of which is the empty set. Removing the empty set from such a family presents no problem (cf., e.g., the proof of Theorem 7); Theorem 16 is then deduced with the aid of Theorems 14 and 15. ■

In §7 we shall need a somewhat more general proposition, which we now formulate. According to Uspenskiĭ [173], a family B of **re**-sets is called ω-*dense* iff its subfamily of finite sets is strongly enumerable and for every finite subset a of a set $b \in B$, there is a finite set $c \in B$ such that $a \subseteq c \subseteq b$.

Theorem 17: *If* β *is a principal numbering of an* ω-*dense family* B *of* **re**-*sets, and* $C \subseteq B$ *is a subfamily with a* β-*subspecial numbering* γ, *then* C *is also* ω-*dense.*

We present a detailed proof of this theorem, although it is virtually a repetition of Lachlan's proof of the narrower theorem cited above.

By assumption there is a **gr**-function $h(x)$ such that

$$\beta h(0), \beta h(1), \ldots, \beta h(n), \ldots \tag{14}$$

is a strong enumeration of all the finite sets in B, and there is a **pr**-function $g(x)$ such that

$$\beta n \in C \Rightarrow g(n) \text{ defined and } \beta n = \gamma g(n)\,,$$

$$g(n) \text{ defined } \Rightarrow \gamma g(n) \leqslant_\beta \beta n \Rightarrow \gamma g(n) \subseteq \beta n\,. \tag{15}$$

First we show that the family of finite sets in C is strongly enumerable. Since all the finite sets in C appear in the sequence (14), we have only to find those numbers n for which $\beta h(n) \in C$. From (15) we see that

$$\beta h(n) \in C \Leftrightarrow g(h(n)) \text{ defined and } \gamma g(h(n)) = \beta h(n) \Leftrightarrow$$

$$\Leftrightarrow g(h(n)) \text{ defined and } \beta h(n) \subseteq \gamma g(h(n))\,.$$

The set of n satisfying the last conditions in this chain is recursively enumerable because the powers of the sets $\beta h(n)$, as well as π-numbers of these and the sets $\gamma g(h(n))$, are given effectively.

It remains to show that every finite subset U of any set $X \in C$ is included in a finite set V: $X \supseteq V \in C$. Suppose this is false; in fact, suppose U is a finite subset of some $X \in C$ for which there is no V with the desired properties. Since X belongs to the ω-dense family B, we can find $k \in \mathcal{F}_{\text{gr}}$ such that

$$U \subseteq X_0 \subseteq X_1 \subseteq X_2 \subseteq \ldots, \quad \bigcup_i X_i = X,$$

where $X_i = \beta h(k(i))$ for $i \in N$.

Let $t(x)$ be a **gr**-function with nonrecursive range. Put

$$T_i = \{t(0), t(1), \ldots, t(i)\} \quad (i \in N)\,, \quad T = \bigcup_i T_i\,,$$

$$R_i = X_0 \cup \bigcup_{i \notin T_j} X_j \quad (i \in N)\,.$$

We thus have

$$i \notin T \Rightarrow R_i = X\,, \quad i \in T \Rightarrow R_i = X_{m_i} \text{ for some } m_i \in N\,.$$

So $\{R_i\colon i \in N\}$ is a subfamily of B. Moreover, the numbering $i \to R_i$ is π-computable. By Theorem 1, this numbering is also β-computable, i.e., for some **gr**-function $s(x)$ we have $R_i = \beta s(i)$. From the implications

$$i \notin T \Rightarrow \beta s(i) = X \Rightarrow g(s(i)) \text{ defined and } X = \gamma g(s(i)) \Rightarrow$$

$$\Rightarrow g(s(i)) \text{ defined and } U \subseteq \gamma g(s(i))\,,$$

$$i \in T \text{ and } g(s(i)) \text{ defined } \Rightarrow \beta s(i) = X_{m_i} \supseteq \gamma g(s(i)) \Rightarrow$$

$$\Rightarrow U \nsubseteq \gamma g(s(i)) \ ,$$

it follows that

$$i \notin T \Leftrightarrow g(s(i)) \text{ defined and } U \subseteq \gamma g(s(i)) \, .$$

But clearly the set of all i satisfying the right-hand side of this equivalence is recursively enumerable, which contradicts the nonrecursiveness of T. ■

Theorem 17 would be a direct consequence of Theorem 16 were it known that every principal numbering of an ω-dense family is effectively principal. Should this assertion prove untrue, the following generalization of Theorem 16 might be of some interest:

Let β be a precomplete and principal numbering of an ω-dense family B of **re**-*sets. A subfamily $C \subseteq B$ is β-subspecial iff consists of the members of a strongly enumerable family D of finite sets together with all the sets in B that are limits of increasing sequences of members of D.*

The necessity of this condition is deduced from Theorems 11 and 17, while the sufficiency is proved just as for the corresponding theorem of Lachlan [85]. ■

In addition to ω-dense families, so-called τ-closed families are studied in Uspenskiĭ [173]. It is easy to see that for ω-dense families, being τ-closed is equivalent to being closed under limits of computable increasing sequences of **re**-sets. Therefore, Theorem 16 can be reformulated as: *a family of* **re**-*sets is subspecial iff it is ω-dense and τ-closed.*

§7. Totally enumerable families

Let $\langle A, \alpha \rangle$ be a numbered set and let $B \subseteq A$ be a family with α-computable numbering β. According to Lemma 2 in §2, each family of the form $B \cap C$ – where C is an α-**te** family – is a β-**te** subfamily of B. These are called *α-extendable* β-**te** families. Of particular interest are the numberings β for which all β-**te** families are α-extendable. For such a numbering β, the β-topology of B coincides with the topology induced by the basic α-topology on A. Uspenskiĭ described one sort of these numberings ([174], Theorem 5). We shall see that this result is a direct consequence of those obtained in §§5 and 6.

Following Uspenskiĭ, we call a computable numbering β of a family B of **re**-sets *effectively open* iff every nonempty β-**te** subfamily C consists of all possible supersets in B of members of a fixed strongly enumerable family of finite sets belonging to B.

It is clear that every subfamily C of this form is π-extendable. Thus, effectively open numberings belong to the class of numberings whose totally enumerable families are all extendable, but they do not exhaust this class.

Theorem 18: *Suppose the basic numbering α of the set A is upper and complete, the numbering β of the family $B \subseteq A$ is α-principal, and a, b are objects in B for which $a <_\alpha b$. Then every numerical set $M \subseteq N$ satisfying*

$$\beta^{-1}\{a\} \subseteq M, \quad M \cap \beta^{-1}\{b\} = \emptyset \tag{16}$$

is productive.

Let T be a creative set. By Lemma 3 in §2 the family D of all α-superobjects of the object a possesses an α-principal numbering γ complete at a. Let r be a γ-number of b, and let $t(x)$ be the function taking the value r on T and undefined outside of T. By the completeness of γ, there is a **gr**-function $g(x)$ such that

$$\gamma g(n) = \begin{cases} \gamma t(n) = b & \text{if } n \in T, \\ a & \text{if } n \in N \sim T. \end{cases}$$

Consequently, the numbering δ defined by $\delta x = \gamma g(x)$ is an α-computable numbering of the family $C = \{a, b\}$. The family C is included in the α-principally numbered family $\langle B, \beta \rangle$. Therefore, by Theorem 1, δ is β-computable: for some $f \in \mathcal{F}_{gr}$ we have $\gamma g(n) = \beta f(n)$. From (16) we obtain

$$n \in T \Rightarrow \beta f(n) = b \Rightarrow f(n) \in \beta^{-1}\{b\} \Rightarrow f(n) \notin M$$

$$n \in N \sim T \Rightarrow \beta f(n) = a \Rightarrow f(n) \in M\,;$$

that is,

$$n \in \widetilde{T} \Leftrightarrow f(n) \in M\,.$$

Thus the productive set $\widetilde{T}$ is multireducible to M; this means M is productive. ∎

Corollary: *Suppose the numbering α of the set A is upper and complete. Let β be an α-principal numbering of a family $B \subseteq A$. If a is an object in a β-***te*** family $C \subseteq B$, then C contains all α-superobjects of a that belong to B.*

For in the contrary case there is a object $b \in B \sim C$, such that $a <_\alpha b$. Then the set $M = \beta^{-1} C$ satisfies (16), so M is productive, which contradicts its assumed recursive enumerability. ∎

Theorem 19: *For any numbered set* $\langle B, \beta \rangle$, *every nonempty* β-**te** *family* $C \subseteq B$ *is* β-*subspecial.*

By hypothesis the set $\beta^{-1}C$ is nonempty and recursively enumerable and is thus the range of some **gr**-function $h(x)$. The numbering γ of C determined by $\gamma n = \beta h(n)$ is β-computable. We claim it is β-subspecial relative to the **pr**-function $g(x)$ defined by

$$g(n) = \min_z (h(z) = n) .$$

Indeed,

$$\beta n \in C \Rightarrow n \in \beta^{-1}C \Rightarrow g(n) \text{ defined and } n = h(g(n)) \Rightarrow$$

$$\Rightarrow \beta n = \gamma g(n) ,$$

$$g(n) \text{ defined } \Rightarrow \beta n = \beta h(g(n)) = \gamma g(n) . \blacksquare$$

Theorem 20 (Uspenskiĭ [174]): *Any principal numbering* β *of an* ω-*dense family* B *of* **re**-*sets is effectively open.*

Let $C \subseteq B$ be a nonempty β-**te** family. By Theorem 19, C is β-subspecial. Applying Theorem 17, we see that C is ω-dense. Therefore, the family C_{fin} composed of all the finite sets belonging to C is strongly enumerable, and every member of C has at least one subset belonging to C_{fin}. By the corollary of Theorem 18, every set that belongs to B, and includes a set in C, itself belongs to C. Consequently, C is the collection of all those sets in B that have subsets in C_{fin}. ■

In conclusion we indicate a type of normal family consisting of **re**-sets connected with β-**te** families.

Theorem 21: *Suppose the family* B *of* **re**-*sets has a computable numbering* β, *and suppose* $h(x)$ *is a* **gr**-*function such that* $\{\beta h(n): n \in N\}$ *is a family of finite sets from* B *with strongly computable numbering* γ *given by* $\gamma n = \beta h(n)$. *Let* D_γ *denote the family of all* **re**-*sets that have the form* $\beta^{-1}C$, *where either* C *is empty or there exists a* **gr**-*function* $k(x)$ *such that* C *consists of all sets in* B *having at least one subset of the form* $\gamma k(m)$. *Then the family* D_γ *is* π-*normal.*

Let δn be the collection of all β-numbers of those sets in B that have at least one subset of the form γz for $z \in \pi n$. If $\pi n = \emptyset$, we take $\delta n = \emptyset$. It is clear that δ is a π-computable numbering of the family D_γ, and that

$$\pi n \subseteq \pi m \Rightarrow \delta n \subseteq \delta m . \tag{17}$$

Furthermore, $\delta n = \beta^{-1} C_n$, where C_n is the subfamily of B consisting of supersets of members of the family $\gamma(\pi n)$; in particular,

$$\gamma(\pi n) \subseteq C_n \,. \tag{18}$$

Let s_n be a π-number for $\gamma^{-1} C_n$. From (18) we see that $\pi s_n \supseteq \pi n$. Therefore, by (16), $\delta s_n \supseteq \delta n$. On the other hand, every set γx with $x \in \pi s_n$ includes a subset γy for some $y \in \pi n$; hence, $\delta s_n \subseteq \delta n$. Thus

$$\delta s_n = \delta n \,. \tag{19}$$

Since

$$x \in \pi s_n \Leftrightarrow \beta h(x) \in C_n \Leftrightarrow h(x) \in \beta^{-1} C_n \Leftrightarrow x \in h^{-1}(\delta n),$$

we see $\pi s_n = h^{-1}(\delta n)$.

Therefore, for every $m \in N$, there is an $n \in N$ such that

$$\pi m \in D_\gamma \Rightarrow \pi m = \delta n \Rightarrow \pi s_n = h^{-1}(\pi m) \,. \tag{20}$$

Let g be a **gr**-function such that $h^{-1}(\pi x) = \pi g(x)$. From (20) we see that $\pi s_n = \pi g(m)$. Thus in (19) we can take s_n to be $g(m)$; together with (20) this gives $\pi m = \delta g(m)$. Consequently, for every $m \in N$,

$$\pi m \in D_\gamma \Rightarrow \pi m = \delta g(m);$$

in other words, δ is normal. ■

From Theorems 20 and 21 one immediately derives the result mentioned at the end of §5: the normality of the family of all **re**-sets of the form $\pi^{-1} C$, where C is a totally enumerable family of **re**-sets.

NOTES

(^{1}) The empty family must at times be admitted, at times avoided.

(^{2}) The demand that $a \vee \alpha n$ exists and be unique should be a part of this definition. In particular, the α-topology must be T_0.

(^{3}) Not quite. A family consisting of a single object is always normal.

(^{4}) The earlier requirement that αN contain at least two objects has been dropped in this article.

([5]) I.e., the numbering $[i,j] \to T_{i,j}$ is strongly computable; see §6.

([6]) The possibility that a intersects every member of B has been neglected; cf. [85, Lemma 1.2].

([7]) It is necessary to assume also that e is minimal, i.e., has no distinct α-subobjects in A (or, equivalently, that $A \sim \{e\}$ is open in the α-topology). Cf., e.g., the numbering κ/σ following Theorem 7.2 in [XXV].

CHAPTER 28

POSITIVE AND NEGATIVE NUMBERINGS

A *numbering* of a set B of arbitrary objects is a well-defined map β from the set N of all natural numbers onto B. The *number equivalence* of β is the binary relation θ_β on N defined by

$$x\theta_\beta y \Leftrightarrow \beta x = \beta y \qquad (x, y \in N)\ .$$

If the numbering β is (freely) isomorphic to a computable numbering of some family of recursively enumerable sets (**re**-sets) – for terminology see [XXV] and [XXVII] – then there exists a general recursive function (**gr**-function) $F(x, y, u, v)$ such that for all $x, y \in N$,

$$\beta x = \beta y \Leftrightarrow (\forall u \in N)(\exists v \in N)(F(x, y, u, v) = 0)\ .$$

Hence, *the number equivalences of computable numberings of families of* **re**-*sets are of class* $\forall\exists$ *in the Kleene classification.* ■ Below we examine in some detail the properties of numberings β for which θ_β can be expressed in the form $(\exists u \in N)(F(x, y, u) = 0)$ or the form $(\forall u \in N)(F(x, y, u) = 0)$ for an appropriate **gr**-function F. In the first case the set of all pairs $\langle x, y\rangle$ for which $\beta x = \beta y$ is recursively enumerable, and the numbering β is called *positive*. In the second case β is said to be *negative.* When β is negative, the set of all pairs $\langle x, y\rangle$ for which $\beta x \neq \beta y$ is recursively enumerable. A numbering is *decidable* iff it is both positive and negative.

§1. Let $\langle A, \alpha\rangle$ be a numbered set. A numbering β of a subset (family) $B \subseteq A$ is called *α-computable* iff there is a **gr**-function $f(x)$ such that $\beta n = \alpha f(n)$ for all $n \in N$. A numbering β of B is called *uniformizable* iff there exists a 1–1 β-computable numbering of B. A family $B \subseteq A$ is *α-uniform* iff it admits an α-computable 1–1 numbering. A family $B \subseteq A$ is *α-isolated* iff B admits an α-computable numbering, and all such numberings of B are equivalent to one another (i.e., are computable relative to one another). If the fundamental set is the set $\mathcal{W}$ of all **re**-sets, and it is endowed with the usual Post (= Gödel) numbering π, then instead of "π-computable", we shall use "computable", etc.

Theorem 1: *If the basic numbering α of the set A is positive (negative), then every α-computable numbering β of a family $B \subseteq A$ is likewise positive (negative). If α is positive and B is α-computable (i.e., has at least one α-computable numbering), then B is α-isolated.*

The first assertion is a consequence of the definitions. We shall prove the second. By hypothesis, all the pairs $\langle x, y\rangle$ for which $\alpha x = \alpha y$ can be arranged in a recursive sequence

$$\langle x_0, y_0\rangle, \langle x_1, y_1\rangle, \dots, \tag{1}$$

possibly with repetitions. Let β, γ be arbitrary α-computable numberings of B, and suppose $\beta x = \alpha f(x)$, $\gamma x = \alpha g(x)$, where f, g are suitable **gr**-functions. For a given number n, we find the first pair in (1) of the form $\langle f(n), g(m)\rangle$ and put $h(n) = m$. The **gr**-function h so defined satisfies $\beta x = \gamma h(x)$ for all $x \in N$. ■

Corollary: *Uniformizable positive numberings are decidable.*

Indeed, suppose α positively numbers the set A, and let β be a $1-1$ α-computable numbering of the family $B = A$. By the preceding theorem, β and α are equivalent. Since β is decidable, so is α. ■

§**2.** A family B of **re**-sets is called *finitely separable* iff there exists a strongly enumerated sequence ([1]) of finite sets $a_0, a_1, \dots$ such that every set belonging to B includes at least one of these finite sets, but none of the latter is a subset of more than one member of B. E.g., if the family B is a *partition* (i.e., the members of B are nonempty and pairwise disjoint, and their union is N), then B is finitely separated by the sequence $\{\{0\}, \{1\}, \dots\}$.

Theorem 2: *Every positive numbering α of a set of arbitrary objects is isomorphic to a computable numbering of the corresponding partition N/θ_α. Every computable and finitely separable family B of* **re**-*sets is isolated, and every computable numbering of B is positive.*

Let $[r] = \{x: \alpha x = \alpha r\}$. By assumption, for some **gr**-function $F(r, x, y)$ we have

$$\alpha x = \alpha r \Leftrightarrow (\exists y \in N)(F(r, x, y) = 0)$$

for all $x, r \in N$. Consequently, $[r]$ is the domain of definition of the partial recursive function (**pr**-function) $\varphi_r(x)$ given by

$$\varphi_r(x) = \min_y (F(r, x, y) = 0) ;$$

therefore, the map $\alpha_1 : r \to [r]$ is a computable numbering of the partition N/θ_α. Since $\alpha r = \alpha s \Leftrightarrow \alpha_1 r = \alpha_1 s$, the numberings α, α_1 are isomorphic.

To prove the second part, let us consider a strongly enumerated sequence of finite sets $a_0, a_1, \ldots$ that separates B. Let β be an arbitrary computable numbering of B. If we take any natural number n and successively compute the members of the sets $\pi n, a_i, \beta j$ $(i, j = 0, 1, \ldots)$, we can always find a pair $\langle i, j \rangle$ such that $a_i \subseteq \pi n, a_i \subseteq \beta j$. We define a **pr**-function $g(x)$ by letting $g(n)$ be the number j in the first such pair found. It is clear that for all $n \in N$,

$$\pi n \in B \Rightarrow g(n) \text{ defined and } \pi n = \beta g(n).$$

In other words, every computable numbering β of B is *subnormal* [XXVII]. But all subnormal numberings are equivalent ([XXVII]), Theorem 4), so B is isolated. It only remains to show that β is positive. We compute successively the elements of the sets $a_i, \beta j$ $(i, j = 0, 1, \ldots)$, and we mark those pairs $\langle r, s \rangle$ such that for some i, $a_i \subseteq \beta r$ and $a_i \subseteq \beta s$. In so doing, we enumerate all the pairs $\langle r, s \rangle$ for which $\beta r = \beta s$. ■

From Theorem 2 it follows that *if a finitely separable family admits a computable but undecidable numbering, then it is not uniform.* ■

A simple example of such a family is constructed as follows. Let M be a nonrecursive, recursively enumerable set of numbers: $M = \{m_0, m_1, \ldots\}$. We construct sets $M_0, M_1, \ldots$ in stages; at the ith stage we put the number i into M_i and both the numbers $2m_i, 2m_i{+}1$ into M_{2m_i} and M_{2m_i+1}. The mapping $\alpha: n \to M_n$ is a computable positive numbering of the partition $\{M_0, M_1, \ldots\}$. This numbering is not decidable, since $M_{2n} = M_{2n+1} \Leftrightarrow n \in M$.

By setting $\beta n = M_0 \cup M_1 \cup \ldots \cup M_n$, we obtain a computable numbering of a nondecreasing sequence of finite sets:

$$\beta 0 \subseteq \beta 1 \subseteq \ldots, \bigcup_n \beta n = N .$$

The numbering β is isomorphic to α and is, therefore, positive and undecidable. According to the corollary of Theorem 1, the numbering β is not uniformizable. At the same time, from the reasoning of Friedberg [40] it follows (cf. [M9]) that every computable family of **re**-sets that includes a strongly enumerated increasing sequence of finite sets admits a 1–1 computable numbering. Therefore, the family βN is seen to have some 1–1 computable numbering and thus serves as an example of a uniform family admitting a positive computable numbering that cannot be uniformized.

§**3**. Let β be a computable numbering of a family of unary **gr**-functions. This just means that for some **gr**-function $U(n, x)$ we have $(\beta n)(x) = U(n, x)$ for all $n, x \in N$. Computing $U(0,0)$, $U(0,1)$, $U(1,0)$, ... in sequence and marking each pair $\langle r, s\rangle$ such that for some x, $U(r, x) \neq U(s, x)$, we eventually list every pair $\langle r, s\rangle$ for which $\beta r \neq \beta s$. Therefore, *every computable numbering of a family of* **gr**-*functions is negative.* ■

The ordinary numbering of the family of all primitive recursive functions [M9] can serve as a typical example of a negative numbering that is not positive.

Theorem 3: *Every negative numbering* α *of an arbitrary set is isomorphic to a computable numbering of a family of* **gr**-*functions whose values are limited to* 0, 1.

By assumption, the set $\widetilde{\theta}_\alpha$ of all pairs $\langle r, s\rangle$ such that $\alpha r \neq \alpha s$ can be arranged in a recursive sequence

$$\langle r_0, s_0\rangle, \langle r_1, s_1\rangle, \ldots \tag{2}$$

We construct a **gr**-function $U(n, x)$ and auxiliary finite sets $S^m_{k,0}$, $S^m_{k,1}$ of natural numbers by means of the following algorithm. For an arbitrary number m we put

$$U(r_m, m) = 0, \quad U(s_m, m) = 1\ ,$$

$$S^m_{0,0} = \{r_m\}, \quad S^m_{0,1} = \{s_m\}\ .$$

Suppose that for some number p we have already defined finite sets $S^m_{p,0}$, $S^m_{p,1}$ such that

$$i \in S^m_{p,0} \Rightarrow U(i, m) = 0\ , \quad i \in S^m_{p,1} \Rightarrow U(i, m) = 1,$$

and if $i \in S^m_{p,0}$ and $j \in S^m_{p,1}$, then $\langle i, j\rangle \in \widetilde{\theta}_\alpha$. Let z be the smallest natural number not belonging to $S^m_{p,0} \cup S^m_{p,1}$. Choose some pair $\langle r, s\rangle$, where $r \in S^m_{p,0}$ and $s \in S^m_{p,1}$. Since $\alpha r \neq \alpha s$, we know either $\alpha r \neq \alpha z$ or $\alpha s \neq \alpha z$; consequently, on searching through the pairs in (2), we find either $\langle r, z\rangle$ or $\langle s, z\rangle$. Thus $\widetilde{\theta}_\alpha$ contains either all the pairs $\langle r, z\rangle$ $(r \in S^m_{p,1})$. If the first holds, we set

$$U(z, m) = 1, \quad S^m_{p+1,0} = S^m_{p,0}, \quad S^m_{p+1,1} = S^m_{p,1} \cup \{z\};$$

if the first case does not hold, then the second does and we put

$$U(z, m) = 0, \quad S^m_{p+1,0} = S^m_{p,0} \cup \{z\}, \quad S^m_{p+1,1} = S^m_{p,1}.$$

The inductive hypotheses are preserved. This means that $U(z, m)$ can be defined for all $z, m \in N$, and that U is a **gr**-function. Clearly,

$$\alpha r = \alpha s \Leftrightarrow (\forall x \in N)(U(r, x) = U(s, x))$$

for all $r, s \in N$; therefore, the numbering $n \to U(n, x)$ is isomorphic to α. ∎

Among the various numberings of a family, the precomplete ([XXVII], [M9]) are commonly the ones of interest. However, for negative numberings we observe the following

Remark: *No negative numbering α of a set A containing at least two elements can be precomplete.*

For let a, b be distinct elements of A. The set of all α-numbers of all objects not equal to a is recursively enumerable, as is the set of α-numbers of objects different from b. In other words, the families $A \sim \{a\}, A \sim \{b\}$ are totally α-enumerable. They are distinct from A, but their union is A; if α were precomplete, this would contradict Theorem 12 in [XXVII]. ∎

NOTE

(1) I.e., $n \to a_n$ is a strongly computable numbering in the sense of [XXVII], §6.

CHAPTER 29

IDENTICAL RELATIONS IN VARIETIES OF QUASIGROUPS

The purpose of this article is to construct a variety $\mathcal{L}$ of commutative loops that is defined by a finite system of identities in one variable and such that there is no general algorithm whereby for an arbitrary identity in one variable, one can decide whether it is true in every $\mathcal{L}$-loop. This means that in the variety $\mathcal{L}$, the free loop with one free generator is not constructive ([XVIII], §4.1).

In §1 various well-known definitions and facts are recalled. In §2 we produce an auxiliary variety of algebras with two unary operations that has a nonconstructive free algebra. With the aid of this variety we specify in §4 a variety of loops and a variety of commutative loops such that in both, the free loops with one free generator are not constructive. A useful lemma on the completion of partial loops is proved in §1.

§1. The problem of identical relations

Let $\boldsymbol{f}_1, ..., \boldsymbol{f}_s$ be a finite number of symbols, each of which has an associated natural number called its *arity*. We prescribe an *algebra with signature* $\Sigma = \{\boldsymbol{f}_1, ..., \boldsymbol{f}_s\}$ by specifying a nonempty base set A and n_i-ary operations f_i defined on A and taking values in A, where n_i is the arity of the symbol $\boldsymbol{f}_i$ $(i = 1, ..., s)$. The operation f_i is called the *values* of the function symbol $\boldsymbol{f}_i$ in the algebra $\mathfrak{A} = \langle A; f_1, ..., f_s\rangle$. An arbitrary collection of algebras with a fixed signature is called a *class of algebras.*

The notion of a *term of signature* $\Sigma = \{\boldsymbol{f}_1, ..., \boldsymbol{f}_s\}$ in certain individual variables $\boldsymbol{x}_1, ..., \boldsymbol{x}_m$ is defined by recursion:

(I) An expression of the form $\boldsymbol{x}_j$ is a term of signature Σ $(j = 1, ..., m)$;

(II) If $\mathfrak{a}_1, ..., \mathfrak{a}_{n_i}$ are terms of signature Σ (in the $\boldsymbol{x}_j$), then so is the expression $\boldsymbol{f}_i(\mathfrak{a}_1, ..., \mathfrak{a}_{n_i})$ $(i = 1, ..., s)$.

To prescribe a value for the individual variable $\boldsymbol{x}_j$ in the algebra $\mathfrak{A}$ means to associate with the symbol $\boldsymbol{x}_j$ an element of the base A. If $\mathfrak{a}$ is a term of signature Σ in the variables $\boldsymbol{x}_1, ..., \boldsymbol{x}_m$ and values of all these variables are given in the algebra $\mathfrak{A}$, then performing the operations of $\mathfrak{A}$ on these values of the $\boldsymbol{x}_j$ as indicated by the notation of $\mathfrak{a}$ produces an element of A; this element is called the *value* in $\mathfrak{A}$ of the term $\mathfrak{a}$ for the given values of the variables $\boldsymbol{x}_1, \ldots, \boldsymbol{x}_m$.

An expression of the form $\mathfrak{a} \approx \mathfrak{b}$ where $\mathfrak{a}$, $\mathfrak{b}$ are terms of signature Σ in the variables $\boldsymbol{x}_1, \ldots, \boldsymbol{x}_m$ is called an *identity* (or *identical relation*) of signature Σ and rank m. The identity $\mathfrak{a} \approx \mathfrak{b}$ is *true* (or *valid*) in the algebra $\mathfrak{A}$ iff for all values of $\boldsymbol{x}_1, \ldots, \boldsymbol{x}_m$ the values of $\mathfrak{a}$ and $\mathfrak{b}$ in $\mathfrak{A}$ are equal. The identity $\mathfrak{a} \approx \mathfrak{b}$ is *true in a class* $\mathcal{K}$ of algebras with signature Σ iff it is true in every algebra in $\mathcal{K}$ ($\mathcal{K}$-algebra).

By $\mathrm{I}_m(\mathcal{K})$ we denote the collection of all identities of signature Σ in $\boldsymbol{x}_1, \ldots, \boldsymbol{x}_m$ that are true in the class $\mathcal{K}$. The union of all the $\mathrm{I}_m(\mathcal{K})$ ($m = 1, 2, \ldots$) is denoted by $\mathrm{I}(\mathcal{K})$. Since identities are words on the alphabet $\{ \boldsymbol{x}, f_1, \ldots, f_s, \approx, `,), (\}$ – we can code $\boldsymbol{x}_j$ as the word consisting of $\boldsymbol{x}$ repeated j times – we can pose the question of whether the set $\mathrm{I}(\mathcal{K})$ is recursive (or equivalently, whether all the sets $\mathrm{I}_m(\mathcal{K})$ are recursive ([1])). Determining membership in $\mathrm{I}(\mathcal{K})$ is called the *problem of identical relations* (or *identity problem*) for $\mathcal{K}$. It is clear that for any m, if $\mathrm{I}_m(\mathcal{K})$ is not recursive, then neither is $\mathrm{I}(\mathcal{K})$ nor $\mathrm{I}_n(\mathcal{K})$ for $n \geqslant m$.

For every class $\mathcal{K}$ and every set of symbols $\boldsymbol{a}_1, \ldots, \boldsymbol{a}_m$ we can construct a special algebra called the free algebra of rank m (or the algebra with free generators $\boldsymbol{a}_1, \ldots, \boldsymbol{a}_m$) for the class $\mathcal{K}$. This construction is carried out as follows. Suppose $\Sigma = \{f_1, \ldots, f_s\}$ is the signature of $\mathcal{K}$. Let A_m be the set of all terms of signature Σ in the variables $\boldsymbol{a}_1, \ldots, \boldsymbol{a}_m$. Terms $\mathfrak{a}, \mathfrak{b}$ in A_m are said to be *equivalent* iff the identity $\mathfrak{a} \approx \mathfrak{b}$ is true in $\mathcal{K}$ (iff $\mathfrak{a} \approx \mathfrak{b}$ becomes a member of $\mathrm{I}_m(\mathcal{K})$ when the $\boldsymbol{x}_j$ are substituted for the $\boldsymbol{a}_j$). Let E_m be the set of equivalence classes of terms in A_m, and define operations $f_1, \ldots, f_s$ on E_m by setting

$$f_i([\mathfrak{a}_1], \ldots, [\mathfrak{a}_{n_i}]) = [f_i(\mathfrak{a}_1, \ldots, \mathfrak{a}_{n_i})] ,$$

where $[\mathfrak{a}_j]$ is the class of terms equivalent to the term $\mathfrak{a}_j$ from A_m ($j = 1, \ldots, m$; $i = 1, \ldots, s$). The algebra $\mathfrak{F}_m = \langle E_m; f_1, \ldots, f_s\rangle$ is called the $\mathcal{K}$-*free algebra* of rank m with free generators $\boldsymbol{a}_1, \ldots, \boldsymbol{a}_m$. From this construction it is seen that the problem of determining whether two arbitrary words in the generators and operation symbols represent the same element in $\mathfrak{F}_m$ (the *word problem* for $\mathfrak{F}_m$) essentially coincides with the problem of identities of rank m for the class $\mathcal{K}$.

Let $\mathcal{K}$ be a class of algebras with signature Σ, and let S be a system of identities of this signature. By $\mathcal{K}(\mathrm{S})$ we denote the collection of all those $\mathcal{K}$-algebras in which all the identities in S are valid. A class $\mathcal{K}_1$ of algebras is called a *variety* of $\mathfrak{K}$-algebras (or a *$\mathfrak{K}$-variety*, or a *primitive* or *equational class in* $\mathfrak{K}$) iff $\mathfrak{K}_1 = \mathfrak{K}(S)$ for some system S of identities. A class $\mathfrak{K}_1$ is a *finitely defined* variety of $\mathfrak{K}$-algebras iff $\mathfrak{K}_1 = \mathfrak{K}(S)$ for a finite system S. A class $\mathfrak{K}_1$ is a $\mathfrak{K}$-variety *of rank m* iff $\mathfrak{K}_1 = \mathfrak{K}(S)$, where S consists of identities, each involving no more than m variables. Finally, if $\mathfrak{K}$ is the class of all algebras with signature Σ, then a $\mathfrak{K}$-variety is called an *absolute variety* or simply a *variety of algebras* (with the given signature). Examples of finitely defined absolute varieties are the classes of groups, abelian groups, rings, associative rings, Lie rings, lattices, semigroups, etc. An algebra free for a variety belongs to that variety. In all the varieties just listed the free algebras are constructive, i.e., they have algorithmically solvable word problems (cf. [XVIII], §4.1). As far as the author knows, no example has previously been published of a finitely defined absolute variety with a free algebra whose word problem is not thus solvable. (2) Constructed below are finitely defined absolute varieties of rank 1 with very simple signatures such that the word problems for their free algebras of rank 1 are not algorithmically solvable.

§2. Algebras with unary operations

In a certain sense the simplest algebra is one whose signature consists of a single unary operation symbol. According to Ehrenfeucht, the class of all algebras with this signature has a recursively decidable elementary theory; therefore, the identity problem for any finitely axiomatizable class of algebras with one unary operation is recursively solvable. *A fortiori,* the identity problem for a finitely defined variety of algebras with a single unary operation admits an algorithmic solution.

When the algebras have two unary operations, the situation changes radically.

Theorem 1: *There exists a finitely defined, rank 1 variety of algebras with two unary operations for which the problem of identities in one variable is not algorithmically solvable.*

According to the theorem of Post–Markov (see [102]), there is a semigroup $\mathfrak{C}$ presented ([XVIII], §1.4) by generators $\boldsymbol{c}_1, \ldots, \boldsymbol{c}_r$ and defining relations

$$\mathfrak{a}_i \approx \mathfrak{b}_i \quad (i = 1, \ldots, n)\,, \tag{1}$$

where $\mathfrak{a}_i$, $\mathfrak{b}_i$ are certain words in $\boldsymbol{c}_1, \ldots, \boldsymbol{c}_r$, such that the word problem for $\mathfrak{C}$ is not algorithmically solvable. Let $\boldsymbol{a}_1, \boldsymbol{a}_2$ be new individual symbols. In the relations (1) we make the substitutions

$$c_k \leftarrow a_1 a_2 a_1^{k+1} a_2^{k+1} \qquad (k = 1, \ldots, r)\,, \tag{2}$$

replacing each occurrence of the letters c_k with the corresponding word in a_1, a_2. Under these substitutions the relations (1) are turned into relations of the form

$$a_{\alpha(i,1)} a_{\alpha(i,2)} \ldots a_{\alpha(i,p_i)} \approx a_{\beta(i,1)} a_{\beta(i,2)} \ldots a_{\beta(i,q_i)}\,, \tag{3}$$

where $\alpha(i,j), \beta(i,j) = 1, 2$; $i = 1, \ldots, n$. The relations (3) present a semigroup $\mathfrak{D}$ with formal generators a_1, a_2 (let $a_1 = [a_1]$, $a_2 = [a_2]$ be the actual elements – equivalence classes – generating $\mathfrak{D}$). Clearly, (2) determines an isomorphic embedding of $\mathfrak{C}$ in $\mathfrak{D}$ (see [51]). Since the word problem for $\mathfrak{C}$ is not algorithmically solvable, neither is the word problem for $\mathfrak{D}$.

Let us now consider algebras whose signature consists of two unary operation symbols f_1, f_2. Corresponding to each defining relation (3), we write the identity

$$f_{\alpha(i,1)}(f_{\alpha(i,2)} \ldots (f_{\alpha(i,p_i)}(x)) \ldots) \approx f_{\beta(i,1)}(\ldots f_{\beta(i,q_i)}(x)) \ldots) \tag{4}$$

of signature $\{f_1, f_2\}$ in the variable x ($i = 1, \ldots, n$). Let $\mathfrak{V}$ denote the variety of those algebras with signature $\{f_1, f_2\}$ in which all the identities (4) are valid.

An arbitrary, rank 1 identity of signature $\{f_1, f_2\}$ has either the form $x \approx x$, or the form

$$f_{\alpha_1}(f_{\alpha_2} \ldots (f_{\alpha_p}(x)) \ldots) \approx f_{\beta_1}(f_{\beta_2} \ldots (f_{\beta_q}(x)) \ldots) \tag{5}$$

or the form

$$f_{\alpha_1}(f_{\alpha_2}(\ldots (f_{\alpha_p}(x)) \ldots) \approx x\,.$$

It is easy to convince ourselves that *an identity (5) is true in the variety $\mathfrak{V}$ iff the corresponding equation*

$$a_{\alpha_1} a_{\alpha_2} \ldots a_{\alpha_p} = a_{\beta_1} a_{\beta_2} \ldots a_{\beta_q} \tag{6}$$

holds for the elements a_1, a_2 that generate the semigroup $\mathfrak{D}$.

Indeed, let $\mathfrak{A}$ be an arbitrary algebra in $\mathfrak{V}$. Every unary operation defined on $\mathfrak{A}$ (i.e., on its base) is a map from $\mathfrak{A}$ into $\mathfrak{A}$. All such maps from $\mathfrak{A}$ into compose a semigroup under the usual multiplication for mappings. In this semigroup we single out the subsemigroup $\mathfrak{D}^*$ generated by the maps f_1, f_2. The truth of the identities (4) in $\mathfrak{A}$ signifies that in $\mathfrak{D}^*$ the relations (3) are satisfied by giving $\boldsymbol{a}_1, \boldsymbol{a}_2$ the values f_1, f_2. Suppose (6) holds in $\mathfrak{D}$: so the generators of $\mathfrak{D}$ satisfy the relation

$$\boldsymbol{a}_{\alpha_1}\boldsymbol{a}_{\alpha_2} \dots \boldsymbol{a}_{\alpha_p} \approx \boldsymbol{a}_{\beta_1}\boldsymbol{a}_{\beta_2} \dots \boldsymbol{a}_{\beta_q} ;$$

by the definition of $\mathfrak{D}$, this relation is satisfied in any semigroup by any elements that satisfy the relations (3). In particular, we have

$$f_{\alpha_1}f_{\alpha_2} \cdots f_{\alpha_p} = f_{\beta_1}f_{\beta_2} \cdots f_{\beta_q}$$

in $\mathfrak{D}^*$, but this means (5) is true in $\mathfrak{A}$. If, therefore, the equation (6) holds in $\mathfrak{D}$, then the identity (5) is valid in $\mathcal{V}$.

Conversely, suppose (5) is true in the variety $\mathcal{V}$. We adjoin an outside unit element to the semigroup $\mathfrak{D}$, i.e., an element e of which we require that $e \notin \mathfrak{D}$, $ee = e$, and $ex = xe = x$ for all $x \in \mathfrak{D}$. This extension of $\mathfrak{D}$ is denoted by $\mathfrak{D}^e$. On $\mathfrak{D}^e$ we define operations f_1, f_2 by putting

$$f_1(x) = a_1 x, \quad f_2(x) = a_2 x \qquad (x \in \mathfrak{D}^e) . \tag{7}$$

Let us consider the algebra $\mathfrak{A}^e$ with the same base set as $\mathfrak{D}^e$ and with these two unary operations. From the relations (3) it follows that the identities (4) are true in $\mathfrak{A}^e$, so $\mathfrak{A}^e$ belongs to $\mathcal{V}$. The identity (5) is valid in $\mathfrak{A}^e$, being true throughout $\mathcal{V}$ by assumption. When we evaluate (5) in $\mathfrak{A}^e$ by taking $\boldsymbol{x}$ to have the value e and then apply (7), we find that (6) holds in $\mathfrak{D}$. ▲

Thus for the variety $\mathcal{V}$, defined by a finite number of identities of rank 1, the identity problem of the first rank is not algorithmically solvable; this proves Theorem 1. ■

This variety has another property, which we shall need later on.

Lemma. *Suppose $\mathcal{V}$ is the variety above, composed of all algebras with two unary operations f_1, f_2 satisfying the identities (4). Let $\mathcal{V}_0$ be the class consisting of every infinite $\mathcal{V}$-algebra in which there is an element o such that for any elements x, y,*

(α) $f_2^{j+1}(x) = x \Leftrightarrow x = o$,

(β) $f_1 f_2^j(x) = x \Leftrightarrow x = o$,

(γ) $f_1(x) = f_2(y) \Leftrightarrow x = y = o$,

(δ) $f_2(x) = f_2(y) \Leftrightarrow x = y$,

(ϵ) $f_1 f_2^{j+1}(x) = f_1(x) \Leftrightarrow x = o$, $\qquad (j = 0, 1, \ldots)$.

Then every identity of the form (5) that is true in $\mathfrak{V}_0$ is also true in $\mathfrak{V}$. In particular, the identity problem for $\mathfrak{V}_0$ does not have an algorithmic solution.

Suppose an identity (5) is true in every $\mathfrak{V}_0$-algebra. Consider the algebra $\mathfrak{A}^e$ constructed in the course of proving Theorem 1. We adjoin a new element o to it and set $f_1(o) = o$, $f_2(o) = o$. This extended algebra is denoted by $\mathfrak{A}^o$. When $\boldsymbol{x}$ has the value o, (4) holds trivially. Therefore, the identity (4) is true in $\mathfrak{A}$, so $\mathfrak{A}^o$ belongs to $\mathfrak{V}$. We claim that the algebra $\mathfrak{A}^o$ has the properties (α)–(ϵ); the implications from right to left are, of course, trivial in this case.

The elements of $\mathfrak{A}^o$ are e, o, and the elements of the infinite semigroup $\mathfrak{D}$, which can be represented by nonempty words in the letters $\boldsymbol{a}_1, \boldsymbol{a}_2$. Moreover, two of these words are equivalent – i.e., represent the same element in $\mathfrak{D}$ – iff one of them can be converted into the other by means of the elementary transformations derived from the defining relations (3). Let us call a word in $\boldsymbol{a}_1, \boldsymbol{a}_2$ *regular* iff it decomposes (literally) into a concatenation of words of the form $\boldsymbol{a}_1\boldsymbol{a}_2\boldsymbol{a}_1^{k+1}\boldsymbol{a}_2^{k+1}$ $(k > 0)$. A special feature of the relations (3) is that their left-hand and right-hand sides are regular words. Hence it follows, in particular, that a regular word can be equivalent only to another regular word. Moreover, every letter of an arbitrary word can appear in at most one subword of the form $\boldsymbol{a}_1\boldsymbol{a}_2\boldsymbol{a}_1^{k+1}\boldsymbol{a}_2^{k+1}$; therefore, every word breaks into a composition of maximal regular chunks joined by irregular segments containing no subwords of the form $\boldsymbol{a}_1\boldsymbol{a}_2\boldsymbol{a}_1^{k+1}\boldsymbol{a}_2^{k+1}$. Under the elementary transformations, only the maximal regular pieces are subject to change: the irregular segments remain unaltered.

Let us verify (γ). Suppose $f_1(x) = f_2(x)$ and either $x \neq o$ or $y \neq o$. Then both $x \neq o$ and $y \neq o$, so $a_1 x = a_2 y$. The initial letter of a word never changes under elementary transformations. Hence, $a_1 x = a_2 y$ is impossible in $\mathfrak{D}^e$, and thus $x = y = o$.

Suppose $f_2(x) = f_2(y)$. If $x = o$ or $y = o$, then $x = y = o$. If $x \neq o$ and $y \neq o$, then in $\mathfrak{D}^e$ we have $a_2 x = a_2 y$. But the letter $\boldsymbol{a}_2$ standing at the beginning of a word cannot participate in elementary transformations. Therefore, $x = y$; this means (δ) holds in $\mathfrak{A}^o$. Finally, if $f_1(x) = f_1 f_2(x)$ and $x \neq o$, then $a_1 x = a_1 a_2 x$ in $\mathfrak{D}^e$. The left-hand and right-hand sides of the defining relations (3) are words having even lengths. Hence, the difference in length of two equivalent words is an even number. The difference of the lengths of the obvious words representing a_1x, $a_1a_2x \in \mathfrak{D}$ is one, so they cannot be equivalent, i.e., a_1x is not equal to a_1a_2x in $\mathfrak{D}^e$. This proves (ϵ), and the remaining properties are verified similarly.

Thus the algebra $\mathfrak{A}^o$ belongs to the class $\mathcal{V}_0$, and the identity (5) is true in it. Taking $\boldsymbol{x}$ to have the value e and using (7), we obtain the equation (6), which, as we have already shown, implies the identity (5) is valid throughout the variety $\mathcal{V}$. ■

§3. Partial quasigroups

An algebra $\mathfrak{Q}$ with a single binary operation $*$ is called a *quasigroup* iff for any $a, x, y \in \mathfrak{Q}$, the cancellation laws

$$\begin{aligned} a * x = a * y &\Rightarrow x = y\,, \\ x * a = y * a &\Rightarrow x = y\,, \end{aligned} \tag{8}$$

hold, and for every a, $b \in \mathfrak{Q}$ the equations

$$a * x = b, \quad y * a = b$$

are solvable in $\mathfrak{Q}$ for x and y.

An element o of a quasigroup $\mathfrak{Q}$ is called the *neutral element* of $\mathfrak{Q}$ iff

$$x * o = o * x = x \qquad (x \in Q)\,.$$

A system $\mathfrak{Q} = \langle Q, *\rangle$ consisting of a set Q containing a fixed element o and a partial binary operation $*$ on Q will be called a *partial quasigroup* with neutral element o iff in $\mathfrak{Q}$ the products $o * x$, $x * o$ are defined and equal to x for all $x \in Q$, and the cancellation laws (8) hold in $\mathfrak{Q}$, provided that all the products involved are defined.

A quasigroup is called *commutative* iff the identity $\boldsymbol{x} * \boldsymbol{y} \approx \boldsymbol{y} * \boldsymbol{x}$ is true in it. A partial quasigroup is said to be *commutative* iff for any of its elements x, y, whenever $x * y$ is defined, so is $y * x$, and $x * y = y * x$.

Theorem 2: *Let $\mathfrak{Q}_0 = \langle Q, *\rangle$ be a countable and commutative partial quasigroup with neutral element o, and suppose $\mathfrak{Q}_0$ satisfies the conditions:*

(i) For every $a \neq o$, the set of those x in Q for which $x * a$ or $a * x$ is defined is finite;

(ii) For every $r \in Q$, there exist infinitely many $p \in Q$ such that the equation $p * x = r$ has no solutions in Q.

Then the partial operation $*$ can be completed to a totally defined operation that turns Q into a commutative quasigroup with neutral element o.

With no loss of generality we can consider the natural numbers to be the elements of $\mathfrak{Q}_0$, with 0 playing the role of the neutral element. Let us arrange all possible pairs of natural numbers in some definite sequence – e.g., in the order

$$\langle 0,0\rangle, \langle 0,1\rangle, \langle 1,0\rangle, \langle 0,2\rangle, \langle 1,1\rangle, \langle 2,0\rangle, \dots . \tag{9}$$

Let M_0 be the set of all those pairs $\langle x, y\rangle$ for which the product $x * y$ is defined in the partial quasigroup $\mathfrak{Q}_0$. We extend the set M_0 in stages, at each step extending the operation $*$ appropriately, and making sure that the set Q with the new operation will form a commutative partial quasigroup with the properties (i) and (ii).

So suppose that after the nth step we have obtained the set M_n of pairs whose product under $*$ has been defined, and suppose the partial quasigroup $\mathfrak{Q}_n$ with base Q and this extended operation is commutative and satisfies (i) and (ii). Choose the first pair $\langle a, b\rangle$ in (9) that does not appear in M_n. From the commutativity of $\mathfrak{Q}_n$ and the properties of the neutral element it follows that $\langle b, a\rangle \notin M_n$, $a \neq 0$, and $b \neq 0$. According to (i), the set of all $x \in Q$ for which $a * x$ or $b * x$ is defined is finite. Therefore, we can find a number c different from all the extant products $a * x$, $b * x$ $(x \in Q)$. By setting $a * b =$ $= b * a = c$, we construct a commutative partial quasigroup $\mathfrak{Q}'_n$ whose operation $*$ has the domain of definition $M'_n = M_n \cup \{\langle a, b\rangle, \langle b, a\rangle\}$. Clearly, (i) and (ii) hold for $\mathfrak{Q}'_n$.

Next we take the first pair $\langle p, r\rangle$ in (9) for which the equation $p * x = r$ is not solvable in $\mathfrak{Q}'_n$. By virtue of (i) and (ii), there is an element q such that the equation $q * x = r$ is unsolvable and $p * q$ is not defined in $\mathfrak{Q}'_n$. We extend the operation by putting $p * q = q * p = r$. This gives us a new commutative partial quasigroup $\mathfrak{Q}_{n+1}$ whose operation has the set $M_{n+1} = M'_n \cup \{\langle p, q\rangle, \langle q, p\rangle\}$ as its domain. Obviously, $\mathfrak{Q}_{n+1}$ has the properties (i) and (ii).

We thus obtain a sequence of partial quasigroups $\mathfrak{Q}_0$, $\mathfrak{Q}_1$, $\mathfrak{Q}_2$, ..., based on the set of natural numbers, that extend the operation $*$ step by step. The limit algebra $\mathfrak{Q}$ is the commutative quasigroup with neutral element we are seeking. ■

We note that Theorem 2 holds for arbitrary (noncommutative) quasigroups, as well. It is also easily generalized to the uncountable case.

§4. Varieties of quasigroups

In §2 we dealt with algebras with two unary operations. The next simplest algebras, in some sense, are the ones with a single binary operation, i.e., *groupoids.* The situation for groupoids is much the same as for algebras with two unary operations.

Theorem 3: *There exists a finitely defined, rank 1 variety $\mathcal{G}$ of groupoids such that the problem of identical relations of rank 1 is not algorithmically solvable for any subclass of $\mathcal{G}$ that contains every infinite commutative quasigroup with neutral element belonging to $\mathcal{G}$.*

We consider the system of identities (4) mentioned in the proof of Theorem 1. Working by recursion from inside out in each identity (4), we replace expressions of the forms $f_1(\mathfrak{a}), f_2(\mathfrak{a})$ with terms of signature $\{*\}$ according to the scheme:

$$\begin{aligned} f_1(\mathfrak{a}) &\leftarrow (\mathfrak{a} * \mathfrak{a}) * \mathfrak{a}\,, \\ f_2(\mathfrak{a}) &\leftarrow \mathfrak{a} * \mathfrak{a}\,. \end{aligned} \tag{10}$$

E.g., the expression $f_2(f_1(\boldsymbol{x}))$ would be converted into the term

$$((\boldsymbol{x} * \boldsymbol{x}) * \boldsymbol{x}) * ((\boldsymbol{x} * \boldsymbol{x}) * \boldsymbol{x})\,.$$

As a result, we obtain identities

$$\mathfrak{c}_i \approx \mathfrak{d}_i\,, \tag{11}$$

where $\mathfrak{c}_i$, $\mathfrak{d}_i$ are terms of signature $\{*\}$ in the variable $\boldsymbol{x}$ $(i = 1, \ldots, n)$. Let $\mathcal{G}$ be the variety of groupoids defined by the identities (11), and let $\mathcal{G}_0$ be the class of all commutative quasigroups with neutral elements that belongs to $\mathcal{G}$.

Let us take an arbitrary identity of the form (5). The transformation (10) converts it into an identity

$$\mathfrak{c} \approx \mathfrak{d} \tag{12}$$

involving only terms of signature $\{*\}$.

To prove Theorem 3, it is sufficient to show that (i) the truth of the identity (12) in the class $\mathcal{G}_0$ implies the validity of the identity (5) in the class $\mathcal{V}_0$ (from the lemma in §2) of algebras with two binary operations

f_1, f_2, and that (ii) the truth of (5) in the class $\mathcal{V}_0$ implies the truth of (12) in the variety $\mathcal{G}$.

We start by proving the second assertion. Suppose (5) is valid in the class $\mathcal{V}_0$ described in the lemma in §2. Then by that lemma, it is also true in the variety $\mathcal{V}$ defined by the identities (4). Let $\mathfrak{G} = \langle G, *\rangle$ be an arbitrary groupoid in the variety $\mathcal{G}$. We define unary operations f_1, f_2 on G by setting

$$\begin{aligned} f_1(x) &= (x * x) * x\,, \\ f_2(x) &= x * x \end{aligned} \tag{13}$$

for all $x \in G$; the new algebra $\langle G; f_1, f_2\rangle$ is denoted by $\mathfrak{G}^{\#}$. Since the identities (11) are true in $\mathfrak{G}$, the identities (4) are valid in $\mathfrak{G}$. Thus $\mathfrak{G}^{\#}$ belongs to $\mathcal{V}$, so that (5) is true in $\mathfrak{G}^{\#}$; this in turn means – in view of (13) – that (12) is true in the groupoid $\mathfrak{G}$.

It's a bit more complicated to prove the first assertion. Suppose the identity (12) is true in the class $\mathcal{G}_0$; let $\mathfrak{A} = \langle A; f_1, f_2\rangle$ be an arbitrary algebra in the class $\mathcal{V}_0$. So $\mathfrak{A}$ has the properties (α)–(ϵ) listed in the lemma. In particular, $\mathfrak{A}$ has an element o satisfying those conditions. On the set A we define a partial binary operation $*$ by requiring (13) and

$$\begin{aligned} x * (x * x) &= f_1(x)\,, \\ o * x = x * o &= x\,, \end{aligned} \tag{14}$$

to hold for all $x \in A$.

From the properties (α)–(ϵ) it follows that the definitions (13) and (14) are consistent. In fact, the first formula in (13) defines the product of diagonal pairs $\langle x, x\rangle$, and this agrees with the second part of (14) because $f_2(o) = o$. A pair $\langle x * x, x\rangle$ can be diagonal iff $x * x = x$, i.e., just when $f_2(x) = x$. By (α) this implies $x = o$; from (β) we find $f_1(o) = o$. Finally, a pair $\langle x * x, x\rangle$ cannot have the form $\langle o, x\rangle$ for $x \neq o$, again by virtue of (α). Thus, on the base A we have constructed a *partial groupoid* $\mathfrak{Q}_0 = \langle A, *\rangle$. This partial groupoid is commutative and has a neutral element o. We verify the cancellation laws for it. Suppose for some $a, x, y \in A$ we have $a * x = a * y$ in $\mathfrak{Q}_0$. If $a = o$, then $x = y$. So we can assume $a \neq o$ from now on. The product $a * x$ is defined in $\mathfrak{Q}_0$ in the following four cases only:

$$x = o\,,\quad x = a\,,\quad x = a * a\,,\quad a = x * x\,;$$

similarly, $a * y$ is defined only in the cases

$$y = o\,, \quad y = a\,, \quad y = a * a\,, \quad a = y * y\,. \tag{15}$$

We have to check all possible combinations of these cases. If $x = o$, then in the last three cases in (15) we would have

$$a = a * a = f_2(a)\,, \quad a = f_1(a)\,, \quad a = f_1(y) = f_2(y)\,,$$

respectively, and this would mean $y = a = o$. in view of (α)–(γ). Similarly, in all the remaining cases we conclude $x = y$. Therefore, $\mathfrak{Q}_0$ is a commutative partial quasigroup with neutral element o.

It is easy to verify that $\mathfrak{Q}_0$ satisfies the conditions (i) and (ii) of Theorem 2. Indeed, suppose $a \in A$ and $a \neq o$. Then the product $a * y$ is defined only in the cases (15). But by (δ) the equation $y * y = f_2(y) = a$ can have no more than one solution. Therefore, there are no more than four elements y such that $a * y$ is defined in $\mathfrak{Q}_0$. Hence, (i) holds for $\mathfrak{Q}_0$.

Now suppose we are given an arbitrary element r in $\mathfrak{Q}_0$. If $r = o$, then the equation $p * x = r$ is unsolvable in $\mathfrak{Q}_0$ for all $p \neq o$. So suppose $r \neq o$, and consider the elements $f_2(r), f_2^2(r), \ldots$. According to (α), all these elements are different. The equation

$$f_2^{j+1}(r) * y = r \tag{16}$$

can have a solution y only in the cases (15). The first three cases yield

$$f_2^{j+1}(r) = r\,, \quad f_2^{j+2}(r) = r\,, \quad f_1 f_2^{j+1}(r) = r\,;$$

in the fourth case, $f_2(y) = f_2^{j+1}(r)$, which by (α) implies $y = f_2^j(r)$, which implies $f_1 f_2^j(r) = r$. From each of these four relations we conlcude on the basis of (α) or (β) that $r = o$. This contradiction shows that (16) has no solutions in $\mathfrak{Q}_0$ for any natural number j. Hence, (ii) holds for $\mathfrak{Q}_0$.

Since the commutative partial quasigroup $\mathfrak{Q}_0$ satisfies the hypotheses of Theorem 2, it can be completed to a commutative quasigroup $\mathfrak{Q}$ with neutral element o. The identity (4) is true in the algebra $\mathfrak{A}$. Therefore, by (13) all the identities (11) are valid in the quasigroup $\mathfrak{Q}$; hence, $\mathfrak{Q}$ belongs to $\mathcal{G}_0$, and (12) is consequently true in $\mathfrak{Q}$. The relations (13) and the nature of the transformation (10) now guarantee that the identity (5) is true in $\mathfrak{A}$. ■

A *(primitive) loop* is an algebra with a binary operation of multiplication and two binary operations of division (/, \) in which the identities

$$(xy)/y \approx y\backslash(yx) \approx y(y\backslash x) \approx (x/y)y \approx x, \quad x\backslash x \approx y/y$$

are valid.

Commutative loops are those in which the identity $xy \approx yx$ is true.

With respect to its multiplication operation, each loop is a quasigroup with neutral element. Conversely, in every quasigroup with neutral element it is possible to define division operations so that the enriched algebra is a loop. Theorem 3, therefore, immediately implies the

Corollary: *There exists a finitely defined, rank 1 variety $\mathfrak{L}$ of loops such that the problem of identical relations of rank 1 is not algorithmically solvable for any subclass of $\mathfrak{L}$ that contains all the commutative loops in $\mathfrak{L}$.* ∎

In particular, the identity problem of rank 1 has no algorithmic solution for the absolute variety consisting of all the commutative loops belonging to $\mathfrak{L}$.

In addition to the well-known question of the algorithmic solvability of the identity problem for an arbitrary variety of groups, it would be interesting to settle the question of whether there exists a variety of loops, the finite members of which constitute a class of loops with an algorithmically unsolvable identity problem.

NOTES

(1) This should read "uniformly recursive". For every degree of unsolvability there is a variety $\mathfrak{X}$ such that $I(\mathfrak{X})$ has this degree, but for each m, $I_m(\mathfrak{X})$ is recursive. The only examples I know are artificial.

(2) Mal'cev was at the time unaware that Tarski had indicated (*J. Symbolic Logic* 18 (1953), 188) that the finitely defined variety $\mathfrak{X}$ of all relation algebras has nonrecursive $I_1(\mathfrak{X})$. From this it is not hard to construct such a variety consisting of algebras with a single binary operation.

CHAPTER 30

ITERATIVE ALGEBRAS AND POST VARIETIES

Let $P_A^{(m)}$ be the set of all m-ary functions defined on a nonempty set A and taking values in this same set. Let $P_A = \bigcup_m P_A^{(m)}$. The following operations on the finitary functions composing P_A are, in a certain sense, the most elementary.

Consider a term $\mathfrak{a}(\boldsymbol{x}_1, ..., \boldsymbol{x}_n; \boldsymbol{f}_1, ..., \boldsymbol{f}_s)$ written with the help of individual symbols $\boldsymbol{x}_1, ..., \boldsymbol{x}_n$ and function symbols $\boldsymbol{f}_1, ..., \boldsymbol{f}_s$ (with respective arities $m_1, ..., m_s$). Substituting concrete functions $f_1, ..., f_s \in P_A$ for the symbols $\boldsymbol{f}_1, ..., \boldsymbol{f}_s$ in $\mathfrak{a}$ determines a concrete function $f \in P_A$, which has the convenient notation

$$f(x_1, ..., x_n) = \mathfrak{a}(x_1, ..., x_n; f_1, ..., f_s) .$$

The function f is called the *result of the termal operation* $\mathfrak{a}$ applied to the functions $f_1, ..., f_s$. E.g., suppose A is the set of natural numbers and

$$\mathfrak{a} = \boldsymbol{f}_1(\boldsymbol{x}_2, \boldsymbol{f}_2(\boldsymbol{x}_1, \boldsymbol{f}_1(\boldsymbol{x}_1, \boldsymbol{x}_3))) .$$

Then the result of applying $\mathfrak{a}$ to the usual arithmetic functions $+, \circ$ is the function

$$f(x_1, x_2, x_3) = x_2 + x_1(x_1 + x_3) = \mathfrak{a}(x_1, x_2, x_3; +, \circ) .$$

The algebra whose base set is P_A and whose basic operations are all the possible termal operations is called the *algebra of $\overline{\overline{A}}$-valued logic,* where $\overline{\overline{A}}$ denotes the cardinality of the set A.

Thanks chiefly to the fundamental work of E. Post, the properties of the algebra of two-valued logic are now quite well known; this algebra is also called simply the algebra of logic. Because of their intrinsic interest and their connection with many-valued logics and the theory of automata, in recent years algebras of k-valued logics ($k \geqslant 2$) have enjoyed the attention of many authors (see S.V. Jablonskiǐ [65], Yu. I. Janov and A.A. Mučnik [66], A. Salomaa [143], and the literature they cite).

The theory of boolean algebras has developed in parallel to that of the algebra of logic and has become a fully ramified discipline. The first analog of a boolean algebra for the case of k-valued logic was introduced by Post himself. The foundation for the theory of these *Post algebras* was laid in 1942 by P.C. Rosenbloom [142]. In 1960 G. Epstein [29] introduced the more convenient notion of a *Post lattice* and gave a representation theory for these lattices. For each k, the classes of Post algebras and Post lattices are termally (or rationally) equivalent [IX] and merely describe in different languages the same object corresponding to the algebra of k-valued logic.

Until now, apparently, no published work has been devoted to a more detailed study of the relation between algebras of logics and Post lattices (or algebras). The basic purpose of this article is to draw the connections between these important concepts in ordinary algebraic terms and, in particular, to describe the representation theory for algebras of logics on the basis of the theory of representations for Post lattices and with its help to give a complete classification of the subalgebras of the algebra of l-valued logic that are isomorphic to the algebra of k-valued logic.

Since there are infinitely many termal operations, the algebras of logics have infinite signatures. Besides this, termal operations are applicable only to collections of functions with certain fixed arities. Therefore, algebras of logics must be regarded either as algebras with partial operations or as graded algebras, which in any case undesirably complicates their theory. With the purpose of giving a standard algebraic appearance to this theory, various authors (cf., e.g., P.M. Cohn [22], H.J. Whitlock [185]) have proposed limiting the basic operations to only a few of the termal operations. In contrast we propose to take as the basic operations either the five $\zeta, \tau, \Delta, \nabla, *$ defined below or the four $\zeta, \tau, \Delta, *$. With these operations it is easy to express all the termal ones; moreover, they play a fundamental role in modern automata theory. This proposal is probably not new, since it leads, in my opinion, to so many conveniences, beginning with the purely terminological.

As already mentioned, with the algebra of k-valued logic is associated two varieties: the variety of Post algebras of order k, introduced by Rosenbloom, and the variety of Post lattices of order k (our term), studied by Epstein. Using the method whereby these varieties were obtained, one can construct from the algebra of k-valued logic (even for infinite k) a series of other varieties, which we propose to name *Post varieties*. For each k, all of these are termally equivalent to one another and, therefore, represent a single object from an algebraic point of view.

§1. Iterative algebras

The operations $\zeta, \tau, \Delta, \nabla, *$ on the set P_A of functions are defined by the equations

$$(\zeta f)(x_1, x_2, \ldots, x_m) = f(x_2, \ldots, x_m, x_1) ,$$

$$(\tau f)(x_1, x_2, x_3, \ldots, x_m) = f(x_2, x_1, x_3, \ldots, x_m) ,$$

$$(\Delta f)(x_1, x_2, \ldots, x_{m-1}) = f(x_1, x_1, x_2, \ldots, x_{m-1}) ,$$

$$(\nabla f)(x_1, x_2, \ldots, x_{m+1}) = f(x_2, \ldots, x_{m+1}) ,$$

$$(f * g)(x_1, \ldots, x_p, x_{p+1}, \ldots, x_{p+m-1}) = f(g(x_1, \ldots, x_p), x_{p+1}, \ldots, x_{p+m-1}) ,$$

where f, g are arbitrary m-ary and p-ary functions in P_A. If the function f is unary, then we define

$$\zeta f = \tau f = \Delta f = f .$$

The álgebras

$$\mathfrak{P}_A = \langle P_A ; \zeta, \tau, \Delta, \nabla, * \rangle ,$$

$$\mathfrak{P}_A^* = \langle P_A ; \zeta, \tau, \Delta, * \rangle ,$$

we shall respectively call the *iterative algebra* and *pre-iterative algebra* over the set A. The power of A is called the *order* of these algebras. The power of their base P_A is always infinite and is, of course, equal to 2^A when A is infinite.

It is clear that an arbitrary function $f \in P_A$ is representable as the result of a proper term (i.e., not simply an individual symbol $\boldsymbol{x}_i$) applied to functions $f_1, \ldots, f_s \in P_A$ iff f can be obtained from $f_1, \ldots, f_s$ by means of the operations $\zeta, \tau, \Delta, \nabla, *$. For example,

$$f(g_1(x), \ldots, g_m(x)) = \Delta^{m-1}(\zeta(\ldots \zeta(\zeta(f * g_1) * g_2) \ldots * g_m))(x) .$$

The functions $e_i^n \in P_A^{(n)}$ defined by

$$e_i^n(x_1, \ldots, x_n) = x_i \qquad (i \leqslant n; \;\; i, n = 1, 2, 3, \ldots)$$

are called the *selector functions* on A.

The function e_1^1 (or simply e) is also called the *identity function* on A, for $e(x) = x$ for all $x \in A$. The collection of all selector functions forms a subalgebra of $\mathfrak{P}_A$ and $\mathfrak{P}_A^*$. The relations

$$\Delta e_2^2 = e\,, \quad e_2^2 * e_n^n = e_{n+1}^{n+1}\,, \quad \zeta^i e_n^n = e_{n-i}^n\,,$$

show that the function e_2^2 is an element generating this entire subalgebra in both $\mathfrak{P}_A$ and $\mathfrak{P}_A^*$.

From the equation $\nabla f = f * e_2^2$ we see that every termal function $\mathfrak{a}(x_1, ..., x_n, f_1, ..., f_s)$ – barring improper terms – can be obtained from the functions $f_1, ..., f_s, e_2^2$ with the aid of the operations ζ, τ, Δ, $*$ of the pre-iterative algebra. Therefore, if we are only interested in such subalgebras of $\mathfrak{P}_A$ as contain the selector functions, then in place of the iterative algebra $\mathfrak{P}_A$ we can consider the pre-iterative algebra $\mathfrak{P}_A^*$.

§2. Iterative algebras of partial functions

Let $Q_A^{(m)}$ be the set of all partial m-ary functions defined on a subset of A and taking values in A; let $Q_A = \bigcup_m Q_A^{(m)}$. So

$$P_A^{(m)} \subset Q_A^{(m)}\,, \quad P_A \subset Q_A\,.$$

The operations ζ, τ, Δ, $*$ are defined for partial functions $f, g \in Q_A$ by the same equations we used in the case of total functions. In this regard, we take the value of the function $f * g$ to be defined just when the values $g(x_1, ..., x_p)$ and

$$f(g(x_1, ..., x_p), x_{p+1}, ..., x_{p+m-1})$$

are defined. The algebra

$$\mathfrak{Q}_A^* = \langle Q_A; \zeta, \tau, \Delta, * \rangle$$

will be called the *pre-iterative algebra of partial functions* over A.

In order to forge a link between algebras of the forms $\mathfrak{P}_A^*$ and $\mathfrak{Q}_B^*$, we fix an arbitrary element $w \in A$. A function $f \in P_A^{(n)}$ is called a *w-function* iff

$$x_1 = w \text{ or ... or } x_n = w \Rightarrow f(x_1, ..., x_n) = w\,.$$

It is easy to see that the set of all w-functions in P_A forms a subalgebra of

$\mathfrak{P}^*_A$. This subalgebra we denote by $\mathfrak{U}^w_A$ (or simply $\mathfrak{U}^w$) and call the *subalgebra special at w.* Suppose the set $A_w = A \sim \{w\}$ is nonempty. With every function $f \in Q_{A_w}$ we associate the function $\bar{f} \in U^w$ whose values coincide with those of f on the domain of f and which takes the value w where f is not defined.

One can easily verify that the mapping $f \to \bar{f}$ $(f \in Q_{A_w})$ is an isomorphism of the algebra $\mathfrak{Q}^*_{A_w}$ onto the special subalgebra $\mathfrak{U}^w$; hence, studying the structure of $\mathfrak{Q}^*_{A_w}$ reduces to studying the structure of $\mathfrak{P}^*_A$.

If $A = \{0, 1, ..., n-1\}$, then as w we shall always choose the number $n-1$, and the algebras $\mathfrak{P}^*_A$, $\mathfrak{U}^w_A$, $\mathfrak{Q}^*_A$ will also be denoted as $\mathfrak{P}^*_n$, $\mathfrak{U}_n$, $\mathfrak{Q}^*_n$. In particular, the algebras $\mathfrak{U}_n$, $\mathfrak{Q}^*_{n-1}$ are isomorphic.

§3. Congruences on $\mathfrak{P}^*_A$ and $\mathfrak{Q}^*_{A_w}$

On the algebra $\mathfrak{P}_A$ and on all its subalgebras, as on any algebra, there are two trivial congruences θ_O, θ_I, where θ_O coincides with the equality relation, and θ_I coincides with the identically true relation. Besides these, on any subalgebra of $\mathfrak{P}_A$ there exists another congruence $\theta_\#$ called the *arity congruence.* By definition, $f \equiv g \ (mod\, \theta_\#)$ iff the functions f, g have the same arity. The factor algebra $\mathfrak{P}_A/\theta_\#$ is obviously isomorphic to the algebra $\mathfrak{P}_1$.

We introduce another relation θ_w, defined on P_A by taking $f \equiv g \ (mod\, \theta_w)$ iff either $f = g$ or f, g are functions with different arities that take the constant value w on A. It is easy to check that the relation θ_w becomes a congruence on the special subalgebra $\mathfrak{U}^w$.

Finally, if the set A consists of just two elements a, w, then we sort the functions in $\mathfrak{U}^w$ into two classes. In the first we place the n-ary function $t^{(n)}_w$ with constant value w $(n = 1, 2, ...)$; the second class consists of all remaining elements of $\mathfrak{U}^w$. This partition corresponds to an equivalence relation θ_2 on $\mathfrak{U}^w$ that is also a congruence on $\mathfrak{U}^w$.

Theorem 1: *Let* $\mathfrak{A}$ *be an arbitrary subalgebra of* $\mathfrak{P}^*_A$ *that includes the subalgebra* $\mathfrak{U}^w_A$, *but is not equal to it. Then the only congruences on* $\mathfrak{A}$ *are* θ_O, θ_I, $\theta_\#$. *If* $\bar{\bar{A}} \geqslant 3$, *then the only congruences on* $\mathfrak{U}^w_A$ *are* $\theta_O, \theta_I, \theta_\#, \theta_w$. *The algebra* $\mathfrak{U}_2$ *has five congruence relations:* $\theta_O, \theta_I, \theta_\#, \theta_w, \theta_2$. (1)

We divide the proof into four parts.

(I) *Let* θ *be a congruence on an algebra* $\mathfrak{B}$, *where* $\mathfrak{U}^w \subseteq \mathfrak{B} \subseteq \mathfrak{P}^*_A$. *If there are two functions* f_1, f_2 *in* $\mathfrak{B}$ *that are* θ*-congruent and have the same arity, then any two functions in* $\mathfrak{B}$ *that have identical arities are* θ*-congruent.*

By assumption there are elements $a_1, ..., a_m, u_1, u_2$ in A such that $u_1 \neq u_2$ and

$$f_i(a_1, ..., a_m) = u_i \qquad (i = 1, 2)\,.$$

It is clear that the functions q_u, t_u^w, e^w $(u \in A)$ defined by

$$q_u(x) = \begin{cases} u & \text{if } x = u , \\ w & \text{if } x \neq u , \end{cases}$$

$$t_u^w(x) = \begin{cases} u & \text{if } x \neq w , \\ w & \text{if } x = w , \end{cases}$$

$$e^w(x, y) = \begin{cases} y & \text{if } x \neq w , \\ w & \text{if } x = w . \end{cases}$$

all belong to $\mathfrak{U}^w$ and thus to $\mathfrak{B}$, as well. We note that $t_w^w = t_w^{(1)}$.

Of the two distinct elements u_1, u_2, one must be different from w: suppose $u_1 \neq w$. We begin by showing $t_{u_1}^w \equiv t_w^w(\theta)$; there are two cases to be examined.

Case 1: in $\mathfrak{B}$ there is a function h for which $h(w, ..., w) \neq w$. Suppose $h(w, ..., w) = v \neq w$. Then the constant function $t_v = t_v^{(1)}$ satisfies the relation

$$t_v(x) = h(t_w^w(x), ..., t_w^w(x)) \quad (x \in A) ,$$

so it belongs to $\mathfrak{B}$; with t_v the subalgebra $\mathfrak{B}$ contains every unary constant function t_u $(u \in A)$ because

$$t_u(x) = t_u^w(t_v(x)) .$$

From $f_1 \equiv f_2(\theta)$ it follows that the functions t_{u_1}, t_{u_2} are θ-congruent, for

$$t_{u_i}(x) = f_i(t_{a_1}(x), ..., t_{a_m}(x)) \quad (i = 1, 2) .$$

From $t_{u_1} \equiv t_{u_2}$ it follows that $q_{u_1} * t_{u_1} \equiv q_{u_1} * t_{u_2}$, i.e., $t_{u_1} \equiv t_w$. But $t_u^w = \Delta(\zeta(e^w) * t_u)$, so $t_{u_1}^w \equiv t_w^w$.

Case 2: for every h in B, $h(w, ..., w) = w$. From $f_1 \equiv f_2$ it now follows that the functions $t_{u_1}^w$, $t_{u_2}^w$ are θ-congruent, since

$$t_{u_i}^w(x) = f_i(t_{a_1}^w(x), ..., t_{a_m}^w(x)) \quad (i = 1, 2) .$$

From $t_{u_1}^w \equiv t_{u_2}^w$ we get $q_{u_1} * t_{u_1}^w \equiv q_{u_1} * t_{u_2}^w$, i.e., $t_{u_1}^w \equiv t_w^w$.

Thus in both cases, $t^w_{u_1} \equiv t^w_w(\theta)$; therefore, the functions

$$e = \Delta(e^w * t^w_{u_1}), \quad t_w = \Delta(e^w * t^w_w)$$

are also θ-congruent.

Consequently, for any k-ary function $f \in \mathfrak{B}$, we have $e * f = t_w * f$; that is, f is θ-congruent to $t^{(k)}_w$, the k-ary function with constant value w. ▲

(II) *Let θ be a congruence on an algebra $\mathfrak{B}$, where $\mathfrak{U}^w \subseteq \mathfrak{B} \subseteq \mathfrak{P}^*_A$. If there are two θ-congruent functions f, g that have distinct arities m, n $(m < n)$, then all the constant w-functions $t^{(k)}_w$ are θ-congruent.*

From $f \equiv g$ it follows that $t^{(1)}_w * f \equiv t^{(1)}_w * g$, or $t^{(m)}_w \equiv t^{(n)}_w$. Hence, $\Delta^{n-2} t^{(m)}_w \equiv \Delta^{n-2} t^{(n)}_w$, i.e., $t^{(1)}_w \equiv t^{(2)}_w$. From this we get $e^w * t^{(1)}_w \equiv e^w * t^{(2)}_w$, or $t^{(2)}_w \equiv t^{(3)}_w$, etc. ▲

(III) *Again let θ be a congruence on an algebra $\mathfrak{B}$ $(\mathfrak{U}^w \subseteq \mathfrak{B} \subseteq \mathfrak{P}^*_A)$, and suppose that $f, g \in \mathfrak{B}$ are θ-congruent functions with distinct arities m, n. If $f \neq t^{(m)}_w$ and either $\overline{\overline{A}} \geqslant 3$ or $\mathfrak{B} \neq \mathfrak{U}^w$, then $\theta = \theta_I$.*

If under these conditions we have $g = t^{(n)}_w$, then by (II) we also have $t^{(m)}_w \equiv t^{(n)}_w \equiv f$; from (I) and (II) we learn that $\theta = \theta_I$.

Thus we can assume that $f \neq t^{(m)}_w$, $g \neq t^{(n)}_w$, and $m < n$. Let $a_1, \ldots, a_m, v$ be elements of A such that $f(a_1, \ldots, a_m) = v \neq w$. From $f \equiv g$ it follows that the functions

$$q = \Delta^{n-2}\zeta(\ldots\zeta(f * t^w_{a_1}) \ldots * t^w_{a_m}),$$

$$r = \Delta^{n-2}\zeta(\ldots\zeta(g * t^w_{a_1}) \ldots * t^w_{a_m}).$$

are θ-congruent; moreover, $q(x) = v$ if $x \neq w$. From $q \equiv r$ we see

$$\Delta(e^w * q) \equiv \Delta(e^w * r),$$

or $e \equiv e^w * s$, where e is the identity function and s is the unary function Δr. If $s = t^{(1)}_w$, then $e^w * s = t^{(2)}_w$, which brings us to the case considered above. So suppose $s \neq t^{(1)}_w$. From $\tau e = e$ and $\tau e \equiv \tau(e^w * s)$ we obtain $e^w * s \equiv \tau(e^w * s)$. If $e^w * s \neq \tau(e^w * s)$, then by applying (I) and (II) we again find that $\theta = \theta_I$. Therefore, we can assume

$$e^w(s(x), y) = e^w(s(y), x) \quad (x, y \in A). \tag{1}$$

Since $e = \Delta e \equiv \Delta(e^w * s)$, we can also assume

$$e^w(s(x), x) = x \qquad (x \in A) ;$$

consequently, for $x \in A$,

$$x \neq w \Rightarrow s(x) \neq w .$$

If we replace x in (1) with an element $a \in A$ other than w, we find that $y = a$ whenever y is different from w; this is impossible if $\overline{\overline{A}} \geqslant 3$.

Therefore, suppose $\overline{\overline{A}} = 2, A = \{w, a\}$, and $\mathfrak{B} \neq \mathfrak{U}^w$. Let $d \in \mathfrak{B} \sim \mathfrak{U}^w$. Permuting the arguments of d by means of the operations τ, ζ and applying Δ as needed, we derive from d a binary function $h \in \mathfrak{B}$ such that either $h(w, a) = a$ or $h(w, w) = a$ (if d is unary, we can take $h = e^w * d$).

We return to the function $s \in \mathfrak{B}$ appearing in (1). If $s(w) = a$, then by putting $x = w$ in (1), we get $y = w$ for any $y \in A$; this is impossible. This means we must have $s(w) = w$ and $s(a) = a$, since we have assumed $s \neq t_w^{(1)}$. Thus $s = e$, which turns $e \equiv e^w * s$ into $e \equiv e^w$, whence

$$h = h * e \equiv h * e^w, \quad \zeta^2 h \equiv \zeta^2(h * e^w) .$$

Inasmuch as $\zeta^2 h = h$, we have $h * e^w \equiv \zeta^2(h * e^w)$. If

$$h * e^w \neq \zeta^2(h * e^w) ,$$

then in view of (I), (II), and $e \equiv e^w$, we again find that $\theta = \theta_I$. Therefore, it can be assumed that

$$h(e^w(x, y), z) = h(e^w(z, x), y) \qquad (x, y, z \in A) .$$

If we take $x = z = w$ and $y = a$ in this last equation we see that $h(w, w) = a$. Thus the constant function $t_a^{(1)}$ is equal to $(\Delta h) * t_w^{(1)}$ and belongs to $\mathfrak{B}$. Working from $e \equiv e^w$, we obtain

$$t_a^{(1)} = e * t_a^{(1)} \equiv e^w * t_a^{(1)} = e_2^2 .$$

From $\tau t_a^{(1)} = t_a^{(1)}$ we now get $e_2^2 \equiv \tau e_2^2$; that is, $e_2^2 \equiv e_1^2$. Since $e_2^2 \neq e_1^2$, this shows $\theta = \theta_I$ by appeal to (I) and (II). ▲

(IV) *If a subalgebra $\mathfrak{B}$ of $\mathfrak{P}_A^*$ properly includes $\mathfrak{U}^w$, then θ_w is not a congruence on $\mathfrak{B}$.*

Indeed, the closure of $\mathfrak{B}$ under ζ guarantees that for some $n \geqslant 1$, there is an n-ary function $h \in \mathfrak{B}$ and elements $a_2, ..., a_n \in A$ such that $h(w, a_2, ..., a_n) \neq w$. Since the functions $t_w^{(1)}, t_w^{(2)}$ are θ_w-equivalent, the functions $h * t_w^{(1)}$, $h * t_w^{(2)}$ should also be θ_w-equivalent if θ_w is to be a congruence on $\mathfrak{B}$. But they are not, for they have different arities and take a value other than w. ▲

By combining (I)–(IV), we obtain Theorem 1. ■

§4. Automorphisms

Let φ be an arbitrary 1–1 mapping of A onto itself. For every $f \in P_A$ we define a function f^α by setting

$$f^\alpha(x_1, ..., x_n) = f(x_1^{\varphi^{-1}}, ..., x_n^{\varphi^{-1}}) . \tag{2}$$

One easily verifies that $\alpha: f \to f^\alpha$ is an automorphism of the algebra P_A. The relation (2) is written more commonly as

$$f(x_1, ..., x_n)^\varphi = f^\alpha(x_1^\varphi, ..., x_n^\varphi) . \tag{3}$$

An automorphism α of the form (3) is called an *inner automorphism* (the one generated by φ) of the α-invariant subalgebras of $\mathfrak{P}_A$ or $\mathfrak{P}_A^*$. In particular, if $w \in A$ and $w^\varphi = w$, then the mapping α is an automorphism of the special subalgebra $\mathfrak{U}_A^w$, which is isomorphic to the pre-iterative algebra $\mathfrak{Q}_A^*$ of partial functions over A.

Theorem 2: *Let w belong to A, and let $\mathfrak{U}$ be the special subalgebra of $\mathfrak{P}_A^*$ consisting of all w-functions from P_A. Suppose $\mathfrak{A}$ is a subalgebra of $\mathfrak{P}_A^*$ that includes $\mathfrak{U}$. Then all the automorphisms of $\mathfrak{A}$ are inner. In particular, the automorphisms of the full iterative algebra $\mathfrak{P}_A$ and those of the pre-iterative algebra $\mathfrak{Q}_A$ of partial functions are all inner.*

For each automorphism α of the algebra $\mathfrak{A}$, we have to construct a 1–1 mapping φ of A onto itself such that (2) holds for all $x_1, ..., x_n \in A$ and $f \in \mathfrak{A}$. Since the arities of functions are not changed by an automorphism (because it commutes with Δ), α is an automorphism of the semigroup $\mathfrak{A}_0$ of all unary functions in $\mathfrak{A}$ under the operation $*$. Let $\mathfrak{U}_0$ be the semigroup of $\mathfrak{A}_0$ consisting of all unary maps of A into itself that leave the point w fixed. Standard arguments show that α is an inner automorphism of $\mathfrak{A}_0$ in the sense above. To be thorough, we produce a proof.

For $a, v \in A$, let t_a denote the unary function with constant value a, and let s_{va} be the function for which

$$s_{va}(v) = v\,, \quad x \neq v \Rightarrow s_{va}(x) = a\,.$$

We now make a few observations about unary functions.

(A) *Let* $f \in P_A^{(1)}$. *If* $f * g = f$ *for all* $g \in \mathfrak{U}_0$ *(and all the more,* $g \in \mathfrak{A}_0$*), then* f *is constant.*

Indeed, let $c, x \in A$ with $c \neq w$; choose a function g such that $g(w) = w$, $g(c) = x$. Then from $f * g = f$ we get $f(x) = f(c)$. ▲

In particular, since $t_w \in \mathfrak{A}_0$ and since $t_w * g = t_w$ implies $t_w^\alpha * g^\alpha = t_w^\alpha$, we learn from (A) that t_w^α is a constant function: $t_w^\alpha = t_{w'}$ for some $w' \in A$.

(B) *For* $f \in P_A^{(1)}$ *and* $a \in A$,

$$f * t_a = t_a * f \Leftrightarrow f(a) = a. \ ▲$$

Therefore, $\mathfrak{U}_0^\alpha = \mathfrak{U}_0'$, where $\mathfrak{U}_0'$ is the subsemigroup of $\langle P_A^{(1)}, * \rangle$ consisting of all unary w'-functions.

(C) *If* f *is a unary* v*-function that is invertible, then* $s_{va} * f = s_{va}$ *for every* $a \in A$. *Let* $g \in P_A^{(1)}$; *if* $g * f = g$ *for every invertible* v*-function* f, *then* $g = s_{va}$ *for some* $a \in A$.

The proof is obvious. ▲

Since invertible functions are carried onto invertible ones by automorphisms, from (B) and (C) it follows that for every $a \in A$, there exists an $a' \in A$ – necessarily unique – such that $s_{wa}^\alpha = s_{w'a'}$. Let φ denote the mapping $a \to a'$. Comparing the action of the automorphism α^{-1} shows that φ is a 1–1 map of A onto itself.

The mapping φ generates an inner automorphism α_φ of the algebra $\mathfrak{P}_A^*$. We now examine the isomorphism $\beta = \alpha(\alpha_\varphi)^{-1}$ of the subalgebra $\mathfrak{A}$ onto the subalgebra $\mathfrak{A}^\beta$. We first show that β leaves each function in $\mathfrak{A}_0$ fixed.

From the construction of the isomorphism β it is seen that $s_{wa}^\beta = s_{wa}$ for all $a \in A$. Suppose $f \in \mathfrak{U}_0$. From the obvious relation $f * s_{wa} = s_{wf(a)}$ we obtain $f^\beta * s_{wa} = s_{wf(a)}$; that is, $f^\beta(a) = f(a)$ $(a \in A)$. Consequently,

$$f^\beta = f \qquad (f \in \mathfrak{U}_0)\,.$$

Now suppose $f \in \mathfrak{A}_0$, but $f(w) = a \neq w$. Then $t_a = f * t_w$ belongs to $\mathfrak{A}_0$. Given an element c in A, we define a function g by taking $g(a) = c$, $g(x) = x$ $(x \neq a)$. Since $g(w) = w$, g belongs to $\mathfrak{U}_0$, and $t_c = g * t_a$ belongs to $\mathfrak{A}_0$. In other words, when $\mathfrak{A}_0$ properly includes $\mathfrak{U}_0$, it contains every unary constant function t_c. Note that because $t_c = g * (f * t_w)$, $t_c^\beta = t_b$ for some $b \in A$.

We shall see that for each $c \in A$, $t_c^\beta = t_c$. We introduce a function $h \in \mathfrak{U}_0$ by setting $h(c) = h(w) = w$, $h(x) = x$ $(x \neq c)$. Since $h * t_c = t_w$, we know that $h^\beta * t_c^\beta = t_w^\beta$, or $h * t_c^\beta = t_w$. We saw above that $t_c^\beta = t_b$ for an appropriate $b \in A$. The condition $h * t_c^\beta = t_w$ yields $h(b) = w$; consequently, $b = w$ or $b = c$. The former is impossible because $t_w^\beta = t_w$. Therefore, $t_c^\beta = t_c$.

We turn again to the function f. For any $c \in A$ we clearly have $f * t_c = t_{f(c)}$, whence $f^\beta * t_c = t_{f(c)}$, so $f^\beta(c) = f(c)$. Since c was arbitrary, $f^\beta = f$. Thus, β leaves $\mathfrak{A}_0$ pointwise fixed.

This proves an arbitrary automorphism α of $\mathfrak{A}_0$ is inner. To extend this result to $\mathfrak{A}$, we have only to show that the map β leaves the multiplace functions in $\mathfrak{A}$ fixed, too.

E.g., let F be an arbitrary binary function in $\mathfrak{A}$. For any $f \in \mathfrak{A}_0$ we see that $\Delta(F * f) \in \mathfrak{A}_0$, and therefore

$$\Delta(F * f) = (\Delta(F * f))^\beta = \Delta(F^\beta * f)\ ;$$

that is,

$$F(f(x), x) = F^\beta(f(x), x) \quad (x \in A)\,. \tag{4}$$

If $a, b \in A$ and $a \neq w$, then the function g given by

$$g(a) = b\ , \quad g(x) = x \quad (x \neq a)$$

lies in $\mathfrak{U}_0$. Putting such functions in (4) for f gives the result:

$$F(b, a) = F^\beta(b, a) \qquad (a, b \in A;\ a \neq w)\,. \tag{5}$$

Since (5) applies to τF equally well, we learn that

$$F(b, a) = F^\beta(b, a) \quad (a, b \in A;\ b \neq w)\,. \tag{6}$$

Finally, from $(\Delta F)^\beta = \Delta F$ it follows that

$$F(x, x) = F^\beta(x, x) \qquad (x \in A)\,. \tag{7}$$

Combining (5)–(7) gives $F = F^\beta$.

The same method can be used to show $F = F^\beta$ for a function $F \in \mathfrak{A}$ of any arity. ∎

§5. Representations of iterative algebras

By a *representation* of the algebra $\mathfrak{P}_A$ in the algebra $\mathfrak{P}_B$ we mean a homomorphism from $\mathfrak{P}_A$ into $\mathfrak{P}_B$. According to §3, homomorphisms of $\mathfrak{P}_A$ that are not isomorphisms are essentially trivial, so the study of representations of $\mathfrak{P}_A$ reduces to the study of isomorphisms of $\mathfrak{P}_A$ into $\mathfrak{P}_B$. Below we indicate a few obvious isomorphisms, which we shall call *standard isomorphisms.* It will be shown later that an arbitrary isomorphism from $\mathfrak{P}_A$ into $\mathfrak{P}_B$ reduces to a combination of standard ones.

Suppose we are given a 1–1 map $\varphi: A \to B$. Put $C = A^\varphi$. Suppose that we also have a projection $\psi: B \to C$, i.e., a mapping of B into C that leaves each point of C fixed. With every function $f(x_1, ..., x_n)$ in P_A we associate a function $f_{\varphi\psi} \in P_B^{(n)}$, defined by the following requirements:

$$f_\varphi(x_1^\varphi, ..., x_n^\varphi) = f(x_1, ..., x_n)^\varphi \qquad (x_1^\varphi, ..., x_n^\varphi \in C)\,,$$

$$f_{\varphi\psi}(y_1, ..., y_n) = f_\varphi(y_1^\psi, ..., y_n^\psi) \;\; (y_1, ..., y_n \in B)\,.$$

The mapping $f \to f_\varphi$ is the *canonical map* of $\mathfrak{P}_A$ onto $\mathfrak{P}_C$ induced by the 1–1 onto map $\varphi: A \to C$. For every function $g \in P_C^{(n)}$, the function

$$g_\psi(y_1, ..., y_n) = g(y_1^\psi, ..., y_n^\psi)$$

is called the *projective ψ-continuation* of g to the set B. An easy check shows that the operations $\zeta, \tau, \Delta, \nabla$, $*$ are all preserved under projective continuation; therefore, the map $f \to f_{\varphi\psi}$ is an isomorphism from $\mathfrak{P}_A$ onto the subalgebra of $\mathfrak{P}_B$ composed of the projective continuations of the functions f_φ $(f \in P_A)$.

Isomorphisms of the form $f \to f_{\varphi\psi}$ will be called *projective.* If φ maps A onto B, then ψ is the identity map on B and $f \to f_{\varphi\psi}$ is the canonical isomorphism of $\mathfrak{P}_A$ onto $\mathfrak{P}_B$ induced by φ. Theorem 2 shows there are no other isomorphisms from $\mathfrak{P}_A$ onto $\mathfrak{P}_B$.

Suppose we are given a system of representations

$$\beta_\iota: \mathfrak{P}_A \to \mathfrak{P}_{B_\iota} \qquad (\iota \in \mathrm{I})\,.$$

We form the cartesian product

$$B = \prod_{\iota \in \mathrm{I}} B_\iota\,.$$

Each function $f \in P_A^{(n)}$ is mapped onto a function $f^{\beta_\iota} \in P_{B_\iota}^{(n)}$ $(\iota \in \mathrm{I})$. Let f^α

be the n-ary function defined on B whose projection on B_ι is f^{β_ι} ($\iota \in \mathrm{I}$). One easily checks that the map

$$\alpha: f \to f^{\alpha} \qquad (f \in P_A) \tag{8}$$

is a representation of $\mathfrak{P}_A$ in $\mathfrak{P}_B$; it is called the *cartesian product* of the β_ι ($\iota \in \mathrm{I}$).

If all the representations β_ι coincide with a fixed representation β, the (8) gives a representation $\alpha = \beta^{\mathrm{I}}: P_A \to P_{B^{\mathrm{I}}}$ called a *cartesian power* of β. In particular, taking various powers of the identity representation $\epsilon: \mathfrak{P}_A \to \mathfrak{P}_A$, we obtain the series of *power representations*

$$\epsilon^k: \mathfrak{P}_A \to \mathfrak{P}_{A^k} \quad (k = 1, 2, \dots).$$

Suppose A is finite, consisting of n elements. Then the set A^k consists of n^k elements. For $k = 1, 2, \dots$ and for any set B with $\overline{\overline{B}} = n^k$, the representation (9) determines an iterative subalgebra of $\mathfrak{P}_B$ isomorphic to $\mathfrak{P}_n$.

A representation $\alpha: \mathfrak{P}^*_A \to \mathfrak{P}^*_B$ is called a *selector representation* iff it carries the selector function $\overset{2}{\underset{A}{e}}{}_2$ defined on A onto the selector function $\bar{e}_2^2$ defined on B. Since with the help of the operations ζ, τ, Δ, $*$ one can obtain from e_2^2 all of the selector functions e_i^m ($i \leqslant m$; $m = 1, 2, \dots$), all selectors in $\mathfrak{P}^*_A$ are carried by selector representations onto the corresponding selectors in $\mathfrak{P}^*_B$. If we pass from the pre-iterative algebras $\mathfrak{P}^*_A$, $\mathfrak{P}^*_B$ to the iterative algebras $\mathfrak{P}_A$, $\mathfrak{P}_B$ by adding the operation ∇, it becomes obvious that a representation preserves all the selectors iff it maps the identity function e_1^1 in P_A onto the identity function in P_B.

From the definition of the cartesian product it is seen that the product of selector representations is itself a selector representation. In particular, all the power representations of $\mathfrak{P}_A$ are selector maps.

§6. Post varieties

The problem of finding representations for the algebra $\mathfrak{P}_A$ is nicely related to the theory of special varieties that we shall name Post varieties and define as follows. In the set P_A we choose some system of functions

$$f_\iota(x_1, \dots, x_{n_\iota}) \quad (\iota \in \mathrm{I}), \tag{10}$$

and for each $\iota \in \mathrm{I}$ we take an n_ι-ary function symbol $\boldsymbol{f}_\iota$. Now consider the algebra

$$\mathfrak{A} = \langle A; f_\iota : \iota \in \mathrm{I} \rangle$$

with signature $\Sigma = \{f_\iota : \iota \in \mathrm{I}\}$. The smallest variety (cf. [IV], [XXIX]) with signature Σ that contains $\mathfrak{A}$ is denoted by $\hat{\mathfrak{A}}$. If it is possible to obtain every function in P_A from the functions $e_2^2, f_\iota\ (\iota \in \mathrm{I})$ by means of the operations $\zeta, \tau, \Delta, *$ – in other words, if $e_2^2, f_\iota\ (\iota \in \mathrm{I})$ generate the algebra $\mathfrak{P}_A^*$ – then the variety $\hat{\mathfrak{A}}$ is called the *Post variety* associated with the system $\{f_\iota : \iota \in \mathrm{I}\}$. The power of the set A is called the order of the Post variety $\hat{\mathfrak{A}}$. Although to each cardinal $\mathfrak{m} = \overline{\overline{A}}$ correspond many different Post varieties, depending as they do on choices of generators f_ι, it is clear that all Post varieties of one and the same order are *rationally equivalent.* By this we mean that in any algebra in either of the varieties, new operations can be expressed as the results of terms applied to the old basic operations so that the new algebra belongs to the other variety; moreover, the form of the terms depends only on the two varieties and not on the choice of any particular algebra in the given variety, and the varieties are thus put into 1–1 correspondence (cf. [IX]).

The first concrete Post varieties of finite rank ever studied were introduced by Rosenbloom [142] in 1942. He called their members Post algebras. Their definition can be cast in the following form. *Post algebras* of order n $(n \geqslant 2)$ are the algebras in the variety $\hat{\mathfrak{R}}_n$, where

$$\mathfrak{R}_n = \langle \{0, 1, \ldots, n-1\}; \oplus, ' \rangle ,$$

$$x \oplus y = \min(x, y) ,$$

$$(n-1)' = 0 , \quad x' = x + 1 \ \ (x < n-1) .$$

Rosenbloom proved that for each n the variety $\hat{\mathfrak{R}}_n$ is finitely axiomatizable (i.e., defined by some finite system of identities). Hence it follows that all Post varieties of finite order are finitely axiomatizable. Rosenbloom further showed that the power of any finite Post algebra of order n is of the form n^k for some $k \geqslant 0$, and that all finite Post algebras of the same power are isomorphic. This implies every finite Post algebra is isomorphic to a cartesian power of a generating algebra $\mathfrak{R}_n$.

Another type of Post variety of finite order was defined by Epstein [29] in 1960. He took *Post lattices* of order n to be the members of the variety generated by the algebra

$$\mathfrak{L}_n = \langle \{0, 1, \ldots, n-1\}; \vee, \wedge, C_0, C_1, \ldots, C_{n-1} \rangle ,$$

where

$$x \vee y = \max(x, y)\,, \qquad x \wedge y = \min(x, y)\,,$$

$$C_i(i) = n - 1, \quad C_i(x) = i \qquad (x \neq i)\,.$$

Epstein showed that the variety $\hat{\mathfrak{L}}_n$ is characterized by the following simple system of axioms:

L1: *A Post lattice* $\mathfrak{L} \in \hat{\mathfrak{L}}_n$, *relative to the operations* $\vee, \wedge$, *is a distributive lattice with zero O and unit I.*

L2: *For every element x of* $\mathfrak{L}$,

$$C_0(x) \vee C_1(x) \vee \ldots \vee C_{n-1}(x) = I\,,$$

$$C_i(x) \wedge C_j(x) = O \qquad (i \neq j)\,.$$

L3: $\mathfrak{L}$ *contains elements* $d_0, d_1, \ldots, d_{n-1}$ *such that*

(a) $$d_{i-1} \wedge d_i = d_{i-1}\,,$$

(b) $$x \wedge d_1 = O \Rightarrow x = O\,, \qquad (i = 1, \ldots, n-1)$$

(c) $$x \vee d_{i-1} = d_i \Rightarrow x = d_i\,,$$

(d) $$x = (d_1 \wedge C_1(x)) \vee (d_2 \wedge C_2(x)) \vee \ldots \vee (d_{n-1} \wedge C_{n-1}(x))\,.$$

Let a set S be given. A unary function f from S into the set $\{0, 1, \ldots, n-1\}$ is called *n-valued*. For any n-valued functions f, g we define the functions $f \vee g, f \wedge g, C_i(f)$ as follows: for every $x \in S$,

$$\begin{aligned} &(f \vee g)(x) = \max(f(x), g(x))\,, \\ &(f \wedge g)(x) = \min(f(x), g(x))\,, \\ &C_i(f) = \begin{cases} n-1 & \text{if} \quad f(x) = i\,, \\ 0 & \text{otherwise}\,. \end{cases} \end{aligned} \tag{11}$$

Then the following basic theorem holds.

Theorem 3 (Epstein [29]): *Every Post lattice of order n is isomorphic to the algebra whose base consists of all continuous n-valued functions on some totally disconnected, compact Hausdorff space S, and whose operations*

$$\vee, \wedge, C_0, \ldots, C_{n-1}$$

are defined by (11). ■

This tells us that every finite Post lattice is isomorphic to a cartesian power of a fundamental Post lattice $\mathfrak{L}_n$.

§7. Selector representations of pre-iterative algebras

It is easy to turn the representation theorem for Post lattices into a representation theorem for pre-iterative algebras of finite order. In fact, suppose we have a Post lattice

$$\mathfrak{L} = \langle B; \overline{\vee}, \overline{\wedge}, \overline{C}_0, \ldots, \overline{C}_{n-1} \rangle .$$

We define a relation α between functions in P_n (functions defined on the base $A = \{0, 1, \ldots, n-1\}$ of $\mathfrak{L}_n$) and functions in P_B by setting

$$\mathfrak{c}(\vee, \wedge, C_0, \ldots, C_{n-1}, e_2^2)\, \alpha\, \mathfrak{c}(\overline{\vee}, \overline{\wedge}, \overline{C}_0, \ldots, \overline{C}_{n-1}, \overline{e}_2^2) , \tag{12}$$

where $\mathfrak{c}$ is an arbitrary term composed with the aid of symbols for the indicated functions and operation symbols for $\zeta, \tau, \Delta, *$; e_2^2, $\overline{e}_2^2$ are the appropriate selector functions defined on A, B.

Since from the functions $\vee, \wedge, C_0, \ldots, C_{n-1}, e_2^2$ we can get any function in P_n with the help of the operations $\zeta, \tau, \Delta, *$, the relation α defines a mapping (possibly many-valued) from P_n into P_B. We now show that α is single valued.

Suppose $\mathfrak{c}, \mathfrak{d}$ are terms of the above sort such that

$$\mathfrak{c}(\vee, \wedge, C_0, \ldots, C_{n-1}, e_2^2) , \quad \mathfrak{d}(\vee, \wedge, C_0, \ldots, C_{n-1}, e_2^2)$$

are identical functions in P_n. In other words, the formal identity

$$\mathfrak{c}(\mathbf{V}, \mathbf{\Lambda}, \boldsymbol{C}_0, \ldots, \boldsymbol{C}_{n-1}, \boldsymbol{e}_2^2)(\boldsymbol{x}_1, \ldots, \boldsymbol{x}_s)$$
$$\approx \mathfrak{d}(\mathbf{V}, \ldots, \boldsymbol{e}_2^2)(\boldsymbol{x}_1, \ldots, \boldsymbol{x}_s) \tag{13}$$

is valid in $\mathfrak{L}_n$; here, s is the arity of the functions determined by $\mathfrak{c}, \mathfrak{d}$. Therefore, (13) is true in every algebra in $\hat{\mathfrak{L}}_n$. In particular, (13) holds in $\mathfrak{L}$, so

$$\mathfrak{c}(\overline{\vee}, \overline{\wedge}, \overline{C}_0, \ldots, \overline{C}_{n-1}, \overline{e}_2^2) = \mathfrak{d}(\overline{\vee}, \overline{\wedge}, \overline{C}_0, \ldots, \overline{C}_{n-1}, \overline{e}_2^2).$$

Thus α is a single-valued mapping.

From (12) we see that α preserves the operations $\zeta, \tau, \Delta, \nabla, *$ and is, therefore, a selector representation of $\mathfrak{P}_A$ in $\mathfrak{P}_B$.

Suppose, on the other hand, we are given a selector isomorphism of the pre-iterative algebra $\mathfrak{P}_n^*$ into some pre-iterative algebra $\mathfrak{P}_B^*$. Among the functions composing the base of $\mathfrak{P}_n^*$ are found

$$\vee, \wedge, C_0, \ldots, C_{n-1}, e_2^2.$$

Their images in $\mathfrak{P}_B^*$ are certain functions

$$\overline{\vee}, \overline{\wedge}, \overline{C}_0, \ldots, \overline{C}_{n-1}, \overline{e}_2^2,$$

and for any term $\mathfrak{c}$ of the sort described above,

$$\overline{\mathfrak{c}(\vee, \wedge, C_0, \ldots, C_{n-1}, e_2^2)} = \mathfrak{c}(\overline{\vee}, \overline{\wedge}, \overline{C}_0, \ldots, \overline{C}_{n-1}, \overline{e}_2^2). \qquad (14)$$

Any identity valid in the lattice $\mathfrak{L}_n$ can be represented in the form (13). By (14) such an identity must also be valid in the algebra

$$\mathfrak{L} = \langle B; \overline{\vee}, \overline{\wedge}, \overline{C}_0, \ldots, \overline{C}_{n-1} \rangle;$$

therefore, this algebra is a Post lattice of order n.

In place of the variety of Post lattices in these arguments we can clearly take any Post variety of the same order. Nor is the finiteness of the order essential. Thus we have

Theorem 4: *In $\mathfrak{P}_A$ choose a system of generators f_ι ($\iota \in I$), and let $\hat{\mathfrak{A}}$ be the smallest variety containing the algebra $\mathfrak{A} = \langle A; f_\iota : \iota \in I \rangle$.*

A mapping $\alpha: \mathfrak{P}_A^* \to \mathfrak{P}_B^*$ is a selector isomorphism of $\mathfrak{P}_A^*$ in $\mathfrak{P}_B^*$ iff the algebra $\mathfrak{B} = \langle B; f_\iota^\alpha : \iota \in I \rangle$ belongs to $\hat{\mathfrak{A}}$. ([2]) ■

Isomorphisms that do not preserve selectors can be reduced to selector isomorphisms as follows.

Theorem 5: *Suppose α is an arbitrary isomorphism of $\mathfrak{P}_A^*$ into $\mathfrak{P}_B^*$, e is the identity function on A, e^α is its image in P_B, C is the set of values taken*

by e^α on B, and $\bar{f}$ is the restriction of f^α to C ($f \in P_A$). Then the map $\beta: f \to \bar{f}$ is a is a selector isomorphism of $\mathfrak{P}_A^*$ into $\mathfrak{P}_C^*$, the map $e^\alpha: x \to e^\alpha(x)$ is a projection of B onto C, and the original isomorphism α is the projective e^α-continuation of the isomorphism β.

In fact, since $e * e = e$ in $\mathfrak{P}_A^*$, $e^\alpha * e^\alpha = e^\alpha$; therefore, $e^\alpha(e^\alpha(x)) = e^\alpha(x)$ for $x \in B$. Thus $x \to e^\alpha(x)$ is a projection of B onto C.

Similarly, from $f = e * f$ it follows that $f^\alpha = e^\alpha * f^\alpha$, i.e., the values of f^α belong to C ($f \in P_A$). Hence, the map $\beta: f \to \bar{f}$ is a homomorphism of $\mathfrak{P}_A^*$ into $\mathfrak{P}_C^*$. From

$$f(x_1, ..., x_n) = f(e(x_1), ..., e(x_n)) \quad (x_i \in A)$$

we deduce that

$$f^\alpha(y_1, ..., y_n) = \bar{f}(e^\alpha(y_1), ..., e^\alpha(y_n)) \quad (y_i \in B),$$

showing that α is the e^α-continuation of β. In turn, the relations

$$e_2^2(e^\alpha(x), e^\alpha(y)) = (e_2^2)^\alpha (e^\alpha(x), e^\alpha(y)) = e^\alpha(y) \quad (x, y \in B)$$

show that β is a selector homomorphism.

We see from Theorem 1 that if A consists of one element or C contains more than one, then β is an isomorphism. If C has one element, but A has two or more, then β and α, too, are proper homomorphisms, but this contradicts the conditions of the theorem. ■

§8. Subalgebras

Theorems 3–5 lead to a complete description of all those subalgebras of $\mathfrak{P}_s$ that are isomorphic to $\mathfrak{P}_n$ ($n \leqslant s$; n, s finite). Let us say a function h in a subalgebra $\mathfrak{A}$ of an algebra $\mathfrak{P}_B$ is a *unit* in $\mathfrak{A}$ if $\Delta h = h$ (so h is unary) and $f * h = h * f$ for every $f \in \mathfrak{A}$. Certainly, each subalgebra can contain no more than one unit, and if the identity function belongs to a given subalgebra, then it serves as its unit.

The *rank* of a unary function $h \in P_B$ is the power of the set $h(B)$. If a subalgebra $\mathfrak{A} \subseteq \mathfrak{P}_B$ contains a unit, then the rank of the unit is called the *rank of the subalgebra* $\mathfrak{A}$.

Two subalgebras of $\mathfrak{P}_B$ are called *conjugate* iff each is carried onto the other by some automorphism of $\mathfrak{P}_B$. Subalgebras of $\mathfrak{P}_B$ are called *protectively conjugate* iff they are projective continuations of conjugate subalgebras.

Theorem 6: *Any subalgebra of* $\mathfrak{P}_s$ *that is isomorphic to* $\mathfrak{P}_n$ *has rank of the form* n^k *for some* $k \geqslant 1$. ([3]) *For every number of the form* n^k *not greater than s, there exists one and – up to projective conjugacy – only one subalgebra of* $\mathfrak{P}_s$ *that is isomorphic to* $\mathfrak{P}_n$ *and has rank* n^k.

In particular, if $s = n^k$, then $\mathfrak{P}_s$ has only one subalgebra – not counting conjugates – that is isomorphic to $\mathfrak{P}_n$ and contains the identity function in $\mathfrak{P}_s$.

Assume $n \leqslant n^k \leqslant s$. From the set $B = \{0, 1, ..., s-1\}$ we choose n^k arbitrary elements to form the subset D; to the base D we transfer the structure of the kth power $\mathfrak{L}_n^k$ of the fundamental Post lattice $\mathfrak{L}_n$. The result is a Post lattice

$$\mathfrak{L} = \langle D; \overline{\vee}, \overrightarrow{\wedge}, \overline{C}_0, ..., \overline{C}_{n-1} \rangle .$$

By the discussion preceding Theorem 4, this lattice generates an isomorphism β: $\mathfrak{P}_n \to \mathfrak{P}_D$. Let χ be some projection of the set B onto its subset D, and use it to continue β to an isomorphism α: $\mathfrak{P}_n \to \mathfrak{P}_s$. The image of $\mathfrak{P}_n$ under α is the desired rank n^k subalgebra of $\mathfrak{P}_s$ isomorphic to $\mathfrak{P}_n$.

Now suppose $\mathfrak{A}_1, \mathfrak{A}_2$ are subalgebras of $\mathfrak{P}_s$ that have the same rank r and are both isomorphic to $\mathfrak{P}_n$. Let χ_1, χ_2 be their units, and put $\chi_1(B) = D_1$, $\chi_2(B) = D_2$. According to Theorem 5, the given isomorphism α_i: $\mathfrak{P}_n \to \mathfrak{A}_i$ is the χ_i-continuation of a selector isomorphism β_i: $\mathfrak{P}_n \to \mathfrak{P}_{D_i}$ $(i = 1, 2)$.

For $i = 1, 2$, the map β_i induces a Post lattice of order n with base D_i:

$$\mathfrak{L}_i = \langle D_i; \vee_i, \wedge_i, C_0^i, ..., C_{n-1}^i \rangle .$$

The powers of these two lattices coincide. Therefore, by the Rosenbloom-Epstein theorem they are both isomorphic to an appropriate cartesian power $\mathfrak{L}_n^k$ $(k \geqslant 1)$ of the fundamental Post lattice $\mathfrak{L}_n$. In particular, the sets D_1, D_2 have power n^k, and there exists a 1–1 correspondence φ: $D_1 \to D_2$ that is an isomorphism of $\mathfrak{L}_1$ onto $\mathfrak{L}_2$. Extend φ in any agreeable fashion to a 1–1 map ψ of B onto B. This map induces an inner automorphism α_ψ of the algebra $\mathfrak{P}_B$. It is clear that the automorphism α_ψ maps the algebra $\mathfrak{A}_1$ onto an algebra that is a projective continuation of $\mathfrak{P}_n^{\beta_2}$. ([4]) ∎

NOTES

([1]) Note that $\mathfrak{U}_1$ coincides with $\mathfrak{P}_1$; by part (II) of the proof, its only congruences are θ_O, θ_I.

(^{2}) That e_2^2, f_ι ($\iota \in I$) generate $\mathfrak{P}_A^*$ is a weaker hypothesis; α should be a many-valued correspondence satisfying conditions analogous to (12). Alternatively, we could take the f_ι alone to generate $\mathfrak{P}_A^*$ and α to be a homomorphism. The set B must have more than one element if A does.

(^{3}) The proof of this does not require $n \leqslant s$; thus for $n > s$, $\mathfrak{P}_n$ has only trivial representations in $\mathfrak{P}_s$.

(^{4}) See [M15] for weak conditions under which a subalgebra of $\mathfrak{P}_n$ is *complete*, i.e., coincides with $\mathfrak{P}_n$.

CHAPTER 31

A FEW REMARKS ON QUASI VARIETIES OF ALGEBRAIC SYSTEMS

Varieties (primitive classes) and quasivarieties (quasiprimitive classes) of algebras have been studied before [IV], [22], but varieties, etc. of other structures have apparently not been examined in the previous literature. The first section of this paper features natural definitions of varieties and quasivarieties of *algebraic systems,* structures whose signatures can contain predicate symbols as well as function symbols. All the common properties of the usual varieties, etc. carry over to the new sorts without change, except for the theory of totally characteristic congruences: such congruences come to be replaced with totally characteristic factor systems.

In §2 with the aid of filtered products we establish structural characterizations and relations for quasivarieties that are in some measure analogous to those known from the theory of varieties of algebras.

§1. Identities and quasidentities

Let Σ be a set of predicate and function symbols, each having a fixed finite arity. Let Σ_P be the part of Σ containing all the former symbols, and Σ_f the latter. The equality sign $\approx$ is not included among the symbols in Σ. Formulas of first-order predicate logic (FOPL) of the form

$$\boldsymbol{P}(\mathfrak{a}_1(\boldsymbol{x}_1, \ldots, \boldsymbol{x}_n), \ldots, \mathfrak{a}_s(\boldsymbol{x}_1, \ldots, \boldsymbol{x}_n)) , \tag{1}$$

$$\boldsymbol{P}_1(\mathfrak{a}_1^1, \ldots, \mathfrak{a}_{s_1}^1) \;\&\; \ldots \;\&\; \boldsymbol{P}_m(\mathfrak{a}_1^m, \ldots, \mathfrak{a}_{s_m}^m) \rightarrow \boldsymbol{P}(\mathfrak{a}_1, \ldots, \mathfrak{a}_s) , \tag{2}$$

– where $\boldsymbol{P}, \boldsymbol{P}_i \in \Sigma_P \cup \{\approx\}$, the $\boldsymbol{x}_j$ are individual variables, and the $\mathfrak{a}_k, \mathfrak{a}_l^i$ are terms of signature Σ_f in the $\boldsymbol{x}_j$ (Σ_f-words [22]) – are respectively called *identities* and *quasidentities* of signature Σ. An identity (1) or a quasidentity (2) is said to be *true* (or *valid*) in an algebraic system $\mathfrak{A}$ with signature Σ iff the indicated formula is true in $\mathfrak{A}$ for all choices of values for the variables $\boldsymbol{x}_j$ in $\mathfrak{A}$ (i.e., in the base set of $\mathfrak{A}$).

An arbitrary nonempty collection $\mathfrak{K}$ of algebraic systems with a fixed signature Σ is called an *(abstract) class* of systems iff $\mathfrak{K}$ contains all systems isomorphic to each of its members.

For any collection $\mathfrak{K}$ of systems with signature Σ, by $\mathbf{I}(\mathfrak{K})$ and $\mathbf{Q}(\mathfrak{K})$ we denote the sets of all identities and of all quasidentities of signature Σ true in every system in $\mathfrak{K}$ ($\mathfrak{K}$-system). If S is a set of identities of quasidentities of signature Σ, then $\mathbf{K}_\Sigma(\mathrm{S})$ (or just $\mathbf{K}(\mathrm{S})$, if the signature is understood) is the class of all systems with signature Σ in which every member of S is valid.

A class $\mathfrak{K}$ is called a *variety* (*quasivariety*) iff there exists a set S of identities (quasidentities) such that $\mathfrak{K} = \mathbf{K}(\mathrm{S})$. In other words, a class $\mathfrak{K}$ is a variety iff $\mathfrak{K} = \mathbf{KI}(\mathfrak{K})$, and $\mathfrak{K}$ is a quasivariety iff $\mathfrak{K} = \mathbf{KQ}(\mathfrak{K})$.

For any class $\mathfrak{K}$ of systems we let $\mathbf{S}(\mathfrak{K})$, $\mathbf{P}(\mathfrak{K})$, $\mathbf{F}(\mathfrak{K})$, $\mathbf{H}(\mathfrak{K})$ be the respective classes of all subsystems of $\mathfrak{K}$-systems, all direct products of $\mathfrak{K}$-systems, all filtered (reduced) products of $\mathfrak{K}$-systems, and all homomorphic images of $\mathfrak{K}$-systems.

According to Birkhoff's theorem [9], a class of algebras is a variety iff the following three conditions all hold:

(a) $\mathbf{S}(\mathfrak{K}) = \mathfrak{K}$ ($\mathfrak{K}$ is hereditary),

(b) $\mathbf{P}(\mathfrak{K}) = \mathfrak{K}$ ($\mathfrak{K}$ is multiplicatively closed),

(c) $\mathbf{H}(\mathfrak{K}) = \mathfrak{K}$ ($\mathfrak{K}$ is homomorphically closed).

These same conditions are necessary and sufficient for a class $\mathfrak{K}$ of algebraic systems to be a variety.

Suppose $\mathfrak{K}$ is a class of systems with signature Σ and $\mathfrak{A}$ is an arbitrary system with the same signature, but not necessarily a member of $\mathfrak{K}$. A set M of elements of $\mathfrak{A}$ is called *$\mathfrak{K}$-independent* iff every mapping α of M into an arbitrary $\mathfrak{K}$-system $\mathfrak{C}$ can be extended to a homomorphism $\alpha^*\colon \mathfrak{M}^* \to \mathfrak{C}$, where $\mathfrak{M}^*$ is the subsystem generated in $\mathfrak{A}$ by M. From this definition readily follows (cf. [22]).

Corollary 1: *If the set M is $\mathfrak{K}$-independent, then M is also* $\mathbf{HSP}(\mathfrak{K})$*-independent.* ■

A set $M \subseteq \mathfrak{A}$ is called a *$\mathfrak{K}$-free basis* for $\mathfrak{A}$ iff M generates $\mathfrak{A}$ and is $\mathfrak{K}$-independent. The system $\mathfrak{A}$ is a *free system of rank* $\mathfrak{n}$ *in the class* $\mathfrak{K}$ iff $\mathfrak{A}$ belongs to $\mathfrak{K}$ and has a $\mathfrak{K}$-free basis of power $\mathfrak{n}$. $\mathfrak{A}$ is *free (in itself) of rank* $\mathfrak{n}$ iff it has an $\{\mathfrak{A}\}$-free basis of power $\mathfrak{n}$.

A system $\mathfrak{E}$ with signature Σ is said to be a *unit* iff it contains only one element e, and for all $P \in \Sigma_{\mathrm{P}}$, $P(e, ..., e)$ is true in $\mathfrak{E}$. A class $\mathfrak{K}$ is called *unitary* iff it consists solely of unit systems (necessarily isomorphic to each other). A class $\mathfrak{K}$ is *total* iff it consists of all systems with a fixed signature. The family of all varieties with a given signature is lattice ordered by set-theoretic inclusion;

the unitary and total classes are the smallest and largest varieties, respectively.

It is easy to verify that in every variety containing a system with more than one element there are free systems of every rank.

For any collection $\mathfrak{K}$ of similar algebraic systems, the class **KI**($\mathfrak{K}$) is the smallest variety including $\mathfrak{K}$. From Birkhoff's theorem we deduce the relation

$$\mathbf{KI}(\mathfrak{K}) = \mathbf{HSP}(\mathfrak{K}) .$$

Suppose $\mathfrak{A}$ is an algebraic system with signature Σ and base set A. By A/θ we denote the collection of equivalence classes into which A is partitioned by an equivalence relation θ. An algebraic system $\mathfrak{A}/\theta$ with signature Σ and base A/θ is called a *factor system* for $\mathfrak{A}$ relative to θ iff the canonical map $a \to a\theta$ ($a\theta$ is the equivalence class containing $a \in A$) is a homomorphism of $\mathfrak{A}$ onto $\mathfrak{A}/\theta$. We observe that, relative to a given equivalence θ, $\mathfrak{A}$ will have many different factor systems. The collection of all factor systems for $\mathfrak{A}$ is partially ordered (see [V]): we take $\mathfrak{A}/\theta \leqslant \mathfrak{A}/\eta$ iff $a\theta \subseteq a\eta$ ($a \in A$) and the map $a\theta \to a\eta$ is a homomorphism of $\mathfrak{A}/\theta$ onto $\mathfrak{A}/\eta$.

A factor system $\mathfrak{A}/\theta$ is called *totally characteristic* iff for any endomorphism φ: $\mathfrak{A} \to \mathfrak{A}$ we have

$$P(a_1\theta, ..., a_n\theta) \Rightarrow P((a_1^{\varphi})\theta, ..., (a_n^{\varphi})\theta) \quad (\boldsymbol{P} \in \Sigma_{\mathrm{P}} \cup \{\approx\};\ a_i \in A) .$$

In place of the theorem on the correspondence between varieties of algebras and totally characteristic congruences, we have the following assertion concerning varieties of systems.

Theorem 1: *Let $\mathfrak{F}$ be a free system of countable rank in a variety $\mathfrak{K}$ of algebraic systems. The map*

$$\mathfrak{F}/\theta \to \mathbf{KI}(\{\mathfrak{F}/\theta\})$$

is an antisomorphism of the lattice of all totally characteristic factor systems for $\mathfrak{F}$ onto the lattice of all subvarieties of $\mathfrak{K}$. ∎

If $\mathfrak{K}$ is a variety of algebras, then the factor systems are in 1–1 correspondence with the congruence relations, and Theorem 1 reduces to the result mentioned on totally characteristic congruences.

§2. Quasivarieties of algebraic systems

The signature Σ of an algebraic system $\mathfrak{A}$ consists in general of a function (or operation) part Σ_f and a predicate part Σ_P. With each basic operation $f(x_1, \ldots, x_m)$ of $\mathfrak{A}$ ($f \in \Sigma_f$), we associate the predicate $P_f(x_1, \ldots, x_m, x)$ that is true just in case $f(x_1, \ldots, x_m) = x$. Let

$$\Sigma^* = \Sigma_P \cup \{P_f : f \in \Sigma_f\}.$$

The model $\mathfrak{A}^* = \langle A; \Sigma^* \rangle$ with signature Σ^* so obtained from $\mathfrak{A} = \langle A; \Sigma \rangle$ is called the *model corresponding to the system* $\mathfrak{A}$, or simply the *model of the system* $\mathfrak{A}$. Submodels of $\mathfrak{A}^*$ are called *submodels of the system* $\mathfrak{A}$. If $\Sigma_1 \subseteq \Sigma$, then the Σ_1-*impoverishment* of $\mathfrak{A}$ is the system $\mathfrak{B} = \langle A; \Sigma_1 \rangle$ in which the signature symbols from Σ_1 designate the same basic predicates and operations on A as they do in $\mathfrak{A}$. When Σ_1 is finite, $\mathfrak{B}$ is said to be a *finite impoverishment.*

From theorems of Tarski-Łoś and Horn we easily deduce

Theorem 2: *A class $\mathfrak{K}$ of algebraic systems is a quasivariety iff the following three conditions hold:*

(a) $\mathfrak{K}$ contains a unit system;

(b) $\mathfrak{K}$ is multiplicatively closed;

(c) for any system $\mathfrak{A}$ with the same signature as $\mathfrak{K}$, if each finite impoverishment of each finite submodel of $\mathfrak{A}$ is isomorphically embeddable in a $\mathfrak{K}$-system, then $\mathfrak{A}$ belongs to $\mathfrak{K}$. ∎

By applying Theorem 1.15 from [38] and using the properties of filtered products, we can avoid the not wholly "algebraic" condition (c) in Theorem 2:

Theorem 3: *In order for a class $\mathfrak{K}$ of systems to be a quasivariety, it is necessary and sufficient that the following all be in force:*

(i) $\mathfrak{K}$ contains every filtered product of its systems;

(ii) $\mathfrak{K}$ is hereditary;

(iii) $\mathfrak{K}$ contains a unit. ∎

For any collection $\mathfrak{K}$ of systems, the class $\mathbf{KQ}(\mathfrak{K})$ is called the *quasiprimitive closure* of $\mathfrak{K}$, or the quasivariety *generated* by $\mathfrak{K}$. Theorem 3 now yields

Corollary 2: *For every class $\mathfrak{K}$ of systems,*

$$\mathbf{KQ}(\mathfrak{K}) = \mathbf{SF}(\mathfrak{K}_u), \tag{3}$$

where $\mathfrak{K}_u$ is the class that results from $\mathfrak{K}$ on adding the unit systems. ∎

Suppose we are given a class $\mathfrak{K}$ of systems with signature Σ and an arbitrary system $\mathfrak{A}$ with this signature but not necessarily in $\mathfrak{K}$. An epimorphism

$\alpha: \mathfrak{A} \to \mathfrak{A}_{\mathcal{K}}$ is called a $\mathcal{K}$-*morphism* iff $\mathfrak{A}_{\mathcal{K}} \in \mathcal{K}$ and for every homomorphism $\gamma: \mathfrak{A} \to \mathfrak{C} \in \mathcal{K}$, there exists a homomorphism $\xi: \mathfrak{A}_{\mathcal{K}} \to \mathfrak{C}$ such that $\gamma = \alpha\xi$. If $\mathfrak{A}$ admits a $\mathcal{K}$-morphism, then the map α and system $\mathfrak{A}_{\mathcal{K}}$ satisfying these conditions are unique up to equivalence; $\mathfrak{A}_{\mathcal{K}}$ is called a $\mathcal{K}$-*replica* of $\mathfrak{A}$ (cf. [VIII]). The class $\mathcal{K}$ is called *replica-complete* (or *replete*) iff every system $\mathfrak{A}$ with signature Σ has a $\mathcal{K}$-replica, i.e., iff every $\mathfrak{A}$ admits a $\mathcal{K}$-morphism. It is easily verified (cf. [VIII]) that a class $\mathcal{K}$ is replete iff it is hereditary and multiplicatively closed and contains a unit ([1]). Hence, for any class $\mathcal{K}$, the class $\mathbf{SP}(\mathcal{K}_u)$ is the smallest replete class including $\mathcal{K}$.

We note that every (first-order) axiomatizable and replete class of systems is a quasivariety. If the class $\mathcal{K}$ is axiomatizable then its replica and quasi-primitive closures coincide, i.e.,

$$\mathbf{SF}(\mathcal{K}_u) = \mathbf{SP}(\mathcal{K}_u)\,.$$

The usual theory of defining relations holds in replete classes and in them only [VIII]. In particular, among the axiomatizable classes of systems the quasivarieties alone admit a full theory of defining relations. Any partial algebra with signature Σ can be viewed as a Σ^*-model; a variety or quasi-variety of partial algebras can thus be viewed as a quasivariety of the corresponding models. So the theory of defining relations for partial algebras is but a special case of the theory for algebraic systems.

For any collection of algebraic systems $\mathfrak{A}_\iota$ $(\iota \in I)$ with signature Σ and any class $\mathcal{K}$ of systems with this same signature, a *(free)* $\mathcal{K}$-*composition* of the systems $\mathfrak{A}_\iota$ is any $\mathcal{K}$-system $\mathfrak{A}$ for which there exist homomorphisms $\alpha_\iota: \mathfrak{A}_\iota \to \mathfrak{A}$ $(\iota \in I)$ with these properties:

(a) $\mathfrak{A}$ is generated by the set $\bigcup_{\iota \in I} A_\iota^{\alpha_\iota}$ (A_ι is the base of the system $\mathfrak{A}_\iota$);

(b) for any $\mathcal{K}$-system $\mathfrak{C}$ and any homomorphisms

$$\gamma_\iota: \mathfrak{A}_\iota \to \mathfrak{C} \quad (\iota \in I)\,,$$

if $\mathfrak{C}$ is generated by $\bigcup_{\iota \in I} A_\iota^{\gamma_\iota}$, then there exists a homomorphism $\xi: \mathfrak{A} \to \mathfrak{C}$ such that $\gamma_\iota = \alpha_\iota \xi$ $(\iota \in I)$.

Such a composition of systems $\mathfrak{A}_\iota$ is called *injective* iff all the homomorphisms α_ι are monomorphisms.

In the words of Łoś, a collection of systems $\mathfrak{A}_\iota$ $(\iota \in I)$ is said to be *(simultaneously) embeddable* in a $\mathcal{K}$-system iff for some $\mathcal{K}$-system $\mathfrak{A}$ there are embeddings $\mathfrak{A}_\iota \to \mathfrak{A}$ $(\iota \in I)$. From the compactness theorem for FOPL it follows that if a class $\mathcal{K}$ is axiomatizable and any pair of $\mathcal{K}$-systems is embeddable in in a $\mathcal{K}$-system, then every collection of $\mathcal{K}$-systems is embeddable in a $\mathcal{K}$-system. This in turn yields

Corollary 3: *Let $\mathcal{K}$ be an arbitrary quasivariety. Then every $\mathcal{K}$-composition of $\mathcal{K}$-systems is injective iff every pair of $\mathcal{K}$-systems can be embedded in a $\mathcal{K}$-system.* ■

From the relation (3) and these observations we get

Theorem 4: *A quasivariety $\mathcal{K}$ is generated by a single system iff every pair of $\mathcal{K}$-systems in embeddable in a $\mathcal{K}$-system.*

Suppose $\mathcal{K}$ has this embeddability property. For every quasidentity $\Phi \notin Q(\mathcal{K})$, choose a $\mathcal{K}$-system $\mathfrak{A}_\Phi$ in which Φ is not valid. From comments above it follows that the collection of systems $\mathfrak{A}_\Phi$ can be embedded in a $\mathcal{K}$-system $\mathfrak{A}$; $\{ \mathfrak{A} \}$ obviously generates $\mathcal{K}$. The converse statement is quickly proved by using (3). ■

NOTE

([1]) By Theorem 5 of [VIII], the class of groups with only multiplication is replete, but it is not hereditary. Subsequent claims thus need modifying (cf. [X], Theorem 6).

CHAPTER 32

MULTIPLICATION OF CLASSES OF ALGEBRAIC SYSTEMS

A collection (possibly empty) of algebraic systems with signature Σ that is closed under isomorphisms is called a *class of systems with signature* Σ. Let F be some collection of subclasses of a given class $\mathfrak{X}$ of algebraic systems. F is partially ordered by the usual relation of inclusion. In many important cases the partially ordered structure $\langle F, \subseteq \rangle$ will be completely lattice-ordered; in such cases we can speak of the *lattice* F. As a rule, the structure of such a lattice is complicated and rarely admits an explicit description.

If $\mathfrak{X}$ is the class of all groups and F is the collection of all its subvarieties, then besides the lattice operations on F there is an operation of multiplication of varieties, introduced by H. Neumann; with respect to this operation, F becomes a free semigroup with zero and identity elements (see [115], [114], [153]). Thus, while the structure of the lattice F remains unknown, the structure of the semigroup F (apart from its power) is entirely clear.

In this article we define, by analogy with the multiplication of varieties of groups, the product of any two subclasses $\mathcal{A}, \mathcal{B}$ of an arbitrary fixed class $\mathfrak{X}$ of systems. This product depends on the choice of the basic class $\mathfrak{X}$, and when $\mathfrak{X}$ is the variety of all groups and $\mathcal{A}, \mathcal{B}$ are subvarieties, the new product coincides with Mrs. Neumann's. Only the general properties of products of classes of systems will be studied here. A short report of the fundamental results was given by the author in his address before the International Congress of Mathematicians, Moscow, 1965 (cf. [XXXIV]). The terminology agrees with that in [XXXI].

§1. The basic definition

Let $\mathfrak{A}$ be an algebraic system in some fixed class $\mathfrak{X}$ and let $\mathfrak{A}/\theta$ be an arbitrary factor system for $\mathfrak{A}$. The elements of $\mathfrak{A}/\theta$ are equivalence classes $a\theta$ ($a \in \mathfrak{A}$), each of which can be viewed as a submodel of the model corresponding to $\mathfrak{A}$. Some of these submodels may happen to be subsystems of the system $\mathfrak{A}$ and belong to $\mathfrak{X}$, but others may not be subsystems, or even if they are, they may not be members of $\mathcal{K}$. For any two subclasses $\mathcal{A}, \mathcal{B}$ of the class $\mathcal{K}$, we define their *$\mathcal{K}$-product* $\mathcal{A} \underset{\mathcal{K}}{*} \mathcal{B}$ by requiring

$$\mathfrak{A} \in \mathcal{A} \underset{\mathcal{K}}{*} \mathcal{B} \Leftrightarrow \mathfrak{A} \in \mathcal{K} \;\&\; (\exists \mathfrak{A}/\theta)(\mathfrak{A}/\theta \in \mathcal{B} \;\&$$

$$\& \; (\forall a \in \mathfrak{A})(a\theta \in \mathcal{K} \Rightarrow a\theta \in \mathcal{A}))\,. \qquad (1)$$

E.g., if $\mathcal{K}$ and $\mathcal{B} \subseteq \mathcal{K}$ are arbitrary classes, and $\mathfrak{A}$ is a $\mathcal{K}$-system for which there is a factor system $\mathfrak{A}/\theta \in \mathcal{B}$, none of whose members is a $\mathcal{K}$-system, then by (1) we see that $\mathfrak{A} \in \mathcal{X} \underset{\mathcal{K}}{*} \mathcal{B}$ for every $\mathcal{X} \subseteq \mathcal{K}$.

We note several consequences of the basic definition (1).

Let $\mathcal{K}$ be an arbitrary class of systems, Σ its signature, and $\mathcal{A}, \mathcal{A}', \mathcal{B}, \mathcal{B}'$ subclasses of the class $\mathcal{K}$.

Corollary 1: *The operation of $\mathcal{K}$-multiplication of subclasses preserves the relation of inclusion, i.e.,*

$$\mathcal{A}' \subseteq \mathcal{A} \;\&\; \mathcal{B}' \subseteq \mathcal{B} \Rightarrow \mathcal{A}' \underset{\mathcal{K}}{*} \mathcal{B}' \subseteq \mathcal{A} \underset{\mathcal{K}}{*} \mathcal{B}\,. \qquad (2)$$

The proof is obvious. ∎

We recall that a subclass $\mathcal{L}$ of a class $\mathcal{K}$ of systems is called *hereditary in* $\mathcal{K}$ iff every $\mathcal{K}$-subsystem of an arbitrary $\mathcal{L}$-system is itself an $\mathcal{L}$-system. A class $\mathcal{K}$ hereditary in the total class $\mathcal{K}_\Sigma$ of all systems with the given signature Σ is said to be *absolutely hereditary,* or simply *hereditary.*

It is clear that the relation of hereditariness is transitive: if $\mathcal{A}$ is a hereditary subclass of $\mathcal{L}$, and $\mathcal{L}$ is a hereditary subclass of $\mathcal{K}$, then $\mathcal{A}$ is also hereditary in $\mathcal{K}$. In particular, every hereditary subclass of an absolutely hereditary class is absolutely hereditary.

We note an obvious property of absolutely hereditary classes. Let $\mathcal{K}$ be a hereditary class, $\mathfrak{A}$ a $\mathcal{K}$-system, and $\mathfrak{A}/\theta$ some factor system for $\mathfrak{A}$. Then *if some equivalence class $a\theta$ is included in a $\mathcal{K}$-subsystem $\mathfrak{B}$ of the system $\mathfrak{A}$, then $a\theta \in \mathcal{K}$*. Indeed, let $b \in \mathfrak{B}$, $x_1, \ldots, x_n \in a\theta$, $f^{(n)} \in \Sigma$. Since $\mathfrak{B}$ is a subsystem of $\mathfrak{A}$, $f(b, \ldots, b) \in \mathfrak{B}$; consequently,

$$f(x_1, \ldots, x_n)\,\theta = f(b, \ldots, b)\,\theta = a\theta\,.$$

This means $a\theta$ is a subsystem of $\mathfrak{A}$. Because $\mathcal{K}$ is hereditary, we can conclude $a\theta \in \mathcal{K}$. ∎

Corollary 2: *If $\mathcal{L}$ is a hereditary subclass of a class $\mathcal{K}$ of systems and $\mathcal{A}, \mathcal{B} \subseteq \mathcal{L}$, then*

$$\mathcal{A} \underset{\mathcal{K}}{*} \mathcal{B} = (\mathcal{A} \underset{\mathcal{K}}{*} \mathcal{B}) \cap \mathcal{L}. \tag{3}$$

For suppose $\mathfrak{A} \in \mathcal{A} \underset{\mathcal{K}}{*} \mathcal{B}$ and $\mathfrak{A}/\theta$ is a factor system satisfying the conditions in (1). Then $\mathfrak{A} \in \mathcal{K}$, $\mathfrak{A}/\theta \in \mathcal{B}$, and for any $a \in \mathfrak{A}$ we have

$$a\theta \in \mathcal{K} \Rightarrow a\theta \in \mathcal{K} \ \&\ a\theta \subseteq \mathfrak{B} \in \mathcal{L} \Rightarrow a\theta \in \mathcal{L} \Rightarrow a\theta \in \mathcal{A}.$$

because $\mathcal{L}$ is hereditary in $\mathcal{K}$. This shows

$$\mathcal{A} \underset{\mathcal{L}}{*} \mathcal{B} \subseteq (\mathcal{A} \underset{\mathcal{K}}{*} \mathcal{B}) \cap \mathcal{L}.$$

The converse inclusion

$$(\mathcal{A} \underset{\mathcal{K}}{*} \mathcal{B}) \cap \mathcal{L} \subseteq \mathcal{A} \underset{\mathcal{L}}{*} \mathcal{B}$$

obviously holds for an arbitrary (not necessarily hereditary) subclass $\mathcal{L}$. Thus (3) holds for hereditary $\mathcal{L}$. ∎

Let $\mathcal{E}_\Sigma$ (or just $\mathcal{E}$) denote the unit class with signature Σ, i.e., the class consisting of the one-element systems in which all the signature predicates are true.

Corollary 3: *Assume $\mathcal{E} \subseteq \mathcal{K}$; for every $\mathcal{A} \subseteq \mathcal{K}$,*

$$\mathcal{A} \underset{\mathcal{K}}{*} \mathcal{E} = \mathcal{A}. \tag{4}$$

For every hereditary subclass $\mathcal{A}$ of a class $\mathcal{K}$,

$$\mathcal{A} \underset{\mathcal{K}}{*} \mathcal{A} \supseteq \mathcal{A}. \tag{5}$$

For the proof it suffices to note that among the factor systems of any system we can count the system itself and the unit system. ∎

Let $\mathcal{E}_0$ be the collection of all one-element $\mathcal{K}$-systems. We similarly convince ourselves that for every $\mathcal{B} \subseteq \mathcal{K}$,

$$\mathcal{E}_0 \underset{\mathcal{K}}{*} \mathcal{B} \supseteq \mathcal{B}. \tag{6}$$

In particular, from (2), (4) and (6) we get

$$\mathcal{E} \subseteq \mathcal{B} \Rightarrow \mathcal{A} * \mathcal{B} \supseteq \mathcal{A}, \tag{7}$$

$$\mathcal{E}_0 \subseteq \mathcal{A} \Rightarrow \mathcal{A} * \mathcal{B} \supseteq \mathcal{B}. \tag{8}$$

It is easy to see that in (6) inclusion cannot in general be replaced with equality. Suppose, e.g., that $\mathcal{K}$ is the class of all semigroups, $\mathcal{B}$ is the class of all commutative semigroups, and $\mathfrak{A}$ is the semigroup consisting of the pairs $\langle a, b\rangle$ $(a, b = 1, 2, 3, \dots)$ with the multiplication

$$\langle a, b\rangle\langle c, d\rangle = \langle a+c,\ a+bc+d\rangle\,;$$

let $\mathfrak{G}$ be the additive semigroup of positive integers. Under the homomorphism $\pi\colon \langle a, b\rangle \to a$ mapping $\mathfrak{A}$ onto $\mathfrak{G}$, none of the members of $\mathfrak{A}/\pi$ are semigroups, yet $\mathfrak{A}/\pi \cong \mathfrak{G} \in \mathcal{B}$. Consequently, $\mathfrak{A} \in \mathcal{E} * \mathcal{B}$ but $\mathfrak{A} \notin \mathcal{B}$.

Let $\mathfrak{A}$ be an arbitrary algebraic system, and let $\mathfrak{A}/\theta$, $\mathfrak{A}/\eta$ be any two of its factor systems. We put $\mathfrak{A}/\theta \leqslant \mathfrak{A}/\eta$ iff $a\theta \subseteq a\eta$ for every $a \in \mathfrak{A}$, and $a\theta \to a\eta$ $(a \in \mathfrak{A})$ is a homomorphism of $\mathfrak{A}/\theta$ onto $\mathfrak{A}/\eta$ (cf. [XXXI], §1). This partially orders the set of all factor systems of $\mathfrak{A}$. A factor $\mathfrak{A}/\rho$ is called the *$\mathcal{A}$-replica* of $\mathfrak{A}$ iff $\mathfrak{A}/\rho$ is the smallest of the factor systems of $\mathfrak{A}$ belonging to $\mathcal{A}$.

Corollary 4: *Let $\mathcal{K}$ be an absolutely hereditary class with subclasses $\mathcal{A}, \mathcal{B}$. If a $\mathcal{K}$-system $\mathfrak{A}$ has a $\mathcal{B}$-replica $\mathfrak{A}/\rho$, and $\mathcal{A}$ is a hereditary subclass, then $\mathfrak{A} \in \mathcal{A} \underset{\mathcal{K}}{*} \mathcal{B}$ iff $a\rho \in \mathcal{K} \Rightarrow a\rho \in \mathcal{A}\ (a \in \mathfrak{A})$.*

We have only to verify that

$$\mathfrak{A} \in \mathcal{A} \underset{\mathcal{K}}{*} \mathcal{B} \Rightarrow (\forall a \in \mathfrak{A})(a\rho \in \mathcal{K} \Rightarrow a\rho \in \mathcal{A}).$$

Let $\mathfrak{A}/\theta$ be a factor system satisfying the conditions in (1), and suppose for some $a \in \mathfrak{A}$ we have $a\rho \in \mathcal{K}$. Since $a\rho \subseteq a\theta$ and $a\rho$ is a subsystem of $\mathfrak{A}$, $a\theta$ is also a subsystem of $\mathfrak{A}$. By hypothesis $\mathcal{K}$ contains all subsystems of $\mathfrak{A}$. Thus $a\theta \in \mathcal{K}$, and so $a\theta \in \mathcal{A}$. But $\mathcal{A}$ is hereditary in $\mathcal{K}$. Therefore, $a\rho \in \mathcal{A}$. I.e.,

$$(\forall a \in \mathfrak{A})(a\rho \in \mathcal{K} \Rightarrow a\rho \in \mathcal{A})\,. \blacksquare$$

Theorem 1: *In a hereditary class $\mathcal{K}$ the product of two hereditary subclasses $\mathcal{A}, \mathcal{B}$ is a hereditary subclass.*

Let $\mathfrak{A} \in \mathcal{A} * \mathcal{B}$ and have a factor $\mathfrak{A}/\theta$ satisfying the requirements in (1), and let $\mathfrak{A}_1 \subseteq \mathfrak{A}$, $\mathfrak{A}_1 \in \mathcal{K}$. Let $\mathfrak{A}_1/\theta_1$ denote the corresponding factor system for $\mathfrak{A}_1$ that is a subsystem of $\mathfrak{A}/\theta$.

Since $\mathcal{B}$ is absolutely hereditary, and $\mathfrak{A}_1/\theta_1$ is a subsystem of the $\mathcal{B}$-system $\mathfrak{A}/\theta$, we learn that $\mathfrak{A}_1/\theta_1 \in \mathcal{B}$. Suppose for some $a_1 \in \mathfrak{A}_1$ the equivalence class $a_1\theta_1$ is a $\mathcal{K}$-system. Since $a_1\theta_1 \subseteq a_1\theta$, $a_1\theta$ is a subsystem of $\mathfrak{A}$. Thus $a_1\theta \in \mathcal{K}$, and so $a_1\theta \in \mathcal{A}$. Inasmuch as $\mathcal{A}$ is hereditary and $a_1\theta_1$ is a subsystem of the $\mathcal{A}$-system $a_1\theta$, we see that $a_1\theta_1 \in \mathcal{A}$, and therefore, $\mathfrak{A}_1 \in \mathcal{A} * \mathcal{B}$. Thus

$$\mathfrak{A} \in \mathcal{A} * \mathcal{B} \ \ \& \ \ \mathfrak{A}_1 \subseteq \mathfrak{A} \ \ \& \ \ \mathfrak{A}_1 \in \mathcal{K} \Rightarrow \mathfrak{A}_1 \in \mathcal{A} * \mathcal{B} :$$

the class $\mathcal{A} * \mathcal{B}$ is hereditary. ∎

Let $\mathbf{F} = \langle \mathrm{I}, F \rangle$ be a *filter over* I, i.e., a nonempty collection F of subsets of a nonempty set I with the following properties:

$$\text{(a)} \qquad \emptyset \notin \mathbf{F},$$

$$\text{(b)} \qquad \mathrm{A} \in \mathbf{F} \ \& \ \mathrm{A} \subseteq \mathrm{B} \Rightarrow \mathrm{B} \in \mathbf{F}, \qquad (\mathrm{A}, \mathrm{B} \subseteq \mathrm{I})$$

$$\text{(c)} \qquad \mathrm{A}, \mathrm{B} \in \mathbf{F} \Rightarrow \mathrm{A} \cap \mathrm{B} \in \mathbf{F}.$$

When F consists of the single set I, then the filter is called *cartesian.* A filter $\mathbf{F}$ is an *ultrafilter* iff for every subset $\mathrm{A} \subseteq \mathrm{I}$, either A or its complement $\mathrm{I} \sim \mathrm{A}$ belongs to F. By $\prod_{\iota \in \mathrm{I}} \mathfrak{A}_\iota/\mathbf{F}$ we denote the $\mathbf{F}$-*product* (i.e., the filtered or reduced product with respect to $\mathbf{F}$) of the algebraic systems $\mathfrak{A}_\iota$ $(\iota \in \mathrm{I})$.

A class $\mathcal{K}$ of systems is called $\mathbf{F}$-*closed* iff all the $\mathbf{F}$-products of arbitrary $\mathcal{K}$-systems belong to $\mathcal{K}$. $\mathcal{K}$ is *multiplicatively closed* iff it contains every cartesian product of its systems. $\mathcal{K}$ is *ultraclosed* iff every ultraproduct of $\mathcal{K}$-systems is a $\mathcal{K}$-system. A subclass $\mathcal{A}$ of a class $\mathcal{K}$ is said to be $\mathbf{F}$-*closed in* $\mathcal{K}$ iff every $\mathcal{K}$-system isomorphic to an $\mathbf{F}$-product of $\mathcal{A}$-systems belongs to $\mathcal{A}$. We relativize the other two closure properties similarly.

Theorem 2: *Let $\mathcal{A}, \mathcal{B}$ be subclasses of a hereditary class $\mathcal{K}$ of algebraic systems; suppose the class $\mathcal{A} \subseteq \mathcal{K}$ is multiplicatively closed in $\mathcal{K}$, and the class $\mathcal{B} \subseteq \mathcal{K}$ is (absolutely) multiplicatively closed. Then the class $\mathcal{A} \underset{\mathcal{K}}{*} \mathcal{B}$ is multiplicatively closed in $\mathcal{K}$.*

In particular, in a hereditary, multiplicatively closed class $\mathcal{K}$, the $\mathcal{K}$-product of any two multiplicatively closed subclasses is a multiplicatively closed subclass.

Suppose $\mathfrak{A} \in \mathcal{X}$, $\mathfrak{A} \cong \prod_{\iota \in I} \mathfrak{A}_\iota$, and $\mathfrak{A}_\iota \in \mathcal{A} \underset{\mathcal{X}}{*} \mathcal{B}$ $(\iota \in I)$. For certain factor systems $\mathfrak{A}_\iota/\theta_\iota$, we have for each $\iota \in I$

$$\mathfrak{A}_\iota/\theta_\iota \in \mathcal{B}\,, \quad b\theta_\iota \in \mathcal{X} \Rightarrow b\theta_\iota \in \qquad (b \in \mathfrak{A}_\iota)\,. \tag{9}$$

Consider the canonical homomorphism $\mathfrak{A} \to \prod_{\iota \in I} (\mathfrak{A}_\iota/\theta_\iota)$ and let $\mathfrak{A}/\theta$ be the corresponding factor system. Since

$$\mathfrak{A}/\theta \cong \prod_{\iota \in I} (\mathfrak{A}_\iota/\theta_\iota)\,, \qquad \mathfrak{A}_\iota/\theta_\iota \in \mathcal{B},$$

and the class $\mathcal{B}$ is multiplicatively closed, we see $\mathfrak{A}/\theta \in \mathcal{B}$. Let a_ι be the projection on $\mathfrak{A}_\iota$ of an arbitrary element $a \in \mathfrak{A}$; then we have

$$a\theta \cong \prod_{\iota \in I} a_\iota \theta_\iota \,, \tag{10}$$

where $a\theta$, $a_\iota\theta_\iota$ are viewed as submodels (relational substructures) of the models corresponding to $\mathfrak{A}$, $\mathfrak{A}_\iota$. Suppose that $a\theta \in \mathcal{X}$ – in particular, we assume $a\theta$ is an algebraic system. Then all of its projections $a_\iota\theta_\iota$ are also algebraic systems. Since $\mathcal{X}$ is absolutely hereditary and $a_\iota\theta_\iota$ is a subsystem of $\mathfrak{A}_\iota$, $a_\iota\theta_\iota$ belongs to $\mathcal{X}$ and, by (9), to $\mathcal{A}$ as well. From the decomposition (10) and the multiplicative closedness of $\mathcal{A}$ in $\mathcal{X}$ we find that $a\theta \in \mathcal{A}$. Since a was arbitrary, $\mathfrak{A} \in \mathcal{A} \underset{\mathcal{X}}{*} \mathcal{B}$. ■

Theorem 3: *Suppose $\mathcal{A}, \mathcal{B}$ are subclasses of a hereditary class $\mathcal{X}$ of algebraic systems whose signature Σ contains only a finite number of operation symbols, and* $\mathbf{F} = \langle I, F \rangle$ *is an arbitrary ultrafilter. If $\mathcal{A}$ is* **F**-*closed in $\mathcal{X}$ and includes $\mathcal{E}$, and if $\mathcal{B}$ is (absolutely)* **F**-*closed, then the subclass $\mathcal{A} \underset{K}{*} \mathcal{B}$ is* **F**-*closed in $\mathcal{X}$.*

Suppose $\mathfrak{A} \in \mathcal{X}$ and $\mathfrak{A} \cong \prod_{\iota \in I} \mathfrak{A}_\iota/\mathbf{F}$, where $\mathfrak{A}_\iota \in \mathcal{A} \underset{\mathcal{X}}{*} \mathcal{B}$ $(\iota \in I)$; suppose the factor systems $\mathfrak{A}_\iota/\theta_\iota$ satisfy the conditions (9). As before, we consider the canonical homomorphism $\mathfrak{A} \to \Pi (\mathfrak{A}_\iota/\theta_\iota)/\mathbf{F}$ and let $\mathfrak{A}/\theta$ be the corresponding factor system. Letting a_ι be the cartesian projection of an arbitrary element $a \in \mathfrak{A}$ on the factor $\mathfrak{A}_\iota$ $(\iota \in I)$, we find

$$a\theta \cong \prod_{\iota \in I} (a_\iota\theta_\iota)/\mathbf{F}\,.$$

Assume $a\theta \in \mathcal{X}$. Since the signature Σ contains only a finite number of operation symbols, the assumption that $a\theta$ is a model corresponding to an algebraic system with signature Σ can be expressed by a closed formula (or *sentence, axiom*) Φ of first-order predicate logic (FOPL) involving predicate symbols corresponding to the function symbols in Σ. We write $\mathfrak{M} \vdash \Psi$ to mean that the FOPL sentence Ψ is true in the model $\mathfrak{M}$. Since $\mathbf{F}$ is an ultrafilter,

$$a\theta \vdash \Phi \Leftrightarrow \{\iota : a_\iota \theta_\iota \vdash \Phi\} \in F .$$

Let

$$\mathrm{I}_0 = \{\iota : a_\iota \theta_\iota \vdash \Phi\}, \qquad F_0 = \{\mathrm{X} \cap \mathrm{I}_0 : \mathrm{X} \in F\}, \quad \mathbf{F}_0 = \langle \mathrm{I}_0, F_0 \rangle .$$

Then $\mathrm{I}_0 \in F$, $\mathbf{F}_0$ is an ultrafilter, and

$$\prod_{\iota \in \mathrm{I}} (a_\iota \theta_\iota)/\mathbf{F} \cong \prod_{\kappa \in \mathrm{I}_0} (a_\kappa \theta_\kappa)/\mathbf{F}_0 .$$

Taking

$$\mathfrak{B}_\kappa = a_\kappa \theta_\kappa \quad (\kappa \in \mathrm{I}_0), \qquad \mathfrak{B}_\iota \in \mathcal{E} \quad (\iota \in \mathrm{I} \sim \mathrm{I}_0) ,$$

we get

$$a\theta \cong \prod_{\kappa \in \mathrm{I}_0} (a_\kappa \theta_\kappa)/\mathbf{F}_0 \cong \prod_{\iota \in \mathrm{I}} \mathfrak{B}_\iota/\mathbf{F} . \tag{11}$$

Each factor $\mathfrak{B}_\kappa = a_\kappa \theta_\kappa$ is a subsystem of the $\mathcal{X}$-system $\mathfrak{A}_\kappa$; therefore, it belongs to $\mathcal{X}$– and to $\mathcal{A}$ by (9). The remaining factors $\mathfrak{B}_\iota$ are unit systems and also belong to $\mathcal{A}$. The subclass $\mathcal{A}$ is $\mathbf{F}$-closed in $\mathcal{X}$; therefore, by (11) we know $a\theta \in \mathcal{A}$. Thus

$$(\forall a \in \mathfrak{A})(a\theta \in \mathcal{X} \Rightarrow a\theta \in \mathcal{A}) . \blacksquare$$

In this argument we need to know $\mathcal{E}$ is included in $\mathcal{A}$ in order to conclude that $\Pi(a_\kappa \theta_\kappa)/\mathbf{F}_0$ belongs to $\mathcal{A}$, knowing it belongs to $\mathcal{X}$. But $\mathbf{F}_0$ is an ultrafilter, so this will hold automatically if $\mathcal{A}$ is ultraclosed in $\mathcal{X}$. Consequently, if $\mathcal{A}$ is an ultraclosed subclass of the hereditary class $\mathcal{X}$, $\mathfrak{B}$ is an absolutely ultraclosed subclass of $\mathcal{X}$, and the signature of $\mathcal{X}$ contains a finite number of operation symbols, then the subclass $\mathcal{A} \underset{\mathcal{X}}{*} \mathcal{B}$ is ultraclosed in $\mathcal{X}$.

A class $\mathcal{X}$ of algebraic systems with signature Σ is called *replica-complete* (or *replete*) iff it is hereditary and multiplicatively closed and contains a unit system. A replete class $\mathcal{X}$ contains a $\mathcal{X}$-replica of each system with its signature (cf. [VIII]). From Theorems 1–3 we immediately get

Corollary 5: *If a class $\mathcal{K}$ is replete, then the $\mathcal{K}$-product of any of its replete subclasses is replete. If the signature of a hereditary, ultraclosed class $\mathcal{K}$ contains only a finite number of operation symbols, then the $\mathcal{K}$-product of any two ultraclosed subclasses of $\mathcal{K}$ is ultraclosed.* ■

Observe that the corollary still does not let us speak of the groupoid of replete subclasses of a given replete class $\mathcal{K}$, for this collection cannot be a set with a definite cardinality, except in trivial cases. This shortcoming is easy to fix by considering not all subclasses of $\mathcal{K}$, but only subclasses subject to some stronger conditions: e.g. cardinality limitations or axiomatizability.

§2. Products of axiomatizable classes

Let Γ be some type of sentences of FOPL (with equality and function symbols). In particular, Γ could be the class I of all identities, the class Q of all quasidentities (cf. [XXXI]), or the class $\forall$ of all universal sentences. If we are given some class $\mathcal{K}$ of algebraic systems with signature Σ, then by $\Gamma(\mathcal{K})$ we denote the set of all sentences of the type Γ and signature Σ that are true in every $\mathcal{K}$-system. The set $\Gamma(\mathcal{K})$ is called the Γ-*theory* of the class $\mathcal{K}$. Conversely, if we are given a set S of FOPL sentences of signature Σ, then $\mathbf{K}_{\Sigma}(S)$ (or just $\mathbf{K}(S)$) denotes the class of all systems with signature Σ in which all the sentences in S are valid. The class of all FOPL sentences is denoted by **E**.

A subclass $\mathcal{L}$ of a class $\mathcal{K}$ is called a Γ-*subclass* of $\mathcal{K}$ iff $\mathcal{L} = \mathcal{K} \cap \mathbf{K}\Gamma(\mathcal{L})$.

A class $\mathcal{K}$ is called an *(absolute)* Γ-*class* (or Γ-axiomatizable) iff $\mathcal{K} = \mathbf{K}\Gamma(\mathcal{K})$. E-axiomatizable classes are simply called *(first-order) axiomatizable.* I-classes and Q-classes are respectively called *varieties* and *quasivarieties*; $\forall$-classes are called *universal classes* (or *universals*).

For any type Γ, the intersection of any family of Γ-subclasses of a given class $\mathcal{K}$ is itself a Γ-subclass of $\mathcal{K}$. Therefore, the family of all Γ-subclasses of a given class $\mathcal{K}$ of systems can be viewed as a complete lattice relative to the usual relation of inclusion. This lattice is denoted by $\mathfrak{L}_{\Gamma}(\mathcal{K})$. Since the $\mathcal{K}$-product of two Γ-subclasses might not be a Γ-subclass, the family of all Γ-subclasses of $\mathcal{K}$ together with the $\mathcal{K}$-multiplication will be a partial groupoid, and only for special $\mathcal{K}$, Γ will it be an ordinary groupoid with a totally defined operation of multiplication. This partial groupoid is denoted by $\mathfrak{G}_{\Gamma}(\mathcal{K})$;

We shall show that the family of all subvarieties of a given variety need not be a (total) groupoid. Let $\mathcal{K}$ be the variety of all semigroups with identity element e (which is viewed as a distinguished element designated by a 0-ary operation symbol in the signature); let $\mathcal{A}$ be the subvariety of all commutative semigroups in $\mathcal{K}$. It is clear that $\mathcal{K} \neq \mathcal{A} \underset{\mathcal{K}}{*} \mathcal{A}$. E.g., let $\mathfrak{A}_5$ be the semigroup of all even permutations of the numbers 1, ..., 5, i.e., the multiplicative semigroup of the alternating group of degree 5. The only factor algebras $\mathfrak{A}_5$ has are itself and the unit semigroup. But $\mathfrak{A}_5 \notin \mathscr{A}$, so $\mathfrak{A}_5 \notin \mathscr{A} * \mathscr{A}$.

On the other hand, let $\mathfrak{F}$ be the $\mathcal{K}$-free semigroup with free generators a, b. The factor semigroup $\mathfrak{F}/\theta$ by the congruence θ defined by

$$a^{m_1} b^{n} \dots a^{m_k} b^{n_k} \ \theta \ a^{p_1} b^{q_1} \dots a^{p_l} b^{q_l} \Leftrightarrow m_1 + \dots + m_k = p_1 + \dots + p_l$$

is an abelian semigroup with the single generator $a\theta$; therefore, $\mathfrak{F}/\theta \in \mathscr{A}$. Since e is distinguished in $\mathcal{K}$, the only congruence classes $x\theta$ belonging to $\mathcal{K}$ are those that contain $e \in \mathfrak{F}$. But there is only one such class: $e\theta = \{e, b, b^2, \dots\}$, and it is a commutative semigroup. Therefore, $\mathfrak{F} \in \mathscr{A} * \mathscr{A}$.

If the class $\mathscr{A} * \mathscr{A}$ were a subvariety of $\mathcal{K}$, then all the factor semigroups of $\mathfrak{F}$ would have to belong to $\mathscr{A} * \mathscr{A}$. But the semigroup $\mathfrak{A}_5$ mentioned above is generated by two of its elements and is, therefore, isomorphic to a factor semigroup of $\mathfrak{F}$ that cannot belong to $\mathscr{A} * \mathscr{A}$.

The next theorems show that, in contrast to varieties, quasivarieties and universals behave more regularly.

Theorem 4: *Let $\mathcal{K}$ be a universal class whose signature contains only a finite number of function symbols. Then the $\mathcal{K}$-product of any two universal subclasses $\mathscr{A}, \mathscr{B}$ of $\mathcal{K}$ is universal. Thus $\mathfrak{G}_\forall(\mathcal{K})$ is a groupoid.*

For inasmuch as the classes $\mathcal{K}, \mathscr{A}, \mathscr{B}$ are universal, they are hereditary and ultraclosed. So the class $\mathscr{A} \underset{\mathcal{K}}{*} \mathscr{B}$ is hereditary by Theorem 1 and ultraclosed by Theorem 3. This implies it is universal. ■

Theorem 5: *For every quasivariety $\mathcal{K}$ whose signature contains but a finite number of operation symbols, the partial groupoid $\mathfrak{G}_Q(\mathcal{K})$ of the subquasivarieties of $\mathcal{K}$ is a groupoid.*

Suppose $\mathscr{A}, \mathscr{B}$ are subquasivarieties of $\mathcal{K}$. By the preceding theorem the class $\mathscr{A} \underset{\mathcal{K}}{*} \mathscr{B}$ is universal; according to Theorem 2 this class is multiplicatively closed; it also contains a unit system. Any universal, multiplicatively closed class that contains unit systems is a quasivariety (cf. [XXXI], §2); hence, $\mathscr{A} \underset{\mathcal{K}}{*} \mathscr{B}$ is a quasivariety. ■

These proofs of Theorems 4 and 5 are based on properties of ultrafilters. These theorems, however, are easy to prove without recourse to ultrafilters. Let $\mathcal{K}$ be a class of algebraic systems with signature Σ, and let $\Sigma_1 \subseteq \Sigma$. Viewing the systems in $\mathcal{K}$ as systems with signature Σ_1 gives us a class $\mathcal{K}^{\Sigma_1}$ with signature Σ_1, called the Σ_1-*projection* (or Σ_1-*impoverishment*) of the class $\mathcal{K}$. Projections of axiomatizable classes are called *projective classes.*

Theorem 6: *If $\mathfrak{K}$ is a finitely axiomatizable class of models, then the $\mathfrak{K}$-product of any two axiomatizable subclasses $\mathcal{A}$, $\mathfrak{B}$ of $\mathfrak{K}$ is a projective class of models.*

By hypothesis the signature Σ of these classes of models consists of predicate symbols $\boldsymbol{P}_\xi$ with corresponding arities n_ξ ($\xi \in \Xi$). We associate with each $\boldsymbol{P}_\xi$ a new predicate symbol $\boldsymbol{P}_\xi^*$ of the same arity as $\boldsymbol{P}_\xi$; Σ^* is the signature composed of these new symbols. We introduce an auxiliary binary predicate symbol $\boldsymbol{\theta}$; we form the set $\mathrm{S_h}$ from the following axioms (initial universal quantifiers suppressed):

$$\boldsymbol{x\,\theta x} \,\&\, (\boldsymbol{x\,\theta y} \rightarrow \boldsymbol{y\,\theta x}) \,\&\, (\boldsymbol{x\,\theta y} \,\&\, \boldsymbol{y\,\theta z} \rightarrow \boldsymbol{x\,\theta z}) \,,$$

$$\boldsymbol{P}_\xi(\boldsymbol{x}_1, \ldots, \boldsymbol{x}_{n_\xi}) \rightarrow \boldsymbol{P}_\xi^*(\boldsymbol{x}_1, \ldots, \boldsymbol{x}_{n_\xi}) \,, \qquad (\xi \in \Xi)$$

$$\boldsymbol{P}_\xi^*(\boldsymbol{x}_1, \ldots, \boldsymbol{x}_n) \,\&\, \boldsymbol{x}_1 \boldsymbol{\theta} \boldsymbol{y}_1 \,\&\, \ldots \,\&\, \boldsymbol{x}_{n_\xi} \boldsymbol{\theta} \boldsymbol{y}_{n_\xi} \rightarrow \boldsymbol{P}_\xi^*(\boldsymbol{y}_1, \ldots, \boldsymbol{y}_{n_\xi}) \,.$$

The significance of the axiom system $\mathrm{S_h}$ is the following: a model

$$\mathfrak{A} = \langle A; \Sigma, \Sigma^*, \theta \rangle$$

satisfies all the axioms in $\mathrm{S_h}$ iff the Σ^*-projection of $\mathfrak{A}$ is in an obvious sense a θ-factor system of the Σ-projection of $\mathfrak{A}$.

By assumption of the class $\mathfrak{K}$ can be characterized by a single axiom Φ, while the classes $\mathcal{A}, \mathfrak{B}$ can be characterized by certain axiom systems $\mathrm{S}_{\mathcal{A}}$, $\mathrm{S}_{\mathfrak{B}}$, possibly infinite. Construct Φ^*, S^* by replacing the symbols $\boldsymbol{P}_\xi$, $\approx$ in the corresponding axioms with the respective symbols $\boldsymbol{P}_\xi^*$, θ. Let $\Psi_{\boldsymbol{x}}$ be the relativization (specialization) of the axiom Ψ to the formula $\mathrm{R}(\boldsymbol{y}) = \boldsymbol{y\,\theta x}$. Consider the system S consisting of Φ, Φ^*, the members of $\mathrm{S_h}$ and $\mathrm{S}^*_{\mathfrak{B}}$ and all the axioms

$$(\boldsymbol{x})(\Phi_{\boldsymbol{x}} \rightarrow \Psi_{\boldsymbol{x}}) \quad (\Psi \in \mathrm{S}_{\mathcal{A}}) \,.$$

A model $\mathfrak{A} = \langle A; \Sigma, \Sigma^*, \theta \rangle$ satisfies S iff the Σ-projection of $\mathfrak{A}$ belongs to $\mathcal{A} \underset{\mathfrak{K}}{*} \mathfrak{B}$. Thus this class is a projection of the axiomatizable class **K(S)**. ■

Corollary 6: *Let $\mathfrak{K}$ be a universal class of algebraic systems whose signature Σ contains only a finite number of operation symbols. Then the $\mathfrak{K}$-product of any axiomatizable subclasses $\mathcal{A}, \mathfrak{B}$ of the class $\mathfrak{K}$ is a projective class of systems.*

Consider the total class $\mathcal{K}_\Sigma = \mathbf{K}_\Sigma(\emptyset)$ of all algebraic systems with signature Σ. According to Corollary 2,

$$\mathcal{A} \underset{\mathcal{K}}{*} \mathcal{B} = (\mathcal{A} \underset{\mathcal{K}_\Sigma}{*} \mathcal{B}) \cap \mathcal{K}.$$

The intersection of an axiomatizable class and a projective class is a projective class. Therefore, it suffices to verify the projectiveness of $\mathcal{A} \underset{\mathcal{K}_\Sigma}{*} \mathcal{B}$.

Every algebraic system with signature Σ, which contains operation symbols $f_\xi^{(n_\xi)}$ $(\xi \in \Xi)$, can be viewed as a model satisfying the universal axioms

$$\left.\begin{aligned} &(\exists y) P_\xi(x_1, \ldots, x_{n_\xi}, y), \\ &P_\xi(x_1, \ldots, x_{n_\xi}, y) \;\&\; P_\xi(x_1, \ldots, x_{n_\xi}, z) \to y \approx z, \end{aligned}\right\} \quad (\xi \in \Xi) \qquad (12)$$

which ensure that the predicates the P_ξ designate are in fact functions. Since Ξ is finite, there are a finite number of axioms (12). Thus the class $\mathcal{K}$, viewed as a class of models, is finitely axiomatizable. Applying Theorem 6 proves that $\mathcal{A} \underset{\mathcal{K}}{*} \mathcal{B}$ is projective. ■

From Corollary 4 we can immediately deduce Theorem 4, and with it Theorem 5. Indeed, by this corollary the hypotheses of Theorem 4 imply that the class $\mathcal{A} \underset{\mathcal{K}}{*} \mathcal{B}$ is projective. By Theorem 1 this class is hereditary. It is known that hereditary projective classes are universal.

We shall see that the conclusion of projectiveness cannot in general be strengthened to axiomatizability. Consider the class $\mathcal{K}$ of algebras whose signature consists of two unary operation symbols f, g and which satisfy the identities

$$f(g(x)) \approx g(f(x)) \approx x. \qquad (13)$$

Let a be an element of an arbitrary $\mathcal{K}$-algebra $\mathfrak{A}$. We set $a^0 = a$ and

$$a^n = f^n(a) = \underbrace{f \ldots f}_{n}(a), \qquad a^{-n} = g^n(a) \qquad (n = 1, 2, \ldots).$$

By virtue of the axioms (13) we have

$$a^i = b^j \Leftrightarrow a = b^{j-i} \qquad (i, j = 0, \pm 1, \pm 2, \ldots; a, b \in \mathfrak{A}).$$

We say that elements $a, b \in \mathfrak{A}$ belong to the same *cycle* iff $a = b^i$ for some integer i. Clearly, the algebra $\mathfrak{A}$ splits into disjoint cycles, each of which is a subalgebra of $\mathfrak{A}$.

Let $\mathscr{A}$ be the class of those $\mathcal{X}$-algebras that satisfy the sentences

$$(\exists x)(x^1 \approx x)\,, \tag{14}$$

$$(\exists x)(x^1 \not\approx x)\,, \tag{15}$$

$$x^1 \approx x \,\&\, y^1 \approx y \rightarrow x \approx y\,, \tag{16}$$

$$x^i \approx x \rightarrow x^1 \approx x \qquad (i = 2, 3, \dots)\,; \tag{17}$$

the significance of (14)–(17) is that every $\mathscr{A}$-algebra decomposes into one one-element cycle and some (at least one) infinite cycles.

It is easy to verify that arbitrary $\mathcal{X}$-algebras $\mathfrak{A}$, $\mathfrak{B}$ are isomorphic iff for every $n = 1, 2, \dots, \omega$, the powers of the sets of n-element cycles in $\mathfrak{A}$ and $\mathfrak{B}$ are the same. Consequently, any two $\mathscr{A}$-algebras that have the same uncountable power are isomorphic. Since all $\mathscr{A}$-algebras are infinite, by Vaught's test the class $\mathscr{A}$ is a minimal axiomatizable class. In particular, if it turns out that

$$\mathscr{A} \underset{\mathcal{X}}{*} \mathscr{A} \subseteq \mathscr{A}\,, \qquad \mathscr{A} \underset{\mathcal{X}}{*} \mathscr{A} \neq \mathscr{A}\,, \tag{18}$$

then $\mathscr{A} \underset{\mathcal{X}}{*} \mathscr{A}$ is certainly not axiomatizable.

We now prove (18). Let $\mathfrak{A} \in \mathscr{A} \underset{\mathcal{X}}{*} \mathscr{A}$, and let θ be a congruence on $\mathfrak{A}$ for which

$$\mathfrak{A}/\theta \in \mathscr{A}, \quad a\theta \in \mathcal{X} \Rightarrow a\theta \in \mathscr{A} \quad (a \in \mathfrak{A})\,. \tag{19}$$

According to (14), in $\mathfrak{A}/\theta$ there is a one-element cycle $c\theta = (c\theta)^1 = c^1\theta$. This shows that the congruence class $c\theta$ is a subalgebra of $\mathfrak{A}$; hence, by (19) $c\theta \in \mathscr{A}$. Consequently, there exists a one-element cycle $\{e\} = \{e^1\}$ in $c\theta$ – and in $\mathfrak{A}$, too. If in $\mathfrak{A}$ there were another finite cycle $\{a, a^1, \dots, a^m\}$, then the set $\{a\theta, a^1\theta, \dots, a^m\theta\}$ would form a finite subalgebra of $\mathfrak{A}/\theta$. But $\mathfrak{A}/\theta$ is an $\mathscr{A}$-algebra, so its only finite subalgebra is $\{e\theta\}$; hence $\{a, a^1, \dots, a^m\} \subseteq e\theta \in \mathscr{A}$, so $a = e$. Thus the algebra $\mathfrak{A}$ contains only one finite cycle, namely $\{e\}$. On the other hand, since $\mathfrak{A}/\theta$ belongs to $\mathscr{A}$, $\mathfrak{A}$ is infinite. Therefore, $\mathfrak{A}$ is an $\mathscr{A}$-algebra, proving the first condition in (18).

Consider now the $\mathscr{A}$-algebra $\mathfrak{C}$ that splits into a one-element cycle $\{e\}$ and one infinite cycle $\{\dots, a^{-1}, a, a^1, \dots\}$. Let θ be an arbitrary congruence for which $\mathfrak{C}/\theta \in \mathscr{A}$. For any $i \neq j$, if $a^i\,\theta\, a^j$ or $a^i\,\theta\, e$, then the algebra $\mathfrak{C}$ is finite, contradicting a property of $\mathscr{A}$-algebras. Therefore, the congruence θ is just equality, $\{e\} = e\theta \in \mathcal{X}$, but $e\theta \notin \mathscr{A}$, so $\mathfrak{C} \notin \mathscr{A} \underset{\mathcal{X}}{*} \mathscr{A}$.

Thus, (18) holds and the class $\mathscr{A}$ can serve as an example of an axiomatizable subclass of a finitely axiomatizable variety $\mathfrak{X}$ such that the class $\mathscr{A} \underset{\mathfrak{X}}{*} \mathscr{A}$ is not axiomatizable ([1]).

We shall further show that the $\mathfrak{X}$-product of finitely axiomatizable subvarieties of $\mathfrak{X}$ need not be finitely axiomatizable, although it must be a quasivariety by Theorem 5. Let $\mathscr{A}$ be the variety of $\mathfrak{X}$-algebras satisfying the identity $\boldsymbol{x}^2 \approx \boldsymbol{x}$. Suppose $\mathfrak{A} \in \mathscr{A} \underset{\mathfrak{X}}{*} \mathscr{A}$, and suppose θ is a congruence on $\mathfrak{A}$ satisfying (19). Assume that for some $a \in \mathfrak{A}$, $m > 0$ we have $a^{2m+1} = a$. Since $(a\theta)^2 = a\theta$, we successively conclude

$$a^2 \theta a, \quad a^{2m} \theta a, \quad a = a^{2m+1} \theta a^1, \quad a\theta \in \mathfrak{X}.$$

Whence, in view of (19), we get

$$a^2 = a, \quad a^{2m} = a, \quad a = a^1 .$$

In other words, in the class $\mathscr{A} \underset{\mathfrak{X}}{*} \mathscr{A}$ the quasidentities

$$\boldsymbol{x}^{2m+1} \approx \boldsymbol{x} \rightarrow \boldsymbol{x}^1 \approx \boldsymbol{x} \qquad (m > 0) \tag{20}$$

hold.

Conversely, suppose that in some $\mathfrak{X}$-algebra $\mathfrak{A}$ all the quasidentities (20) hold; so the algebra $\mathfrak{A}$ decomposes into certain sets (possibly empty) of one-element cycles $\{a_\alpha\}$, finite cycles $\{b_\beta, b_\beta^1, \ldots, b_\beta^{2m_\beta - 1}\}$ of even order, and infinite cycles $\{\ldots, c_\gamma^{-1}, c_\gamma, c_\gamma^1, \ldots\}$. We introduce a binary relation $\hat{\theta}$ on $\mathfrak{A}$ by setting

$$b^i \hat{\theta}\, b^{i+2m}, \quad c^i \hat{\theta}\, c^{i+2m} \qquad (i, m = 0, \pm 1, \pm 2, \ldots).$$

Let θ be the equivalence relation generated by $\hat{\theta}$; θ is clearly a congruence on $\mathfrak{A}$ and $\mathfrak{A}/\theta \in \mathscr{A}$. Since for every $a \in \mathfrak{A}$,

$$a\theta \in \mathfrak{X} \Rightarrow a\theta a^1 \Rightarrow (\exists \alpha)(a = a_\alpha) \Rightarrow a^2 = a,$$

we have $\mathfrak{A} \in \mathscr{A} \underset{\mathfrak{X}}{*} \mathscr{A}$.

Thus the class $\mathscr{A} \underset{\mathfrak{X}}{*} \mathscr{A}$ is characterized in $\mathfrak{X}$ by the infinite system of quasidentities (20). This system is equivalent in $\mathfrak{X}$ to no finite part of itself, since for any $n > 0$, the $\mathfrak{X}$-algebra consisting of a single cycle of length $2n + 1$ satisfies those quasidentities (20) in which $m < n$, but not the one with $m = n$.

§3. Multiplication in special classes of systems

In studying the partial groupoids $\mathfrak{G}_\Gamma(\mathcal{K})$, we naturally encounter problems of the form: for which $\Gamma, \mathcal{K}$ does the partial groupoid $\mathfrak{G}_\Gamma(\mathcal{K})$ have this or that property, e.g., is total, associative, or commutative, etc. We stated above simple conditions for the class $\mathcal{K}$ that guarantee $\mathfrak{G}_\forall(\mathcal{K})$ and $\mathfrak{G}_Q(\mathcal{K})$ are groupoids. Now we indicate conditions ensuring that $\mathfrak{G}_I(\mathcal{K})$ is a groupoid, and then others securing the associativity of $\mathfrak{G}_\Gamma(\mathcal{K})$.

An element a is called an *idempotent* relative to an operation $f(x_1, ..., x_n)$ iff $f(a, ..., a) = a$. An element a of an algebraic system $\mathfrak{A}$ is called an *idempotent of* $\mathfrak{A}$ iff it is an idempotent with respect to every signature operation on $\mathfrak{A}$. In particular, if the signature designates 0-ary operations, then an idempotent of $\mathfrak{A}$ and the elements distinguished as the constant values of these operations must all coincide.

A *polar operation* (or *support operation,* or *polar*) on a system $\mathfrak{A}$ is a constant, termal, unary operation whose single value is an idempotent of $\mathfrak{A}$. The value of a polar on $\mathfrak{A}$ is called a *polar element* (or *support element,* or *pole*). A unary term that defines a polar operation in every system in a class $\mathcal{K}$ is called a *polar* of $\mathcal{K}$ ($\mathcal{K}$*-polar*). A class is called *polarized* iff it admits at least one polar. It is obvious that every pole of a system $\mathfrak{A}$ forms a one-element subsystem, and every subsystem of $\mathfrak{A}$ contains all the poles of $\mathfrak{A}$. Therefore, no system has more than one pole. If $\mathrm{t}(\boldsymbol{x})$ is a polar of a class $\mathcal{K}$, then it determines a polar on each $\mathcal{K}$-system $\mathfrak{A}$ whose value $p_\mathfrak{A}$ is an idempotent of this system. The mapping $\mathfrak{A} \to p_\mathfrak{A}$ has the following obvious property: for any homomorphism φ of a $\mathcal{K}$-system $\mathfrak{A}$ into a $\mathcal{K}$-system $\mathfrak{B}$,

$$\varphi(p_\mathfrak{A}) = p_\mathfrak{B} \,. \qquad (21)$$

We shall show that if a class $\mathcal{K}$ contains a $\mathcal{K}$-free system $\mathfrak{F}$ of rank 1, and if in every $\mathcal{K}$-system $\mathfrak{A}$ an idempotent $p_\mathfrak{A}$ can be chosen so that (21) holds, then the class $\mathcal{K}$ is polarized.

For suppose v is a free generator of $\mathfrak{F}$. Then $p_\mathfrak{F} = \mathrm{t}^\mathfrak{F}(v)$, where t is an appropriate formal term, and $\mathrm{t}^\mathfrak{F}$ is the operation it defines in $\mathfrak{F}$. Suppose $\mathfrak{A} \in \mathcal{K}$ and $a \in \mathfrak{A}$. By hypothesis there is a homomorphism φ: $\mathfrak{F} \to \mathfrak{A}$ with $\varphi(v) = a$. From (21) it follows that $\varphi(p_\mathfrak{F}) = \mathrm{t}^\mathfrak{A}(a) = p_\mathfrak{A}$. Thus the term t determines the pole $p_\mathfrak{A}$ in each $\mathfrak{A} \in \mathcal{K}$.

We make one more obvious remark. Suppose the system $\mathfrak{A}$ has the pole p, and suppose $\mathfrak{A}/\theta$ is a factor system for $\mathfrak{A}$. Then among the equivalence classes forming $\mathfrak{A}/\theta$ the class $p\theta$ is the only one that is a subsystem of $\mathfrak{A}$. In particular, if $\mathcal{A}, \mathcal{B}$ are subclasses of a polarized class $\mathcal{K}$, then $\mathfrak{A} \in \mathcal{A} \underset{\mathcal{K}}{*} \mathcal{B}$ iff there exists a factor system $\mathfrak{A}/\theta$ such that $\mathfrak{A}/\theta \in \mathcal{B}$ and $p\theta \in \mathcal{A}$, where p is the pole of $\mathfrak{A}$.

Varieties of loops and groups are polarized. The pole of a loop or group is its identity element. The variety of all nonassociative rings with signature $\{-, \cdot\}$ is polarized by the term $\boldsymbol{x}-\boldsymbol{x}$; the pole is always the zero element. By contrast, the varieties of all lattices and all semigroups have no polars.

Theorem 7: *If $\mathcal{K}$ is a polarized variety of algebras and all the congruences on each $\mathcal{K}$-algebra commute, then the $\mathcal{K}$-product of any two subvarieties $\mathcal{A}, \mathcal{B} \subseteq \mathcal{K}$ is a variety.*

Thus for such $\mathcal{K}$ the partial groupoid $\mathfrak{G}_I(\mathcal{K})$ is a groupoid with identity $\mathcal{E}$ and zero $\mathcal{K}$.

To prove the theorem we note that by Theorems 1 and 2 the class $\mathcal{A} \underset{\mathcal{K}}{*} \mathcal{B}$ is hereditary and multiplicatively closed. By Birkhoff's theorem we have only to show that this calss contains all factor algebras of every algebra $\mathfrak{A} \in \mathcal{A} \underset{\mathcal{K}}{*} \mathcal{B}$. Let θ be a congruence on $\mathfrak{A}$ satisfying the two requirements

$$\mathfrak{A}/\theta \in \mathcal{B}, \quad a\theta \in \mathcal{K} \Rightarrow a\theta \in \mathcal{A} \quad (a \in \mathfrak{A}),$$

and let p be the pole of $\mathfrak{A}$.

Consider an arbitrary factor algebra $\mathfrak{A}/\sigma$. From the relations

$$(\mathfrak{A}/\sigma)/\sigma\theta \cong (\mathfrak{A}/\theta)/\sigma\theta \in \mathcal{B},$$

$$(p\sigma)\sigma\theta = (p\sigma\theta)/\sigma \cong (p\theta)/(\sigma \cap \theta) \in \mathcal{A}$$

and the remark made above we immediately see that $\mathfrak{A}/\sigma \in \mathcal{A} \underset{\mathcal{K}}{*} \mathcal{B}$. ■

It is known that all congruences on a group, ring, or loop commute, and that classes of any of these algebras are polarized. Therefore, for any variety $\mathcal{K}$ of groups, rings, or loops, the partial groupoid $\mathfrak{G}_I(\mathcal{K})$ is a groupoid, but not as a rule associative. In order to formulate sufficient conditions for its associativity, we introduce a few definitions.

A congruence σ on an algebra $\mathfrak{A}$ is called *characteristic* iff every automorphism of $\mathfrak{A}$ naturally induces an automorphism of $\mathfrak{A}/\sigma$. A congruence σ on $\mathfrak{A}$ is called *verbal* (*quasiverbal*) iff there exists a variety (quasivariety) such that σ is the smallest among the congruences on $\mathfrak{A}$ whose factor algebras belong to $\mathcal{A}$.

Let $\mathcal{K}$ be a fixed class of algebraic systems, and let $\mathfrak{A}$ belong to $\mathcal{K}$ and $\mathfrak{A}/\theta$ be one of its factor systems belonging to $\mathcal{K}$. The factor system $\mathfrak{A}/\theta$ is called *characteristic* (*totally characteristic*) iff for every automorphism (endomorphism) φ: $\mathfrak{A} \rightarrow \mathfrak{A}$ and every basic predicate P of $\mathfrak{A}/\theta$ – including equality and the signature operations –

$$P(x_1\theta, \ldots, x_n\theta) \Rightarrow P(\varphi(x_1)\theta, \ldots, \varphi(x_n)\theta) \quad (x_i \in \mathfrak{A}) .$$

The factor system $\mathfrak{A}/\theta$ is called *verbal* (*quasiverbal*) iff there is a variety (quasivariety) $\mathscr{A}$ such that $\mathfrak{A}/\theta$ is an $\mathscr{A}$-replica of $\mathfrak{A}$.

Let $a_\mu\theta$ $(\mu \in \mathrm{M})$ be those equivalence classes in $\mathfrak{A}/\theta$ that are $\mathfrak{K}$-systems, and suppose for each class $a_\mu\theta$ we have a factor system $a_\mu\theta/\eta_\mu \in \mathfrak{K}$. The collection $\{a_\mu\theta/\eta_\mu : \mu \in \mathrm{M}\}$ is called a *partial $\mathfrak{K}$-subfactor* for the factor system $\mathfrak{A}/\theta$. A partial subfactor is called (*totally*) *characteristic* iff it consists of (totally) characteristic factor systems. A partial subfactor is called *verbal* (*quasiverbal*) iff the factors of which it consists are $\mathscr{A}$-replicas in an appropriate variety (quasivariety) $\mathscr{A}$.

A partial subfactor $\{a_\mu\theta/\eta_\mu : \mu \in \mathrm{M}\}$ is said to be *$\mathfrak{K}$-extendable* iff all of its members can be simultaneously extended to some factor system $\mathfrak{A}/\eta \in \mathfrak{K}$ $(\mathfrak{A}/\eta \leqslant \mathfrak{A}/\theta)$, i.e., iff

$$x\eta = x\eta_\mu \qquad (x \in a_\mu\theta, \mu \in \mathrm{M}) ,$$

and for each $\mu \in \mathrm{M}$, the map $x\eta \to x\eta_\mu$ $(x \in a_\mu\theta)$ is an isomorphism of $a_\mu\theta/\eta$ onto $a_\mu\theta/\eta_\mu$.

A system $\mathfrak{A} \in \mathfrak{K}$ is called *transcharacteristic* (*totally transcharacteristic*) in $\mathfrak{K}$ iff every (totally) characteristic partial $\mathfrak{K}$-subfactor of each (totally) characteristic factor $\mathfrak{K}$-system of $\mathfrak{A}$ is $\mathfrak{K}$-extendable.

Similarly, a system $\mathfrak{A} \in \mathfrak{K}$ is *transverbal* (*transquasiverbal*) in $\mathfrak{K}$ iff every (quasi)verbal partial $\mathfrak{K}$-subfactor of an arbitrary (quasi)verbal factor system of $\mathfrak{A}$ in $\mathfrak{K}$ is $\mathfrak{K}$-extendable.

A class $\mathfrak{K}$ is called *transverbal* (*transquasiverbal*), (*transcharacteristic*) iff every $\mathfrak{K}$-system is transverbal (transquasiverbal) (transcharacteristic) in $\mathfrak{K}$.

Since for every factor system $\mathfrak{A}/\theta$,

$$\mathfrak{A}/\theta \text{ verbal} \Rightarrow \mathfrak{A}/\theta \text{ quasiverbal} \Rightarrow \mathfrak{A}/\theta \text{ characteristic,}$$

we see that for any class $\mathfrak{K}$ and any $\mathfrak{A} \in \mathfrak{K}$,

$$\mathfrak{A} \text{ transcharacteristic in } \mathfrak{K} \Rightarrow \mathfrak{A} \text{ transquasiverbal in } \mathfrak{K} \Rightarrow$$

$$\Rightarrow \mathfrak{A} \text{ transverbal in } \mathfrak{K};$$

an analogous observation holds for classes.

A subclass $\mathcal{L} \subseteq \mathcal{K}$ is called *homomorphically closed* in $\mathcal{K}$ iff for every factor system $\mathfrak{A}/\theta$ of a system $\mathfrak{A} \in \mathcal{L}$,

$$\mathfrak{A}/\theta \in \mathcal{K} \Rightarrow \mathfrak{A}/\theta \in \mathcal{L} .$$

From these definitions it follows that every homomorphically closed subclass of a transverbal class is transverbal.

Suppose the system $\mathfrak{A}$ has a pole p, and $\mathcal{K}$ is a class of systems containing $\mathfrak{A}$. A subsystem $\mathfrak{C}$ of $\mathfrak{A}$ is called *$\mathcal{K}$-normal in* $\mathfrak{A}$ iff there exists a factor system $\mathfrak{A}/\sigma$ in $\mathcal{K}$ such that $p\sigma = \mathfrak{C}$. Normal subsystems corresponding to verbal (quasiverbal) (characteristic) factor systems are called *verbal (quasiverbal) (characteristic) subsystems.* Since among the equivalence classes in $\mathfrak{A}/\sigma$ only the class $p\sigma$ is a subsystem: if the class $\mathcal{K}$ is hereditary, then the polarized system $\mathfrak{A}$ is transverbal (transquasiverbal) (transcharacteristic) iff every verbal (quasiverbal) (characteristic) subsystem of each verbal (etc.) subsystem of $\mathfrak{A}$ is $\mathcal{K}$-normal in $\mathfrak{A}$.

A characteristic subgroup of a normal divisor of any group is a normal divisor of that group. Therefore, a variety of groups is transcharacteristic; hence, it is transverbal.

It becomes clear that the variety of all associative rings is not transverbal. For let $\mathcal{R}_6$ be the variety of all associative rings in which the identity $\boldsymbol{x_1x_2x_3x_4x_5x_6} \approx \boldsymbol{0}$ is valid, and let $\mathcal{A}$ be the variety of all commutative and associative rings. Let $\mathfrak{A}$ be the $\mathcal{R}_6$-free ring with free generators a_1, a_2. The $\mathcal{A}$-verbal ideal $\mathfrak{J}$ in $\mathfrak{A}$ consists of integral linear combinations of members of the form $\mathfrak{a}(a_1a_2 - a_2a_1)\mathfrak{b}$ where $\mathfrak{a}, \mathfrak{b}$ are monomials. The $\mathcal{A}$-verbal ideal of the ring $\mathfrak{J}$ consists of elements that can be written in the form

$$(m+n)\,ca_ic - ma_ic^2 - nc^2a_i \;,$$

where $i = 1, 2$; $m, n = 0, \pm 1, \pm 2, \ldots$; $c = a_1a_2 - a_2a_1$. These cannot form an ideal in $\mathfrak{A}$.

Theorem 8: *In a hereditary and trans(quasi)verbal class $\mathcal{K}$ of algebraic systems, any hereditary subclass $\mathcal{A}$ and any absolute sub(quasi)varieties $\mathcal{B}$, $\mathcal{C}$ satisfy the law of associativity:*

$$\mathcal{A} \circ \mathcal{B}\mathcal{C} = \mathcal{A}\mathcal{B} \circ \mathcal{C} .$$

First of all, it is easy to see that for any hereditary class $\mathcal{K}$, any subclasses $\mathcal{A}, \mathcal{B}, \mathcal{C} \subseteq \mathcal{K}$ satisfy

$$\mathcal{A} \circ \mathcal{B}\mathcal{C} \subseteq \mathcal{A}\mathcal{B} \circ \mathcal{C} .$$

Indeed, suppose $\mathfrak{A} \in \mathcal{A} \cdot \mathcal{B}\mathcal{C}$; thus $\mathfrak{A} \in \mathcal{K}$ and for some appropriate factor system $\mathfrak{A}/\theta$,

$$\mathfrak{A}/\theta \in \mathcal{B}\mathcal{C}, \qquad a\theta \in \mathcal{K} \Rightarrow a\theta \in \mathcal{A} \quad (a \in \mathfrak{A}) . \tag{22}$$

That $\mathfrak{A}/\theta \in \mathcal{B}\mathcal{C}$ just means that $\mathfrak{A}/\theta \in \mathcal{K}$ and for some factor system $(\mathfrak{A}/\theta)/\rho \cong \mathfrak{A}/\rho$ we have $\mathfrak{A}/\rho \in \mathcal{C}$ and

$$(a\theta)\rho = a\rho/\theta \in \mathcal{K} \Rightarrow a\rho/\theta \in \mathcal{B} , \tag{23}$$

for every $a \in \mathfrak{A}$.

Suppose $a\rho \in \mathcal{K}$. We want to show that $a\rho \in \mathcal{A}\mathcal{B}$. Since $a\rho$ is a subsystem of $\mathfrak{A}$, its image $a\rho/\theta$ in $\mathfrak{A}/\theta$ is a subsystem of $\mathfrak{A}/\theta$. Inasmuch as $\mathfrak{A}/\theta \in \mathcal{K}$ and the class $\mathcal{K}$ is hereditary, we learn that $a\rho/\theta \in \mathcal{K}$. In view of (23), this shows us that $a\rho/\theta \in \mathcal{B}$; (22) can then be applied, giving $a\rho \in \mathcal{A}\mathcal{B}$.

The proof of the converse inclusion

$$\mathcal{A} \cdot \mathcal{B}\mathcal{C} \supseteq \mathcal{A}\mathcal{B} \cdot \mathcal{C}$$

will be based on the strict hypotheses of Theorem 8. Let $\mathfrak{A}$ be a system in $\mathcal{A}\mathcal{B} \cdot \mathcal{C}$ and let $\mathfrak{A}/\gamma$ be the $\mathcal{C}$-replica of $\mathfrak{A}$. Since the class $\mathcal{A}\mathcal{B}$is hereditary and $\mathfrak{A} \in \mathcal{A}\mathcal{B} \cdot \mathcal{C}$, by Corollary 4 in § 1 we know

$$a_\mu\gamma \in \mathcal{A}\mathcal{B} \qquad (\mu \in \mathrm{M}) , \tag{24}$$

where $a_\mu\gamma$ $(\mu \in \mathrm{M})$ are those equivalence classes in $\mathfrak{A}/\gamma$ that are $\mathcal{K}$-subsystems of $\mathfrak{A}$. Let $a_\mu\gamma/\beta_\mu$ be the $\mathcal{B}$-replica of the system $a_\mu\gamma$ $(\mu \in \mathrm{M})$. Because $\mathcal{A}$ is hereditary, it follows from (24) that

$$x\beta_\mu \in \mathcal{K} \Rightarrow x\beta_\mu \in \mathcal{A} \qquad (x \in a_\mu\gamma,\ \mu \in \mathrm{M}) . \tag{25}$$

The class $\mathcal{K}$ and the factor system $\mathfrak{A}/\gamma$ are trans(quasi)verbal and $\mathcal{B}$ is a (quasi)variety, so all the $\mathcal{B}$-replicas $a_\mu\gamma/\beta_\mu$ must have a common extension $\mathfrak{A}/\beta \in \mathcal{K}$ with $\mathfrak{A}/\beta \leqslant \mathfrak{A}/\gamma$ and $a_\mu\gamma/\beta = a_\mu\gamma/\beta_\mu$ $(\mu \in \mathrm{M})$; moreover,

$$x \in \mathfrak{A} \ \&\ x\beta \in \mathcal{K} \Rightarrow (\exists \mu \in \mathrm{M})(x \in a_\mu\gamma \ \&\ x\beta = x\beta_\mu) . \tag{26}$$

In view of (26) and (25) the desired relation $\mathfrak{A} \in \mathcal{A} \cdot \mathcal{B}\mathcal{C}$ will be proved if we manage to show that $\mathfrak{A}/\beta \in \mathcal{B}\mathcal{C}$. But

$$(\mathfrak{A}/\beta)/\gamma \cong \mathfrak{A}/\gamma \in \mathcal{C} ,$$

so we only need the implication

$$x\gamma/\beta = (x\beta)\gamma \in \mathfrak{X} \Rightarrow (x\beta)\gamma \in \mathfrak{B} \quad (x \in \mathfrak{A}) .$$

Suppose $x\gamma/\beta \in \mathfrak{X}$. Since $x\gamma$ is the full preimage of the subsystem $x\gamma/\beta \subseteq \mathfrak{A}/\beta \in \mathfrak{X}$ under the homomorphism $\mathfrak{A} \to \mathfrak{A}/\beta$, $x\gamma$ must be a subsystem of the system $\mathfrak{A} \in \mathfrak{X}$. The hereditariness of $\mathfrak{X}$ implies that $x\gamma \in \mathfrak{X}$, and thus $x\gamma = a_\nu\gamma \in \mathcal{A}\mathfrak{B}$ for some $\nu \in \mathrm{M}$, whence

$$x\gamma/\beta = a_\nu\gamma/\beta_\nu \in \mathfrak{B} \ . \blacksquare$$

Corollary 7: *For every transquasiverbal quasivariety $\mathfrak{X}$ of algebraic systems, the groupoid $\mathfrak{G}_\mathrm{Q}(\mathfrak{X})$ is a semigroup.* ■

Combining Theorems 7 and 8 gives us

Corollary 8: *For every polarized transverbal variety $\mathfrak{X}$ of algebras with commuting congruences, the partial groupoid $\mathfrak{G}_\mathrm{I}(\mathfrak{X})$ is a semigroup with zero $\mathfrak{X}$ and identity $\mathcal{E}$.* ■

As mentioned already, every variety of groups is polarized and transverbal, and congruence relations on any group commute. Hence, from Corollary 8 we derive (cf. H. Neumann [115]): *for every variety $\mathfrak{X}$ of groups, $\mathfrak{G}_\mathrm{I}(\mathfrak{X})$ is a semigroup with zero and identity.* ■

According to a theorem of the Neumanns [114] and Šmel'kin [153], if $\mathcal{G}$ is the variety of all groups, then $\mathfrak{G}_\mathrm{I}(\mathcal{G})$ is a free semigroup with zero and identity elements. For other group varieties $\mathfrak{X} \subseteq \mathcal{G}$, the structure of $\mathfrak{G}_\mathrm{I}(\mathfrak{X})$ can be more complex.

§4. Additional observations

Besides the operation of multiplication of classes we can introduce an operation of (right) division of classes that in a certain sense is inverse to multiplication. Suppose $\mathcal{C}$ is an arbitrary subclass and $\mathfrak{B}$ is a replete subclass of a class $\mathfrak{X}$. The *$\mathfrak{X}$-quotient* $\mathcal{C} /_\mathfrak{X} \mathfrak{B}$ is the class consisting of every $\mathfrak{X}$-system that can be embedded isomorphically in a $\mathfrak{X}$-system that is an equivalence class belonging to the $\mathfrak{B}$-replica of a $\mathcal{C}$-system.

From this definition it is immeidately seen that the subclass $\mathcal{C}/_\mathfrak{X}\mathfrak{B}$ is always hereditary. Furthermore, if $\mathcal{L}$ is a hereditary subclass of $\mathfrak{X}$ and $\mathcal{C},\mathfrak{B}$ are subclasses of $\mathcal{L}$, then $\mathcal{C} /_\mathfrak{X} \mathfrak{B} = \mathcal{C} /_\mathcal{L} \mathfrak{B}$. If $\mathfrak{X}$ is fixed, we shall write $\mathcal{C} / \mathfrak{B}$ instead of $\mathcal{C} /_\mathfrak{X} \mathfrak{B}$.

Theorem 9: *Suppose $\mathcal{K}$ is a hereditary class of algebraic systems and $\mathcal{A}$, $\mathcal{B}$, $\mathcal{C}$ are hereditary subclasses with $\mathcal{B}$ replete. Then*

$$\mathcal{C}/\mathcal{B} \subseteq \mathcal{C} \subseteq \mathcal{C}/\mathcal{B} * \mathcal{B}\,, \tag{27}$$

$$(\mathcal{A} * \mathcal{B})/\mathcal{B} \subseteq \mathcal{A}\,, \tag{28}$$

$$(\mathcal{A} * \mathcal{B})/\mathcal{B} * \mathcal{B} = \mathcal{A} * \mathcal{B}\,. \tag{29}$$

According to the definition, from $\mathfrak{A} \in \mathcal{C}/\mathcal{B}$ it follows that $\mathfrak{A}$ is a $\mathcal{K}$-subsystem of some $\mathcal{C}$-system. Since $\mathcal{C}$ is hereditary, $\mathfrak{A}$ is a $\mathcal{C}$-system. So $\mathcal{C}/\mathcal{B} \subseteq \mathcal{C}$. In addition, if $\mathfrak{C} \in \mathcal{C}$ and $\mathfrak{C}/\theta$ is the $\mathcal{B}$-replica of $\mathfrak{C}$, then $\mathfrak{C}/\theta \in \mathcal{B}$ and

$$c \in \mathfrak{C} \;\&\; c\theta \in \mathcal{K} \Rightarrow c\theta \in \mathcal{C}/\mathcal{B}$$

by definition, i.e., $\mathfrak{C} \in \mathcal{C}/\mathcal{B} * \mathcal{B}$. This proves (27).

We turn to the proof of (28). Suppose $\mathfrak{A} \in (\mathcal{A} * \mathcal{B})/\mathcal{B}$. Then $\mathfrak{A}$ is a $\mathcal{K}$-subsystem of some system of the form $c\theta$, where $c \in \mathfrak{C} \in \mathcal{A} * \mathcal{B}$ (and $\mathfrak{C}/\theta$ is the $\mathcal{B}$-replica of $\mathfrak{C}$). But the classes $\mathcal{K}$, $\mathcal{A}$ are hereditary, so $c\theta \in \mathcal{A}$, and thus $\mathfrak{A} \in \mathcal{A}$. Hence, (28) is true.

By (27)

$$\mathcal{A} * \mathcal{B} \;\dot{\subseteq}\; (\mathcal{A} * \mathcal{B})/\mathcal{B} * \mathcal{B}\,,$$

and by multiplying both sides of (28) by $\mathcal{B}$ we get

$$(\mathcal{A} * \mathcal{B})/\mathcal{B} * \mathcal{B} \subseteq \mathcal{A} * \mathcal{B}\;;$$

therefore, (29) is also true. ■

For an arbitrary class $\mathcal{L}$ of systems, let $\mathcal{L}^{\forall}$ be the smallest $\forall$-class including $\mathcal{L}$, i.e., $\mathcal{L}^{\forall} = \mathbf{K}\,\forall(\mathcal{L})$. It can also be defined as

$$\mathcal{L}^{\forall} = \mathbf{SU}(\mathcal{L})\,,$$

where $\mathbf{S}(\mathcal{K})$, $\mathbf{U}(\mathcal{K})$ respectively denote the class of all subsystems of $\mathcal{K}$-systems and the class of all isomorphs of ultraproducts of $\mathcal{K}$-systems.

Let $\mathcal{L}^{\mathrm{I}}$, $\mathcal{L}^{\mathrm{Q}}$ be the variety and quasivariety generated by $\mathcal{L}$.

In §1 we saw a variety $\mathcal{K}$ with a subvariety whose square is not a variety. In this example, $\mathfrak{G}_{\mathrm{I}}(\mathcal{K})$ is only a partial groupoid. However, by trading the operation $\underset{\mathcal{X}}{*}$ for the operation $\underset{\mathcal{X}}{*}\mathrm{I}$, defined thus:

$$\mathcal{A}\underset{\mathcal{X}}{*}\mathrm{I}\mathcal{B} = (\mathcal{A}\underset{\mathcal{X}}{*}\mathcal{B})^{\mathrm{I}},$$

we turn the collection of all subvarieties of any variety $\mathcal{X}$ into a groupoid $\mathfrak{G}_{\mathrm{I}}^{\#}(\mathcal{X})$ with a totally defined operation $*\,\mathrm{I}$.

In a similar fashion we specialize the division / of classes to division operations /I, /Q by defining

$$\mathcal{C}/\Delta\mathcal{B} = (\mathcal{C}/\mathcal{B})^{\Delta} \quad (\Delta = \mathrm{I}, \mathrm{Q}).$$

For $\Delta = \mathrm{I}, \mathrm{Q}$ and for any Δ-class $\mathcal{X}$, in addition to the groupoid $\mathfrak{G}_{\Delta}^{\#}(\mathcal{X})$ we have a quasigroup

$$\mathfrak{Q}_{\Delta}(\mathcal{X}) = \langle \mathfrak{L}_{\Delta}(\mathcal{X}); *\Delta, /\Delta \rangle \quad (*\mathrm{Q} = *).$$

Since whenever $\mathcal{A}$ belongs to $\mathfrak{Q}_{\Delta}(\mathcal{X})$ it equals $\mathcal{A}^{\Delta}$, from (27) and (28) we get

$$\mathcal{C}/\Delta\mathcal{B} \subseteq \mathcal{C} \subseteq \mathcal{C}/\Delta\mathcal{B}*\Delta\mathcal{B},$$

$$(\mathcal{A}*\Delta\mathcal{B})/\Delta\mathcal{B} \subseteq \mathcal{A}, \tag{30}$$

whence

$$(\mathcal{A}*\Delta\mathcal{B})/\Delta\mathcal{B}*\Delta\mathcal{B} = \mathcal{A}*\Delta\mathcal{B}.$$

We note that inclusion in (30) can be strengthened to equality iff the corresponding groupoid $\mathfrak{G}_{\Delta}^{\#}(\mathcal{X})$ satisfies the law of right cancellation.

An algebraic system $\mathfrak{A}$ is called *$\mathcal{L}$-decomposable* iff there exists a factor system $\mathfrak{A}/\sigma$ with at least two elements that belongs to the class $\mathcal{L}$. In the contrary case, $\mathfrak{A}$ is *$\mathcal{L}$-indecomposable* [160]. In particular, if the class $\mathcal{L}$ is replete, then a system $\mathfrak{A}$ is $\mathcal{L}$-indecomposable iff its $\mathcal{L}$-replica has only one member. A system $\mathfrak{A}$ is called *$\mathcal{L}$-attainable* iff it has a factor system belonging to $\mathcal{L}$ whose every member (an equivalence class) that is a subsystem of $\mathfrak{A}$ is $\mathcal{L}$-indecomposable. If $\mathcal{L}$ is replete, then $\mathfrak{A}$ is $\mathcal{L}$-attainable iff its $\mathcal{L}$-replica has the property described.

Finally, a subclass $\mathcal{A}$ of a hereditary class $\mathcal{X}$ is called *attainable* in $\mathcal{X}$ iff every $\mathcal{X}$-system is $\mathcal{A}$-attainable.

Theorem 10: *If a replete subclass $\mathcal{A}$ of a hereditary class $\mathcal{X}$ is attainable in*

$\mathcal{K}$, *then for any hereditary class* $\mathfrak{X} \subseteq \mathcal{K}$,

$$(\mathfrak{X} * \mathcal{A}) * \mathcal{A} = \mathfrak{X} * \mathcal{A} \ . \tag{31}$$

Indeed, every replete class contains a unit system. Therefore, $\mathcal{E} \subseteq \mathcal{A}$, and by (7)

$$\mathfrak{X}\mathcal{A} \subseteq \mathfrak{X}\mathcal{A} \cdot \mathcal{A} \ .$$

Conversely, suppose $\mathfrak{A} \in \mathfrak{X}\mathcal{A} \cdot \mathcal{A}$ and let $\mathfrak{A}/\theta$ be the $\mathcal{A}$-replica of the system $\mathfrak{A}$; then we have by Corollary 4:

$$a\theta \in \mathcal{K} \Rightarrow a\theta \in \mathfrak{X}\mathcal{A} \quad (a \in \mathfrak{A}) \ .$$

Suppose $a\theta \in \mathcal{K}$. Then letting $a\theta/\sigma$ be the $\mathcal{A}$-replica of the system $a\theta$, from the relation $a\theta \in \mathfrak{X}\mathcal{A}$ we derive

$$x\sigma \in \mathcal{K} \Rightarrow x\sigma \in \mathfrak{X} \quad (x \in a\theta) \ .$$

By hypothesis the system $\mathfrak{A}$ is $\mathcal{A}$-attainable, so $a\theta/\sigma$ has but one element; thus $a\theta = a\sigma \in \mathfrak{X}$.

In other words,

$$a\theta \in \mathcal{K} \Rightarrow a\theta \in \mathfrak{X} \quad (a \in \mathfrak{A}) \ ,$$

and $\mathfrak{A} \in \mathfrak{X}\mathcal{A}$. ∎

In an arbitrary groupoid an element a satisfying $a^2 = a$ is called an *idempotent,* and an element a satisfying $(xa)a = xa$ for all x is called a *right idempotent.* It is clear that in a groupoid with a left identity every right idempotent is an idempotent. In an associative groupoid (semigroup) the converse is true: every idempotent is a right idempotent.

We observe that *if a replete subclass* $\mathcal{A}$ *is attainable in a hereditary class* $\mathcal{K}$ *and the* $\mathcal{A}$*-replica* $\mathfrak{A}/\theta$ *of an arbitrary* $\mathcal{K}$*-system* $\mathfrak{A}$ *satisfies the requirement*

$$(\forall a \in \mathfrak{A})\,(a\theta \in \mathcal{K} \Rightarrow a\theta \in \mathcal{E}_0) \Rightarrow \mathfrak{A}/\theta \cong \mathfrak{A} \ , \tag{32}$$

where $\mathcal{E}_0$ *is the class of one-element* $\mathcal{K}$*-systems. Then* $\mathcal{A} \underset{\mathcal{K}}{*} \mathcal{A} = \mathcal{A}$.

For (32) guarantees that $\mathcal{E}_0\mathcal{A} = \mathcal{A}$, so by taking $\mathfrak{X} = \mathcal{E}_0$ in (31), we get $\mathcal{A}^2 = \mathcal{A}$. ∎

E.g., the condition (32) is certainly satisfied by any $\mathcal{K}$-factor $\mathfrak{A}/\theta$ for any algebra $\mathfrak{A}$ in any hereditary class $\mathcal{K}$ of quasigroups, groups, or rings. Therefore, every replete, attainable subclass $\mathcal{A}$ of such a class is equal to its square.

Theorem 11: *If a sub(quasi)variety $\mathcal{A}$ of a hereditary and trans(quasi)verbal class $\mathcal{K}$ satisfies $\mathcal{A} \underset{\mathcal{K}}{*} \mathcal{A} = \mathcal{A}$, then $\mathcal{A}$ is attainable in $\mathcal{K}$.*

Let $\mathfrak{A}/\theta$ be the $\mathcal{A}$-replica of an arbitrary $\mathcal{K}$-system $\mathfrak{A}$. For $a \in \mathfrak{A}$, if the equivalence class $a\theta$ is a subsystem, let $a\theta/\sigma_a$ be its $\mathcal{A}$-replica. Since the system $\mathfrak{A}$ is trans(quasi)verbal, there is a factor system $\mathfrak{A}/\sigma \leqslant \mathfrak{A}/\theta$ such that

$$a\theta \in \mathcal{K} \Rightarrow a\theta/\sigma_a = a\theta/\sigma \qquad (a \in \mathfrak{A}) .$$

Inasmuch as $(\mathfrak{A}/\sigma)/\theta \in \mathcal{A}$ and for $a \in \mathfrak{A}$,

$$(a\sigma)\theta \in \mathcal{K} \Rightarrow (a\sigma)\theta \in \mathcal{A},$$

we see that $\mathfrak{A}/\sigma \in \mathcal{A} \cdot \mathcal{A}$, but $\mathcal{A}^2 = \mathcal{A}$, so $\mathfrak{A}/\sigma \in \mathcal{A}$.

Since $\mathfrak{A}/\theta$ is the $\mathcal{A}$-replica of $\mathfrak{A}$, we have $\mathfrak{A}/\theta \leqslant \mathfrak{A}/\sigma$, which together with $\mathfrak{A}/\sigma \leqslant \mathfrak{A}/\theta$ gives $\mathfrak{A}/\theta = \mathfrak{A}/\sigma$; hence,

$$a\theta \in \mathcal{K} \Rightarrow a\theta/\sigma = a\theta/\theta \in \mathcal{E}_0 \qquad (a \in \mathfrak{A}) .$$

This means the class $\mathcal{A}$ is attainable in $\mathcal{K}$. ∎

As already noted, every group is a transquasiverbal algebra, each of whose factors satisfies (32). Therefore, in any quasivariety $\mathcal{K}$ of groups, the attainable quasivarieties are just the idempotents of the semigroup $\mathfrak{G}_Q(\mathcal{K})$.

A few examples. According to the Neumann-Šmel'kin theorem, for the variety $\mathcal{G}$ of all groups, the semigroup $\mathfrak{G}_I(\mathcal{G})$ is a free semigroup with zero and identity. In such a semigroup the zero and identity are the only idempotent elements. Applying the remarks above, we see that the variety $\mathcal{G}$ has no nontrivial attainable subvarieties. Tamura [160] proved this assertion by other methods.

Let $\mathcal{N}_k$ be the variety of all k-step nilpotent groups. The structures of the lattice $\mathfrak{L}_Q(\mathcal{N}_1)$ (see [182]) and the lattices $\mathfrak{L}_I(\mathcal{N}_3)$ (see [125], [67]) are known explicitly. Each subvariety of the variety $\mathcal{N}_1$ of all abelian groups is definable in $\mathcal{N}_1$ by a single identity of the form

$$\boldsymbol{x}^m \approx \boldsymbol{1} \qquad (m = 0, 1, 2, \dots) ;$$

denote these subvarieties by $\mathcal{A}_m$ $(m \geqslant 0)$. Then we have

$$\mathcal{A}_m \underset{\mathfrak{N}_1}{*} \mathcal{A}_n = \mathcal{A}_{mn} \qquad (m, n = 0, 1, 2, \dots),$$

so the semigroup $\mathfrak{G}_{\mathrm{I}}(\mathfrak{N}_1)$ is isomorphic to the multiplicative semigroup of natural numbers.

The subvarieties of the variety $\mathfrak{N}_2$ are in 1–1 correspondence with the pairs of identities

$$x^m \approx 1, \quad (x^{-1}y^{-1}xy)^n \approx 1 \quad (n \mid m;\ m, n = 0, 1, 2, \dots)$$

that define them in $\mathfrak{N}_2$; let $[m, n]$ denote such a subvariety. A simple computation shows that

$$[r, s] \underset{\mathfrak{N}_2}{*} [m, n] = [rm, (\frac{rm}{(2, rm)}, r^2 n)] ,$$

where (u, v) is the greatest common divisor of the numbers u, v. From this formula we see that the semigroup $\mathfrak{G}_{\mathrm{I}}(\mathfrak{N}_2)$ is not commutative and does not obey either cancellation law.

As a last example, we consider the class $\mathfrak{K}_\Sigma$ of all algebras with a given signature Σ and take $\mathcal{P}$ to be the variety of algebras with this signature defined by the identities

$$f(x, x, \dots) \approx x \qquad (f \in \Sigma). \tag{33}$$

It is easy to convince ourselves that

$$\mathcal{P} \underset{\mathfrak{K}_\Sigma}{*} \mathcal{P} = \mathcal{P}.$$

For let $\mathfrak{A}/\theta$ be the $\mathcal{P}$-replica of an arbitrary algebra $\mathfrak{A} \in \mathcal{P}^2$. Then for elements $x_1, x_2, \dots$ chosen from an arbitrary congruence class $a\theta$, we see that

$$f(x_1, x_2, \dots)\,\theta = f^\theta(x_1\theta, x_2\theta, \dots) = f^\theta(a\theta, a\theta, \dots) = a\theta ,$$

where f is an arbitrary signature operation of $\mathfrak{A}$, and f^θ is the corresponding operation of $\mathfrak{A}/\theta$. Thus each member $a\theta$ of $\mathfrak{A}/\theta$ is a subalgebra of $\mathfrak{A}$. Since $\mathfrak{A}$ belongs to $\mathcal{P}^2$, $a\theta$ belongs to $\mathcal{P}$ for all $a \in \mathfrak{A}$. Thus the identities (33) are all valid in $\mathfrak{A}$, and $\mathfrak{A}$ itself belongs to $\mathcal{P}$.

It is easy to verify that if Σ contains a non-unary symbol, then for a suitable hereditary $\mathcal{X}$ we have $\mathcal{X}\mathcal{P} \cdot \mathcal{P} \neq \mathcal{X}\mathcal{P}$, so $\mathcal{P}$ is not attainable in $\mathfrak{K}_\Sigma$. If Σ contains only unary operation symbols, then the variety $\mathcal{P}$ is attainable in $\mathfrak{K}_\Sigma$. Indeed, a factor $\mathfrak{A}/\sigma$ of an arbitrary $\mathfrak{K}_\Sigma$ algebra $\mathfrak{A}$ belongs to $\mathcal{P}$ iff all the congruence classes $a\sigma$ ($a \in \mathfrak{A}$) are subalgebras; moreover, any partition of $\mathfrak{A}$ into disjoint subalgebras determines a factor algebra, and this factor belongs to $\mathcal{P}$. From this it follows that each congruence class belonging to the $\mathcal{P}$-replica of $\mathfrak{A}$ admits no proper partition into subalgebras and thus is $\mathcal{P}$-indecomposable.

NOTE

(1) It is necessary (and easy) to show $\mathcal{A} * \mathcal{A}$ is nonempty. In fact, every $\mathcal{A}$-algebra not isomorphic to $\mathfrak{C}$ belongs to $\mathcal{A} * \mathcal{A}$.

CHAPTER 33

UNIVERSALLY AXIOMATIZABLE SUBCLASSES OF LOCALLY FINITE CLASSES OF MODELS

A subclass $\mathcal{L}$ of a class $\mathcal{K}$ of models is said to be *universally axiomatizable in* $\mathcal{K}$ iff there exists a collection S of universal, prenex, closed formulas of first-order predicate logic (FOPL) such that $\mathcal{L}$ consists of just those $\mathcal{K}$-models in which all the closed formulas (or sentences, axioms) in S are true. If S can be chosen to be finite (independent), then the subclass $\mathcal{L}$ is said to be *finitely* (*independently*) $\forall$-axiomatizable in $\mathcal{K}$. We similarly define the notions of finite and independent Γ-axiomatizability for any other type Γ of FOPL sentence. This article indicates several simple tests for the finite or independent $\forall$-axiomatizability of subclasses of a locally finite class $\mathcal{K}$. E.g., it is shown in §3 that there are continuum many different universally axiomatizable subclasses of the class of all nonoriented graphs of degree $\leqslant 2$, and just as many $\forall$-axiomatizable subclasses of partially ordered sets of fixed dimension. The latter problem arose because the class of all linearly ordered sets and the class of all boolean algebras each have only a countable number of universally axiomatizable subclasses. In §1 we recall the well-known criterion – modified in form – of Tarski and Łoś for $\forall$-axiomatizability.

§1. Conditions for universal axiomatizability

Suppose we are given a universal sentence

$$\Phi = (\forall \boldsymbol{x}_1 \dots \boldsymbol{x}_n)\, \Psi(\boldsymbol{x}_1, \dots, \boldsymbol{x}_n)$$

of signature $\Sigma = \{\boldsymbol{P}_0, \dots, \boldsymbol{P}_s\}$; we assume that Σ contains only predicate symbols and that $\boldsymbol{P}_0$ is the equality sign $\approx$. The subformula $\Psi(\boldsymbol{x}_1, \dots, \boldsymbol{x}_n)$ of the formula Φ is assumed to be a $\{\&, \vee, \neg\}$-polynomial in atomic formulas of the form $\boldsymbol{P}_j(\boldsymbol{x}_{\alpha_1}, \dots, \boldsymbol{x}_{\alpha_{nj}})$ $(j = 0, \dots, s)$. Suppose $\Omega_1, \dots, \Omega_r$ are all the possible atomic formulas of this form. Then Ψ is equivalent to the conjunction of certain disjunctions, each of the form

$$\sigma_1\Omega_1 \vee \sigma_2\Omega_2 \vee \ldots \vee \sigma_r\Omega_r \quad (\sigma_i = \neg\,,\ \Lambda\,(\text{the empty string}))\ ;$$

therefore, the sentence Φ is equivalent to a conjunction of sentences of the form

$$\Phi_k = \neg(\exists \boldsymbol{x}_1 \ldots \boldsymbol{x}_n)(\tau_1\Omega_1 \,\&\, \ldots \,\&\, \tau_r\Omega_r) \quad (\tau_i = \neg\,\sigma_i)\,. \tag{1}$$

Up to isomorphism there is no more than one model $\mathfrak{M}_k$ with signature Σ, the elements of which can be designated by the symbols $\boldsymbol{x}_1, \ldots, \boldsymbol{x}_n$ so that the formula

$$\tau_1\Omega_1 \,\&\, \ldots \,\&\, \tau_r\Omega_r \tag{2}$$

is satisfied in $\mathfrak{M}_k$. If there is no such model, then the sentence (1) is identically valid, and this conjunct can be dropped from Φ. Suppose the model $\mathfrak{M}_k$ exists. Then the truth of (1) in some model $\mathfrak{M}$ with signature $\Sigma_0 \supseteq \Sigma$ means that the model $\mathfrak{A}_k$ is not (isomorphically) embeddable in the model $\mathfrak{M}$ (symbolically, $\mathfrak{A}_k$ $\overline{emb}$ $\mathfrak{A}$); the original sentence Φ, equivalent to $\Phi_1 \,\&\, \ldots \,\&\, \Phi_t$, is true in $\mathfrak{A}$ iff none of the models $\mathfrak{M}_1, \ldots, \mathfrak{M}_t$ is embeddable in $\mathfrak{A}$.

Conversely, suppose we are given a finite collection of finite models $\mathfrak{M}_1, \ldots, \mathfrak{M}_t$ with some finite signature Σ. Let n be the greatest of the powers of these models. Then the elements of any model $\mathfrak{M}_k$ can be designated by the symbols $\boldsymbol{x}_1, \ldots, \boldsymbol{x}_n$ (possibly with repetitions). For each k we construct a diagram (2) of $\mathfrak{M}_k$ in terms of the $\boldsymbol{x}_i$. Taking the conjunction of the corresponding sentences (1) leads us to a universal sentence Φ whose truth in a model is equivalent to the nonembeddability in that model of the models $\mathfrak{M}_1, \ldots, \mathfrak{M}_t$. We thus arrive at the following proposition, various versions of which are well known (cf. [163]).

Theorem 1.1: *For every universal sentence Φ of finite signature Σ that is not identically valid, there exists a finite sequence of models $\mathfrak{M}_1, \ldots, \mathfrak{M}_t$ with signature Σ such that the truth of Φ in a model $\mathfrak{A}$ with signature $\Sigma_0 \supseteq \Sigma$ is equivalent to the nonembeddability in $\mathfrak{A}$ of every model $\mathfrak{M}_1, \ldots, \mathfrak{M}_t$. Conversely, if $\mathfrak{M}_1, \ldots, \mathfrak{M}_t$ are models with finite signature Σ whose powers do not exceed the finite number n, then there is a universal sentence Φ with n quantifiers whose truth in an arbitrary model $\mathfrak{A}$ with signature $\Sigma_1 \supseteq \Sigma$ is equivalent to the nonembeddability in $\mathfrak{A}$ of each and every of the models $\mathfrak{M}_1, \ldots, \mathfrak{M}_t$.* ■

This theorem speaks of truth in an arbitrary model with given signature. In applying it, not to all models with this signature, but to the members of some special class $\mathcal{K}$, we run into the unpleasantness that those $\mathfrak{M}_1, \ldots, \mathfrak{M}_t$

mentioned in the theorem might not belong to $\mathcal{K}$. To avoid this hangup, we introduce the following

Definition: *A class $\mathcal{L}$ of models is called locally $\mathcal{K}$-finite iff in every $\mathcal{L}$-model every finite set of elements is included in a finite $\mathcal{K}$-submodel. If $\mathcal{L}=\mathcal{K}$, we say $\mathcal{L}$ is locally finite.*

If $\mathcal{K}$ is some class of models with signature Σ and S is a set of sentences of signature Σ, then $\mathcal{K}(S)$ denotes the class of all $\mathcal{K}$-models in which every sentence in S is valid. A class of the form $\mathcal{K}(S)$ where S is a set of $\forall$-sentences is called a *universally axiomatizable subclass* of the class $\mathcal{K}$, or a *universal* of $\mathcal{K}$-models. From Theorem 1.1 we immediately deduce the following modification of the well-known Tarski-Łoś theorem [163]:

Theorem 1.2: *For every subuniversal $\mathcal{A}$ of a class $\mathcal{K}$ of models with signature Σ, there exists a collection $\{\mathfrak{M}_\xi : \xi \in \Xi\}$ of finite models $\mathfrak{M}_\xi$ with finite signatures $\Sigma_\xi \subseteq \Sigma$ such that a $\mathcal{K}$-model $\mathfrak{A}$ belongs to $\mathcal{A}$ iff none of the models $\mathfrak{M}_\xi$ is embeddable in $\mathfrak{A}$. For any such collection $\{\mathfrak{M}_\xi : \xi \in \Xi\}$, the subclass $\mathcal{A}$ so defined is a subuniversal of $\mathcal{K}$.* ■

In general, the models mentioned in this formulation cannot be taken to belong to $\mathcal{K}$, for it could be that $\mathcal{K}$ contains no finite models at all. But if $\mathcal{K}$ contains "sufficiently many" finite models, such a provision can be made.

Theorem 1.3: *For every subuniversal $\mathcal{A}$ of a locally finite class $\mathcal{K}$ of models, there exists a collection $\{\mathfrak{N}_\zeta : \zeta \in Z\}$ of finite $\mathcal{K}$-models such that a $\mathcal{K}$-model $\mathfrak{A}$ belongs to $\mathcal{A}$ iff none of the $\mathfrak{N}_\zeta$ is embeddable in $\mathfrak{A}$.*

Suppose $\{\mathfrak{M}_\xi : \xi \in \Xi\}$ is the set of finite models whose existence is asserted by Theorem 1.2. We shall show that the demands of Theorem 1.3 are satisfied by the set $\{\mathfrak{N}_\zeta : \zeta \in Z\}$ of those finite $\mathcal{K}$-models in which at least one of the models $\mathfrak{M}_\xi$ is embeddable. Indeed, if the $\mathcal{K}$-model $\mathfrak{A}$ belongs to $\mathcal{A}$, then no $\mathfrak{M}_\xi$ is embeddable in it; hence, neither can any $\mathfrak{N}_\zeta$ be embedded in $\mathfrak{A}$. Conversely, if $\mathfrak{A} \in \mathcal{K}$ but $\mathfrak{A} \notin \mathcal{A}$, then for some $\nu \in \Xi$ there is an embedding $\varphi: \mathfrak{M}_\nu \to \mathfrak{A}$. By the local finiteness of $\mathcal{K}$ we can find a finite $\mathcal{K}$-submodel $\mathfrak{N}$ of $\mathfrak{A}$ that includes the set $\varphi(\mathfrak{M}_\nu)$. Thus for some $\eta \in Z$, $\mathfrak{N} = \mathfrak{N}_\eta$, so not all the members of $\{\mathfrak{N}_\zeta : \zeta \in Z\}$ fail to be embeddable in $\mathfrak{A}$. ■

§2. Independent axiomatizability

Consider some type Γ of FOPL sentences, e.g., identities (I-sentences), quasidentities (Q-sentences), or universal sentences ($\forall$-sentences). A subclass $\mathcal{L}$ of some class $\mathcal{K}$ of models with signature Σ is called a *Γ-axiomatizable* *subclass* of $\mathcal{K}$ (or a Γ-*subclass in* $\mathcal{K}$) iff there exists a set S of Γ-sentences of signature Σ such that $\mathcal{L} = \mathcal{K}(\mathrm{S})$. A subclass $\mathcal{L} \subseteq \mathcal{K}$ is *finitely* Γ-*axiomatizable* in $\mathcal{K}$ iff there is a finite set S of Γ-sentences for which $\mathcal{L} = \mathcal{K}(\mathrm{S})$. We let **E** denote the class of all FOPL sentences. A Γ-subclass of the total class of all models with a given signature is called simply a Γ-*axiomatizable class.* E-axiomatizable classes are simply called *(first-order) axiomatizable.* Note the compactness theorem implies that if the base class $\mathcal{K}$ is axiomatizable, then any Γ-subclass is finitely Γ-axiomatizable iff it is finitely axiomatizable.

A set S of sentences of signature Σ is called *independent* relative to a class $\mathcal{K}$ of models with this signature iff $\mathrm{S}_1 \subset \mathrm{S}$ implies $\mathcal{K}(\mathrm{S}_1) \supset \mathcal{K}(\mathrm{S})$. A subclass $\mathcal{L} \subseteq \mathcal{K}$ is called *independently* Γ-*axiomatizable* in $\mathcal{K}$ iff $\mathcal{L} = \mathcal{K}(\mathrm{S})$ for some $\mathcal{K}$-independent system S of Γ-sentences. It is clear that every finitely Γ-axiomatizable subclass is also independently Γ-axiomatizable.

We present an example of a Q-subclass that is not independently Q-axiomatizable.

Let $\mathcal{K}$ be the class of algebras with signature $\{\boldsymbol{0}, \boldsymbol{f}, \boldsymbol{g}\}$ (where $\boldsymbol{0}$ is an individual constant symbol, and $\boldsymbol{f}, \boldsymbol{g}$ are unary function symbols) defined by the quasidentities (the universal quantifiers have been dropped for clarity):

$$f(0) \approx 0\,,$$

$$f(x) \approx x \rightarrow x \approx 0\,,$$

$$f(g(x)) \approx g(f(x)) \approx x\,, \tag{3}$$

$$f(x) \approx f(y) \rightarrow x \approx y\,. \tag{4}$$

Let $\mathcal{L}$ be the subclass characterized in $\mathcal{K}$ be the quasidentities

$$f^n(x) \approx x \rightarrow x \approx 0 \quad (n = 1, 2, 3, \ldots)\,. \tag{5}$$

We shall show that $\mathcal{L}$ is not independently Q-axiomatizable in $\mathcal{K}$.

We introduce the abbreviations

$$x^0 = x, \quad x^k = f^k(x)\,, \quad x^{-k} = g^k(x) \quad (k = 1, 2, \ldots)\,.$$

By using these and (3), we can rewrite an arbitrary quasidentity of signature $\{\boldsymbol{0}, \boldsymbol{f}, \boldsymbol{g}\}$ in the variables $x_1, \ldots, x_l$ in the equivalent (in $\mathcal{K}$) form

$$u_1^{m_1} \approx v_1^{n_1} \,\&\, \ldots \,\&\, u_r^{m_r} \approx v_r^{n_r} \rightarrow u^m \approx v^n\,, \tag{6}$$

where $\boldsymbol{u}, \boldsymbol{u}_i, \boldsymbol{v}, \boldsymbol{v}_i \in \{\boldsymbol{0}, \boldsymbol{x}_1, \ldots, \boldsymbol{x}_l\}$ and m, m_i, n, n_i are integers. From (4) it follows that

$$\boldsymbol{x}^j \approx \boldsymbol{y}^k \leftrightarrow \boldsymbol{x} \approx \boldsymbol{y}^{j-k}$$

is valid in $\mathcal{K}$. Therefore, the quasidentity (6) can be reduced over $\mathcal{K}$ to the form

$$\boldsymbol{x}^{p_1} \approx \boldsymbol{x} \,\&\, \ldots \,\&\, \boldsymbol{x}^{p_s} \approx \boldsymbol{x} \,\&\, \boldsymbol{y}^{q_1} \approx \boldsymbol{y} \,\&\, \ldots \,\& \\ \& \, \boldsymbol{y}^{q_t} \approx \boldsymbol{y} \rightarrow \boldsymbol{x}^p \approx \boldsymbol{z}\ , \tag{7}$$

where $\boldsymbol{z}$ is $\boldsymbol{x}, \boldsymbol{y}$, or $\boldsymbol{0}$. By substituting $\boldsymbol{0}$ for $\boldsymbol{x}$ and $\boldsymbol{y}$ separately, we see that (7) is equivalent over $\mathcal{K}$ either to a quasidentity of the form

$$\boldsymbol{x}^{m_1} \approx \boldsymbol{x} \,\&\, \ldots \,\&\, \boldsymbol{x}^{m_k} \approx \boldsymbol{x} \rightarrow \boldsymbol{x}^m \approx \boldsymbol{x}\ , \tag{8}$$

or to one or two quasidentities of the form

$$\boldsymbol{x}^{m_1} \approx \boldsymbol{x} \,\&\, \ldots \,\&\, \boldsymbol{x}^{m_k} \approx \boldsymbol{x} \rightarrow \boldsymbol{x} \approx \boldsymbol{0}\,. \tag{9}$$

But in $\mathcal{K}$ we have the equivalence

$$\boldsymbol{x}^{m_1} \approx \boldsymbol{x} \,\&\, \ldots \,\&\, \boldsymbol{x}^{m_k} \approx \boldsymbol{x} \leftrightarrow \boldsymbol{x}^d \approx \boldsymbol{x}\,,$$

where d is the greatest common divisor of the numbers $m_1, \ldots, m_k$. Therefore, (8) and (9) are respectively equivalent in $\mathcal{K}$ to quasidentities of the form

$$\boldsymbol{x}^d \approx \boldsymbol{x} \rightarrow \boldsymbol{x}^m \approx \boldsymbol{x}\,, \quad \boldsymbol{x}^d \approx \boldsymbol{x} \rightarrow \boldsymbol{x} \approx \boldsymbol{0}\,.$$

Thus the matter reduces to the question: can't the class $\mathcal{L}$ be characterized in $\mathcal{K}$ by a $\mathcal{K}$-independent system $\{\Phi_1, \Phi_2, \ldots\}$, each axiom Φ_i having the form ([1])

$$(\boldsymbol{x}^{m_1} \approx \boldsymbol{x} \rightarrow \boldsymbol{x} \approx \boldsymbol{0}) \,\&\, \ldots \,\&\, (\boldsymbol{x}^{m_s} \approx \boldsymbol{x} \rightarrow \boldsymbol{x} \approx \boldsymbol{0}) \,\& \\ \& \, (\boldsymbol{x}^{p_1} \approx \boldsymbol{x} \rightarrow \boldsymbol{x}^{q_1} \approx \boldsymbol{x}) \,\&\, \ldots \,\&\, (\boldsymbol{x}^{p_t} \approx \boldsymbol{x} \rightarrow \boldsymbol{x}^{q_t} \approx \boldsymbol{x})? \tag{10}$$

The meaning of (10) is easy to picture. Every algebra $\mathfrak{A}$ in the class $\mathcal{K}$ is the union of minimal subalgebras of certain of the following three forms: a cycle of length 1 whose only element is 0; a cycle $\{x_0, x_1, ..., x_{n-1}\}$ of finite length n with

$$f(x_i) = x_{i+1}, \quad g(x_{i+1}) = x_i, \quad (i = 0, ..., n-2)$$

$$f(x_{n-1}) = x_0, \quad g(x_0) = x_{n-1};$$

an infinite cycle $\{..., x_{-1}, x_0, x_1, ...\}$, where

$$f(x_{i-1}) = x_i = g(x_{i+1}) \quad (i \text{ an integer}).$$

Any set of disjoint cycles of these forms that contains a single cycle of length 1 determines a $\mathfrak{X}$-algebra up to isomorphism.

Clearly, the quasidentity $\boldsymbol{x}^m \approx \boldsymbol{x} \rightarrow \boldsymbol{x} \approx \boldsymbol{0}$ is valid in a $\mathfrak{X}$-algebra $\mathfrak{A}$ iff $\mathfrak{A}$ includes no cycle whose length is a divisor of $m > 1$. On the other hand, since

$$\boldsymbol{x}^p \approx \boldsymbol{x} \;\&\; \boldsymbol{x}^q \approx \boldsymbol{x} \leftrightarrow \boldsymbol{x}^d \approx \boldsymbol{x} \quad (d = \gcd(p, q)),$$

is valid in $\mathfrak{X}$, the quasidentity

$$\boldsymbol{x}^p \approx \boldsymbol{x} \rightarrow \boldsymbol{x}^q \approx \boldsymbol{x}$$

is equivalent in $\mathfrak{X}$ to the quasidentity

$$\boldsymbol{x}^{ld} \approx \boldsymbol{x} \rightarrow \boldsymbol{x}^d \approx \boldsymbol{x} \quad (ld = p), \tag{11}$$

the truth of which in the algebra $\mathfrak{A}$ is equivalent to the total absence from $\mathfrak{A}$ of cycles whose lengths are greater than 1, divide ld, and are distinct from d ([2]).

Thus each sentence Φ_i asserts in the algebra $\mathfrak{A}$ that it includes no cycles of certain lengths $\alpha_1, ..., \alpha_s > 1$. By (5) the $\mathfrak{X}$-algebras in the class $\mathfrak{L}$ have no finite cycles besides $\{0\}$. Therefore, among the axioms $\Phi_1, \Phi_2, ...$ there must be an axiom Φ_j asserting that $\mathfrak{A}$ has no cycles of length $a = (\alpha_1 \,...\, \alpha_s)^2$ (so $j \neq i$). This can happen only if the conjunction (10) for Φ_j contains a member of at least one of the forms

$$\boldsymbol{x}^{ma} \approx \boldsymbol{x} \rightarrow \boldsymbol{x} \approx \boldsymbol{0}, \qquad \boldsymbol{x}^{ma} \approx \boldsymbol{x} \rightarrow \boldsymbol{x}^q \approx \boldsymbol{x},$$

the latter reduced as in (11). If the first appears, the axiom Φ_j also asserts the absence from $\mathfrak{A}$ of cycles of lengths $\alpha_1, ..., \alpha_s$ (as well as others), i.e.,

Φ_j implies Φ_i in $\mathcal{K}$: the system $\{\Phi_1, \Phi_2, \ldots\}$ is not $\mathcal{K}$-independent. If only the second occurs: for Φ_j not to imply Φ_i, q must be equal to one of the numbers $\alpha_1, \ldots, \alpha_s$, say α_1. Thus the axiom Φ_j asserts that $\mathfrak{A}$ lacks certain lengths from among $\alpha_2, \ldots, \alpha_s, \beta_1, \ldots, \beta_t$. Repeating the argument shows that among the axioms $\Phi_1, \Phi_2, \ldots$ there has to be a sentence Φ_k whose truth in $\mathfrak{A}$ implies the lack of cycles of length

$$b = (\alpha_1 \alpha_2 \ldots \alpha_s \beta_1 \ldots \beta_t)^2 .$$

We can pass directly to the case when Φ_k contains a conjunct of the form

$$x^{nb} \approx x \rightarrow x^r \approx x ,$$

and r is some member of $\{\alpha_1, \ldots, \alpha_s\}$. But if $r = \alpha_1$, then $\Phi_k \Rightarrow_{\mathcal{K}} \Phi_j$, and if $r \neq \alpha_1$, $\{\Phi_k, \Phi_i\} \Rightarrow_{\mathcal{K}} \Phi_j$; either way, this contradicts the $\mathcal{K}$-independence of $\{\Phi_1, \Phi_2, \ldots\}$.

Thus the quasivariety $\mathcal{L}$ cannot be independently Q-axiomatized in $\mathcal{K}$. The following general proposition shows that $\mathcal{L}$ is not even independently $\forall$-axiomatizable in $\mathcal{K}$.

Theorem 2.1: *Suppose the class $\mathcal{K}$ of models and its subclass $\mathcal{L}$ are multiplicatively closed and contain a unit model (a one-element model, all of whose basic predicates are true). If $\mathcal{L}$ is independently $\forall$-axiomatizable in $\mathcal{K}$, then it is independently Q-axiomatizable in $\mathcal{K}$.*

Suppose $\{\Phi_1, \Phi_2, \ldots\}$ is a system of $\forall$-sentences, independent relative to $\mathcal{K}$, that defines $\mathcal{L}$ in $\mathcal{K}$. Since $\mathcal{K}$ is multiplicatively closed, each axiom Φ_i is equivalent in $\mathcal{K}$ to a conjunction of simple Horn sentences $\Psi_{i1}, \ldots, \Psi_{is_i}$. Since these sentences have to be valid in the unit model in $\mathcal{K}$, none can be a purely negative disjunction; hence, they can be viewed as Q-sentences. Thus, $\mathcal{L}$ is characterized in $\mathcal{K}$ by the union of the groups

$$\{\Psi_{11}, \ldots, \Psi_{1s_1}\}, \{\Psi_{21}, \ldots, \Psi_{2s_2}\}, \ldots$$

of Q-axioms; moreover, each group is $\mathcal{K}$-independent from the union of the remaining groups. We now check each axiom Ψ_{ij} successively and throw out any that is a consequence of the axioms remaining at the time of checking. Since each of the groups is independent and finite, the axioms remaining at the completion of this procedure ([3]) form the desired $\mathcal{K}$-independent system of Q-sentences characterizing $\mathcal{L}$ in $\mathcal{K}$. ■

Theorem 2.2: *Every* $\forall$*-subclass* $\mathcal{L}$ *of a locally finite class* $\mathcal{K}$ *of models with finite signature is independently* $\forall$*-axiomatizable in* $\mathcal{K}$.

According to Theorem 1.3, $\mathcal{L}$ consists of those $\mathcal{K}$-models in which none of the fixed finite models $\mathfrak{N}_1, \mathfrak{N}_2, \ldots$ can be embedded; we can assume $\mathfrak{N}_1, \mathfrak{N}_2, \ldots$ are pairwise nonisomorphic. Suppose $\mathfrak{N}_{\alpha_1}, \mathfrak{N}_{\alpha_2}, \ldots$ are those models among $\mathfrak{N}_1, \mathfrak{N}_2, \ldots$ in which no other model in this sequence can be embedded. Clearly, $\mathcal{L}$ consists of those $\mathcal{K}$-models in which none of finite models $\mathfrak{N}_{\alpha_1}, \mathfrak{N}_{\alpha_2}, \ldots$ can be embedded. Let Υ_i be an $\forall$-sentence expressing the nonembedability of $\mathfrak{N}_i$. We claim the system $\{\Upsilon_{\alpha_1}, \Upsilon_{\alpha_2}, \ldots\}$ defines $\mathcal{L}$ in $\mathcal{K}$ and is $\mathcal{K}$-independent. E.g., the system $\{\Upsilon_{\alpha_2}, \Upsilon_{\alpha_3}, \ldots\}$ defines a subclass $\mathcal{L}_1$ in $\mathcal{K}$ that contains the model $\mathfrak{N}_{\alpha_1}$ since none of the models $\mathfrak{N}_{\alpha_1}, \mathfrak{N}_{\alpha_3}, \ldots$ can be embedded in this model. But $\mathfrak{N}_{\alpha_1} \notin \mathcal{L}$, so $\mathcal{L}_1 \neq \mathcal{L}$. ∎

Theorem 2.3: *In order that every subuniversal of a locally finite class* $\mathcal{K}$ *of models with a finite signature be finitely axiomatizable in* $\mathcal{K}$, *it is necessary and sufficient that there exist no infinite system* $\{\mathfrak{N}_\iota : \iota \in I\}$ *of finite* $\mathcal{K}$*-models, none of which is embeddable in any other.*

For suppose $\{\mathfrak{N}_\iota : \iota \in I\}$ is an infinite set of finite $\mathcal{K}$-models, none of which can be embedded in another. Let Υ_ι be a universal axiom expressing the nonembeddability of $\mathfrak{N}_\iota$. From the argument above it follows that the system $\{\Upsilon_\iota : \iota \in I\}$ is independent relative to $\mathcal{K}$. Therefore, distinct subsystems of this system determine distinct subuniversals in $\mathcal{K}$, and thus the power of the set of all subuniversals of $\mathcal{K}$ is equal to the power of the continuum. The set of finitely axiomatizable subuniversals of $\mathcal{K}$ has no more than countable power. Consequently, there are subuniversals in $\mathcal{K}$ that are not finitely axiomatizable.

Conversely, suppose the subuniversal $\mathcal{A}$ is not finitely axiomatizable in $\mathcal{K}$. According to Theorem 1.3, there is a set $\{\mathfrak{N}_\iota : \iota \in I\}$ of pairwise nonembeddable finite $\mathcal{K}$-models such that a $\mathcal{K}$-model $\mathfrak{A}$ belongs to $\mathcal{A}$ iff none of the models $\mathfrak{N}_\iota$ is embeddable in $\mathfrak{A}$. If the set I were finite, the $\forall$-subclass would be defined in $\mathcal{K}$ by the finite axiom system $\{\Upsilon_\iota : \iota \in I\}$ (where Υ_ι says $\mathfrak{N}_\iota$ is not embeddable), but this is impossible. ∎

§3. Graphs of finite degree

As an example we consider the universal $\mathcal{P}_r$ of all partially ordered sets (**po**-sets) whose dimensions do not exceed the given number r (see [11]). How many different subuniversals does the universal $\mathcal{P}_r$ have?

Theorem 3.1: *In the universal* $\mathcal{P}_r$ $(r \geqslant 2)$ *there are subuniversals that cannot be finitely axiomitized; thus* $\mathcal{P}_r$ *has continuum many distinct subuniversals.*

According to Theorem 2.3, we have only to indicate an infinite sequence of finite **po**-sets of dimension 2 such that none is embeddable in any other. The sequence

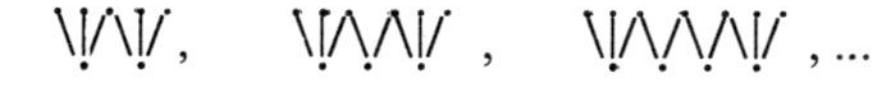

obviously has the desired property. Indeed, under an embedding of one graph in another – even when they are viewed as certain **po**-sets – the degree of a vertex cannot be decreased. In each of the graphs in the suggested sequence there are only two vertices of degree 3. So if an embedding were possible, these vertices would have to be mapped onto one another. The broken lines joining them would also have to coincide under the embedding, but this is impossible inasmuch as the length of this line differs from graph to graph. ∎

The reasoning is also valid when the sequence displayed is regarded as a set of nonoriented graphs, rather than **po**-sets. Therefore, *the universal of nonoriented graphs of degree* $\leqslant r$ $(r \geqslant 3)$ *has continuum many different sub-universals.* ∎

By considering the sequence of regular polygons with increasing numbers of vertices, we readily convince ourselves that the universal class of nonoriented graphs of degree $\leqslant 2$ also includes continuum many subuniversals.

It is clear that the universal of all linearly ordered sets has but a countable set of subuniversals. A subtler example of such a universal is the class ([4]) of convergent **po**-sets, which is characterized by the axioms for partial order plus these two:

$$(xyz)(x \leqslant z \,\&\, y \leqslant z \rightarrow x \leqslant y \vee y \leqslant x)\,,$$

$$(xy)(\exists z)(z \leqslant x \;\&\; z \leqslant y)\,.$$

Finite convergent **po**-sets are just finite trees. According to a theorem of Kruskal [79]), there is no infinite sequence of pairwise nonembeddable finite trees. By Theorem 2.3, this means that every subuniversal of convergent **po**-sets is finitely axiomatizable, so there are only a countable number of such subuniversals.

§4. Uniformly locally finite classes

Refining the notion of local finiteness guides us to the following definition: a class $\mathfrak{L}$ of models is called *uniformly locally* $\mathfrak{X}$*-finite* ($\mathfrak{X}$**-ulf**) iff there exists a function $\lambda\colon N \rightarrow N$ $(N = \{1, 2, \ldots\})$ such that for any $\mathfrak{L}$-model $\mathfrak{A}$, any $m \in N$, and any elements $a_1, \ldots, a_m$ in $\mathfrak{A}$, the elements $a_1, \ldots, a_m$ are contained in a $\mathcal{K}$-submodel $\mathfrak{A}' \subseteq \mathfrak{A}$ whose power does not exceed $\lambda(m)$. A class $\mathcal{L}$ is called *uniformly locally finite* (**ulf**) iff it is $\mathcal{L}$-**ulf**.

Obviously, every $\mathcal{L}$-**ulf** class is locally $\mathcal{K}$-finite ($\mathcal{K}$-**lf**). The converse is not generally true, but the following theorem holds:

Theorem 4.1: *Suppose $\mathcal{K}$ is an arbitrary class of models, and $\mathcal{L}$ is an axiomatizable class, both with finite signature Σ. If $\mathcal{L}$ is locally $\mathcal{K}$-finite, then it is uniformly locally $\mathcal{K}$-finite.*

Suppose $\mathcal{L}$ is not uniformly locally $\mathcal{K}$-finite. This means there is a positive integer m such that for any number $n \geqslant m$ we can find an $\mathcal{L}$-model $\mathfrak{A}$ and elements $a_1, \ldots, a_m \in \mathfrak{A}$ such that no submodel $\mathfrak{A}' \subseteq \mathfrak{A}$ that contains $a_1, \ldots, a_m$ and has no more than n elements can be a $\mathcal{K}$-model. In order to express this property by means of a **FOPL** formula, we introduce individual symbols $\boldsymbol{x}_1, \ldots, \boldsymbol{x}_n$ and let $\Delta_1^n, \Delta_2^n, \ldots$ be all possible diagrams of all possible models with signature Σ and no more than n elements, taking $\boldsymbol{x}_1, \ldots, \boldsymbol{x}_n$ to designate the elements in all possible combinations. Suppose $\Delta_1^n, \ldots, \Delta_{s_n}^n$ are all those diagrams that correspond to models not belonging to $\mathcal{K}$. Now the property above can be formulated as: for any $n \geqslant m$, there exists an $\mathcal{L}$-model $\mathfrak{A}_n$ in which the sentences

$$(\exists \boldsymbol{x}_1 \ldots \boldsymbol{x}_m)(\forall \boldsymbol{x}_{m+1} \ldots \boldsymbol{x}_n)(\Delta_1^n \vee \ldots \vee \Delta_{s_n}^n) \quad (^5) \tag{12}$$

is valid.

We supplement the signature Σ with individual symbols $\boldsymbol{a}_1, \ldots, \boldsymbol{a}_m$ and denote by $\mathcal{K}^*, \mathcal{L}^*$ the classes of those models with the new signature $\Sigma^* = \Sigma \cup \{\boldsymbol{a}_1, \ldots, \boldsymbol{a}_m\}$ that are obtained respectively from $\mathcal{K}$- and $\mathcal{L}$-models by supplementing them with arbitrary distinguished elements as values for the symbols $\boldsymbol{a}_1, \ldots, \boldsymbol{a}_m$. By hypothesis the class $\mathcal{L}$ is characterized by some system S of FOPL sentences. The class $\mathcal{L}^*$ is defined by the same system **S**. By the assumed property of $\mathcal{L}$, there is a model $\mathfrak{A}_n^* \in \mathcal{L}^*$ in which the sentence

$$\Phi_n = (\forall \boldsymbol{x}_{m+1} \ldots \boldsymbol{x}_n)\,(\Delta_1^n(\boldsymbol{a}_1, \ldots, \boldsymbol{a}_m, \boldsymbol{x}_{m+1}, \ldots, \boldsymbol{x}_n) \vee \ldots$$

$$\ldots \vee \Delta_{s_n}^n(\boldsymbol{a}_1, \ldots, \boldsymbol{a}_m, \boldsymbol{x}_{m+1}, \ldots, \boldsymbol{x}_n))$$

is true. This means the system $\mathrm{S} \cup \{\Phi_n\}$ is consistent. From the sense of the sentence Φ_n it follows that if Φ_{n+k} is true in some model $\mathfrak{D}^*$ with signature Σ^*, then Φ_n is also true in $\mathfrak{D}^*$. Thus every finite part of the system $\mathrm{S} \cup \{\Phi_m, \Phi_{m+1}, \ldots\}$ is consistent. By the compactness principle, this whole system is consistent. Let $\mathfrak{B}^*$ be a model satisfying this infinite system. Then $\mathfrak{B}^*$ belongs to $\mathfrak{L}^*$ and satisfies the sentences Φ_n $(n \geqslant m)$; the latter means that for $n \geqslant m$ the set $\{a_1, ..., a_m\}$ is included in no $\mathfrak{K}^*$-submodel of $\mathfrak{B}^*$ of power not greater than n. In other words, the set $\{a_1, ..., a_m\}$ is included in no finite $\mathfrak{K}$-submodel of $\mathfrak{L}$, but this contradicts the supposed local $\mathfrak{K}$-finiteness of $\mathfrak{L}$. ∎

The condition that Σ be finite in Theorem 4.1 cannot be dropped. We give an example of an **lf** universal of algebras with infinite signature that is not **ulf**.

Example: Let the signature Σ consist of the unary function symbols $f_1, f_2, ...$. Let $\mathfrak{L}$ be the universal of algebras with signature Σ defined by the axioms

$$f_i(x) \not\approx x \rightarrow f_j(x) \approx y \quad (i \neq j; i, j \in N),$$

$$f^{\pi_i}(x) \approx x \quad (i \in N),$$

where π_i is the ith prime number. Then $\mathfrak{L}$ is **lf**, but not **ulf**.

Let $a_1, ..., a_m$ be elements of an $\mathfrak{L}$-algebra $\mathfrak{A}$. If the sentences $(x)(f_i(x) \approx x)$ $(i \in N)$ are true in $\mathfrak{A}$, then the set $\{a_1, ..., a_m\}$ is itself a finite $\mathfrak{L}$-subalgebra of $\mathfrak{A}$ containing the given elements. Suppose, rather, that there is an element $a \in \mathfrak{A}$ and an index $i \in N$ such that $f_i(a) \neq a$. Then for $j \neq i$, $(x)(f_j(x) \approx x)$ is valid in $\mathfrak{A}$; hence, the set

$$\{a_1, f_i(a_1), ..., f_i^{\pi_i - 1}(a_1), ..., a_m, f_i(a_m), ..., f_i^{\pi_i - 1}(a_m)\}$$

is a finite subalgebra containing the elements $a_1, ..., a_m$. Thus the class $\mathfrak{L}$ is **lf**.

For $i \in N$ we construct an $\mathfrak{L}$-algebra $\mathfrak{B}_i$ by taking $b_1, ..., b_{\pi_i}$ as its elements and defining the operations by the equations

$$f_j(b_k) = b_k \quad (j \neq i; k = 1, ..., \pi_i),$$

$$f_i(b_l) = b_{l+1} \quad (l = 1, ..., \pi_i - 1),$$

$$f_i(b_{\pi_i}) = b_1 .$$

Each element of an algebra $\mathfrak{B}_i$ generates the whole algebra. It follows that $\mathfrak{L}$ cannot be **ulf**. ∎

Remark: *Every* **lf** *quasivariety* $\mathfrak{L}$ *of algebras with any signature is* **ulf**.

Indeed, for any $m \in N$, $\mathfrak{L}$ contains a free algebra $\mathfrak{F}_m$ with m free generators. Because $\mathfrak{L}$ is locally finite, the power $|\mathfrak{F}_m|$ of $\mathfrak{F}_m$ is finite. Suppose $a_1, ..., a_m$ are in an arbitrary $\mathfrak{L}$-algebra $\mathfrak{A}$. By mapping the free generators of

$\mathfrak{F}_m$ onto $a_1, ..., a_m$, we obtain a homomorphism $\psi: \mathfrak{F}_m \to \mathfrak{A}$; the subalgebra $\psi(\mathfrak{F}_m) \subseteq \mathfrak{A}$ belongs to $\mathcal{L}$ and contains $a_1, ..., a_m$. Therefore, we can take $\lambda(m) = |\mathfrak{F}_m|$. ∎

Using the notion of uniform local finiteness, we can formulate the following appendix to Theorem 1.3:

Theorem 4.2: *For every finitely $\forall$-axiomatizable subclass $\mathcal{L}$ of a uniformly locally finite class $\mathcal{K}$ of models with finite signature Σ, there exists a finite set $\mathcal{N}$ of finite $\mathcal{K}$-models such that a $\mathcal{K}$-model $\mathfrak{A}$ belongs to $\mathcal{L}$ iff no member of $\mathcal{N}$ can be embedded in $\mathfrak{A}$.*

According to Theorem 1.1, if a subclass $\mathcal{L} \subseteq \mathcal{K}$ is finitely $\forall$-axiomatizable in $\mathcal{K}$, then there is a finite set $\{\mathfrak{M}_1, ..., \mathfrak{M}_t\}$ (possibly empty) of finite models with signature Σ (not necessarily members of $\mathcal{K}$) that has the property indicated in Theorem 4.2. Let m be the greatest of the powers of the $\mathfrak{M}_i$, and let λ be the function guaranteed by the uniform local finiteness of $\mathcal{K}$. Since Σ is finite, there are but a finite number of pairwise nonisomorphic $\mathcal{K}$-models whose powers do not exceed $\lambda(m)$. Let these be $\mathfrak{N}_1, ..., \mathfrak{N}_r$. Of these let $\mathfrak{N}_{\alpha_{11}}, ..., \mathfrak{N}_{\alpha_{ir_i}}$ be the models in which $\mathfrak{M}_i$ is embeddable ($i = 1, ..., t$). We claim the set

$$\mathcal{N} = \{\mathfrak{N}_{\alpha_{11}}, ..., \mathfrak{N}_{\alpha_{1r_1}}, ..., \mathfrak{N}_{\alpha_{t1}}, ..., \mathfrak{N}_{\alpha_{tr_t}}\}$$

of finite $\mathcal{K}$-models has the property required in the theorem. For suppose $\mathfrak{A} \in \mathcal{L}$. Then no model $\mathfrak{M}_i$ and *a fortiori* no model $\mathfrak{N}_{\alpha_{ij}}$ can be embedded in $\mathfrak{A}$. Conversely, suppose $\mathfrak{A} \in \mathcal{K}$, but $\mathfrak{A} \notin \mathcal{L}$. Then some model $\mathfrak{M}_k$ admits an embedding $\varphi: \mathfrak{M}_k \to \mathfrak{A}$. Since $|\varphi(\mathfrak{M}_k)| \leqslant m$, there is a $\mathcal{K}$-submodel $\mathfrak{A}' \subseteq \mathfrak{A}$ such that $|\mathfrak{A}'| \leqslant \lambda(m)$ and $\varphi(\mathfrak{M}_k) \subseteq \mathfrak{A}'$. Because $\mathfrak{A}' \in \mathcal{K}$ and $|\mathfrak{A}'| \leqslant \lambda(m)$ and $\mathfrak{M}_k$ *emb* $\mathfrak{A}'$, the model $\mathfrak{A}'$ must be isomorphic to some model $\mathfrak{N}_{\alpha_{kl}}$; hence, a member of $\mathcal{N}$ is embeddable in $\mathfrak{A}$. ∎

NOTES

(1) We can assume the exponents are positive. Zero exponents yield conjuncts either tautologous or clearly not valid in $\mathcal{L}$.

(2) Alas, the lengths must also not *divide* d; the author's proof not only breaks down at this point, but the quasivariety $\mathcal{L}$ actually has a $\mathcal{K}$-independent Q-axiomatization: let q_n be the product of the first n primes; then

$$\boldsymbol{x}^{\mathrm{lcm}(n,q_n)} \approx \boldsymbol{x} \to \boldsymbol{x}^{q_{n-1}} \approx \boldsymbol{x} \quad (n \geqslant 3),$$

$$\boldsymbol{x}^2 \approx \boldsymbol{x} \to \boldsymbol{x} \approx \boldsymbol{0}$$

are $\mathcal{K}$-independent axioms for $\mathcal{L}$ in $\mathcal{K}$. However, those algebras in $\mathcal{K}$, each of whose finite cycles has length not divisible by the square of any prime, form a subquasivariety of $\mathcal{K}$ that is not even $\mathcal{K}$-independently $\forall$-axiomatizable.

([3]) This procedure is inadequate. In fact, let I, J be two infinite, disjoint sets of prime numbers; let $\mathcal{K}$ be the quasivariety defined in the example preceding Theorem 2.1; let $\mathcal{L}$ be the class of all $\mathcal{K}$-algebras $\mathfrak{A}$ such that no finite cycle in $\mathfrak{A}$ has length divisible by any member of I or the square of any prime in J. Then $\mathcal{L}$ is a subquasivariety of $\mathcal{K}$ that is not $\mathcal{K}$-independently Q-axiomatizable, but $\mathcal{L}$ can be $\mathcal{K}$-independently characterized by sentences that are each the conjunction of two quasidentities.

([4]) This is not a universal class, but it *is* locally finite.

([5]) It would be more accurate, but not really necessary, to conjoin

$$x_1 \not\approx x_2 \ \&\ x_1 \not\approx x_3 \ \&\ \dots\ \&\ x_{m-1} \not\approx x_m$$

to the matrix in (12) – and in Φ_n below, changing x_i to a_i.

CHAPTER 34

PROBLEMS ON THE BORDER BETWEEN ALGEBRA AND LOGIC (*)

In this report I want to survey some results and problems in a mathematical discipline which has arisen in the last decades on the boundary between mathematical logic and classical abstract algebra, and which to date has no generally accepted name. It is most frequently called *model theory* or *universal algebra,* or sometimes *general algebra.*

The basic mathematical structures studied in this general algebra are *algebraic systems,* i.e., sequences consisting of a nonempty set and a certain number of operations and predicates of various finite arities defined on it. As a typical example of an algebraic system, we take an ordered ring $\langle A; -, \cdot; \leqslant: \upsilon\rangle$, consisting of the *base set* A of elements of the ring (also called its *carrier*), the symbols $-$, $\cdot$ for the binary operations of subtraction and multiplication, the symbol $\leqslant$ for the relation of order, and the mapping υ that associates with the symbols $-$, $\cdot$, $\leqslant$ those concrete operations and relations designated by these symbols in the given concrete ring. The collection of symbols $-$, $\cdot$, $\leqslant$ together with their arities 2, 2, 2 is called the *signature of an ordered ring.*

In the general case, the *signature* is a pair of nonintersecting sets Σ_f, Σ_P and a map α: $\Sigma_f \cup \Sigma_P \to N$ of their union into the set N of natural numbers 0, 1, 2, ... The elements of Σ_f are called the *function* (or *operation*) *signature symbols,* and the elements of Σ_P are called the *predicate signature symbols.* From now on, the signature and the set of all signature symbols will be denoted by the same letter Σ. The natural number αs is called the *arity* of the symbol $s \in \Sigma$.

A sequence

$$\mathfrak{A} = \langle A; \Sigma: \upsilon\rangle$$

is called an *algebraic system with signature* Σ iff A is a nonempty set and υ is a mapping that assigns to each $f \in \Sigma_f$ a function f^υ: $A^{\alpha f} \to A$ and to each

$\boldsymbol{P} \in \Sigma_{\mathrm{P}}$ a predicate $\boldsymbol{P}^{v} \subseteq A^{\alpha \boldsymbol{P}}$. The functions $\boldsymbol{f}^{v}$ (or f) and the predicates $\boldsymbol{P}^{v}$ (or P) are called the *values* designated by the signature symbols $\boldsymbol{f}, \boldsymbol{P}$ in the system $\mathfrak{A}$, and the map v is called the *valuation* of $\mathfrak{A}$. Mention of the valuation is usually dropped from the notation, and $\langle A; s\colon \boldsymbol{s} \in \Sigma \rangle$ is written instead of $\langle A; \Sigma\colon v \rangle$. The algebraic system $\mathfrak{A}$ is called an *algebra* when Σ_{P} is empty, and called a *model* when Σ_{f} is empty. $\mathfrak{A}$ is *finite* when its base A is finite.

In the usual way we define the concepts of subsystem of a given algebraic system and of isomorphic and homomorphic mappings of an algebraic system into another system with the same signature (cf. [22]). An arbitrary collection of algebraic systems with the same signature Σ is called a *class of systems with signature* Σ. A class of systems is called *abstract* iff together with each of its members it contains all systems isomorphic to it.

An inkling of the necessity for studying algebras with arbitrary signatures was seen at the close of the last century, but this idea underwent no real development for more than three decades. Instead, deep theories of particular classes of algebras – fields, rings, groups, lattices – were founded on the one hand; in mathematical logic, on the other hand, broad research was conducted on the simpler formal languages. In the late thirties it was noted that the unification of the ideas of algebraic systems and first-order languages permitted the formulation of propositions whose specializations to the classical systems (fields, groups) not only yielded nontrivial theorems already known in the theories of groups and fields, but also answered certain questions in group theory open at that time ([1]). Thus at the junction of classical abstract algebra and mathematical logic burgeoned a new discipline, general algebra; here, in contrast to classical algebra, a prominent position is held by problems on the bond between structural properties of classes of algebras and the properties of the formal languages which can be used to define the classes.

Research in general algebra reached full maturity in the postwar years. Particularly significant advances were completed in the late fifties and sixties. The creation of the theories of filtered products and complete classes comes to mind. Because detialed reviews of these theories have already appeared in the journals, I shall focus on other trends of research.

We recall a few more concepts. Suppose a signature Σ is given. Obeying the usual rules, we combine signature symbols, parentheses, comma, symbols $\boldsymbol{x}_1, \boldsymbol{x}_2, \ldots$ for individual variables, logical symbols &, $\vee$, $\neg$, $\rightarrow$, $\approx$, and quantifiers

$\forall \boldsymbol{x}_i$ – "for every element x_i in the carrier A of the system",

$\exists \boldsymbol{x}_i$ – "there exists an element $x_i \in A$ such that", to form certain finite strings of symbols called *formulas* of signature Σ of first-order predicate logic (FOPL). E.g., the strings

$$(\forall x_1)(\forall x_2)(x_1 + x_2 \approx x_2 + x_1), \tag{1}$$

$$(\forall x_1)(\exists x_2)(x_2 \leqslant x_1 \;\&\; x_2 \not\approx x_1).$$

– with the common abbreviations – are closed FOPL formulas (or sentences, axioms) of any signature that contains the symbols $+, \leqslant$.

If we are given an algebraic system $\mathfrak{A}$ with signature Σ and a FOPL sentence Φ of the same signature, then the semantic valuation of the signature symbols occurring in Φ determines a truth value for Φ in the system $\mathfrak{A}$ (cf. [161]). For example, if

$$\mathfrak{A} = \langle \{0, 1, 2, \ldots\}; +; \leqslant \rangle,$$

where the symbols $+, \leqslant$ have the usual arithmetic values, then the sentence (1) is true in $\mathfrak{A}$, but (2) is false.

Let E denote the class of all FOPL sentences, and let E_Σ be the set of FOPL sentences of signature Σ. In addition to the class E, we shall need several subclasses of sentences of special forms. Recall that formulas in whose notation appear function, individual variable, and punctuation symbols only are named *terms* (or *polynomials*) in these variables. E.g., if $+, \wedge$ are binary operation symbols, then the expressions

$$x + (x \wedge y), \quad (x + y) + (x + x)$$

are both terms of signature $\{+, \wedge\}$ in x, y (even though neither $\wedge$ nor y occurs in the second).

We introduce the following special classes of sentences:

I – the class of *identities*, i.e., sentences of the form

$$(\forall x_1) \ldots (\forall x_n) P(\mathfrak{a}_1, \ldots, \mathfrak{a}_r),$$

where r is the arity of P, which is either a predicate symbol or the equality sign, and $\mathfrak{a}_1, \ldots, \mathfrak{a}_r$ are terms in $x_1, \ldots, x_n$;

Q – the class of *quasidentities*, those sentences of the form

$$(\forall x_1) \ldots (\forall x_n)(P_1(\mathfrak{a}_1^1, \ldots \mathfrak{a}_{r_1}^1) \;\&\; \ldots$$

$$\ldots \;\&\; P_q(\mathfrak{a}_1^q, \ldots, \mathfrak{a}_{r_q}^q) \rightarrow P(\mathfrak{a}_1, \ldots, \mathfrak{a}_r)),$$

where the $\boldsymbol{P}_i, \boldsymbol{P}$ are predicate symbols or the equality sign, and the $\mathfrak{a}^i_j$, $\mathfrak{a}_k$ are terms in $\boldsymbol{x}_1, \dots, \boldsymbol{x}_n$;

$\forall$ – the class of *universal sentences,* i.e., those of the form

$$(\forall \boldsymbol{x}_1) \dots (\forall \boldsymbol{x}_n)\, \Psi ,$$

where Ψ is a FOPL formula without quantifiers;

$\forall\exists$ – the class of *Skolem sentences,* which have the form

$$(\forall \boldsymbol{x}_1) \dots (\forall \boldsymbol{x}_m)\, (\exists \boldsymbol{x}_{m+1}) \dots (\exists \boldsymbol{x}_n)\Psi ,$$

where Ψ is a quantifier-free FOPL formula;

D – the class of *diophantine sentences,* closed formulas of the form

$$(\exists \boldsymbol{x}_1) \dots (\exists \boldsymbol{x}_n)\, (\mathfrak{a}^1_1, \dots, \mathfrak{a}^1_{r_1})\, \& \dots \& \, \boldsymbol{P}_q(\mathfrak{a}^q_1, \dots, \mathfrak{a}^q_{r_q})) ,$$

where the $\boldsymbol{P}_i$ are predicate symbols or the equality sign, and the $\mathfrak{a}^i_j$ are terms in $\boldsymbol{x}_1, \dots, \boldsymbol{x}_n$.

Let Γ be a type of FOPL sentence, e.g., one of the classes I, Q, $\forall$, $\forall\exists$, D specified above. Let $\mathcal{K}$ be a class of algebraic systems with signature Σ. Then the Γ-*theory* of the class $\mathcal{K}$ is the set $\Gamma(\mathcal{K})$ of all those sentences from $\Gamma \cap \mathbf{E}_\Sigma$ that are true in every system in $\mathcal{K}$. On the other hand, for any class Γ of FOPL sentences, we let $\mathbf{K}_\Sigma(\Gamma)$ denote the class of all algebraic systems with signature Σ in which all Γ-sentences of signature Σ are true. In particular, $\mathbf{K}_\Sigma(\emptyset)$ is the *total class* of all algebraic systems with signature Σ (we denote it briefly by $\mathcal{K}_\Sigma$). Let $\mathcal{K}^{\text{fin}}$ be the class of all finite $\mathcal{K}$-systems.

The set $\mathbf{E}_\Sigma$ of FOPL sentences of signature Σ is a subset of the set W_Σ of all finite strings of symbols from among the signature symbols, parentheses, comma, the logical signs &, ..., $\forall$, $\exists$, and the symbol $\boldsymbol{x}$ (assuming $\boldsymbol{x}_n = (\boldsymbol{x}\boldsymbol{x} \dots \boldsymbol{x})$). That is to say, E_Σ is a subset of the set of *words* on the alphabet consisting of the symbols indicated. The collection of all words on a fixed alphabet bears the structure of an inductive algebra, which lets us define recursive and recursively enumerable sets of words, etc. For finite alphabets, such objects are well explored in the theory of algorithms, but recently, R. Peter, F. Schwenkel [147], and others studied their properties for infinite alphabets as well, so that now we can speak of the recursiveness or nonrecursiveness of theories of classes of algebraic systems with not only finite signatures, but also infinite ones. The following two programs thus naturally present themselves:

(1) to discover the algorithmic nature of the theory $\Gamma(\mathcal{K})$ for the most

important classes $\mathcal{K}$ of systems and the most interesting types Γ of sentences;

(2) to study the general algebraic properties of the classes $\mathbf{K}_\Sigma(\Gamma_1)$ $(\Gamma_1 \subseteq \Gamma)$ for the most interesting types Γ.

Classes of the form $\mathbf{K}_\Sigma(\Gamma_1)$ $(\Gamma_1 \subseteq \Gamma)$ are called Γ-*classes.* I-classes are called *varieties,* Q-classes *quasivarieties,* and $\forall$-classes *universally axiomatizable classes* (or *universals*) of algebraic systems.

Let us briefly review the new results and open problems contributing to these programs.

§1. The algorithmic nature of theories

§1.1. E-*theories and theories of total classes*

The question whether the theory $\mathbf{E}(\mathcal{K}_\Sigma)$ is recursive is known as the decision problem for first-order predicate logic with signature Σ. For sufficiently rich signatures, it was answered negatively by A. Church [19] in 1939. The question of recursiveness – or decision problem – for the theories $\Gamma(\mathcal{K}_\Sigma)$, $\Gamma(\mathcal{K}_\Sigma^{\text{fin}})$ for various types Γ and signatures Σ has attracted the attention of many authors, e.g., B.A. Trahtenbrot [168]. One of the more remarkable results along these lines was obtained by Wang Hao [184], who proved the nonrecursiveness of $\forall^1\exists^1\forall^1(\mathcal{K}_\Sigma)$ when Σ consists of a single binary and infinitely many unary predicate symbols ($\forall^1\exists^1\forall^1$-sentences, of course, have the form $(\forall x)(\exists y)(\forall z)\Psi$ with Ψ quantifier-free). Using Wang's methods, Ju. Š. Gurevič [50] last year obtained definitive results in this direction. To present these, we let

$$\Pi_1 = \{\forall^m\exists^n : m, n = 0, 1, 2, \ldots\},$$

$$\Pi_2 = \{\forall^m\exists^i\forall^n : i = 1, 2; \;\; m, n = 0, 1, 2, \ldots\},$$

$$\Upsilon = \{\boldsymbol{P} : \boldsymbol{P} \text{ is a unary predicate symbol}\}.$$

Let Π be any set of words on the alphabet $\{\forall, \exists\}$; then we take Γ^Π to be the set of all prenex FOPL sentences containing no occurrences of $\approx$ whose quantifier prefixes belong to Π, and we define $\tilde{\Gamma}^\Pi$ similarly, now admitting the equality sign. Let Σ be any signature containing no function symbols. Then Gurevič proved:

$$\begin{aligned} Rec\, \Gamma^{\Pi}(\mathcal{K}_{\Sigma}) &\Leftrightarrow Rec\, \tilde{\tilde{\Gamma}}^{\Pi}(\mathcal{K}_{\Sigma}) \\ &\Leftrightarrow Rec\, \Gamma^{\Pi}(\mathcal{K}_{\Sigma}^{\mathrm{fin}}) \\ &\Leftrightarrow Rec\, \tilde{\tilde{\Gamma}}^{\Pi}(\mathcal{K}_{\Sigma}^{\mathrm{fin}}) \\ &\Leftrightarrow \neg\, Creat\, \Gamma^{\Pi}(\mathcal{K}_{\Sigma}) \\ &\Leftrightarrow \Sigma \subseteq \Upsilon \vee \Pi \subseteq \Pi_1 \cup \Pi_2 \vee \\ &\quad \vee (Fin\, \Sigma\ \&\ Fin(\Pi \sim (\Pi_1 \cup \Pi_2))) , \end{aligned}$$

where *Rec M, Creat M, Fin M* respectively signify that the set *M* is recursive, creative, finite.

By tradition these investigations are counted as research in "pure logic". On the contrary, when $\mathcal{K}$ is a special class such as the class of groups, finite groups, etc., the decision problems for $\mathbf{E}(\mathcal{K})$, $\mathbf{Q}(\mathcal{K})$, $\mathbf{I}(\mathcal{K})$ more often involve the theory of the class $\mathcal{K}$ itself (group theory, the theory of finite groups, etc.). Early examples are the theorems that first-order arithmetic $\mathbf{E}(\langle N; +, \cdot\rangle)$ is not recursive (Rosser), but that $\mathbf{E}(\langle N; +\rangle)$ is recursive (Presburger).

The significance of this domain of inquiry because especially clear after A. Tarski [162] showed $\mathbf{E}(\langle C; +, \cdot\rangle)$ and $\mathbf{E}(\langle R; +, \cdot\rangle)$ to be recursive, and J. Robinson [134] showed the opposite for $\mathbf{E}(\langle Q; +, \cdot\rangle)$, where *C, R, Q* are the sets of complex, real, and rational numbers. The well-known book [166] of Tarski et al. summed up the early development of the new field. For many important classes $\mathcal{K}$, however, the algorithmic nature of the *elementary theory* $\mathbf{E}(\mathcal{K})$ was still unknown. During the following decade, the nonrecursiveness of the elementary theories of many classes was demonstrated; in particular, the elementary theory $\mathbf{E}(\mathcal{K}^{\mathrm{fin}})$ of the finite part of many a well-known class $\mathcal{K}$ proved not to be (recursively) decidable. Also found were a number of classes $\mathcal{K}$ for which the theory $\mathbf{E}(\mathcal{K})$ is recursive. Eršov et al. [33] have reviewed the results obtained through 1964.

In recent years, interest has sharply increased in theories of the form $\Gamma(\mathcal{K})$ for sundry types Γ. In this case, it is natural to ask not only whether a theory $\Gamma(\mathcal{K})$ is recursive, but also what its degree of unsolvability is. Another engaging question: for which $\mathcal{K}$, $\mathcal{L}$, Γ is $\Gamma(\mathcal{K})$ equal to $\Gamma(\mathcal{L})$?

To investigate the structure of theories a number of general methods have been developed. In particular, the workhorse for proving undecidability of theories is the method of interpretations. For establishing recursiveness, besides the direct method of eliminating quantifiers, A. Robinson's technique of model completeness [132] has become a valuable tool. The quest for coinciding elementary theories also can be implemented now with several techniques: shuttling, ultraproducts, or strategies. Nevertheless, the algorithmic natures of many important theories are still completely unknown.

I shall list below a series of important new results, obtained since the last Congress, and quote a few of the related open problems.

§**1.2.** *Number theory*

One of the most significant problems still unsolved remains *Hilbert's tenth problem:*

(A) *Is the theory* $\mathrm{D}(\langle N; -, \cdot\rangle)$ ([2]) *recursive?*

Also unanswered is the closely related question:

(B) *For every recursively enumerable set M of natural numbers, does there exist a polynomial* $F(x_1, ..., x_n)$ *with integer coefficients such that for all* $x \in N$,

$$x \in M \Leftrightarrow (\exists y_1 \ldots y_n \in N)(F(x, y_1, ..., y_n) = 0)\ ?$$

The following variant of Hilbert's problem is of interest:

(C) *Are there natural numbers s,* $t > 0$ *and polynomials* $F_{1i}(x), ..., F_{si}(x)$ $(i = 1, ..., t)$ *with integer coefficients such that those* $n \in N$ *for which the equation*

$$\sum_{i=1}^{t} \pm x_1^{F_{1i}(n)} x_2^{F_{2i}(n)} \ldots x_s^{F_{si}(n)} = 0$$

is solvable form a nonrecursive set? If they do exist, what are the least values for s, t, and the degrees of the F_{ki}*?*

Although these problems are still open, the similar issue of the existence of an algorithm for determining from the integral coefficients of an arbitrary polynomial $F(x_1, ..., x_m, y_1, ..., y_n)$ whether the exponential equation

$$F(x_1, ..., x_m, 2^{x_1}, ..., 2^{x_m}) = 0$$

is solvable has been settled negatively in the remarkable paper [24] of Davis et al. It remains to find the simplest form of polynomials F for which there is no solvability algorithm.

§**1.3.** *Field theory*

Somewhat akin to the Hilbert problem are the decision problems for elementary theories of classes of fields. Since the report to the present Congress by Ju. L. Eršov is devoted to this area, I shall limit myself to a few general remarks. For almost fifteen years, essentially only two classes of fields were known to have decidable elementary theories: the class of algebraic fields of fixed characteristic and the class of real closed fields. Meanwhile, the number of classes of fields known to have nonrecursive theories grew steadily. Finally in 1964–65 the elementary theories of the field of p-adic numbers and of certain other fields were shown to be recursive by Ax and Kochen [5], who used ultraproducts, and independently by Eršov [31], [32], who applied model completeness. We may regard these papers as laying the foundation for a theory of fields with decidable elementary theories. Using the subtle apparatus of model theory, these authors simultaneously proved several conjectures of Lang and Artin on forms.

The work on Hilbert's problem and the work of Ax, Kochen and Eršov is of special interest because it opens the way for model theory to "invade" classical number theory – and algebraic geometry, possibly.

In spite of the great advances in studying the elementary theories of classes of fields, many problems quite easy to formulate are still unsolved in the domain. Among these at present are the decision problems for the elementary theories of

(a) *the class of all finite fields (J. Robinson);*

(b) *the field of rational functions in the variables* $x_1, \ldots, x_m$ *over an arbitrary coefficient field (A.I. Mal'cev);*

(c) *the field of those complex numbers which can be constructed with ruler and compass, and its subfield of real numbers (A. Tarski).*

In number theory itself, theorems like Thue's theorem offer the interesting challenge of being made effective; model-theoretic methods may help meet it.

§**1.4.** *The theory of groups and semigroups*

The undecidability of the elementary theory of the class $\mathcal{G}$ of all groups was established by A. Tarski. The stronger result that $\mathrm{Q}(\mathcal{G})$ is not recursive was obtained by P.S. Novikov and W.W. Boone. From 1960 on, many classes of groups were seen to have undecidable elementary theories; among them are the class of all finite groups, the class $\mathcal{S}_k$ of all k-step solvable groups ($k \geqslant 2$), and so on. (3) Open questions:

(I) *For* $k \geqslant 2$, *is the theory* $\mathrm{Q}(\mathcal{S}_k)$ *recursive? (The answer is affirmative iff the word problem for finitely presented* $\mathcal{S}_k$*-groups is recursively solvable.)*

(II) *Are the theories* $\mathrm{D}(\mathfrak{F}_n)$, $\exists(\mathfrak{F}_n)$, $\mathrm{E}(\mathfrak{F}_n)$ *recursive?* ($\mathfrak{F}_n$ *is the free group of rank n for* $n \geqslant 2$.)

A.D. Taĭmanov recently showed that

$$\forall\exists\forall\exists(\mathfrak{F}_m) = \forall\exists\forall\exists(\mathfrak{F}_n) \qquad (m, n \geqslant 2)\,,$$

but the following assertions are still unproved:

(III) $\mathrm{E}(\mathfrak{F}_m) = \mathrm{E}(\mathfrak{F}_n)$ $(m, n \geqslant 2)$;

(IV) $\mathrm{E}(\mathfrak{G}) = \mathrm{E}(\mathfrak{G}')$ & $\mathrm{E}(\mathfrak{H}) = \mathrm{E}(\mathfrak{H}') \Rightarrow \mathrm{E}(\mathfrak{G} * \mathfrak{H}) = \mathrm{E}(\mathfrak{G}' * \mathfrak{H}')$,

where $\mathfrak{G}$, $\mathfrak{G}'$, $\mathfrak{H}$, $\mathfrak{H}'$ are arbitrary groups and $\mathfrak{G} * \mathfrak{H}$, $\mathfrak{G}' * \mathfrak{H}'$ are the corresponding free products.

Ju. I. Merzljakov has shown that $\mathfrak{F}_n$ has no nonabelian subgroups definable by positive formulas with one free variable. It is still unclear whether or not

(V) $\mathfrak{F}_n$ *has proper nonabelian formular subgroups* $(n \geqslant 2)$.

In 1949, W. Szmielew established the recursive decidability of the elementary theory of the class of all abelian groups. Ju. Š. Gurevič [49] proved in 1964 that the E-theory of the class of ordered abelian groups is also decidable; at the same time, he found conditions under which the E-theories of two ordered abelian groups coincide. A.I. Kokorin and N.G. Hisamiev [76] studied the class $\mathcal{O}_{\mathrm{la}}$ of lattice-ordered abelian groups, finding conditions for the equality of the E-theories of two such groups with finite numbers of filets. Hisamiev [58] then showed $\forall(\mathcal{O}_{\mathrm{la}})$ is recursive. Using standard methods, Gurevič soon proved $\mathrm{E}(\mathcal{O}_{\mathrm{la}})$ is not recursive.

The class of abelian semigroups is more complicated than that of abelian groups. Taĭclin and Tarski showed that the E-theory of the class of cancellative abelian semigroups is not recursive. Taĭclin [156] next found a series of classes of abelian semigroups with decidable E-theories. In particular, he discovered that the E-theory of each individual finitely generated abelian semigroup is recursive.

Of tremendous interest for abelian semigroup theory is this *isomorphism problem:*

(VI) *Is there an algorithm whereby for every two finite systems of defining relations for semigroups, one can tell whether or not they define isomorphic semigroups in the class of all abelian semigroups?*

Such an algorithm is known for relations with two generators (cf. Rédei [124]). Taĭclin has described an algorithm corresponding to four generators. For the class of cancellative abelian groups, the isomorphism problem was solved positively by E.A. Halezov. Taĭclin found a number of other important classes of abelian semigroups whose isomorphism problems have positive solutions. The isomorphism problem (VI) is a consequence of Hilbert's tenth problem: does the converse hold?

According to P.S. Novikov, the answer to the isomorphism problem for the variety of all groups is negative. For the variety of abelian groups, the answer is affirmative; it is also affirmative for a few other varieties of groups in which all the finitely generated groups are finite. Nothing is known for other group varieties; in particular, the isomorphism problem for varieties of polynilpotent groups is open, even for the variety of metabelian groups.

§**1.5.** *The identity problem*

For a variety $\mathfrak{X}$, the theory $\mathbf{I}(\mathfrak{X})$ is recursive iff the free algebras of finite rank of the class $\mathfrak{X}$ admit constructive descriptions ([4]) (cf. [XXIX], §1; [XVIII], §4.1). In many cases such descriptions have been found – e.g., when $\mathfrak{X}$ is the class of all rings, associative rings, Lie rings, or lattices, or any variety of polynilpotent groups. But there is a finitely axiomatizable variety of commutative loops whose I-theory is not recursive ([XXIX], §4). It would be pleasing to know whether

(A) *there is a finitely axiomatizable variety of groups with a nonrecursive* I-*theory,*

(B) *there is a finitely axiomatizable variety of associative (or Lie) rings with a nonrecursive* I-*theory.*

It is not yet known even whether

(C) *every variety of groups is finitely axiomatizable (B. Neumann),*

(D) *every variety of associative (Lie) rings is finitely axiomatizable (Specht).*

§**1.6.** *Degrees of unsolvability of theories*

From Gödel's completeness theorem it follows that for every recursively axiomatizable class $\mathfrak{X}$ of systems, the theories $\mathbf{E}(\mathfrak{X})$, $\mathbf{I}(\mathfrak{X})$, $\mathbf{Q}(\mathfrak{X})$, $\forall(\mathfrak{X})$ are certainly recursively enumerable. What degrees of unsolvability can these theories have? What happens when $\mathfrak{X}$ is a finitely axiomatizable variety? It's not at all hard to construct an infinitely axiomatizable class $\mathfrak{X}$ whose theory $\mathbf{E}(\mathfrak{X})$ has an arbitrary given recursively enumerable degree of unsolvability. Hanf [52] obtained the analogous result for finitely axiomatizable varieties. Nevertheless, for all the natural (i.e., not inspired solely by this problem) finitely axiomatizable classes $\mathfrak{X}$, whenever the unsolvability degree of $\mathbf{E}(\mathfrak{X})$ is known, it turns out to be either 0 (and $\mathbf{E}(\mathfrak{X})$ is recursive) or $0'$ (and $\mathbf{E}(\mathfrak{X})$ is creative).

It is thus of special interest to ask:

(i) *What degree of unsolvability has the theory*

$$\mathrm{E}(\langle R; +, \cdot, exp\rangle),$$

where R is the set of real numbers and $exp(x, y) = x^y$? *(Grzegorczyk).*

Let $B_{\mathfrak{T}}$ be the set of all subsets of a topological space $\mathfrak{T}$, and let $\cup, ', {}^{-}$ be the operations of union, complement, and $\mathfrak{T}$-closure defined on $B_{\mathfrak{T}}$. Consider the theory

$$\mathrm{T} = \mathrm{E}(\langle B_{\mathfrak{T}}; \cup, ', - \rangle).$$

If $\mathfrak{T}$ is the open euclidean unit square $(0,1) \times (0,1)$, then according to Grzegorczyk [48], the theory **T** is nonrecursive. But

(ii) *What is the unsolvability degree of* **T** *if* $\mathfrak{T}$ *is the open interval* (0,1)? *(Grzegorczyk).*

Rewarding tasks might be to catalog the unsolvability degrees attained in the isomorphism problem for finitely axiomatizable varieties, and to apply the metric theory of algorithms to find the degree of complexity of theories $\mathrm{E}(\mathfrak{K})$, $\mathrm{I}(\mathfrak{K})$ when they are recursive.

§2. Varieties and quasivarieties

§2.1. *Lattices of subvarieties*

Suppose Γ is a fixed type of sentence. The collection of all Γ-subclasses of an arbitrary Γ-class $\mathfrak{K}$ is completely lattice-ordered relative to set-theoretic inclusion. We denote this lattice by $\mathfrak{L}_\Gamma(\mathfrak{K})$.

The atoms of $\mathfrak{L}_\Gamma(\mathfrak{K}_\Sigma)$ are called Γ-*minimal* or Γ-*complete* classes. Clearly, a Γ-class $\mathfrak{K}$ is Γ-minimal iff the lattice $\mathfrak{L}_\Gamma(\mathfrak{K})$ has only two elements. Note that the smallest element in $\mathfrak{L}_\Gamma(\mathfrak{K})$ may be the empty class. The creation of the theory of E-complete classes (see [180]) was, in my opinion, one of the major events in general algebra in recent years.

Equally momentous, from a purely algebraic point of view, would be the creation of a theory as detailed as possible of I-, Q-, and $\forall$-classes, the simplest classes with respect to the logical language used for axiomatization. Although convenient set-theoretic characterizations of these classes are well known (cf. [XXXI], [XXXIII]), they serve only as a starting point for this study. ([5]).

We may regard the article [9] by G. Birkhoff (1935) as the origin of the general theory of varieties of algebras, and the article [112] by B. Neumann (1937) as the source of group variety theory. The effort dedicated to both these aspects of variety theory increased sharply during the fifties. The obvious task – to describe the lattice $\mathfrak{L}_I(\mathfrak{X})$ explicitly for the classical varieties $\mathfrak{X}$ – turns out to be most difficult and has been completed for very simple varieties only. E.g., let $\mathfrak{N}_k$ be the variety of all k-step nilpotent groups, and let $\mathfrak{S}_k$ be the variety of all k-step solvable groups ($k = 1, 2, 3, \ldots$; $\mathfrak{N}_1 = \mathfrak{S}_1$ = the class of abelian groups). The lattice $\mathfrak{L}_I(\mathfrak{N}_1)$ has long been known to consist of the subvarieties $\mathfrak{A}_m$ defined in $\mathfrak{N}_1$ by the identities $x^m \approx 1$ and ordered according to:

$$\mathfrak{A}_m \supseteq \mathfrak{A}_n \Leftrightarrow m \mid n \quad (m, n = 0, 1, 2, \ldots) .$$

The structure of the lattice $\mathfrak{L}_Q(\mathfrak{N}_1)$ was recently discovered by A.A. Vinogradov [182]; $\mathfrak{L}_I(\mathfrak{N}_1)$ is countable, but $\mathfrak{L}_Q(\mathfrak{N}_1)$ was found to have the power of the continuum. The structures of $\mathfrak{L}_I(\mathfrak{N}_2)$, $\mathfrak{L}_I(\mathfrak{N}_3)$ are now fully known [125], [67], and $\mathfrak{L}_I(\mathfrak{S}_2)$ has been partially described. In connection with certain conjectures on the structure of such lattices, it is important to

(a) *complete the characterization of* $\mathfrak{L}_I(\mathfrak{S}_2)$ *(B. Neumann and H. Neumann, and determine the structure of* $\mathfrak{L}_I(\mathfrak{N}_4)$.

Little is understood of the structure of $\mathfrak{L}_I(\mathfrak{L})$, where $\mathfrak{L}$ is the variety of all lattices. Its best-known elements are the variety $\mathfrak{M}$ of modular lattices and the variety of distributive lattices; the latter is the only atom in $\mathfrak{L}_I(\mathfrak{L})$. Other varieties of lattices have been pointed out by Iqbalunnisa [60] and H. Löwig [93]. There is still hope, I feel, that $\mathfrak{L}_I(\mathfrak{L})$ or $\mathfrak{L}_I(\mathfrak{M})$ is not too complicated, and that someone will manage to

(b) *describe* $\mathfrak{L}_I(\mathfrak{L})$ *or* $\mathfrak{L}_I(\mathfrak{M})$.

General arguments show that every variety includes at least one I-complete subvariety and every quasivariety includes at least one Q-complete subquasivariety. In particular, every minimal variety includes a minimal quasivariety; the converse, however, fails. Therefore, the number of minimal subquasivarieties of an arbitrary variety is greater than or equal to the number of its minimal subvarieties.

For many of the classical varieties $\mathfrak{X}$, Tarski and Kalicki described the atoms in $\mathfrak{L}_I(\mathfrak{X})$ explicitly. Kalicki also demonstrated that the variety of all groupoids includes continuum many minimal subvarieties. This result was recently strengthened by Bol'bot (Novosibirsk), who showed that the variety of groupoids defined by the identities $x \cdot xy \approx yx \cdot x \approx x$ also has continuum many subvarieties.

From the compactness theorem it follows that the lattices $\mathfrak{L}_E(\mathfrak{X})$, $\mathfrak{L}_I(\mathfrak{X})$, $\mathfrak{L}_Q(\mathfrak{X})$ cannot have arbitrary form; we may ask:

(c) *Which lattices can be represented in the form* $\mathfrak{L}_I(\mathcal{K})$ (*or* $\mathfrak{L}_Q(\mathcal{K})$) *for an appropriate variety (or quasivariety)* $\mathcal{K}$?

§2.2. *Groupoids of quasivarieties*

In 1956, H. Neumann [115] found a new approach for studying group varieties; she introduced an associative operation of multiplication on these varieties and proposed concentrating on the resulting semigroup $\mathfrak{G}_I(\mathcal{G})$ instead of the lattice $\mathfrak{L}_I(\mathcal{G})$ of group varieties. Guided by the work cited, B. Neumann et al. [114] and A. Šmel'kin [153] independently proved that $\mathfrak{G}_I(\mathcal{G})$ is a free semigroup with zero and identity elements. But this semigroup's power (clearly infinite but not greater than that of the continuum) is still unknown ([6]).

It is probably to good purpose to introduce an analogous operation of multiplication on the subclasses of an arbitrary class $\mathcal{K}$ of algebraic systems. Namely, for any $\mathcal{A}, \mathcal{B} \subseteq \mathcal{K}$, take $\mathfrak{A}$ to belong to $\mathcal{A} \underset{\mathcal{K}}{*} \mathcal{B}$ iff $\mathfrak{A}$ belongs to $\mathcal{K}$ and there exists a factor system $\mathfrak{A}/\theta \in \mathcal{B}$ such that every equivalence class $a\theta$ ($a \in \mathfrak{A}$) that is a $\mathcal{K}$-subsystem of $\mathfrak{A}$ belongs to $\mathcal{A}$ (cf. [XXXII]). It is easy to prove that if $\Delta = \forall, Q$ and $\mathcal{K}$ is a Δ-class with finite signature, then the $\mathcal{K}$-product of any two of its Δ-subclasses is again a Δ-subclass; thus in addition to the lattice $\mathfrak{L}_\Delta(\mathcal{K})$, we have a groupoid $\mathfrak{G}_\Delta(\mathcal{K})$ to study. It is worth noting that even when $\mathcal{K}$ is a variety of semigroups, the $\mathcal{K}$-product of two subvarieties may not be a subvariety. But if $\mathcal{K}$ is a variety of algebras, if all the congruence relations on each $\mathcal{K}$-algebra commute, and if there is a term that defines a one-element subalgebra in every $\mathcal{K}$-algebra, then the $\mathcal{K}$-product of subvarieties will be a subvariety of $\mathcal{K}$; hence, if the signature of $\mathcal{K}$ is finite, $\mathfrak{G}_I(\mathcal{K})$ will be a subgroupoid of the groupoid $\mathfrak{G}_Q(\mathcal{K})$. One can also state conditions under which the groupoids $\mathfrak{G}_Q(\mathcal{K})$, $\mathfrak{G}_I(\mathcal{K})$ become associative (see [XXXII], §3).

Suppose $\mathcal{K}$ is a quasivariety with finite signature and $\mathcal{L} \in \mathfrak{G}_Q(\mathcal{K})$. Clearly, the $\mathcal{L}$-product of two subquasivarieties of $\mathcal{L}$ does not, in general, coincide with their $\mathcal{K}$-product, so that $\mathfrak{G}_Q(\mathcal{L})$ may not be a subgroupoid of $\mathfrak{G}_Q(\mathcal{K})$. But if $\mathcal{L} \underset{\mathcal{K}}{*} \mathcal{L} = \mathcal{L}$, then $\mathfrak{G}_Q(\mathcal{L})$ will be a subgroupoid of $\mathfrak{G}_Q(\mathcal{K})$. This suggests that, apart from determining the overall structure of the groupoid $\mathfrak{G}_Q(\mathcal{K})$, it would be valuable to identify its idempotent elements.

Closely related to the latter problem is T. Tamura's search [160], for so-called *attainable* subquasivarieties of $\mathcal{K}$. Indeed, the attainability of a subquasivariety $\mathcal{A} \subseteq \mathcal{K}$ implies that

$$(\mathcal{K} \underset{\mathcal{K}}{*} \mathcal{A}) \underset{\mathcal{K}}{*} \mathcal{A} = \mathcal{K} \underset{\mathcal{K}}{*} \mathcal{A}$$

for every $\mathfrak{X} \in \mathfrak{G}_Q(\mathfrak{X})$. Under certain conditions, the converse also holds.

From earlier remarks it follows that if $\mathfrak{X}$ is a variety of groups, rings, or loops, then the subvarieties of $\mathfrak{X}$ form a groupoid relative to $\mathfrak{X}$-multiplication. It is possible the structure of some of these groupoids is not too complex.

NOTES

(*) Although the translation [XXXIV(ET)] represents a later and shorter version of this report, it incorporates no substantial changes.

(1) Please compare [I] and [II].

(2) What algebra is this? The D-theories of $\langle N; +, \cdot \rangle$, $\langle N; \dot{-}, \cdot \rangle$, and $\langle N; 0, +, \cdot \rangle$ are trivially decidable. But Hilbert's tenth problem is equivalent to the decision problem for the D-theory of $\langle N; ', +, \cdot \rangle$, or $\langle N; 1, +, \cdot \rangle$, or even $\langle N; ', \cdot \rangle$ (where $'$ is the successor operation). These theories were finally shown to have the highest recursively enumerable degree of unsolvability by Ju. V. Matijasevič (*Doklady ANSSSR* 191 (1970), 279–282).

(3) The reader is referred to [XIV], [XV], and [XIX] – [XXI].

(4) The descriptions should be uniform in the number of generators. Please see Note 1 in Chapter 29.

(5) In [M16], Mal'cev's last publication (appearing soon in English), there is a detailed exposition of the current theory of varieties and quasivarieties.

(6) There are now known to be continuum many group varieties; see, e.g., M.R. Vaughn-Lee's article, *Bull. London Math. Soc.* 2 (1970), 280–289.

BIBLIOGRAPHY

The consolidated bibliographies from the 34 papers here translated, augmented with other useful references, are presented below in six parts. In order, these comprise the titles of:

[I–XXXIV], the articles from which this book has been translated;

[XII(ET)–XXXIV(ET)], previously published translations of the corresponding articles above;

[M1–M16], other works of Mal'cev that he or the editor cites;

[R1–R7], reviews cited by the editor;

[B1–B3], biographical material on Mal'cev and comprehensive lists (in Russian and English) of his published works;

[1–188], publications of other authors.

Russian titles are given in a convenient transliteration. The journals, *Trudy Matematičeskogo Instituta imeni V.A. Steklova* and *Algebra i Logika Seminar,* are both published by the Academy of Science USSR, the former in Moscow; the latter, issued by the Academy's Siberian Division in Novosibirsk, was founded by A.I. Mal'cev, who chaired this important seminar. The works [XXXIII, XXXIV, M15, M16] were published posthumously under the editorship of several of Mal'cev's students and colleagues.

PART I: The articles translated in this collection

with date of submission and review references (JSL = J. Symbolic Logic, MR = Math. Reviews, RŽ = Referativnyi Žurnal Matematika); papers marked (ET) *have been translated previously (see Part II)*

[I] Untersuchungen aus dem Gebiete der mathematischen Logik, *Mat. Sbornik* (nov. ser.) **1** (43) (1936), 323–336 (with Russian summary). (Rcvd. 16/XII/35; *JSL* **2**, 84.)

[II] Ob odnom obščem metode polučenija lokal'nyh teorem teorii grupp, *Učenye Zapiski Ivanov. Ped. Inst. (Fiz-mat. Fakul'tet)* **1** (1941), no. 1, 3–9. (*JSL* **24**, 55, *MR* 17–823.)

[III] O predstavlenijah modeleǐ, *Dokl. Akad. Nauk SSSR* **108** (1956), 27–29. (Rcvd. 3/I/56; *JSL* **24**, 55, *MR* 18–370, *RŽ* 57#1999.)

[IV] Kvaziprimitivnye klassy abstraktnyh algebr, *Dokl. Akad. Nauk SSSR* 108 (1956), 187–189. (Rcvd. 6/I/56; *JSL* **24**, 57, *MR* 18–107, *RŽ* 59#174.)

[V] Podprjamye proizvedenija modeleǐ, *Dokl. Akad. Nauk SSSR* **109** (1956), 264–266. (Rcvd. 13/III/56; *JSL* **24**, 57, *MR* 19–240, *RŽ* 58#179.)

[VI] O proizvodnyh operacijah i predikatah, *Dokl. Akad. Nauk SSSR* **116** (1957), 24–27. (Rcvd. 14/III/57; *MR* 20#1647, *RŽ* 59#2232.)

[VII] O klassah modeleǐ s operacieǐ poroždenija, *Dokl. Akad. Nauk SSSR* **116** (1957), 738–741. (Rcvd. 15/VI/57; *MR* 20#2271, *RŽ* 59#1299.)

[VIII] Opredeljajuščie sootnošenija v kategorijah, *Dokl. Akad. Nauk SSSR* **119** (1958), 1095–1098. (Rcvd. 29/I/58; *MR* 20#3805, *RŽ* 59#5634.)

[IX] Strukturnaja harakteristika nekotoryh klassov algebr, *Dokl. Akad. Nauk SSSR* **120** (1958), 29–32. (Rcvd. 6/II/58; *MR* 20#5154, *RŽ* 59#5635.)

[X] O nekotoryh klassah modeleǐ, *Dokl. Akad. Nauk SSSR* **120** (1958), 245–248. (Rcvd. 20/II/58; *MR* 20#5155, *RŽ* 59#8883.)

[XI] Model'nye sootvetstvija, *Izv. Akad. Nauk SSSR* (ser. mat.) **23** (1959), 313–336. (Rcvd. 27/XI/58; *JSL* **33**, 299, *MR* 22#10909, *RŽ* 60#12554.)

[XII] Reguljarnye proizvedenija modeleǐ, *Izv. Akad. Nauk SSSR* (ser. mat.) **23** (1959), 489–502. (Rcvd. 2/I/59; *MR* 23#A1536, *RŽ* 60#12555.) (ET)

[XIII] O malyh modeljah, *Dokl. Akad. Nauk SSSR* **127** (1959), 258–261. (Rcvd. 18/IV/59; *MR* 21#5553, *RŽ* 60#6091.)

[XIV] O svobodnyh razrešimyh gruppah, *Dokl. Akad. Nauk SSSR* **130** (1960), 495–498. (Rcvd. 11/XI/59; *JSL* **30**, 99, *MR* 22#8056, *RŽ* 61#3A191.) (ET)

[XV] Ob odnom sootvetstvii meždu kol'cami i gruppami, *Mat. Sbornik* (nov. ser.) **50** (92) (1960), 257–266. (Rcvd. 17/IX/59; *JSL* **30**, 393, *MR* 22#9448, *RŽ* 61#2A155.) (ET)

[XVI] O nerazrešimosti ǝlementarnyh teoriǐ nekotoryh poleǐ, *Sibir. Mat. Ž.* **1** (1960), 71–77. (Rcvd. 19/II/60; *JSL* **30**, 395, *MR* 23#A3094, *RŽ* 61#10A273.) (ET)

[XVII] Zamečanie k stat'e "O nerazrešimosti ələmentarnyh teoriĭ nekotoryh poleĭ", *Sibir. Mat. Ž.* **2** (1961), 639. (Rcvd. 2/II/61; *JSL* **30**, 395, *RŽ* 62#6A243.) (ET)

[XVIII] Konstruktivnye algebry (I), *Uspehi Mat. Nauk* **16** (1961), no. 3 (99), 3–60. (Rcvd. 6/I/61; *JSL* **31**, 647, *MR* 27#1362, *RŽ* 62#11A229.) (ET)

[XIX] Nerazrešimost' ələmentarnoĭ teoriĭ konečnyh grupp, *Dokl. Akad. Nauk SSSR* **138** (1961), 771–774. (Rcvd. 27/II/61; *JSL* **30**, 394, *MR* 27#3550, *RŽ* 62#5A288.) (ET)

[XX] Ob ələmentarnyh svoĭstvah lineĭnyh grupp, in: *Nekotorye Problemy Matematiki i Mehaniki.* Novosibirsk, Akad. Nauk SSSR (Sibir. Otdel.), 1961, pp. 110–132. (*RŽ* 63#1A279.)

[XXI] Əffektivnaja neotdelimost' množestva toždestvenno istinnyh i množestva konečno oproveržimyh formul nekotoryh ələmentarnyh teoriĭ, *Dokl. Akad. Nauk SSSR* **139** (1961), 802–805. (Rcvd. 17/IV/61; *JSL* **30**, 394, *MR* 25#17, *RŽ* 62#5A82.) (ET)

[XXII] Strogo rodstvennye modeli i rekursivno soveršennye algebry, *Dokl. Akad. Nauk SSSR* **145** (1962), 276–279. (Rcvd. 9/IV/62; *JSL* **31**, 649, *MR* 26#1254, *RŽ* 64#2A109.) (ET)

[XXIII] Aksiomatiziruemye klassy lokal'no svobodnyh algebr nekotoryh tipov, *Sibir. Mat. Ž.* **3** (1962), 729–743. (Rcvd. 28/II/62; *JSL* **32**, 278, *MR* 26#59, *RŽ* 63#7A211.)

[XXIV] O rekursivnyh abelevyh gruppah, *Dokl. Akad. Nauk SSSR* **146** (1962), 1009–1012. (Rcvd. 6/VII/62; *JSL* **31**, 649, *MR* 27#1363, *RŽ* 64#1A231.) (ET)

[XXV] Polno numerovannye množestva, *Algebra i Logika Sem.* **2** (1963), no. 2, 4–29. (Rcvd. 25/II/63; *RŽ* 64#3A56.)

[XXVI] Nekotorye voprosy teorii klassov modeleĭ, in: *Trudy 4-go Vsesojuznogo Mat. S"ezda (Leningrad 1961),* vol. 1. Leningrad, Akad. Nauk SSSR, 1963, pp. 169–198. (*MR* 27#5693, *RŽ* 64#2A356.) (ET)

[XXVII] K teorii vyčislimyh semeĭstv ob"ektov, *Algebra i Logika Sem.* **3** (1964), no. 4, 5–31. (Rcvd. 25/VIII/64; *RŽ* 65#4A67.)

[XXVIII] Positivnye i negativnye numeracii, *Dokl. Akad. Nauk SSSR* **160** (1965), 278–280. (Rcvd. 19/X/64; *RŽ* 65#6A56.) (ET)

[XXIX] Toždestvennye sootnošenija na mnogoobrazijah kvazigrupp, *Mat. Sbornik* (nov. ser.) **69** (111) (1966), 3–12. (Rcvd. 19/II/65; *MR* 34#61, *RŽ* 66#6A184.) (ET)

[XXX] Iterativnye algebry i mnogoobrazija Posta, *Algebra i Logika Sem.* **5** (1966), no. 2, 5–24. (Rcvd. 26/III/66; *MR* 34#7424.)

[XXXI] Neskol'ko zamečaniĭ o kvazimnogoobrazijah algebraičeskih sistem, *Algebra i Logika Sem.* **5** (1966), no. 3, 3–9. (Rcvd. 10/V/66; *MR* 34#5728, *RŽ* 67#2A243.)

[XXXII] Ob umnoženii klassov algebraičeskih sistem, *Sibir. Mat. Ž.* **8** (1967), 346–365. (Rcvd. 7/X/66; *MR* 35#4140, *RŽ* 67#11A267.) (ET)

[XXXIII] Universal'no-aksiomatiziruemye podklassy lokal'no konečnyh klassov modeleĭ, *Sibir. Mat. Ž.* **8** (1967), 1005–1014. (Rcvd. 3/V/67; *MR* 36#6275, *RŽ* 68#6A132.) (ET)

[XXXIV] O nekotoryh pograničnyh voprosah algebry i metamatičeskoĭ logiki, in: *Proc. Int. Congress of Math. (Moscow 1966).* Moscow, Mir, 1968, pp. 217–231. (*MR* 38#2072, *RŽ* 67#12A295.) (ET)

PART II: Previous English translations

[XII(ET)] Regular products of models, *Amer. Math. Soc. Transl.* (2) **39** (1964), 139–206. Translated by J.N. Whitney.

[XIV(ET)] On free soluble groups, *Soviet Math.* **1** (1960), 65–68. Translated by K.A. Hirsch.

[XV(ET)] On a correspondence between rings and groups, *Amer. Math. Soc. Transl.* (2) **45** (1965), 221–231. Translated by F. Albrecht.

[XVI/XVII(ET)] On the undecidability of elementary theories of certain fields, Remark on the paper "On the undecidability of elementary theories of certain fields", *Amer. Math. Soc. Transl.* (2) **48** (1965), 36–44. Translated by J.M. Danskin.

[XVIII(ET)] Constructive algebras (I), *Russian Math. Surveys* **16** (1961), no. 3, 77–129. Translated by K.A. Hirsch.

[XIX(ET)] Undecidability of the elementary theory of finite groups, *Soviet Math.* **2** (1961), 714–717. Translated by A.J. Lohwater.

[XXI(ET)] Effective inseparability of the set of identically true from the set of finitely refutable formulas of certain elementary theories, *Soviet Math.* **2** (1961), 1005–1008. Translated by Elliott Mendelson.

[XXII(ET)] Strongly related models and recursively complete algebras, *Soviet Math.* **3** (1962), 987–991. Translated by Elliott Mendelson.

[XXIV(ET)] On recursive abelian groups, *Soviet Math.* **3** (1962), 1431–1434. Translated by Lisa Rosenblatt.

[XXVI(ET)] Some problems of the theory of classes of models, *Amer. Math. Soc. Transl.* (2) **83** (1969), 1–48. Translated by Andrew Yablonsky.

[XXVIII(ET)] Positive and negative numerations, *Soviet Math.* **6** (1965), 75–77. Translated by Elliott Mendelson.

[XXIX(ET)] Identical relations on varieties of quasigroups, *Amer. Math. Soc. Transl.* (2) **82** (1969) 225–235. Translated by K.A. Hirsch.

[XXXII(ET)] Multiplication of classes of algebraic systems, *Siberian Math. J.* **8** (1967), 254–267.

[XXXIII(ET)] Universally axiomatizable subclasses of locally finite classes of models, *Siberian Math. J.* **8** (1967), 764–770.

[XXXIV(ET)] Some questions bordering on algebra and mathematical logic, *Amer. Math. Soc. Transl.* (2) **70** (1968), 89–100. Translation provided by the author.

PART III: Other works of A.I. Mal'cev

cited by the author in the present collection:

[M1] O vklučenii associativnyh sistem v gruppy, *Mat. Sbornik* (nov. ser.) **6** (48) (1959), 331–336; (II), *Ibid.* **8** (50), 251–264.
[M2] O douporjadočenii grupp, *Trudy Mat. Inst. im. Steklova* **38** (1951), 173–175.
[M3] *Osnovy Lineĭnoĭ Algebry*, 2nd edition. Moscow, Gostehizdat, 1956.
[M4] Zamečanie o častično uporjadočennyh gruppah, *Učenye Zapiski Ivanov. Ped. Inst.* **10** (1956), 3–5.
[M5] Svobodnye topologičeskie algebry, *Izv. Akad. Nauk SSSR* (ser. mat.) **21** (1957), 171–198.
[M6] O gomomorfizmah na konečnye gruppy, *Učenye Zapiski Ivanov. Ped. Inst.* **28** (1958), 49–60.
[M7] Ob odnom sootvetstvii meždu kol'cami i gruppami, *Uspehi Mat. Nauk* **14** (1959), no. 5 (89), 208–209.
[M8] Ob əlementarnyh teorijah lokal'no svobodnyh universal'nyh algebr, *Dokl. Akad. Nauk SSSR* **138** (1961), 1009–1012.
[M9] *Algoritmy i Rekursivnye Funkcii.* Moscow, Nauka, 1965

cited for their logical, metamathematical, or universal-algebraic appeal (in addition to [M1, M2, M4, M5, M9] *):*

[M10] Ob algebrah s toždestvennymi opredeljajuščimi sootnošenijami, *Mat. Sbornik* (nov. ser.) **26** (68) (1950), 19–33.
[M11] Ob odnom klasse algebraičeskih sistem, *Uspehi Mat. Nauk* **7** (1953), no. 1 (53), 165–171.
[M12] K obščeĭ teorii algebraičeskih sistem, *Mat. Sbornik* (nov. ser.) **35** (71) (1954), 3–20.
[M13] Ob uravnenii $zxyx^{-1}y^{-1}z^{-1} = aba^{-1}b^{-1}$ v svobodnoĭ gruppe, *Algebra i Logika Sem.* 1 (1962), no. **5**, 45–50.
[M14] O standartnyh oboznačenijah i terminologii v teorii algebraičeskih sistem, *Algebra i Logika Sem.* **5** (1966), no. 1, 71–77.
[M15] Ob odnom usilenii teorem Słupeckiego i Jablonskogo, *Algebra i Logika Sem.* **6** (1967), no. 3, 61–75 (with English summary).
[M16] *Algebraičeskie Sistemy*. Moscow, Nauka, 1970.

PART IV: Reviews cited in editor's notes

[R1] Review of [II] and [III] (with historical notes concerning [I]) by L. Henkin and A. Mostowski, *J. Symbolic Logic* 24 (1959), 55–57.
[R2] Review of [IV] by A. Mostowski, *J. Symbolic Logic* 24 (1959), 57
[R3] Review of [XIV] by V.H. Dyson, *J. Symbolic Logic* 30 (1965), 99.
[R4] Review of [XVI] and [XVII] by B.F. Wells III, *J. Symbolic Logic* 30 (1965), 395–397.
[R5] Review of [XVIII] by V.H. Dyson, *J. Symbolic Logic* 31 (1966), 647–649.
[R6] Review of [XXVI] by A. Mostowski, *Math. Reviews* 27 (1964), #5693.
[R7] Review of [157] by P.G. Hinman, *J. Symbolic Logic* 30 (1965), 253–254.

PART V: Obituaries with bibliographies

providing detailed information on the life and work of A.I. Mal'cev

[B1] Obituary by P.S. Aleksandrov, Ju.L. Eršov, M.I. Kargapolov, E.N. Kuz'min, D.M. Smirnov, A.D. Taĭmanov, and A.I. Širšov, *Uspehi Mat. Nauk* 23 (1968), no. 3 (141), 159–170. (An English translation by Haya Freedman appears in *Russian Math. Surveys* 23 (1961), no. 3, 157–168.)

[B2] *Sibir. Mat. Ž.* 8 (1967), no. 4, i–vi. (English translation: *Siberian Math. J.* 8 (1967), 541–546.)

[B3] *Algebra i Logika Seminar* 6 (1967), no. 4, i–xi.

PART VI: General references

a list of other publications cited by the author and the editor

[1] W. Ackermann, Untersuchungen über das Eliminationsproblem der mathematischen Logik, *Math. Ann.* **110** (1935), 390–413.

[2] W. Ackermann, *Solvable Cases of the Decision Problem* (Studies in Logic). Amsterdam, North-Holland Publ. Co., 1954.

[3] K.I. Appel, Horn sentences in identity theory, *J. Symbolic Logic* **24** (1959), 306–310.

[4] M. Auslander and R.C. Lyndon, Commutator subgroups of free groups, *Amer. J. Math.* 77 (1955), 929–931.

[5] J. Ax and S. Kochen, Diophantine problems over local fields (I, II), *Amer. J. Math.* 87 (1965), 605–630, 631–648; (III), *Ann. of Math.* (2) 83 (1966), 437–456.

[6] R. Baer, Nilpotent groups and their generalizations, *Trans. Amer. Math. Soc.* **47** (1940), 393–434.

[7] E. Beth, On Padoa's method in the theory of definition, *Indag. Math.* **15** (1953), 330–339.

[8] K. Bing, On arithmetical classes not closed under direct union, *Proc. Amer. Math. Soc.* **6** (1955), 834–846.

[9] G. Birkhoff, On the structure of abstract algebras, *Proc. Cambridge Phil. Soc.* **31** (1935), 433–454.

[10] G. Birkhoff, Subdirect unions in universal algebras, *Bull. Amer. Math. Soc.* **50** (1944), 764–768.

[11] G. Birkhoff, *Teorija Struktur.* Moscow, Izd. In. Lit., 1952. (Russian translation of *Lattice Theory,* Colloq. Publ. vol. 25, New York, Amer. Math. Soc. [rev. edition: 1948].)

[12] J.R. Büchi, Weak second-order arithmetic and finite automata, *Z. math. Logik und Grundl. Math.* 6 (1960), 66–92.

[13] J.R. Büchi, On a problem of Tarski, *Notices Amer. Math. Soc.* 7 (1960), 382.

[14] S.N. Černikov, Beskonečnye special'nye gruppy, *Mat. Sbornik* (nov. ser.) **6** (48) (1939), 199–214.

[15] S.N. Černikov, Beskonečnye lokal'no razrešimye gruppy, *Mat. Sbornik* (nov. ser.) 7 (49) (1940), 35–64.

[16] S.N. Černikov, O gruppah s silovskim množestvom, *Mat. Sbornik* (nov. ser.) **8** (50) (1940), 377–394.

[17] C.C. Chang, On unions of chains of models, *Proc. Amer. Math. Soc.* **10** (1959), 120–127.

[18] C.C. Chang and A.C. Morel, On closure under direct product, *J. Symbolic Logic* 23 (1958), 149–154.

[19] A. Church, A note on the Entscheidungsproblem, *J. Symbolic Logic* **1** (1936), 40–41, 101–102.

[20] A. Church, Application of recursive arithmetic in the theory of computers and automata, *Advanced Theory of the Logical Design of Digital Computers.* University of Michigan, Ann Arbor, 1958.

[21] J.P. Cleave, Creative functions, *Z. math. Logik und Grundl. Math.* 7 (1961), 205–212.

[22] P.M. Cohn, *Universal Algebra.* New York, Harper & Row, 1965.

[23] W. Craig, Linear reasoning. A new form of the Herbrand-Gentzen theorem, *J. Symbolic Logic* 22 (1957), 250–268.

[24] M. Davis, H. Putnam, and J. Robinson, The decision problem for exponential diophantine equations, *Ann. of Math.* (2) 74 (1961), 425–436.

[25] J.C.E. Dekker and J. Myhill, Some theorems on classes of recursively enumerable sets, *Trans. Amer. Math. Soc.* 89 (1958), 25–59.

[26] A. Ehrenfeucht, On theories categorical in power, *Fund. Math.* 44 (1957), 241–248.

[27] A. Ehrenfeucht and A. Mostowski, Models of axiomatic theories admitting automorphisms, *Fund. Math.* 43 (1956), 50–68.

[28] S. Eilenberg and S. MacLane, General theory of natural equivalences, *Trans. Amer. Math. Soc.* 58 (1945), 231–294.

[29] G. Epstein, The lattice theory of Post algebras, *Trans. Amer. Math. Soc.* **95** (1960), 300–317.

[30] Ju.L. Eršov, Nerazrešimost' nekotoryh poleĭ, *Dokl. Akad. Nauk SSSR* **161** (1965), 27–29.

[31] Ju. L. Eršov, Ob əlementarnyh teorijah lokal'nyh poleĭ, *Algebra i Logika Sem.* **4** (1965), no. 2, 5–30.

[32] Ju.L. Eršov, Ob əlementarnoĭ teoriĭ maksimal'nyh normirovannyh poleĭ, *Algebra i Logika Sem.* 4 (1965), no. 3, 31–70; (II), *Ibid.* 5 (1966), no. 1, 5–40.

[33] Ju.L. Eršov, I.A. Lavrov, A.D. Taĭmanov, and M.A. Taĭclin, Əlementarnye teorii, *Uspehi Mat. Nauk* 20 (1965), no. 4 (124), 37–108.

[34] T. Evans, Embeddability and the word problem, *J. London Math. Soc.* 28 (1953), 76–80.

[35] C.J. Everett and S. Ulam, Projective algebra (I), *Amer. J. Math.* 68 (1946), 77–88.

[36] S. Feferman and R.L. Vaught, The first order properties of products of algebraic systems, *Fund. Math.* 47 (1959), 57–103.

[37] A.L. Foster, Generalized "Boolean" theory of universal algebras (I), *Math. Zeitschrift* 58 (1953), 306–336.

[38] T.E. Frayne, A.C. Morel, and D.S. Scott, Reduced direct products, *Fund. Math.* 51 (1962), 195–228.

[39] T.E. Frayne and D.S. Scott, Model-theoretical properties of reduced products, *Notices Amer. Math. Soc.* 5 (1958), 675.

[40] R.M. Friedberg, Three theorems on recursive enumeration, *J. Symbolic Logic* 23 (1958), 309–316.

[41] A. Fröhlich and J.C. Shepherdson, Effective procedures in field theory, *Phil. Trans. Royal Soc. London* (A) 248 (1956), 407–432.

[42] L. Fuchs, On subdirect unions (I), *Acta Math. Acad. Sci. Hungar.* 3 (1952), 103–119.

[43] T. Fujiwara, Remarks on the Jordan-Hölder-Schreier theorem, *Proc. Japan Acad.* 31 (1955), 135–140.

[44] V.M. Gluškov, Nekotorye problemy sinteza cifrovyh avtomatov, *Ž. Vyčislit. Mat. i Mat. Fiz.* 1 (1961), 371–411.

[45] V.M. Gluškov, Abstraktnaja teorija automatov, *Uspehi Mat. Nauk* 16 (1961), no. 5 (101), 3–62.

[46] K. Gödel, Die Vollständigkeit der Axiome des logischen Funktionenkalküls, *Monatsh. Math. und Physik* 37 (1936), 349–360.

[47] G. Grätzer, *Universal Algebra* (University Series in Higher Math.). Princeton, Van Nostrand, 1968.

[48] A. Grzegorczyk, Undecidability of some topological theories, *Fund. Math.* 38 (1951), 137–152.

[49] Ju.Š. Gurevič, Ɜlementarnye svoĭstva uporjadočennyh abelevyh grupp, *Algebra i Logika Sem.* 3 (1964), no. 1, 5–39.

[50] Ju. Š. Gurevič, Ob əffektivnom raspoznavanii vypolnimosti formul UIP, *Algebra i Logika Sem.* 5 (1966), no. 2, 25–55.

[51] M. Hall, The word problem for semi-groups with two generators, *J. Symbolic Logic* 14 (1949), 115–118.

[52] W. Hanf, Model-theoretic methods in the study of elementary logic, in: *The Theory of Models (Proc. Int. Symposium, Berkeley, 1963)* (Studies in Logic). Amsterdam, North-Holland Publ. Co., 1965, pp. 132–145.

[53] L. Henkin, The completeness of the first-order functional calculus, *J. Symbolic Logic* 14 (1949), 159–166.

[54] L. Henkin, Some interconnections between modern algebra and mathematical logic, *Trans. Amer. Math. Soc.* 74 (1953), 410–427.

[55] L. Henkin, On a theorem of Vaught, *Indag. Math.* 17 (1955), 326–328.

[56] D. Hilbert and W. Ackermann, *Osnovy Teoretičeskoĭ Logiki.* Moscow, Izd. In. Lit., 1947. (Russian translation of *Grundzüge der theoretischen Logik.* Berlin, J. Springer, 2nd edition: 1938.)

[57] D. Hilbert and P. Bernays, *Grundlagen der Mathematik,* vol. 2. Berlin, J. Springer, 1939.

[58] N.G. Hisamiev, Universal'naja teorija strukturno uporjadočennyh abelevyh grupp, *Algebra i Logika Sem.* 5 (1966), no. 3, 71–76.

[59] A. Horn, On sentences which are true of direct unions of algebras, *J. Symbolic Logic* 16 (1951), 14–21.

[60] Iqbalunnisa, On types of lattices, *Fund. Math.* 59 (1966), 97–102.

[61] J.R. Isbell, Some remarks concerning categories and subspaces, *Canad. J. Math.* 9 (1957), 563–477.

[62] J.R. Isbell, Adequate subcategories, *Illinois J. Math.* 4 (1960), 541–552.

[63] J.R. Isbell, Two set-theoretic theorems in categories, *Fund. Math.* 53 (1963), 43–49.

[64] J.R. Isbell, Subobjects, adequacy, completeness and categories of algebras, *Rozprawy Matematyczne* 36 (1964).

[65] S.V. Jablonskiĭ, Funkcional'nye postroenija v k-znacnoĭ logike, *Trudy Mat. Inst. im. Steklova* 51 (1958), 5–142.

[66] Ju. I. Janov and A.A. Mučnik, O suščestvovanii k-značnyh zamknutyh klassov, ne imejuščih konečnogo bazisa, *Dokl. Akad. Nauk SSSR* 127 (1959), 44–46.

[67] B. Jónsson, Varieties of groups of nilpotency three, *Notices Amer. Math. Soc.* 13 (1966), 488.

[68] C. Jordan, Mémoire sur les équations différentielles linéaires à intégrale algébrique, *J. reine und angew. Math.* 84 (1878), 89–215.

[69] H.J. Keisler, Theory of models with generalized atomic formulas, *J. Symbolic Logic* 25 (1960), 1–26.

[70] H.J. Keisler, Ultraproducts and elementary classes, *Indag. Math.* 23 (1961), 477–495.

[71] S.K. Kleene, *Vvedenie v Metamatematiku.* Moscow, Izd. In. Lit., 1957. (Russian translation of *Introduction to Metamathematics,* Princeton, Van Nostrand, 1952.)

[72] S. Kochen, Ultraproducts in the theory of models, *Ann. of Math.* (2) 74 (1961), 221–261.

[73] S.R. Kogalovskiĭ, Universal'nye klassy modeleĭ, *Dokl. Akad. Nauk SSSR* 124 (1959), 260–263.

[74] S.R. Kogalovskii, Ob odnom obščem metode polučenija strukturnyh harakteristik aksiomatiziruemyh klassov, *Dokl. Akad. Nauk SSSR* 136 (1961), 1291–1294.

[75] S.R. Kogalovskii, O mul'tiplikativnyh polugruppah kolec, *Dokl. Akad. Nauk SSSR* 140 (1961), 1005–1007.

[76] A.I. Kokorin and N.G. Hisamiev, Ǝlementarnaja klassifikacija strukturno uporjadočennyh abelevyh grupp s konečnym čislom niteĭ, *Algebra i Logika Sem.* 5 (1966), no. 1, 41–50.

[77] A.N. Kolmogorov and V.A. Uspenskiĭ, K opredeleniju algoritma, *Uspehi Mat. Nauk* 13 (1958), no. 4 (82), 3–28.

[78] P.G. Kontorovič, Gruppy s bazisom rasščeplenija (III), *Mat. Sbornik* (nov. ser.) 22 (64) (1948), 79–100.

[79] J.B. Kruskal, Well-quasi-ordering, the tree theorem, and Vazsonyi's conjecture, *Trans. Amer. Math. Soc.* 95 (1960), 210–225.

[80] A.G. Kuroš, *Teorija Grupp,* 2nd edition. Moscow, Gostehizdat, 1953.

[81] A.G. Kuroš and S.N. Černikov, Razrešimye i nil'potentnye gruppy, *Uspehi Mat. Nauk* 2 (1947), no. 3 (19), 18–59.

[82] A.G. Kuroš, A.H. Lifšic, and E.G. Šul'geĭfer, Osnovy teorii kategoriĭ, *Uspehi Mat. Nauk* 15 (1960), no. 6 (96), 3–52.

[83] A.V. Kuznecov, O problemah toždestva i funkcional'noĭ polnoty dlja algebraičeskih sistem, in: *Trudy 3-go Vsesojuznogo Mat. S''ezda (Moscow 1956)*, vol. 2. Moscow, Akad. Nauk SSSR, 1956, pp. 145–146.

[84] A.V. Kuznecov, Algoritmy kak operacii v algebraičeskih sistemah, *Uspehi Mat. Nauk* 13 (1958), no. 3 (81), 240–241.

[85] A.H. Lachlan, Standard classes of recursively enumerable sets, *Z. math. Logic und Grundl. Math.* 10 (1964), 23–42.

[86] J. Łoś, On the existence of linear order in a group, *Bull. Acad. Polon. Sci.* (3) 2 (1954), 21–23.

[87] J. Łoś, On the categoricity in power of elementary deductive systems and some related problems, *Colloq. Math.* 3 (1954), 58–62.

[88] J. Łoś, Quelques remarques, théorèmes et problèmes sur les classes définissables d'algèbres, in: *Mathematical Interpretations of Formal Systems* (Studies in Logic). Amsterdam, North-Holland Publ. Co., 1955, pp. 98–113.

[89] J. Łoś, On the extending of models (I), *Fund. Math.* 42 (1955), 38–54.

[90] J. Łoś and C. Ryll-Nardzewski, Effectiveness of the representation theory for Boolean algebras, *Fund. Math.* 41 (1954), 49–56.

[91] J. Łoś and R. Suszko, On the extending of models (II), *Fund. Math.* 42 (1955), 343–347; (IV), *Ibid.* 44 (1957), 52–60.

[92] L. Löwenheim, Über Möglichkeiten im Relativkalkül, *Math. Ann.* 76 (1915), 447–470.

[93] H. Löwig, On the importance of the relation $[(A,B),(A,C)] < (A,[(B,C),(C,A),(A,B)])$ between three elements of a structure, *Ann. of Math.* (2) **44** (1943), 573–579.

[94] R.C. Lyndon, Existential Horn sentences, *Proc. Amer. Math. Soc.* **10** (1959), 994–998.

[95] R.C. Lyndon, An interpolation theorem in the predicate calculus, *Pacific J. Math.* 9 (1959), 129–142.

[96] R.C. Lyndon, Properties preserved under homomorphisms, *Pacific J. Math.* 9 (1959), 143–154.

[97] R.C. Lyndon, Properties preserved under subdirect products, *Pacific J. Math.* 9 (1959), 155–164.

[98] J.C.C. McKinsey, The decision problem for some classes of sentences without quantifiers, *J. Symbolic Logic* 8 (1943), 61–76.

[99] D.H. McLain, Local theorems in universal algebras, *J. London Math. Soc.* **34** (1959), 177–184.

[100] S. MacLane, Duality for groups, *Bull. Amer. Math. Soc.* **56** (1950), 485–516.

[101] E. Marczewski, Sur les congruences et les propriétés positives d'algèbres abstraites, *Colloq. Math.* **2** (1951), 220–228.

[102] A.A. Markov, Teorija algorifmov, *Trudy Mat. Inst. im. Steklova* 42 (1954), 1–375.

[103] A. Morel, D.S. Scott, and A. Tarski, Reduced products and the compactness theorem, *Notices Amer. Math. Soc.* **5** (1958), 674.

[104] A. Mostowski, On models of axiomatic systems, *Fund. Math.* **39** (1952), 133–158.

[105] A. Mostowski, On direct products of theories, *J. Symbolic Logic* **17** (1952), 1–31.

[106] A. Mostowski, On a system of axioms which has no recursively enumerable arithmetic model, *Fund. Math.* 40 (1953), 56–61.

[107] A. Mostowski, Sovremennoe sostojanie issledovaniĭ po osnovanijam matematiki, *Uspehi Mat. Nauk* 9 (1954), no. 3 (61), 3–38.

[108] A. Mostowski, Concerning a problem of H. Scholz, *Z. math. Logik und Grundl. Math.* **2** (1956), 210–214.

[109] A.A. Mučnik, Izomorfizm sistem rekursivno-perečislimyh množestv s эffektivnymi svoĭstvami, *Trudy Moskov. Mat. Obšč.* **7** (1958), 407–412.

[110] J. Mycielski, A characterization of arithmetical classes, *Bull. Acad. Polon. Sci.* (3) 5 (1957), 1025–1027.

[111] J. Myhill, Creative sets, *Z. math. Logik und Grundl. Math.* **1** (1955), 97–108.

[112] B.H. Neumann, Identical relations in groups (I), *Math. Ann.* **114** (1937), 506–525.

[113] B.H. Neumann, An embedding theorem for algebraic systems, *Proc. London Math. Soc.* (3) 4 (1954), 138–153.

[114] B.H. Neumann, H. Neumann, and P.M. Neumann, Wreath products and varieties of groups, *Math. Zeitschrift* 80 (1962), 44–62.

[115] H. Neumann, On varieties of groups and their associated near-rings, *Math. Zeitschrift* **65** (1956), 36–69.

[116] P.S. Novikov, Ob algoritmičeskoĭ nerazrešimosti problemy toždestva slov v teorii grupp, *Trudy Mat. Inst. im. Steklova* **44** (1955), 1–143.

[117] P.S. Novikov, *Эlementy Matematičeskoĭ Logiki.* Moscow, Fizmatgiz, 1959.

[118] A. Oberschelp, Über die Axiome produkt-abgeschlossener arithmetischer Klassen, *Archiv. math. Logik und Grundlagenforschung* 4 (1958), 95–123.

[119] R. Peter, *Rekursivnye Funkcii,* Moscow, Izd. In. Lit., 1954. (Russian translation of *Rekursive Funktionen.* Budapest, Akadémiai Kiadó, 1951.)

[120] G. Pickert, Direkte Zerlegungen von algebraischen Strukturen mit Relationen, *Math. Zeitschrift* **57** (1953), 395–404.
[121] E.L. Post, Recursively enumerable sets of positive integers and their decision problems, *Bull. Amer. Math. Soc.* **50** (1944), 284–316.
[122] M.O. Rabin, Arithmetical extensions with prescribed cardinality, *Indag. Math.* **21** (1959), 439–446.
[123] H. Rasiowa and R. Sikorski, A proof of the completeness theorem of Gödel, *Fund. Math.* **37** (1950), 193–200.
[124] L. Rédei, *Theorie der endlich erzeugbaren kommutativen Halbgruppen* (Hamburger math. Einzelschriften 41). Würzburg, Physica-Verlag, 1963.
[125] V.N. Remeslennikov, Dva zamečanija o 3-stupenno nil'potentnyh gruppah, *Algebra i Logika Sem.* **4** (1965), no. 2, 59–65.
[126] H.G. Rice, Classes of recursively enumerable sets and their decision problems, *Trans. Amer. Math. Soc.* **74** (1953), 358–366.
[127] H.G. Rice, On completely recursively enumerable classes and their key arrays, *J. Symbolic Logic* **21** (1956), 304–308.
[128] A. Robinson, *On the Metamathematics of Algebra* (Studies in Logic). Amsterdam, North-Holland Publ. Co., 1951.
[129] A. Robinson, Note on an embedding theorem for algebraic systems, *J. London Math. Soc.* **30** (1955), 249–252.
[130] A. Robinson, A result on consistency and its application to the theory of definition, *Indag. Math.* **18** (1956), 47–58.
[131] A. Robinson, Note on a problem of L. Henkin, *J. Symbolic Logic* **21** (1956), 33–35.
[132] A. Robinson, *Complete Theories* (Studies in Logic). Amsterdam, North-Holland Publ. Co., 1956.
[133] A. Robinson, Obstructions to arithmetical extension and the theorem of Łoś and Suszko, *Indag. Math.* **21** (1959), 489–495.
[134] J. Robinson, Definability and decision problems in arithmetic, *J. Symbolic Logic* **14** (1949), 98–114.
[135] J. Robinson, General recursive functions, *Proc. Amer. Math. Soc.* **1** (1950), 703–718.
[136] J. Robinson, The undecidability of algebraic rings and fields, *Proc. Amer. Math. Soc.* **10** (1959), 950–957.
[137] J. Robinson, The decision problem for fields, in: *The Theory of Models (Proc. Int. Symposium, Berkeley 1963)* (Studies in Logic). Amsterdam, North-Holland Publ. Co., 1965, pp. 299–311.
[138] R.M. Robinson, Primitive recursive functions, *Bull. Amer. Math. Soc.* **53** (1947), 925–942.
[139] R.M. Robinson, Undecidable rings, *Trans. Amer. Math. Soc.* **70** (1951), 137–159.
[140] R.M. Robinson, The undecidability of pure transcendental extensions of real fields, *Z. math. Logik und Grundl. Math.* **10** (1964), 275–282.
[141] H. Rogers, Gödel numberings of partial recursive functions, *J. Symbolic Logic* **23** (1958), 331–341.
[142] P.C. Rosenbloom, Post algebras I. Postulated and general theory. *Amer. J. Math.* **64** (1942), 167–188.
[143] A. Salomaa, On basic groups for the set of functions over a finite domain, *Ann. Acad. Sci. Fenn.* (A.1) **338** (1963).

[144] A. Schmidt, Über deduktive Theorien mit mehreren Sorten von Grunddingen, *Math. Ann.* **115** (1938), 485–506.

[145] A. Schmidt, Die Zulässigkeit der Behandlung mehrsortiger Theorien mittels der üblichen einsortigen Prädikatenlogik, *Math. Ann.* **123** (1951), 187–200.

[146] I. Schur, Über Gruppen periodischer linearer Substitutionen, *S.-B. preuss. Akad. Wiss.* (1911), 619–627.

[147] F. Schwenkel, Rekursive Wortfunktionen über unendlichen Alphabeten, *Z. math. Logik und Grundl. Math.* **11** (1965), 133–147.

[148] D. Scott, Equationally complete extensions of finite algebras, *Indag. Math.* 18 (1956), 35–38.

[149] N. Shapiro, Degrees of computability, *Trans. Amer. Math. Soc.* 82 (1956), 281–299.

[150] R. Sikorski, Products of abstract algebras, *Fund. Math.* 39 (1952), 211–228.

[151] T. Skolem, Logisch-kombinatorische Untersuchungen über die Erfüllbarkeit oder Beweisbarkeit mathematischer Sätze nebst einem Theoreme über dichte Mengen, *Skrifter utgit av Videnskapsseskapet i Kristiania* (1) no. 4 (1920), 1–36.

[152] T. Skolem, Über die Nicht-charakterisierbarkeit der Zahlenreihe mittels endlich oder abzählbar unendlich vieler Aussagen mit ausschliesslich Zahlenvariablen, *Fund. Math.* **23** (1934), 150–161.

[153] A.L. Šmel'kin, Polugruppa mnogoobrazii̮ grupp, *Dokl. Akad. Nauk SSSR* **149** (1963), 543–545.

[154] R.M. Smullyan, *Theory of Formal Systems* (Ann. of Math. Studies no. 47). Princeton, Princeton University, 1961.

[155] W. Szmielew, Elementary properties of Abelian groups, *Fund. Math.* **41** (1955), 203–271.

[156] M.A. Taĭclin, Ob əlementarnyh teorijah kommutativnyh polugrupp, *Algebra i Logika Sem.* **5** (1966), no. 4, 55–89.

[157] A.D. Taĭmanov, O klasse modeleĭ, zamknutyh otnositel'no prjamogo proizvedenija, *Izv. Akad. Nauk SSSR* (ser. mat.) **24** (1960), 493–510.

[158] A.D. Taĭmanov, Harakteristika aksiomatiziruemyh klassov modeleĭ (I, II), *Izv. Acad. Nauk SSSR* (ser. mat.) **25** (1961), 601–620, 755–764.

[159] A.D. Taĭmanov, Harakteristika konečno-aksiomatiziruemyh klassov modeleĭ, *Sibirsk. Mat. Ž.* **2** (1961), 759–766.

[160] T. Tamura, Attainability of systems of identities on semigroups, *J. Algebra* **3** (1966), 261–276.

[161] A. Tarski, Der Wahrheitsbegriff in den formalisierten Sprachen, *Studia Philosophica* **1** (1936), 261–404.

[162] A. Tarski, *A Decision Method for Elementary Algebra and Geometry*, 2nd edition, rev. Berkeley, University of California, 1951.

[163] A. Tarski, Contributions to the theory of models (I, II), *Indag. Math.* **16** (1954), 572–581, 582–588; (III), *Ibid.* **17** (1955), 58–64.

[164] A. Tarski, An extension of the Löwenheim-Skolem theorem to a second-order logic (abstract), in: *Short Comm. Int. Congress of Math. (Edinburgh, 1958)*. Cambridge, University Press, 1960, p. 10.

[165] A. Tarski, The elementary undecidability of pure transcendental extensions of real closed fields, *Notices Amer. Math. Soc.* **10** (1963), 355.

[166] A. Tarski, A. Mostowski, and R.M. Robinson, *Undecidable Theories* (Studies in Logic). Amsterdam, North-Holland Publ. Co., 1953.

[167] A. Tarski and R.L. Vaught, Arithmetical extensions of relational systems, *Comp. Math.* 13 (1957), 81–102.

[168] B.A. Trahtenbrot, Nevozmožnost' algoritma dlja problemy razrešenija v konečnyh klassah, *Dokl. Acad. Nauk SSSR* 70 (1950), 569–572.

[169] B.A. Trahtenbrot, O rekursivnoĭ neotdelimosti, *Dokl. Akad. Nauk SSSR* 88 (1953), 953–956.

[170] B.A. Trahtenbrot, Sintez logičeskih seteĭ, operatory kotoryh opisany sredstvami isčislenija predikatov, *Dokl. Akad. Nauk SSSR* 118 (1958), 646–649.

[171] B.A. Trahtenbrot, Nekotoryc postroenija v logike odnomestnyh predikatov, *Dokl. Akad. Nauk SSSR* 138 (1961), 320–321.

[172] B.A. Trahtenbrot, Konečnye avtomaty i logika odnomestnyh predikatov, *Sibir. Mat. Ž.* 3 (1962), 103–131.

[173] V.A. Uspenskiĭ, O vyčislimyh operacii, *Dokl. Akad. Nauk SSSR* 103 (1955), 773–776.

[174] V.A. Uspenskiĭ, Sistemy perečislimyh množestv i ih numeracii, *Dokl. Akad. Nauk SSSR* 105 (1955), 1155–1158.

[175] V.A. Uspenskiĭ, K teoreme o ravnomernoĭ nepreryvnosti, *Uspehi Mat. Nauk* 12 (1957), no. 1 (73), 99–142.

[176] V.A. Uspenskiĭ, Neskol'ko zamečaniĭ o perečislimyh množestvah, *Z. math. Logik und Grundl. Math.* 3 (1957), 157–170.

[177] R.L. Vaught, Applications of the Löwenheim-Skolem-Tarski theorem to problems of completeness and decidability, *Indag. Math.* 16 (1954), 467–472.

[178] R.L. Vaught, Remarks on universal classes of relational systems, *Indag. Math.* 16 (1954), 589–591.

[179] R. Vaught, On sentences holding in direct products of relational systems, in: *Proc. Int. Congress of Math. (Amsterdam, 1954)*, vol. 2. Amsterdam, North-Holland Publ. Co., 1954, pp. 409–410.

[180] R. Vaught, Models of complete theories, *Bull. Amer. Math. Soc.* 69 (1963), 299–313.

[181] A.A. Vinogradov, Častično-uporjadočennye lokal'no-nil'potentnye gruppy, *Učenye Zapiski Ivanov. Ped. Inst.* 4 (1953), 3–18.

[182] A.A. Vinogradov, Kvazimnogoobrazija abelevyh grupp, *Algebra i Logika Sem.* 4 (1965), no. 6, 15–19.

[183] B.L. van der Waerden, *Gruppen von linearen Transformationen.* Berlin, J. Springer, 1935.

[184] H. Wang, Dominoes and the $\forall\exists\forall$ case of the decision problem, in: *Proc. Symposium Math. Theory of Automata (New York, 1962).* Brooklyn, Polytechnic Institute of Brooklyn, 1963, pp. 23–25.

[185] H.I. Whitlock, A composition algebra for multiplace functions, *Math. Ann.* 157 (1964), 167–178.

[186] D.A. Zaharov, K teoreme Łosia-Suszki, *Uspehi Mat. Nauk* 16 (1961), no. 2 (98), 200–201.

[187] A.A. Zykov, Problema spektra v rasširennom isčislenii predikatov, *Izv. Akad. Nauk SSSR* (ser. mat.) 17 (1953), 63–76.

[188] Editor's note concerning results of A. Tarski, *Fund. Math.* 23 (1934), 161 . (German)

TOPIC TABLE

Chapter		I	II				III	IV	V
			G	*R*	*F*	*O*			
1		x							
2			x						
3		x							
4									x
5		x							
6		x							x
7		x							
8		x							
9		x							x
10		x							x
11		x	x			x			x
12		x					x		
13		x							
14			x				x		
15			x	x			x		
16					x		x		
17					x		x		
18	Survey	x					x	x	x
19			x	x			x		
20			x		x		x		
21			x	x			x		
22			x	x				x	
23		x				x	x		
24			x					x	
25								x	
26	Survey	x							
27								x	
28								x	
29						x	x		x
30						x			x
31									x
32		x	x						x
33									x
34	Survey	x	x	x	x	x	x		x

I = Metamathematics and the general theory of algebraic systems.
II = Metamathematical applications and results in specific algebraic theories (*G* = groups, *R* = rings, *F* = fields, *O* = other).
III = Decidability results.
IV = Contributions to a theory of numbered sets and constructive algebraic systems.
V = Varieties, quasivarieties, or universals.

INDEX

n following a page number means the reference is to an editor's note on that page

algebra, 27, 152, 314, 461
 absolutely free, 160
 finitely generated, 156, 201
 finitely presented by conditional identities, 159
 free, 28, 263, 385
 iterative, 398
 locally free, 160, 263, 266
 of logic, 396
 partial, 30, 152
 pre-iterative, 398, 399
 simple, 85, 211
 with defining relations, 28, 55, 159
 with signature Σ, 262, 384
algebraic system, 62, 152, 315, 460
 abstract (unnumbered), 187
 constructive, 191, 206, 282
 free, 417
 general recursive, 191
 numbered, 149, 187
 positive or negative, 190
 primitive recursive, 192
 with signature Σ, 460
application of a formula, 8
arity [*see* rank]
axiom [*see* sentence]

base (of an algebraic system) (= carrier), 68, 152, 314, 460
basic notions of an algebraic system, 68, 152, 314

category, 51, 326
category of structures, 52
 additive, 57
 bounded, 53
 homomorphically closed, 56
 multiplicatively closed, 53
 (quasi) free, 59
 (quasi) primitive, 62
 R-complete, 54
 regular, 53
chain, 34
class of algebras (or algebraic systems, or models), 27, 68, 152, 262, 315
 Γ-class, 429
 I-class (variety), 429
 Q-class (quasivariety), 429
 ∀-class (universal), 429
 abstract, 32, 52, 68, 263, 315
 attainable, 442
 axiomatizable [*see* class, (first-order) axiomatizable]
 decidable [*see* theory, (recursively) decidable]
 decomposable, 442
 finitely axiomatizable, 262, 319, 450
 (first-order) axiomatizable, 71, 262, 319, 450
 free, 28, 59
 hereditary, 423
 homomorphically closed, 56, 327
 minimal, 98, 470

class of algebras (*continued*)
multiplicatively closed, 53, 263, 331, 426
primitive (equational) [*see also* variety], 27, 62, 157, 386
projective, 71, 344, 430
pseudoaxiomatizable (= quasiaxiomatizable), 44, 77, 99, 334, 352*n*
quasiprimitive [*see also* quasivariety], 28, 62, 157
quasiuniversal, 24, 84, 349
recursively axiomatizable, 319
replica-complete (= R-complete, replete) [*definitions vary*], 54, 420, 428
total, 417, 463
ultraclosed, 340, 426
universally axiomatizable (= universal), 62, 68, 263, 447, 464
commutator sequence, 19
composition, direct, 51
free, 51, 160, 420
conditional equation [*see* quasidentity]
configuration, 4
consistent and complete, 5
congruence relation, 64, 196
consistent set of propositions, 1
consistent set of sentences, 16
constructive description [*see* numbering, constructive]
correspondence of models, 70, 96
axiomatizable, 70, 97
projective, 70
splitting, 97

defining relations, 28, 52, 159, 160
degree of unsolvability, 164
diagram, 45, 69
distinguished elements, 68, 152
domain (of a model), 5
of a function, 151

elementary class [*see* class, (first-order) axiomatizable]
elementary property, 16, 221
elementary subsystem, 120, 238
elementary theory [*see* theory, elementary]
embedding (isomorphic), 30
elementary, 337

enrichment, 132
inessential, 133
equality, relative and absolute, 6
relativized, 7
strong, 151, 157
equation [*see* identity]
equivalence, elementary (arithmetic), 238, 278, 325
rational, 59, 409
structural, 59
syntactical, 123, 244

factor (= quotient) algebra (or algebraic system, or model), 33, 418
family of objects, 293, 353
totally enumerable, 293, 356
finite subdiagram, 45, 69
first-order predicate logic (FOPL), 317
formula
closed [*see* sentence]
of signature Σ (Σ-formula), 262, 462
open, 199
positive, 199
formulatable property, 15
function, 151
characteristic, 164
(general) recursive, 162
partial, 151
partial recursive, 162
primitive recursive, 162
fundamental operation, predicate, set, etc. [*see* basic notions]

generalized Jordan cell, 222
generating set, 44, 55, 153, 156
free and dense, 56
generation, natural, 44
group, 158
abelian, 158
enriched (= with fixed elements), 126
free solvable, 119
freely orderable, 92
metabelian, 123*n*, 137*n*
p-group, 131
R-group, 120
with fixed elements, 121
groupoid, 157, 263, 265, 392

homomorphic image, 32
homomorphism of algebraic systems, 32, 51, 152, 196, 314
  canonical, 197
  strong, 315
homomorphism of numberings, 292
homomorphism of partial algebras, 30

identity (= identical relation), 157, 385, 416, 462
  conditional [*see* quasidentity]
identity problem, 385, 469
impoverishment, 134, 419, 430
indecomposable element, 268
independent set of axioms, 417, 450
individual constant, 156
interpretation (of a theory in another theory), 134, 256
  exact, 134
isomorphism of algebraic systems, 51, 152, 314
isomorphism of numberings, 292, 354

Jordan form, 223

Kleene universal function, 183, 289

limit, direct, 63, 338
local embeddability, 65
local property, 347
loop, 394

mapping, 151, 152
model, 2, 14*n*, 29, 152, 314, 461
  correspondence [*see* correspondence of models]
  corresponding to an algebra, 314
  decidable, 112
  multibase (many-sorted), 68, 344
  small, 114, 321
  with defining relations, 55
multiplication of classes, 423

number set, 165
numbering, 149, 165, 288
  complete [*old definition* = precomplete], 182, 291, 365
    [*new definition*], 288, 354
  computable, 305, 354, 379

numbering (*continued*)
  constructive, 191, 256, 282
  decidable, 166, 379
  Kleene, 183, 289
  positive or negative, 166, 188, 379
  Post, 185, 289
  precomplete [*see* numbering, complete]
  principal, 306, 354
  simple, 165, 288
  standard, 164, 193, 202
    extended, 203
  steadfast, 256
  (sub)normal, 360
  (sub)special, 369
  trivial, 165

operation, partial, 30, 151
  termal, 37, 155, 199, 396
  total, 151, 313
order of a model (or class), 113, 321

polar, 435
Post algebras, lattices, and varieties, 397, 409
power of a model, 316
predicate, 67, 151, 313
  formular (= elementary), 39, 81, 138, 223, 318
  fundamental (or basic), 68, 152
  invariant, 39, 42
  persistent, 39
problem of identical relations [*see* identity problem]
product, direct, 33, 53, 100, 331
  filtered, 426
  free, 160
  $\mathcal{K}$-product, 423
  regular, 102
    proper, 103
  subdirect, 33, 72, 333
projection of a system (or class) [*definitions vary*], 344, 430
propositional calculus (PC), 1

quantifiers, relativization of, 76
  relativized, 67
  unification of, 75

quasidentity (= conditional identity), 416, 157, 462
quasifree closure, 63
quasigroup, 390
quasiprimitive closure, 419
quasivariety, 28, 157, 417, 429, 464
  strict (finite), 28, 157

R- [*stands for any of* primitive recursive, general recursive, *or* partial recursive *and is prefixed to the following terms*]
  equivalence, 166, 188
  function, 162, 169
  isomorphism, 167, 172, 188
  map, 166
  monomorphism, 166, 188
  multiequivalent (= **m**-equivalent), 172
  multireducible (= **m**-reducible, reducible), 171, 185, 291
  numbering, 187
    Gödel, 204
  operation, 169
  predicate, 166, 169
  set, 164
  stable, 181
  subset, 165
  subsystem, 192
  uniequivalent, 172
  unimorphism, 167, 171, 188, 292
  unireducible, 171
range of a function, 151
rank (= arity), 151, 153, 460
reducible [*see* **R**-multireducible]
relation [*see* predicate]
  defining, 28, 52, 159, 160
replica, 53, 420
representation of a model, 22
  direct, 23
  predicate, 23
representing function, 149
ring, 158
  associative, 159
  of characteristic p (= char p ring), 131
rring, 131

second-order predicate logic (SOPL), 342
Segre characteristic, 222
selector function, 398
semigroup, 157, 386
  cancellative, 158
sentence (= axiom, closed formula), 16 37, 68, 318
  finitely refutable, 248
  Horn, 331
  in Skolem form, 84, 319, 463
  universal, 68, 318, 463
set, numbered, 165
  primitive recursive, 164
  recursive, 164
  recursively enumerable, 164
signature, *Foreword,* 156, 315, 460
similarity type, *Foreword,* 68, 151, 314
spectrum, 279, 342
structure, 52
subclass, finitely axiomatizable, 133, 450
  Γ-subclass, 450
subdiagram, finite, 45, 69
  realizable, 69
subdirect indecomposability, 33, 333
submodel, 68, 319
  B-submodel, 68
  elementary, 119
subobject, 357
subproduct, B-, 113*n*
substructure, 52
Sylow sequence, 18
symbol, individual, 153
  operation (= function), 153, 460
  predicate, *Foreword,* 70, 460
  signature, 460
symmetry conditions, 266

term, 154, 462
  of signature Σ (Σ-term), 263, 384, 462
theorems
  compactness theorem for FOPL, 14*n*, 16, 22, 76, 320
  compactness theorem for PC, 1

theorems (*continued*)
local theorem
of logic [*see* compactness theorem for FOPL]
extrinsic, 77, 88, 93
intrinsic, 85, 92, 349
Löwenheim-Skolem theorem, 1, 12, 78, 320
Rogers' theorem generalized, 183, 184, 292
Tarski-Łoś theorem on universal subclasses, 69, 449
theory
Γ-theory, 429, 463
categorical, 322
complete, 98, 322
elementary (= E-theory), 122, 138, 248, 279, 316, 465
essentially undecidable, 133
(recursively) decidable, 122, 133, 138, 279, 463
undecidable, 133
type of algebraic system [*see* similarity type]
type of regular product, 103

ultrafilter, 335, 426
ultralimit, 339
ultrapower, 336
ultraproduct, 336, 426
unit system (or structure), 53, 417
universal, 429, 464

validity of a formula, 157
value of a term (or symbol), 154, 315
variety, 27, 157, 385, 386, 417, 429, 464

word problem, 205, 385